Fruit and Nut Production

Brenda Olcott-Reid, Ph.D.
William Reid, Ph.D.

ISBN 978-1-58874-575-0

Published by

Stipes Publishing L.L.C.
204 W. University Ave.
Champaign, Illinois 61820

Captions for Cover Photos

Front cover photo: Author Brenda Olcott-Reid harvests peaches from her Kansas orchard.

Back cover photo: This fruit market in Ontario illustrates the diversity of fruit grown locally in North America.

All photos by authors unless otherwise noted.

Preface

This book is designed for students, growers, researchers and Extension workers who deal with temperate-zone, perennial fruit and nut crops. The book emphasizes ecological and economical production methods. Part I, the basics of fruit and nut production, introduces principles of establishing and maintaining a planting that apply to all fruit and nut crops.

The remaining sections provide specific instructions for growing the individual fruit and nut crops. Part II covers the principle small fruits, meaning those that grow as bushes, vines, or herbaceous perennials rather than as trees. Part III details the major tree fruits, and Part IV covers the tree nuts. Part V introduces some lesser-grown fruit crops that offer great potential for certain areas, and Part VI details the use of fruit plants in home landscapes.

In some cases we mention specific pest control or other products used in growing fruits and nuts. We mention these to inform readers of available choices, and such mention does not imply endorsement of any products, nor criticism of products not mentioned.

Many fruit and nut researchers, Extension workers and growers have provided information for this book, either via materials they have written or in communications with us. We have cited many of them throughout the book, but some others are not cited, and we thank them all. Any errors, however, are completely ours.

We welcome corrections, new cultivar or product announcements, and updates for future editions of this book. Send these to the address below.

Drs. Brenda Olcott-Reid and William Reid
10331 SW 95th St.
Chetopa, KS 67336-8567
e-mail: reidfarm@ckt.net

Table of Contents

Chapter 1

Choosing Types of Fruits and Nuts, Cultivars and Rootstocks

Fruits and nuts are exceptional crops. Often regarded as dessert crops, they have a special appeal to our taste buds. Fruits (the fleshy tissues surrounding seeds) are high in natural sugars and nuts (the seeds) are high in oils, and both often are used in dessert recipes. Yet unlike many desserts, these are healthy foods. Medical researchers are discovering more and more health benefits from a diet high in fruits and nuts, such as preventing many types of cancers and keeping arteries unclogged.

Fruits and nuts are perennial crops, borne on plants that keep producing for many years. That makes them excellent soil-holding crops, especially on marginal land that is prone to soil erosion. Homeowners will find that many fruit plants are attractive, long-lived additions to the landscape that can add value to a property. However, many types of fruit and nut plants take several years of care before they begin to bear, and even more years before the producer begins to profit from his investment.

Once plants bear, fruits and nuts are high-value crops. They command excellent prices, often producing 10 to 100 times the profit per acre as do agronomic crops. Even greater profits can be achieved through value-added processing or marketing efforts. Homeowners who grow their own fruits can save much more money than they can by growing annual vegetables.

Consider fruit and nut plantings as long-term investments that require a few years to produce dividends, but then yield great returns. Because these plantings will continue to produce for several to many years, pay careful consideration to which types of fruit or nut plants to grow and to choosing the site.

CLIMATE CONSIDERATIONS

Some type of fruit or nut crop can be grown almost anywhere in the temperate zones. The first step in deciding which types of fruit and nut plants to produce is to study your climate. Ask your county Extension Service agent for data for your area, and compare that with Tables 1-1 to 1-3. The climate factors outlined below dictate where each type of crop can be grown profitably.

Winter temperatures. All of the fruit and nut plants covered in this book are adapted to the temperate zone, meaning they adjust to colder temperatures in fall and can withstand winter temperatures below freezing. Various fruit and nut species differ greatly in their degree of cold hardiness; some can withstand temperatures down to -40 F (-40 C), while others are injured at temperatures below 10 F (-12 C). Within each species, the various cultivars (cultivated varieties) also differ greatly in cold hardiness. Fruit cultivars are propagated asexually, so each cultivar is a clone, identical to all other plants of the same cultivar name, in hardiness and other plant attributes.

Most fruit and nut nurseries include a copy of the U.S. Dept. of Agriculture's Hardiness Zone Map (Figure 1-1). The USDA developed this map based on average minimum winter temperatures, and it is now widely used to rate the cold hardiness of individual cultivars of perennial plants. The map is generally accurate except in mountainous areas,

Table 1-1.

CLIMATE ADAPTATION OF SMALL FRUITS

Kind of Fruit	Hardiness Zones of Adaptaion	Chill Hours Needed	Summer Temperature Adaptation *	Bearing Season	Fruit Bud Hardiness
Blackberry					
Erect	5 to 8	200 to 600	warm to hot	June - August	sensitive
Trellised	6 to 9	700 to 1100	warm to hot	June - August	very sensitive
Blueberry					
Highbush	4 to 10	200 to 900	warm	June - July	mod. hardy
Half-high	3 to 5	700 to 1000	cool	July - August	hardy
Rabbiteye	7 to 9	350 to 650	hot	June - August	sensitive
Currant	2 to 8	800 to 1500	cool to warm	June - July	very hardy
Gooseberry	2 to 8	800 to 1500	cool to warm	June - July	very hardy
Grape					
Bunch	3 to 10	100 to 1500	warm to hot	August - Sept.	mod. hardy
Muscadine	7 to 10	300 to 700	long, hot	August - Sept.	sensitive
Raspberry					
Red & Yellow	3 to 10	800 to 1700	cool to warm	July - October	very hardy
Black & Purple	4 to 9	800 to 1200	cool to hot	June - July	hardy
Strawberry					
Spring-bearing	3 to 10	200 to 300	cool to hot	May - June	hardy
Everbearing	3 to 10	200 to 300	cool	May - Sept.	hardy

* Summer heat:
"Cool" areas have high temperatures mostly in the 70's and low 80's F.
"Warm" areas have high temperatures mostly in the 80's and low 90's F.
"Hot" areas often have high temperatures above 90 F.

Note: The values listed are approximations and may vary with cultivars and locations.

where altitude becomes a greater determinant of temperature minimums than latitude.

A fruit or nut grower should know his hardiness zone, and choose species and cultivars adapted to that zone. Fruit breeders have expanded the range of many fruit crops (for examples, highbush blueberries and peaches) but only a few cultivars are adapted to areas at either edge of a crop's normal hardiness range. A cultivar that does well in zone 4 won't do well in zone 9, and vice versa.

Over most of the country, winter temperature fluctuations place more stress on plants than the minimum temperatures. "Super-hardy" cultivars, rated hardy to the lowest zone numbers, are essential in the coldest regions but often don't do as well in milder zones. These "super-hardy" cultivars are often adapted to prolonged, steady cold but not to wider winter temperature swings. Growers in milder zones should choose cultivars that are hardy enough for their zone, but also adapted to the other climate factors below.

Amount of winter chilling. Temperate-zone fruit and nut plants all need a period of winter chilling, with temperatures between 32 F and 45 F, before they will bloom and grow in the spring. The number of chilling hours needed varies with the

Table 1-2.

CLIMATE ADAPTATION OF FRUIT TREES

Kind of Fruit	Hardiness Zones of Adaptaion	Chill Hours Needed	Summer Temperature Adaptation *	Bearing Season	Fruit Bud Hardiness
Apple					
Dwarf	4 to 9	300 to 1000	cool to hot	July - Oct.	hardy
Semi-dwarf	3 to 9	300 to 1000	cool to hot	July - Oct.	hardy
Standard	2 to 9	300 to 1000	cool to hot	July - Oct.	hardy
Apricot	3 to 9	350 to 1000	warm to hot	July	very sensitive
Sweet Cherry	5 to 7	650 to 1300	cool to warm	June	very sensitive
Tart Cherry					
Genetic dwarf	3 to 8a	800 to 1200	cool to warm	June	hardy
Stardard	4 to 8a	800 to 1200	cool to warm	June	hardy
Peach & Nectarine					
Genetic dwarf	6 to 8	600 to 900	warm to hot	July - Sept.	sensitive
Standard	5 to 9	150 to 1100	warm to hot	July - Sept.	sensitive
Pear					
Dwarf	5b to 9	600 to 900	cool to hot	Aug. - Oct.	hardy
Standard	3 to 9	600 to 1100	cool to hot	Aug. - Oct.	hardy
Asian Pear	6 to 9	300 to 850	warm to hot	Aug. - Sept.	sensitive
Plum					
European	4 to 8	750 to 1100	cool to warm	Aug. - Sept.	mod. hardy
Japanese	5 to 9	500 to 1200	warm to hot	July - Aug.	very sensitive
American	2 to 8	500 to 900	cool to hot	Aug. - Sept.	hardy

* Summer heat:
"Cool" areas have high temperatures mostly in the 70's and low 80's F.
"Warm" areas have high temperatures mostly in the 80's and low 90's F.
"Hot" areas often have high temperatures above 90 F.

Note: The values listed are approximations and may vary with cultivars and locations.

type of fruit, and between cultivars of the same fruit species. Figure 1-2 shows the average number of chilling hours provided in the United States.

Growers in warm-winter zones 9 and 10 should choose "low-chill" cultivars that need about as many chilling hours as their area provides. Most cultivars of most temperate-zone fruits and nuts have higher chilling requirements, and would fail to bloom or bloom very erratically if planted in the Deep South or the mild-winter West Coast.

In hardiness zones 5 through 8, where winters are colder and spring frosts are troublesome, cultivars with high chill requirements do best. High-chill cultivars bloom a week or so later than cultivars with average chilling requirements, which may help these plants escape frosts during bloom

Table 1-3.

CLIMATE ADAPTATION OF NUT TREES

Kind of Nut Tree	Hardiness Zones of Adaptaion	Chill Hours Needed	Summer Temperature Adaptation*	Bearing Season	Fruit Bud Hardiness
Almond	7 to 9	300 to 500	warm to hot	August	very sensitive
Chinese Chestnut	5 to 9	500 to 700	cool to hot	Sept. - Oct.	hardy
Hazelnut	4 to 8	800 to 1200	cool to warm	Sept. - Oct.	sensitive
Pecan	5b to 9	500 to 900	long, hot	Oct. - Nov.	hardy
Pistachio	8 to 9	950 to 1050	long, hot	Sept. - Oct.	very sensitive
Walnut					
Black	3 to 8	600 to 1100	cool to hot	Sept. - Oct.	hardy
Persian	6 to 9	500 to 1000	warm to hot	Sept. - Oct.	sensitive

* Summer heat:
"Cool" areas have high temperatures mostly in the 70's and low 80's F.
"Warm" areas have high temperatures mostly in the 80's and low 90's F.
"Hot" areas often have high temperatures above 90 F.

Note: The values listed are approximations and may vary with cultivars and locations.

when the open flower buds are most vulnerable to cold damage.

In the coldest areas, zones 3 and 4, plant hardiness to minimum winter temperatures is more important than resistance to spring frosts. Very cold winters slow flower bud development, so adapted plants usually escape spring frosts even if they have low chilling requirements. In fact, very winter-hardy cultivars often have low chilling requirements, so they don't do as well in the middle hardiness zones.

Summer temperatures. Certain fruit species grow poorly where summers are very hot. Temperatures above 90 F increase plants' respiration rates, causing more sugars to be burned off instead of accumulating in fruits. This reduces yields and quality of certain fruits, such as red raspberries and many "northern" cultivars of apples.

Other fruit and nut species need hot summers to ripen properly. The plants themselves may grow well in areas with cooler summers, but produce few or no crops of good quality. Some pecan cultivars, for example, are hardy to zone 5 and produce well in Midwestern parts of zone 5 with hot summers, but the kernels don't fully fill the shells in Eastern parts of zone 5 or 6 with cooler summers. Both George Washington and Thomas Jefferson planted pecans on their Virginia estates; the 200-year-old trees are still alive but have rarely produced good crops.

The American Society for Horticultural Science has developed a heat tolerance map, but it documents plant survival, not fruit production. This map is used mainly for ornamental plants.

Length of growing season. The growing season--the number of frost-free days--becomes a concern in winter-hardiness zones 3, 4 and 5. The season in these northern or mountainous areas is too short for some late-maturing cultivars of fall-ripening fruits and nuts, such as apples, pears, grapes, fall red raspberries, and pecans. In these areas, choose cultivars that will ripen in the number of days your season provides or less.

Late-ripening fruits may also fail to mature in coastal areas with cool summers, because they take

Figure 1-1.
USDA HARDINESS ZONE MAP

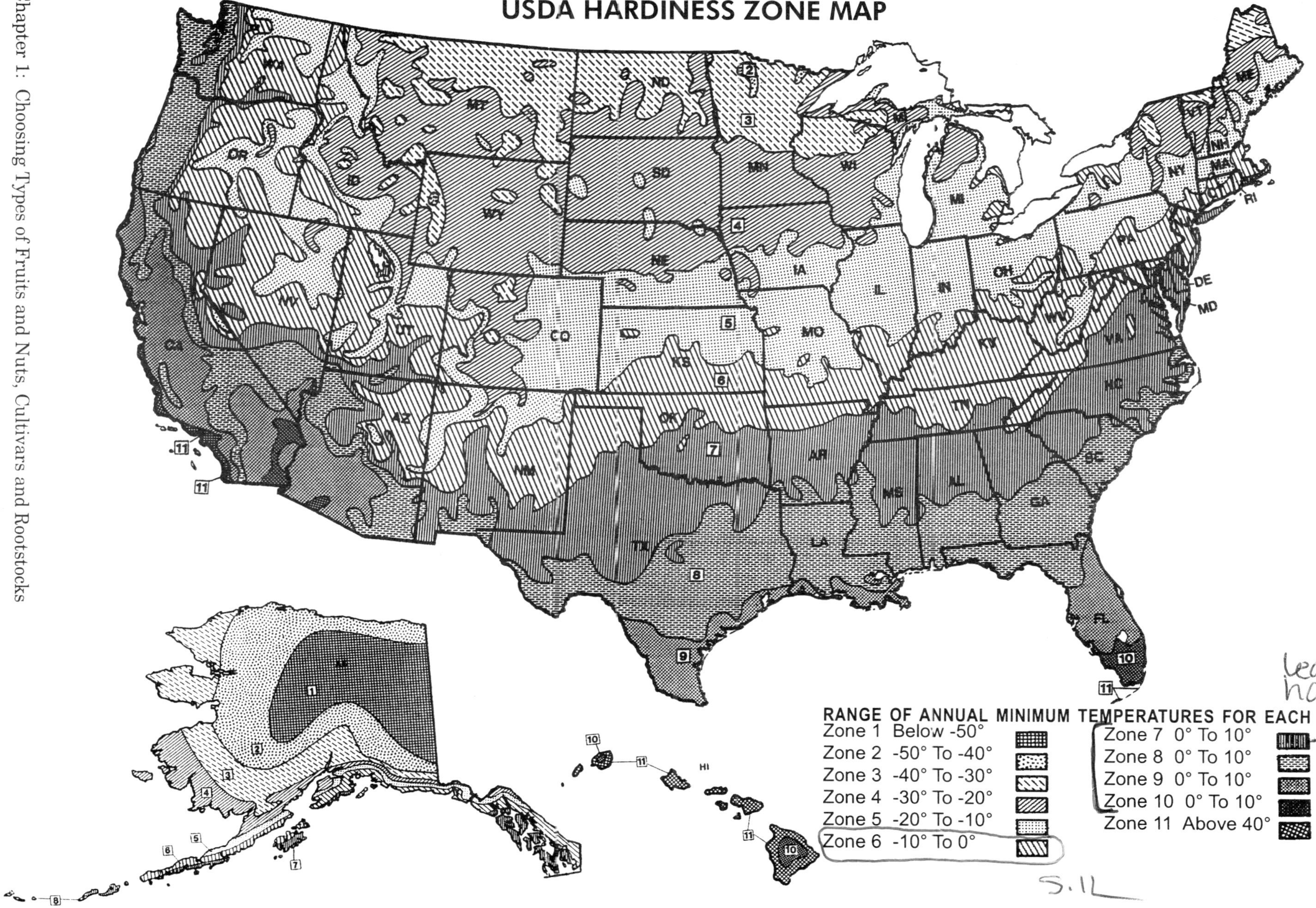

Figure 1-2.
APPROXIMATE MINIMUM HOURS OF CHILLING

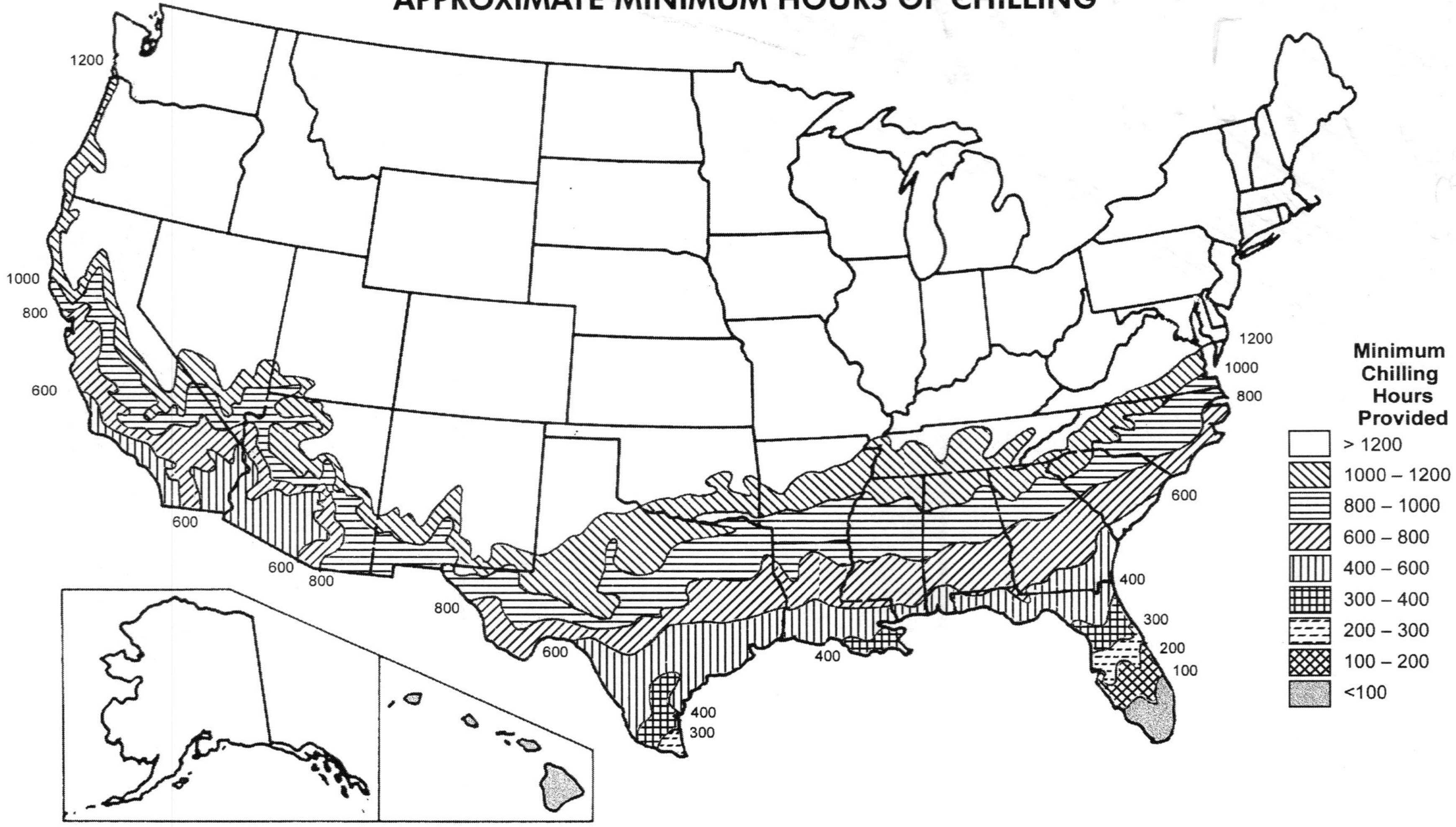

Figure 1-2. This map represents minimum total hours of winter tempratures below 45° and above 32°F. Chilling hours may be much greater during some seasons. The map is generalized and many local variations exist, particularly in mountainous areas.

Drawn by Brenda Olcott-Reid based on maps and data provided by Dr. George Ray McEachern of Texas A&M University and Dr. William R. Okie of USDA.

more days to ripen at cooler temperatures. Extension agents can help growers to determine cultivars' growing-season needs in these areas.

Where growing seasons are longer and warmer, later-ripening cultivars often do better. Early-maturing cultivars of fall-ripening fruits and nuts would ripen during the heat and drought of late summer in these zones. Late-ripening cultivars of these same species take advantage of the longer season, and develop better flavor and yields during the cooler temperatures and higher moisture levels of fall.

Rainfall and humidity. Fruit crops, in general, need about one inch of rainfall or irrigation per week. Eastern areas generally provide enough rainfall, but periodic droughts and the use of irrigation systems for frost protection make irrigation highly advisable for commercial production even in the East. Western areas often require irrigation, which increases production costs.

Western areas also offer lower humidity, however, which reduces problems with many diseases and some insect pests. Growers in Eastern areas can greatly reduce their need for fungicide sprays by choosing cultivars resistant to the major diseases of each fruit crop in their area (Photo 1-1). Even more importantly, choose cultivars resistant to certain bacterial diseases, such as fire blight of apple and pear or bacterial canker of peaches, plums and cherries in humid areas. Such bacterial diseases can kill susceptible plants, and bacteriocides are much more expensive than fungicides.

Certain cultivars develop better quality in areas of lower humidity. 'Golden Delicious' apple, for example, develops a rusty-colored "russet" on the skin of the fruit in Eastern areas, while producing a longer and smooth-skinned yellow apple in Western areas. Eastern apple producers do better with 'Smoothee Golden Delicious', a strain of the cultivar that resists russeting (Photo 1-2). Even better, plant cultivars resistant to russeting and several diseases, such as 'Gold Rush'.

PROFIT POTENTIAL

Fruit and nut crops offer the greatest profit potential of any farm products. However, profit level can vary widely between operations, based on many factors including type of marketing, degree of processing, demand for the types of fruits and nuts grown, location, advertising, production costs and yields. Tables 1-4 through 1-6 document the general costs and profit potentials of many fruit and nut crops.

Photo 1-1. Fruit and nut growers can reduce pesticide costs by choosing cultivars resistant to the diseases prevalent in their area. These 'Jonafree' apples resist apple scab, fire blight and powdery mildew and need no sprays for those diseases.

Choose your market first. Decide whether you will sell retail or wholesale or both before choosing which types of fruits and nuts to grow, as the choice of market will affect the setup of your entire operation. Retail growers sell directly to the public, receiving the entire price paid for their products, but devoting much labor time to customer service. Wholesale growers sell to buyers who process, package and resell their product, receiving only a percentage of the final price, but concentrating on the growing rather than the selling. Some operations are both retail and wholesale, but most growers find it easier to gear an operation to one marketing strategy or the other.

Retail growers find that proximity to a strong market is crucial. Generally, a retail grower should be located within 30 miles of at least 20,000 people. Pick-your-own sales (Photo 1-3) will be strongest if the farm lies within ½ mile of a paved road, and

Photo 1-2. 'Smoothee Golden Delicious' resists the russeting that plagues other strains of 'Golden Delicious' in humid areas. 'Goldrush' is another yellow apple that resists apple scab, fire blight, cedar-apple rust and powdery mildew as well as russeting.

Table 1-4.

PLANTING ECONOMICS OF SMALL FRUITS

Kind of Fruit	Plant Spacing (ft)	Plants per Acre	Start-up costs[x] ($/Acre)	First Crop Year	First Profit Year	Yield at Maturity (Tons/Acre)	Potential Net Profit[y] ($/Acre)	Years of Expected Life
Blackberry								
Erect	10 x 4	1089	1600	2nd	3rd	3 to 4	4500	10 to 12
Trellised	12 x 6	605	2800	2nd	4th	3 to 5	5000	10 to 12
Blueberry								
Highbush	10 x 4	1089	3000	3rd	5th	3 to 5	4500	25 to 30
Half-high	10 x 3	1452	3900	4th	6th	2 to 4	3000	20 to 25
Rabbiteye	12 x 6	605	2400	3rd	5th	4 to 6	4500	25 to 30
Currant	10 x 4	1089	1600	2nd	3rd	4 to 6	3000	12 to 15
Gooseberry	10 x 4	1089	1600	2nd	3rd	3 to 5	2500	12 to 15
Grape								
Bunch	10 x 8	544	3800	3rd	5th	4 to 10	2000	25 to 30
Muscadine	12 x 20	182	2800	3rd	5th	4 to 7	1800	25 to 30
Raspberry								
Red & Yellow	10 x 2	2178	1600	1st	2nd	1 to 3	4000	8 to 10
Black & Purple	10 x 3	1452	1900	2nd	3rd	1 to 2	2500	8 to 10
Strawberry								
Spring-bearing	4 x 2	5445	2100	2nd	3rd	4 to 5	3500	3 to 5
Everbearing	4 x 1.5	7260	2400	1st	2nd	1 to 2	1800	2 to 4

Pollination note: All small fruits are self-fertile, so only one cultivar is needed for pollination, EXCEPT half-high and rabbiteye blueberries, which need at least two cultivars.

[x]Start-up costs include plants, planting and care labor, materials and equipment use costs until first harvest. Land charges, fumigation (if needed), overhead costs and irrigation system establishment are not included.

[y]Net profit potential is the approximate average return after all production, harvesting, depreciation and overhead costs have been subtracted.

Note: The values listed are approximations and may vary with cultivars, locations, and care given to fruit plants. Economic data is based on reports from the University of Illinois and other state Extension Services, and the USDA National Agricultural Statistics Service.

Photo 1-3. Choose your marketing system before deciding which types of fruit or nuts to grow, and how much of each. Pick-your-own marketing minimizes labor costs and maximizes grower returns, if the fruit planting is located near population centers and close to a paved road.

Photo 1-4. Value-added processing, such as pressing apples into cider, can utilize smaller or disfigured fruit and greatly increase grower returns. However, processing requires more start-up capital for equipment and means following strict health and safety procedures.

such operations require about one acre of parking area for each five acres in crops to be picked. Bringing harvested fruit to customers at city-sponsored farmers' markets or by supplying restaurants can also be very profitable, especially for more out-of-the-way growers. Growing several different types of fruits and nuts, to provide products throughout the growing season, increases profit potential for all of these retail ventures.

Wholesale growers generally concentrate on fewer types of fruits and nuts. Larger acreages allow the grower to purchase more specialized equipment for efficient pest control and harvesting of each crop. Wholesale production requires much greater capital investments and risks greater profit fluctuation from year to year, since the grower relies on a smaller number of crops. Yet wholesale production of unique crops can be very profitable, as the market size is greatly magnified.

Value-added processing also requires more capital investment, but can pay off greatly. Consumers often pay ten to twenty times as much per unit for fruits or nuts that have been processed into pies, jams, candies, or other products as for the raw crops. These preserved products also offer a longer shelf life than raw fruits or nuts. Commercial kitchen facilities and health inspections are required for any value-added processing, as in Photo 1-4.

Certified organic production is increasing, as more people are willing to pay premium prices for pesticide-free produce. Organic growers usually must receive higher prices to make up for lower yields and higher input costs for organic fertilizers and pest control products. Organic growers should plan on using a portion of the crop for processing, to utilize the greater amount of fruit with superficial diseases or insect damage. Organic certification requires allowing inspectors from the state Board of Agriculture or an organic growers' association to determine that all production practices meet organic guidelines.

PRODUCTION COSTS

Small fruit – those that grow on bushes, vines or herbaceous perennial plants rather than on trees--will produce crops sooner than tree fruits and nuts, thus recovering the grower's investment sooner. Small fruits, except for grapes, also generally require fewer pesticide sprays and less worker-hours than tree fruits or nuts, and less expensive equipment, further increasing their profit potential. Many beginning growers find it economical to start with small fruits, often just one acre of strawberries, raspberries or blackberries, or a combination of these.

Irrigation is essential for small fruits, however, as their root systems are shallower than tree fruits or nuts. Many small fruit plantings also must be replaced on new ground after five to ten years.

Tree fruit and nut plantings last longer than

Table 1-5.

PLANTING ECONOMICS OF FRUIT TREES

Kind of Fruit	Pollinizer Needs [x]	Tree Spacing (ft)	Trees per Acre	Start-up costs ($/Acre) [y]	First Crop Year	First Profit Year	Yield at Maturity (Tons/ Acre)	Potential Net Profit ($/Acre) [z]	Years of Expected Life
Apple									
Dwarf	SS or PSF	14 x 8	389	5600	3rd - 4th	6th - 7th	30 to 40	4000	15 to 25
Semi-dwarf	SS or PSF	18 x 14	173	4800	4th - 5th	7th - 8th	20 to 25	2400	25 to 30
Standard	SS or PSF	22 x 20	99	6000	5th - 7th	9th - 10th	10 to 15	1200	30 to 40
Apricot	PSF	20 x 18	121	3000	4th - 5th	8th - 9th	4 to 7	400	20 to 30
Sweet Cherry									
Semi-dwarf	SS	18 x 16	151	3200	4th - 5th	8th - 9th	4 to 8	2500	30 to 40
Standard	SS	20 x 18	121	3000	4th - 5th	9th - 10th	3 to 6	1800	30 to 40
Tart Cherry									
Genetic dwarf	SF	14 x 8	389	3400	3rd	5th - 6th	4 to 6	300	12 to 15
Stardard	SF	18 x 16	151	2400	4th	7th - 8th	3 to 5	200	15 to 20
Peach & Nectarine									
High Density	SS	14 x 8	389	3100	3rd	5th	10 to 12	2000	8 to 10
Standard	SS	18 x 16	151	2100	3rd	6th	6 to 8	1200	10 to 15
Pear									
Dwarf	SS	16 x 10	272	4500	3rd - 4th	7th - 8th	20 to 24	2000	10 to 15
Standard	SS	20 x 20	109	6000	5th - 7th	9th - 10th	10 to 16	1000	30 to 40
Asian Pear	PSF	20 x 12	181	4000	3rd - 4th	6th - 7th	18 to 22	1800	30 to 40
Plum									
Dwarf		18 x 12	202	2500	3rd - 4th	6th - 7th	6 to 8	1000	10 to 15
Standard		20 x 18	121	2600	4th - 5th	8th - 9th	4 to 6	800	15 to 20
European	PSF								
Japanese	SS								
American	SF or PSF								

[x] Pollination key:

SF = self-fertile: only one cultivar needed for pollination.

PSF = partially self-fertile: generally sets heavier fruit crop with two or more cultivars planted nearby.

SS = self-sterile. Two cultivars needed for pollination; occasional exceptions with some cultivars.

[y] Start-up costs include plants, planting and care labor, materials and equipment use costs until first harvest. Land charges, fumigation (if needed), overhead costs and irrigation system establishment are not included.

[z] Net profit potential is the approximate average return after all production, harvesting, depreciation and overhead costs have been subtracted.

Note: The values listed are approximations and may vary with cultivars, locations, and care given to fruit plants. Economic data is based on reports by Pennsylvania State University and other states' Extension Services, and the USDA National Agricultural statistics Service.

Table 1-6.

PLANTING ECONOMICS OF NUT TREES

Kind of Nut Tree	Initial Tree Spacing (ft)[x]	Trees per Acre	Start-up costs ($/Acre)[y]	First Crop Year	First Profit Year	Yield at Maturity (Tons/Acre)	Potential Net Profit ($/Acre)[z]	Years of Expected Life
Almond	20 x 20	109	2600	4th	10th - 11th	1 to 1.5	400	20 to 30
Chinese Chestnut	25 x 25	70	2000	3rd	8th - 9th	1.5 to 2.5	300	60 to 100
Hazelnut	18 x 10	242	2600	3rd - 4th	9th - 10th	0.5 to 1.5	200	40 to 50
Pecan	35 x 35	35	2100	7th - 8th	12th - 13th	0.5 to 1.5	400	60 to 100
Pistachio	24 x 12	151	3600	5th - 6th	13th -14th	1.5 to 2.5	350	40 to 50
Walnut								
Black	30 x 30	48	2400	6th - 7th	13th -14th	0.5 to 1.0	100	60 to 100
Persian	20 x 20	109	3100	5th - 6th	12th - 13th	2.5 to 3	300	40 to 50

Pollination note: Most cultivars of all nut species are self-sterile, so at least two cultivars are needed for pollination. Often, one cultivar must be protandrous or male, and the other protogynous or female.

[x] Most nut orchards today are planted at these higher-density spacings, and when trees start to crowd after 10-12 years, alternate trees are removed, except in almonds.

[y] Start-up costs include plants, planting and care labor, materials and equipment use costs until first harvest. Land charges, fumigation (if needed), overhead costs and irrigation system establishment are not included.

[z] Net profit potential is the approximate average return after all production, harvesting, depreciation and overhead costs have been subtracted.

Note: The values listed are approximations and may vary with cultivars, locations, and care given to plants. Economic data is based on reports by the University of California and other states' Extension Services, and the USDA National Agricultural Statistics Service.

Photo 1-5. Small fruit plantings, such as strawberries, produce crops and profits sooner than do tree fruit or nut plantings. Pick-your-own harvesting, as here, reduces labor costs to growers and guarantees freshness and reduced price to customers.

small fruit plantings, and may become just as profitable after ten years or so. The smaller the trees, the more can be planted per acre and the quicker they come into bearing. Naturally smaller trees such as peaches and tart cherries are planted closer than standard-sized apples and pears. Most apple and pear growers are now planting dwarf trees to increase production and reduce time until trees bear. Nut trees grow the largest and take the longest time to begin bearing.

Harvesting methods significantly affect production costs, and should be determined before planting. Fruit crops are labor-intensive, and labor costs can become the greatest production expense. Small retail growers can greatly reduce labor costs with a pick-your-own operation (Photo 1-5). Harvested fruit can also be offered for sale at a price that covers the additional cost of harvest labor, generally 67% to 75% more than the pick-your-own fruit.

Mechanical harvesting equipment is available for many fruit and nut crops. Many harvesters can be adjusted to harvest several types of similar crops, such as blackberries and blueberries, or peaches and plums. Growers must harvest enough acreage to justify the high equipment costs. Smaller growers often custom-harvest neighbors' plantings to help recoup equipment expenses.

Mechanical harvesting of fruits works well with a processing operation, but some types of mechanical harvesters are too rough on the fruit for fresh marketing. Mechanical nut harvesters do not damage the nut kernels, protected in their hard shells, and are standard equipment in nut orchards.

POLLINATION NEEDS

Fruit and nut growth begins with the pollination of flowers. Bees and other insects transfer pollen from the anthers to the stigmas in most fruit plants. These fruit plants have showy petals and nectar to attract the insects, and heavy, sticky pollen that adheres to the insects' legs. Most nut trees are pollinated by wind, and so have no petals or nectar in their flowers, but produce large quantities of dry, light-weight pollen that is easily carried by gusts of wind to the large, feathery stigmas.

Flowers of most fruit plants contain both male and female parts, while nut trees usually have separate male and female flowers on each tree. Fruit and nut cultivars fall into one of three pollination classes:

1. Self-fertile cultivars set a good fruit crop when pollinated with pollen from the same plant, so only one cultivar is needed. These include almost all berries, grapes, peaches, tart cherries and some European plums. For the longest harvest season of these fruits, growers may want to plant several cultivars of each that ripen at different times, but each cultivar can be planted in its own block.

2. Partially self-fertile cultivars set some fruit alone, but will bear larger crops if a different cultivar provides pollen. These include most Asian pears, apricots and European plums, along with a few cultivars of apples, European pears, sweet cherries, and walnuts. These cultivars should each be planted in smaller blocks, to allow some cross-pollination.

3. Self-infertile cultivars include most cultivars of apples, European pears, sweet cherries, Japanese and American plums, pecans and walnuts. These need pollen from a different cultivar in order to set fruit. Growers should intermix at least two different cultivars of the same fruit species in each block. Bloom times of the two cultivars must overlap, so don't rely on a late bloomer to provide pollen for an early bloomer. Also make sure the cultivars are cross-compatible, as some cultivars are too closely related to cross-pollinate each other, and some cultivars have sterile pollen.

The male and female flowers of most nut trees, including pecans and walnuts, mature at different times. Some nut cultivars are protandrous, shedding pollen before their female flowers are receptive, while others are protogynous, with female flowers receptive before male flowers shed pollen. Choose at least one protandrous variety and at least one protogynous variety per block to ensure best pollination of nut trees.

TASTE AND STORAGE CONSIDERATIONS

Within the group of fruit and nut plants that are adapted to your area's climate and disease problems, you can choose from many different flavors. Many nursery catalogs glowingly describe all cultivars they sell, so taste or get advice on the flavor of various cultivars from the individual fruit chapters in this book, University trials, other fruit growers, roadside markets, and your county Extension agent. Offering unique flavors can boost sales, especially among retail markets.

Determine when the cultivars you're considering ripen in your area. Planting several types of fruit, and several cultivars of each which ripen at different times, can provide you with a succession of fruit ripening continuously from early summer to late fall, an advantage for retail markets. Wholesale growers who mechanically harvest may prefer a more concentrated ripening season.

Some cultivars are better than others for cooking, canning, or freezing. Some peaches and nectarines turn brown within minutes when cut, while others stay fresh-looking for hours. Some apples and pears store well in a cooler or refrigerator for months, while others keep only a few weeks. Consider these factors, your potential customers' needs, and your storage facilities when choosing cultivars.

STANDARD VERSUS DWARF FRUIT TREES

Tree densities have been increasing in tree fruit and nut plantings for the past 30 years. Planting more trees per acre increases establishment costs, but greatly increases returns in the early years and throughout the life of the planting.

Most nut trees grow very large, so higher-density nut orchards require later removal of half to two-thirds the trees as they begin to crowd each other. In contrast, tree fruit producers maintain all trees permanently smaller in high-density plantings. Some full-size fruit trees can be kept smaller by pruning, but most high-density fruit orchards rely on dwarf trees.

A standard fruit tree is a cultivar grafted onto a rootstock of the same or a closely-related species. Standard fruit trees grow 15 to 35 feet tall and wide, depending on the species. The taller species of standard fruit trees--apple, pear, or sweet cherry--require ladders for hand pruning and harvesting. The smaller species--peaches, nectarines, plums and tart cherries--can be kept low enough by pruning to do most work from the ground.

Dwarf fruit trees grow about 8 to 12 feet tall and wide, and semidwarfs grow 12 to 20 feet tall and wide, depending on the species. These trees produce less fruit per tree than standards, but they can be planted at higher densities and produce much more per acre, and start bearing a year or two sooner. Dwarf trees take less time to prune and harvest per amount of fruit, and allow most work to be done without ladders, greatly reducing labor costs and liability.

All dwarf fruit trees are not created equal. Most are kept small by a dwarfing rootstock, but many different dwarfing rootstocks are used, each with its own assets and liabilities. The individual fruit chapters provide information on rootstocks commonly used for each tree fruit and nut, and the appendix details the rootstocks used by many fruit and nut nurseries.

Choose rootstocks as carefully as cultivars. Some dwarfing rootstocks cause early tree death, others aren't as winter hardy as standard rootstocks for that type of fruit, and others are very susceptible to certain diseases or soil insect pests. Many dwarfing rootstocks have shallow root systems and require staking or trellising, extra fertilizer, and irrigation during drought. A dwarfing interstem grafted between the cultivar and a standard rootstock can prevent these problems, but such three-part trees cost nurseries more to produce and so are hard to find.

A few dwarf fruit trees are genetic dwarfs. These cultivars are naturally small-growing and are grafted onto a standard rootstock of the same or a closely related species. The standard rootstock provides better tree anchorage and water and nutrient uptake than most dwarfing rootstocks, and the tree will probably live as long as full-size trees if the cultivar is adapted to the area where it's planted. The problem with genetic dwarfs is that cultivar selection is limited, and many aren't very winter hardy or disease resistant.

ORDERING NURSERY PLANTS

Always buy fruit plants from a reputable nursery whose plants are state-inspected and guaranteed to be free of diseases when sold. Certified virus-free plants are advisable for the types of fruit most troubled by viruses: raspberries, strawberries, cherries and peaches. A good nursery will also guarantee its plants to be true-to-name, and will replace any plants that fail to grow despite receiving proper care.

Order plants from the fruit nursery as far in advance as possible, to be sure of getting the cultivars and rootstocks desired in the quantities you need

HOW FRUIT AND NUT PLANTS ARE PRODUCED

Fruit and nut cultivars are not grown from seed because these plants possess a great deal of genetic variability. Seeds from a selected plant each grow into a plant producing fruit different from its sister plants and from the parent plant. So to produce plants identical to a selected cultivar, vegetative propagation is used. Strawberries are produced from runners, blackberries and raspberries are grown from tip layers or cuttings, and blueberries and grapes are propagated from cuttings.

Fruit and nut trees are propagated by grafting a small stem piece from the desired cultivar onto a young tree called a rootstock (Photo 1-6). Once the graft "takes", and the cultivar starts to grow, the top of the rootstock is cut off above the graft. The desired cultivar then becomes the top half of the tree, and produces fruit identical to the tree from which it was taken. The rootstock provides the bottom half of the tree, its root system, and can affect tree size, cold hardiness and disease resistance.

An interstem, a stem piece of a third type of tree, is sometimes grafted between the cultivar and the rootstock (Photo 1-7). The interstem may be used between a rootstock and a cultivar that do not graft together well because of genetic differences, or it may help dwarf the tree or increase its cold hardiness.

Photo 1-6. Fruit trees are produced in the nursery by grafting a small piece of the desired cultivar, in this case just a chip containing a bud, onto a small tree called a rootstock. The rootstock will later be cut off above the bud, so that the cultivar will form the top half of the tree and the rootstock supplies the bottom half.

Photo 1-7. Some fruit trees are grafted twice, so that an interstem piece lies between the cultivar and the rootstock, providing dwarfing or cold tolerance. In this case the dwarfing interstem has overgrown the cultivar and rootstock, making its boundaries (marked by arrows) easy to see.

at the proper shipping date. Requests for specific cultivar/rootstock combinations should be made a year in advance.

Most orchardists order one-year-old fruit and nut trees, which have a two-year-old root system and a stem that grew one year after grafting. They are graded by height and caliper, from 3-4 feet tall with a 3/8-inch caliber to 6-7 feet tall with a 3/4" caliber, with the more vigorous trees costing more. Usually, the less vigorous trees, 3 to 5 feet tall, are the best buy. Extra height will be cut off after planting anyway, and less vigorous trees are easier to train to a high-density system. You don't want very low-vigor trees, though, so avoid trees less than 3 feet tall.

One-year-old apples, pears, sweet cherries, European plums, and nut trees are usually whips: single stems with no branches. Many one-year-old peaches, Japanese plums, tart cherries and apricots have branches, but many of these are not positioned well enough to become scaffold branches for the tree, and must be cut off at planting.

"Feathered" trees, with a large caliber and four or more strong branches arising 18 inches or more above the graft union, are often available for a higher price, especially for apples. Planting "feathered" trees can bring an orchard into production a year sooner, but they require better soil preparation and more irrigation to succeed.

Two-year-old trees are sometimes sold at a higher price. These have bigger tops and more branches, but the roots are dug at the same depth, putting more stress on the roots in transplanting. Again, many of those branches will have to be cut off anyway because they'll arise from the trunk at too narrow an angle.

Grapes, blueberries, currants and other bush fruits are usually ordered as two-year-old plants grown from stem cuttings. Brambles are sold as one-year-old plants grown from suckers, root cuttings, layers or tissue culture, and strawberries as one-year-old plants grown from runners or tissue culture.

The nursery should carefully dig all bare-root plants at a depth that preserves most of the feeder roots, gently wash them to remove loose soil, and pack them in a material that keeps roots moist but not soggy. Such nursery plants should be kept in cold storage until sold.

Beware of "bargain"-priced fruit plants. You may get a cultivar that's mislabeled, highly disease susceptible, or not at all suited to your climate. The initial cost of fruit plants is minimal compared to investment of labor and materials to grow them and the value of the fruit produced. High-quality plants, of well-adapted cultivars and rootstocks, will produce the best fruit the soonest and are, in the long run, the true bargain.

Further Reading

Each state's Extension Service offers many bulletins on fruit and nut growing. A Web site that allows access to the Extension Services of all states is:
http://www.hcs.ohio-state.edu/fs/list.html

ATTRA, Appropriate Technology Transfer for Rural Areas, offers printed and downloadable materials on fruit growing, including organic certification. Experts also prepare customized answers to questions and connect people to sources of products and information. ATTRA, P.O. Box 3657, Fayetteville, AR, 72702 Phone: 1-800-346-9140 Fax: 501-442-9842 Web site:
http://www.attra.org

Chapter 2
Site Selection and Preparation

Sites for fruit and nut crops should be carefully chosen and prepared, as site decisions can make or break a planting. Some aspects of soils and topography can not be changed, so choosing a site conducive to fruit plantings is crucial. Other soil factors can be improved inexpensively before planting, while after planting such improvement measures are more difficult or costly.

SOIL SELECTION

Get a copy of your county's soils map from the local Natural Resources Conservation Service office, and use it to compare the soils of all potential sites. This resource contains information on each soil type's drainage, depth, texture, erodability and many other pertinent attributes.

Water drainage and soil depth. Soils for fruit plants need not be especially fertile, but must be well-drained and fairly deep. Root rots and plant death occur where soils remain saturated with water too long. This means not only to avoid low areas where water accumulates during rains, but also to check the soil's internal drainage.

First consult your county's soils map, to give you an idea of where to look for deep, well-drained soil. Then check the existing vegetation on your proposed sites. Short, weak growth may indicate shallow areas in the field, and patches of water-loving plants such as smartweed, barnyardgrass or sedges can indicate poorly drained areas.

Check soil depth at several sites, using a soil auger or post-hole digger. Check down at least two feet deep for small fruits, and four feet deep for tree fruits, observing soil color and texture and how these change as you dig. Deep, well-drained soils are generally uniform in color with no streaks, with root growth throughout the soil profile (Photo 2-1) and no restrictive layers (heavy clay, hardpan, bedrock or water table). Poorly drained soils have grayish or bluish streaks at some point before you've dug three feet deep (Photo 2-2). A soil may be poorly drained because it is mostly tightly-packed clay without enough air spaces, or it may be looser near the surface but have a clay layer beneath that restricts drainage below that point.

To be sure soil drainage is adequate, fill a two-foot-deep hole with water and let it all drain out. Then fill the hole with water again and check how long it takes to empty again. If the water is gone in

Photo 2-1. Choose a site with deep, well-drained soil, as evidenced by deep plant roots and no mottling.

Photo 2-2. Avoid a poorly-drained site, as evidenced by rusty-colored mottling in the subsoil. This site also contains a plowpan, seen as the gray-brown line beween the upper, light brown plow layer and the darker brown soil beneath. Break up a plowpan with a subsoiler before planting.

Photo 2-3. Build raised beds if the soil is only moderately well drained, to prevent root rots. This bedmaker was built by modifying a disk harrow.

less than 10 hours, soil drainage is good. If water remains in the hole after 10 hours, the site is only moderately well-drained and should be improved as outlined below. If water still remains after 24 hours, the soil is poorly drained and you should find another spot.

To improve drainage, you can install tile drains or plant on raised mounds or beds, six to eight inches high and as wide as the mature plant's spread. Many growers contruct raised beds using a moldboard plow, modified disk harrow or other equipment to move topsoil from the future row-middles into the row centers (Photo 2-3). You'll also need to build beds or mounds for fruit trees if your soil isn't at least three feet deep for trees on dwarfing rootstocks, or at least five feet deep for standard-size trees. Be sure to construct raised beds so that surface water drains along the row middles and out of the field. Plantings on raised beds will also require irrigation systems, because the beds dry out quicker during summer droughts even though the site drains water slowly during wet winters or springs.

Soil Texture. Soil texture--the percentage of sand, silt and clay in a soil--isn't as important as good drainage, but does influence the growth of fruit plants. Sandy soils have excellent drainage but also dry out quickly during summer droughts, making irrigation a necessity. Sandy soils also hold fewer soil nutrients, so require higher fertilizer levels. Clay soils, with their finer particles that pack tightly, can easily become compacted and restrict root growth and water drainage. The fine clay particles store many nutrients but also hold water tightly during a drought, releasing less of it to plants than the intermediate-sized silt particles.

Generally, the best soils for fruit crops are loams or silt loams, mixtures of 10-50% sand, 30-80% silt, and 10-25% clay. Sandy loams and sandy clay loams, with a sand percentage up to 70%, or clay loams and silty clay loams, with a clay percentage up to 40%, are also acceptable. Clay and silty clay soils are not advisable except for a few crops such as pecans, while sand and loamy sand soils are suited to only a few crops such as blueberries.

Certain fruit plants and rootstocks tolerate clayey soils well and other withstand sandy soils better. In general, blueberries, peaches, almonds, sweet cherries, Japanese plums, and apricots do better in sandier soils. Grapes, apples, pears, tart cherries, European plums, walnuts and pecans do better in heavier clay soils. Strawberries and raspberries do best in loamy soil.

Adding organic matter is the best way to improve both sandy soil and clayey soils. Organic matter additions help sandy soils hold more water, and loosen up tightly-packed clay soils. High organic matter levels in any soil type make soil nutrients, especially micronutrients, more available to plants, and stimulate naturally-occurring organisms that help to control potential plant diseases. The best way to increase soil organic matter is to grow green manure crops before planting the fruit plants, as outlined below under Site Preparation.

Soil pH and nutrients. Soil pH refers to the acidity or alkalinity of the soil solution (Figure 2-1).

In a neutral soil, the concentration of acid hydrogen ions [H^+] equals the concentration of alkaline hydroxyl ions [OH^-], so they cancel each other out. The pH measurement refers to the negative exponent of the logarithm of the concentration of hydrogen ions [H^+]. In a neutral soil, [H^+] = [OH^-] = 1/10,000,000 water molecules (H_2O) = 10^{-7}, so pH = 7.0.

A pH value less than 7.0 indicates an acid soil, while a pH value greater than 7.0 indicates an alkaline, or basic, soil. The logarithmic nature of the scale means that small pH value differences can indicate great changes in acidity or alkalinity. A soil with a pH of 6.0 has ten times more hydrogen ions and ten times less hydroxyl ions than a neutral soil, and a soil with a pH of 5.0 has 100 times more and less of each (Figure 2-2).

The pH has a large effect on nutrient availability and plant growth (Figure 2-3). Most fruit plants grow best with a slightly acid soil, pH 6.2 to 6.8. This pH range maximizes the availability of most plant nutrients, and reduces the availability of certain metals (especially aluminum and manganese) which may be needed in small quantities but are toxic in large quantities. Many fruits can also be grown in neutral to slightly alkaline soils, up to pH 8.0 or 8.5, but will require fertilization with certain nutrients such as zinc or boron that are less available at higher pH's.

Blueberries, cranberries and lingonberries are native to acid, sandy, peaty soil with lots of water, so they need a pH of 4.2 to 4.8. These plants have developed a high need for iron, which is more available at lower pH's, and show severe iron deficiency when grown in near-neutral soils. These acid-loving plants are grown commercially mostly in areas that naturally provide the soil conditions they need. Other soils can be modified to suit them, but the process is expensive.

Soil pH can be adjusted before planting, if needed, as described below under Site Preparation.

Current and previous vegetation. Choosing a site where no woody plants grew for the previous five years can avoid certain soil-borne diseases that can't be cured. Oak root rot (also known as mushroom root rot or Armillaria root rot) attacks many woody plants and can survive on pieces of roots or wood in the soil, to later attack fruit plants. Newly-cleared oak forest land is the most prone to infection. Verticillium wilt and crown gall also attack many woody plants and can be devastating to fruit plants.

Verticillium wilt, fusarium wilt, and root knot nematodes can also build up on susceptible annual

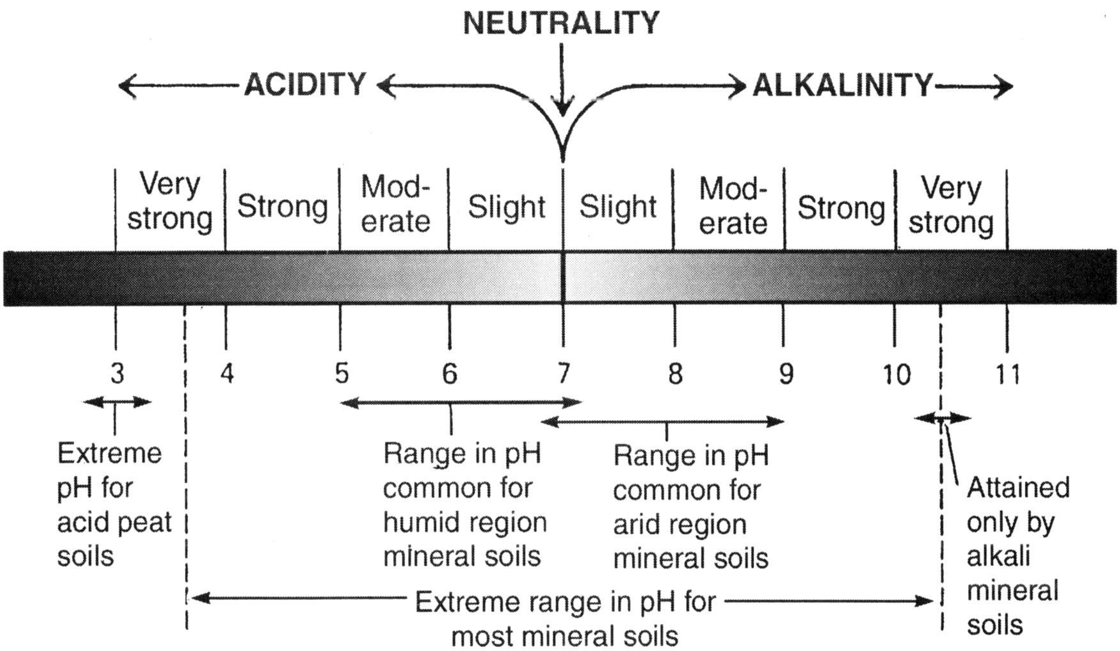

Figure. 2-1. Soil pH scale. Taken from *The Nature and Properties of Soils, 8th Edition*, by Nyle C. Brady, MacMillan Publishing Co.

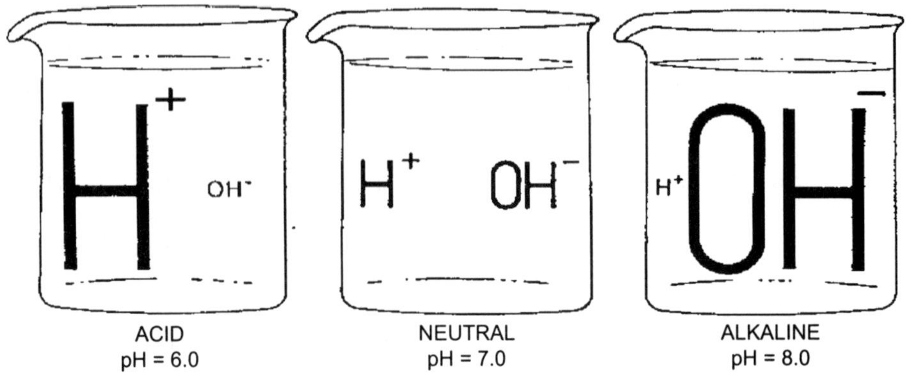

Diagrammatic representation of neutrality, acidity, and alkalinity. At neutrality (center) the H ions and OH ions of a solution are balanced, their respective numbers being the same (pH7). (Left) At pH 6, the H ions are dominant, being 10 times greater, while the OH ions have decreased proportionately, being only 1/10 as numerous. The solution, therefore, is acid, there being 100 times more H ions than OH inons present. (Right) At pH 8, the exact reverse is true, the OH ions being 100 times more numerous than the H ions. hence, the solution is alkaline. This mutually invrse relationship must always be kept in mind using pH data.

Figure 2-2. Diagramatic representation of soil acidity, neutrality and alkalinity. Taken from *The Nature and Properties of Soils*, 8th Edition, by Nyle C. Brady, MacMillan Publishing Co.

vegetables crops, including melons and peppers, and on common broadleaf weeds such as lambsquarters, pigweed and nightshade. Old orchard sites have probably built up several viruses and nematodes, and are best rotated out of fruit plants for several years.

The ideal site was planted for several years to a non-invasive grass pasture and hay crop, ideally fescue, or to grass crops such as wheat, sorghum or corn. These grass crops reduce levels of harmful nematodes, fungi, bacteria and viruses in the soil, and increase organic matter and helpful microorganisms.

TOPOGRAPHY AND AIR MOVEMENT

The topography of the land greatly affects the local climate, or microclimate. Carefully choosing a favorable site can enable growers to produce crops that won't do well in the surrounding region.

Air drainage. Sites for most fruit and nut crops must provide good air drainage as well as water drainage. This means that colder air flows out of the area into a lower area. Avoid "frost pockets", where cold air settles (Figure 2-4), as these cause more winter damage and spring frost damage to fruit buds. Also avoid windswept ridgetops. Gentle to moderate slopes are generally best (Photo 2-4).

Fruit and nut crops are a good choice for sloping land, since they are perennials grown with a sod groundcover that holds the soil against erosion. Gentle slopes of up to 10% pose no problem, although slopes greater than 33% make equipment operation difficult. Even such steep slopes can still be utilized for trees if they are terraced and

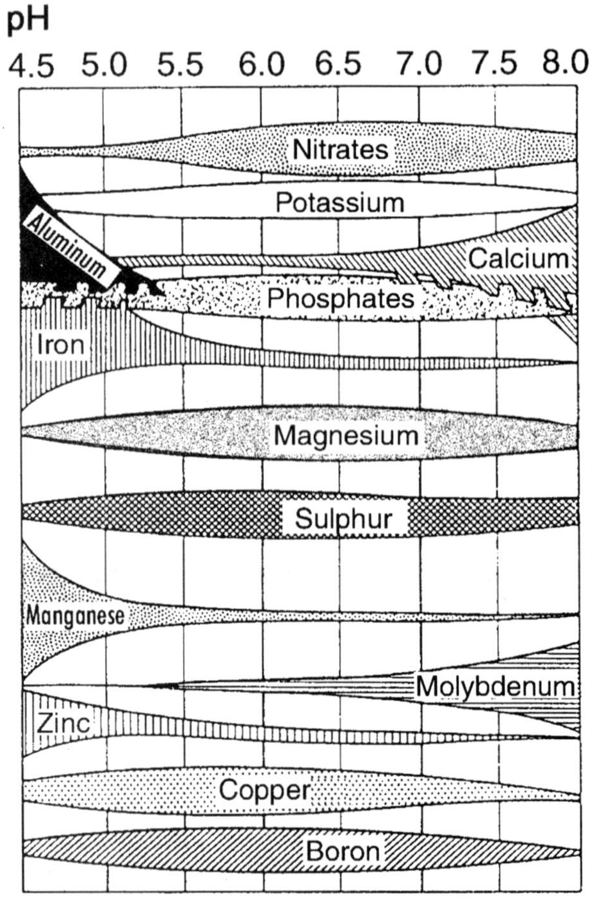

Figure 2-3. Relative availability of 12 essential plant nutrients, and the potential toxin aluminum, in mineral soils in temperate regions as influenced by soil pH. Much phosphate is tied up by aluminum and iron at low pH values, and by calcium at high pH values. Courtesy of Univ. of Kentucky.

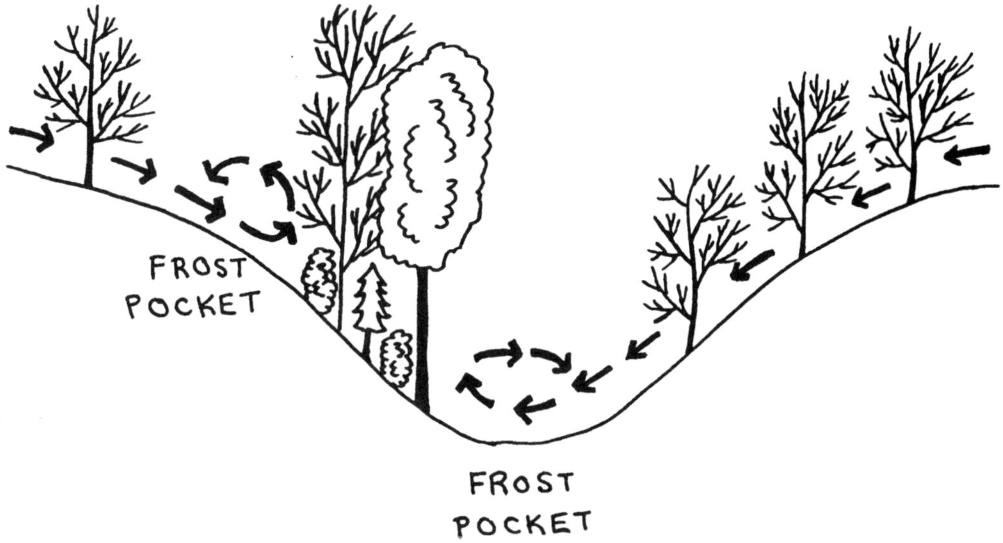

Fig. 2-4. Avoid "frost pockets", where cold air settles, for fruit planting sites.

the trees are planted on the contour across the slopes.

Slope direction. A north-facing slope is a good site for early-blooming, frost-sensitive crops such as peaches, Japanese plums, sweet cherries, and apricots. The shadow cast by the slope in the winter keeps the plant cooler, which delays bud development and bloom in the spring, while the higher summer sun angle provides enough light during the growing season. Fruit ripening will also be delayed on a north slope.

A south-facing slope receives more intense light and heat, and so hastens both bloom and harvest. In areas with short or frequently cloudy growing seasons, late-blooming and late-ripening varieties of apples, pears, European or American hybrid plums, or grapes do better on a south-facing slope. Where summers are very hot and dry, avoid southwestern or western slopes, or give the fruit plants plenty of irrigation. East-facing slopes can be beneficial because the morning sun will dry dew off quickly, minimizing diseases, while the late afternoon heat can be less intense.

Photo 2-4. A site with a gentle to moderate slope offers the best air drainage, important in avoiding frost damage.

Nearby vegetation and bodies of water. In areas with strong winds, choose a site protected by existing plantings, or plant a fast-growing windbreak (see sidebar). Windbreak trees can also serve as border plantings or screens, provide shelter and nesting sites for insect-eating birds, and may lure pest birds away from fruit trees. Avoid areas where existing trees impede air drainage out of the proposed fruit site.

Bodies of water, such as farm ponds and lakes, are slower to cool in fall and to warm in spring than the air. These water bodies moderate the temperatures of the land over which the wind blows off the water, producing prime sites for fruit plantings. Such areas will have warmer winter temperatures, less spring frosts, and cooler and more humid summers. The larger the body of water, the greater the effect, and the greater the choice of excellent fruit-growing sites.

SITE PREPARATION

Once you've chosen your site, have the soil tested for pH and levels of important nutrients, preferably the season before you intend to plant. Your Extension agent can provide information on procedures for taking soil samples and sending them to a soil testing lab. If necessary, add ground limestone (calcium carbonate) to raise the pH or

Planting a Windbreak

Strong winds increase fruit plants' moisture needs and can whip trees around in their planting holes, weakening them and leaving them vulnerable to pest attack. Heavy spring winds can interfere with pollination by desiccating blossoms and hindering bee movement. Strong summer or fall winds can knock fruit off trees, while winter winds can increase cold damage.

In many parts of the United States, fall and winter winds usually blow from the northwest, and spring and summer winds usually blow from the south. If winds become strong during any season in your area, consider planting a windbreak on the side from which the worst winds come, to reduce the brunt of the windiest season. In most areas, plant a summer windbreak to the south of your fruit planting, as many feet away as the windbreak trees grow tall, to avoid shading the fruit planting.

A summer windbreak for fruit trees can be just one row of fast-growing but long-lived, drought-tolerant, deciduous trees (Photo 2-5). Small (15- to 25-foot) windbreak trees, such as redbud, Russian olive, white (Chinese) or black mulberry, or serviceberry, will provide enough protection for small fruits or dwarf fruit trees. Medium-size (35- to 50-foot) windbreak trees, such as thornless honey locust, red mulberry, or lacebark (true Chinese) elm, will be needed for standard-size fruit trees. Cottonwoods also provide a fast-growing windbreak, but must be planted farther from the fruit trees as they grow taller.

Winter windbreaks, planted to the north and west of the fruit planting in most areas, should contain a row of evergreens, usually pines or spruce. Again, plant the windbreak as many feet away from the fruit planting as the windbreak trees grow tall. Don't box in fruit plantings with windbreaks on all sides; that could stop air movement, which can aggravate frost and disease problems.

Avoid planting trees in the windbreak that could spread diseases in your fruit planting, such as Eastern red cedar (alternate host of cedar-apple rust), scab- or fire-blight-susceptible crabapples, hawthorn, mountain ash, or ornamental cherries. Also avoid overly competitive, brittle-wooded, or short-lived trees such as Lombardy poplar, hackberry, green or white ash, silver maple, catalpa, or Siberian elm (often erroneously called Chinese elm).

Photo 2-5. Where summer winds are strong, plant a single row of fast-growing but long-lived deciduous trees as a windbreak. These poplars, behind the trellised pears,are starting to die out; longer-lived lacebark elm or red mulberry would do better.

sulfur to lower the pH, in amounts specified by your soil test results. Tables 2-1 and 2-2 show approximate amounts of lime or sulfur to add, based on the current and desired pH and the soil texture. Add phosphate or potash fertilizers if phosphorus or potassium levels are low. All these materials move slowly through the soil, so are best added, and worked into the soil to a depth of 6 to 10 inches, the season before planting (Photo 2-6).

Photo 2-6. Add phosphorous and potassium, along with lime or sulfur if needed to adjust pH, the season before planting fruit crops, and work these ammendments in to a depth of 6 to 10 inches.

Table 2-1
LIME NEEDS OF VARIOUS SOILS

Topsoil texture	Current topsoil pH			
	4.5 to 4.9	5.0 to 5.4	5.5 to 5.9	6.0 to 6.4
Sands	2.5	2	1.5	1
Loamy sands	3	2.5	2	1
Sandy loams	4	3	2.5	1.5
Loams & clay loams	5	4	3	2
Clays & silty clays	6	5	4	2

Approximate tons of lime, with a calcium carbonate equivalent of 90%, to add per acre to raise soil pH to 6.8. Less pure lime requires greater amounts; for example, lime with a calcium carbonate equivalent of 60% requires 1.5 times the tons listed.

Table 2-2
SULFUR NEEDS OF VARIOUS SOILS

Desired soil pH	Current topsoil pH						
	5	5.5	6	6.5	7	7.5	8
4.5 – blueberries							
sandy	174	349	523	654	828	1002	
loamy	523	1046	1526	2000	2529	3008	
6.0 – brambles							
sandy					349	523	654
loamy					1046	1526	2000
clays					1134	1613	2093
7.0 – tree fruits							
sandy						174	349
loamy						523	1046
clays						610	1134

Approximate number of pounds of sulfur to add per acre to lower soil pH to the level needed.

Control any perennial weeds on the site, such as johnsongrass or bermudagrass, before planting. These invasive grasses can be killed with herbicides, or with two years of tilling in weed-suppressive cover crops. The most effective weed-smothering crops in Oklahoma trials included sorghum-sudangrass hybrids for a summer crop and rye with hairy vetch for a winter crop. The vetch, a legume, also adds nitrogen to the soil. Dr. Marvin Pritts found that no-till planting of strawberries in New York, into a rye cover crop killed with glyphosate (Roundup™), effectively controlled an infestation of nutsedge, which is extremely difficult to control after planting.

Another way to add nutrients and organic matter to the soil is to turn under composted organic wastes. Some states are starting to subsidize farmers for applying composted municipal wastes to their soil. Test both the soil and the compost before applying it, to be sure the compost isn't too high in salts or metals.

After turning under cover crops or existing vegetation and soil ammendments, shape the soil into raised beds before planting most small fruit crops in many areas, to ensure adequate soil drainage. Before or after planting the fruit or nut crop, plant a less competitive groundcover between rows of all fruit and nut plants except strawberries. Creeping red fescue or Kentucky-31 tall fescue are widely recommended, and can be combined with white clover or subterranean clover for nitrogen production if pest bugs are managed (see chapter 5).

Where a suitable sod, preferably fescue, already exists on the site and raised beds aren't needed, simply kill the grass sod in a four-foot wide strip down each row, using glyphosate (Roundup™) herbicide. Then plant the fruit or nut crop directly into the killed sod (Photo 2-7). The killed grass roots decompose, providing organic matter and leaving small pores in the soil that aid aeration and water infilatration. The killed grass can trigger up to 45% more growth on young fruit trees than planting in bare soil, researchers have found.

NEMATODE AND DISEASE MANAGEMENT STRATEGIES.

If plant-pathogenic nematodes (microscopic soil worms) are a problem in your area and your planting site contained broad-leaved weeds or plants, test the soil for nematodes before planting. If levels are high, follow one of the programs below. Also follow one of these programs if replanting an old orchard site or a site previously infected with Verticillium, Armillaria or other soil pathogens.

Soil fumigation. Chemical fumigants, injected into the soil to kill plant-parasitic nematodes and fungi, have been widely used in site preparation where previous fruit crops grew on the site or where known pathogens occurred. These chemicals are being used less and less, due to the following problems:

- Many of the chemical fumigants contain halogens, chlorine or bromine, which are highly volatile and accumulate in the upper atmosphere, where they can damage the ozone layer. Consequently, these chemicals are being withdrawn from use or restricted.
- Chemical fumigants are broad-spectrum biocides, killing many beneficial soil microorganisms as well as plant pathogens. Studies have found that levels of harmful nematodes and fungi rebound to higher levels three or four years after fumigation than on land that was not fumigated, due to the decline of beneficial organisms that competed with or helped control the pests.
- Fumigants also kill beneficial mycorrhizae that help plants absorb water and nutrients, leading to nutritional deficiencies and greater susceptibility to drought.
- Fumigants are extremely toxic to humans and dangerous to applicators.
- Application of a fumigant adds $850 to $2200 per acre to the cost of establishing a fruit planting, and can only be done when temperature and moisture conditions are ideal, which is a narrow time period in many locations.

Photo 2-7. Planting into a killed sod, when possible, allows minimal soil disturbance and leads to greater tree growth. Fewer weed seeds grow when the soil isn't disturbed, and the killed sod acts as a mulch, further supressing weed growth and conserving water.

Soil fumigants are most used in plasticulture strawberries grown as annuals and replanted year after year on the same site, in California and several southern states. Soil fumigants are also still widely used in fruit nurseries, to produce disease-free plants for growers.

Several newer, safer chemicals are under test to replace methyl bromide and other fumigants being banned. These include metam sodium and sodium tetrathiocarbonate (STTC, trade name Enzone), which have proved effective at reducing pathogenic soil fungi and nematodes.

Soil solarization. High soil temperatures can kill many disease-causing organisms and weed seeds, yet research has shown they do not kill many beneficial soil microorganisms. Such high temperatures can be achieved by applying sheets of 0.6-mil clear plastic film to cover the soil for six to eight weeks during the hottest part of the summer before fall or spring planting. Remove the plastic before planting most fruit crops.

Soil solarization by itself does not reduce weed populations as well as it reduces plant disease agents. Increase weed kill during solarization by working in manure and organic materials before applying the plastic. These materials encourage the buildup of beneficial microbes which begin to compost the materials, causing it to heat the soil to higher temperatures and release methane gas. The methane, trapped in the soil by the plastic, kills weed seeds.

Plastic application is especially feasible for strawberry or vegetable growers already equipped to apply rolls of plastic to the soil (Photo 2-8). For

Photo 2-8. Growers who are equipped to apply rolls of plastic can reduce soil-borne diseases by solarizing the soil for several summer months before planting. (Photo from Univ. of California.)

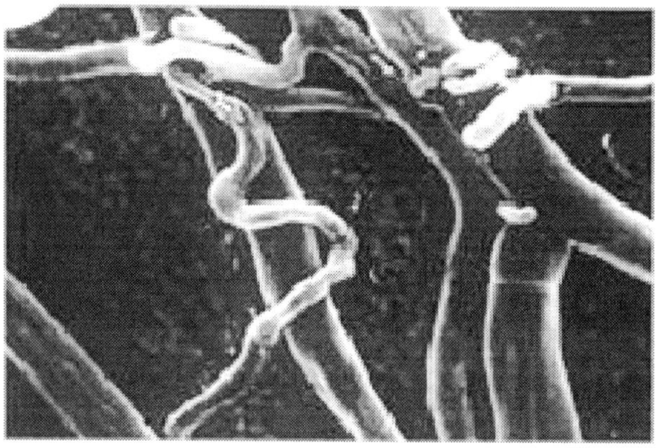

Photo 2-9. Certain beneficial soil fungi, including species of *Gliocladium* and *Trichoderma*, parasitize plant pathogenic fungi by strangling them and releasing antibiotics into the soil. This *Trichoderma* coils around the plant pathogen *Rhizoctonia solani*. Commercial products containing these beneficial fungi are now available. (Photo by Ilan Chet, Weizmann Institute of Science, Israel.)

a plasticulture strawberry grower, preparing the planting beds by late spring can allow soil solarization over the summer, before installing black plastic over the clear plastic and planting into the plastic in early fall.

Most plasticulture strawberry growers treat strawberries as annuals, and remove the planting after the bearing period ends in early summer. Where growers wish to keep the planting for several years and summer temperatures reach high levels, clear or black plastic sheets must be painted white or covered with several inches of organic mulch to avoid damaging plants.

Reducing soil pathogens with biocontrol agents. Researchers are discovering, and developing as pest control products, more and more natural enemies of soil-borne plant pathogens. One such product is DiTerra from Abbott Laboratories (1-800-323-9597; www.abbott.com), composed of proteins, sugars and lipids produced by Myrothecium verrucaria, a hyphomycete fungus. DiTerra kills plant-parasitic nematodes, in the soil and in plant roots, on contact, when applied at 60 to 100 pounds per acre and incorporated at least one week before planting. It has no effect on free-living nematodes that feed on soil microorganisms such as plant pathogenic fungi. DiTerra effectively controls many nematodes troublesome to fruit growers, including lesion, root knot and sting nematodes. Currently, it is only registered for use on grapes and berry crops among the deciduous fruit crops.

Other biocontrol products inhibit fungi that can cause root rots and plant death. SoilGard biofungicide, a formulation of the beneficial fungus Gliocladium virens strain GL-21 from Thermo Trilogy, Inc. (1-800-847-5620), reduces populations of several plant pathogenic fungi, including Pythium and Rhizoctonia. Root Shield and T-22 Planter Box are both formulations of Trichoderma harzianum strain T-22, and reduce Fusarium, Pythium and Rhizoctonia. Mycostop, a formulation of the actinomycete Streptomyces griseoviridis strain K61, inhibits all of the pathogens listed plus Alternaria, Botrytis, Phomopsis and Phytophthora. BioWorks (www.bioworksbiocontrol.com) makes RootShield and T-22 Planter Box, and Geiger, Inc. (1-800-443-4437; www.ecgeiger.com) distributes these plus Mycostop. All of these benefical fungi produce antibiotics that kill pathogens, and Gliocladium and Trichoderma also wrap their strands around the plant-pathogenic fungi (Photo 2-9), strangling them and siphoning nutrients from them so that they can't attack plants.

These biocontrol products cost more than chemical fumigants, but do not harm the environment and may persist in the soil, especially if organic matter levels are raised with cover crops, offering extended control of soil pathogens. Adding composted organic wastes is another way to both build up organic matter and provide beneficial microorganisms that can inhibit soilborne plant pathogens. Ohio plant pathologist Harry Hoitink has found that apple growers can prevent Phytophthora collar rot for five years after planting by applying ½ to 1 bushel of composted hardwood bark per tree at planting.

Choosing rootstocks or cultivars resistant to soil pathogens. Rootstocks that resist harmful nematodes or fungi have been developed for several fruit and nut tree species, and certain cultivars of small fruits also resist soil-borne diseases. Where such diseases present problems, choose resistant plants in addition to modifying or treating the soil to reduce the pathogen. 'Guardian', 'Nemaguard' and 'Nemared' peach rootstocks resist root-knot nematodes, but 'Lovell' peach rootstock is more tolerant of lesion and ring nematodes, which do more damage than root-knot nematodes in some areas. Several clones of paradox walnut rootstocks, and several apple rootstocks, such as MM111, resist Phytophthora collar rot.

Some red raspberry cultivars resist Phytophthora root rot, a common disease which can kill susceptible plants, especially on less well-drained soils. Many strawberry cultivars resist red stele and Verticillium wilt. Planting on raised beds, in addition to choosing resistant cultivars, helps for both raspberries and strawberries (Photo 2-10).

Using cover crops to reduce soil pathogens. Adding organic matter to a soil can enhance beneficial microorganisms and reduce plant pathogens, in addition to improving soil fertility and structure, moisture-holding capacity and drainage of excess water. Turning under a green manure crop or two before planting helps a great deal.

Root-rotting soil fungi such as Cylindrocarpon, Phytophthora, Pythium and Rhizoctonia often have built up in the soil if it is poorly drained or was in fruit plants for many years. Dr. Mark Mazzola, a USDA plant pathologist, found that these fungi are the primary cause of apple replant disease in Washington state, rather than lesion nematodes as was previously thought. Growing three quick cover crops of 'Eltan' soft white winter wheat, for three weeks each, before planting apples in replant sites significantly increased the young trees' growth and suppressed root rot.

Using small grain cover crops, and building raised beds, before planting raspberries or strawberries can reduce similar root rots. Connecticut researchers Wade Elmer and James La Mondia demonstrated that a turning under a cover crop of 'Saia' oats before planting strawberries reduced lesion nematodes and black root rot severity. A cover crop of 'Triple S' sorgho-sudangrass reduced the nematodes only, and 'Garry' oats reduced the number of roots blackened by Rhizoctonia fragariae only, while 'Saia' oats reduced both. New England strawberry growers often rotate strawberries with oats to provide a source of mulch, so such cover crops fit easily into the overall farm plan.

Photo 2-10. This cultivar of strawberry, 'Earliglow', resists red stele and Verticillium wilt, two common soilborne diseases. Raised beds further reduce incidence of these diseases.

Some small grains and other grass crops, if grown before planting fruit crops, also reduce harmful nematodes. Dr. John Potter at the Ag Canada Research Station in Vineland, Ontario has shown that 'Kentucky 31' tall fescue and 'Roland 21' red fescue reduced lesion nematodes the most, while wheat, barley, corn, clovers, buckwheat and other cover crops allowed lesion nematode populations to increase. Using a wheat-sorghum double-crop rotation for a year before planting peaches greatly reduced the number of ring nematodes in the soil in Georgia, which in turn reduces cold damage, fungal and bacterial cankers, bacterial spot, and other diseases in the peaches. A wheat or other grass intercrop between peaches or other fruit trees does not reduce harmful nematode populations, so the grass crop must be grown for at least a year before planting the fruit crop.

Where the soil was already in grain crops for several years, a good approach is to plant canola (rapeseed) in August, then French marigolds or black-eyed susan after turning under the canola in April. Then plant the fruit crop, with a creeping red fescue or tall fescue groundcover for all crops except strawberries, after turning under the marigolds the second fall. Research has demonstrated that canola, marigolds and black-eyed susan all produce compounds toxic to damaging nematodes. Broccoli, a relative of rapeseed, also reduces nematode populations and Verticillium wilt, so vegetable growers could plant fruits following a broccoli crop.

On old orchard sites, sites that have tested high in nematodes, or fruit nurseries, several years of certain cover crops can reduce nematodes and other pathogens as well as a fumigation treatment. Follow this sequence, developed by Dr. Paul Steiner, Extension Fruit Pathologist, University of Maryland:

1) October, year 1: Remove old fruit plants and roots. Sample soil and apply lime (or sulfur) if needed. Plant winter wheat, barley, rye or oats two and a half years before spring planting or two to three years before fall planting.
2) April, year 2: Disk under the winter cover crop, fertilize with 50 lb./A actual nitrogen and plant sudex, a sorghum-sudan grass hybrid that quickly produces a large amount of biomass and roots that penetrate four to six feet deep (Photo 2-11).
3) July & August, year 2: Mow the sudex, fertilize with ammonium sulfate (75 to 100 lbs./A) for additional nitrogen and sulfur for the next cover crop. Mow again a month later, add another 50 to 75 lb./A ammonium sulfate, then disk under. Two weeks later, plant 'Dwarf Essex' rapeseed (canola) or another brassica that produces sulfur compounds toxic to nematodes.
4) April, year 3: Mow and disk under the rapeseed in small sections, so each section is turned under within two hours of mowing. Otherwise, the volatile compounds effective against nematodes escape into the air. Two weeks later, add another 50 to 75 lb./A ammonium sulfate and plant a second rapeseed crop. Tests have shown that two successive canola crops are equivalent to a chemical fumigation treatment.
5) August, year 3: Collect and submit soil samples. Mow and immediately disk under the canola as before. Two weeks later, add any lime and nutrients needed for fruit crop, plus 15 to 20 pounds actual nitrogen, NOT in ammonium sulfate form.
6) Sept. year 3 to April, year 4: If planting tree fruits or nuts or grapes, plant Kentucky-31 tall fescue or creeping red fescue between future rows in September to further suppress nematodes and serve as permanent groundcover, along with winter wheat or winter oats as a nurse crop.

Plant trees in October to November or March to April. For small fruits, form raised beds, if needed, and plant fruits in fall or spring.

Photo 2-11. Cover crops of sudex, a sorghum-sudangrass hybrid, produce a great deal of biomass, which helps to build populations of beneficial microorganisms that break it down and suppress many soil pathogens. (Photo from Univ. of Florida.)

Further Reading

Natural Resources Conservation Service offices in each county offer free soil survey maps for that county.

Blok, W.J., J.G. Lamers, A.J. Temorshuizen, and G.J. Bollen. 2000. Control of soilborne plant pathogens by incorporating fresh organic amendments followed by tarping. *Phytopathology* 90:253-259.

Cook, R.J. 1982. Use of pathogen-suppressive soils for disease control. p.51-88. In: R.W. Schneider (ed.). *Suppressive soils and plant disease.* APS Press, St. Paul, MN.

Elmer, W.H. and J.A. LaMondia. 1999. Influence of ammonium sulfate and rotation crops on strawberry black root rot. *Plant Dis.* 83:119-123.

Nyczepir, A.P., P.F. Bertrand, and B.M. Cunfer. 1996. Suitability of a wheat-sorghum, double-crop rotation to manage Criconemella xenoplax in peach production. *Plant Dis.* 80:629-632.

Soil solarization, a nonchemical method for controlling diseases and pests. University of California Cooperative Extension leaflet no. 21377.

Steiner, P.W. 1999. Orchard site bio-renovation. *Small Farm*, Nov.-Dec., 44-46.

Sztejnberg, A., S. Freeman, I. Chet and J. Katan. 1987. Control of Rosellinia necatrix in soil and in apple orchard by solarization and Trichoderma harzianum. *Plant Dis.* 71:365-369.

Widmer, T.L., J.H. Graham and D.J. Mitchell. 1998. Composted municipal waste reduces infection of citrus seedlings by *Phytophthora nicotianae. Plant Dis.* 82:683-688.

Wilcox, W.F., M.P. Pritts and M.J. Kelly. 1999. Integrated control of Phytophthora root rot of red raspberry. Plant Dis. 83:1149-1154.

Chapter 3
Establishing New Plantings

Fruit plants can be planted in fall or early spring, as dormant plants. Planting in late fall, when nursery plants are available, enables roots to become better established before new leaves start requiring water in spring, and avoids the common problem of frequent spring rains leaving the soil too wet to plant. However, fall planting risks winter injury to the not-yet-settled plants. Fall planting works well for the hardier fruits and nuts, including apples, pears, pecans, black walnuts, raspberries, currants and gooseberries, in zones 6 and warmer. More cold-tender fruits and nuts, such as stone fruits, Persian walnuts, grapes, strawberries, blackberries and blueberries, may be planted in fall in zones 8 and warmer.

LAYOUT OF THE PLANTING

Small fruits are generally planted in hedgerows, with an in-row spacing just wide enough to allow plants to fill out well and touch, forming a solid hedge (Photo 3-1). The width of tractors and equipment for mowing and spraying determines the between-row spacing. Traditionally this has been 10 to 12 feet wide, but newer compact equipment may enable narrowing that distance to 9 or even 8 feet in some cases. The closer the between-row spacing, the greater the number of plants that can be planted per acre, and the greater the yield potential. However, if irrigation water is limited, increase the spacing between rows, and slightly increase the in-row spacing. Trellised or staked dwarf fruit trees, and some smaller freestanding trees, are also planted in hedgerows, with in-row spacings as close as five to ten feet apart.

Lay out hedgerows so the rows run north to south, to optimize light interception and consequent yields of both small fruits and dwarf fruit trees. A north-south row orientation ensures that when shadows are longest, early and late in the day, such shadows fall between the rows rather than over them. When the sun is due south in the middle of the day, its higher angle minimizes in-row shading.

Photo 3-1. Plant small fruits, such as these pick-your-own blueberries, in hedgerows, so the plants within the row touch. Between-row spacing should be wide enough to accommodate equipment.

Standard and semidwarf fruit trees are usually planted in a square or rectangular pattern. Larger trees, including most nut trees, are often planted in a square pattern, because equipment will be run under the trees. Medium-sized trees with lower fruiting branches are generally planted close enough within the row so that trees will touch, and about six feet further apart between rows to allow room for equipment.

On erodable slopes, orient rows along the contour. Very steep slopes may require constructing a terrace for each row.

Figure the number of plants to buy as shown.

$$\text{\# plants per acre} = \frac{43{,}560}{\text{(ft. between rows) X (ft. between plants in row)}}$$

Table 3-1 illustrates planting densities at some common spacings.

BED PLANTING SYSTEMS

Bed plantings group two or more rows of plants into a bed which will not be serviced by larger equipment, with wider spacings between the beds to allow such equipment access. Bed plantings allow a greater number of plants per acre, thus increasing yields per acre.

Strawberries planted for same-season harvest (which includes most California plantings as well as many plantings in the Southeast) are often planted in two-row beds, and sometimes four-row beds. Matted-row strawberry plantings, the traditional system in most areas for harvest over several years starting the year after planting, utilize runner plants to form a bed the width of two to three rows.

Orchards can also be laid out in beds of two to three rows at narrow spacings, with wider spacings between beds to accommodate sprayers and bin-handling equipment (Figure 3-1). Spacing between rows in the bed can be the same distance as that between plants in the row, or just a few feet wider. Such bed systems can increase tree density per acre (Table 3-2), if the spraying equipment can provide good penetration of the bed. Two sets of some equipment may also be needed, to accommodate the varying row widths within and between beds. Smaller-width mowers will likely be needed to mow between rows within the beds, in addition to wider-width mowers to mow between the beds.

HIGH-DENSITY ORCHARD SYSTEMS

Higher-density tree plantings require greater establishment costs and greater manipulation of plants, but greatly increase returns, especially in

Table 3-1

NUMBER OF PLANTS PER ACRE

	Within rows																	
Between rows	2'	3'	4'	5'	6'	7'	8'	9'	10'	12'	14'	16'	18'	20'	22'	24'	30'	35'
4'	5445																	
6'	3630	2420																
8'		1815	1361	1089	908	777	680											
9'		1613	1210	968	807	691	605	537										
10'		1452	1089	870	726	622	544	484	435									
11'			990	792	660	565	495	440	396									
12'			908	726	605	518	453	403	363	302								
14'				622	518	444	388	345	311	259	222							
16'				544	453	388	340	302	272	226	194	170						
18'					403	345	302	268	242	201	172	151	134					
20'							272	242	217	181	155	136	121	109				
22'									198	165	141	124	110	99	90			
24'										151	130	113	101	91	83	76		
26'											120	105	93	84	76	70		
28'											111	97	86	78	71	65		
30'													81	73	66	61	48	
35'																	41	36

Table 3-1. Number of plants per acre at various spacings

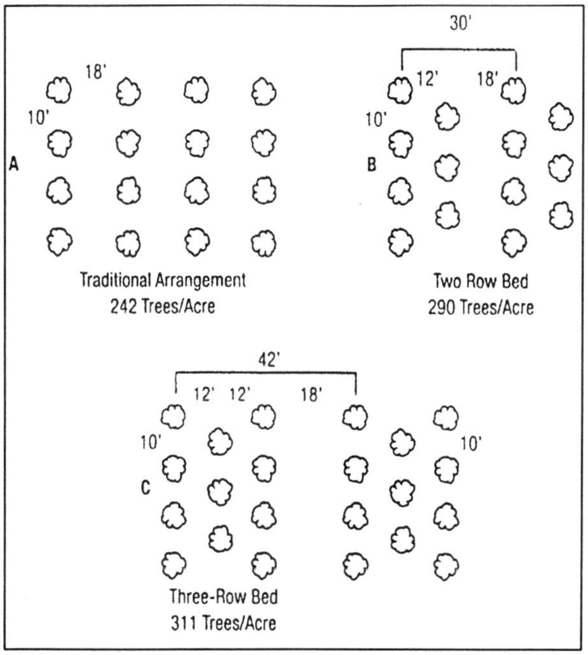

Figure 3-1. Traditional orchard tree arrangement, with wide spacing between all rows to allow equipment access, and possible arrangement of two- and three-row beds, with larger equipment restricted to the wide spacing between beds. The narrower spacing between rows within beds allows greater tree density and yields per acre. Taken from *Pennsylvania Tree Fruit Production Guide*, 1998-1999.

the early life of the orchard. High-density orchards also outyield mature low-density orchards, because large trees become shaded and unproductive at their centers, producing fruit mainly around the outside perimeter of the tree. Smaller trees put more energy into fruit production and less into growth of wood, resulting in reduced pruning costs per amount of fruit produced, as well as higher yields of higher-quality fruit. Many high-density systems also aim to arrange branches so as to maximize light interception and distribution within the orchard.

High-density orchards require:

- less vigorous species and cultivars and/or dwarfing rootstocks to keep trees smaller, which must be adapted to your climate and soil type
- often a support system of trellises or stakes
- rigorous training of tree branches into optimum positions
- increased irrigation

Some of the most popular high-density orchard planting systems are outlined below. Training and pruning for each system is covered in chapter 4.

Slender Spindle (Photo 3-2): This system was developed for high-density dwarf trees with a strong central leader, mainly apples. Closely-spaced trees, five to six feet apart in the row, are each staked or trellised. Encourage strong side branches by cutting back the leader each year, and cut back the side branches each year once they've developed. Maintain trees at five to six feet tall and wide.

Vertical Axis Trellis (Photo 3-3): This system was also designed for dwarf trees with a strong central leader, especially apples. Support young trees, spaced closely at five to six feet apart in the row, with a single-wire trellis 7.5 feet high. Let the trees rapidly grow to about 10 feet tall within three years, developing a tall, narrow tree with weak-growing fruiting branches rather than strong side branches. Each tree's leader must be tied to

Table 3-2

TREE DENSITY: CONVENTIONAL VS BEDS

Conventional planting			Bed planting specifications					
In-row tree spacing	Row spacing	Trees/A	In-row tree spacing	Wide row spacing	In-bed row spacing	Rows/ bed	Trees/A	Percent increase in trees/A
10	18	242	10	18	14	2	272	12%
			10	18	14	3	284	17%
			10	18	12	2	290	20%
			10	18	12	3	311	29%
14	22	141	14	22	18	2	156	11%
			14	22	18	3	161	14%
			14	22	16	2	164	16%
			14	22	16	3	173	23%

Table 3-2. Increase in tree density made possible using beds. Taken from Pennsylvania Tree Fruit Production Guide, 1998-1999.

Photo 3-2. Hedgerow planting systems, such as this slender spindle system for apples, are also used on other dwarf fruit trees.

Photo 3-3. The vertical axe system involves rapidly growing dwarf fruit trees, such as these apples, to bearing size by minimizing early pruning.

a bamboo pole attached to the wire, as cropping on the leader will bend it down and limit tree height. Light penetration is maintained as trees grow by removing all large upper branches.

V-System (Photo 3-4): This system works best for naturally spreading trees and has been utilized for peaches and other stone fruits. Plant trees at the usual between-row spacing of 20 to 24 feet, but just five to six feet apart in the row. Train each tree to only two scaffolds, perpendicular to the tree row and parallel to the neighboring trees' scaffolds, forming a V-shape down the row. These closely-spaced trees fill the row more quickly than the open-center system, and the V-shape maximizes light interception. Some versions of this system use a V-trellis for easier training of branches into position, but the system works with freestanding trees and is less costly to establish and remove that way.

Temporary Trees (Photo 3-5): This system is utilized for large-growing tree species, especially nut trees, when no dwarfing rootstock or genetic dwarf cultivars are available. Plant trees at an initial spacing of about half the distance they'll need at maturity. This means planting about four times as many trees per acre as the number that will be permanent trees, and producing four times as much yield in early years. When trees start to crowd after 10 to 12 years, remove every other row. After another 10 years or so, remove every other tree in the remaining rows. Many nut and fruit trees produce valuable hardwood lumber if they are

Photo 3-4. The V-trellis system provides excellent light penetration for naturally spreading trees, such as these peaches.

Photo 3-5. Temporary trees, as in this Persian walnut planting, yield crops until the planting starts to crowd, when they are removed.

pruned to at least a ten-foot straight trunk, so the trees to be removed can often be sold.

PLANTING PROCEDURES

Start planting your fruit plants the day you get them, if possible. If weather conditions delay planting, keep the roots moist until you can plant. Before planting, soak the roots of bare-root plants in buckets of water—a few hours for fruit and nut trees, a half hour for small fruits—and keep them in water until planting .

The planting rows are often already indicated by raised beds, terraces or killed sod. If not, many growers first stretch a taut string between two fenceposts, and align that string, using a compass or triangulation, so it runs north-south or parallel to the other rows. The planting spot for each tree can then be measured off with a tape measure, and each planting hole marked with a flag. Since small fruits are more closely spaced, planting spots can be conveniently determined while planting, using a spacing bar cut to the desired distance between plants in the row (Photo 3-6).

Tractor-mounted augers are commonly used to dig holes for many fruit and nut plants. In soils high in clay content, auger-drilled holes have slick sides that can trap roots in the hole like a clay pot. Avoid this by breaking up the sides of the planting hole with a shovel before setting in plants.

Larger small-fruit growers often use transplanting machines to set strawberries, grapes and other small fruits. Such machines are expensive and require larger acreages to justify their cost, but reduce labor costs.

Be sure planting holes are a few inches deeper and wider than the roots' spread. Cut off any broken roots and any extra-long roots, rather than bending them, and spread the roots out as widely as possible when planting.

Photo 3-6. When planting by hand, as with these blackberries, use a spacing bar to ensure even spacing of plants within the row.

Do not add any nitrogen fertilizers, such as chemical fertilizer mixes or fresh manures, at planting time, since they can burn young roots. Nitrogen fertilizer should be applied to the soil surface before or four to six weeks after planting.

PREVENTING CROWN GALL

Crown gall is a common disease of fruit trees, grapes, brambles and many other woody plants. Stone fruits and blackberries are particularly susceptible. Caused by a bacterial pathogen (*Agrobacterium tumefaciens*) common in most soils, it produces swollen galls on the plants' roots and crown (Photo 3-7) that limit plant growth and productivity. Two similar diseases, cane gall of brambles and hairy root of apples, are caused by other bacteria in this genus (*A. rubi* and *A. rhizogenes*, respectively). Once plants are infected, there is no cure. Planting time offers the only opportunity to prevent these diseases.

You ordered certified disease-free plants from a reputable nursery, but occasionally infected plants slip through the cracks and get shipped. Inspect all plants carefully, and remove any plants with galls before soaking plant roots in water. Call the nursery to ask if they want you to return the infected plants for a replacement or refund or destroy them. If the

Photo 3-7. Prevent crown gall, such as on these blackberries, by buying certified disease-free plants from a reputable nursery and by spraying or dipping the roots with the biocontrol agent Galltrol™ just before planting.

nursery doesn't want the infected plants returned, burn them. Just removing the galls from the plants will not completely remove the pathogen.

Prevent healthy plants from getting crown gall, cane gall or hairy root after planting by inoculating their roots with a biological control agent just before planting. This biocontrol agent, sold as Galltrol by AgBioChem, Inc. of Orinda, CA (925-254-0789), is a nonvirulent species (*Agrobacterium radiobacter* strain 84) of the same bacterial genus as the pathogens that cause the diseases. When applied to plant roots before planting, it colonizes them and uses up sugary exudates from the roots which would otherwise attract the pathogens, and also produces a material toxic to the pathogens. The biocontrol agent causes no damage to plants and prevents the later attack of the pathogens.

Order Galltrol™ several months before planting, to be shipped a week before planting, and store it in a refrigerator until planting time. Remember, this biocontrol contains living bacteria which can be killed and rendered ineffective if subjected to heat or drying. Mix the material with water in a small hand sprayer immediately before planting, and mist each plant's roots thoroughly just before you set it in the planting hole. Alternatively, mix the material in a bucket of water and dip the roots dipped into it just before planting (Photo 3-8).

Galltrol™ will not cure nursery-infected plants, but will prevent future crown gall infections from the soil when planting clean stock.

PLANTING DEPTHS

Set small fruit plants at the same depth that they grew in the nursery. Set fruit and nut trees an inch or two higher than you ultimately want them, so you can mound the soil up around the trunk to reduce problems with crown rot. The planting depth for fruit trees depends on the type of rootstock, but shouldn't be more than four inches deeper than the trees grew in the nursery.

Set trees on standard rootstocks (including genetic dwarf cultivars) with the graft union a few inches below soil level to reduce root suckering. (If the graft union is four inches or more above the soil line, just set the tree a few inches deeper than it grew in the nursery.) Set trees on dwarfing rootstocks with the graft union several inches above the soil; if the union on these trees is buried, the cultivar could root and overcome the rootstock's dwarfing effect. Set interstem trees with the interstem piece half above and half below the soil. Set all fruit trees with the curved part above the graft facing into the direction of prevailing summer winds.

Photo 3-8. Keep plant roots soaking in a bucket of water, as the blackberry plants in the larger bucket at left, until planting each one. Just before planting, each of these plants was dipped in a Galltrol™ solution, in the smaller bucket, to prevent crown gall.

Spread out the roots in the hole and press the soil around them firmly as you fill the hole. Air pockets around the roots can kill the new plants. After filling the hole around a fruit tree, mound up additional soil to about six inches above soil level. Make a watering moat in this mound a few inches away from the tree's trunk. Water each new plant enough to thoroughly soak the soil.

PRUNE AFTER PLANTING

Cut back the new plants immediately if planting in spring, but wait until early next spring to prune plants set in the fall. This pruning adjusts the top to the reduced root system, which was pruned in transplanting, and avoids stressing roots too much. Pruning also stimulates buds to grow just below the cuts, so proper pruning helps the new plant develop the best shape for fruiting.

Cut fruit trees off at about 30 inches from the ground, just above a bud. If the tree has branches, remove any that are less than 18 inches off the ground, unless training to a V-trellis for which you want branches to arise as low as possible. If any strong branches above this height arise from the trunk at a wide angle and in desired positions, leave them on, but cut them back by one-half. Cut off all branches arising at a narrow angle, less than 45 degrees, since these would develop into structurally weak branches, likely to split under heavy crop loads and very susceptible to winter injury. Don't worry if you're left with few or no side branches—buds on the trunk will grow into side branches that you can train to wider angles.

Trim back the branches of blueberries, currants, gooseberries, saskatoons, bush cherries and other bush fruits by about half. Cut new grapevines back to a single stem with two live buds. Remove the old canes of raspberries and blackberries and the old, fully expanded leaves of strawberries, leaving the new bramble buds or shoots and developing strawberry leaves.

CARE FOR NEWLY-SET FRUIT PLANTS

Fruit trees on some dwarfing rootstocks are poorly anchored and should be tied to sturdy stakes or trellised so the wind doesn't blow them over. Fenceposts or two-inch-thick wooden posts, four or five feet tall, make good stakes. Drive the posts a foot into the ground, six to twelve inches away from each tree. Tie the tree loosely to the stake with soft material in a figure-eight loop, so you don't girdle the tree. Trees on standard and semi-dwarf rootstocks do not require staking.

Protect tree trunks from sunscald, borer insects and rodents by wrapping white plastic tree guards around them (Photo 3-9), sinking the guards an inch into the soil. Remove these when the tree's branches have grown enough to shade the trunk. Another way to prevent sunscald and borer damage is to paint the trunk with white interior or exterior latex paint diluted half-and-half with water. A 14-inch tall wire mesh guard around each tree, sunk into the ground two inches, can also protect against rodents.

A thick layer of mulch or killed sod around each fruit tree or down the hedgerow will eliminate the need for most weed control. Keep a foot-wide radius around each tree in bare soil, however, to avoid harboring rodents. Preemergence herbicides are often used to control weeds in the tree row. When glyphosate (Roundup™) postemergence herbicide is used for spot treatments, be sure that all root suckers are pruned off and that the glyphosate does not contact any leaves on the trees.

Straw mulch provides a good deal of weed control within rows of strawberries, brambles and other bush fruits. Sawdust, pine bark or other acidic mulches are often used for blueberry rows. Several preemergence herbicides are registered for small fruits.

Newly-set fruit plants need frequent irrigation or rainfall during the first growing season. Inspect plantings twice a week during the first few weeks to determine need for irrigation, pest control or additional cultural practices.

Further Reading

Blizzard, S.H., S. Singha, T.A. Baugher, and B.D. Cayton. 1988. Yield and fruit quality of apple trees under three high density management systems. *Fruit Var. J.* 42:67-72.

Grossman, Y.L. and T.M. DeJong. 1998. Training and pruning system effects on vegetative growth potential, light interception, and cropping efficiency in peach trees. *J. Amer. Soc. Hort. Sci.* 123:1058-1064.

Jackson, J.E. 1980. Light interception and utilization by orchard systems. *Hort. Reviews* 2:208-267.

Photo 3-9. White plastic tree guards protect the trunks of young trees from sunscald and rodent injury. Grass clippings should be directed away from the trees when mowing, or raked away from the trunk.

Chapter 4
TRAINING AND PRUNING FRUIT PLANTS

Newly-set fruit and nut plants must be trained to grow a certain way in order to mature into healthy and productive individuals. *Training* a fruit plant involves selecting strong shoots or branches to become the main bearing framework of the plant, and positioning these shoots or branches so that they'll receive enough sunlight. Training begins at planting time and includes pruning branches, pinching new growth, and branch spreading or tying.

The aim is to develop a branch structure that's strong enough to support heavy crops of fruit and open enough to mazimize light intake and minimize disease problems. Well-trained fruit plants bear heavier crops at an early age. Most training is accomplished during the first two years after planting.

Once fruit plants are trained, they need yearly *pruning* (cutting off excess branches) to remain healthy and keep producing good crops of high-quality fruit. Unpruned fruit trees grow too tall and dense, bearing fruit only around the outside of the tree where sunlight is adequate. Unpruned brambles and grapes become choked with dead canes, while unrenovated strawberry patches suffer from too many competing strawberry plants. With all fruits and nuts, getting rid of dead, diseased, and unproductive growth will prevent disease spread and channel the plants' energies toward better fruit production.

WHEN TO PRUNE

Prune fruit and nut plants twice each year: late in the dormant season and again in summer. Pruning just before or soon after the plants break dormancy, when undesirable branches are easier to see and remove, shapes the plant and encourages new growth. Pruning young plants again in early summer channels that new growth in desired directions, and pruning mature plants in mid- to late summer controls plant size and allow better light penetration.

Wait until late winter or early spring to perform dormant pruning, as plants pruned in early winter are more susceptible to damage from subzero temperatures. Wait until after buds swell or bloom to prune trees with winter-tender fruit buds or those which bloom early.

In the summer, prune the current season's shoot growth, but generally don't cut into older wood. The later in the growing season you prune, the less regrowth will occur. In areas with cold winters, prune no later than midsummer to give new growth a chance to harden before winter.

PRUNING TOOLS

Proper tools, kept sharp, are essential to pruning. A tool that is dull or not strong enough to make a certain cut can tear the bark or leave a jagged surface that heals slowly or becomes diseased. Fruit growers need the following tools:

1. Hand pruners: scissors type, used to prune small branches up to 3/8 inch in diameter. The scissors-type pruner (Photo 4-1) leaves a much cleaner cut than the anvil-type pruner, which often smashes branches.

2. Loppers: long-handled, again scissors-type, used for cuts of 3/8 to 3/4 inch in diameter (Photo 4-2). Pneumatic pole pruners (loppers on very long

Photo 4-1. Use hand pruners to remove small branches less than about 3/8 inch in diameter.

Photo 4-2. Use hand loppers, as shown here on peaches, or pneumatic loppers to remove branches between 3/8 inch and 3/4 inch in diameter.

handles, powered by air compression) are often used for faster pruning in commercial fruit plantings.

3. Ladders: 10- or 12-foot stepladders are safer than ladders that are leaned against trees. Dwarf trees eliminate the need for ladders, greatly reducing pruning costs and potential liability. To prune the tops of taller trees, always have workers stand on ladders. Standing on tree limbs to prune can damage the bark.

4. Pruning saws: teeth set so saw cuts on pull rather than push stroke, making it easier to cut from below; used for cuts larger than ½ inch. Commercial orchardists often use pneumatic pruning saws (Photo 4-3).

5. Orange shellac: for painting cuts larger than two inches immediately after pruning (available at hardware stores). Shellac helps stop borers from invading the cut surfaces, yet allows air exchange to minimize fungal growth. Do NOT use asphalt-based or oil-based wound dressings, as they do not allow air exchange and actually encourage trapped fungi to decay the cut surface.

With most fruit tree training systems, you should remove undesirable branches while they're still small, eliminating most needs for a saw and shellac. Pruning small fruit plants requires only the hand pruners and loppers.

Some larger orchards use tractor-pulled mechanical pruning implements (Photo 4-4) to reduce labor costs for pruning. Such systems work best for height control of trees, as shown. Overly dense or diseased branches will still need thinning by hand or pneumatic tools. Also, mechanical pruning will tear some branches and leave stubs that

Photo 4-3. Use pneumatic pruning saws, as shown here on peaches, or hand saws to remove branches larger than ¾ inch.

Photo 4-4. Mechanical pruning, such as this summer mowing of peaches to reduce tree height, can reduce pruning labor costs but requires follow-up hand pruning of damaged wood and stubs.

require follow-up hand pruning, to avoid diseases colonizing the wounds and spreading throughout the trees. Summer mechanical pruning of young shoots is less damaging than dormant mechanical pruning of branches.

MAKING PRUNING CUTS

There are only two types of pruning cuts: thinning and heading. *Thinning*, or thinning out, means cutting excess branches off completely where they arise from the larger branch (photo 4-5). *Heading*, or heading back, means cutting a branch back to a bud or lateral branch but not removing it entirely (Photo 4-6). *Pinching*, removing part of a new shoot in summer while it's still growing, is a form of heading (Photo 4-7).

Thinning cuts reduce the number of new shoots produced in that part of the plant, while heading cuts encourage more new shoots to grow (Figure 4-1). Late-summer thinning cuts have a strong dwarfing effect, whereas dormant-season heading cuts invigorate the tree. When deciding whether to remove a branch or just cut it back, consider whether you want to reduce or stimulate new shoot production in that part of the plant.

Make thinning cuts as close as possible to the bark branch collar (see **Pruning Terms** text box and Figure 4-2), but do not cut through the collar.

Photo 4-5. A thinning cut removes a branch entirely, by cutting it off at its base, resulting in less regrowth.

Photo 4-6. A heading cut removes only the end of a branch, by cutting just above a bud, and stimulates buds below the cut to grow, resulting in more regrowth.

Photo 4-7. Pinching is a form of heading done with the fingers on new shoots in spring or summer, to induce branching below the pinch.

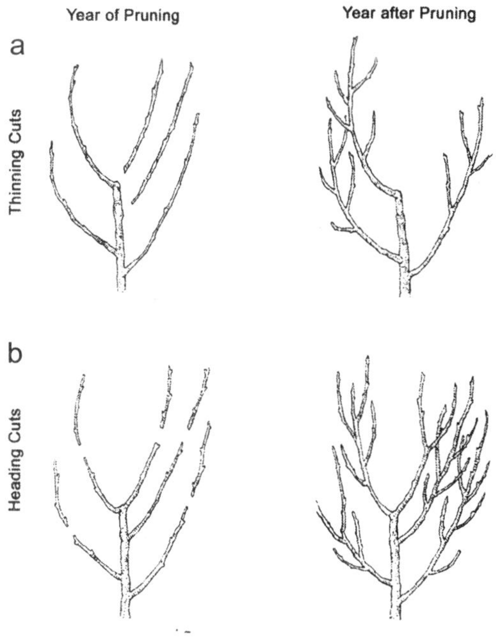

Figure 4-1. Thinning cuts (a) reduce the number of new shoots arising the following year, while heading cuts (b) increase the number of new shoots that grow the next year.

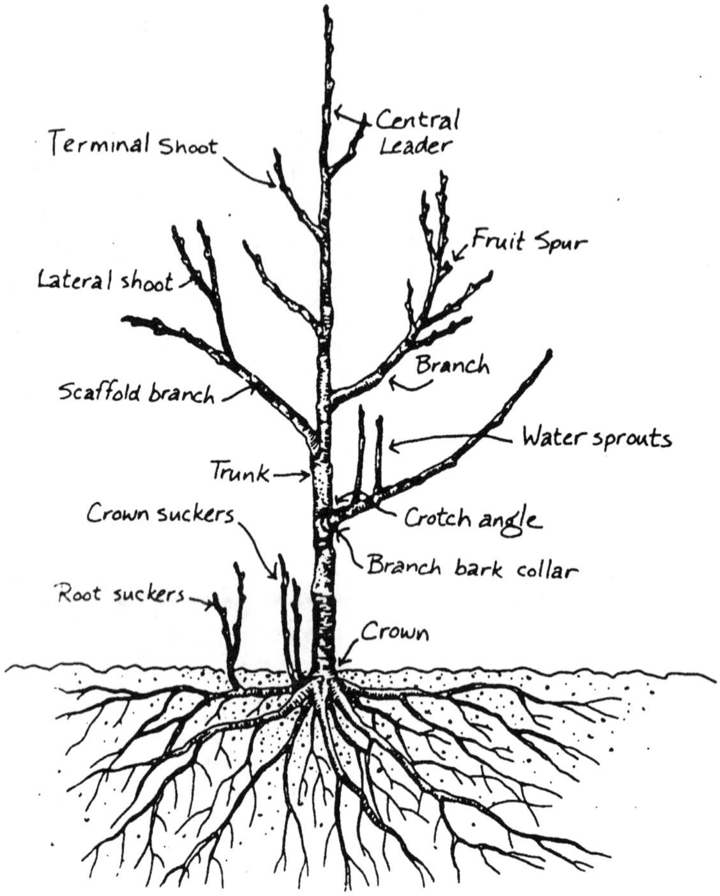

Figure 4-2.

Pruning Terms

Shoot: new growth developing during the current growing season.

Branch: a mature shoot that has passed through one or more dormant seasons.

Terminal shoot or branch: growth from the end of a branch.

Lateral shoot or branch: growth from the side of a branch.

Trunk: the main central stem of the tree.

Limbs: matured main structural branches arising from the trunk, after several years' growth.

Leader: an upright-growing branch in the center of the tree, selected as the continuation of the trunk.

Scaffold branch or limb: one of the main branches (later limbs) arising from the trunk and growing outward, forming the tree's basic framework.

Crotch angle: the angle at which a branch arises from the trunk or scaffold branch.

Branch bark collar: the raised ridge of tissue in the crotch of a branch, where it comes out from the larger branch.

Fruit spurs: short, thick growth upon which flowers and fruit develop on many fruit trees, growing on branches two years old or older, and continuing to produce fruit for many years.

Watersprout: vigorous, unfruitful shoot or branch growing straight up from a larger branch.

Crown: the juncture of the top of the tree and the root system; the part of the trunk right around the soil line.

Sucker: shoot or branch growing from the roots or crown of the rootstock.

If you leave a stub or cut through the collar, the branch will heal poorly and may be invaded by disease and insects. Making heading cuts just above a bud or short lateral branch growing in the direction you want the new terminal to grow.

TRAINING SYSTEMS FOR SMALL FRUITS

The goal of small fruit training is to keep branches spread out so they compete less with each other for light. Annual pruning of all small fruits is very important to thin out excess branches, keeping the plants open enough to receive adequate light and minimize disease. Pruning of small fruits also removes excess fruit buds, which is crucial to preventing overcropping and subsequent small, poor-quality fruit. Fruit thinning is not practiced on small fruits, so pruning is the main way to control crop load and avoid cycles of too-heavy crops one year followed by sparse crops the next.

Specific training and pruning systems for each small fruit crop are outlined in the individual fruit chapters.

TRAINING SYSTEMS FOR FRUIT AND NUT TREES

The branches of not-yet-bearing fruit and nut trees tend to grow too upright without the weight of fruit pulling them down. Such branches develop into limbs that are susceptible to winter injury at their bases and structurally weak, often breaking under heavy crop loads (Photo 4-8). Spreading upright branches (Photo 4-9) enables them to develop into a strong bearing framework, reduces the amount of pruning needed later, and promotes the formation of flower buds. Early fruiting in turn takes some of the tree's energy away from vegetative growth, so is a powerful dwarfing agent.

Keeping the branches growing outward instead of upward is important to all fruit tree training systems. Beyond that, fruit trees can be trained into many different systems.

Free-standing fruit and nut trees are trained to one of two basic systems, the central leader or the open center, or to the modified leader which is a compromise between the two. The ideal central leader tree (Photos 4-9 and 4-10) is shaped like a Christmas tree: a dominant central trunk continuing straight up to the top of the tree, with six to eight main scaffold branches radiating out from it like steps in a spiral staircase. The uppermost branches are pruned back closer to the trunk than the lower branches to allow light to reach the lower branches.

Photo 4-8. Allowing narrow-angled branches to develop as scaffolds often results in limb breakage later under heavy crop loads. Prevent this by removing or spreading narrow-angled branches.

Photo 4-9. Spreading branches while they are young trains the branches into a stronger framework that can bear heavier crops without breaking.

Photo 4-10. Central leader trees are shaped like Christmas trees, with branches radiating out from the dominant central trunk in a spiral pattern.

An *open center* tree (Photos 4-3 and 4-11) is shaped like a vase: the central trunk is cut off above the point where three or four scaffold limbs arise from it. These scaffold limbs grow upward and outward, forming the walls of the vase, while the center is kept free from branches to allow better light penetration.

Training for a *modified leader* (Photo 4-12) begins like that for a central leader, but after five or six main scaffold limbs have formed, the central leader is cut back above the top outward-growing scaffold limb. The center is kept open from that point up.

The training system to use depends partly on the type of fruit or nut tree. Train dwarf trees to a central leader for best light penetration and yields. Dwarf apples, pears, European plums, and cherries are easy to train to this system, since one branch tends to grow dominantly over the others. Semi-dwarf or standard-size trees of these fruits, apricots, and nut trees also yield best when trained to a central leader, but growers usually train them to a modified leader because that's easier to maintain with a larger tree. Peaches, nectarines, almonds and Japanese plums are more easily trained to an open center because they don't have a dominant upward-growing branch.

Many high-density training systems for fruit trees are scaled-down versions of these traditional systems. Slender spindle trees are like miniature central leaders, while the perpendicular V system creates open-center trees with just two scaffolds instead of three. The vertical axe system borrows techniques from espaliered fruit trees (those grown in a flat plane) to keep the central leader vigorous and keep side branches weak vegetatively and more fruitful.

A fruit tree's basic framework should be in place by the third or fourth season, when it starts bearing. Young bearing trees need little pruning. Remove diseased or damaged branches, those which grow upright, watersprouts (new growth from old pruning cuts), and root suckers (shoots from the roots) at the base of the tree. As the tree fills its allotted space, prune to keep it in bounds, thin out crowded branches, and maintain and renew fruiting wood by cutting back older, non-productive branches to new growth.

Photo 4-11. Open center trees have three main scaffolds that extend outward, keeping the center open for better light penetration.

Photo 4-12. Modified leader trees start off like central leader trees, then have the leader cut back above a certain point like open center trees.

CENTRAL LEADER TRAINING AND PRUNING

These instructions are for training and pruning dwarf fruit trees to a central leader (Figure 4-3). Standard and semidwarf trees can also be trained to a central leader. Allow a couple more scaffolds to develop on these larger trees, and space them farther apart along the trunk.

Planting time: Most growers using this system plant branchless whips; simply head these back at 36 inches off the ground. With "feathered" trees, remove any narrow-angled branches and those arising below 20 inches off the ground. Keep well-spaced, wide-angled branches, but head the branches back by 1/3, and head the leader back to 12 inches above the top branch.

Early summer of first year: Choose the most vigorous upright shoot arising from one of the uppermost buds on the trunk as the central leader. Prune off any trunk stub above this shoot. Remove the next two shoots below the central leader so they don't compete with it.

Remove any shoots arising within 20 inches off the ground. If possible, select several strong-growing side shoots for scaffold branches. The lowest branch should be 20 to 22 inches off the ground, and each successive branch should arise five to eight inches above the one below and grow out in a different direction. When these selected shoots are three to four inches long, spread them with clothes-pins or other branch spreaders so they develop a wide crotch angle.

Other shoots should be kept weak growing, so they don't compete with the selected ones, and encouraged to bear fruit. One way to do this is to repeatedly pinch these shoots back to about four inches long, once a month during June, July, August, and (where growing seasons are long) September. Pinching induces the formation of fruit buds and fruit-bearing spurs. As an alternative you can severely bend down these shoots in mid- to late summer, when they are long enough, by looping them down and tying them to the trunk.

Early spring of second year: Make final selections of lower scaffold branches, and pull these down to an almost horizontal positions by tying them to the trunk or to stakes in the ground. Head back the central leader at about 16 or 18 inches above the top scaffold branch to induce more scaffold branches to develop above the first set. Without this severe heading back of the leader, lower vegetative buds will not grow and the next set of scaffolds will be too far above the first. Make the cut just above a bud.

Head back the scaffold branches until the tree conforms to a Christmas tree shape, wider at the base and narrowing to the central leader tip. If the scaffolds' growth was thin and weak, head them back severely; cut only a few inches off strong-growing scaffolds. Makes cuts to a bud on the side of the branch; cutting back to a top bud will produce a shoot that grows straight up, and heading to a bottom bud leads to weak growth. Heading the scaffolds back will strengthen them and induce more secondary branches to form. This is necessary to develop a strong tree framework, but it delays fruiting on the scaffolds.

Encourage early fruiting and reduced vegetative growth on nonscaffold branches arising from the trunk. If not already pinched back or tied down, bend these down to below horizontal and tie to the trunk. Do not head these branches back during the dormant season; that would encourage vegetative growth. Thin, weak branches can be left alone.

Early summer of second year: Again choose the strongest-growing upright shoot near the top of the tree as the central leader, and prune off any trunk stub above this shoot. Remove the next two shoots below the central leader.

Similarly, choose a single shoot to continue the outward growth of each scaffold branch chosen last year, and remove the next two shoots below this shoot. When the terminal shoot of each scaffold is about 10 to 12 inches long, pinch out the growing tip to encourage more lateral branching.

Select well-spaced, strong-growing side shoots on the trunk for the next set of scaffolds, and spread them with clothespins. Pinch back or tie down other side shoots arising from the main trunk and overvigorous shoots arising from the scaffold branches chosen last year. Remove any root suckers.

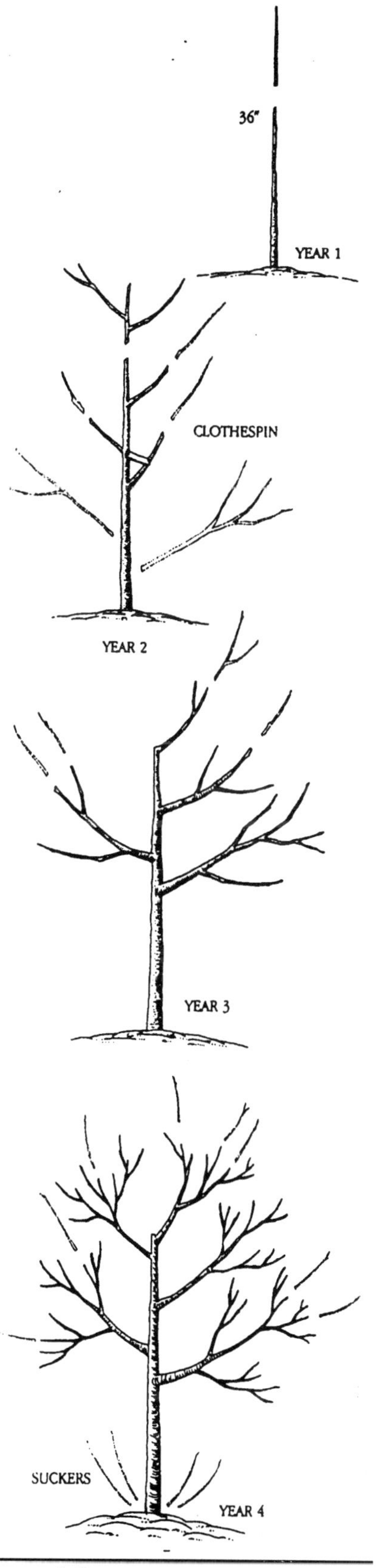

Figure 4-3.

Early spring of third year: Follow the same instructions as for the spring of the second year. Very little, if any, thinning of secondary branches on the scaffolds is needed at this time; remove only branches that are crossing, diseased, or damaged. Tie down secondary branches that grow vigorously or upright.

Early summer of third year: Follow the same instructions as for the summer of the second year. Your trees may bear some fruit this year. If the fruit load starts to bend the central leader, support the leader in an upright position with a temporary stake (½-by-½ inch will work) tied to the trunk and extending to the top of the leader.

Early spring of fourth and subsequent years: Continue this training until the tree has developed six to eight scaffold limbs and reached the desired height. At that time, during early spring of the fourth or fifth year, cut the leader back only slightly to a weak lateral branch. When the scaffolds reach the desired length, cut them back slightly to weak laterals also. Retain the Christmas tree shape: the higher the branches are in the tree, the shorter they should be.

Continue to remove dead, diseased, or damaged wood each year. Remove vigorous, upright branches or head them back to laterals. Thin out crossing or crowded branches, paying special attention to the upper and outer parts of the tree to admit adequate light into the center.

Branches that have borne crops for several years will be pulled down under the weight. Do not remove all drooping branches or you will lose fruit production in the lower part of the tree. Head drooping branches after they have fruited for three or four years, cutting them back to an outward-growing side branch.

Summer of fourth and subsequent years: Continue to select a new leader and scaffolds, and pinch or tie down other shoots, each year until the tree is trained. Continue to remove root suckers and watersprouts each year in midsummer; this produces much less regrowth than sucker removal during the dormant season. Retain the established tree shape by pinching back shoots that are growing too vigorously.

After the scaffold branches begin bearing heavy crops in the fourth or fifth season, remove the weak temporary branches or cut them back to fruit-bearing spurs in midsummer. When the tree has filled its allotted space, keep it in bounds by pinching or cutting about ½ inch off of all new shoot growth in midsummer. This pinching also allows more light into the tree canopy, increasing formation of spurs and flower buds for next year's crop, and producing larger, sweeter fruit with better color. Do not summer-pinch trees that have lost many leaves to diseases.

OPEN CENTER TRAINING AND PRUNING

Naturally spreading fruit trees are often trained to an open center (Figure 4-4).

Planting time: Head the trees back at 30 inches off the ground. Remove any narrow-angled branches, and branches arising less than 18 inches off the ground. Keep well-spaced, wide-angled branches arising from 18 to 26 inches off the ground.

Summer of first year: By early June most buds on the tree will have developed into leaves or growing shoots. The most vigorous shoots, arising from the top two or three buds, will have narrow crotch angles that are unsuitable for scaffold limbs.

Cut these vigorous shoots back to two or three buds. Below these there should be wide-angled shoots, which you want to encourage. If you have no lower shoots, this severe pruning will usually cause lower buds to grow. If the new shoots from these lower buds start growing too upright, spread them with clothespins before the crotch angles harden.

About a month later, check the trees and cut the top shoots back again if necessary. As soon as the tree has three wide-angled branches spaced about equally around the trunk and arising from the trunk at 2- to 4-inch intervals, 18 to 26 inches off the ground, cut other branches back to a few inches of growth. If one of these scaffold branches is overly vigorous, cut it back to bring it into balance with the other two.

Early spring of second year: Choose three scaffold branches if you didn't select them the summer before. If only two good scaffold branches have developed, you can train a third from a strong side branch coming from the base of one of the scaffolds. If one scaffold has grown much more than the other two, cut it back to bring it into balance with the others. Otherwise, do not head back the scaffolds; doing so will delay the onset of fruiting.

Cut the branches remaining in the center of the tree, above the three scaffolds, back to short stubs. A small cluster of shoot growth will develop from these stubs; this will help keep the scaffold branches growing outward and the center of the tree open.

Completely remove all growth arising from the trunk below the scaffold branches and any growth

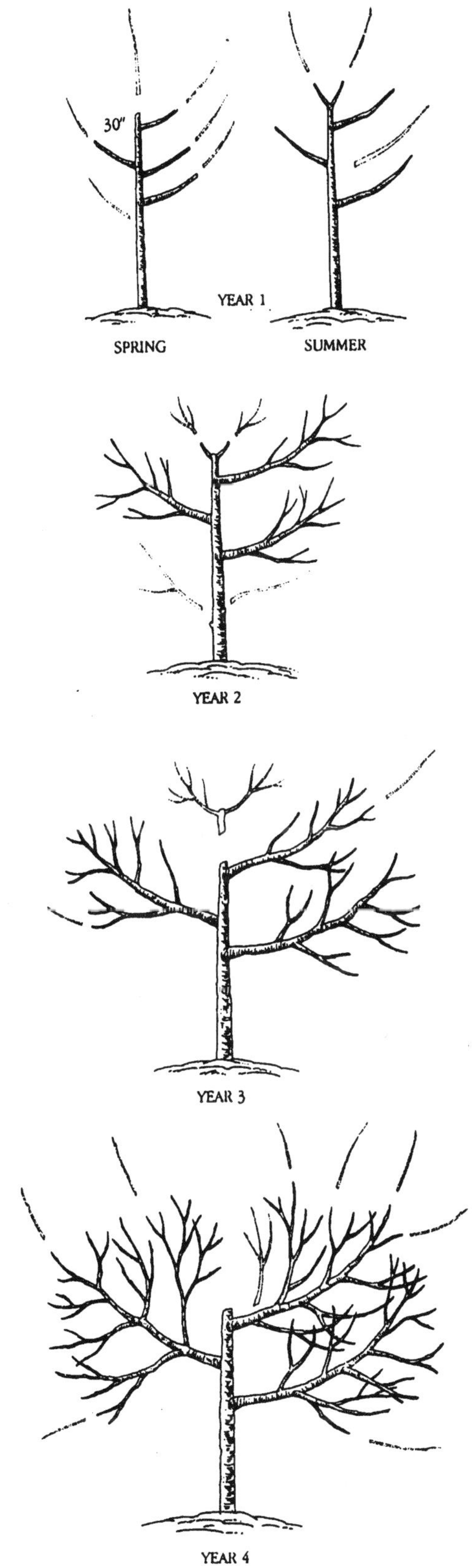

Figure 4-4.

on the scaffold branches within six inches of the trunk. Remove any broken side branches and those with poor crotch angles, winter injury, or insect damage. Allow all other side branches on the scaffolds to remain.

Early summer of second year: Cut back the shoots arising from the stubby growth above the scaffold branches to three to five inches. Remove any growth arising from the trunk below the scaffold branches. If necessary, lightly prune the scaffold branches to maintain tree balance.

Early spring of third and subsequent years: Completely remove the stubby growth above the three scaffold limbs. The shape of the tree should now be well established.

The tree should have some flower buds by this time. Prune just before, during, or just after bloom, so you can assess the amount of winter damage to the flower buds, prune off branches with dead buds, and keep those with live blossoms. If bloom is light, leave as many blossoms as possible, but if it is heavy, cut longer branches back to lighten the crop load.

Continue to cut out dead, diseased, or damaged wood each year. Remove upright growth or cut it back to outward-growing branches. Thin out crowded branches and remove crossing or dropping branches.

As the tree grows, head back older branches to encourage the growth of new fruiting wood close to the trunk. Otherwise you will end up with long, unfruitful scaffold limbs bearing fruit only around the edges of the tree.

Summer of third and subsequent years: If the tree is vigorous and leaf cover is dense, pinch or cut the new growth back by an inch or so once or twice in mid-summer. This will keep the tree from putting too much energy into leaf growth instead of into developing flower buds, and will help keep the tree smaller. Be sure to leave enough leaf cover to prevent sunburn of newly exposed fruit. If a tree is weak or has lost many of its leaves to disease, don't cut back new growth.

MODIFIED LEADER TRAINING AND PRUNING

Modified leader training begins like central leader training, with well-spaced scaffold limbs radiating out in spiral-staircase fashion from a central trunk, instead of just three scaffolds arising from close to the same point as in the open center. In the modified leader system you will allow four to six scaffold limbs to develop, then cut back the leader to the top scaffold branch, rather than to a weak lateral branch as in the central leader system. This leaves the center of the tree open above the scaffolds.

Follow the same instructions as for central leader training, but lightly head back the scaffold branches each spring to encourage lateral branching, and keep them all about the same length. Once four to six well-spaced scaffolds have developed, cut the leader off above the top one.

When the scaffolds reach the desired length, cut them back to a weak lateral branch as in the central leader system. Subsequent pruning is the same as for the central leader system, except that you keep the center open. Undesirable branches can again be temporarily retained, severely pinched back, or tied down to provide some fruit while the scaffolds are developing.

VERTICAL AXIS TRAINING AND PRUNING

The vertical axis training system works for upright dwarf trees such as apples, pears, European plums and tart cherries. It originated in France and has many variations and names, including French Axe, Axe, Triple Axe and Slender Pyramid. This system can induce bearing as early as the second season, high yields, and high-quality fruit with improved size, color and soluble solids due to better light penetration into the trees.

The vertical axe system requires a single-wire trellis to support the trees to a height of 10 feet, and a 10-foot pole attached to the trellis next to each tree (Photo 4-13), but the increased establishment cost of installing the trellis are offset by reduced labor costs for pruning. Basically, the trees are pruned as little as possible while young to induce rapid growth in height and early fruiting, then as trees mature a few large limbs are removed each year rather than making many detailed pruning cuts. The following instructions are adapted from those of Stephen Hoying and Terence Robinson of Cornell University.

Planting time: Remove any branches below 22 inches off the ground. If the tree has only one or two branches, remove them, and head back such trees and branchless trees to 32 inches. Keep well-spaced, wide-angled branches on feathered trees with three or more branches, but head the branches back by 1/3 and head the leader at 12 inches above the top branch.

Attach trees to trellis poles with permanent tree ties, leaving a 2-inch-diameter loop to allow for trunk growth. Rub off the second and third buds below the chosen leader shoot to eliminate competition with the leader. Remove any flowers. Spread new side shoots with clothes pins when 2 to 4 inches long, removing the clothespins after shoots harden in mid-summer. Tie the developing leader to the support system.

Early spring of second year: Do NOT head the leader or prune trees while dormant. If some trees haven't developed enough side branches, score above dormant trunk buds at the desired spots for scaffolds, to induce these buds to grow.

Photo 4-13. The vertical axe system involves minimal early pruning so trees achieve the desired height and produce heavy yields within just three years. Once the tree framework is set, one upper limb is removed each year to maintain light penetration and renew fruiting wood.

When lateral shoots are 4 to 6 inches long, pinch back those laterals in the top ¼ portion of last year's leader growth, to encourage the leader to grow taller. Remove the last 4 to 5 leaves of each lateral pinched.

Summer of second year: About 6 weeks and 12 weeks later, in early and mid-summer, pinch back the laterals in the top ¼ portion of last year's leader growth again. Remove any fruit on 1-year-old wood, and thin remaining fruit to 6 inches apart.

Tie the tree to the trellis pole at 6 feet, and tie the developing leader to the support system. Tie down 4 to 5 permanent scaffold branches to a horizontal position, and tie down other vigorous upright shoots to *below* horizontal.

Early spring of third year: Do NOT head the leader or prune trees while dormant. Tie down any additional vigorous upright shoots to *below* horizontal.

Again, pinch back lateral shoots in the top ¼ portion of last year's leader growth, when they are 4 to 6 inches long.

Summer of third year: Again, about 6 weeks and 12 weeks later, in early and mid-summer, pinch back the laterals in the top ¼ portion of last year's leader growth. Thin fruit to 4 inches apart.

Tie the tree to the trellis pole at 8 feet, and tie the developing leader to the support system. Prune back any permanent scaffold branches that aren't strong enough to support the developing crop. Tie down other vigorous upright shoots to *below* horizontal.

Early spring of fourth year: Do NOT head the leader. Remove any limbs that are overly vigorous.

Summer of fourth year: Tie the tree to the trellis at 10 feet, the top of the pole. Lightly prune the scaffold branches to allow light penetration and maintain the pyramidal tree shape.

Early spring of fifth and sixth years: Do NOT head the leader. Each year, remove one of the least desirable lower-tier scaffolds until only 4 remain. Head back the remaining lower-tier scaffolds to a side branch where needed to allow equipment movement and maintain light penetration to preserve fruit quality.

If the upper scaffolds have grown vigorously, begin removing one upper scaffold limb with a diameter of 1 inch or greater per year, to allow light penetration into the tree and regrowth of a new scaffold limb from a bud below the cut. Head back drooping branches to the horizontal portion of the branch.

Summer of fifth and sixth years: Pinch shoots or head back branches as needed to maintain a pyramidal tree shape with good light penetration.

Early spring of seventh and later years: Cut the leader back to a fruitful side branch at the desired height of 10 to 12 feet. Keep the remaining 4 lower-tier scaffolds permanently, and each year head them back to a side branch to allow equipment movement and maintain light penetration.

Remove one to two vigorous upper scaffold limbs per year. This maintains only smaller-caliper fruiting wood in the top of the tree, producing more fruit and less shading of the limbs below. Preserve all weak fruiting wood, heading back drooping branches where needed.

Summer of seventh and later years: Pinch shoots or head back branches as needed to maintain a pyramidal tree shape with good light penetration.

PERPENDICULAR V TRAINING AND PRUNING

The perpendicular V system works for peaches, Japanese plums and other trees traditionally trained to the open center system. By reducing the scaffolds from 3 to 2 and keeping the scaffolds perpendicular to the tree row, trees can be planted as close as 5 to 6 feet apart (Photo 4-14). This requires planting 3 to 4 times as many trees, but can triple yields during the early years of the orchard.

Planting time: With this system, scaffold branches should arise from just a few inches off the ground to form the V. Head the trees lower, at just 20 inches off the ground. Remove any narrow-angled branches, and branches arising less than 3 inches off the ground, as those could be easily hit with herbicides. If available, keep two oppositely-spaced, wide-angled branches arising from 4 to 12 inches off the ground, and plant the tree so these two branches are perpendicular to the tree row.

Summer of first year: Pinch or cut the top few vigorous shoots back to two or three buds. Choose two lower shoots perpendicular to the tree row to be scaffolds, and spread them with clothespins before the crotch angles harden.

Pinch or cut the top shoots back again a month later, if necessary, and cut other non-scaffold branches back to a few inches of growth. If one of the scaffold branches is overly vigorous, cut it back to bring it into balance with the other one.

Early spring of second year: If one scaffold has grown much more than the other, cut it back to bring it into balance with the other. Otherwise, do not head back the scaffolds; doing so will delay the onset of fruiting.

Cut the branches remaining in the center of the tree, above the two scaffolds, back to short stubs. A small cluster of shoot growth will develop from these stubs; this will help keep the scaffold branches growing outward.

Completely remove all growth arising from the trunk below the scaffold branches and any growth on the scaffold branches within six inches of the trunk. Remove any broken side branches and those that are strongly upright or with poor crotch angles,

Photo 4-14. The perpendicular V system is similar to the open center system, but with two main scaffolds instead of three. With the two scaffolds perpendicular to the tree row, trees can be spaced just 5 to 6 feet apart, potentially tripling early yields. A trellis is not essential.

winter injury, or insect damage. Allow all other side branches on the scaffolds to remain.

Early summer of second year: Cut back the shoots arising from the stubby growth above the scaffold branches to three to five inches. Remove any growth arising from the trunk below the scaffold branches. If necessary, lightly prune the scaffold branches to maintain tree balance and keep just one leader at the end of each scaffold. Remove excessively vigorous shoots arising from the scaffolds that threaten to dominate the tree's center.

Early spring of third and subsequent years: Completely remove the stubby growth above the tree scaffold limbs. The V shape of the tree should now be well established.

Prune just before, during, or just after bloom, so you can assess the amount of winter damage to the flower buds, prune off branches with dead buds, and keep those with live blossoms. If bloom is light, leave as many blossoms as possible, but if it is heavy, cut longer branches back to lighten the crop load.

Continue to cut out dead, diseased, or damaged wood each year. Remove upright growth or cut it back to outward-growing branches. Thin out crowded branches and remove crossing or dropping branches. As the tree grows, head back older branches to encourage the growth of new fruiting wood close to the trunk.

Summer of third and subsequent years: If the tree is vigorous and leaf cover is dense, pinch or cut the new growth back by an inch or so once or twice in mid-summer. This will keep the tree from putting too much energy into leaf growth instead of into developing flower buds, and will help keep the tree smaller. Be sure to leave enough leaf cover to prevent sunburn of newly exposed fruit. If a tree is weak or has lost many of its leaves to disease, don't cut back new growth.

RENOVATING NEGLECTED MATURE TREES

Fruit trees that have not been pruned in several years are overly dense, unproductive throughout much of the tree canopy, too tall and wide, and probably contain many diseased and dead branches. You can reinvigorate these trees by careful dormant-season and summer pruning. If the tree has been neglected for many years or if many large cuts are needed, complete rejuvenation may take two or three years of rather severe pruning.

Up to half of the total wood can be removed form peaches or Japanese plums, so you can often renovate these trees in a single year. One the other hand, you should remove no more than one-third of the total wood during one pruning season of apples, cherries, apricots, or European plums. Don't remove more than one-tenth of the total wood of pears each year to avoid stimulating vigorous new growth susceptible to fire blight.

To renovate a tree by dormant-season pruning:

1. Remove all root suckers and watersprouts from the lower trunk to enable you to see the tree better.
2. Remove undesirable large limbs from the center of the tree, if necessary. Limbs that are entirely dead or diseased, or those which have split or have very narrow crotch angles, should be removed. Up to four large limbs may be removed in one year.
3. Remove large secondary branches with these same problems, and thin out crowded branches.
4. Lower the height of the tree where necessary. Up to 4 or 5 feet of growth can be removed in one year. Make cuts to outward-growing lateral branches.
5. Reduce the tree's spread by heading to lateral branches. Prune upper branches to shorter lengths than lower branches.
6. Remove all dead, diseased, and broken branches.
7. Remove most watersprouts and upright-growing branches entirely, or if they are needed to replace removed branches, head them back to a lateral branch or outer bud. Work from the outside of the tree in for better visibility.
8. Head back very weak drooping branches, and those which grow too long, to a strong lateral branch. Do not remove or head back all drooping branches, or the tree's fruiting zone will remain far off the ground. Drooping branches will become productive once they are now longer shaded from above, and are willowy enough to resist breaking under fruit loads.
9. If the remaining branches are still too dense, thin out those which are too close or cross each other. Make more thinning cuts in the outer parts of the tree to permit adequate light penetration into the center.

This heavy pruning will stimulate new growth, including root suckers and watersprouts. Remove these in midsummer. Also pinch or cut back the growing tips of all shoots around the outside of the tree. Continue renovating the tree the following dormant season, if needed.

Further Reading

Barden, J.A. and R.P. Marini. 1998. Growth and cropping of young apple trees as affected by heading and support of the central leader. *HortScience* 33:39-41.

Forshey, C.G., D.C. Elfving, and R.L. Stebbins. 2000. *Training and pruning of apple and pear trees.* ASHS Press, Alexandria, VA. 166 pages.

Guimond, C.M., G.A. Lang, and P.K. Andrews. 1998. Timing and severity of summer pruning affects flower initiation and shoot regrowth in sweet cherry. *HortScience* 33:647-649.

Ito, A., H. Yaegaki, H. Hayama, S. Kusaba, I. Yamaguchi and H. Yoshioka. 1999. Bending shoots stimulates flowering and influences hormone levels in lateral buds of Japanese pear. *HortScience* 34:1224-1228.

Marini, R.P. and J.A. Barden. 1987. Summer pruning of apple and peach trees. *Hort Reviews* 9:351-375.

Mika, A. 1986. Physiological responses of fruit trees to pruning. *Hort Reviews* 8:337-378

Chapter 5
Groundcovers, Mulches and Weed Control

Most fruit and nut growers today grow a *groundcover* between the rows of their plantings. This groundcover is usually a perennial planting (Photo 5-1), although in some cases a succession of annual cover crops or intercrops can be grown. The groundcover's most important functions include:

- protecting the soil from erosion by water or wind
- providing a firm surface for harvesting, spraying and other fruit or nut planting operations
- crowding out harmful weeds, yet not competing too much with the fruit or nut plants and needing only minimal mowing
- helping manage water in the planting, by growing quickly and absorbing excess water during periods of heavy rain, and going dormant so as to need less water during droughts
- ncreasing soil organic matter through the decomposition of old roots and of mown top growth

Certain groundcovers can also provide other benefits. The roots of legume plants fix nitrogen in the soil, reducing the need for nitrogen fertilizer application, but legumes require extra management to avoid increasing problems with stinkbugs and plant bugs. Certain groundcover plants can increase populations of beneficial insects in the planting, which can greatly aid control of fruit insect pests.

Within the rows of a fruit planting, all other vegetation must be controlled, to minimize competition for water and nutrients and make harvesting easier. Mulches, organic or inorganic materials laid on top of the ground within the fruit planting rows, preserve soil water by reducing evaporation and reduce weed growth. Certain organic mulches can also stimulate beneficial soil microorganisms that help prevent diseases of fruit plants. Growers also use herbicides or mechancial weed control to prevent weed problems.

GROUNDCOVER MANAGEMENT

The topography of the land greatly influences the choice of groundcover. Protect sloping land from soil erosion with a permanent, perennial groundcover and rows laid out on the contour.

Photo 5-1. Most fruit growers use perennial groundcovers between rows of fruit plants, and keep the within-row area vegetation-free with herbicides and/or mulches.

Land with little erosion potential can be managed with either a perennial groundcover, annual cover crops, or intercrops.

The method of harvest is the other major factor that largely determines the choice of groundcover. Hand-harvested crops, especially pick-your-own plantings, must provide a clean, dry surface for pickers. This generally means mulching within the rows of small fruit plantings (and also between the rows of strawberries) and herbicide strips within tree rows.

Hand-harvested crops also generally require a perennial groundcover between the rows, kept mown fairly often during harvest periods. Small fruits and tree fruits which are harvested mechanically, using a shake-and-catch type of harvester, offer more leeway for groundcover management. Nut crops, generally harvested using a shake-and-sweep approach, require a smooth orchard floor free of debris during harvest, to enable the nuts to be retrieved off of the ground.

Choosing perennial groundcovers. Choose a groundcover that grows well in your climate, will persist for many years, and smothers weed competition without competing much with the fruit crop. Groundcover plants that put on most of their growth during the cooler seasons complement the fruit or nut planting well.

Kentucky-31 tall fescue has been recommended for many years in many areas because it is a cool-season grass that goes dormant in summer, competing little with the fruit plants when they most need water and nutrients. Tall fescue increases its water and nutrient absorption in fall, helping fruits plants to harden for winter, and helps absorb extra water during winter and spring, when excess rainfall often occurs. Creeping red fescue is another cool-season grass with the same advantages of tall fescue, plus the advantage of lower growth that needs less frequent mowing.

Recent research, however, shows many benefits from incorporating several different species in the groundcover, rather than just one grass. Including cool-season perennial legumes, such as white or alsike clover, provides slow-release nitrogen. Including plants with flowers attractive to beneficial insects provides pest control, as detailed below.

Using annual cover crops or intercrops. Quick-growing annual crops can be tilled into the soil to increase organic matter in young orchards with little soil erosion potential. Such tillage would damage the roots too much in older orchards and in small fruit plantings.

Cool-season annual legumes, such as Austrian winter peas or subterranean clover, are often planted in fall to turn under in late spring and increase soil nitrogen. Fast-growing summer annuals such as buckwheat or brassicas might be grown during summer in humid areas. Growers in drier areas generally leave the orchard fallow during summer, to conserve moisture for the fruit or nut crop.

When using cool-season cover crops, seed heavily enough and promptly to get a good stand. Sparse stands can allow winter annual weed growth, which leads to increased problems with tarnished plant bug, stinkbugs and other bug pests. These bugs are especially numerous on flowering winter annual weeds such as chickweed, wild geranium and henbit. These pest bugs also build up on winter wheat cover crops, so avoid these.

Young orchards on non-erodable land also offer the opportunity for intercropping: growing a marketable crop between the tree rows for a few years until the trees reach mature size. Annual intercrops such as pumpkins could be grown for 2 or 3 years; after that, discontinue the annual tillage and plant a perennial groundcover to avoid damaging the tree roots. Short-lived perennial intercrops which don't require annual tillage, such as strawberries (Photo 5-2), could be cropped for 3 to 5 years before replacing them with a perennial groundcover.

GROUNDCOVER MANAGEMENT FOR ARTHROPOD PEST CONTROL

Arthropod pests of fruit or nut crops (insects and mites) have natural enemies that can help control them, if the population of natural enemies is large enough at the time that control is needed. Groundcovers can increase populations of benefi-

Photo 5-2. Short-lived perennial intercrops, such as these strawberries between peaches, can produce crops for 3 to 4 years before tilling them under and planting perennial groundcovers.

cial arthropods (insects, mites and spiders) in two basic ways.

Groundcovers can provide alternate prey for predators. *Predators*, arthropods that eat smaller arthropods, often require an *alternate prey species* to feed on before the fruit or nut pest that they attack becomes numerous, or after the pest population is reduced. The predator population can increase while feeding on the alternate prey, so that the predators are numerous enough to control the pest when its population builds up. Without the alternate prey, predator populations would be insufficient to control the pest. Groundcover plants can provide such alternate prey, which does not hurt the fruit or nut crop.

Cool-season perennial legumes such as white clover, for example, can serve as reservoirs of pea aphids, allowing lady beetle predator populations to increase. Do not mow the clover in early spring, or the lady beetles will disperse from the orchard. When pest aphids, such as walnut aphids on walnuts or rosy apple aphids on apples, build up on the trees in mid-spring, mowing the clover groundcover can force the lady beetles into the trees, where they control the pests. Keep the clover closely mowed through early summer to minimize problems with pest bugs. Cool-season legumes such as white clover naturally go dormant during summer drought, which helps minimize pest bug problems, but may need renewed close mowing in fall. Drought and continued close mowing can cause the clover to die out, necessitating reseeding in mid-fall.

Annual cool-season cover crops can provide even greater populations of alternate prey and beneficial arthropods. Winter cover crops of rye and the legume hairy vetch in Georgia pecan orchards provided a succession of aphids and thrips as alternate prey in spring, allowing lady beetle populations to increase to six times the levels of unmown resident grasses and weeds and 87 times the levels of mown grasses and weeds, although no increase in biocontrol of pecan aphids was shown. If the annual legume crop, whether hairy vetch or crimson clover or arrowleaf clover (Photo 5-3), does not die naturally before pest bugs become numerous, disk it under early. This will also prevent a buildup of spider mites which often occurs on vetches in orchards in early summer, and the resultant dispersal of the spider mites to the trees when the vetch dies.

Cool-season legumes also provide important overwintering sites and early-season feeding sites for predator mites, which can completely control pest mites if the predator population is high enough. Such legumes also support many other helpful predators, such as assassin bugs, trash bugs, lacewings and spiders, which help control pests.

Avoid warm-season legumes in fruit and nut plantings. Annual warm-season legumes, such as cowpeas and sesbania, and perennial warm-season legumes such as alfalfa, do support high densities of alternate prey and beneficial insects, but also harbor large numbers of pest bugs: stinkbugs, leaf-footed bugs and plant bugs. Attempts to use these warm-season legumes as trap crops near fruit or nut plantings, to attract the pest bugs away from the fruits or nuts and spray insecticide only on the trap crops, have not been successful.

Groundcovers can produce nectar and pollen for adult parasitoids. *Parasitoids* lay their eggs in arthropods large than themselves, and the eggs hatch into larvae that eat the pest from the inside out. The adult parasitoids feed on nectar and pollen. Groundcover plants that provide such nectar and pollen can enable adult parasitoids to live up to five times longer than they would in fruit or nut plantings with little nectar or pollen available. Consequently, the longer-lived adult parasitoids lay many more eggs, that parasitize a much greater percentage of pests.

Photo 5-3. Arrowleaf clover in the groundcover of a pecan orchard provides nitrogen and supports beneficial predator insects that help control aphids. Wildflowers such as the clasping coneflower provide nectar for adult parasitoids of pests.

The majority of these beneficial parasitoids are tiny wasps and flies that attack insect pests; they do not sting humans. Flowers with many small nectaries fit their tiny mouthparts and help them the most. Such flowers include those in the carrot family, such as Queen Anne's lace (wild carrot), dill, caraway, and wild parsnip, and those in the sunflower family with tiny flowers, such as yarrows, goldenrods, fleabanes, asters, blazing stars, and coneflowers (Photo 5-4). Many legumes with extrafloral nectaries, such as showy partridge pea and sweetclover, also help, as do wild mustards, milkweeds, buckwheat and the extrafloral nectaries of smartweeds.

Photo 5-4. Small flowers in the carrot family, such as these dill flowers, feed beneficial adult parasitoids, enabling them to live up to five times longer and lay many more eggs in pest insects than if such flowers are not present in the groundcover. The parasitoids' eggs hatch into larvae that eat the pests from the inside out, helping to control them.

Apple orchard understories rich in such flowering plants showed greater parasitism of codling moth larvae and tent caterpillar larvae in studies in Ontario, Russia and California, and of leafroller larvae in Hungary and New Zealand. Several seed companies now offer seed mixtures of plants shown to aid beneficial insects, with names such as "Insectary Blend" or "Good Bug Blend". Unfortunately, many of these seed mixes contain both plants that need to be left unmown to flower and provide nectar and pollen, and legumes which require close mowing to prevent problems with pest bugs. Native wildflower seed mixes often provide a better option for flowers that aid parasitoids, as they contain many of the plants listed above.

Some growers have planted such seed mixtures between rows of trees instead of a traditional groundcover. A prune grower in northern California planted two different mixtures in alternating alleys, and manages them by high-mowing every third alley every two weeks beginning in late June. Such reduced mowing permits some plants always to be flowering, allows self-seeding, has caused beneficial insects to persist in the groundcover and has avoided the previously annual outbreaks of spider mites that came with close mowing.

Where a more traditional grass-and-clover groundcover is needed near the fruit or nut plants to provide a firm, clean surface for pick-your-own harvests or cultural operations, growers could plant parasitoid-aiding plants only in a narrow strip in the center of each alley. That strip could be mowed less often than the grass and clover on either side of it. Alternatively, plants that increase parasitoids could be planted between widely-spaced trees within the tree row.

MULCHES

Small fruit growers commonly use mulches to preserve moisture and minimize weeds within the rows of their plantings, because small fruit plants are spaced so close together. Tree fruits and nuts are generally not mulched, because the greater spacing distance would require a much greater amount of mulch, and because organic mulches attract voles, mouse-like rodents which can kill trees by gnawing away the bark. Mulch also increases problems with *Phytophthora* crown rot on certain plants, such as susceptible apple rootstocks and certain red raspberry cultivars. Where voles and crown rot aren't severe problems, mulch can be applied within orchard tree rows if it is kept at least one foot away from tree trunks, especially in winter.

Organic mulches. Organic mulches include any natural material that can be obtained in sufficient quantity to apply to the soil surface, and that will eventually decompose, adding organic matter to the soil. Because organic mulches break down, they need replenishing every year to maintain a depth of four to six inches. Organic mulches keep the soil cooler in summer and warmer in winter, minimizing temperature fluctuations that damage fruit plants. Organic mulches are commonly placed on top of strawberry plants to protect them over the winter, as well as between all the small fruit plants (Photo 5-5).

Wheat or oat straw is a common mulch in regions where these grains are grown, as are sorghum or sudangrass residues, and fescue or other grass hay. All of these materials, as they decompose, exude allelopathic chemicals that inhibit weed seed germination and growth. Be sure to obtain straw or hay cut from a clean field, to minimize weed seeds in the mulch. Some grain or grass seeds will always

Photo 5-5. Mulch from the straw of grain crops or grass hay is commonly used between small fruit plants such as blackberries and strawberries, and is placed over strawberry plants during winter to protect them from cold injury.

Photo 5-6. Sawdust mulch, especially pine sawdust, or pine woodchips make good mulches for blueberries because they acidify the soil as they break down.

come with the straw or hay, but these can be easily controlled with a herbicide such as sethoxydim (Poast) that kills only grasses. Volunteer grass seedlings can also be minimized by allowing the straw or hay to sit out through a couple of rains, which sprout the seeds, before spreading it.

Mowing the groundcover alleys within orchards or between small fruit rows can provide mulch material. Blow or rake the mulch into the fruit rows.

Blueberries benefit from an acidifying mulch such as sawdust (Photo 5-6), especially pine sawdust, or pine bark. Avoid using such acidifying mulches on other fruit crops.

Cottonseed meal and alfalfa hay are two mulches that also provide significant nitrogen to the fruit crop. Apply these only in late winter, so they will become available for early spring growth and not continue releasing nitrogen into the fall when the plants' growth should slow to prepare for winter.

Plasticulture and polyethylene mulch. Fruit growers use plastic mulch mainly when growing strawberries in high-density plantings cropped for just one season. Clear plastic mulch (Photo 5-7) has been widely used in conjunction with methyl bromide treatments, which kill weed seeds. Black plastic mulch eliminates the need for herbicides or fumigation against weeds, because it blocks the light needed for weed seed germination and development. Black plastic also keeps the soil warmer, which accelerates spring growth

Photo 5-7. Clear plastic mulch has been widely used in conjunction with methyl bromide fumigation for strawberry production. Now that methyl bromide is being phased out, black or opaque white plastic mulches may be substituted. (Photo courtesy of University of California Cooperative

and bloom, necessitating greater frost protection. Plants are established in early fall, cropped in late spring, and then removed, avoiding the heat of the summer which would otherwise fry plants in black plastic mulch.

A few blueberry growers have substituted black woven polyethylene (landscape fabric weed barrier), available in rolls of various widths, for an organic mulch. The initial cost is high, but the material does not need replacing for 10 to 12 years. Squares of this material with an X cut in the center are also available for placing around trees after planting, to reduce weeds and conserve moisture and soil warmth. Some studies have demonstrated faster tree growth using such squares, while others have not. The squares must be anchored to the soil with metal pins, which is very time consuming, and the pins are difficult to remove if a tree needs to be replaced. Weeds also grow freely through the center X, reducing the squares' effectiveness for weed control.

CHEMICAL AND MECHANICAL WEED CONTROL

The area within the rows of fruit or nut plantings should be kept free of weeds for a width of 2½ to 3 feet. Some weeds, such as chickweed and dandelions, may harbor viruses that can spread to the fruit planting. Other weeds, including many winter annual broadleaf weeds, attract pest insects that can damage the fruit. Some weeds, such as dandelions, bloom during the fruit plants' bloom period and can entice many pollinating insects away from the fruit blossoms. Uncontrolled weed growth under trees also leads to vole damage.

Many growers, however, have kept herbicide strips too wide. Ian Merwin, integrated weed management specialist at Cornell University in New York state, has found that apples on dwarfing rootstocks grew just as well in herbicide strips 2½ /2 feet wide as in strips 4 or 6 feet wide, saving chemical costs.

Preemergent herbicides. Many fruit and nut growers apply preemergent herbicides once or twice per year, to kill newly-germinated weed seeds before or soon after they emerge from the soil. Such preemergent herbicides must be applied about two to four weeks before most weed seeds will germinate, in early spring for summer weeds and in late fall for winter weeds, and must be incorporated into the soil by rainfall or irrigation.

Only a few preemergent herbicides are registered for bearing fruit and nut plantings, although several more are registered for non-bearing young orchards. These herbicides bind to soil clay particles and organic matter, so soils with greater percentages of clay and organic matter require higher application rates, while some products should not be used on very sandy soils low in organic matter. Always check the label to be sure the herbicide you plan to use is suited to your soil type and percent organic matter, and registered for your crop at its current growth stage.

Be sure to check nozzles and calibrate your sprayer to avoid overapplying the herbicide, since that could injure the crop. Overapplication could also contaminate surface or groundwaters, since these preemergent herbicides persist in the soil for several years.

Any time you apply a pesticide, record the date of application, material and formulation used, rate, area treated, volume of water, growth stage of crop and weeds, temperature, cloud cover, and time until the next ½-inch rainfall or irrigation. Such records, in addition to being required for many pesticides, enable growers to evaluate the effectiveness of pesticide applications and adjust application rates or timing if needed.

A few weed species often escape each type of preemergent herbicide, so it's a good idea to rotate the preemergent herbicides used from year to year. For example, when simazine (Simazine, Princep or Caliber) herbicide was repeatedly applied to peach tree rows in North Carolina, narrowleaf vetch was not affected and came to dominate the orchard floor vegetation. The vetch harbored spider mites, which migrated to the peaches and caused damage when the vetch matured and died in mid-summer. Napropamide (Devrinol) has little activity against annual morningglory, black nightshade, jimsonweed, prickly sida, or velvetleaf, so these weeds can become problems. Continued use of other preemergent herbicides may lead to buildups of other weeds.

Corn gluten meal is a natural preemergent herbicide that has recently been discovered and marketed. A byproduct of milling corn, it effectively inhibits root formation of germinating seeds of a wide variety of weed species. This could be useful for certified organic growers, but unfortunately costs about ten times as much as synthetic preemergent herbicides to treat the same area. The website http://www.hort.iastate.edu/gluten/cframe.html lists corn gluten meal sources.

Postemergent herbicides. Growers apply postemergent herbicides to kill weeds that have already emerged. Two advantages of such applications over preemergent herbicide applications are that only the weeds need be treated rather than the entire soil surface, and that these postemergent herbicides do not persist in the soil or contaminate surface or groundwaters.

Selective postemergent herbicides may be sprayed over an entire planting to kill only certain weeds. For example, sethoxydim (Poast) is often used to kill grassy weeds in strawberries or within rows of brambles. This herbicide affects only grasses so won't injure the crop, but won't control broadleaf weeds either. 2,4-D controls broadleaf weeds, and is best applied to orchards in early spring or fall when trees are dormant to avoid damaging them. Do not use 2,4-D in or near vineyards, as many grape cultivars are extremely sensitive to it.

Nonselective postemergent herbicides, such as glyphosate (Roundup) or glufosinate (Rely), must be applied selectively. Orchardists commonly use these two herbicides to kill weeds within the tree rows, after first removing from the trees all root sucker growth that could take up these herbicides and lead to tree damage. A similar natural product is pelargonic acid, a fatty acid found in geraniums and other plants, sold as Scythe.

Cultivation and flaming. Most fruit and nut growers no longer cultivate the soil to control weeds, except possibly in young orchards when cover crops are grown. Disking or other methods of cultivation damage the fruit or nut crops' feeder roots near the soil surface, which absorb most of the water and nutrients. Cultivation close to trees can also damage the trunks. Cultivation stirs up weed seeds which were buried too deeply to grow, leading to increased weed germination, and leaves the soil open to erosion.

Where cultivation must be practiced within the rows of fruit plants, as in some certified organic fruit plantings, use mechanical hoes (also known as grape hoes) such as Green Hoe™ or Weed Badger™. These tractor-mounted hoes just scrape the soil surface and bring up fewer weed seeds and do less damage to fruit plants than other forms of cultivation.

Flame weeding has been tried in some orchards and small fruit plantings. It requires a special machine that produces flames directed beneath the fruit plants, and a water spray to shield the crop. This process is dangerous and expensive.

Hoeing, hand-pulling weeds, and weeder geese. Some weeds will always escape other treatments, so a limited amount of hoeing and hand-pulling will always be needed. The high cost of labor relative to herbicide costs prohibits most growers from using these methods extensively, although certified organic growers who receive higher prices for their fruit may rely on them.

Some small fruit growers have utilized weeder geese to eliminate most of the grassy weeds and many tender broadleaf weeds from a planting. Geese will also eat ripe fruit and new shoots of the fruit crop, so they must be excluded from the plantings during early spring and harvest. Movable electric fencing allows rotational, intensive grazing of each area. Geese do compact the soil, so shouldn't be kept in any one area too long.

Further Reading

Aylsworth, J.D. 2000. IPM 2000, part 1: IPM techniques for weed management. *Amer. Fruit Grower,* Feb. issue, p. 6-10.

Brown, M.W. and D.M. Glenn. 1999. Ground cover plants and selective insecticides as pest management tools in apple orchards. *J. Econ. Entomol.* 92:899-905.

Bugg, R.L. and C. Waddington. 1994. Using cover crops to manage arthropod pests of orchards: a review. *Agric. Ecosyst. Environ.* 50:11-28.

Clark, M.S., S.H. Gage, L.B. DeLind, and M. Lennington. 1995. The compatibility of domestic birds with a nonchemical agroecosystem. *Amer. J. Alternative Agric.* 10:114-120.

Merwin, I.A., J.A. Ray, and P.D. Curtis. 1999. Orchard groundcover management systems affect meadow vole populations and damage to apple trees. *HortScience* 34:271-274.

Olcott-Reid, B. 1989. Use beneficial insects to fight pests. *Amer. Fruit Grower,* Jan. issue, p. 28-31.

Skroch, W.A. and J.M. Shribbs. 1986. Orchard floor management: an overview. *HortScience 21:390-394.*

Chapter 6
Mineral Nutrition and Fertilizers

Over 90% of the weight of most plant tissues, including fruits, consists of water. Plants produce 90% of their dry matter—sugars, starches and other carbohydrates—from carbon dioxide they obtain from the air and from water they take up from the soil. Thus the most important compound that can limit plant growth is water. Providing adequate water for fruit growth is discussed in Chapter 7, Irrigation.

The remaining 10% of plant dry weight consists of *mineral nutrients* taken up by plants when the nutrients dissolve in the soil water. These nutrients, though needed in far smaller quantities than water and carbon dioxide, are nonetheless essential to plant growth. Often several of these nutrients become limiting and must be added, to the soil or directly to the plants, as fertilizers.

Excessive amounts of these nutrients can cause problems, too. Figure 6-1 illustrates the effects of increasing amounts of a plant nutrient in a crop. At extremely low levels of the nutrient, fruit yield and quality suffer and the crop displays *deficiency* symptoms. At higher but still not adequate levels of the nutrient, "*hidden hunger*" conditions exist,

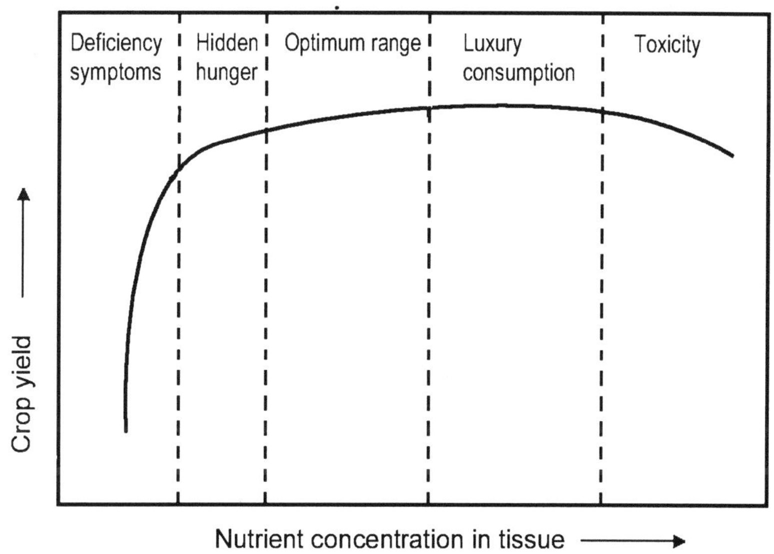

Figure 6-1. The effects of increasing amounts of a plant nutrient in a crop. Based on data originally published by Smith, P.F. 1962. Mineral analysis of plant tissues. *Ann. Rev. Plant Physiol. 13:81-108.*

reducing yield although no obvious symptoms occur. Levels of the nutrient within the optimum range constitute a *sufficiency* of the nutrient that produces optimum yield and quality. Higher levels than are needed result in *luxury consumption* of that nutrient, which means money was wasted on fertilizer for that nutrient, and the excess may interfere with uptake of other needed nutrients. Very high levels of the nutrient can even induce *toxicity* symptoms.

MACRONUTRIENTS AND MICRONUTRIENTS

Macronutrients are the mineral nutrients needed in fairly large quantities. The three macronutrients needed in the greatest amounts—carbon (C), oxygen (O) and hydrogen (H)—are supplied by carbon dioxide from the air and soil water, as mentioned. The other macronutrients are commonly added to the soil as fertilizers or other soil ammendments.

Micronutrients, though essential, are needed in such tiny quantities that they do not usually need to be added as fertilizers. However, some fruit plants have an especially high need for certain micronutrients, and some soils are low in certain micronutrients and require additions.

Fertilizer, or primary, macronutrients. The three macronutrients most commonly added as fertilizers are nitrogen (N), phosphorous (P) and potassium (K). Fertilizer materials must specify an *analysis* of the percent of these three minerals in the fertilizer. This analysis is denoted by three numbers separated by dashes, for example 10-5-8. This example indicates that 10% of the fertilizer's weight is N, the P level is equivalent to 5% P_2O_5, and the K level is equivalent to 8% K_2O. The actual amount of P is 43% that of the P_2O_5 number, and the actual amount of K is 87% that of the K_2O number.

A fertilizer containing all three of the primary macronutrients is called a *complete* fertilizer. Most fruit crops on most soils do not require a complete fertilizer every year. Growers should sample their soils to determine P and K levels, and their plants to determine N levels, before choosing a fertilizer material.

pH-regulating, or secondary, macronutrients. The other three macronutrients—calcium (Ca), magnesium (Mg) and sulfur (S)—usually occur in sufficient quantities in soils or are added in materials to change the soil's pH. Crushed limestone, the most common material added to an acidic soil to raise its pH, consists of calcium carbonate. Dolomitic limestone or dolomite consists partly of calcium carbonate and partly of magnesium carbonate, and is added to acid soils when magnesium is lacking. Sulfur is added to alkaline soils to lower the pH.

Occasionally, soil pH is within the correct range but Ca, Mg, or S are deficient. Then these elements can be included with fertilizers supplying the primary macronutrients. Many fertilizer analyses now contain a fourth number, which tells the percent of S in the fertilizer. If Ca is deficient, calcium nitrate can be chosen as a nitrogen source. If Mg is lacking, magnesium ammonium phosphate supplies that element as well as N and P.

Micronutrients. The six micronutrients sometimes needed in fertilizers include iron (Fe), manganese (Mn), zinc (Zn), copper (Cu), boron (B) and molybdenum (Mo). Several other micronutrients are essential to plants, but always occur in sufficient quantities in soils.

Soil pH has a large affect on availability of micronutrients (see Figure 2-3), and adjusting the pH is often the most economical way to provide adequate levels of needed micronutrients. For example, the acid-loving crops (blueberries, cranberries and lingonberries) have a higher need for iron than most other fruit crops, because Fe is more available at the low pH range (4.0 to 5.0) to which these plants are native. Acidifying the soil with S can prevent Fe deficiency in these crops.

When supplying a needed micronutrient directly to the crop is more economical, growers often use foliar sprays rather than soil ammendments. This allows the plants to immediately take the nutrient into their leaves, where it is needed. Foliar sprays also avoid having much of the micronutrient tied up in the soil and unavailable to plants, which is why these soils show deficiencies to that nutrient.

SOIL AND LEAF TISSUE ANALYSIS

Most state Extension services provide both soil and tissue testing services for a small fee. Growers collect the samples and deliver them to their county Extension agent, who will mail them to the University testing labs.

Soil tests let growers know the soil pH, the levels of P and K, and the percent organic matter. Soil tests for N are available, but are much less accurate than soil P and K tests, because excess N either gets tied up in organic matter or leaches out of soils, rather than adhering to mineral particles as do P and K.

Leaf analysis provides information on foliar levels of N, P, K, Ca, Mg, Fe, Mn, Cu, B and Zn. Leaf analysis is especially useful for checking the actual amounts of N and micronutrients that the plants have absorbed. Foliar analysis can also warn of an approaching deficiency or toxicity problem before the plants show any visible symptoms of distress. Table 6-1 details the desired nutrient ranges for each fruit and nut crop.

How to sample soil. Take soil samples the year before planting each fruit crop to allow time for any needed ammendments to break down, and every three to four years thereafter to be sure the pH, P, K and organic matter stay within desired levels. Using a shovel or soil auger, collect soil from at least five or six different locations throughout the field. At each location, collect a small amount of soil from about four to six inches depth. Place all the soil in a small paper bag and mix together, creating one composite sample to be tested.

Wherever soil conditions change dramatically, take a separate sample. This includes areas with a different soil type, drainage, slope, or cropping history, and any areas that received different soil ammendments from the rest of the field.

How to sample leaves. Collect leaf samples during mid-summer before the first cropping year, and about every three years thereafter, or whenever any symptoms of possible nutrient deficiency or excess show up. The period from mid-July to mid-August is the best time to sample because nutrient levels are most stable then. Early in the growing season, leaves expand rapidly and nutrients flow into the leaves from the roots, while later in the season leaves begin to senesce and nutrients flow out of them into the fruits. To avoid pesticide contamination, take samples as long as possible after the last spray or just before the next spray application.

Sample healthy, fully expanded new leaves toward the middle of current-season shoots. Collect 50 to 60 leaves per sample, taking them from as many plants as possible in the sample area. Sample different cultivars and rootstocks separately, and collect separate samples from areas where the soil varies from the rest of the field. Two plantings of different ages should also be sampled separately.

If fruit development or plant growth appears abnormal and a nutritional deficiency or excess is suspected, take paired samples: one set from healthy-looking plants and one set from abnormal plants.

After collecting leaf samples, clip the petioles or leaf stalks from the leaves with shears or scissors. Then air-dry the leaves until brittle before placing them in a small paper bag.

FERTILIZING

Reports generated by soil and leaf tissue analysis let growers know which nutrients need adjusting. Then growers must decide which soil ammendments or foliar sprays to apply and when to apply them. The following guidelines detail, for each nutrient, the best application timing and some of the most common fertilizer materials used.

Nitrogen (N). Check the color and amount of growth on your fruit plants each mid-summer, in addition to taking leaf samples for foliar analysis, to guide you in fertilizing with N the following spring and early summer. Leaves should be dark green, and the percent N should fall within the range listed in Table 6-1. Fruit tree terminals—shoots from the ends of branches—should grow 16 to 24 inches during each of the first three years, and about 12 to 15 inches per year during succeeding years. Grape and bramble shoots should grow three or more feet, blueberry shoots about a foot, and strawberry plants should be vigorous with a good amount of runners. If plants received adequate water but leaves are yellowish and shoots didn't grow that long, add more N fertilizer next year. If shoots grow excessively long, cut down on N.

Growers can't rely solely on soil N tests to provide a fertilizer recommendation, as they can with soil P and K tests. That's because most of the soil N is tied up in soil organic matter, while only the small amount of soluble N in the nitrate form is tested. Growers will gain a feel for the amount of N to add based on experience with crop growth and yield goals, foliar analysis, and previous cropping history.

Apply N fertilizer 4 to 6 weeks after spring planting, or in early spring after fall planting. During the second and following growing season, make two fertilizer applications: one when growth begins in spring and one in early summer. In orchards and vineyards, apply about 0.01 to 0.04 pounds of actual N per tree or vine per year of age, up to 0.3 pounds actual N per mature tree or vine. For small fruits, apply about 30 to 40 pounds of actual N per acre each time. Specific amounts of N to add will vary with the soil type, irrigation and disease management, and other factors.

Make the second N application at least six weeks before fruit ripen or wait until after harvest, to avoid inducing overly-large and watery fruit that's more prone to rot. Because strawberries ripen so early, skip the first fertilizer application on them to avoid increasing fruit rot. An after-harvest N application to strawberries and raspberries

Table 6-1.

FRUIT PLANT FOLIAGE NUTRIENTS

	Macronutrients: % dry weight					Micronutrients: ppm dry weight				
Crop	**N**	**P**	**K**	**Ca**	**Mg**	**Fe**	**B**	**Zn**	**Mn**	**Cu**
Almond	2.00-2.60	0.10-0.25	0.70-1.80	1.00-2.00	0.30-0.70	50-200	30-85	20-100	30-150	5-12
Apple	1.70-2.50	0.15-0.30	1.20-1.90	1.50-3.00	0.25-0.35	50-250	20-60	15-100	25-150	5-12
Apricot	2.30-3.00	0.15-0.30	2.00-3.00	1.50-3.00	0.50-0.95	50-200	25-75	20-100	30-150	5-10
Blueberry	1.80-2.40	0.10-0.22	0.50-0.85	0.40-0.85	0.10-0.28	100-350	32-50	20-60	25-170	12-20
Brambles	2.65-3.20	0.25-0.40	1.50-2.50	0.75-2.50	0.36-0.68	50-200	30-60	20-60	50-150	4-25
Cherry	2.30-3.30	0.16-0.40	1.00-3.00	1.60-2.60	0.40-0.90	50-250	20-65	15-70	20-150	5-25
Grape	2.40-2.60	0.21-0.24	1.20-1.40	2.50-3.50	0.24-0.27	100-250	26-40	30-150	30-200	6-25
Hazelnut	1.50-2.80	0.10-0.40	0.60-1.30	0.80-2.40	0.20-0.60	50-250	35-75	28-85	24-120	8-12
Peach	2.50-3.80	0.15-0.40	1.50-2.50	1.90-3.50	0.25-0.60	100-200	20-80	15-60	20-150	6-15
Pear	1.80-2.40	0.12-0.25	1.00-2.00	1.30-3.50	0.25-0.90	100-300	20-60	20-60	20-170	6-20
Plum	2.10-3.00	0.10-0.25	2.00-3.50	1.80-3.80	0.20-0.60	50-300	30-56	18-100	25-150	5-10
Pecan	2.50-3.00	0.12-0.18	0.90-1.30	0.80-2.50	0.30-0.65	50-300	60-80	30-300	100-300	7-30
Strawberry	2.35-2.95	0.25-0.40	1.50-2.50	0.75-1.70	0.28-0.50	50-150	24-90	20-75	50-150	6-18
Walnut	2.20-2.60	0.12-0.30	1.00-3.00	1.25-2.50	0.20-0.90	50-250	35-80	20-200	25-220	5-20

Optimum concentrations of plant nutrients in the foliage of fruit plants during July and August.

is important, to stimulate growth of new plants that will produce next year's harvest. Similarly, an early-summer or mid-fall N boost to fruit trees will stimulate greater fruit bud production for next year's crop. Avoid applying N in late summer or early fall, as that could stimulate late-season growth that's more susceptible to winter injury.

Established fruit plants thrive on regular additions of organic matter and just enough N to produce dark green leaves and a reduced amount of shoot growth. Bearing trees that produce crops on one-year-old branches (peaches, apricots, Japanese plums and nuts) should make 15 to 18 inches of terminal growth per season. Those which bear mainly on spurs (apples, pears, cherries, and European plums) should make no more than 10 to 14 inches of terminal growth per season. New canes of brambles, grapes, and blueberries should be mostly pencil-thickness, rather than too thick and vigorous or spindly and weak. Strawberries should form thick crowns with large leaves, and should runner well but not excessively.

Common sources of N include nitrate fertilizers, ammonium fertilizers, and urea, a high-N material that breaks down into ammonium. Urea and all the ammonium-containing fertilizers make the soil more acid as they break down in the soil, because they release hydrogen ions. For most fruit crops, several years of ammonium fertilization will necessitate lime additions to bring the pH back up to ideal levels. Alkaline soils or acid-loving crops such as blueberries benefit from ammonium sulfate or ammonium phosphate fertilization.

Ammonium nitrate is not as acidifying as other ammonium fertilizers, and other nitrate fertilizers, such as calcium nitrate and potassium nitrate, slightly increase the soil's pH. Avoid applying large amounts of nitrate fertilizer at once, especially before a rain, because the highly soluble nitrate easily leaches out of soils and has become a severe pollutant of surface waters and groundwater.

Organic matter in the soil releases N slowly over time as proteins decompose. Growers can use a new computer model, the Nitrogen Decision Aid, to determine how much nitrogen is being released naturally by microbial breakdown of the organic matter. This will help prevent application of excess nitrogen, which would waste money and could worsen plant diseases or pollute the environment.

The model, developed by soil scientist Alan Olness at the USDA's North Central Soil Conservation Research Laboratory in Morris, MN, is available at www.infolink.morris.mn.us/~lwink/products/products.htm. The model is designed for Northern climates that have some frost in the soil during the winter. Growers need to know the clay content and organic matter content of their soils, and have some data from a field weather station, to use the model.

Organic fertilizers contain lower percentages of N, but it is stably bound by the organic matter and much less subject to leaching out of soils. Cottonseed meal, spent mushroom compost, feather meal and blood meal are good organic fertilizers that are fairly high in N. Animal manures are less desirable for fruit crops: they have much less nitrogen and may contain excessive amounts of phosphorus, which can cause zinc or iron deficiency in some fruit crops.

Apply organic fertilizers only in early spring. If applied in summer, they would still release N into early fall, when reduced nitrogen levels are desired for fruit ripening and plant hardening for winter.

Growing a legume groundcover between rows of fruit plants can provide part to all of the N that the fruit crops need, but unfortunately legumes also attract stinkbugs and plant bugs, which can damage the fruit. Chapter 6 describes ways to manage a legume groundcover to minimize harmful bugs.

Phosphorus (P). The best time to apply P, if the soil needs it, is before planting when it can be worked into the soil, because P fertilizers move little in the soil and become available slowly. Raise the available soil phosphorus level to 100 pounds per acre.

Established fruit plants often don't benefit from P applications in most soils, and excess P can cause zinc or iron deficiency. Use leaf analysis every few years to determine if P levels are adequate for that crop.

P fertilizers include ammonium phosphate (which also provides N), superphosphate, and calcium phosphate. Organic growers use finely-ground rock phosphate (which is extremely slowly available), manures or bone meal.

Potassium (K). K may not be needed, but may be deficient on alkaline soils high in Ca and Mg, or in soils that have been heavily fertilized with nitrogen. Add K before planting, if needed, according to soil test results.

Foliage levels of K over 1.2 to 1.3%, from the leaf analysis performed every few years, generally indicate no additional K is needed. Avoid excess K, which can inhibit uptake of Ca and Mg.

The most common K fertilizer is potassium chloride. Potassium nitrate, which also provides nitrogen, is sometimes used. Wood ashes and bark composts are high in K.

Calcium (Ca). Naturally alkaline soils have plenty of Ca, while lime additions provide Ca for acid soils. The soil test before planting, and every few years thereafter, will determine the need for lime. Use calcium carbonate lime unless the soil test indicates a need for magnesium also, which would call for dolomite.

Some fruit crops may also benefit from foliar-applied Ca. Calcium chloride sprays are recommended for apples to prevent corkspot and bitter pit, two physiological disorders resulting from low levels of Ca in the fruit. These disorders show up during cold storage of the fruit, greatly reducing storage life. Prevent them by adding calcium chloride to six to eight of the summer pesticide cover sprays.

Magnesium (Mg). Alkaline soils are often high in Mg, and it is generally adequate in acid soils. Do a soil test for magnesium before planting if you suspect your soil may be low in Mg. Do not use dolomite (high-magnesium lime) on an acid soil unless the soil test indicates a need for Mg, because excess Mg inhibits calcium uptake.

If foliar analysis indicates a lack of Mg and a quick addition is needed, use magnesium sulfate sprays. This is a temporary solution until dolomite can be added to the soil.

Sulfur (S). Organic matter releases S along with nitrogen over time, as proteins decompose. S does not generally need to be added to soils, except when acidifying alkaline soils or soils for acid-loving crops. In those cases, add ground S according to the pH test and the desired pH, before planting and every few years thereafter.

Many fertilizers for other nutrients provide S also, including ammonium sulfate, superphosphate, and magnesium sulfate. The most commonly-used micronutrient sprays are metal sulfates: zinc sulfate, copper sulfate, etc. Avoid excessive use of fertilizers containing S on acid soils for most fruits, because they will make the soil even more acid, necessitating more lime.

S is widely used as a fungicide, often combined with lime, which renders the mixture pH-neutral. Do not apply S fungicides when temperatures exceed 85 F, as it can burn plant foliage under such heat.

Micronutrients. Boron (B), zinc (Zn), and sometimes copper (Cu) are often deficient on alkaline soils in the West and in some eastern soils, and are best applied as foliar sprays in spring or in fall after harvest. Base applications on leaf analysis.

Trees take up foliar-applied B, such as Borospray or Solubor, better than soil applications of borax, although annual soil applications may suffice for trees more than three years old. Spray zinc sulfate for Zn deficiency and copper sulfate for Cu deficiency. Apply two to four sprays in early spring for long-season nut crops such as pecans, or in fall after harvest while leaves are still green for many fruit crops. Where copper or zinc fungicides are used for disease control, they provide the necessary amounts of these micronutrients.

Symptoms of Nutrient Deficiency or Toxicity in Fruit and Nut Crops

Nitrogen (N)

Deficiency: Short, spindly shoots have pale green to yellow basal leaves. In stone fruits, leaves redden as deficiency worsens and "shothole" spots may fall out. Fruits are smaller and mature earlier.

Toxicity: Shoots grow excessively long with dark green leaves which drop later in fall. Such excess growth is more prone to fire blight and winter injury. Fruits develop less color and mature later, with less flavor and storage life. Fruits are larger but more prone to calcium-deficiency problems and fruit rots.

Phosphorus (P)

Deficiency: Fruit plants rarely exhibit deficiencies severe enough to produce symptoms. If symptoms do occur, terminal shoots appear slender, and young expanding leaves look unusually dark green and leathery, then purplish or reddish, especially on the undersides. Fruit skins are redder and may crack, fruits mature early and flavor is unpleasant.

Toxicity: Plants usually show deficiencies to one or more of the metallic micronutrients: zinc, iron, copper, or manganese.

Potassium (K)

Deficiency: Older leaves at the base of shoots are small and bluish-green with yellowed or dying edges. Scorched leaf margins may curl upward, especially in stone fruits. Fruits exhibit reduced size, red color, sugar content, and storage life.

Toxicity: Calcium or magnesium deficiency can occur.

Calcium (Ca)

Deficiency: Fruit symptoms often show up before leaf symptoms. Apples and pears can exhibit cork spot or bitter pit. Other fruits can be less firm and more prone to rot. Severe deficiency can cause the margins of young leaves to cup upward, expanding leaves to turn uniformly yellow, and new shoots to die back.

Toxicity: No known symptoms occur.

Magnesium (Mg)

Deficiency: Older leaves lose green color between the veins, progressively from the leaf tip to its base, causing a "herringbone" pattern of green veins to stand out. As the season progresses, faded areas turn brown and die, older leaves drop and symptoms develop on progressively younger leaves. Fruit may drop early.

Toxicity: Calcium or potassium deficiency can occur.

Sulfur (S)

Deficiency: Younger leaves turn uniformly yellow, without the green veins of iron deficiency. This is similar to nitrogen deficiency, but young leaves rather than older leaves are affected first, leaves aren't as stunted and may be twisted. Veins may turn reddish. Shoots are spindly and soft, similar to an *excess* of nitrogen.

Toxicity: Excesses in the soil are rare. Sulfur dioxide pollution from industrial sources, deposited by wind on leaves or petals moist from dew, may cause marginal burning of leaves or reduced fruit set if during bloom. If sulfur fungicide is applied during high temperatures, it may burn leaves, shoots and young fruits.

Iron (Fe)

Deficiency: The very young leaves lose green color between the veins, while veins remain dark green. Interveinal areas can turn bright yellow or even white. Iron deficiency is common with a high soil pH.

Toxicity: This occurs rarely; if it does, symptoms are similar to manganese deficiency.

Manganese (Mn)

Deficiency: The older leaves lose green color from the margins inward, although the main veins remain green. These symptoms can be distinguished from iron deficiency because the youngest leaves are not affected at first, and the finest veins do not remain green. Symptoms differ from magnesium deficiency in that faded areas do not turn brown.

Toxicity: The bark of young branches may blister, crack and peel, as in the "measles" or "internal bark necrosis" disorder of apples. This occurs often when the pH is too low.

Zinc (Zn)

Deficiency: Newly-developing young leaves remain smaller and narrower than normal, in tight clusters due to reduced shoot elongation, causing a "rosette" effect.

Toxicity: Symptoms are rare.

Copper (Cu)

Deficiency: Younger leaves appear stunted or misshapen, narrow with wavy margins. Rapidly-growing shoot terminals may develop necrotic lesions and die back. The next season, branching occurs below the dead area, creating "witch's broom" growth.

Toxicity: Soil levels are rarely excessive, but may cause symptoms resembling zinc deficiency. Copper fungicides may burn shoots, leaves, and flower clusters if used at excessive rates. Do not use copper sprays during bloom, as copper deposited on the stigmas will prevent pollen germination and thus fertilization and fruit growth.

Boron (B)

Deficiency: Fruit symptoms often show up before leaf symptoms. Apples and pears produce gnarled, misshapen fruit underlaid by hard corky tissue. Plums develop brown sunken areas in the flesh, and the flesh of peaches next to the pit becomes dry and corky. The tips of terminal shoots can die, causing growth of many short side shoots into a "witch's broom".

Toxicity: Twigs can die back, older leaves can yellow along the midrib and large lateral veins, and leaf margins may crinkle. Fruit may mature and drop early, apples and pears can develop internal breakdown, and stone fruits may develop split pits. Fruit bud formation and set may be reduced.

Chlorine (Cl)

Toxicity: Though an essential micronutrient, chlorine is never deficient in soils, but is often excessive on irrigated soils in the West. Mild toxicity causes reduced vigor without foliar symptoms, while severe toxicity causes a marginal leaf scorch and eventual leaf drop.

Sodium (Na)

Toxicity: Sodium is also an essential micronutrient, never deficient in soils, and often excessive on irrigated soils in the West. Toxicity causes a marginal leaf burn that is striated, unlike that of chlorine. Excess sodium also tends to accumulate in the roots and wood rather than the leaves, and can kill fruit and nut plants if it builds up.

Photo 6-1. Nitrogen deficiency, peach. Note small, yellow-green leaves.

Photo 6-2. Severe potassium deficiency, apple. Note very small leaves with necrotic margins. (Photo by Andrea Clarke, Ontario Min. of Agric.)

Photo 6-3. Potassium deficiency, bunch grape. Note leaves with yellowed margins, which will later turn brown. (Photo by Tom Zabadal, Michigan State Univ.)

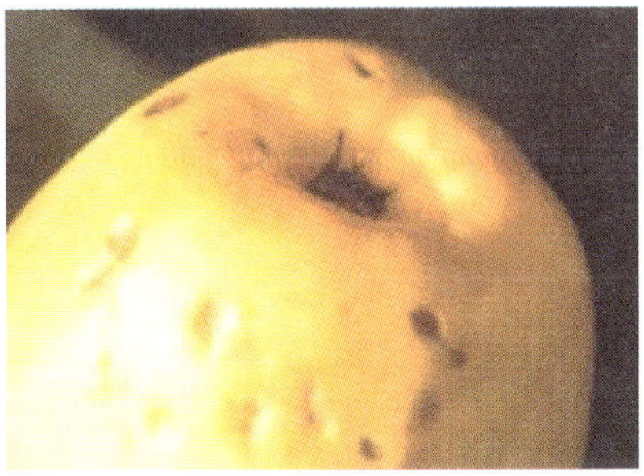

Photo 6-4. Calcium deficiency, apple, causes "bitter pit" of fruit. Note sunken corky spots on this 'Golden Delicious' apple. (Photo from British Columbia Min. of Agric.)

Photo 6-5. Magnesium deficiency, highbush blueberry. Note interveinal chlorosis of older leaves.

Photo 6-6. Magnesium deficiency, muscadine grape. Note interveinal chlorosis and necrosis of older leaves.

Photo 6-7. Iron deficiency, highbush blueberry, common in less-acid soils. Note bright yellow interveinal areas of younger leaves.

Photo 6-8. Iron deficiency, bunch grape. Note interveinal yellowing, especially in young leaves.

Photo 6-9. Manganese deficiency, strawberry. Note very small leaves with interveinal chlorosis. (Photo by Warren Bendixen, Univ. of California.)

Photo 6-10. Severe zinc deficiency, peach. Note leaves curled upward at margins. (Photo by Jerome Frecon, Rutgers Univ., NJ.)

Photo 6-11. Severe zinc deficiency, pecan, common in alkaine soils. Note necrotic shoot tips.

Photo 6-12. Boron deficiency, apple. Note fruit corking throughout flesh, not just near skin as with calcium deficiency. (Photo by Andrea Clarke, Ontario Min. of Agric.)

Photos by William Reid unless noted otherwise.

Fe deficiency often shows up when the pH is too high for a crop, especially acid-loving crops such as blueberries. A high pH renders iron less available. The least expensive solution is to add S to the soil to make it more acid, but an iron sulfate or chelated iron spray can provide immediate iron while waiting for the sulfur to lower pH.

A pH too low for the crop can produce Mg toxicity. The solution here is to raise the pH with lime. Mg deficiency may occasionally occur on alkaline soil; if so, use manganese sulfate sprays or fungicides containing manganese.

Further Reading

Jones, J. B. Jr. 1985. Soil testing and plant analysis: guides to the fertilization of horticultural crops. *Hort. Reviews* 7:1-68.

Shear, C.B. and M. Faust. 1980. Nutritional ranges in deciduous tree fruits and nuts. *Hort. Reviews* 2:142-163.

Stiles, W.C. and W.S. Reid. 1991. Orchard nutrition management. Information Bull. 219. Cornell Cooperative Extension.

Swietlik, D. and M. Faust. 1984. Foliar nutrition of fruit crops. *Hort. Reviews* 6:287-356.

Chapter 7
Irrigation

An irrigation system is essential for growing fruit crops commercially in arid and semi-arid areas, and remains the best form of crop insurance in many humid areas. Irrigating young plantings promotes vigorous growth and earlier production, allowing growers to recoup their investment sooner, and can prevent much first-season death of plants and the cost of replacing them. Irrigating bearing fruit plantings increases fruit size, increasing both quality and yield, often producing a two- to ten-fold return on the investment in irrigation equipment and water. Irrigation during drought also ensures flower bud initiation and development for next year's crop, helping to reduce the biennial bearing tendency of some fruit and nut crops.

DEVELOPING AN IRRIGATION WATER SOURCE

Surface waters or groundwater can be tapped for irrigation. Most states regulate the use of surface and subsurface water, so contact the appropriate state agency for proper permits. Always check on surface water and groundwater rights before buying or leasing land for a fruit planting.

Surface water costs depend on access to existing bodies of water or construction costs for ponds, piping distance, and the cost of increased filtering needed for low-volume irrigation systems. In the eastern half of North America where surface waters are plentiful, diverting runoff water from a stream or building an irrigation pond usually costs less than drilling a well for groundwater. Along the West Coast of the United States, the government has diverted much surface water into dammed reservoirs, canals and other water projects, and set up irrigation water management districts to allocate the water to various farms.

Where aquifers occur near the surface, wells can be economical, and often offer clearer water for irrigation. Well water costs depends on depth to water, well flow rate, and possible problems with minerals dissolved in the groundwater.

Before developing an irrigation system, test all potential water sources for any of these factors that could pose a problem in your area: pH, suspended solids, dissolved solids, calcium, magnesium, sodium, carbonate, bicarbonate, sodium absorption ratio, chloride, sulfate, boron, nitrate, iron, manganese, and bacteria.

Determining Water Source Needs. To determine the size of the irrigation water source you'll need, first determine the amount of water needed by the crop over a growing season, and the period for which natural rainfall will be inadequate to meet this demand. As a general rule, fruit crops require 1 inch of rainfall per week during the spring and fall, increasing to 2 inches per week during July and August. In many eastern and Midwestern areas, spring and fall rainfall is more than adequate (sometimes excessive) to meet water needs, so the irrigation season coincides with the time of greatest water needs in July and August. Western fruit regions often require irrigation throughout the growing season.

Soil type, type of plant, rootstock and cultivar, and plant spacing all influence the amount of irrigation water needed. Most soils will store adequate moisture in the soil profile for 3 to 4 weeks after the last spring rains, but sandy soils will dry out sooner unless they have been enriched with organic matter. Clay soils store more water, but less of that water is available to plants than in loams. Table 7-1 details the available water storage of various soil textures, both in inches of water per foot of soil depth and in gallons per acre-foot. The lower the amount of soil water storage, the more frequently growers will need to irrigate.

The shallower the crop's root system, the sooner it will require irrigation. Strawberries and blueberries will probably need irrigation just 2 weeks after the last spring rain, and require weekly water applications during droughts. Fruit trees on dwarfing rootstocks require more frequent irrigation than those on semidwarfing or standard rootstocks. Later-ripening cultivars often require more irrigation to achieve good fruit size. High-density plantings require more irrigation than the same plants on the same rootstocks at wider spacings.

After you've estimated the gallons of water used per plant for your crop, rootstock, soil type and spacing, multiply that amount by the number of plants per acre to determine the gallons needed per acre. Multiply that number by the number of days during the irrigation season to determine the gallons of water supply needed. For example, each mature peach tree at 18 x 24-foot spacing requires 30 gallons of water per day during summer. The 100 trees per acre at this spacing require 3000 gallons per day. An irrigation season of 92 days, from mid-June to mid-September, requires 276,000 gallons of water per acre of orchard.

This amount could be supplied by 10 inches of water depth from a one-acre pond (because one acre-foot of water = 325,851 gallons), or a 30-gallon-per-minute well running 5 hours every third day. Thus 60 acres of peaches would require a 5-acre pond 10 feet deep, or 5 60-gallon-per-minute wells running 10 hours every day. When you need to develop a water source, compare the construction costs of ponds vs. wells to provide the amount of water you'll need.

Irrigation Ponds and Other Surface Waters. Ponds can be created by embankment, excavation or both. Growers with land containing a sloping stream can construct an embankment pond at lower expense, by building a dam across the stream where the depression is deep enough to provide a water depth of at least 6 feet. Growers with flatter topography can construct an excavated pond in a low spot by digging a pit and piling up the excavated soil onto a berm all the way around the pond. A terrace can be constructed to channel runoff water into an excavated pond. Contact your county's Natural Resources Conservation Service agent to survey the land and determine the best

Table 7-1.
WATER-HOLDING CAPACITY OF VARIOUS SOIL*

	Available Water-Holding Capacity	
Soil Texture	**Inches/Foot of Depth**	**Gallons/Acre-Foot**
Sand	0.24 - 0.72	6,534 - 19,602
Fine Sand	0.60 - 0.96	16,292 - 26,136
Loamy Sand	0.72 - 1.20	19,602 - 32,670
Loamy Fine Sand	0.84 - 1.32	22,809 - 35,719
Sandy Loam	1.20 - 1.68	32,670 - 45,738
Fine Sandy Loam	1.32 -1.80	35,719 - 48,787
Very Fine Sandy Loam	1.56 - 2.28	42,253 - 61,911
Sandy Clay Loam	1.44 - 2.04	39,204 - 55,321
Silty Clay	1.32 - 2.16	35,843 - 58,653
Silty Clay Loam	1.68 - 2.28	45,619 - 61,911
Clay Loam	1.80 - 2.40	48,877 - 65,170
Silt Loam	1.80 - 2.64	48,877 - 71,687
Loam	2.28 - 2.88	61,911 - 78,204
Peats & Mucks	2.40 - 3.00	65,170 - 81,462

*Adapted from NRCS soil survey data.

Water-Volume Equivalents

1 acre-foot = 325,851 gallons
1 acre-inch = 27,154 gallons
1 acre-inch in 1 hour = 450 gallons per minute

site for a pond, the estimated runoff from the recharge area, and the recommended method of construction.

Surface water sources will require a coarse screen filter installed on the pump inlet (suction) line to block trash and debris. Locate the pump inlet two feet below the water surface to avoid floating debris. To filter out organic material and grit, install a sand filter or wafer filter and backwash it periodically. Surface water high in soluble solids may require construction of a settling pond or basin to allow the particles to settle out before using the water for irrigation. When a fast-moving stream is used for the water source, a settling pond is recommended to slow velocity of the water, allowing many particles to settle out. Where a settling pond is not feasible, install a hydrocyclone separator to remove coarse materials.

Algae and bacteria can grow in surface waters and pass through filters, causing biological plugging of trickle irrigation emitters. Certain bacteria precipitate iron and manganese compounds, so emitter plugging problems occur commonly when using water with high levels of both biological activity and these minerals. Ponds may require treatment with copper sulfate to prevent growth of algae and bacteria, or the irrigation system might require periodic treatment with chlorine.

Wells. An irrigation well should penetrate the water-bearing rock formations quite deeply while keeping drilling costs within economical limits. Local contractors and Natural Resources Conservation Service agents can help growers determine the best drilling depth and diameter for their irrigation needs. The well must be sufficiently deep to avoid running dry during drought or other heavy drawdown periods, such as for frost protection during spring frosts. In coastal areas, however, saline water underlies the fresh water and salt-water incursion can occur with deeper wells, limiting growers to shallower wells.

Salinity also poses a problem with groundwater in many arid areas. Know the type of aquifer that you are drilling into, and have this water source analyzed for pH, total soluble salts, sodium absorption ratio, and bicarbonate and carbonate content before drilling the well. If the groundwater is high in any of these measurements, try to find a different water source. If saline water must be used, it could be worthwhile to invest in some method of reducing the salinity, such as reverse-osmosis or ion exchange filters.

Chemical plugging of irrigation systems can result from groundwater high in calcium, magnesium, iron, or manganese. When groundwater is pumped to the surface, the temperature, pressure and pH often change, which can result in precipitation of these minerals with carbonates or bicarbonates to form scale on the inside of the irrigation lines. If the groundwater is also high in hydrogen sulfide, as is much Eastern coastal plain groundwater, its acidity will minimize this scale precipitation. Otherwise, injecting acids or acidifying fertilizers through the irrigation system can lower the pH of the irrigation water and prevent scale buildup. Adding scale inhibitors containing polyphosphates will keep mineral precipitates sub-microscopic in size, preventing chemical clogging.

After a well is developed, perform a pumping test to determine potential well yield and select an efficient pump. Many states also require an aquifer test to determine the effects of newly-installed wells on existing wells, the radius of influence of the well, and the drawdowns in the well for different discharge rates. These tests provide information needed for proper well spacing and groundwater management.

METHODS OF IRRIGATION

The choice of irrigation method makes a large difference in the amount of water needed to achieve optimum yields and profits, as well as the cost and additional uses of the irrigation system. The best irrigation system will vary with the crop, climate, soil type and grower objectives.

Microirrigation: Trickle or Drip Irrigation. Microirrigation involves low-pressure, low-volume application of water on, just above or below the soil surface. Trickle irrigation, also known as drip irrigation, was developed by Israeli scientists to greatly reduce the amount of water needed for irrigation applications. Since its development some 40 years ago, trickle irrigation has become the method of choice in most new irrigation installations. Recent modifications of trickle irrigation have included burying the lines for subsurface irrigation, and using entire lateral lines or microsprinklers as water emitters.

Flexible polyethylene tubes (*lateral lines*) run down each row of the fruit planting, with one to several *emitters* placed at each plant to allow water to slowly trickle or drip onto the soil around the plant (Photo 7-1). Trickle irrigation wets only the root zone around the fruit plants, not the entire orchard floor or groundcover, and water does not accumulate on the soil surface as it does with furrow or flood irrigation, so almost no water is lost to evaporation. This can create water-use efficiencies of 90 to 95% and can cut the amount of irrigation water needed per year, as compared to furrow or flood irrigation, almost in half.

Trickle irrigation also requires 20-30% less water than sprinkler irrigation, because the water is applied directly to the soil, rather than to the fruit plants where much of it evaporates. Avoiding wetting the foliage can reduce many foliar diseases, as well. However, trickle systems lack the capacity of sprinkler systems to provide spring frost protection and evaporative cooling in summer, which are important for some fruit crops in certain climates.

Trickle irrigation systems are easily clogged, so require a filter to remove particles in the water (Photo 7-2). Surface water sources may require more filtration than groundwater. Trickle systems also require a very slight slope of just 2 to 3%; trickle systems can be laid out on the contour for steeper slopes. Keep lateral lines 250 to 300 feet long or less, to minimize pressure differences from one end of the line to the other.

Various styles of emitters are available, each emitting a certain amount of water per hour in a specific pattern. Match the emitter style and number to the crop and soil texture. For example, small fruits such as blueberries or brambles often have two emitters per plant, both arising directly from the lateral line (main tube) within the row, one on each side of each plant. Young fruit trees are treated similarly, while mature fruit trees might have four to six emitters per plant, with each emitter consisting of a smaller tube that extends outward from the lateral line, creating a ring of wetted soil around each tree.

On sandy soils, water tends to move straight down, so choose emitters with lower flow rates, and place more emitters around each plant, so that the water spreads out more horizontally around the plants. Use fewer emitters with higher flow rates and run the system less often on loamy soils.

The size of the wetted soil zone will determine root distribution in semi-arid areas, so be sure to provide enough emitters per plant and wet as wide and deep an area as each plant normally uses. Insufficient trickle irrigation can limit root spread, which means plants require even more frequent irrigation. Trickle-irrigated crops do require irrigation more often than crops irrigated by other methods, usually every 1 to 4 days during the summer as opposed to every 7 to 14 days.

On trellised plantings, trickle irrigation tubes can be tied to trellis wires to help prevent damage from equipment and animals (Photo 7-3). Other variations of trickle irrigation include the following.

Subsurface Microirrigation involves burying the trickle irrigation lines a few inches to two foot below the soil surface, depending on the main rooting depth of the crop. This minimizes damage to the

Photo 7-1. This trickle-irrigated almond orchard has a drip tube running down each side of the herbicide strip in each row. This irrigates only the tree row, not the groundcover, resulting in a 90-95% water-use efficiency.

Photo 7-2. A trickle irrigation system requires a filter to remove small particles that can cause clogging. A surface water source may require more filtration than groundwater.

lines from equipment and many animals, and can aid water penetration in soils with less than ideal permeability. Keeping the water below the soil surface reduces germination of shallow weed seeds, and minimizes fruit rots of crops such as strawberries, which fruit on the ground. The expense of burying the lines increases installation costs, but the system can last 12 to 15 years. Lateral lines for subsurface microirrigation can be either polyethylene tubes with built-in emitters set at certain intervals along the tube, or porous tubes.

Porous tubes are lateral lines that exude water along their entire length like a garden soaker hose, eliminating the need for separate emitters (Photo 7-4). They can be installed above or below ground, and can be applied under plastic mulch by modifying the mulch-laying implement to hold a reel of tubing. Porous tubes are useful for closely spaced crops that require continuous irrigation along the row, such as strawberries and blueberries. Some orchardists use porous tubes on both sides of tree rows, to provide a continuous ribbon of water available to the root system.

Drip tape has flat emitters built into the lateral line every 6 to 24 inches, and can be machine-installed like porous tubes. Drip tape has the smallest initial cost but also the shortest useful life of trickle irrigation options. It's used mainly for strawberries and vegetable crops.

Microsprinklers are low-volume systems that spray water in 5- to 12-foot circular patterns a few inches above the ground. Microsprinkler systems use flexible polyethylene lateral lines like those of trickle irrigation, and mini-sprinkler heads replace the emitters. Microsprinklers are often preferred to other forms of trickle irrigation on sandy soils, because their larger wetting pattern reaches a larger percentage of the root system.

Although microsprinklers loose a modest amount of water to evaporation and operate at about 90% water-use efficiency, these systems can also provide frost protection or evaporative cooling to strawberries and the lower branches of blueberries. Microsprinklers are economical on strawberries and blueberries, which requires frequent summer irrigation due to shallow roots, as well as spring frost protection.

Sprinkler Irrigation. Sprinkler systems require larger-volume, rigid lateral lines and high-pressure pumps to spray out water onto the plants as well as the soil. Sprinklers use more water and energy than trickle systems, with a 75 to 85% water use efficiency, but wetting the plants can provide spring frost protection for open blossoms as well as evaporative cooling of foliage in the summer. Sprinkler systems can be used on sloping land as well as near-level land. Researchers have demonstrated that sprinkler irrigation reduces spider mite damage in orchards and vineyards, but wetting the foliage can spread certain diseases.

Sprinklers are most economical where a crop requires spring frost protection or summer evaporative cooling, or both, in addition to irrigation. Late-blooming crops that withstand summer heat well can be trickle irrigated with less water and energy, unless the land is more than gently sloping..

Sprinkling for frost protection is discussed in chapter 8. Evaporative cooling irrigation is activated by thermocouples in the fruit when temperatures exceed a certain level, and deactivated when temperatures fall below that level. Water evaporating from the leaves and fruit surfaces cools the

Photo 7-3. On trellised plantings such as these dwarf apples, trickle irrigation lines can be tied to the trellis to reduce damage from equipment and animals. Some growers bury drip lines to achieve the same purpose.

Photo 7-4. Some Western growers eliminate all summer vegetation on the orchard floor to eliminate competition for water and nutrients. This almond orchard uses porous tubing, a type of trickle irrigation in which the lateral lines ooze water continuously along their length.

plants, preventing heat stress. Excess heat causes higher rates of respiration, leaving fruit smaller and less sweet with less red color in the skin, and can sunburn the fruit. Evaporative cooling increases fruit size, sugar content and red color, and can allow earlier harvest, which can be very important due to higher early-season prices.

Sprinkler systems fall into three basic groups.

Under-tree sprinklers perform well in orchards as well as small fruit plantings, where they are simply called sprinklers (Photo 7-5). These are small, low-angle sprinklers covering 30 to 90 feet per set, usually spaced 20 to 60 feet apart on the sprinkler lateral. This close spacing and the low trajectory of water from the nozzles minimize distortion of the wetting pattern, a common problem with taller sprinklers, especially in windy areas. The short riser pipes and light weight of these systems makes them easy to move by hand.

Under-tree sprinklers can provide frost protection or evaporative cooling to small fruits, dwarf trees or the lower parts of medium-sized trees. They can provide the lower water application rates needed for frost protection, about 0.1 to 0.3 inches per hour. They minimize summer wetting of upper foliage, which can reduce disease spread.

Under-tree sprinkler systems require more lateral pipe and sprinkler sets per acre than do taller sprinklers, due to the closer spacing of the sprinklers. This higher initial cost per acre is offset by the high portability of under-tree sprinklers. Many growers purchase only enough pipe and sprinklers to cover perhaps one-third of their acreage, then move the pipe as needed to irrigate all the fields in succession.

This system works especially well for frost protection of a mix of crops or cultivars with different bloom times. However, labor costs of moving pipe during a long summer drought can become expensive, unless the labor is otherwise not fully utilized during those dry periods, so some growers permanently set small sprinklers. Under-tree sprinkler systems require only 35 to 70 psi of water pressure, so offer the lowest operating costs of sprinkler systems.

Through-the tree sprinklers are medium-sized sprinklers, normally spaced 60 to 90 feet apart on the lateral line and covering a slightly larger width (Photo 7-6). These sprinklers require less pipe and risers per acre than under-tree sprinklers, but operate at higher water pressure, which increases the size of pump needed and operating costs. They can be moved by hand but are heavier, and the taller risers hamper movement between tree rows, so they are usually set permanently in place.

Through-the tree sprinklers can provide greater frost protection and evaporative cooling to medium and large trees than do under-tree sprinklers. Greater summer foliage wetting can increase diseases, although this can be minimized if evaporative cooling is done only during the hottest parts of the day and irrigation is avoided in mornings and evenings. Wind distorts the irrigation pattern more than that of under-tree sprinklers.

Traveling irrigation guns (Photo 7-7) are another, portable form of through-the tree sprinklers. These guns can be towed by tractor to various locations, so can cover more acreage in succession at less cost. A gun is placed at one end of each planting, with the hose unrolled, and the water pressure

Photo 7-5. Movable low sprinklers, or under-tree sprinklers, can be hand-moved to different fields in rotation, reducing initial costs of irrigation equipment but increasing labor costs. Such low sprinklers can provide frost protection of low crops such as these blueberries or strawberries, and partial frost protection of low or dwarf fruit trees.

Photo 7-6. Medium-sized sprinklers, or through-the-tree sprinklers, cover larger areas than low sprinklers and provide greater frost protection for tree crops, but require higher pressures and pumping costs and are subject to greater wind distortion.

during irrigation rolls the hose up again, pulling the sprinkler gun to the other side of the field as it does so. Thus the single gun irrigates an area the width of the sprinkler's throw diameter and the length of the hose, within the few hours that it takes the hose to roll up.

Over-tree sprinklers are large, high-pressure sprinklers mounted high enough to throw water over the treetops (Photo 7-8). Each sprinkler set can cover one to three acres. Over-tree sprinklers often require a tractor to be moved, so most growers who install them fix them permanently in place. These growers usually bury the main lines and laterals so they don't interfere with orchard operations.

Over-tree sprinklers are the most expensive to install and pumping costs are the highest, so they are only justified for taller orchards requiring frost protection or evaporative cooling as well as irrigation. Over-tree sprinklers don't work well in windy areas, where distortion of the sprinkler pattern can leave large dry spots in the orchard.

Furrow Irrigation. Some growers irrigate the furrows or aisles between raised beds of small fruit crops, such as grapes, strawberries, raspberries or blueberries. This method can be quite economical to establish if water can be pumped from the water source into an irrigation channel which flows into the upper end of a field with a 2 to 3% slope, then flows down the furrows. However, furrow irrigation uses a great deal of water, with only 50 to 70% efficiency, and wets most the aisles that need water the least.

This method is most suited to row crops and, of the fruit crops, to small fruits with their narrower spacing between rows. Furrows are sometimes constructed down both sides of tree rows specifically for irrigation, but such furrows interfere with tractor movement and collect debris, requiring periodic cleaning out.

Furrow irrigation should be reserved for loamy soils. On sandy soils, too much water quickly percolates down below the rooting zone, where the crop can't use it. On soils high in clay, water infiltrates slowly, and muddy aisles can keep workers and pickers out of a field for days after a furrow irrigation.

Growers can increase water-use efficiency by shutting off the sluice gates to every other furrow during an irrigation, then irrigating the other furrows during the next scheduled irrigation. Although this requires a greater number of irrigations, the total water use can be greatly reduced.

Flood Irrigation. Flood irrigation involves covering the entire orchard floor with water, between cross dikes (usually the tree rows) that impound the water (Photo 7-9). The orchard must be leveled with a gentle, 2 to 3% slope, and the soil must be high in organic matter with a good water infiltration rate. Land preparation costs can be quite expensive. Flood irrigation wastes the most water due to evaporation from the entire flooded surface. Flooding is used mainly on alkaline soils prone to salt buildup, as it can leach damaging salts below the root zone. Cranberry growers also use flood irrigation, because cranberries are water-harvested and can withstand flooding, while many weeds cannot.

Laser leveling of flood-irrigated orchards has become common in the past few years. This allows

Photo 7-7. Traveling guns are another form of through-the-tree sprinklers, that can be moved to cover larger acreage at less cost. Here, a traveling gun irrigates pecans.

Photo 7-8. Over-tree sprinklers cover the largest area, require the greatest pressure and pumping cost, and suffer the most wind distortion of sprinkler systems. They are most useful for tall crops, such as these apples on standard rootstock, that require spring frost protection and summer evaporative cooling as well as irrigation.

Photo 7-9. Flood-irrigated orchards, such as this pecan planting, cover the entire orchard with water between the raised dikes along the tree rows. Laser leveling can improve the efficiency of flood irrigation from the usual 50% up to 80%.

"dead-level" irrigation, where the grower applies only the water needed to refill the root zone, minimizing water waste. The grower floods the orchard as rapidly as possible to minimize evaporation and differences in water intake between one end of the orchard and the other. Dead-level irrigation requires border distances shorter than 300 to 500 feet, depending on soil type, and at least 3 inches of water application per irrigation. Laser leveling and dead-level irrigation can increase the efficiency of flood irrigation from the usual 50% up to about 80%.

Managers of flood-irrigated orchards have further reduced water use in recent years by applying water to every other tree row middle instead of to every middle. The next irrigation will apply water to the middles left dry by the previous irrigation, and close off water access to the middles irrigated previously. This method produces less tree stress than doubling the time between irrigating the entire orchard.

IRRIGATION SCHEDULING

The total amount of irrigation water a particular crop requires will vary with the amount of rainfall, sun intensity and evapotranspiration during a particular season. The best way to distribute that total amount of water—the irrigation rate and frequency—depends on the type of irrigation system and water supply, the size of the crop's root zone, and such soil factors as infiltration rate, water-holding capacity, aeration and leaching requirements.

The polyethylene pipe costs of trickle irrigation systems favor smaller application rates, and thus more frequent applications, to minimize pipe sizes. Shallow-rooted crops such as strawberries and blueberries also require more frequent irrigation, applying less water each time. Sandy soils or soils low in organic matter, with reduced water-holding capacity, and wells with low output rates would similarly need frequent irrigation at low rates.

The fixed costs associated with each water application using sprinkler, furrow or flood irrigation encourage less frequent irrigation of greater amounts of water, so as to minimize the number of applications. Deep-rooted fruit and nut crops on standard or drought-tolerant rootstocks, and drought-tolerant small fruits such as blackberries, do well with more widely-spaced irrigation. Soils high in silt or clay content and organic matter, with high water-holding capacity, respond well to fewer applications of more water each time. Alkaline soils prone to salt buildup require at least one very heavy irrigation per season, to leach salts out of the root zone.

Growers determine when to irrigate by estimating crop water consumption, usually by measuring the evaporation of free water or soil water status, or by calculating the depletion of a crop water budget. No matter which of these methods, described below, is used, the grower must make allowances for the efficiency of the irrigation system. For example, if a crop needs 1.0 inch of water and the grower uses a trickle system that is 90% efficient, the application rate is 1.0 inch/0.90 = 1.11 inches, or 30,141 gallons per acre. If the grower uses a sprinkler system that is 75% efficient, he'll need to apply 1.0 inch/0.75 = 1.33 inches, or 36,115 gallons per acre. A furrow system that's 50% efficient would require an application rate of 1.0/0.50 = 2.00 inches, or 54,308 gallons per acre.

Determine how many minutes to run the irrigation system by dividing the total gallons needed for the entire acreage by the flow rate in gallons per minute. For the trickle irrigation example where 30,141 gallons per acre are needed, if a 20-acre block is irrigated at one time and the flow rate is 1600 gallons per minute, run the system for (20 A)(30,141 gallons per A)/1600 gallons per minute = 377 minutes = 6 hours and 17 minutes.

Install a flow meter to accurately measure irrigation water, to verify that the correct amount was applied. Meters are available that register in gallons or acre-inches.

Evaporation Measurements. The rate of evapotranspiration, the total amount of water transpired into the atmosphere by the crop as well as evaporated from the soil over a period of time, is highly correlated with the evaporation rate of water from a pan. Evaporation measurements can therefore be used to estimate evapotranspiration, which equals water use for a field. In arid areas and during droughts in humid regions, growers simply supply an amount of water equal to evapotranspiration to maximize yields and quality. The amount of any significant rainfall should be subtracted from the amount of irrigation water to add.

The U.S. Weather Bureau Class-A evaporation pan is widely used to measure evaporation. Any other rigid metal or plastic container having a minimum diameter of 10 inches and a maximum depth of 12 inches can be used, so long as its sides are vertical. Install the container in an open spot, level and elevated to prevent use by animals. Fill the container to within 3 inches of the top, measure the depth with a ruler to the nearest 1/8 or 1/10 of an inch, and record this depth in a notebook. Measure the depth again at a set time each week (or as often as you usually irrigate), then add water to the original depth. The new measurement subtracted from the original water depth is the inches of water evaporated over the period.

Another way to measure evaporation is with an atmometer, an instrument that measures the amount of water evaporated to the atmosphere from a wet, porous ceramic surface covered with green canvas. Install an atmometer in an open part of the planting, placing the instrument in a vertical position with the top of the ceramic cup 3.3 feet above the soil surface. Read the water evaporated, in inches, periodically, and refill the reservoir with distilled water when the water level drops to below two-thirds of the reservoir.

Convert this evaporation data to gallons of water lost per acre by multiplying the inches of water lost by 27,154 gallons per acre-inch. For example, if 5/8 inch of water evaporated, that is (0.625 inches)(27,154 gallons per acre-inch) = 16,971 gallons per acre lost. Multiply this figure by an estimated crop water-use value (k) that usually ranges between 0.5 and 1.0, to determine the gallons of water per acre needed. Find the k value for your crops from your Extension agent, or at the Web site http://www.irrigation.org with its ET (evapotranspiration) connections to states.

The k value increases with closer plant spacing, and decreases with less mature plants or greater drought tolerance of a crop such as grapes or apples on MM111 rootstock. For the example above on grapes with a k value of 0.7, the crop will need 11,880 gallons of water per acre to replace water lost over the period measured. Divide this number by the irrigation system's efficiency to determine how much water to apply.

Crop Water Budgets. Extension specialists in many states or regions have determined the amount of water needed by each type of fruit crop during each week of the growing season, based on average temperatures. This data provides an estimate of evapotranspiration without having to measure evaporation. Growers can use these basic crop water budgets, adjusted for their particular cultivars, rootstocks and plant spacing, to determine when and how long to irrigate.

For example, if a planting requires 3600 gallons of water per acre per day during the first week of August and no rain has fallen since the last irrigation, that entire amount must be supplied by irrigation. If the planting is trickle-irrigated every 4 days, it needs 14,400 gallons of water per acre per irrigation, divided by the irrigation efficiency of 0.95, so apply 15,158 gallons. If ¼ inch of rain fell within the past 4 days, that provided (0.25)(27,154 gallons per acre-inch) = 6788 gallons of water, so irrigation need supply only 8370 gallons. Alternatively, delay the next irrigation by 2 days and apply the 8370 gallons plus two days' worth of water, 7579 gallons, for a total of 15,949 gallons.

Contact your Extension agent to find out what crop water budgets have been developed for your region, or check the Web site http://www.irrigation.org for its ET connections to states. Some Extension services also offer weather monitoring and evapotranspiration data to help determine more exact water budgets for the current season.

Photo 7-10. Soil tensiometers, each with a porous cup permanently set in the ground and its vacuum gauge indicating a moisture tension reading, can help growers schedule irrigation. Deep-rooted crops need tensiometers at two depths, and this especially deep-rooted pecan grove has tensiometers at three soil depths.

Soil water status. Measuring the amount of water available to plants in the soil is another method of determining irrigation needs, especially useful in humid areas where rainfall is frequent even when inadequate. Soil tensiometers (Photo 7-10) are instruments placed in the soil that measure the tension with which soil water is held. Low tension indicates plenty of water available to plants, and tension above a certain point, which varies with the fruit crop and the soil texture, indicates a need to irrigate.

Typical tensiometers measure a scale of 0 to 100 centibars soil water tension. Growers adjust the vacuum gauge on top to read 0 when the tensiometer's porous cup is immersed in a pan of water, then place the cup in the soil in a hole made with an auger or soil probe.

Readings of 0 to 10 indicate soil saturation following a drenching rain or irrigation. Readings of 10 to 30 indicate adequate water and adequate aeration of the soil. Begin irrigating when readings reach 30 to 60, depending on the crop and soil type. In general, irrigate loamy sands and sandy loams when readings reach 30 to 40, silt loams when readings reach 40 to 50, and silty clay loams and silty clays when readings reach 50 to 60. Calculate the approximate length of time to irrigate using the crop water budget method, then check tensiometers the day after the irrigation, when they should read 10 to 15. Tensiometers in trickle-irrigated plantings should always remain between 10 and 20 to 30, depending on soil type, until irrigation stops before harvest.

Growers can purchase tensiometers in lengths ranging from 6 to 48 inches, to sample soil water content at various depths corresponding to the greatest concentration of roots for that crop. Normally, orchardists and vineyardists use two lengths, usually 18 inches and 3 feet for sandier soils, or 2 feet and 4 feet for loamy or clay soils. Strawberry growers need only 6-inch tensiometers, while other small fruit growers use only 12-inch or 18-inch tensiometers. Place the tensiometers permanently in several representative locations in a fruit planting, each on the southwest side of a plant, using one tensiometer or pair of tensiometers for every 20 acres. Growers can monitor them by eye several times a week, or can hook them up to an electronic data logger and computer program that will warn them when it's time to irrigate each block of each crop.

Several other instruments also measure soil moisture. The Gro-Point™ soil moisture sensor operates on a principle similar to radar, measuring the speed of electromagnetic waves, which are highly influenced by the amount of soil moisture surrounding the probe. Like tensiometers, this instrument is placed permanently in the soil and can be hooked up to a data logger and computer program. The Web site http://www.irrigation.org offers an irrigation product locator and irrigation software links.

REDUCED IRRIGATION SCHEDULING

Fruit scientists in Australia, where water is the greatest limiting factor in fruit production, have developed two new methods of scheduling irrigation that can further reduce water use.

Regulated deficit irrigation eliminates any irrigation during the early season, from bud break to fruit set and the period during which fruit remain small. Water deficits during this time will reduce vegetative growth, which is often desirable to keep fruit plants smaller and minimize pruning needs, but will not adversely affect fruit growth.

The length of this period of no irrigation will vary with the ripening time of the crop, from just 6 weeks of no irrigation for early-ripening crops like cherries and blueberries, to 2 months for mid-season peaches and Japanese plums, to 3 months for mid-season apples and pears, to 4 months for pecans and walnuts.

Once plants enter the stage of rapid fruit expansion, irrigate as much as needed, which will maximize yields and provide enough water for the plants to grow as needed and set fruit buds for next year. Early-ripening crops will also need some irrigation after harvest, to ensure plant survival and fruit-bud set for the following year. Late-ripening crops will usually need no irrigation after harvest.

Partial root drying irrigates only one side of each row of fruit plants during each irrigation. This can be achieved by running trickle irrigation or microsprinkler lines down both sides of each row, and turning off one line per row during each irrigation. During the next irrigation 5 to 7 days later, run the lines which were turned off during the last water application, and turn off the lines which ran last time. Furrow irrigation can also produce partial root drying, by applying water to every other furrow during each irrigation.

The half of the root system that dries produces abscisic acid due to the stress, which is transported through the plant to all the leaves. There, abscisic acid induces the leaves' stomata, the pores that allow water vapor loss and the gas exchange needed for photosynthesis, to stay closed longer. The closed stomata minimize water loss and vegetative growth, reducing the total need for irrigation. By watering the stressed side of the plants during the next irrigation within 7 days, root death is minimized and the production of abscissic acid is prolonged.

Partial root drying works very well for crops that suffer reduced fruit quality when provided all the water they can take up, such as wine or raisin grapes. It also works well on fruit trees that tend to grow too vegetatively, such as apples and pears. It has not been tested and probably wouldn't work well on small fruit with limited root systems, such as strawberries and blueberries.

SOIL MANAGEMENT TO REDUCE PLANTS' NEED FOR IRRIGATION

Irrigation is expensive, and growers can reduce the need to irrigate in several ways. Choose a cool-season groundcover, that doesn't compete much with the fruit crop for water during the growing season, and control weeds within the fruit rows. Some growers in arid areas, with level sites not prone to soil erosion, control all vegetation in the orchard during the dry season. Increasing the soil's organic matter content, through cover crops added before planting or grown over the winter, and mulches used between the fruit crops, increases the soil's ability to store water.

Adding gypsum (calcium sulfate) to an alkaline soil high in sodium can improve water penetration and retention. Sample the soil and have it tested by a commercial laboratory to determine how much gypsum to add. Usually, growers apply 3 to 5 tons of gypsum per acre in the fall, incorporating it into the soil as deeply and thoroughly as possible.

Gypsum can also improve infiltration of very low-salt irrigation water into loams or fine sandy loams. Some Western growers rely on irrigation water that originated as snow melt and is very pure, with less than 250 ppm soluble salts, and moves very slowly into the soil. Such growers have increased infiltration rates up to threefold by applying 1 to 2 tons of gypsum per acre in the spring, spreading it on the soil surface under drip irrigation lines or in furrows without incorporating it. Bill Peacock, farm advisor to Tulare County, CA, also found that adding calcium continuously to a drip irrigation system tripled the water infiltration rates under such conditions. Adding gypsum or calcium will not improve water infiltration of a heavy clay or compacted soil, however.

Some growers have added a polyacrylamide gel product to the soil at planting time, to absorb excess water and give it back during droughts. The product is granular, but once in the soil each granule absorbs up to 500 times its weight in water, swelling into a gelatinous chunk about 1/2 inch in diameter. This product is expensive to apply and needs further research, but has greatly reduced irrigation needs in arid areas in some trials. However, some studies have shown that adding this gel product to the soil increased mortality of plants in the first year.

FERTIGATION

Water-soluble fertilizers can be injected through trickle and sprinkler systems, a process that is termed fertigation. The injection point must be upstream of the filters, and the irrigation system must be given time to reach normal operating pressure before injecting fertilizer. N and K are the most commonly injected fertilizers.

Fertigation reduces fuel use by eliminating one or two tractor passes through a fruit planting per year to apply fertilizer, and minimizes leaching loss of N by adding small amounts more frequently. Generally, add fertilizer to the irrigation between March and July. When no spring irrigation are needed, fertilizers must be applied with a fertilizer spreader as usual. Western growers who must irrigate longer will recover the additional cost of fertigation equipment quicker than Eastern growers will.

The type of N to add depends on the soil pH and salinity of the irrigation water. When urea, ammonium nitrate or other ammonium forms of N are added through a trickle system, soil pH decreases in the zones wetted by emitters. In eastern North America where soils are slightly acid, this lowered pH can cause higher availability of toxic aluminum, which reduces root density, water absorption and yields. Lowered pH due to fertigation has also increased availability of Fe, Mn, Zn and sometimes P, and reduced availability of Ca, Mg and K.

Therefore, Eastern growers who fertigate should generally use calcium nitrate or potassium nitrate. If the water source is high in bicarbonates, use potassium nitrate, since calcium nitrate will cause clogging as the calcium combines with the bicarbonate and precipitates. Dry potassium nitrate requires a great deal of water and time to dissolve due to its low solubility, so it's best to use a more expensive liquid formulation labeled for irrigation injection.

In Western areas where soils are alkaline, use acidifying forms of N, particularly ammonium thiosulfate. This will increase availability to plants of nutrients such as Zn and P, which are often tied up in such soils. Fertigating with acidifying N forms can improve marginally saline irrigation water, by freeing soil calcium ions to exchange with some sodium ions in the irrigation water. Lowering soil pH can also prevent cotton root rot, to which many fruit trees and pecans are susceptible.

Phosphoric acid is sometimes injected both to lower pH and provide P. This should not generally be done when the water source is high in calcium, as it can cause calcium carbonate to precipitate at the interface between the water source and injection, resulting in clogging. Scale inhibitors containing polyphosphates keep mineral precipitation in the sub-microscopic range so it doesn't cause clogging, and provide small amounts of P to plants. P may not be needed each year on many soils, and if it is a single spring soil application should suffice.

Fertigation with only N has led to K deficiencies even when adequate soil K levels exist, especially when ammonium forms of N are used. When fertigating, provide both N and K, even when soil K is adequate.

When plantings require fertilization with a micronutrient, growers usually apply sprays so the foliage will absorb the micronutrient quickly. If micronutrients are injected, use chelated forms rather than sulfate forms in the East, to avoid lowering pH, and wherever the water contains high amounts of calcium or magnesium that would precipitate out the sulfate.

Positive displacement injection pumps provide better control of fertilizer injection rates, so are generally preferred over venturi or pressure differential pumps. Choose an injection pump size that will allow the maximum fertilizer application during a season to be completed in 90 minutes or less. After each fertigation, flush all fertilizer from the lateral lines before shutting the irrigation system down, to avoid plugging of emitters or nozzles from the dissolved minerals.

Always install safety backflow prevention devices, in both the main system line and the injection pump line, to prevent accidental flow of chemicals back into surface or groundwater. Most states require this by law. Do not inject fertilizers in combination with chlorine (for killing bacteria in the lines) or pesticides.

SALINITY MANAGEMENT

Crop-damaging salts often accumulate in irrigated arid soils. Use of saline irrigation water causes especially rapid salt accumulation, but even irrigation water with low salt content will eventually cause soil salinity in arid areas, due to the lack of rainfall to wash the salts down below the root profile.

A salt is simply a positively charged ion combined with a negatively charged ion, which dissolves in water and breaks into these two components. Common table salt, sodium chloride, dissolves into positively-charge sodium ions and negatively charged chloride ions. These two ions are among the most toxic to fruit and nut plants, but high concentrations of calcium, magnesium or sulfate ions can also be damaging.

Chemical fertilizers are salts, and animal manures are high in salts, so overapplication of these types of fertilizer can worsen a salinity problem. However, adequate nutrient levels and good management promote vigorous growth that dilutes salt concentrations in plants, minimizing damage. Underfertilization can cause weak growth, resulting in higher salt concentrations in plant tissues and greater salt damage.

Tipburn and marginal burn of leaves usually indicates chloride or sodium toxicity, as described in chapter 6. Collect affected leaves for leaf analysis to be sure; sodium content above 0.2% or chlorine content above 0.5% confirms this diagnosis. Some rootstocks or cultivars resist damage by absorbing less sodium and/or chloride, so growers with salinity problems should try to plant such salt-resistant crops.The frequency and method of irrigation can influence the amount of salt that fruit plants take up. A given level of soil salinity will damage fruit crops less when growers irrigate more frequently. .

After an irrigation, crops absorb water from the soil but take up very little of the salt dissolved in the water, so the remaining soil water becomes saltier. Evaporation of water from the soil further increases the salinity. Frequent irrigation keeps diluting this salinity, resulting in less damage. Irrigate saline soils when tensiometer readings reach 20 to 33, rather than waiting until readings reach 40 to 60 on a non-saline soil. Continue irrigating until tensiometer readings reach 10; shallow irrigation causes salt buildup in the root zone.

Trickle, microsprinkler or low sprinkler irrigation, that doesn't wet the leaves, causes the least salinity damage. Leaves can absorb sodium or chloride directly from irrigation water sprinkled on them, and be damaged by sprinkled water that contains as little as 70 parts per million sodium or 100 parts per million chloride, which would not damage plants if applied to the soil. Fortunately, strawberries absorb relatively little salt through their foliage, so may be safely irrigated with sprinklers or microsprinklers.

Salt accumulates in the ridges of furrow-irrigated plantings, so avoid furrow irrigation wherever salinity poses a problem. Subsequent rainfall can wash this high salt concentration into the root zone, causing extreme damage.

Excess water beyond that needed for crop use must be added to saline soils, to leach salts below the root zone. That excess water may be added with each irrigation by irrigating more frequently as described above, or the field may be periodically flooded, perhaps once per year. The extra water required for leaching varies with the crop's salt tolerance, but usually means that 17 to 20% more water must be added per season.

Further Reading

Bernstein, L. 1980. Salt tolerance of fruit crops. USDA Agric. Info. Bull. 292.

Bravdo, B. and E.L. Proebsting. 1993. Use of drip irrigation in orchards. *HortTechnology* 3:44-49.

Elfving, D.C. 1982. Crop response to trickle irrigation. *Hort Reviews* 4:1-48.

Haynes, R.J. and R.S. Swift. 1987. Effect of trickle fertigation with three forms of nitrogen on soil pH, levels of extractable nutrients below the emitter and plant growth. *Plant & Soil* 102:211-221.

Neilsen, G, P. Parchumchuk, M. Meheriuk, and D. Neilsen. 1998. Development and correlation of K-deficiency in drip-irrigated apple. *HortScience* 33: 258-261.

Several states' Extension Services offer numerous helpful publications on irrigation, including many that can be downloaded off the Internet. Some good sites:

Alabama: http://www.aces.edu/department/irrig

Colorado: http://www.colostate.edu/Depts/CoopExt/PUBS/CROPS

Florida: http://edis.ifas.ufl.cdu/MENU_AE Florida also offers an Irrigation Tool Box CD-ROM for $30. It includes programs to design and operate irrigation systems and calculate evapotranspiration. See http://it.ifas.ufl.edu/software/toolbox-cd.html

Washington: http://sis.prosser.wsu.edu This site has a link to the Public Agriculture Weather System, and the Washington Irrigation Scheduling Expert software can be downloaded at no charge.

Chapter 8
Preventing Winter and Frost Damage

Fruit plants are notorious for suffering damage from winter cold or losing crops to spring frosts, but proper selection of type of crop, cultivar and site, along with some timely measures, can prevent most problems.

PROTECTING FRUIT PLANTS FROM WINTER DAMAGE

Temperate fruit crops can withstand very cold temperatures, as long as the plants prepared for the cold. Lack of plant acclimation and severe fluctuations in winter temperatures cause more cold damage than do extremely low minimum temperatures.

Encouraging plant hardening for winter. Each fall, fruit plants begin to harden in preparation for winter. Growth slows and metabolic changes take place that enable plants to withstand much lower temperatures without damage. Help the plants harden by allowing their growth to slow. Don't apply too much fertilizer or apply it past midsummer, because high nitrogen levels would cause continued growth. Don't irrigate from early fall on. Mow the groundcover less often in early fall to allow the grass to compete and help slow tree growth.

Stop summer pruning at least two months before your average first frost date, so as not to stimulate regrowth that would delay plant hardening. Wait until late winter or early spring to begin dormant pruning, as plants pruned in late fall or early winter are more susceptible to cold damage.

During the winter, desiccation can cause as much injury to plants as cold temperatures. If fall weather has been dry, irrigate fruit plants well after they are dormant, but before the ground freezes. Some fruit plants also benefit from a light nitrogen application after they are dormant, in areas where winters are mild and roots keep growing. Nitrogen-deficient plants are more subject to winter damage.

Cover strawberries with clean straw mulch six inches thick about a month after the first frost. This will protect the fruit buds during winter and increase next year's yields. Remove straw mulch in early spring as new leaves start to grow, several weeks before the last spring frost, placing the mulch between the aisles (Photo 8-1).

Photo 8-1. Protect strawberries from winter heaving of the crowns which is caused by repeated soil freezing and thawing. Cover them with 6 inches of clean straw after the first hard freeze of fall, and remove the straw a few weeks before the last spring frost, as here.

Preventing sunscald and damage from mammals. Paint tree trunks white in late fall to prevent winter sunscald. This damage, often worse than that from absolute cold, occurs when the unshaded trunks heat up to high temperatures during the day (often 25 to 40 degrees F above air temperatures), then rapidly cool to sub-freezing temperatures at night. Combine one part interior white latex paint with two to three parts water, and spray or brush this mixture on trunks up to the first main limb.

Rabbit and rodent repellents can be mixed into the paint to prevent these animals from feeding on the bark in the winter, as they can severely damage trunks and kill trees (photo 8-2). Or place 1/4-inch mesh hardware-cloth cages around trees, making sure the guards reach above the expected snow level (Photo 8-3). On young trees, white plastic tree guards can prevent both sunscald and rodent damage (Photo 8-4). Hoe any tall vegetation or mulch away from tree trunks in the fall to avoid attracting rodents.

Deer can greatly damage fruit trees in winter by feeding on tender twigs and branches. Electric fencing provides the most permanent solution, but several commercial deer repellent sprays are available. The sprays must be applied at monthly intervals through the winter.

SPRING FROST PROTECTION

Choosing fruit crops adapted to your area's expected spring frosts and a site with good air drainage, as mentioned in chapters 1 and 2, is the best and least expensive way to prevent losses to spring frosts. Even well adapted crops on the best sites, however, may lose a portion of their fruit to frosts almost every year (Photo 8-5). Several frost protection methods may reduce these losses, but in many cases the expense of frost protection can be greater than the value of the fruit losses. Therefore, estimate as accurately as possible the extent of economic losses from spring frost damage in your area, before investing in frost protection equipment.

The degree of damage to flower buds caused by a given low temperature depends mainly on the stage of bud development, as shown in Tables 8-1 and 8-2 and Figures 8-1 and 8-2. As buds open, the protective covering of petals and sepals unfurls, leaving the pistils more vulnerable to frost damage. Choosing later-blooming species and cultivars can reduce frost damage greatly, as the likelihood of bud-killing temperatures occurring during full bloom decreases the later the blossoms open.

Frost versus freeze. Two distinct spring weather patterns threaten fruit crops. Most commonly, a *radiation frost* occurs on clear nights with very little wind, as heat on the earth radiates out

Photo 8-2. Rodents can strip tree bark in winter, seriously damaging or killing trees. Remove mulch or dead vegetation from around the trunks to avoid attracting them.

Photo 8-3. Wire mesh guards around young trees, such as these trellised dwarf apples, can prevent rodent damage.

Photo 8-4. White plastic tree guards, or white paint with a rodenticide mixed in, can prevent both sunscald and rodent damage to tree trunks in winter.

Photo 8-5. Spring frosts during bloom can kill many blossoms, preventing fruit formation. Only one bloom of these two pear clusters has survived the frost.

into space with no cloud cover to bounce it back to earth. The warmer air that radiates out often accumulates above the mass of cold air, 30 to 200 feet thick, near the earth. This creates a temperature inversion, a reversal of the normal condition of temperature declining with increasing altitude.

An *advective or windborne freeze* occurs when winds of 5 mph or more bring in a uniformly cold Arctic air mass that's 500 to 5000 feet thick. Conditions may be cloudy or clear, and no temperature inversion occurs. Other than bloom delay, many of the protective measures outlined below are ineffective during an advective freeze, but fortunately this is much less common in spring than a radiation frost.

The best defense against an advective freeze is to choose an elevated site with good air drainage that's protected by a windbreak to the northwest, and ideally a pond to the southeast, and to choose later-blooming species and cultivars. In areas where advective freezes are common, south slopes make better sites than north slopes, because the dry northwest winds of an advective freeze will desiccate buds as well as open blooms on the north slope, while the south slope is protected. However, in most areas where radiation frosts are more common, the bloom delay caused by north slopes make them valuable for certain early-blooming crops.

Monitoring low temperatures. Local weather forecasts and Internet weather services can provide advance warning as to when frosts or freezes may occur, but growers must monitor each planting site to determine if and when frost protection measures are needed. Microclimates can differ enough within an area, even within a farm, so that frost damage occurs on some sites but not on others. Growers may need to start frost protection equipment on some sites hours before it is needed on other sites.

Monitor temperatures at several locations, such as on slopes facing different directions, a slope above a pond, sites protected by winter windbreaks, level fields, low spots, and different soil types. Place a minimum/maximum thermometer at a 5-foot height in a wood shelter at each spot, to record the highest and lowest temperature reached each day and night.

Ideally, growers should record minimum and maximum temperatures during winter and spring of potential planting sites for several years before planting a fruit or nut crop. This can enable them to pick the best sites, that will require the least frost protection, and to choose the best crop species and cultivars for each site. Such temperature data can accurately determine when each cultivar's winter chilling requirement would be satisfied, and when it would bloom on each site. For example, darker-colored soils retain more heat than lighter-colored soils, often fulfilling the fruit buds' heat requirement after chilling sooner, and causing earlier bloom. Growers can then choose sites that promote later bloom for the earlier-blooming species and cultivars, and later-blooming species and cultivars for the earlier sites.

On sites needing frost protection equipment, wire the equipment to a thermocouple attached to a flower bud, or to a standard thermometer, that will start the equipment when the temperature drops below a certain level. That threshold temperature will vary with the crop and bloom stage, as shown in Table 8-1.

Soil management and pruning. Growers who raise winter cover crops between fruit plantings should till them under a few weeks before the expected spring frost season. The bare soil will give off more heat than ground covered with vegetation,

Table 8-1.

CRITICAL BUD-KILLING TEMPERATURES FOR TREE FRUITS AT VARIOUS BUD DEVELOPMENT STAGES*

Bud Stage	10% kill (degrees F)	90% kill (degrees F)	Bud Stage	10% kill (degrees F)	90% kill (degrees F)
Apples			**Pears**		
Silver tip	15	2	Swollen bud	15	0
Green tip	18	10	Bud burst	20	6
1/2" green	23	15	Green cluster	24	15
Tight cluster	27	21	First white bud	25	19
First pink	28	24	Full white bud	26	22
Full pink	28	25	First bloom	27	23
Bloom & post-bloom	28	25	Full & post-bloom	28	24
Peaches			**Apricots**		
Swollen bud	18	1	Swollen bud	15	-10
1/2" green	21	5	Tip separates	20	0
Calyx red	23	9	Calyx red	22	9
Pink	25	15	White bud	24	14
First bloom	26	21	First bloom	25	19
Full bloom	27	26	Full bloom	27	22
Petal fall	28	26	In shuck	27	24
Fruit set	29	26	Green fruit	28	25
Sweet cherries			**Tart cherries**		
Swollen bud	17	5	Swollen bud	15	0
Bud burst	22	9	Bud burst	24	10
Green tip	25	14	Green tip	26	22
Tight cluster	26	17	Tight cluster	26	24
Open cluster	27	21	Open cluster	28	24
White bud	27	24	White bud	28	24
Bloom & post-bloom	28	25	Bloom & post-bloom	28	24
European plums					
Swollen bud	14	0			
Bud burst	17	3			
Green cluster	20	7			
Tight cluster	24	16			
White bud	26	22			
First bloom	27	23			
Full & post-bloom	28	23			

*Data provided by several states' Extension Services.

which can provide a few degrees of frost protection. Similarly, where agronomic or vegetable crops are grown in fields bordering fruit plantings, tilled soil will release heat, providing some frost protection to the adjacent fruit crop.

In fruit plantings with permanent sod row middles, mowing the sod to a height of two inches or less before expected spring frosts will greatly increase the soil's heat storage and release during frosts. Maintain the area beneath the fruit plants free of vegetation, so the bare soil will absorb more heat and radiate it back to the plants at night. This may require an extra herbicide treatment to control winter annual weeds, which begin vigorous growth just before spring frost season. Heavy mulches, like living vegetation, keep the soil cooler because they

Table 8-2.

CRITICAL BUD-KILLING TEMPERATURES FOR SMALL FRUITS AT VARIOUS BUD DEVELOPMENT STAGES*

Bud Stage	90% kill (degrees F)	Bud Stage	90% kill (degrees F)
Strawberries		**Grapes**	
Tight bud	22	Bud swell	26
Early bloom	28	2-4" shoots	28
Full bloom	31	8-10" shoots	31
Post-bloom	28	Bloom	29
Highbush Blueberries		**Rabbiteye Blueberries**	
Swollen bud	20	Swollen bud	21
Flowers separate	25	Flowers separate	25
First bloom	26	First bloom	28
Full bloom	28	Full bloom	31
Post-bloom	27	Post-bloom	28
Red Raspberries		**Black Raspberries**	
Tight bud	25	Tight bud	26
Full green bud	27	Full green bud	28
Full bloom	29	Full bloom	30
Post-bloom	27	Post-bloom	28

*Data provided by several states' Extension Services.

reflect radiation. Wait to replenish mulches until after the threat of spring frosts has passed.

Moist soil stores and releases much more heat than dry soil. Where sprinkler irrigation is not available for frost protection, irrigating the soil before a freeze can provide about 2 degrees F of protection.

Providing enough nitrogen the previous summer to keep fruit plants vigorous helps them to produce healthy buds that resist frost better. Nitrogen-deficient plants tend to bloom a few days earlier than those with adequate nitrogen.

Delaying dormant pruning of frost-sensitive fruit species, until budbreak or later, can reduce winterkill of fruit buds and delay flowering by 2 to 4 days, which can significantly reduce frost risk. Where growers have not invested in special frost protection equipment, they should wait to prune the upper terminal branches of trees until after the threat of frosts has passed, for following radiation frosts, any fruits remaining on the trees will be on the upper branches.

Sprays to reduce frost damage. Researchers have discovered that frost crystals usually form around bacteria commonly present on the flowers. If such bacteria are present in lower numbers, the blossoms often supercool without ice formation, and can withstand lower temperatures without damage. Spraying Frostgard™, a cryoprotectant, on plants a day or two before an expected light frost can inhibit the bacteria that cause ice nucleation, reducing frost damage to buds by 20 to 60%. This level of protection does not approach that of the methods listed below, but can be useful when a grower has not invested in special frost protection equipment. One Frostgard™ spray can partially protect buds for up to two weeks from temperatures down to 25 or 26 F.

Certain growth-regulator sprays may increase plant resistance to frost somewhat. Fall applications of ethephon have increased flower bud hardiness and/or delayed spring bloom in sweet cherry, which has increased yields in years with late spring frosts. Late fall ethephon applications to peaches have delayed flowering by 4 to 7 days, reducing damage to both dormant winter buds and to developing blooms in spring. Treated trees must have good vigor, as ethephon treatments to peach trees stressed by drought or inadequate nutrition has caused severe gummosis. The benefits of fall ethephon applications on other *Prunus* species and on apple are usually outweighed by other adverse reactions, such as flower bud abscission and reduced fruit set.

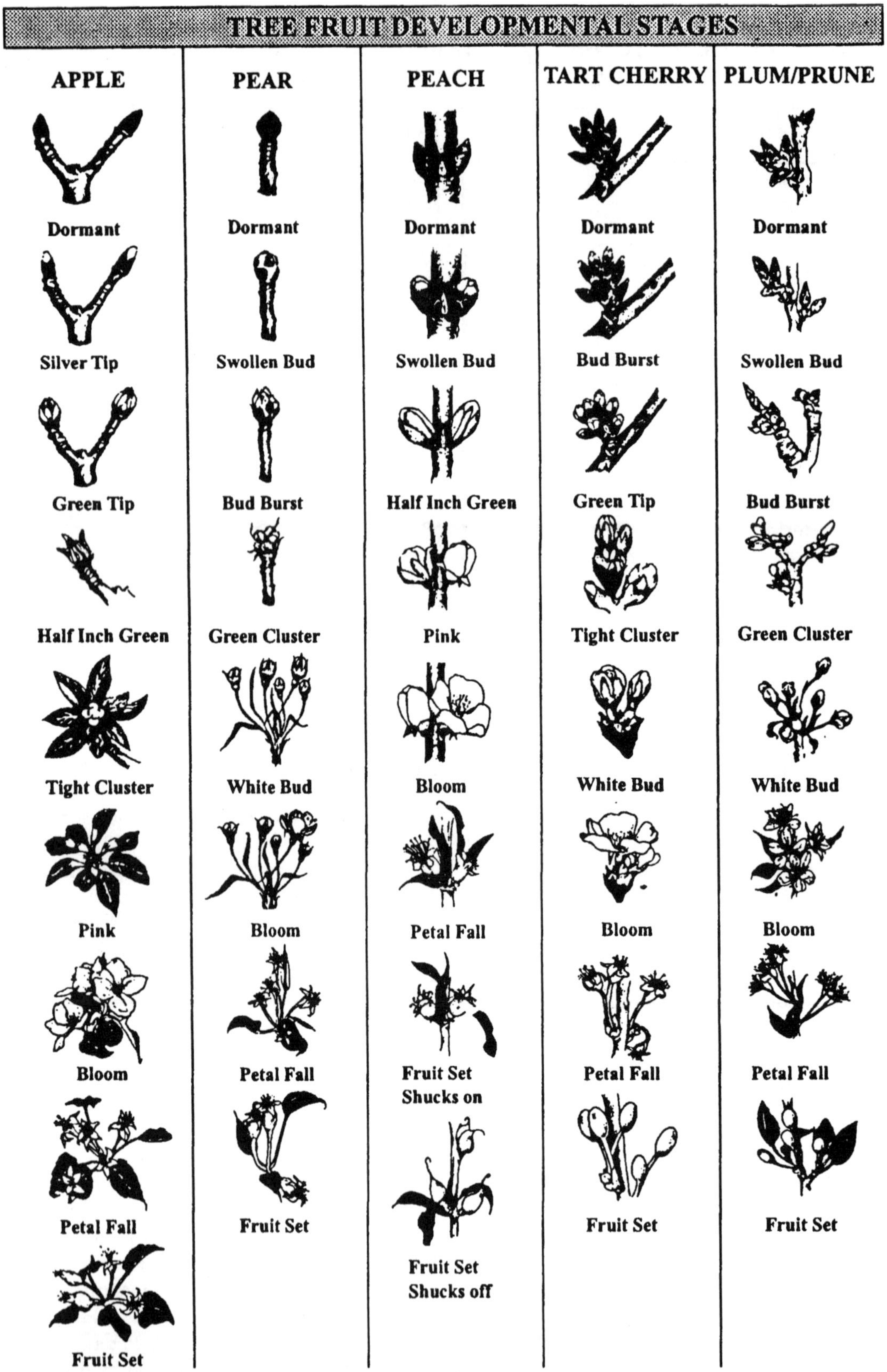

Figure 8-1. Bud development stages of fruit trees. The further along the blooms develop, the more susceptible they become to frost. Taken from *Managing Pests in Home Fruit Plantings*, Purdue University Cooperative Extension Service.

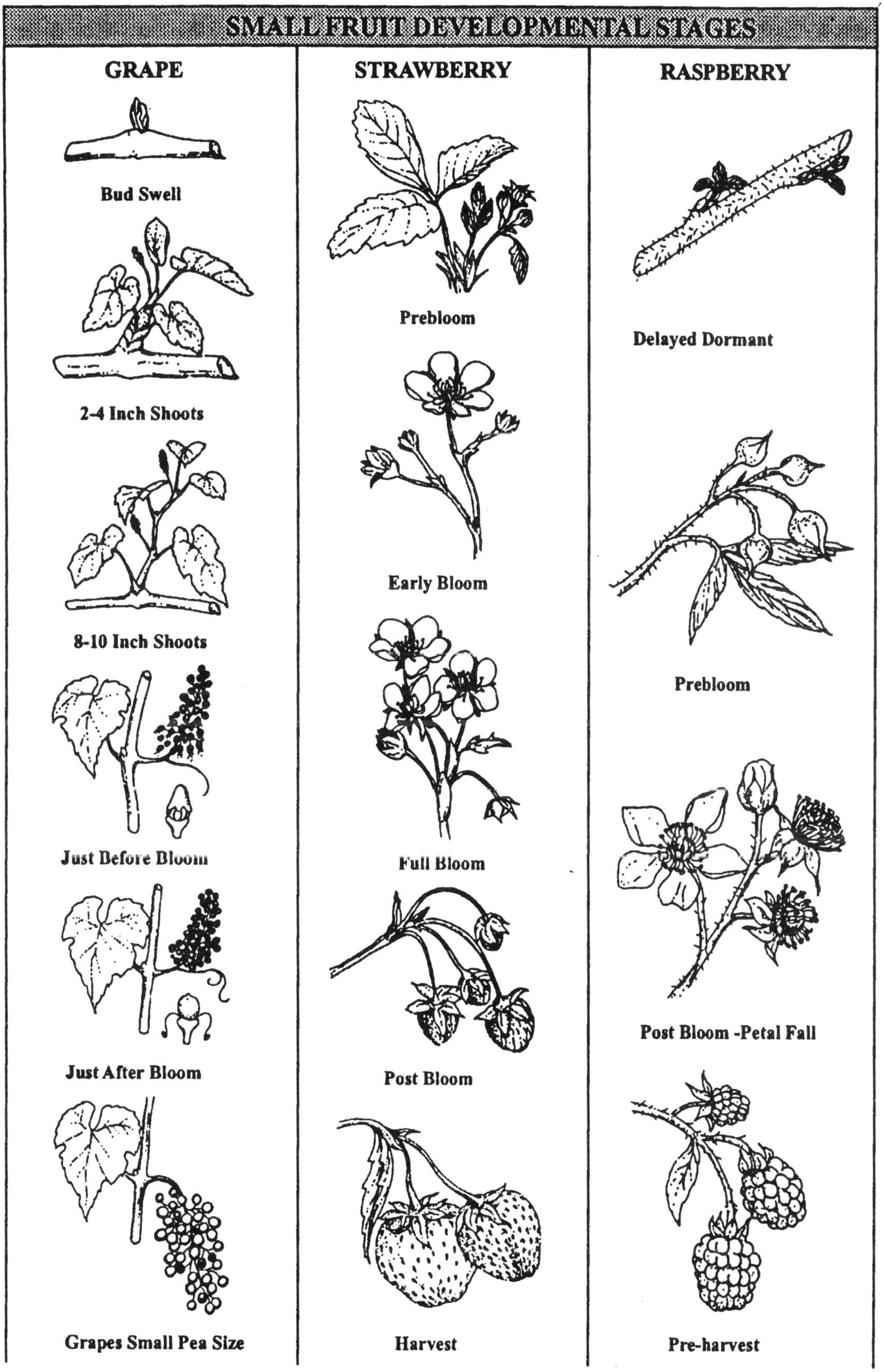

Figure 8-2. Bud development stages of small fruits. Open blossoms are most susceptible to frost. Taken from *Managing Pests in Home Fruit Plantings*, Purdue University Cooperative Extension Service.

Gibberellic acid (GA), sprayed on peaches just before defoliation in fall, has delayed spring bloom and increased yields, although also thinning some flowers. However, fall GA sprays on sweet cherry and almond caused excessive flower bud thinning and reduced yields. GA application to dormant peaches, after the chilling requirement had been partially satisfied, reduced bud hardiness or accelerated bud development, and GA applications to pears 10 days before a frost increased flower bud injury.

Some companies claim that seaweed sprays, which contain cytokinens, have provided a few degrees of frost protection when applied a few days before a frost. Antitranspirant sprays, which inhibit water loss, have not provided freeze protection to flower buds or blooms of fruit trees. A new product to be marketed as Bud Shield™, a byproduct of ethanol distillation containing fermented plant products, has provided several degrees of protection from frosts in trials.

Plant covers. A few growers use agricultural row covers, made of woven or spun-bonded polypropylene, to protect low crops against frost. Plasticulture strawberry growers use mechanical rollers to apply and remove the covers. Covers have also provided outstanding protection for smaller blueberry plants, when anchored with sandbags, irrigation pipes or other objects. Thinner, lightweight covers provide 2 to 3 degrees F protection, but for most situations it's worth the extra investment for thicker covers (0.9 to 1.0 oz. per square yard) which can provide 6 to 9 degrees F protection.

Plant covers must be removed after each frost or freeze event. If left on, they accelerate plant growth, causing buds to lose cold hardiness quicker, and interfere with insect pollination.

A few strawberry growers in northern areas have used row covers for winter protection instead of straw mulch, applying them in late fall and removing them when plants start to bloom. These strawberry plantings bloom about 10 days earlier than straw-mulched plantings, thus requiring more frost protection during bloom, either by repeatedly reapplying the covers at night or by sprinkling. Plasticulture strawberries also bloom about two weeks earlier than strawberries mulched with straw, so plasticulture strawberry growers are advised to use sprinklers for frost protection.

Sprinkler systems. Applying water to fruit blossoms during most frosts can effectively prevent damage down to 20 to 22 F, if sprinkling begins as soon as the temperature drops below 34 F and continues all night until the temperature goes above 32F and the ice melts. So long as water continues to be applied, that water releases heat as it freezes, keeping the blooms near 32 F, above the critical temperature needed to kill the pistils (Photo 8-6).

Sprinkler systems used for irrigation can be adapted for frost protection by simply changing the nozzles to ones that produce a greatly reduced flow rate, to avoid saturating the soil after sprinkling all night. A sprinkler system is especially economical for low crops that need frequent summer irrigation anyway, such as strawberries (Photo 8-7). Some strawberry growers choose microsprinklers, a form of trickle irrigation that uses less water than traditional sprinkler systems yet sprays out enough water to protect strawberry blossoms from frost.

Photo 8-6. These peach flowers in full bloom were protected from frost by the heat released from water freezing into ice continually over the night.

Photo 8-7. Movable sprinkler irrigation systems can protect low fruit crops, such as strawberries, from frost damage. Use frost protection nozzles that put out less water than irrigation nozzles, turn the water on before the temperature drops below 32 F, and leave it running until the temperature rises above 32 F.

Growers of other fruits must decide whether summer irrigation or frost control, or both, is most important. If damaging frosts rarely occur during bloom of a particular crop in that area, trickle irrigation systems can cut summer irrigation costs in half as compared to sprinklers. If spring frosts are often troublesome and only sporadic summer irrigation is needed, a sprinkler system would be a better investment. Some growers who need both frost protection and frequent summer irrigation might invest in both trickle irrigation and a sprinkler system.

Over-tree (or overhead) sprinkler systems consist of fixed pipes and tall risers that can cover the plants (Photo 8-8). These systems are expensive, and growers have found that under-tree sprinkling can protect dwarf and semidwarf fruit trees almost as well at a reduced investment cost for irrigation equipment. Under-tree sprinkler equipment can be moved around to protect different fruit crops that bloom at different times. Furthermore, over-tree sprinkling sometimes caused excessive ice buildup and subsequent limb breakage, which rarely happens with under-tree sprinkling because of lower water usage. However, when under-tree sprinkler irrigation is used for frost protection of medium-sized trees, only the lower limbs may have fruit.

Photo 8-8. Fixed over-tree or overhead sprinkler irrigation systems can protect taller fruit crops from frost damage, such as these grapes, but under-tree systems can provide almost as much control for dwarf to medium-size crops at lower cost. Such sprinklers can also apply water during dormancy, as here, to delay bloom.

Sprinkling for frost protection is tricky. Water application rates must increase the lower the temperature drops, the lower the humidity, and the faster the wind speed. Table 8-3 details the water application rate needed under various temperatures and wind speeds, when the critical bud-killing temperature is 28 F and relative humidity is 75%. If relative humidity is lower or the critical temperature is higher (such as 31 F for strawberries), increase the water application rate. Growers must obtain accurate forecasts for these weather conditions, then pick the appropriate nozzle size (from a range of nozzles they have stockpiled) for the predicted conditions. Excess water application can stimulate root rots and other diseases and build up ice that can damage trees, but inadequate water application can cause blossom temperatures to drop even below the air temperature due to evaporation.

Heating the water to 70 to 150 F before applying it can reduce the amount of water needed. A commercial water heater can accomplish this, or

Table 8-3.
SPRINKLER APPLICATION RATES FOR FROST PROTECTION*

Temperature (degrees F)	Wind speed, mph 1	2	5	10
30	0.05	0.05	0.09	0.12
28	0.07	0.10	0.15	0.16
26	0.10	0.13	0.20	0.24
24	0.12	0.17	0.26	0.32
22	0.14	0.20	0.31	0.40
20	0.17	0.24	0.37	0.47

*Sprinkler application rates for frost protection, inches per hour, when critical bud temperature is 28 F and relative humidity is 75%. Taken from *Pennsylvania Tree Fruit Production Guide, 1998-1999.*

Photo 8-9. Wind machines can protect taller trees, such as these sweet cherries, from a temperature inversion caused by a radiation frost, but don't work under very windy conditions or advective frosts caused by a uniform cold air mass.

Photo 8-10. Wind machines protect this vineyard from frost, but the orchard in the background, on higher ground, needs no frost protection. Choosing sites with good air drainage and late-blooming cultivars can greatly reduce costs for frost protection.

if shallow ponds or wells provide irrigation water, sunny spring days may already have heated that water to 70 or 80 F, so water application rates can be reduced.

Sprinkler systems do not work well under very windy conditions of 10 mph or more or temperatures below 20 F, which are generally caused by an advective freeze. Some growers have reduced damage with sprinkling even when temperatures dropped to 16 F and winds reached 25 mph, however, by applying a grease product to the nozzles to keep them from freezing up, and walking the lines throughout the night to knock ice off nozzles and regrease them. Dr. E. Barclay Poling of North Carolina State University's Small Fruit Center uses Cycle ™ white grease spray with Teflon at a rate of 4 cans per acre per night for strawberries.

Another use of sprinkler systems in combating frost damage is evaporative cooling of plants during winter dormancy (Photo 8-8). This technique requires monitoring winter temperatures and, once the chilling requirement of each cultivar has been satisfied, misting the plants daily whenever the air temperature exceeds 45 F. This keeps plants cooler and delays the accumulation of heat units needed for buds to begin swelling. This can delay bloom for several days to a couple of weeks, hopefully until after the worst threat of spring frosts has passed. Evaporative cooling during dormancy works best on sandier soils that drain quickly, in areas or during seasons of low winter and early spring rainfall. Excess soil water in early spring can greatly reduce cropping.

Wind displacement devices. Many frosts result from a temperature inversion: a layer of cold air sliding underneath warmer air which is radiating up due to calm, clear conditions. These radiation frosts can be corrected by wind machines (Photos 8-9, 8-10), tall spires topped by propeller blades that mix the stratified air up, bringing warmer air back to the orchard floor. Each wind machine can protect about 10 acres of orchard, by bringing down air from 50 or 60 feet that is 6 to 8 degrees F warmer. Wind machines are expensive to purchase and consume expensive fuel to run the motors, but use only 5 to 10% of the energy consumed by orchard heaters.

Wind machines do not work when an advective freeze is caused by uniformly cold air with no temperature inversion, or under windy conditions with winds more than 5 mph. Wind machines may be useful for taller trees on sites with good air drainage but prone to calm, clear nights that cause temperature inversions.

Some growers hire helicopters to achieve the same purpose. During a strong temperature inversion, helicopters can bring down even warmer air from 100 to 200 feet, which can be 8 to 14 degrees F warmer than air on the ground. Each helicopter can protect 50 to 100 acres.

Orchard heating. Burning combustible materials on the orchard floor during a frost can effectively keep temperatures above critical levels, even during an advective freeze. However, this method costs the most and produces the most environmental damage due to air pollution from the smoke. Most growers who heat orchards use heaters that burn fuel oil or propane (Photo 8-11), which have become increasingly expensive. Check local air pollution regulations before investing in orchard heaters, as some areas limit the number of hours which they may be utilized, which can render them uneconomical. Propane or other natural-gas systems produce less pollution than oil heaters do.

Where frosts pose only an occasional problem and regulations do not prohibit burning wood, late-winter prunings from an orchard or vineyard can be

Photo 8-11. Orchard heaters have been widely used to protect trees from frosts, but become less economical as energy costs rise and pollution restrictions increase.

stored in piles between the plants, and lit as needed during an especially frosty night. This still causes air pollution, but costs only the grower's time.

Further Reading

Anderson, J.L. and S.D. Seeley. 1993. Bloom delay in deciduous fruits. *Hort Reviews* 15:97-144.

Arthur, T. 1989. Frost protection in a spray. *Amer. Fruit Grower* Feb. issue:8,44.

Evans, R.G. 1995. Fight frost with under-tree irrigation. *Amer. Fruit Grower* Jan. issue:10-11.

Perry, K.B. 1998. Basics of frost and freeze protection for horticultural crops. *HortTechnology* 8:10-15.

Reiger, M. 1989. Freeze protection for horticultural crops. *Hort Reviews* 11:45-109.

Warmund, M.R. and J.T. English. 1998. Ice nucleation, freezing injury, and colonization of 'Totem' strawberry flowers with ice-nucleation-active (INA) bacteria. *J.Amer. Soc. Hort. Sci.* 123:234-238.

Chapter 9

Managing Pest Arthropods, Diseases, Mammals and Birds

Fruit and nut crops are notorious for having many pest problems, and once chemical pesticides were developed, have been among the most heavily sprayed crops. Such heavy pesticide use allowed growers to produce high-quality crops, but increased costs and caused problems such as damage to beneficial organisms and pollution of ground and surface waters. The good news is that in recent decades, researchers have developed many non-chemical methods of controlling pests, as well as safer chemicals that selectively stop pests without harming beneficial organisms or posing any risk to farm workers or consumers.

Most fruit and nut growers today practice some form of integrated pest management (IPM), a system that minimizes pesticide use by utilizing all economical methods of pest control. IPM also emphasizes monitoring of pest populations, so that pesticides or other control methods are applied only when needed and at the best possible times for control of the pest. IPM labeling programs are now available in several states, including Massachusetts and New York, that let consumers know that the fruit has been grown using certified IPM methods.

Organic certification programs recognize the growers who use only naturally occurring fertilizer and pest control materials. Consumer demand for certified organic fruits and nuts is growing, allowing organic growers to charge premium prices to make up for the greater crop damage they sustain. However, some naturally occurring, and therefore organic, pest control products, such as nicotine sprays made from tobacco, are much more harmful to beneficial organisms and humans than most synthetic chemicals used in IPM programs.

DIRECT VS. INDIRECT PESTS

Arthropods, disease agents or vertebrate pests that attack the fruit or nut itself are called *direct pests*, because they directly reduce the marketable crop. Examples include Oriental fruit moth and plum curculio of pome and stone fruits (Photo 9-1), anthracnose fruit rot, starlings, and squirrels. (Specific management guidelines for each of these pests are covered in the individual fruit and nut chapters.) Direct pests also include those which can

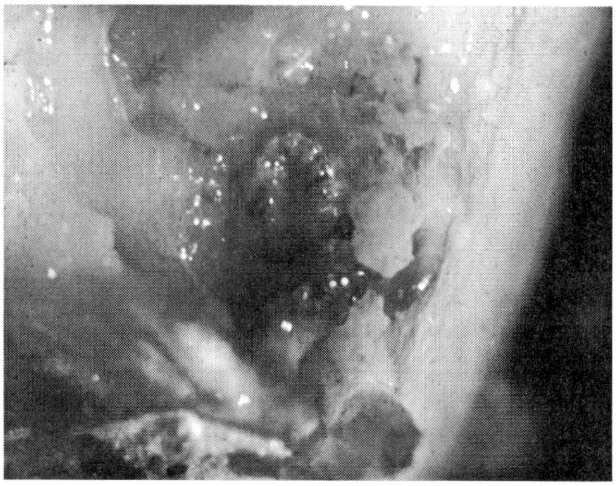

Photo 9-1. Plum curculio in peach and other stone fruits is a direct pest, since it attacks the plant part that is sold.

kill the entire plant, such as peachtree borer, fire blight, voles, and deer. *Indirect pests* feed on the foliage, reducing photosynthesis and yields. These include aphids (photo 9-2), mites, and leafspot diseases.

The *economic threshold* is the level of damage at which control measures become economical (Figure 9-1). Below the threshold level, pest control measures would cost more than the increased yields of marketable fruit or nuts that they produced, so are not justified. Direct pests have very low economic thresholds, usually at about 2 to 4% damage to fruits, nuts or plants. Indirect pests have higher economic thresholds, measured in terms of number of pests per leaf or percent defoliation. Indirect pests can therefore be managed without pesticide applications more easily than can direct pests.

ARTHROPOD PEST MANAGEMENT

Arthropod pests of fruit and nut crops include many species of insects and mites. Insects have three body parts and six legs as adults, while mites have two body parts and eight legs like their

Photo 9-2. Green apple aphid on apple leaves is an indirect pest, potentially reducing yields but not attacking the salable crop.

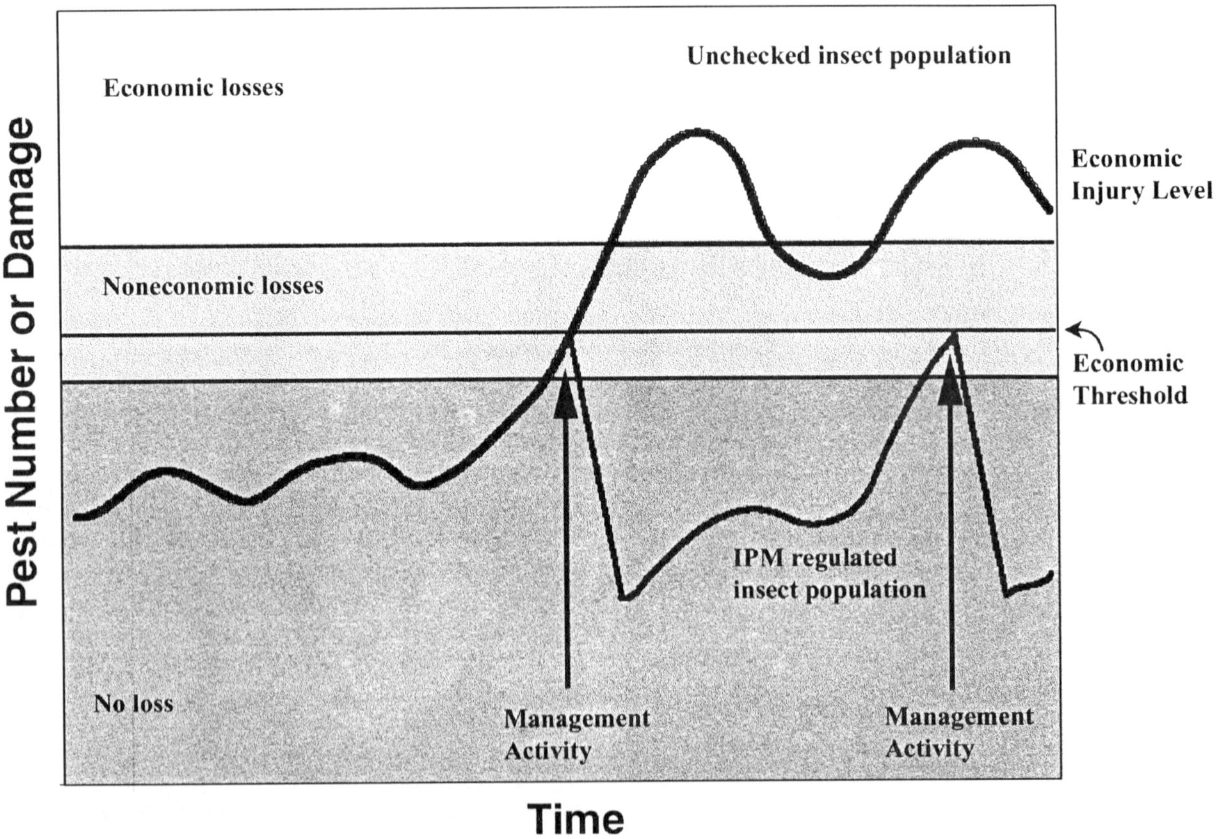

Figure 9-1. The economic threshold for a pest is the population level at which control measures become economically justified.

relatives, ticks and spiders. Mites, as the name implies, are tiny, often requiring a hand lens to see. Slugs and snails belong to the gastropods, a group of organisms related to the arthropods and managed similarly.

Plant vigor and genetic resistance to arthropods. Maintaining plants in good but not excessive nutrition and water status can reduce indirect pest problems. Vigorous plants can produce enough new growth to offset some damage, while drought-stressed plants attract some pests such as mites and grasshoppers. However, excessive nitrogen or water levels lead to very succulent new growth, which is more susceptible to aphids as well as certain diseases.

Some cultivars of certain fruit and nut crops also possess genetic resistance to certain indirect pests. Several pecan cultivars support lower populations of one or more species of aphids or phylloxera than do other cultivars. 'Royalty' purple raspberry repels raspberry aphids, which transmit raspberry mosaic virus (Photo 9-3). Planting such resistant cultivars generally eliminates the need to control that particular pest.

Biocontrol of arthropods. Biocontrol of arthropods involves aiding the pests' arthropod natural enemies, the larger predators that eat them and the smaller parasitoids that lay their eggs inside them. Biocontrol also includes conserving or applying formulations of arthropod disease agents, such as beneficial fungi, bacteria or nematodes, or toxins derived from them.

Fruit plantings contain many native species of beneficial arthropods (Photo 9-4), so the first step in biocontrol is to conserve them by avoiding use of pesticides that kill them. Avoid *broad-spectrum* pesticides that kill a wide variety of organisms, such as soil fumigants and chlorinated-hydrocarbon insecticides. Instead, use the most *selective* pesticides possible (Table 9-1). For example, phosmet (Imidan™) insecticide effectively controls many direct pests while doing little damage to beneficial arthropods, and a selective miticide such as abamectin (Avid™) controls pest mite outbreaks while doing little harm to the predator mites that help control them (Figure 9-2).

Another way to aid biocontrol agents is to provide alternate food for them, to stabilize their populations when pests are present in low numbers. Such alternate food could include non-pest insects or mites that feed on groundcover plants, or nectar and pollen for adult parasitoids. Chapter 5 discusses groundcover management to provide such alternate food for biocontrol agents.

Many fruit growers hire pest management consultants to monitor populations of beneficial and destructive arthropods. Such consultants can advise growers when beneficials are abundant enough to control certain pests, and when pest numbers rise to the point that they need control.

Sometimes growers must control imported pests that hurt the native biocontrol agents. For example, the invasion of red imported fire ants into southeastern pecan orchards upset the natural

Photo 9-3. 'Royalty' raspberry is non-attractive to the aphid vector of raspberry mosaic virus, thus avoiding the disease, raspberry mosaic.

Photo 9-4. Lady beetle larvae, here on a young apple, are valuable predators of aphids.

Table 9-1.
TOXICITY OF COMMON PESTICIDES*

Pesticide		Toxicity to Predators				
Common name	Trade name(s)	*Stethorus punctum*	*Amblyseius fallacis*	*Zetzellia mali*	*Aphidoletes*	General aphid predators*
Insecticides and miticides						
abamectin	Agri-Mek	XX	XX	XX	nd	nd
azinphosmethyl	Guthion, Sniper	X	X	X	XXX	XX
Bt	Dipel, Agree, etc.	0	0	0	0	0
carbaryl	Sevin	XXX	XX	XX	XXX	XX
chlorpyrifos	Lorsban	X	XX	XX	nd	XX
diazinon	Diazinon	X	nd	nd	XXX	XX
dicofol	Kelthane	X	XX	X	X	X
endosulfan	Thiodan, Phaser	XX	X	nd	XX	XX
fenbutin-oxide	Vendex	X	X	XXX	X	X
imidachloprid	Provado	XX	X	X	nd	XX
methomyl	Lannate	XX	XXX	XX	XXX	XXX
methyl parathion	Penncap-M	X	X	X	X	XX
phosmet	Imidan	X	X	X	XX	X
pyrethrin	Permethrin	XXX	XXX	XX	X	XX
Fungicides						
benomyl	Benlate	X	XX	XX	X	X
captan	Captan	X	X	X	X	0
fenarimol	Rubigan	X	X	nd	nd	nd
mancozeb	Dithane, Manzate	X	X	X	nd	0
metiram	Polyram	X	X	X	nd	0
myclobutanil	Nova	X	X	X	nd	0
sulfur	wettable sulfur	X	X	nd	nd	nd
thiophanate-methyl	Topsin-M	X	XX	XX	X	0
thiram	Thiram	X	X	nd	nd	nd
triadimefon	Bayleton	X	X	nd	nd	nd
triforine	Funginex	X	X	nd	nd	nd
ziram	Ziram	X	X	X	nd	0

0 = non-toxic, X = slightly toxic, XX = moderately toxic, XXX = highly toxic, nd = no data.
* General aphid predators include lady beetles, lacewings, syrphid fly larvae, and minute pirate bugs.

*Toxicity of common pesticides, applied at recommended rates, to mite and aphid predators. Adapted from *Pennsylvania Tree Fruit Production Guide, 1998-1999.*

system of aphid control by native predators. The fire ants tend the aphids, protecting them from predators, and also prey on the eggs of aphid predators. Controlling the fire ants allows the natural biocontrol to resume.

When native predators and parasitoids are insufficient to control pests, growers may augment them by releasing beneficial arthropods that were raised and sold by insectary companies. Buying beneficial arthropods to release generally costs more than applying an insecticide or miticide, but may prove economical if the arthropods remain in the planting and increase in population enough to provide season-long control.

To help ensure that this occurs, manage the groundcover and pesticide applications to help and not hurt the biocontrol agents. Releases that have partially controlled fruit insect pests have included green lacewing releases to reduce grape mealybug, and *Trichogramma* wasp releases to reduce codling moth.

Chemicals applied to plantings to control fruit or nut diseases may also affect arthropod pest populations. Beneficial fungal diseases can control certain insect pests in humid climates where fungicides are rarely used, whereas dry air limits these beneficial fungi in arid areas and frequent fungicide sprays can kill them in humid areas. In humid areas where growers limit fungicide sprays by growing cultivars resistant to the major fungal diseases, such as pecan scab or apple scab, the beneficial fungi can proliferate and control certain insect pests, such as aphids.

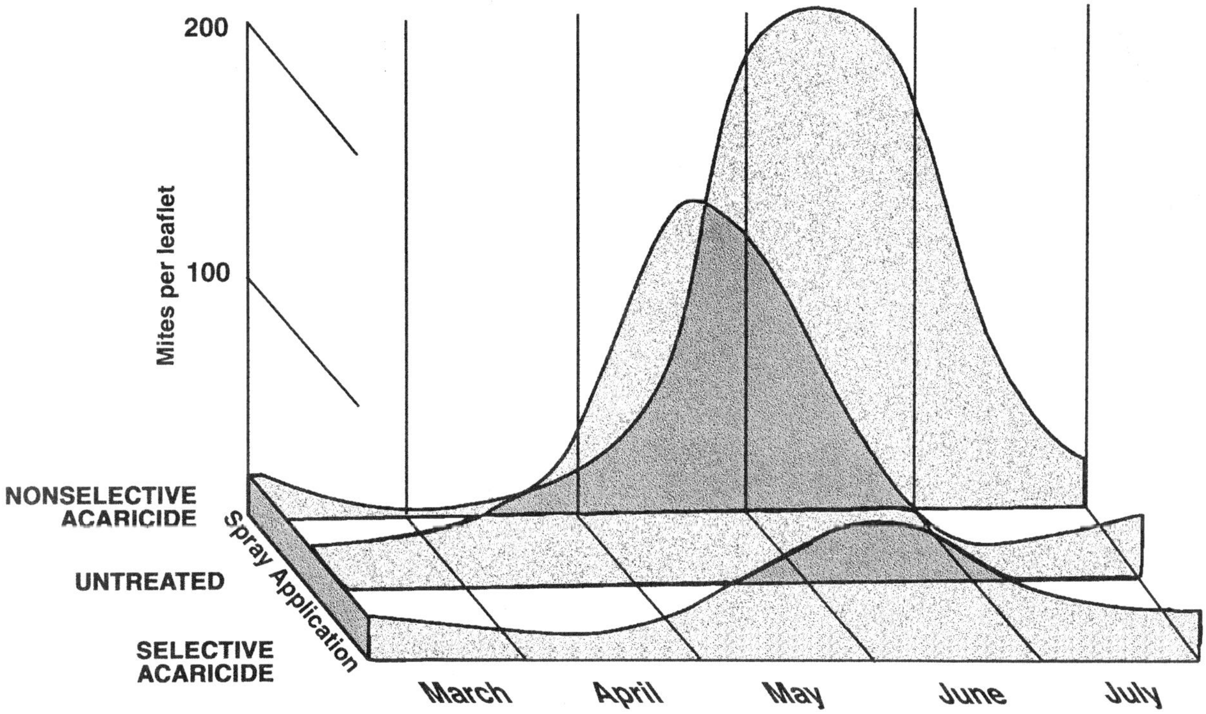

Figure 9-2. Applying a nonselective pesticide, such as the miticide (acaricide) shown here, kills natural enemies of the pest as well as the pest, allowing pest populations to resurge to higher levels than on untreated plants. A Selective pesticide kills the best but not the natural enemies, keeping pest populations below damaging levels. Taken from *Integrated Pest Management for Strawberries*, Univ. of California pub. 3351.

Several bacterial, actinomycete, fungal and nematode pathogens of insects have also been formulated and marketed as biocontrol products. Such products usually cost more than chemical insecticide applications, and their effectiveness depends greatly on weather factors. The oldest, most consistently effective and most economical of the insect disease products is the endotoxin produced by *Bacillus thuringiensis* or Bt for short, sold under numerous trade names including Dipel™ and Javelin™. Bt kills only caterpillars, the larvae of moth pests, so it's completely harmless against beneficial insects, other animals or people.

However, Bt only works on caterpillars that injest the toxin, and it's difficult to apply Bt at the precise time after egg hatch and before fruit worms (such as codling moth and Oriental fruit moth) or destructive moth borers (such as peach tree borer) are protected from the toxin inside the fruit or trunk. Nevertheless, growers who use mating disruption (outlined below) to control key moth pests, such as codling moth on apples and pears, are increasingly using Bt products to control secondary moth pests such as leafrollers.

Two other effective biocontrol products for arthropods are spinosad and abamectin. Spinosad (SpinTor™ or Success Naturalyte™), produced by the actinomycete *Saccharopolyspora spinosa* under anaerobic fermentation conditions, effectively kills caterpillars, sawflies, and leaf-feeding beetles, and is non-toxic to humans, mammals, birds and most beneficial insects and mites. Abamectin (Avid™) miticide is a synthesized version of the same chemical produced by another actinomycete, *Streptomyces avermitilis*, and works very well against pest mites while doing little harm to predator mites or other organisms.

The fungus *Beauveria bassiana* (Mycotrol™ or Naturalis-T™) has provided 30 to 40% control of tarnished plant bug and pecan weevil in trials, not enough for commercial control. Several species of beneficial nematodes attack trunk borers and insects that pupate and/or overwinter in the soil, such as plum curculio and pecan weevil. These beneficial nematodes, sold under many trade names including Ecomask™ and Biosafe™, have provided some control in trials but not at a commercial level.

Monitoring and forecasting arthropod populations. More sophisticated methods of determining just when arthropod populations are building to a level requiring control have greatly reduced the number of pesticide treatments applied in fruit and nut plantings. Before the 1970's, early-season pesticide sprays were timed based on crop flower bud stages, but later sprays, called "cover sprays", were applied every 10 to 14 days on a calendar basis. Newer ways of detecting when each generation of each pest needed control allowed growers to reduce the number of insecticide applications dramatically, such as from 15 or 20 per season down to just 3 or 4 per season in many long-season fruit and nut crops.

The destructive phase of many insect pests is the juvenile stage, either a nymph or a larvae ("worm"). The time to control such pests is a few days after the peak flight of the adults, when most of the eggs have been laid and are just starting to hatch into larvae or nymphs. Two common ways to predict such times are to monitor temperatures and use them to predict when peak flight and egg laying will occur, and to count the adults to see when more of them are present in the planting.

Temperature monitoring involves adding up degree-days. Each insect pest has a threshold temperature below which it remains dormant, and an upper temperature limit above which it doesn't survive. (See "Computing Degree-Days," below.) Between those limits, the warmer the temperature, the faster the insect develops, and the fewer days between generations. When the threshold temperatures for a pest are known, growers can keep track of each day's high and low temperatures to determine the average temperature and the degree days accumulated for each pest. Then the grower will know when to watch for the pest and when to apply control measures.

The insect pests themselves can be monitored in several ways. For indirect pests such as mites or aphids, randomly choose ten leaves per plant and count all the pests on each leaf. Monitor bug pests, such as lygus bugs or stink bugs, by beating the flower stalks of a ten-foot length of preferred weed hosts over a white beating sheet or paper, and counting the bugs. Monitor beetle pests, such as plum curculio, by shaking fruit plants over a white beating sheet spread around the base of the plant. Some beetles, such as pecan weevil, can be trapped in a wire cone trap attached to the north side of the tree, since about 10% of the population crawls

Computing Degree-Days

1. Look up the threshold temperature for the pest in question. Begin tracking temperatures and scouting for nymphs in weeds in March and for adults in strawberries in April.
2. Determine average temperature for each day by adding minimum temperature and maximum temperature and dividing the sum by 2.
3. Determine the degree days (DD) for that day by subtract threshold temperature from average temperature.
4. Keep running total of DD accumulated until treatment time.

For example, for the lygus bug on strawberry (fig. 9-3), the threshold temperature is 54 F. Therefore –

On a day which the minimum temperature is 30 F and the maximum temperature is 70 F, the average temperature is 50F, which is less than 54 F, so 0 DD are accumulated that day.

On a day which the minimum temperature is 35 F and the maximum temperature is 75 F, the average temperature is 55F, which is 1 more than 54 F, so 1 DD is accumulated that day.

On a day which the minimum temperature is 40 F and the maximum temperature is 80 F, the average temperature is 60F, which is 6 more than 54 F, so 6 DD are accumulated that day.

The first generation of nymphs will appear after 252 degree days have accumulated since adults were first seen in strawberries. The second generation of nymphs will appear after 799 degree days have accumulated since nymphs were first seen on weeds. The third generation of nymphs will appear after 799 degree days have accumulated since the first generation of nymphs appeared in strawberries. Begin scouting for the pest at each predicted appearance of nymphs, and treat if levels reach 1 lygus bug per 20 plants.

up a tree trunk after they emerge from pupation in the ground.

Sticky traps of various colors and shapes attract some pests. Monitor apple maggot flies with red spheres (that resemble apples) coated with the sticky material Tanglefoot™. Monitor tarnished plant bug adults and European apple sawfly adults with sticky white rectangles. Sticky yellow rectangles attract aphids, thrips and leafminers.

Monitor moth pests with pheromone traps, sticky traps containing a pheromone dispenser (Photo 9-5). A pheromone is a chemical synthesized to imitate the natural attractant scent that female moths exude to attract male moths. Each moth species produces a slightly different scent, which humans can't detect, so each pheromone attracts only male moths of that species. Therefore, growers need to buy pheromone dispensers for each moth pest species they encounter in their area. For example, apple growers might buy pheromone dispensers for codling moth, Oriental fruit moth, redbanded leafroller, obliquebanded leafroller, and tufted apple bud moth.

Photo 9-5. Pheromone traps, sticky traps containing a pheromone dispenser that attracts male moths of a particular species, help growers monitor population of moth pests, such as grape berry moth.

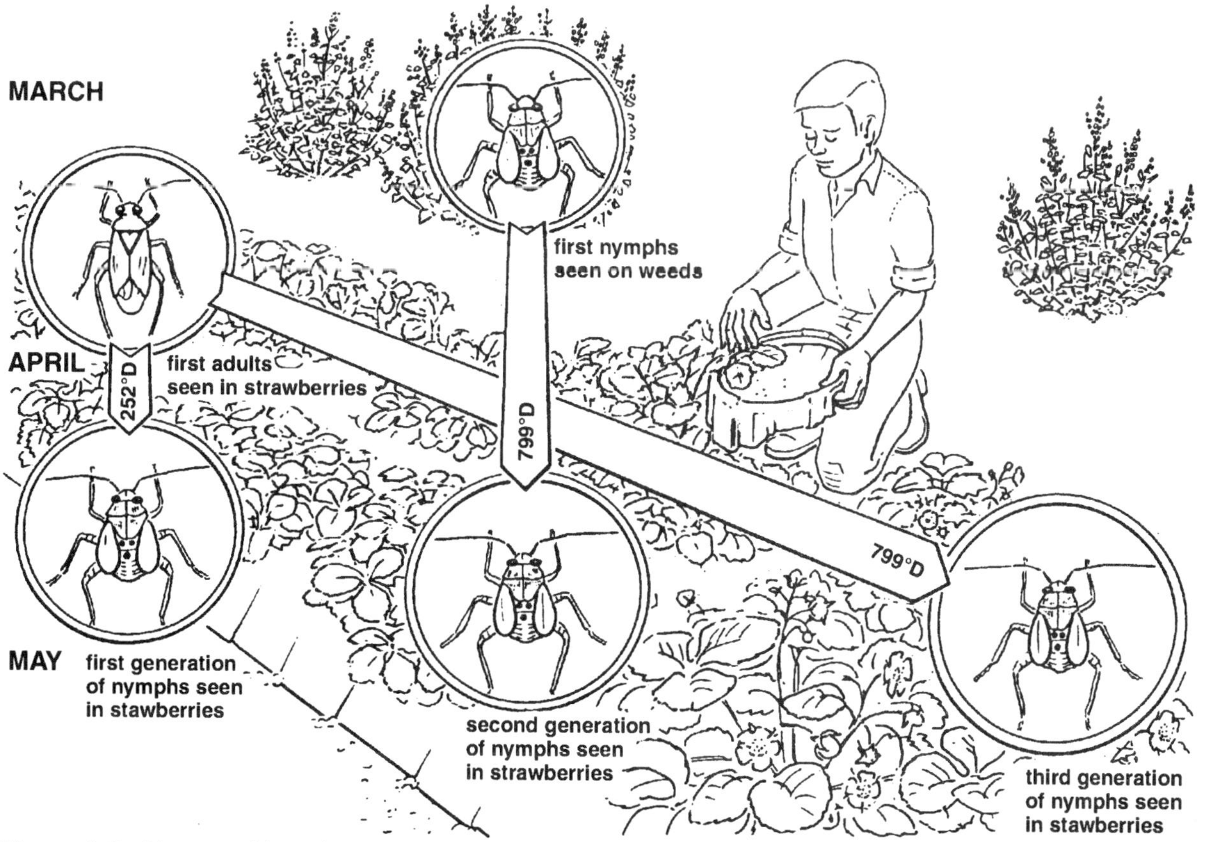

Figure 9-3. Diagram of lygus bug movement into strawberries and accumulations of degree-days above 54 F that predict lygus bug generations in strawberries. Taken from *Integrated Pest Management for Strawberries*, Univ. of California pub. 3351.

Not all of these pests occur in all areas, so only monitor the key pests in your area. For example, apple maggot doesn't occur in the southern half of the U.S., and European apple sawfly is currently a problem only in New England, New York and eastern Canada.

All of these traps help growers determine when to apply control treatments. Attempts at mass trapping, using many traps to reduce the pest population, do not generally reduce damage. Pheromones are being used to disrupt mating of moth pests, as described below.

Mating disruption of arthropods. Mating disruption utilizes pheromones to greatly reduce or completely eliminate insecticide sprays for moth pests in some fruit and nut crops. More and more orchardists and vineyardists have adopted the practice over the last 20 years, and it is now widely used.

The insect pheromone is impregnated in plastic twist-ties similar to those used as bread wrappers, or is also newly available in sprayable form. Workers hand-apply the twist-ties to trees (one per dwarf tree, up to three per large tree) or grapevines (one every third vine) at the beginning of the season. Long-season crops may require replacement twist-ties or sprays in mid-season, although newer formulations can last all season. The dispensers or spray material flood the air with the pheromone, confusing the male moths so that they can't locate the females and mate (Figure. 9-4). No mating results in no egg laying and no larvae hatching (Photo 9-6) to damage the fruit or bore into the tree.

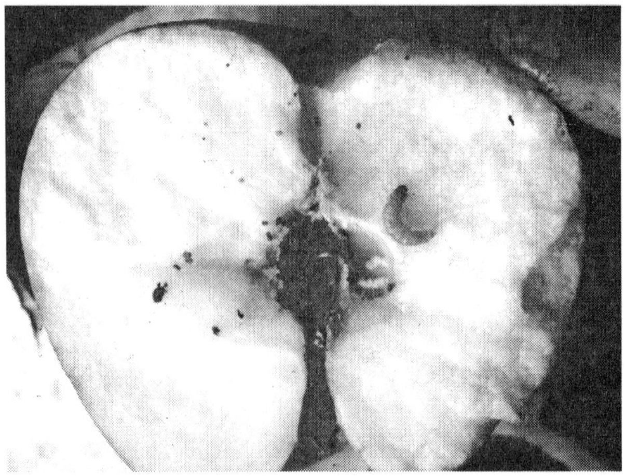

Photo 9-6. Mating disruption uses pheromone dispensers to flood the fruit planting with the pheromone, so that male moths can't locate the females to mate, which practically eliminates egg-laying and thus worm damage, such as this codling moth in apple.

Mating disruption has produced results comparable to or better than insecticide sprays, when used on fruit and nut plantings of at least five acres in size. The only fruit or trunk damage usually occurs around the edges of the planting, if already-mated female moths fly in from other plantings not using mating disruption. Fruit or trunk damage is generally less than 1%, and the cost of applying

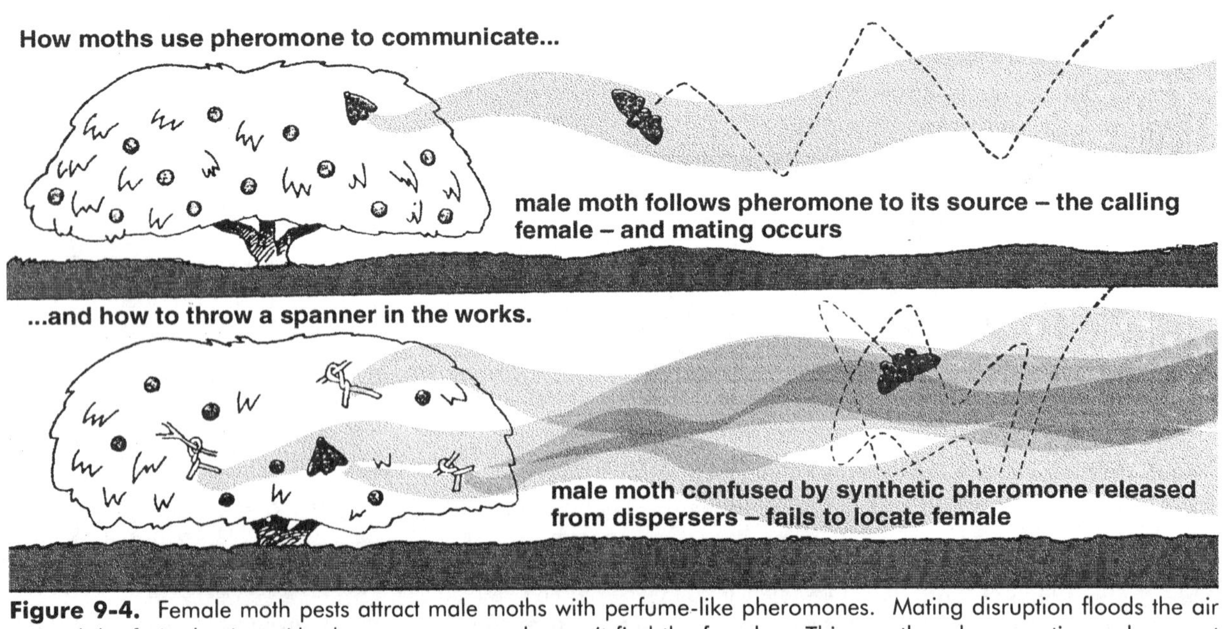

Figure 9-4. Female moth pests attract male moths with perfume-like pheromones. Mating disruption floods the air around the fruit planting with pheromones, so males can't find the females. This greatly reduces mating, subsequent egg-laying, and resulting worms in the fruit or tree trunks. Diagram courtesy of C.S.I.R.O., Australia.

the pheromone dispensers is about the same as that of applying insecticides. Mating disruption offers these advantages over insecticide spraying:

- no difficulty in applications due to windy, rainy or cold weather
- no toxicity to field workers, no safety equipment needed, no re-entry interval
- no residues on fruit
- no harm to beneficial arthropods or wildlife
- use for over three consecutive years greatly reduces the pest population, which in turn reduces the number of dispensers needed per acre and can cut control costs in half

To reduce costs of mating disruption, many growers now apply dispensers at half the recommended rate, and spray an insecticide once in summer, if needed, to keep the pest population down. Other growers apply an insecticide only around the perimeter of the fruit planting, to kill any already-mated female moths that fly in. Certain processors, such as baby food companies, now require growers who raise crops for them to use mating disruption with no organophosphate insecticide sprays, to avoid even the possibility of pesticide residues in the product.

Mechanical barriers and arthropod removal. Particle film technology has developed into a powerful, safe new method of controlling arthropods. Particle film contains purified kaolin clay, a white, nonabrasive, fine-grained aluminosilicate mineral that disperses in water. When sprayed on plants, it forms a thin coating over leaves and fruit (Photo 9-7) that impairs pest recognition of the host

Photo 9-7. Particle film, containing purified kaolin clay, may be sprayed on fruit plants such as the apple row at left. It forms a thin white coating over leaves and fruit that stops some insect pests with a physical barrier, reduces fruit sunburn, and increases fruit color due to reflected light. (Photo from Engelhard Corp.)

as well as pest movement, feeding and egg-laying. Researchers have demonstrated that particle films provide 80 to 98% control of many pests, including mites, pear psylla, thrips, plum curculio, codling moth, apple maggot, Japanese beetle, stink bugs, leafrollers and leafhoppers.

Because particle film stops pests with a physical barrier rather than killing them, it may require more frequent spraying than with insecticides. To control plum curculio, for example, instead of applying 2 insecticide sprays at petal fall and 10 days later, apply 4 particle film sprays at petal fall and every 5 to 7 days thereafter. Rain will eventually wash off the particle film coating, so humid areas probably require the more frequent treatments.

A commercial particle film, Surround™, and its spreader-sticker, M03, are now registered for use on pome fruits, stone fruits, walnuts, wine grapes, processing blueberries and brambles. Surround™, developed by USDA researchers Michael Glenn and Gary Puterka in cooperation with the manufacturer Engelhard Corp. of Iselin, NJ (www.engelhard.com or 877-240-0421), has been approved for organic use and has been widely used in Washington state since 1997. New formulations of the product, with other minerals added, have shown promise in controlling several diseases of fruit crops, including leaf spots, powdery mildew and apple scab. The product has also shown potential for reducing frost damage in trials.

Particle films leave a white residue on plant surfaces that looks unsightly but is completely harmless to humans, bees, lady beetles, green lacewings and most other beneficial organisms. The film does adversely affect predator mites as well as pest mites, so continued use might require occasional treatment with a selective miticide. To avoid leaving noticeable residues at harvest, discontinue spraying Surround™ 6 to 8 weeks before harvest, or wash the fruit after harvest. Many target pests, such as plum curculio and Japanese beetle, mainly require control early in the season anyway. The particle film residue does not block needed light from getting to plants; instead, it actually increases net photosysnthesis by reflecting excess light and reducing plant temperatures on hot days, and provides secondary benefits of fruit sunburn control, smoother fruit finish, better fruit color and higher yields.

Some large strawberry growers use huge field vacuum cleaners, such as Bug Vac™, to suck lygus bugs off the plants. The machines are expensive, but some large growers found that a machine paid for itself in a year by reducing pesticide costs by

75%. The machine removes only insects on the upper surface of the leaves, so isn't useful for pests that prefer the underside, but also preserves beneficial arthropods on the underside of leaves. Bug vacuums have not yet been developed or used for other fruit crops.

Selective insecticides and miticides. Most fruit and nut growers still need to apply some insecticides and miticides, for outbreaks of pests which don't produce pheromones and aren't adequately controlled by particle films, cultivar resistance or biocontrol agents. Use the most selective materials available, to kill only the pest(s) you need to and not beneficial arthropods (Table 9-1). This avoids the creation of secondary pests by killing off the natural enemies that normally control them (Figure 9-5).

It's also a good idea to rotate insecticides or miticides whenever possible, using ones with different modes of action every other application. This avoids continued selection pressure on the pest population for pests that resist a particular pesticide (Figure 9-6). If such resistance developed in a pest population and the insecticide that induced it was repeatedly used, within a few years the insecticide would no longer be effective against the new pest population that had developed.

DISEASE MANAGEMENT

The majority of plant diseases are caused by fungi, which grow as long microscopic filaments that need high humidity to keep from drying out, reproduce by spores, and generally cause the most damage in humid areas. Other plant diseases are caused by single-celled bacteria, by viruses, by nematodes, or by non-living causes such as air pollution.

Whether or not a plant shows symptoms of a disease depends on three factors, called the disease triangle (Figure. 9-7). First, the plant must be genetically susceptible to the pathogen, the cause of the disease. Fruit and nut growers can eliminate many diseases by choosing cultivars and rootstocks resistant, rather than susceptible, to common diseases of their area. Secondly, the pathogen must be present and virulent, capable of causing disease. Starting with disease-free plants and employing cultural practices that destroy overwintering pathogens can greatly reduce disease incidence, as do pesticide sprays and biocontrol agents. Thirdly, the environment must be conducive to disease development, with temperatures within a certain range and generally high levels of humidity and moisture. Proper management of the plant canopy,

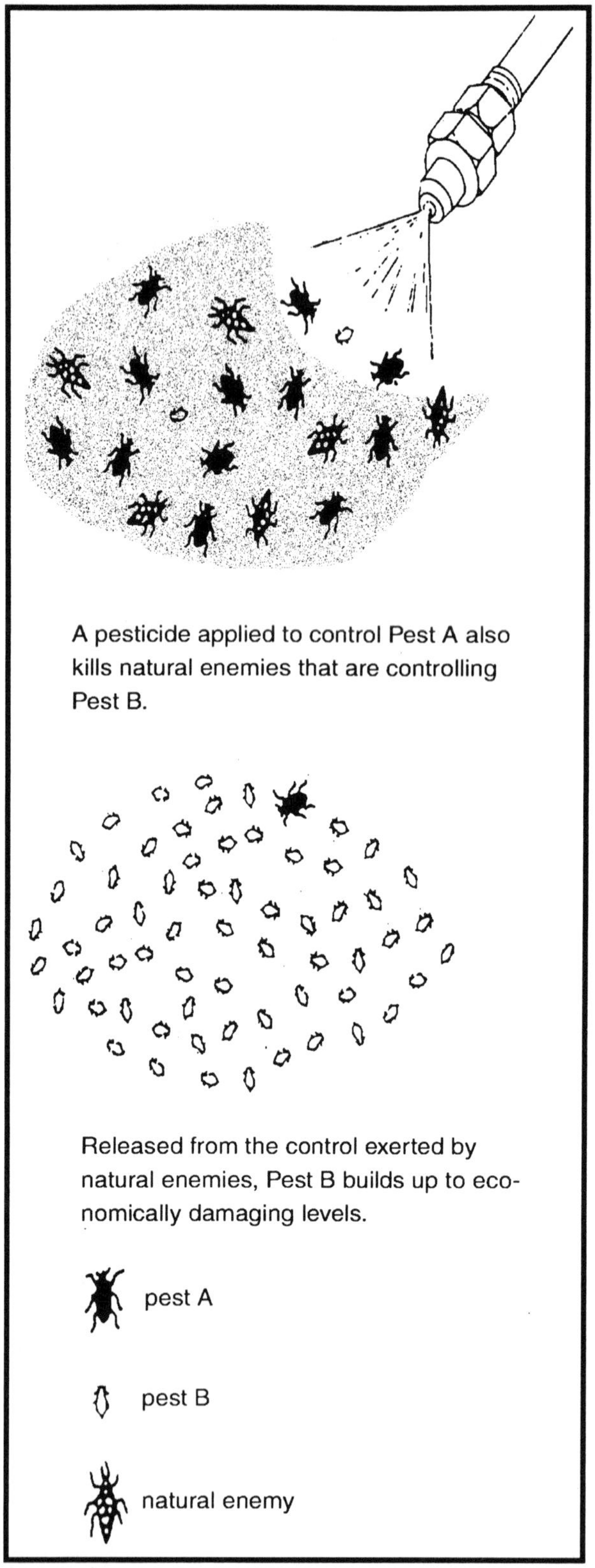

Figure 9-5. Secondary pests like pest B are created by broad-spectrum pesticide applications aimed at primary pests like pest A, that also kill natural enemies that formerly kept pest B in check. Taken from *Integrated Pest Management for Strawberries*, Univ. of California pub. 3351.

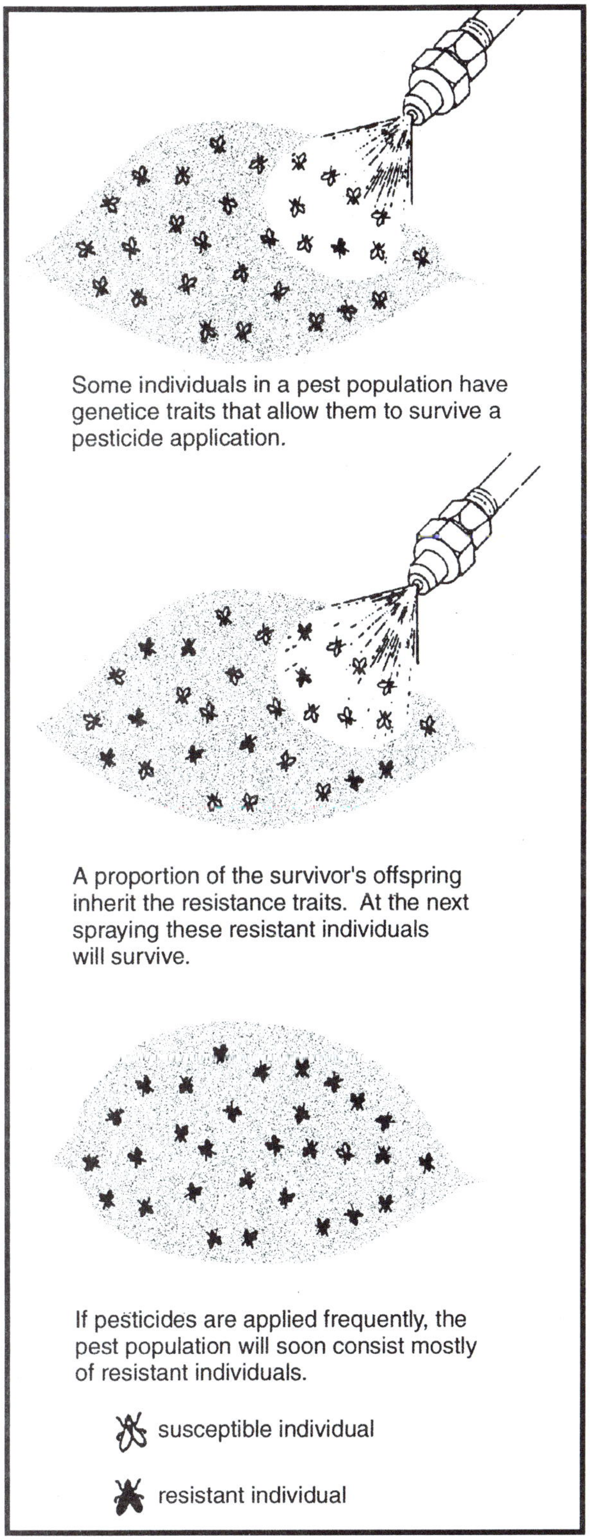

Figure 9-6. Repeated application of the same pesticide may select for resistance to that pesticide within the pest population. Therefore, integrate pesticides with other control measures and rotate types of pesticides used. Taken from *Integrated Pest Management for Strawberries*, Univ. of California pub. 3351.

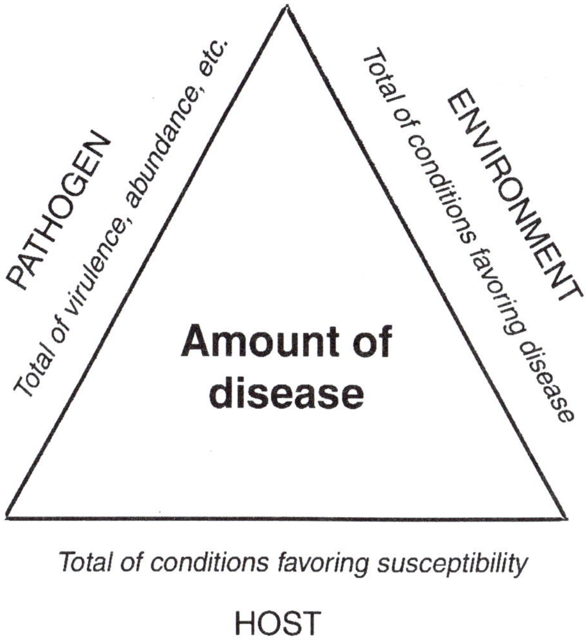

Figure 9-7. The disease triangle illustrates that disease occurs only when all three elements are present: a virulent pathogen, a susceptible host, and an environment conducive to the disease. Excluding or suppressing the pathogen, planting cultivars resistant to the disease, or modifying the environment can prevent disease.

groundcover, mulch and irrigation can alter this environment and further reduce disease incidence.

Genetic resistance to diseases and cultural practices. Fruit and nut breeders have developed many cultivars and rootstocks that resist one or more serious diseases. Growers should focus on disease resistance as one of their top criteria for choosing cultivars and rootstocks. Be careful, though, to look for resistance only to the diseases troublesome in your area, rather than choosing cultivars or rootstocks resistant to the greatest total number of diseases. There's no need to choose plants resistant to a disease that isn't serious in your area, and such plants may be more prone to other diseases or problems. For example, where root-knot nematodes are not serious, don't plant peach rootstocks resistant to these nematodes, since they are more susceptible to lesion nematodes and cold damage.

Strong resistance is widely available to many foliar diseases and diseases that can kill the plants, but moderate resistance is the best available to fruit rots. Rapidly growing, succulent plant tissue is more susceptible to many diseases, so avoiding excess nitrogen and excess irrigation can reduce fruit rots, fire blight and other diseases. Applying urea or another high-nitrogen fertilizer to old leaves after plants have already hardened for winter, or shredding the fallen leaves with a flail mower, will help the leaves decompose faster, destroying many overwintering plant pathogens. Urea applications to apple leaf litter in November reduced sporulation of the apple scab pathogen by 50% in New Hampshire trials, while April applications reduced sporulation by 66%. Shredding the fallen leaves with a flail mower in either November or April reduced sporulation by 80 to 90%. Applying urea in conjunction with flail mowing could reduce disease incidence even further.

Many root rot diseases can be prevented by choosing a well-drained soil and planting on raised beds or mounds, as described in chapter 2. Mowing the groundcover and proper pruning to keep plants open allows the wind to dry off dew quicker, reducing the wetting period and subsequent foliar and fruit diseases. Summer pruning may reduce diseases more effectively than winter pruning. For example, summer pruning consistently reduced flyspeck of apples by 50% where no fungicides were applied, while dormant pruning had no effect on flyspeck during 2 years and increased it during a third year, probably because it stimulated greater regrowth (Photo 9-8).

Virus-free plant propagation and weed control. Virus infections weaken plants, usually causing them to produce low yields of small, misshapen fruit. Viruses are systemic pathogens that move from plant cell to plant cell, infecting the entire plant except for the newest cells at the growing tips. In recent years, nurseries have utilized *meristem culture*, a form of tissue culture that propagates from these few cells at the growing tip, to produce plants free of viruses (fig. 9-8).

Growers should always order certified virus-free plants of the fruit crops most prone to viruses, including strawberries, raspberries, cherries and peaches. After starting with clean plants, keep the planting clean by eliminating weeds that could harbor the viruses, and the aphids or leafhoppers that vector viruses from the weeds to the fruit plants. For example, bramble growers should eliminate wild brambles on their land, which are often virus infected, and all fruit growers should eliminate virus-prone weeds such as dandelions.

Weather monitoring and disease forecasting. Temperature, rainfall and humidity have a large effect on plant disease development, so weather monitoring can help determine when control treatments are needed and when they aren't, reducing many pesticide applications. Some states' Extension services offer disease forecast

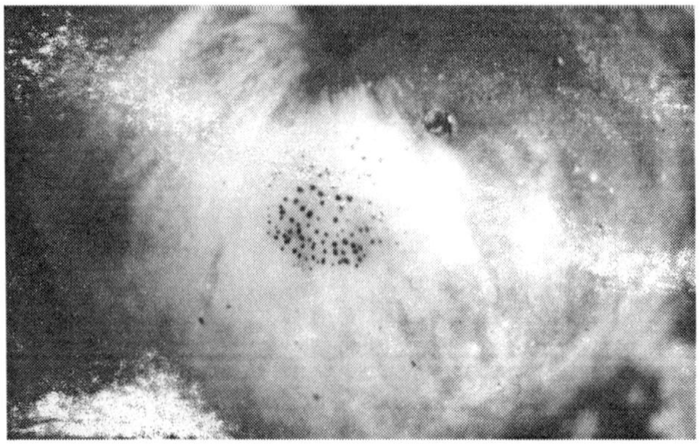

Photo 9-8. Summer pruning can reduce certain fruit diseases, such as flyspeck of apples, by opening plants up to breezes that reduce humidity.

alerts or computer programs that help determine when treatments are needed. Such programs determine "infection periods" for each disease, based on the temperature, rainfall and humidity needs of each disease. Spore traps may also be used to corroborate that the disease agent is present, but for common diseases it's often assumed that it is and the weather determines when disease outbreaks will occur.

A manually checked weather station can be easily and inexpensively set up in a fruit planting. It should contain a U-type max/min thermometer, placed in a small instrument shelter on a 6- to 7-foot post so it's 5 feet off the ground, and a clear plastic rain gauge with 6-inch capacity set on a 4- to 5-foot post so it's 3 feet off the ground. Set up the station at an elevation representative of the entire planting's terrain, at least 100 feet from nonvegetated areas (gravel roads, concrete or paved areas) or buildings, and at least twice as far away from trees as their height.

Automated weather monitors are more expensive, but automatically record data in formats that can be easily analyzed by computer programs. Adcon Telemetry (www.adcon.com) offers relatively inexpensive solar-powered stations that measure soil as well as climate factors, along with software providing irrigation and fertilizer recommendations.

Alternatively, e-mail weather forecasts for your latitude, longitude and elevation are available by subscription to several companies, such as SkyBit E-Weather (800-454-2266). These require no field monitoring but may vary from your actual site conditions at times. SkyBit E-Weather also provides 1- to 7-day predictions for several diseases and insect pests, including apple scab, fire blight, summer diseases, apple maggot, codling moth, Oriental

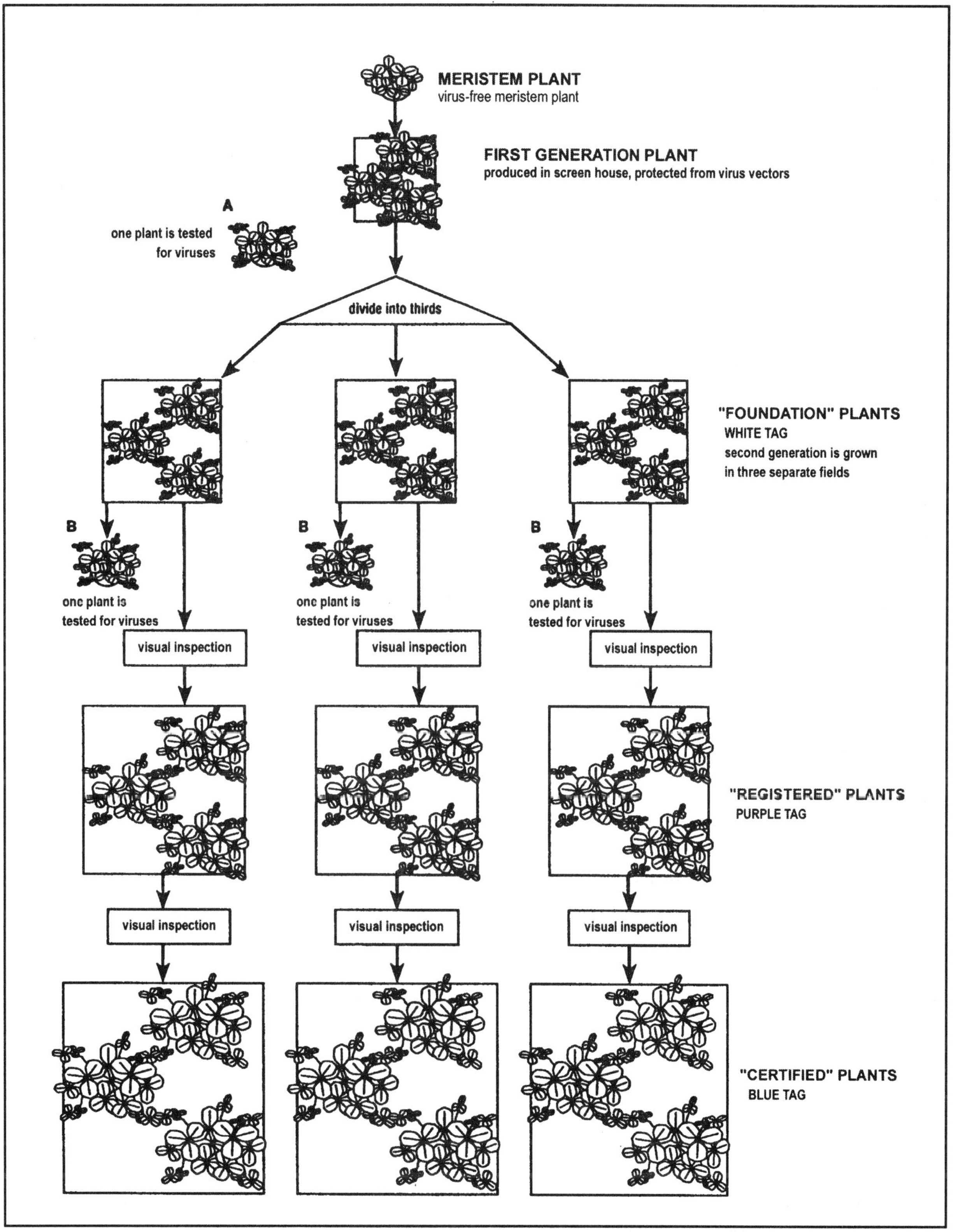

Figure 9-8. Growers can eliminate viruses in new fruit plantings by ordering certified virus-free plants from a nursery. This diagram details how certified strawberry plants are produced. Taken from Integrated Pest Management for Strawberries, Univ. of California pub. 3351.

fruit moth and tufted apple bud moth.

Biocontrol of plant diseases. Beneficial fungi, actinomycetes, bacteria, nematodes and viruses have been developed into biological pesticides that can be sprayed on crops to fight disease agents. Some of these biopesticides simply contain the toxins produced by the beneficial organisms, and these can be applied like any other pesticide. Other biopesticides contain the beneficial organisms themselves, and these, like the pathogens, are highly dependent on favorable temperature and humidity conditions to work properly. Consequently, these latter biopesticides may produce variable results. Growers who use them must pay more attention to timing sprays to take advantage of prime environmental conditions, or must modify environmental conditions or biopesticide delivery systems.

Antibiotics, such as streptomycin (Agrimycin™, Bac-Master™) and terramycin (Myco-Shield™), are some of the oldest biocontrols for plant diseases. Soil actinomycetes, such as several *Streptomyces* species, produce these compounds, which kill bacteria. Fruit growers use them to control bacterial diseases such as fire blight of pome fruits and bacterial spot of stone fruits. Antibiotics are expensive and have been used so much in some areas, however, that bacterial pathogens have developed resistance to them. Therefore, rotate them with other types of bacteriocides to avoid developing pest resistance.

A newer biofungicide, produced by *Streptomyces griseoviridis*, has been marketed as Mycostop™ by the Geiger Company (800-443-4437; www.ecgeiger.com). Mycostop™ has been found effective against these fungal pathogens: *Alternaria, Botrytis, Fusarium, Phomopsis, Pythium and Phytophthora*. Topshield™, a new formulation of the beneficial fungus in Rootshield™ (*Trichoderma harzianum* strain T-22), is registered for foliar application against powdery mildews and *Botrytis* of pome fruit, stone fruit, berries and nuts.

Bacillus subtilis (Serenade™) helps prevent many diseases caused by fungi or bacteria, by stopping fungal spores from germinating, disrupting germ tubes and mycelial growth, and producing an inhibition zone that stops pathogens from attaching to the plant or growing.

Serenade™, manufactured by AgraQuest (530-750-0150; www.agraquest.com), also helps induce the crop plants' natural systemic resistance to certain pathogens. Trials have found it effective against many fruit diseases, including fire blight, powdery mildews, apple scab, and pear scab.

AQ10™ biofungicide, made by Ecogen (800-220-2135; www.ecogeninc.com), has proven as effective as chemical fungicides against powdery mildew on grapes, pome fruits, stone fruits and strawberries. AQ10™ contains spores of the fungus *Ampelomyces quisqualis*, a parasite of the powdery mildew fungus. The powdery mildew fungus must be present in the planting when AQ10™ is applied for it to work, and thorough wetting of all exposed leaf and fruit surfaces during application is important. Apply AQ10™ in the early morning or late evening when humidity levels are highest, so the parasite doesn't dessicate and die before the spores germinate and invade the powdery mildew host.

Chemical fungicides and bacteriocides. Fruit growers have used certain minerals to control fungal and bacterial diseases for centuries. These minerals are considered "organic" in the sense that they occur naturally, although they are inorganic (non-carbon-containing) chemicals. Sulfur sprays control several fungal diseases, notably powdery mildew, but can damage plants if applied at temperatures greater than 85 F or if used within 2 weeks before or after an oil spray, and leave unsightly residues on the fruit. Some fruit processors, such as wineries and baby food manufacturers, do not want growers to apply sulfur after mid-season and will not accept fruit with residues. Copper sprays, such as copper sulfate, control some bacterial as well as fungal diseases. Copper sulfate is usually applied with hydrated spray lime as a safener to prevent foliar damage to plants from the copper; this combination is called Bordeuax mixture because it was originally applied to a vineyard in Bordeaux, France in 1882.

Several companies are now marketing *bicarbonate compounds* as effective and very safe fungicides, especially for control of powdery mildew, leaf spots and gray mold. Some growers have used sodium bicarbonate, also known as baking soda, but potassium bicarbonate is a better choice because it provides needed potassium and avoids a buildup of sodium, which can damage plants (see chapter 6). Both work by creating an alkaline pH around the fruit or leaves, which fungi can't survive.

Potassium bicarbonate is sold as Kaligreen™ by the Monterey Chemical Co. (559-499-2100; www.montereychemical.com). Kaligreen™ is registered for most fruit crops and has received organic approval.

Another new chemical bactericide/fungicide that has received organic approval for use on apples, grapes and hazelnuts is Oxidate™, a combination of hydrogen peroxide and peroxyacetic acid with surfactants that keep it from damaging plant tissues. This powerful, broad-spectrum compound kills many disease organisms and inactivates dormant spores on contact, yet is safe for people and has a 0 hour re-entry time after treatment. Oxidate™ is manufactured by BioSafe Systems (888-273-3088; www.biosafesystems.com).

Elexa 4™ from SafeScience (www.safesci.com), a more selective compound, contains chitosan, a natural organic material that inhibits attachment and penetration of fungi causing downy mildew and powdery mildew. Elexa 4™ is registered for pome fruits, stone fruits, grapes and strawberries.

Other chemical companies have developed safer fungicides in recent years that specifically target fungal enzymes that don't occur in other organisms. One group, known as *sterol inhibitors*, inhibit the fungal enzyme ergesterol. These include myclobutanil (Nova™) and fenarimol (Rubigan™). Another group, known as *strobilurins*, was developed from a natural anti-fungal compound that occurs in a small mushroom in the genus *Strobilurus*, that grows on fallen pine cones. Strobilurins, which include azoxystrobin (Abound™), trifloxystrobin (Flint™) and kresoxim-methyl (Sovran™), provide excellent inhibition of spore germination and secondary sporulation of pathogenic fungi, but have very low toxicity to humans and beneficial organisms and break down quickly in soil. These fungicides are very safe and effective, but a fungicide within each group should be rotated with other fungicides with a different mode of action, to avoid inducing fungal pathogens to develop resistance to these controls. Rotating chemical and biological disease control methods can prolong the usefulness of both.

When choosing fungicides, another factor to consider is that certain fungicides can reduce populations of mycorrhizal fungi, the beneficial soil organisms that live on or in plant roots and help plants to better absorb water, phosphorous and micronutrients. Foliar fungicide sprays will generally not harm mycorrhizae because they don't come in contact with the soil unless sprayed to the point of runoff. However, systemic foliar fungicides such as benomyl (Benlate™) and triadimefon (Bayleton™), which are translocated throughout the plant including the roots, have inhibited mycorrhizae. Several fungicides applied to the soil to control root rot diseases, including metalaxyl (Ridomil™ or Subdue™) and fosetyl-Al (Aliette™), also inhibit mycorrhizae. Try to choose non-systemic foliar fungicides that don't hurt mycorrhizae, and prevent many root rot problems by choosing disease-resistant cultivars and rootstocks and planting on raised beds instead of using soil fungicides.

MANAGING PEST MAMMALS AND BIRDS

Mammals and birds can cause severe losses on many fruit and nut crops. Voles, mouse-like rodents that tunnel underground and strip the bark off lower trunks and roots, can kill many fruit trees if not managed. Rabbits smoothly strip bark from lower trunks in winter (Photo 9-9), while deer jaggedly strip bark from tree trunks and branches during fall antler rubbing or winter feeding, also killing trees. Pest birds such as starlings can dam-

Photo 9-9. Rabbits can kill fruit trees if not controlled.

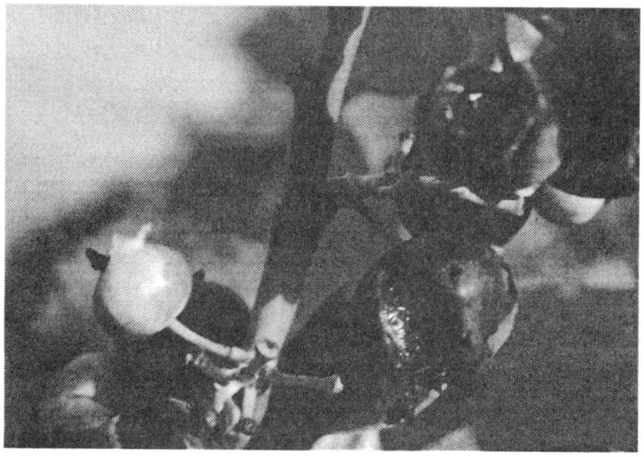

Photo 9-10. Birds can take or damage much of certain fruit and nut crops, such as these blueberries, if not prevented.

age over half the fruit of susceptible crops such as grapes and blueberries (Photo 9-10), and birds and squirrels can take a significant amount of pecans and other nut crops.

Excluding mammals and birds. The best way to deal with most vertebrate pests is to exclude them from the fruit planting or from the plants. *Plastic or hardware cloth guards* around the trunks of trees in winter stop voles, rabbits and other animals from stripping the bark, which would kill the tree. Where borer insects are also a problem, wrap trees instead with sections of 12- to 18-inch-wide metal, fiberglass or nylon window screening, securing it at top and bottom with twist ties, and loosening and replacing these at least once a year. White plastic guards (Photo 9-11) work well on young trees because they also prevent the young trunks, with little protective bark built up and little shade from limbs, from heating up too much in winter and developing sunscald. Alternatively, growers who paint tree trunks white to prevent sunscald can mix a rodenticide/insecticide such as thiram into the paint. Keeping an area at least two feet out from each tree trunk free from vegetation and mulch will also reduce vole and rabbit problems.

Deer and elk must be excluded from the whole orchard. A *slanted, 5-foot high and 8-foot wide electric fence* (Figure 9-9) works best where deer pressure is high, since deer can jump even a 10-foot high non-electric fence. Gallagher Power Fence (Texas office 800-531-5908, Virginia office 800-492-9617) developed this 30-degree-slant fence design with

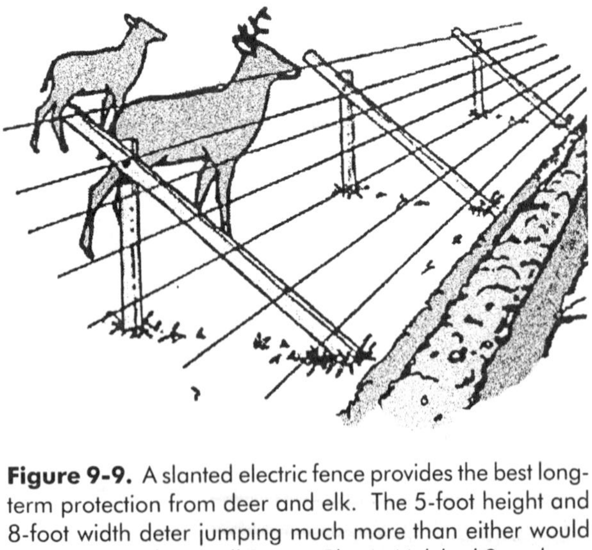

Figure 9-9. A slanted electric fence provides the best long-term protection from deer and elk. The 5-foot height and 8-foot width deter jumping much more than either would alone. Taken from *Hill System Plastic Mulched Strawberry Production Guide for Colder Areas*, Virginia Cooperative Extension pub. 438-018.

Photo 9-11. White plastic tree guards prevent rodent damage and sunscald.

a charger that conditions deer without harming them, and materials cost less than $1 per running foot of perimeter. Some states' Wildlife and Parks Departments offer cost-share payments to farmers to install electric fencing to protect against deer and elk damage. Solar cells are available to power the fence. Each fall, paint the outermost fence wire at intervals with a mixture of equal parts peanut butter and vegetable oil, so the deer get a small shock when they lick it; after that, they'll stay away from the fence. Post signs warning people not to touch the fence.

Where deer don't pose a problem during daylight hours, the fence charger can be set on a timer to operate only from dusk to dawn. Electric fences serve as behavioral barriers, not true physical barriers, so permanent fences must be kept charged all year. If the power is turned off during summer and deer test the fence and don't get shocked, they may find they can jump the fence, and it will no longer keep them out once it's turned on for winter.

Where deer pressure is low to moderate, some growers use *lower-cost, temporary electric fences* that they put up each fall and take down each spring, to avoid keeping a fence charged all summer. A two-strand fence of polywire (white polyethylene braided with the wire strands) or polytape (like polywire but wider and stronger) at 30" and 18" off the ground, between posts spaced 30 to 60 feet apart, can effectively deter low populations when baited with peanut butter. Polywire or polytape costs about $0.18 per running foot of perimeter and lasts 5 to 7 years.

Some crops subject to severe bird predation, such as blueberries and grapes, have been economically protected on small acreages by erecting a frame for *bird netting*, and attaching sections of plastic netting together around and over the entire planting. Other small fruit growers run single strands of smooth wire over the top and sides of each row to support the netting and keep it away from the berries, and use machinery with a front end lift to apply large rolls of netting.

The increased yields in an Illinois blueberry field paid 80% of the cost of netting within the first season, and recouped the remaining cost early in the second season. Retail growers with just a few tart cherries, small trees highly prone to bird predation, have protected their crop by individually netting each tree (Photo 9-12). The netting has a lifespan of about 10 years.

Frightening devices. Fruit growers who need to control birds on larger acreages often use scare tactics. These include automatic cannons, tapes of bird distress calls, bird scare-eye balloons, and Mylar™ reflective tape. All of these devices work for just a few days until the birds get used to them, so it's best to rotate between at least two different scare methods, and move them every few days to extend their effectiveness. Further increase the effectiveness of scare devices by using them in conjunction with repellents or trapping.

Automatic propane exploder cannons require one cannon per 10 acres of fruit crops to be protected. Place each cannon on a platform higher than tree or bush height, and move it every 3 days. Run the cannons from sunrise to sunset, with a varied interval between explosions. Attachments that rotate the cannon with each explosion and provide several different multi-bang versions of explosion will increase the effectiveness of cannons. A motion detector attachment that triggers the cannon only when birds are in the field, rather than continually, keeps cannons effective longer.

Photo 9-12. Netting can prevent bird damage, and is economical on small plantings of dwarf trees, such as tart cherries, or on small blocks of blueberries or grapes.

Firing shotgun shells over the fruit planting whenever birds are present can also scare them away. This works best when firing from a concealed position so birds do not associate explosions with a vehicle or person.

Broadcasting bird distress calls works best when using mobile speakers, such as mounting them on trucks. Broadcast alarm and distress cries only during daylight when birds are in the fruit planting, and alternate these sounds with the sounds of urban traffic, static, steam locomotives, ocean surf, hard rock music and other sounds that may frighten birds.

Bird scare balloons require at least one balloon for every 3 to 5 acres of fruit crops to be protected. Suspend each balloon from a pole or line between posts, at varying heights at least 3 feet above the trees or bushes, and use tethers at least 2 feet long so the balloons will move (Photo 9-13). Move the balloons and vary their heights every 3 to 5 days.

Bird scare tape is red on one side and silver on the other. Stretch the tape between posts at the ends of fruit crop rows, positioning the tape at least 18 inches above the crop. Twist the tape several times before attaching it to the posts, so the visible interval of red/silver is about 16 inches. Tape may not be needed over every row. In some cases, just one or two tape lines per acre have proved ef-

Photo 9-13. Bird scare-eye balloons can repel birds for a few days in larger plantings such as these grapes, but must be moved every few days as the birds get used to them.

fective, so start with that and add more tape as necessary.

Repellent substances. Repellent substances, developed for deer and some birds, work best when integrated with other methods of control. Apply repellents at the first sign of damage to prevent the animals or birds from establishing a feeding pattern in the fruit planting.

Repellents fall into two categories, area or contact. Area repellents ward off the pests by odor, so may be applied only around the borders of the fruit planting to reduce costs. Area repellents for deer include tankage (putrified meat scraps), bags of human hair (obtained from hairdressers), blood meal, and ammonium soaps. Contact repellents work by taste, so must be applied to all plants in the orchard. Contact repellents include putrescent egg solids and hot pepper sauce for deer, and the flavor of 'Concord' grapes for birds.

Apply deer repellents monthly during the dormant season, on dry days when temperatures are above freezing. Repellents work best when the deer density is low and other food sources are available, so keeping part of your land in wildlife habitat and hunting will complement the use of repellents. University of Wisconsin tests found that the area repellent Hinder™ (800-247-8711), containing ammonium soaps of fatty acids that give off an ammonia odor, reduced deer damage by 46%, while bags of human hair curtailed damage by 34%. Contact repellents Deer Away™ (800-468-2472) and Deer Off Big Game Repellent™ (www.medpest.com/deeroff.htm), both made from putrescent egg solids, reduced deer damage by 43%, while hot pepper sauce cut damage by only 15%.

One bird repellent chemical, methyl anthranilate, is a component of 'Concord' grapes that is used to flavor candy, soft drinks, gums and medicines. The sprayable products Bird Shield™ (www.bsrc.com or 800-359-1594) and ReJex-it™ (800-423-2473) contain this chemical and are registered for use on cherries, blueberries and table grapes. The chemical repels birds that taste it, and is effective for about a week, usually reducing bird damage to less than 1% on cherries and blueberries and less than 5% on grapes. It works best if applied when birds first appear, and weekly thereafter, rather than starting just two weeks before harvest. Sunlight and high temperatures break down the active ingredient, so that no flavor remains on harvested fruit.

Biological control. Hawks, owls, shrikes, snakes, raccoons, foxes, weasels, opossums and cats all feed on voles and rabbits. Encourage such predators by leaving some uncultivated woodland and meadow or prairie areas for them as habitat, and do not kill such beneficial animals. Some farmers raise domestic cats, which can be quite effective in killing voles and scaring birds when patrolling fruit plantings.

A British Columbia vineyardist used 10 to 12 falcons to protect 500 acres of grapes from starlings, with each falcon covering 80 to 100 acres at two shifts per day. The falcons do not control robins or some other pest birds, however, and renting falcons often costs more than hiring people to shoot the pest birds.

Hunting, trapping and poison baits. Contact your local Wildlife Conservation office to check your state's regulations before doing any hunting, trapping or baiting. Farmers are generally limited as to the number of animals they can kill and the seasons.

Orchardists can open their land to hunters during deer season to reduce damage. Outside of deer season, most states allow farmers to kill any deer *witnessed* to be causing or about to cause damage to the orchard, but often you must contact your local Wildlife Conservation office first.

Some counties' pest control boards have instituted trapping programs for starlings, crows and other wildlife pests. If these pest birds have become more numerous in your area, contact your local Wildlife Conservation office to request that they begin or step up such a program. Some counties also use the bird poison chloro-ptoluidine (DRC 1339 or Starlicide™) in the areas where starlings or crows group on their way to and from their roosting site. This poison can be used near, but not in, fruit or other food crop plantings, and each state has its own label with varying restrictions.

Poison baits are widely used to control voles within orchards, and most rodenticides can only be used during the dormant season when trees are not bearing fruit. Use pelleted baits designed for rodents, such as zinc phosphide pellets, rather than grain baits, as the latter are more hazardous to ground-feeding birds and other non-target wildlife. Further protect non-target wildlife by placing the pellets in bait stations, rather than broadcasting them throughout the planting. Bait stations can be made from discarded beverage cans, and marked with flags so they can be relocated every few days to cover the entire area.

PESTICIDE APPLICATION TECHNOLOGY

Many pest repellents and pesticides, including biopesticides, require special application equipment. Fruit and nut growers use sprayers to apply liquid materials, and granular applicators to apply dry formulations.

Types of sprayers. Herbicides can be applied with low-pressure boom sprayers or small backpack sprayers for spot applications, but the liquid pesticides and repellents discussed in this chapter generally require a large, high-pressure sprayer. Orchardists usually use an *air-blast sprayer* (Photo 9-14) to reach to the tops and middles of trees with the pesticide. Air blast sprayers use a high-speed, fan-driven air stream to break the spray into fine droplets, which the air stream can carry upward 20 to 150 feet, depending on the model design for different sized trees. Air blast sprayers have traditionally produced a lot of off-target drift, but newer models charge the spray droplets with an electrostatic charge that is attracted to foliage, thus minimizing drift.

Small fruit rows can be sprayed with smaller air-blast sprayers, but *over-the-row sprayers* (Photo 9-15) can reduce pesticide waste even more. Over-the-row sprayers straddle the row, applying the pesticide from both sides, and any material that doesn't stick to the plants recirculates back into the sprayer.

Testing water for pesticide sprays and spray combinations. Test your water source for pH, with a pH meter or test kit, before using it to mix pesticides. Some water sources vary in pH over the seasons, so retest pH before each pesticide application until you know your water's seasonal variations. Many pesticides, particularly insecticides, undergo alkaline hydrolysis if mixed with water with a pH greater than 7.0, which destroys their effectiveness.

A pH in the range of 4.0 to 6.5 is recommended for many pesticide sprays. This can be achieved by adding an acidifying material such as ammonium sulfate to the tank mix, and/or by adding a buffer to keep the pH within a certain range. A few pesticide materials, including those containing copper or lime, should never be acidified, since an acid pH causes these materials to damage plants. Always read and follow the pesticide label carefully as to pH of the spray solution and materials to mix with the pesticide.

Some pesticides should not be combined due to adverse chemical reactions. For example, do not mix certain insecticides requiring an acid pH with disease-controlling chemicals containing copper or lime. Each pesticide's label lists any other chemicals that should not be combined with it in a single spray. Always follow all safety precautions listed on the label for mixing and applying pesticides.

Photo 9-14. Orchardists use air-blast sprayers to reach the tops of trees with insecticides and fungicides. Dwarf fruit trees can utilize a less powerful, thus less expensive, air blast sprayer than taller trees such as these pecans.

Photo 9-15. Grape growers and other small fruit growers often use an over-the-row sprayer, which captures and recirculates most excess spray, greatly reducing drift.

Applying the correct amount of pesticide spray. Fruit and nut growers must apply enough pesticide to provide good pest control, but not excess amounts, which would waste money and could damage the environment. Since fruit and nut plants vary much more in height and width than do field crops, pesticides can be applied much more accurately if amounts are figured on the basis of tree-row volume rather than ground surface area (acres). *Tree-row volume* (TRV) is simply the linear

Adjusting spray rates with tree-row volume

1. Determine the number of linear feet per acre, (A) = L.

   L = 43,560 square ft per A/feet between rows

   For example, trees spaced 20 ft apart have L = 43,560/20 = 2178 ft

2. Measure average tree height, H, and width, W.

   For example, H measures 12 ft and W measures 14 ft

3. Tree-row volume, TRV = (L x H x W) cu. ft per A (Fig. 9-10)

   In this example, TRV = (2178)(12)(14) = 365,904 cu. ft per A

4. Gallons per acre to apply, dilute basis = (TRV x 0.7 gal)/1000 cu. ft

   In this example, dilute application = (365,904 cu. ft per A x 0.7 gal)/1000 cu. ft = 256 gal per A

   For 2x concentrate spraying, apply 128 gal per A
   For 4x concentrate spraying, apply 64 gal per A
   For 8x concentrate spraying, apply 32 gal per A

5. Ounces of pesticide to apply per A = TRV x pesticide rate per 10,000 cu. ft

   For example, Nova 40 WP: rate = 0.1 oz per 10,000 cu. ft.
   Ounces of Nova to apply per A = (365,904 cu. ft per A)(0.1 oz per 10,000 cu. ft.) = 3.66 oz per A

Note: for concentrate spraying, rate of pesticide applied per A remains the same, only the amount of water applied per acre decreases.

feet of fruit or nut planting per acre, multiplied by the average height and width of the row (Figure 9-10). Compare the TRV, in cubic feet per acre, to a pesticide recommendation based on cubic feet per acre to determine how much pesticide to apply per acre (see sidebar, next page).

Low-volume or concentrate spraying has become popular in recent decades, as it reduces the number of times the sprayer tank must be refilled, and thus reduces the labor requirement for spraying. Low-volume spraying applies about the same amount of pesticide per acre, but with a reduced amount of water carrier (see "Adjusting spray rates with tree-row volume," next page). Generally, dilute spraying requires 350 to 400 gallons of water per acre, while concentrate spraying needs just 20 to 100 gallons. Low-volume spraying eliminates pesticide runoff from the fruit plants to the ground, and eliminating this waste can enable a grower to reduce pesticide rates by about 20% per acre.

Concentrate spray applicators must be even more careful than dilute applicators to calibrate the sprayer correctly and keep a consistent speed, as concentrate spraying magnifies the effects of small discrepancies in nozzle output or spray coverage time per plant. Some pesticides, such as dormant oil, bactericides, and growth regulators, must still be applied dilute or at no less than 100 gallons per acre to obtain even coverage. For most other chemicals, concentrating the spray from 400 to 50 gallons of water per acre produces great savings, but further concentrating sprays to less than 50 gallons of water per acre doesn't produce much additional cost savings, and adds more risk of uneven treatment than is justified.

Alternate-middle row spraying involves driving down every other aisle with an air blast sprayer, since some material will be pushed through to the other side of each row of plants. On the next spray date, the applicator drives down the aisles that he did not drive on last time, and skips the aisles that he did drive on last time. This technique cuts the amount of pesticide applied per acre each time in half, but some growers offset this savings by spraying twice as often, say every week instead of every two weeks.

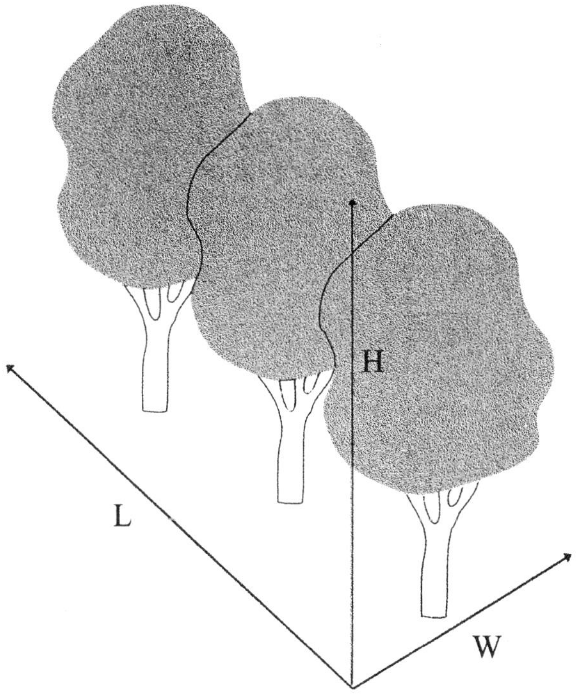

Figure 9-10. Determine tree-row volume by multiplying the number of linear feet of row per acre (L) by average tree height (H) and width (W).

That works well for pesticides which are short-lived on the tree, but many times a compromise can be reached, for example spraying alternate aisles every 10 days, which can produce good control and still reduce the overall amount of pesticide applied. Alternate-middle row spraying requires an air-blast sprayer with sufficient air volume and velocity to reach at least 90% of the fruit or nut plants with at least light spray coverage during each application.

Nozzle selection and sprayer calibration. Each sprayer can be used with several different nozzles, each containing a slightly different opening size and releasing a different amount of spray. Sprayer operating pressure, tractor speed, and how worn the nozzle is also influence the amount of spray released by each nozzle. To ensure applying the correct amount of pesticide, choose nozzle sizes carefully, calibrate the sprayer for the operating pressure and speed the operator will use, and keep speed consistent.

To select the best nozzle size, first consult the pesticide label and decide the gallons of spray to apply per acre, based on concentrate or dilute spraying of that particular pesticide, and tree-row volume.

Table 9-2.
TOTAL AIRBLAST SPRAYER OUTPUT*

Feet between rows	Gallons of spray/A									
	20	30	40	50	60	80	100	200	300	400
36	3.6	5.5	7.3	9.1	10.9	14.5	18.2	36.4	54.6	72.7
34	3.4	5.2	6.9	8.6	10.3	13.8	17.2	34.4	51.6	68.7
32	3.2	4.8	6.5	8.1	9.7	12.9	16.2	32.3	48.5	64.6
30	3.0	4.5	6.1	7.6	9.1	12.1	15.2	30.3	45.5	60.6
28	2.8	4.2	5.7	7.1	8.5	11.3	14.1	28.3	42.4	56.5
26	2.6	3.9	5.3	6.6	7.9	10.5	13.1	26.2	39.4	52.5
24	2.4	3.6	4.8	6.1	7.3	9.7	12.1	24.2	36.3	48.4
22	2.2	3.3	4.4	5.6	6.7	8.9	11.1	22.2	33.3	44.4
20	2.0	3.0	4.0	5.1	6.1	8.1	10.1	20.2	30.3	40.4
18	1.8	2.7	3.6	4.5	5.5	7.3	9.1	18.2	27.3	36.4
16	1.6	2.4	3.2	4.0	4.8	6.5	8.1	16.2	24.2	32.3
14	1.4	2.1	2.8	3.5	4.2	5.7	7.1	14.1	21.2	28.3
12	1.2	1.8	2.4	3.0	3.6	4.8	6.0	12.1	18.1	24.2
10	1.0	1.5	2.0	2.5	3.0	4.0	5.0	10.1	15.1	20.2
Note: Output in gallons per minute should be 20% less at 2.0 mph and 20% more at 3.0 mph. Changing speed without recalibrating results in too much or too little spray applied.										

*Output from both sides, in gallons per minute, at 2.5 mph. Adapted from *Pennsylavania Tree Fruit Production Guide, 1998-1999.*

Table 9-3.
TYPICAL AIR BLAST SPRAYER NOZZLE FLOW*

Nozzle	Pressure, psi					
	60	80	100	150	200	250
D4-25	0.35	0.40	0.45	0.54	0.61	0.68
D4-45	0.43	0.50	0.56	0.68	0.77	0.86
D6-45	0.72	0.83	0.93	1.15	1.32	1.48

*In gallons per minute, of typical air-blast sprayer cone nozzles. Adapted from *Pennsylvania Tree Fruit Production Guide* 1998-1999.

Then decide your tractor speed, which can be 3 to 4 mph for smaller, less thick plants, but only 2 or 2.5 mph for taller, thicker trees. Consult a table such as Table 9-2, which is often provided by the sprayer manufacturer, to determine how much spray needs to be applied in gallons per minute, based on the gallons-per-acre rate and the plant spacing. (If spraying alternate row middles, cut the gallons-per-minute rate in the table in half.)

Divide the gallons-per-minute rate you derived from Table 9-2 by the number of nozzles your sprayer contains, to determine the flow rate you need from each nozzle. Then consult a nozzle flow table such as 9-3, again provided by the sprayer manufacturer, to choose the nozzle size that produces the desired flow rate, and the operating pressure you'll need to maintain to produce that flow rate. (This system of choosing a single nozzle size so that all nozzles produce the same output works well on small fruits, dwarf fruit trees, and tower air-blast spraying of larger trees. Conventional air-blast spraying of tall trees however, requires using nozzles with greater output on the top of the sprayer and nozzles with less output on the sides; the total output of all nozzles should still add up to the desired gallons-per-minute rate.)

Once you've selected the best nozzle size, calibrate the sprayer to be sure the nozzles actually produce the output they are supposed to under your operating conditions. To do this, fill the sprayer tank to a marked level with water only, pressurize the system to the desired level, and spray for 3 minutes. Then measure the gallons of water needed to refill the tank to the same level. Divide this number by 3 to determine the gallons-per-minute rate, and compare that with the flow rate you wanted to achieve. If the flow rate differs from the rate you wanted, adjust pressure or nozzle sizes and repeat this procedure until the correct flow rate is achieved.

Granular pesticide application. Fruit and nut growers use granular applicators to apply slug control granules, rodenticide baits, and certain granular herbicides, as well as granular fertilizers. Granular applicators must be calibrated using the same material that will be applied, under the same temperature and humidity conditions, so run a *check test* each time before applying the material. Do this by disconnecting the spreading mechanism (if one is used) and attaching a catchpan, plastic or paper bag, or other container to catch the granules.

Operate the equipment at the recommended settings and speed to release the desired pounds per acre (determined by consulting the equipment manufacturer's tables), over a measured distance, such as one-tenth of an acre, and measuring the pounds of granules collected in the container. If the amount actually applied differs from the desired amount, adjust settings or ground speed and measure again, until the correct amount is achieved.

Further Reading

Cooley, D.R., J.W. Gamble, and W.R. Autio. 1997. Summer pruning as a method for reducing flyspeck disease on apple fruit. *Plant Dis.* 81:1123-1126.

Epstein, D.L., R.S. Zack, J.F. Brunner, L.Gut, and J.J. Brown. 2000. Effects of broad-spectrum insecticides on epigeal arthropod biodiversity in Pacific Northwest apple orchards. *Environ. Entomol.* 29:340-348.

Howell, J.F. 1997. Organic orchard assistance. *The Grower*, June issue: 14-16.

Klassen, P. and W. Stockwin. 1988. Control insects with pheromones. *Amer. Fruit Grower*, Feb. issue: 26-31.

Pierce, R.A. II and E.P. Wiggers. 1997. Controlling deer damage in Missouri. Agric. pub. MP685. Downloadable from http://muextension.missouri.edu/xplor/miscpubs/mp0685.htm

Puterka, G.J., D.M. Glenn, D.G. Sekutowski, T.R. Unruh, and S.K. Jones. 2000. Progress toward liquid formulations of particle films for insect and disease control in pear. *Environ. Entomol.* 29:329-339.

Reid, W. 1999. Biological pest suppression in native pecan groves. pp 113-122. *In* Charlet, L.D. and G.J. Brewer, (eds.), Biological Control of Native or Indigenous Insect Pests: Challenges, Constraints, and Potential. Thomas Say Publications in Entomology: Proceeedings. Entomol. Soc. of America, Lanham, MD.

Sutton, D.K, W.E. MacHardy, and W.G. Lord. 2000. Effects of shredding or treating apple leaf litter with urea on ascospore dose of *Venturia inaequalis* and disease buildup. *Plant Dis.* 84:1319-1326.

Weakley, C.V., P.Kirsch, and R.E. Rice. 1987. Control of Oriental fruit moth by mating disruption. *Cal. Agric., May-June issue: 7-8.*

Chapter 10
Pollination, Fruit Thinning and Growth Regulators

Flowers must be pollinated for fruit to grow from them. Wind accomplishes this for most nut crops if at least two genetically different cultivars are grown for cross-pollination, as discussed in Chapter 1. Most fruit crops, along with almonds, rely on insects for pollination, whether they are self- or cross-pollinated.

PROVIDING BEES FOR POLLINATION

Native bees, flies and other insects are usually plentiful enough naturally to provide sufficient pollination, particularly if surrounding areas support a wide diversity of vegetation, such as pastures and hedgerows. However, where much of the native vegetation has been replaced by monoculture crops, setting beehives in fruit plantings during bloom may increase fruit set (Photo 10-1). Even in self-pollinating crops such as strawberries, which are often pollinated simply by pollen falling onto the stigmas, providing bees can increase fruit weight by 20 to 25%.

Use of beehives can also provide effective dissemination of certain biocontrol agents to control fruit disease pathogens that attack the flowers. Hive inserts filled with the biocontrol agent dust the bees with the biocontrol as they leave the hive. The bees then transfer the biocontrol agent to each flower they visit.

Researchers at Cornell University in Geneva, NY used this approach to control gray mold of strawberries with spores of a beneficial fungus, *Trichoderma harzianum* strain 1295-22.

It is sold as granular T-22 Planter Box™ or in sprayable form as Topshield™ by BioWorks (www.bioworksbiocontrol.com). Gray mold infected about 20% of the fruit where the biocontrol was not applied, but only 5% of the fruit where honey bees applied the beneficial fungus and 7% of the fruit where bumble bees applied the biocontrol agent. This compared with 4% gray mold in plots sprayed with a fungicide. The bees reduced gray mold better than a single bloom spray of Topshield™, probably because they concentrated more spores on the flowers, where they prevent infection by the pathogen.

Photo 10-1. Setting hives of honeybees in a fruit planting, such as these Japanese plums, can increase fruit set if native pollinators are scarce. Sprays of honeybee-attractant substances have increased pollination and fruit set in some cases, such as very wet weather during bloom, but not in other cases.

Where bees are needed only for pollination, native bees are superior to hives of honeybees, whose populations have been decimated in recent years by mite parasites and Africanized bees. Bumblebees are often present in large numbers, work at lower temperatures than honeybees, and pollinate more flowers. They nest underground, in undisturbed woodlots and meadows, so keeping these habitats preserves them. Early-flowering native plants or groundcovers, especially clovers, attract them to the area. When the fruit crops begin to flower, mow the groundcover to keep them working on the fruit plants. Bumblebee colonies can also be ordered in units of four to provide pollination, at a cost ranging from one-third to about the same cost of buying honeybee hives.

Native mason bees also do a better job of pollination than honeybees. Honeybees visit up to 700 blooms per day, but only pollinate about 5% of the flowers they visit. Mason bees visit up to 1600 blooms per day, pollinating 95 to 99% of them, and work at lower temperatures than honeybees. Unlike honeybees, they prefer the flowers of most fruit plants over groundcover flowers such as dandelions and clover. Thus 300 female mason bees per acre will do a better job of pollination than an entire hive of honeybees (60,000 or so). Another advantage of mason bees is that they're non-aggressive and rarely sting.

Mason bees in the wild nest in brush piles or hollow stems of reed, elderberry, mustards and other plants. These bees are sold commercially when hibernating in 5/16" tubes the size of paper or plastic drinking straws. Nesting receptacles to hold the straws can be bought or built, and placed in fruit plantings where they receive morning sun, to better direct the bees' pollination activities. Mason bees have not yet been tested for disseminating biocontrol agents, but probably such agents could be applied to these nests.

Substances to attract honeybees to an area, such as Bee-Scent™, are now marketed as spray materials. Ohio researchers tested Bee-Scent™ on apples and found that it did attract more honeybees to treated orchards in three of four years, but that it did not dependably increase the number of pollen-collecting honeybees nor fruit set. Bee-Scent™ did increase the amounts of large- and medium-sized fruit and decreased the percentage of small fruit, without affecting total yield, in one year of four, in which continual heavy rains fell during bloom. Washington state studies showed an increase in apple pollination and fruit set due to Bee-Scent™, but New York research showed no effect.

FRUIT DROP AND THINNING

Fruit and nut plants can be *too* fruitful, producing so heavy a crop that the plants are stressed and left more vulnerable to pest damage and winter injury, and likely to produce only a light crop the following year. Pruning out excess branches each year helps reduce this tendency by limiting the number of flower buds, and such pruning provides all the crop reduction needed for small fruits. Tree fruits and nuts often require crop control measures beyond pruning, because if weather conditions are favorable and many pollinating insects are present, far too many fruit may set. This is especially likely with large-fruited trees such as apples, pears, peaches, plums, apricots, almonds and pecans.

Fruit and nut trees may produce ten times more flowers than they need for an optimum crop, as insurance against possible losses from winter cold, spring frosts, poor pollination, and fruit pests. During seasons when losses from these dangers are minimal, fruit set can be very heavy. A tree partially compensates for a heavy fruit set by dropping some of the excess fruit while they are small. This natural fruit drop reduces stress on the tree. A tree with a heavy fruit set can afford to lose up to 90 percent of its tiny fruit.

Most years, growers need to take nature a step further by *thinning* the fruit: removing those that are still too crowded. The fruit remaining on the trees will get bigger and sweeter. Thinning will decrease yields somewhat, preventing a huge harvest of small fruit, but greatly increases profits, because the larger fruit are worth much more than small fruit. Of all the factors influencing fruit size, thinning has the greatest effect (Figure 10-1). Fruit thinning also reduces the number of diseased and insect-damaged fruit.

Fruit thinning reduces branch breakage and stress on the tree, making winter injury less likely. If done early, it also helps prevent biennial bearing, which is a heavy crop of small fruit one year, then very few fruit the next, by stimulating increased fruit bud formation for next year's crop.

Thin fruit early, within three weeks after bloom, for the greatest benefit. The trees will still be dropping some fruit naturally at this time. The ideal thinning method will leave one or two apples or pears per cluster, leave peaches about six to eight inches apart, and leave plums and apricots two to three inches apart. The largest fruit with the most cells should remain, rather than those in precise positions.

Mechanical methods of fruit thinning. *Bloom thinning* involves directing a high-pressure

FACTORS AFFECTING FRUIT SIZE

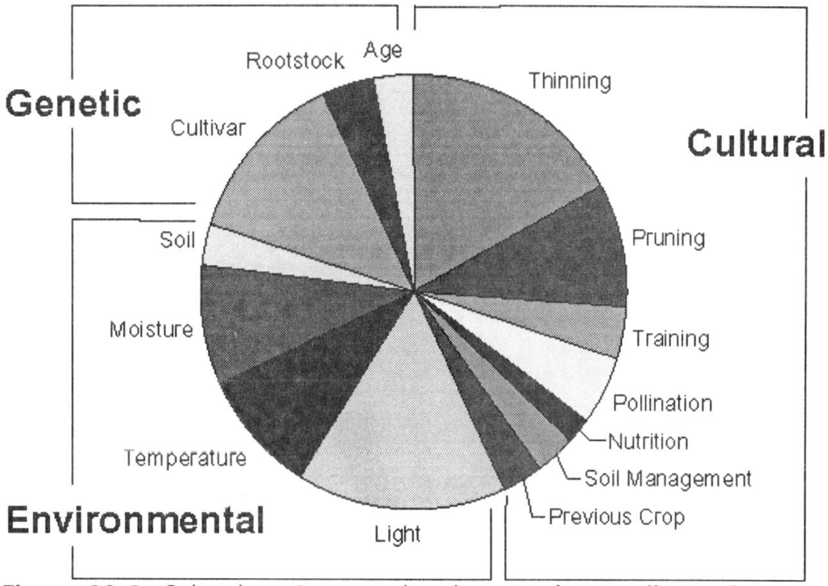

Figure 10-1. Cultural, environmental and genetic factors affecting fruit size. Adapted from David C. Feree's article "What determines fruit size", *Amer. Fruit Grower*, June 2000.

blast of water onto the blossoms a few days after peak bloom. This method works well in areas where few post-bloom frosts occur, but is risky in many areas where frosts can still occur a week or two after bloom, and requires an experienced applicator to prevent overthinning.

Many growers shake excess fruit from limbs or branches of large-fruited stone fruits, two to three weeks after full bloom. Workers can shake the branches with *pneumatic limb shakers* (Photos 10-2, 10-3), knocking off the lightly attached fruit that would probably fall naturally.

Follow-up hand thinning is often helpful following mechanical thinning of stone fruits, to ensure proper spacing and selectively remove pest-infested fruit. When hand-thinning, keep the largest fruit and remove those which are damaged, diseased, or insect-infested, keeping a sharp lookout for the

Photo 10-2. Mechanical methods of fruit thinning, such as limb shaking, remain useful for crops, such as these peaches, that are damaged by many chemical thinning agents.

Photo 10-3. A limb shaker jars loose the fruit that aren't held as tightly.

18"-long, crescent-shaped egg-laying scars of the plum curculio (Photo 10-4), a tough pest that attacks all tree fruits east of the Rockies. Then twist off enough additional fruit so the remaining fruits don't touch, to reduce fruit rots and other diseases. Thinned peaches and nectarines should be spaced 5 to 6 inches apart (Photos 10-5, 10-6), plums and apricots 3 to 4 inches apart.

Growers who utilize *mechanical tree shakers* for harvest (see chapter 11) can also use these machines to thin the trees (Photo 10-7). The operator attaches the shaker to the tree's trunk and shakes the entire tree, in short spurts of 2 to 3 seconds each, until adequate thinning is achieved.

Again, operator experience is important to prevent overthinning. Unfortunately, trunk shaking selectively removes the larger fruit, and removes more fruit from the stiffer areas of the tree. However, trunk shaking is still the best option for mechanically harvested crops that are damaged by chemical thinners and have limbs too high for limb shakers, such as pecans.

Where possible, rake up and burn or otherwise destroy thinned fruit and natural drops to prevent pest problems from proliferating in the orchard. California researchers demonstrated that between 6 and 60% of thinned nectarines developed brown rot and produced spores of the pathogen, and that thinned fruit left on the orchard floor (Photo 10-8) represent a significant source of inoculum for secondary infections of fruit on the trees. Where raking up the thinned fruit is not practical, help them to decompose faster by applying urea to the orchard floor soon after thinning, or disking soon after thinning in a cultivated orchard.

Photo 10-4. Follow-up hand thinning is often helpful after mechanical thinning. Watch for and remove any fruit with these crescent-shaped egg-laying scars of the plum curculio, a major pest. Burn or otherwise destroy such infested fruit.

Chemical fruit thinning. Chemical growth regulators and desiccating materials have been developed that thin excess fruit, and are widely applied to apples and pears. Chemical thinning has produced less success for stone fruits and nuts, as these plants are more sensitive to damage from the chemicals and frequently drop many leaves following growth regulator application.

Photo 10-5. This peach crop before hand thinning is too dense.

Photo 10-6. After hand thinning, peaches should be spaced 5 to 6 inches apart, on average. Space plums and apricots 3 to 4 inches apart.

Photo 10-7. Growers who harvest with mechanical trunk shakers can shake the trees to thin excess fruit, as with these pecans.

Photo 10-8. Thinned fruit left on the orchard floor, such as these thinned peaches, can also be a significant source of fruit diseases, such as brown rot of stone fruits. Rake up and destroy such thinned fruit whenever possible.

Two desiccating or caustic chemicals, Wilthin™ and Thinex™, are currently labeled for blossom thinning of peaches and selected apple cultivars, and ThinRite™ (endothall) is another desiccant newly registered for bloom thinning of apples. These materials are salts or acids that desiccate or burn the flowers' stigmas and styles, preventing fertilization and subsequent fruit growth. Apply these chemicals at 65 to 80% bloom, or when 80 to 90% of the king blooms in apples have been pollinated. These chemicals do not affect flowers that have already been fertilized or haven't opened yet. Chemical blossom thinning, like blossom thinning with water, risks excessive crop loss from late frosts.

Plant growth regulators—synthetic chemicals that mimic naturally occurring hormones in plants—have been developed for post-bloom fruit thinning. After each blossom is fertilized, developing embryos in the young seeds produce growth-promoting hormones that cause fruit tissues to grow around the seeds and keep the fruits on the tree (Photo 10-9). High carbohydrate levels also keep fruit on the tree, while a temporary shortage of carbohydrates can cause young fruits to drop.

Synthetic auxins are among the most commonly used chemical thinners. These include 1-napthalene acetic acid (NAA) and 1-napthalene acetimide (NAD or NAAm; trade name AmidThin™), applied to apples and pears one to three weeks after full bloom at a concentration of 2-25 ppm. Such high amounts of auxins stimulate the production of ethylene, a hormone gas which causes fruit abscission (and sometimes leaf abscission). Ethephon (trade name Ethrel™), a compound which breaks down to release ethylene, is also sometimes used as a thinner, especially on plums and apples. A synthetic cytokinen, 6-benzyladenine (6-BA; trade name Accel™), is another growth regulator registered for thinning apples and increasing fruit size. Do not apply both NAA and Accel™ in the same year, as that can cause excessive formation of pygmy fruit.

The insecticide carbaryl (1-naphthyl N-methyl carbamate; trade name Sevin™) also acts as a potent fruit thinner when applied between 10 and 25 days after full bloom. Carbaryl is highly

Photo 10-9. These young apples, held tightly to the tree because of good pollination and fertilization, should be chemically thinned at this stage.

toxic to bees, so mow the orchard to cut off blooms of wildflowers before applying it, and spray in late evening when bees are no longer foraging. Carbaryl is also toxic to beneficial mite predators and may cause outbreaks of pest mites, so the previously mentioned growth regulators are usually the better choice even though they are more expensive. However, certain cultivars are very difficult to thin chemically, and may require a combination of carbaryl and a growth regulator. 'Fuji' apple, for example, responds only to a combination of carbaryl and Accel™ applied when fruit diameter reaches 10 to 12 mm.

Growth-regulator thinners selectively remove small, weak fruit with fewer cells, thus producing greater size gains in remaining fruit than does mechanical thinning. Chemical thinning also generally costs less than pneumatic limb shaking. However, the degree of thinning can vary greatly depending on orchard conditions. Overthinning may result in these situations:

- young or closely-spaced trees
- high humidity and slow drying conditions
- high maximum temperatures
- frosty nights
- previous heavy crop
- poor pollination

Growers must time applications of chemical thinners not only based on fruit size (in mm), but also on predicted weather conditions from 24 hours before to 72 hours after the applications. They should delay an application for a few days if faced with unusually cold or hot nights or days, or excessively cloudy and humid conditions. Applicators should adjust the concentration of chemical thinner and wetting agent to orchard conditions, and mix the chemicals accurately. A computer program, Thin-it (www.tfrec.wsu.edu/Horticulture/applethin.html), can aid concentration adjustment for chemical thinning of apples. As with all plant growth regulators, a small difference in concentration can produce a large difference in effect on plants.

Ideally, chemical thinning of apples and pears should reduce most fruit clusters to one fruit, the "king" fruit in the center which set first and is generally larger. Some clusters can support two larger fruits (Photos 10-10, 10-11). Limited follow-up hand thinning can be very helpful where chemical thinning left some clusters still full of small fruit.

Photo 10-10. Chemical thinning of apples and pears should remove the less-developed fruit, in the back in this pear cluster, and keep the largest "king" fruit and sometimes one other fruit per cluster.

Photo 10-11. This thinned pear cluster retains its two largest fruit. Most clusters should be thinned to just the one "king" fruit.

OTHER USES OF GROWTH REGULATORS IN FRUIT CROPS

Growth regulators have several other uses in fruit production. These compounds, all synthetic chemicals similar to naturally occurring plant hormones, are expensive, but in some cases produce such large effects that they are widely used.

The same *auxins* (NAA and NAD) that cause excess fruit to drop when applied one to three weeks after full bloom, will instead prevent preharvest fruit drop when applied at about the same concentration (10-20 ppm), three to four weeks before harvest of apples and pears. A much lower concentration of 1 ppm NAA, applied five weeks before harvest, reduces rain cracking in sweet cherries. A very high concentration of 1000 ppm NAA or 2,4-D (another auxin; also a herbicide active against broadleaf herbaceous weeds) controls rootsprouts when applied in early summer to apples, pears, plums, cherries and hazelnuts.

Gibberellins, various form of gibberellic acid (GA; trade names include ProGibb™ and ProVide™), promote cell elongation in young fruits. Grape growers commonly apply GA at full bloom to increase fruit size and cluster size in many seedless grapes, and two to three weeks before full bloom to loosen compact clusters of seeded grapes in order to reduce rot. Sweet cherry growers may apply GA three weeks before harvest to reduce rain cracking, increase fruit size and firmness, and extend the harvest period and fruit storage life. Apple growers may apply GA two to four times from petal fall to up to a month later, to reduce russeting of certain cultivars such as 'Golden Delicious'.

Cytokinens are also growth promoters. The chemical thinner 6-benzyl adenine (6-BA; trade name Accel™) has increased fruit size of certain apple cultivars, including 'Gala' and 'Empire', beyond the effect from thinning, when applied two to three times from petal fall to three weeks later. Promalin™, a mixture of *gibberellins and cytokinen*, stimulates side branching in young fruit trees and is used to aid tree training, as detailed in the training chapter. Promalin™ may also be applied during bloom to lengthen the fruit of certain apple cultivars, particularly 'Delicious'; if used then, do not thin with NAA later, as the combination causes pygmy fruit.

Various synthetic chemicals act as growth inhibitors, by reducing the natural synthesis of gibberellins or other growth promoters. Several growth inhibitors, such as daminozide (Alar™) and chlormequat (CCC), were used on fruit plants several years ago but have been discontinued due to their extended persistence in the plants and questions about their safety to humans. A newly registered compound, prohexadione-calcium (trade name Apogee™), has low toxicity and breaks down completely in plants within one month of application. Apogee™ should prove useful to apple and pear growers, as it has reduced excess vegetative growth and fire blight incidence in apples. The chemical did not affect vegetative growth of peaches.

Ethephon, which breaks down to release ethylene, may be applied to young, nonbearing trees, or to biennially bearing trees, to induce them to bear the following season. Apply ethephon about a month after budbreak to induce bloom the following spring. Ethephon will also advance fruit maturity and cause more concentrated ripening, which facilitates mechanical harvest, when applied 7 to 12 days before anticipated harvest. Since ethephon also promotes fruit abscission, it is often combined with a stop-drop chemical such as the auxins NAA or NAD.

Another stop-drop chemical, ReTain™ (aminoethoxyvinylglycine), inhibits ethylene production and slows maturation of the fruit. Some apple orchardists with large blocks of a few cultivars, all ripening at once, spray some of their acreage with it 4 weeks before harvest, to spread out the harvest so smaller crews can handle the fruit over a longer period. Apples treated with ReTain™ have greater fruit firmness both at harvest and after controlled-atmosphere storage, have slightly lower sugar content at harvest but the same as untreated fruit after 7 months' storage, and develop less watercore and internal browning during storage.

INCREASING FRUIT COLOR AND SIZE BY ADDING LIGHT AND REDUCING HEAT

Fruit require 50 to 70% of full sunlight to develop good skin coloration. Figure 10-1, "Factors Affecting Fruit Size" chart on page 121 shows that the amount of light reaching the leaves around the fruit has the second greatest influence, after fruit thinning, on fruit size, since higher light levels stimulate greater photosynthesis. However, in many areas high light levels also create high heat loads, which increase respiration and thus reduce accumulation of photosynthates, leading to poor fruit size and color.

The development of dwarf trees has greatly increased yields, fruit size and fruit color by allowing light to reach all parts of the trees. Adequate light penetrates only the outer portion of larger trees, leaving the interior shaded and producing few fruit with poor color and size. Planting rows so they run north-south, as mentioned in chapter 3, increases light penetration. Training and pruning systems, discussed in chapter 4, are generally designed to maximize light.

Several newer techniques can increase light levels in fruit plantings, sometimes while also reducing heat levels.

Reflective Mulches. Increasing numbers of apple and stone fruit growers have utilized cloth- or polyethylene-backed aluminum foil mulches under their trees to reflect light back onto the lower fruit. They unroll the foil mulch on the orchard floor 5 to 7 weeks before harvest. The foil has resulted in 30% to 5 times more light reaching the lower canopy, with the greatest effect on a dense canopy, and a 40 to 60% increase in red fruit color in cultivars such as 'Fuji' apple. After harvest, cloth-backed foil is rolled up again and stored; it can last for three seasons. This keeps the expense of cloth-backed foil mulch in line with that of reflective polyethylene mulch, which lasts only one season. Increased profits depend on negotiating a higher price for more highly colored fruits.

Particle Film Sprays. Particle film sprays, originally developed for insect and disease control (see chapter 9), have proven to reduce heat stress and increase fruit color development also. The film's white coating on the fruit (Photo 10-12) and leaves reflects excess heat, reducing the number of sunburned fruit by half and greatly reducing excess respiration of sugars, which would have limited fruit size and color. In one study, particle film reduced skin temperatures of 'Fuji' apple from 122 F to 105 F during the hottest part of the day. The film does not reduce levels of usable light, so photosynthetic rates actually increase up to 30%, which can increase yields and fruit size.

When used late in the season to protect ripening fruit from sunburn and heat stress, particle film residues will remain at harvest and must be washed off after harvest. Growers should use a dump tank with brushes to clean the fruit, add lignin sulfonate to the tank to help disperse and disintegrate the residual particle film, and increase the time that the fruit is in the dump tank.

Photo 10-12. Particle film, containing purified kaolin clay, sprayed on fruit plants such as these 'Red D'Anjou' pears, reduces fruit sunburn by up to 50%, and increases fruit color and size due to reflected light. When applied late in the season, fruit must be washed after harvest to remove particle film. (Photo from Engelhard Corp.)

Further Reading

Andris, H. and C.H. Crisosto. 1996. Reflective materials enhance 'Fuji' apple color. *Calif. Agr.* 50(5):27-30.

Byers, R.E. and K.S. Yoder. 1999. Prohexadione-calcium inhibits apple, but not peach, tree growth, but has little influence on apple fruit thinning or quality. *HortScience* 34:1205-1209.

Ferree, D.C. 2000. What determines fruit size. *Amer. Fruit Grower* 120(6):33-34.

O'Dell, Charlie. 1997. Grow your own disease-resistant pollinators. *Amer. Fruit Grower* 117(2):24.

Hong, C.X., B.A. Holtz, D.P. Morgan, and T.J. Michailides. 1997. Significance of thinned fruit as a source of the secondary inoculum of *Monilinia fruticola* in California nectarine orchards. *Plant Dis.* 81:519-524.

McBride, J. 2000. Whitewashing agriculture. Agric. Res. 48(11):14-17.

Marini, R.P. 1996 . Chemically thinning spur 'Delicious' apples with carbaryl, NAA, and ethephon at various stages of fruit development . *HortTechnology* 6:241-246.

Williams, K.M. and E. Fallahi. 1999. The effects of exogenous bioregulators and environment on regular cropping of apple. *HortTechnology* 9:323-327.

Williams, M.W. 1979. Chemical thinning of apples. *Hort. Reviews* 1:270-300.

Yoder, K.S., S.S. Miller and R.E. Byers. 1999. Suppression of fireblight in apple shoots by prohexadione-calcium following experimental and natural inoculation. *HortScience 34:1202-1204.*

Chapter 11
Harvesting, Storing, Processing and Marketing

Growing high-quality fruits and nuts is only the first step in getting superior products to consumers. That quality must be preserved by harvesting at the optimum time, handling products in the best way to avoid damage, and storing or processing the products to maintain flavor, texture, and nutrition. Finally, growers must be skilled marketers in order to get the best price for their high-quality products.

HARVESTING METHODS

Historically, harvesting represented the most labor-intensive part of fruit and nut production. As labor costs have risen, growers have increasingly turned to customer harvesting and mechanical harvesting. Yet the majority of fruit crops are still hand-harvested by hired workers.

Pick-your-own harvesting by customers. Customer harvest of fruit crops boomed in the 1970's, but has declined since 1980 as more people worked longer hours at jobs and preferred to pay more for pre-harvested fruit, rather than spend time picking it. Many retail growers now market about half their crop pre-picked, and the other half pick-your-own. The growers who continue to succeed with pick- your-own harvesting tend to provide not only high-quality fresh fruit at a good price, but also a "farm experience" that attracts customers who are willing to spend the time for a unique family outing (photo 11-1).

Such growers make sure their farms look attractive, and provide amenities such as plenty of parking space (1 acre of parking per every 5 acres of crops picked at one time), shade trees or structures near the picking area, drinking water, toilet facilities, and picnic tables. Many pick-your-own growers also offer special activities on weekends, such as hay rides, live music, picking contests, or demonstrating how to prepare food items featuring the crop to be harvested.

Pick-your-own farms must be open when customers want to harvest, which usually means weekday evenings and mornings as well as weekends. Growers can educate customers about the value of picking early in the morning to preserve the best fruit quality, but must realize that customers will arrive when it's convenient for them.

Pick-your-own operations must be located within a half-hour's driving distance of at least one large town or city in order to attract enough customers. Each acre of fruit will require about

Photo 11-1. Successful pick-your-own growers provide a unique farm experience for customers as well as high-quality fresh fruit, such as these 'Gala' apples.

350 customer sales, or participation by about 15% of all households in a population of 5000 living within a 20-mile radius. If 50,000 people live within 20 miles of a farm, for example, a grower with no competition could expect to sell up to 10 acres each of several fruit crops that ripen in succession, such as strawberries, blackberries, peaches and apples. As mentioned in chapter 1, pick-your-own operations do best if located less than ½ mile from a paved road.

Most pick-your-own growers provide cardboard flats or baskets for picking into, printed with the farm's name, logo and phone number. Maps can be printed on the bottom or side of the container, so customers can show friends how to get to the farm, or recipes can be printed on the container to give customers ideas of how to use more of the product.

Field attendants should lead customers to assigned rows to pick, using flags to indicate where the last customers stopped picking, so that all parts of the planting get harvested as often as necessary. Workers should show customers how to pick and how to find all the fruit (such as lifting the leaves of strawberry plants). They should instruct customers to pick all ripe fruit, discard any diseased or damaged fruit, and to leave the flag where they stop picking.

Most pick-your-own growers sell their crops by weight, using scales tared to the picking containers. Selling by volume is much less precise, as some customers overfill the containers. Post signs that clearly state the price and units of sale. A few signs can also communicate rules such as "Stay on assigned row," "Pick all ripe fruit," and "Mark your stopping place."

Pick-your-own growers must carry adequate farm liability insurance during the harvest period, to protect against the possibility of customers injuring themselves on the farm and suing for damages.

Hand harvesting by hired workers. Most wholesale fruit growers rely on hired workers to handpick the fruit. Many retail growers do also (Photo 11-2), so they can provide customers with already-picked fruit at on- farm markets, farmers' markets in cities, or for restaurant customers.

Retail growers with smaller acreages may be able to recruit enough harvesters from local people. Growers with crops ripening throughout the summer often hire high school and college students. Fall-ripening crops may present more problems in finding workers. Some retired people like to work as harvesters, or as salespeople in on-farm markets or farmers' markets, to supplement their retirement incomes.

Larger growers often find they can't hire enough local people for work that is seasonal and subject to the elements. Larger U.S. growers therefore contract with individuals or companies that recruit farmworkers, often from Puerto Rico, Mexico or Caribbean countries where unemployment rates are higher, local wages are lower, and people are eager for U.S. farm jobs. Growers must provide housing for such seasonal workers and abide by many government regulations.

Most growers pay their harvesters by the piece, rather than by the hour. Experienced harvesters can easily earn two to four times the minimum wage level when paid by the piece, because they pick so quickly. Crew foremen must train new workers to accurately judge fruit maturity and to pick each type of fruit correctly, and must dock workers who place immature or diseased fruits into picking containers.

Hired workers should harvest fruit early in the day, while temperatures are still cool, but after any surface moisture from dew or rain has dried off. Lower fruit temperatures will reduce the energy costs needed to precool the fruit after harvest, which is done to keep fruit fresh by lowering respiration rates. Keeping fruit dry will greatly reduce postharvest diseases, as will discarding any diseased or damaged fruit in the field.

Photo 11-2. Experienced hand harvesters can earn two to three times the minimum wage, because they pick so rapidly.

Workers should keep their hands clean and fingernails cut short, handle fruit gently to avoid bruising or breaking the fruit skin, and clean and disinfect harvesting containers daily, to further reduce postharvest disease incidence. Workers should place harvested fruit in shaded areas until it is collected for postharvest handling, when the fruit should be moved to the precooling area as soon as possible.

Wholesale growers have harvesters deposit fruit into large bulk bins or boxes, which are moved by forklift to trailers which transport the bins to precooling facilities. Older hardwood bulk boxes provide enough space between boards to allow air circulation, but newer plywood boxes may require drilling many small holes over 5% of the bottom and 8% of the sides. Plastic bulk bins cost more but are easier to clean thus may not harbor as many disease spores, stack more precisely, and fruit cool more quickly due to greater vent space.

Instruct workers not to overfill the boxes, for when overfilled boxes are stacked, many fruit throughout each box become bruised. Using plastic liner inserts in field bins minimizes vibration bruising of fruit as they are moved, and the plastic liners can be discarded after each use to avoid spreading any diseases.

Mechanical harvesting. Nut growers use mechanical harvesters, as do most wholesale fruit growers who grow crops for processing. Mechanical harvesters work well on nut crops because the salable part, the nut, is a hard seed that resists bruising, and it doesn't matter if the shuck or husk (the fruit) surrounding the nut is damaged. Fruit crops, on the other hand, can be easily bruised by many mechanical harvesters, so usually such mechanically harvested fruits must be immediately processed. Several small fruits, however, can be mechanically harvested for fresh market sales.

Mechanical nut harvesting usually requires three machines. Growers first attach a trunk shaker, powered by a tractor, to each tree, shaking the tree for a few seconds to cause the nuts to fall to the ground (Photo 11-3). A nut sweeper then gathers the nuts into windrows in the center of each row (Photo 11-4), where a tractor-pulled or self-propelled nut harvester picks them up (Photo 11-5). The harvester uses rubber paddles to flip the nuts off the ground onto a conveyor belt, which carries them to the hopper while leaves and other light material are blown out.

The main type of mechanical harvester for tree fruits is the shake-and-catch type. This machine shakes the trunks like a nut tree shaker, but has large cushioned pads attached to it to catch the falling fruit to minimize bruising (Photo 11-6). The harvested fruits roll down the inclined pads

Photo 11-3. Mechanically shaking each tree for a few seconds is the first step in harvesting nuts, such as these pecans.

Photo 11-4. A nut sweeper then windrows the pecans that were shaken to the ground in each tree row middle.

Photo 11-5. Finally, a mechanical nut harvester uses rubber paddles to gather windrowed pecans.

into the hopper. Some harvesters shake individual limbs instead of the trunk, which may reduce fruit bruising.

Small fruit mechanical harvesters either shake the bushes to loosen the fruit, or use rubber fingers to pry the fruit loose, then drop the fruit onto a capturing device (Photos 11-7, 11-8). The fruit then moves on a conveyor past an air-blast cleaner, and is finally deposited into field lugs. Most growers use self-propelled, over-the-row harvesters on which the driver sits on top, although some use tractor-pulled harvesters. Cranberries are water harvested with reels that knock the berries from the vines, then the floating berries are corralled with floating booms and taken up conveyor belts into trucks.

Some mechanically harvested fruit crops, such as cherries and blackberries, benefit from spraying the plants with ethephon 7 to 12 days before harvest. Ethephon produces ethylene gas, the fruit-ripening hormone, which induces a greater percentage of the fruit to ripen at once. This reduces the number of mechanical harvests needed, increases fruit color, and reduces fruit acidity.

Mechanical harvesters are expensive, so a grower must raise enough acres of a crop, or several crops that can be harvested with the same machines, to justify the cost of the machines. Alternatively, a group of growers can cooperatively purchase harvesting machines, or a grower who buys a harvester can contract to custom harvest other growers' fields. Mechanical harvesting can reduce labor costs by 30 to 45%, since one machine can replace as many as 80 to 100 manual harvesters.

However, machines harvest many immature, overripe, diseased and damaged fruits that human harvesters would not damage or would sort out in the field, necessitating sorting machines that subject the fruit to additional damage. In areas with frequent rains during the harvest season, fields that could be hand-harvested may often be too wet mechanical harvesters. Mechanically harvested fruit that must be sold for processing returns only a fraction of the price commanded by fresh market fruit. Growers should calculate the economic costs and benefits of mechanical harvesters for their particular crops and market before investing in them.

Growers can minimize the fruit damage that mechanical harvesters cause by harvesting early in the day when the fruit is cool and dry, or at night. Minimize the drop heights on the machine, pad the impact surfaces and eliminate as much vibration as possible. Use shallow field containers with a depth of no more than 4 or 5 inches, handle and dump these containers very gently, and precool the fruit as soon as possible after harvest.

DETERMINING WHEN TO HARVEST

Fruit crops can be harvested at varying stages of maturity. Growers with local retail markets instruct workers or customers to harvest fruit when fully ripe or nearly so, when the fruit has achieved the greatest sweetness and flavor. Wholesale growers who ship fresh fruit usually must harvest the fruit 3 to 7 days before it is fully ripe, while it is still firm enough to resist bruising. Some processed tree fruit crops are harvested early for that reason also. Fruit to be stored may be harvested a few days before it is fully ripe, to reduce deterioration in storage.

Photo 11-6. This shake-and-catch cherry harvester is the most common type of tree fruit mechanical harvester.

Photo 11-7. This blueberry harvester is also a shake-and-catch type.

Most fruit crops ripen over a period of several weeks, necessitating repeated harvests, whether by hand or machine. Most berry crops should be harvested every two to three days during their ripening period, which can extend for a month or two with several cultivars ripening in succession. Several cultivars of a tree fruit, such as peaches or apples, can extend the harvest season for three months or more, with harvests once or twice a week. Some cultivars of mechanically harvested fruits and nuts, however, have been selected for more concentrated ripening, so each plant can be harvested just once or twice.

Fruit maturity can be determined by the following methods. Each is useful for some crops but not others. The "Fruit Maturity Indicators Used for Various Crops" on the next page details which maturity indicators are useful for which crops.

Days or degree-days from full bloom. Growers know when to start checking each fruit or nut cultivar for maturity based on the usual time it ripens each year. Weather conditions vary each year, so the optimum harvest date can vary by up to several weeks. The average number of days from full bloom that a cultivar needs can be a useful rule of thumb, because weather that hastens or delays bloom often hastens or delays harvest as well. Degree-days from full bloom is a more accurate maturity indicator, because it takes into account the heat units accumulated during the growing season.

Both of these indicators merely alert the grower that it's time to sample for other maturity indicators. They do not indicate that the fruit is ready to harvest.

Fruit color. As fruits ripen, several pigments characteristically develop in their skins. Anthocyanins form in many fruits, imparting a red or blue color, while carotenoids provide a yellow or orange color that is more visible as the green chlorophyll of young fruits breaks down. Such color changes coincide with the best harvest time for some fruit crops, such as strawberries, raspberries and cherries (Photo 11-9). Other fruits, however, such as blackberries and blueberries, take on the color of ripeness several days before they reach their peak sugar content and best time to harvest for local sales. Some red-skinned cultivars of apple, pear and plum even develop such red color several weeks before their optimum harvest date.

The skins of tree fruits have both an outer color and a "groundcolor" underneath that. The groundcolor is a much more reliable indicator of maturity than the outer color. When the groundcolor turns

Photo 11-8. The interior of the blueberry harvester shows the arms that shake berries from the bushes onto the pads that catch the falling berries and roll them onto conveyor belts, which deposit the berries into field lugs.

Photo 11-9. Fruit color change accurately indicates harvest maturity for some fruits, including tart cherries.

Fruit Maturity Indicators Used for Various Crops

Almonds. *Fruit stems separate easily* from the branches, as tested by hand-shaking limbs, when the nuts inside are mature. Husks (fruit) do not split open.

Apples. Use *days from full bloom* to decide when to start sampling fruit for more precise maturity indicators. *Fruit groundcolor* turns yellow-green and *fruit stems separate* easily from the spurs, without tearing the spurs, when fruit are mature. *Flesh firmness* falls to between 12 and 20 pounds-force, the *soluble solids* content rises to 12% or more, and the *iodine-starch test* should show 60% or less of the surface area blue-black in color.

Asian pears. *Fruit color* turns yellow-green, or golden-brown for russeted cultivars. *Flesh firmness* declines to 7 to 10 pounds-force, and *soluble solids* readings rise to 11 to 14%, depending on the cultivar.

Apricots. *Fruit groundcolor* turns yellow but *flesh firmness* remains above 10 pounds-force, since apricots are very susceptible to bruising. A *soluble solids* content of 10% or more, and moderate *acidity* of 0.7 to 1.0%, giving a *sugar/acid ratio* of 10.0 to 14.3, confirm fruit maturity.

Blackberries. *Fruit color* becomes fully dull black (a day or two after fruit turn shiny black) and *fruit separates easily* from the plant.

Blueberries. *Fruit color* turns fully blue, without any red or green at the stem end. A *soluble solids* reading of 10 to 15% and a pH of 3.43 to 3.73 confirm fruit maturity.

Cherries. *Fruit color* is the best indicator for tart cherries, which should be bright red. Sweet cherries should be at least light red, preferably red mahogany, and have a *soluble solids* content of 14 to 16% or more.

Grapes. *Soluble solids* readings must reach 14 to 17.5%, depending on the cultivar and production area. A *sugar/acid ratio* of 20 or higher determines maturity for early maturing cultivars in early production areas. Red- and black-skinned cultivars must have significant *fruit color*.

Hazelnuts. Husks (fruit) split open, allowing the nuts to separate easily from the husks.

Peaches and Nectarines. *Fruit groundcolor* turns yellow in yellow-fleshed cultivars and white in white-fleshed cultivars. *Flesh firmness* in pounds-force reaches 10 to 16 for minimum maturity, 6 to 10 for well-mature, or 4 to 5 for "tree ripe" fruit that needs gentler handling to prevent bruising.

Pears. Use *days from full bloom* to decide when to start sampling fruit for *flesh firmness*, the main maturity indicator for European pears. Optimum pounds-force readings vary greatly with cultivar, from 9 to 13 for 'Comice', to 11 to 15 for 'Anjou' and 'Bosc', to 19 to 22 for 'Bartlett'. The *iodine-starch test* can confirm maturity, when 60% or less of the surface area turns blue-black in color. Mature *fruit stems separate easily* from the spurs, without tearing the spurs.

Pecans. Shucks (fruit) split open, allowing the nuts to *separate easily* from the shucks.

Persimmons. *Fruit color* turns orange enough to meet color-chip standards.

Plums. *Fruit color* turns the characteristic shade for each cultivar, and *soluble solids* reach 17% or more. *Flesh firmness* in pounds-force reaches levels like that of peaches: 10 to 16 for minimum maturity, 6 to 10 for well-mature, or 4 to 5 for "tree ripe" fruit. Plums are less susceptible to bruising than peaches at each firmness level.

Raspberries. *Fruit color* turns the characteristic shade for each cultivar, and *fruit separates easily* from the torus, the white cone inside each raspberry.

Strawberries. *Fruit color* turns red or pink over at least 2/3 of the berry surface for minimum maturity, and fully red for field-ripened fruit.

Flesh firmness of the husk (fruit) softens enough so a thumb pressed against the husk will leave an imprint. (Husks do not split open.)

from green to yellow-green, the fruit is ready to harvest (Photo 11-10). Color-chips are available for various fruits, which growers can hold up against the fruit to compare its groundcolor and surface color with the standard.

Fruit firmness. Many fruits soften as they begin to ripen, as pectolytic enzymes form that begin to break down the pectin that cements cell walls together. Growers must let this softening proceed enough for the fruit to become palatable, but not so much that the fruit becomes overly soft, "mushy", and prone to bruising. Certain fruits, such as apples and pears, should still be rather firm and crisp at harvest, while stone fruits and brambles should be softer and need special precautions to prevent bruising.

A pressure tester, a small metal rod that plunges into the fruit and determines the pounds-force of fruit resistance (Photo 11-11), can measure both skin and flesh firmness. Determine skin firmness by testing unpeeled fruit, and flesh firmness by testing a section of the fruit with the peel removed. Flesh firmness is more commonly used as a maturity indicator.

Ease of separation from the plant. Many fruits are difficult to remove from the plants until they are ripe, when they separate easily. Experienced hand harvesters can feel this difference and utilize it to differentiate a ripe blackberry or blueberry from one that has just changed color. Research has demonstrated that mechanical harvesters for these cane fruit crops provide even better quality than hand harvesting, since only the ripe fruit are shaken from the bushes.

Larger fruits, and nuts which remain in their shucks when ripe, may be mechanically shaken from the tree before they are fully ripe, so this method is not a good maturity indicator for mechanically-harvested tree fruits and nuts. However, experienced hand harvesters of tree fruits can judge the ease of fruit separation as they twist the fruit off the tree, using this method as a backup to groundcolor and fruit firmness in choosing which fruit to pick. Growers who will mechanically harvest large fruits or nuts can hand-harvest a sample, or handshake a few limbs, to determine the ease of fruit separation and set the time to harvest.

The fruit (shucks or husks) of some types of nuts, such as pecans and hazelnuts, split open when the nuts inside are mature, allowing the nuts to drop out. The shucks or husks of other nuts, such as almonds and walnuts, do not split open, but the fruit will separate readily from the tree when the nuts inside are mature.

Soluble solids. As fruits ripen, storage starches are broken down into simpler sugars, making the fruit taste sweeter and giving it a less starchy texture. These sugars are soluble in the cells, unlike the more complex starches. A handheld refractometer (Photo 11-12) measures the percent soluble solids, which are mostly sugars, in a few drops of the fruit's juice when it is squeezed onto the device. Soluble solids should rise above a level characteristic to that type of fruit, which can range from 7% for strawberries to 17% for plums.

Iodine-starch test. Dipping fruit slices in an iodine solution for 1 minute will stain the starchy areas blue- black, while the portions where starch

Photo 11-10. A change in groundcolor from green to yellow or white, in peaches and most other tree fruits, pinpoints fruit maturity more reliably than the development of the overcolor or red blush.

Photo 11-11. A pressure tester determines flesh firmness when used on a portion of the fruit with the skin removed. Flesh firmness measurements help determine when to harvest pears and apples.

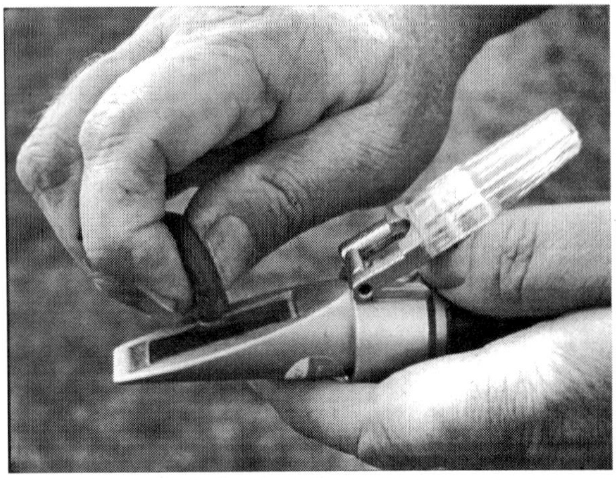

Photo 11-12. A hand-held refractometer measures percent soluble solids, which are mostly sugars, in a few drops of juice squeezed from the fruit onto the instrument. This helps determine maturity of grapes and other fruits.

has already converted into sugars do not stain. This test is sometimes used in conjunction with other tests in apples and pears, where 60% of the cut surface or less should stain blue-black before harvesting commences.

Fruit acidity and pH. Fruit acid concentrations decline and the pH rises as fruit ripen and become less sour. Some fruit growers' associations, such as the Apple Maturity Project in Washington state, test fruit acidity or pH of sampled fruit to provide further evidence of maturity. Test kits are available for measuring pH, although few individual growers do so.

Sugar/acid ratio. The characteristic flavor of many fruits results from both a high sugar content and a moderately high acidity level. Fruits with high sugars and low acidity taste merely insipid. The ratio of the sugar concentration to the concentration of fruit acids provides better correlation to ripeness than either measurement alone. Where growers or their associations test samples for both sugar and acidity, the ratio of these measurements can provide one of the most accurate indicators of the best time to harvest.

Ethylene emissions. Some types of fruit greatly increase their respiration rate as they begin to ripen, as new enzymes are produced that break down starch, pectin and other fruit substances. This burst in respiration just before harvest is called the *climacteric*. Fruits in which this occurs are called *climacteric fruits*, while fruits that produce no burst of respiration during ripening are called *non-climacteric fruits*. Climacteric fruits include most large-fruited tree fruits, such as apples, pears, peaches, plums, apricots, persimmons and figs. Non-climacteric fruits include nuts, the small fruits and small-fruited tree fruits, such as cherries, grapes, berries and nuts.

A burst in fruit ethylene gas emissions immediately precedes and stimulates the respiration increase in climacteric fruits. Sample climacteric fruits can be harvested as the optimum maturity date approaches, placed in sealed containers in a laboratory, and the amount of ethylene they give off measured with a gas chromatograph. Several apple growers' associations measure ethylene emissions, especially for apples to be held in controlled-atmosphere storage, which should be harvested just before fruit respiration starts to rise toward the climacteric peak.

Climacteric fruit can be further ripened after harvest and storage by supplying ethylene at room temperature, usually by packaging the fruit in bags or boxes that trap the ethylene produced by the fruit. In contrast, non-climacteric fruit will not ripen further after harvest, but will merely deteriorate if exposed to ethylene and warm temperatures. Therefore, harvest non-climacteric fruit as ripe as possible and keep such fruit cool the entire time from harvest to sale.

CROP TREATMENT AFTER HARVEST

Postharvest handling is as critical as choosing the best time to harvest. Without proper handling and prompt cooling, a high-quality crop can be converted into a low-quality product in a matter of hours.

Field packing of fruits. Some hand-harvested fruits can be graded as they are picked and packed directly into customer-sized containers for sale or shipping. Highly perishable small fruits, including strawberries, blueberries and raspberries, are commonly handled this way, when picked either by customers or hired workers. Picking such fruits directly into quart or pint sales containers, set in flat cardboard trays, minimizes handling of these easily damaged fruits.

field packing by hired workers requires trained harvesters able to pack only ripe, sound fruit, and additional workers who bring the harvesters flats of containers to pick into and carry off the filled flats. Containers to be shipped are then covered with plastic wrap to retard moisture loss, and containers for either shipping or retail sales are precooled.

Precooling fruits. Placing newly-harvested fruit in some type of cooling system for a few hours until the fruit is packed, shipped or set out for retail sales is called precooling. Precooling removes field heat from the fruit, reducing respiration rates and thus the fruit's softening, shrinking and loss of sugars. The fruit respiration rate at 80 F is nearly 20 times that at 40 F, which means fruit picked at 80 F and cooled to 40 F will remain fresh 20 times longer. Perishable berries might last 10 days instead of ½ day, while larger fruits can retain quality for 2 months instead of 3 days.

In addition to reducing fruit respiration, precooling reduces fruit water loss and inhibits the growth of microorganisms that can cause postharvest fruit rots. Precooling also reduces ethylene production in the fruits that release this gas as they ripen. Ethylene, also known as the ripening hormone, induces greater fruit respiration and softening. It is used commercially to cause ripening in some fruits picked green, such as pears and bananas, but should be minimized in these fruits until ripening is desired, and minimized in all fruits that are harvested when nearly to fully ripe.

Pick-your-own customers should be advised to refrigerate their fruit before preparing or preserving it. Fruit harvested by machines or hired workers should be precooled in one of the following ways. Table 11-1 details the best precooling methods and storage conditions for various types of fruits and nuts.

Room cooling involves placing the fruit in a walk-in cooler, an insulated room equipped with refrigeration units (Photo 11-13). Room coolers are very useful for storing fruits after precooling, but provide only limited precooling capacity. If the room is loaded to capacity with warm fruit that hasn't been precooled some other way, it can take several weeks to cool the fruit to the desired temperature, during which time the fruit deteriorates.

Room coolers may adequately precool fruits with a long storage life, especially apples and pears, if small amounts of fruit are added each day. A room cooler used to precool such fruits will require a larger refrigeration unit than a cooler used only to store fruit that was precooled before loading it in the cooler. More perishable fruits should be cooled more quickly through forced-air cooling or hydrocooling.

Forced-air cooling involves using forced-air fans in conjunction with a room cooler. Such fans pull cool air though the fruit containers, cooling them 4 to 10 times faster than simply placing them in the

Table 11-1.
PRECOOLING METHODS AND STORAGE CONDITIONS FOR FRUIT AND NUT CROPS*

Crop	Precooling methods[1]	Optimum temperature (degrees F)	Optimum humidity	Regular Storage Life	Modifed-atmosphere storage		
					% O2	% CO2	Benefit
Almonds	H, R	0	60 - 75%	1 - 2 years	0.5%	0 - 1%	none
Apples	F, H, R	32	90 - 95%	1 - 8 months	1.5 - 2%	1 - 2%	adds 3 months
Apricots	F, H	32	90 - 95%	1 - 3 weeks	2 - 3 %	2 - 3%	none
Blueberries	F, R	32	90 - 95%	2 weeks	5 - 10%	15 - 20%	adds 1 week
Blackberries	F, R	32	90 - 95%	5 - 10 days	5 - 10%	15 - 20%	adds 2 days
Cherries	F, H	31	90 - 95%	2 - 4 weeks	3 - 10%	10 - 15%	adds 6 weeks
Cranberries	F, H	37	90 - 95%	2 - 4 months	1 -2 %	0.50%	none
Dried Fruits	R	32	55 - 60%	1 year	0.5%	0 - 1%	none
Figs	F	31	85 - 90%	7 - 10 days	5 - 10%	15 - 20%	adds 2 weeks
Grapes	F	31	85 - 95%	2 - 8 weeks	2 - 5%	1 - 5%	adds 3 months
Hazelnuts	R	0	55 - 65%	1 - 2 years	0.5%	0 - 1%	none
Jujubes	R	38	75 - 85%	2 - 3 months	NA	NA	none
Peaches	F, H	32	90 - 95%	2 - 4 weeks	1 - 2%	3 - 5%	adds 4 weeks
Pears	F, H, R	31	90 - 95%	2 - 7 months	1 - 2%	0 - 1%	adds 3 months
Pecans	R	0	55 - 65%	1 - 2 years	0.5%	0 - 1%	none
Persimmons	F, H, R	32	90 - 95%	3 - 5 months	3 - 5%	5 - 8%	adds 1 month
Pistachios	R	0	55 - 65%	1 - 2 years	0.5%	0 - 1%	none
Plums	F, H	32	90 - 95%	2 - 4 weeks	1 - 2%	3 - 5%	adds 4 weeks
Raspberries	F, R	32	90 - 95%	2 - 5 days	5-10%	15-20%	adds 7 days
Strawberries	F, R	32	90 - 95%	5 - 7 days	5 - 10 %	10 - 15%	adds 5 days
Walnuts	R	0	55 - 65%	1 - 2 years	0.5%	0 - 1%	none

[1] F = forced-air cooling, H = hydrocooling, R = room cooling

*Data taken from several states' Extension publications.

Photo 11-13. A room cooler is useful for storing fruit once the field heat has been removed, but forced-air cooling or hydrocooling precool the fruit faster.

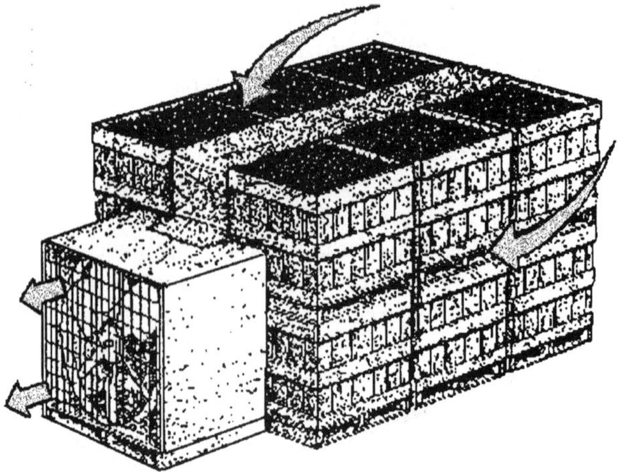

Figure 11-1. Growers can add fans to a room cooler to provide forced-air cooling of large fruits, such as these 600-pound bulk boxes of apples. Taken from North Carolina State University publication AG413-1, Harvest Cooling and Handling of Apples.

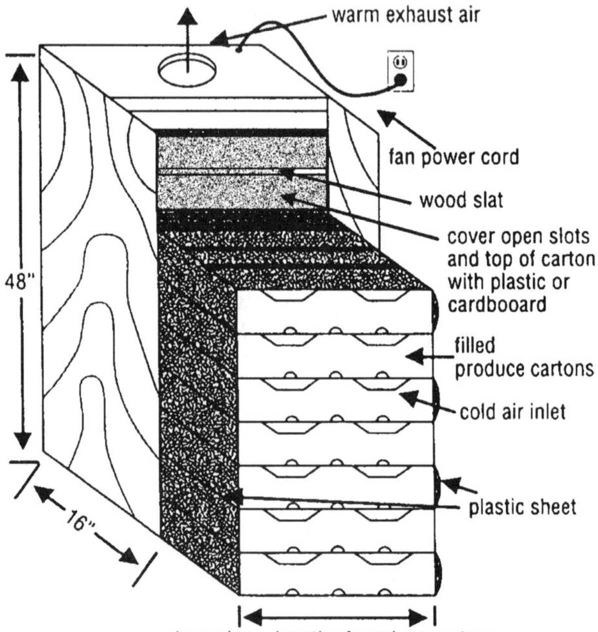

Figure 11-2. Growers can build this simple forced-air precooler for non-wrapped berries in vented cartons. Before constructing the precooler, select and measure cartons and adjust the cooler dimensions to fit the cartons. Taken from *Amer. Fruit Grower*, May 1995 issue, page 31.

room cooler. Use containers with vents to allow the cool air to pass through them, with the vents designed so they retain the stacking strength of the containers.

Adding fans to cool an entire room cooler, and increasing the size of the refrigeration unit to accommodate the additional cooling load, is one way to accomplish forced-air cooling (Figure 11-1). Another method, useful for small fruits that require less space, is building boxes equipped with fans and placing these boxes inside a standard room cooler (Figure 11-2). Workers stack flats of fruit onto the shelves in each box, through which the fan draws cold air. Growers can build these small forced-air coolers themselves, with instructions in the May 1995 *American Fruit Grower* article, "Extending the shelf life of berries", pages 31 and 34.

Choose the correct size of fans and motors for the amount of fruit in each load that will need precooling, so the system will precool an entire load in 2 hours or less. Fans should deliver 1 cubic foot of air per minute for each pound of small fruit to be cooled, or 2 to 3 cubic feet of air per minute for each pound of large fruit, against 0.10" water pressure. Be sure the fans are set to pull air over the fruit, never to blow air across the fruit, which would increase water loss. Equip the fans with a thermostat that turns them off when the fruit reaches the desired temperature (usually 8 to 10 F above the storage temperature), to reduce energy consumption and fruit water loss.

Forced-air cooling is the best way to cool berry crops, since they require fast cooling and are easily damaged by water, so can't be hydrocooled. Forced-air cooling is also useful for tree fruits, as it is 70 to 80% energy efficient.

Hydrocooling involves running bins of tree fruits on a conveyor belt through a machine that causes chilled water to run over the fruit (Figure 11-3). Water removes heat 15 times faster than air, but water also encourages the growth of microorganisms. Do not hydrocool small fruits, which will rot easily if wetted.

Hydrocooling has an energy efficiency of only 20 to 40%, so it consumes 2 to 4 times as much energy as forced-air cooling. Nevertheless, hydrocooling is gaining in popularity for the more perishable tree fruits, such as peaches and plums, because hydrocooling can cool a large amount of fruit rapidly. Due to hydrocooling's high initial investment and operating costs, only large-scale growers or a group of growers, managing at least several hundred acres, find hydrocooling economically viable.

Since hydrocooling water is recirculated, it can easily spread disease from a few infected fruit to all others hydrocooled thereafter. To prevent this, add a chlorine-generating product at 100 ppm to the hydrocooler water. Keep the water's pH at 7.0, because a higher or lower pH requires much more chlorine.

Treatments to control postharvest fruit rots. Rapidly precooling fruit and keeping it dry, or using chlorine in a hydrocooler or fruit washing tank, can prevent many postharvest fruit rots. Some fruits have greater problems with postharvest rots, however, and benefit from additional treatments to prevent them, especially if the fruit will be stored for long periods.

Careful harvesting and handling to prevent wounding the fruit can eliminate much subsequent rot. Most postharvest pathogens cannot invade sound fruit, but many fruits are wounded during harvest and packing, most often by workers' fingernails or fruit stems. Instructing workers to keep their fingernails cut short, and to place fruit gently in the harvest containers or pack so that stems do not puncture other fruits, can greatly reduce subsequent fruit rots.

Fungicide dips of harvested fruit have been commonly used to prevent postharvest fruit rots on pome and stone fruits. Since 1996, fungus resistance to the fungicide benomyl rendered it useless, and the manufacturer of the fungicide iprodione cancelled its use for postharvest treatments. Thiabendazole and dichloran fungicides are still registered for postharvest use and widely used on pome and stone fruits, but many growers and consumers would prefer alternatives to applying a pesticide postharvest, so soon before the fruit is consumed.

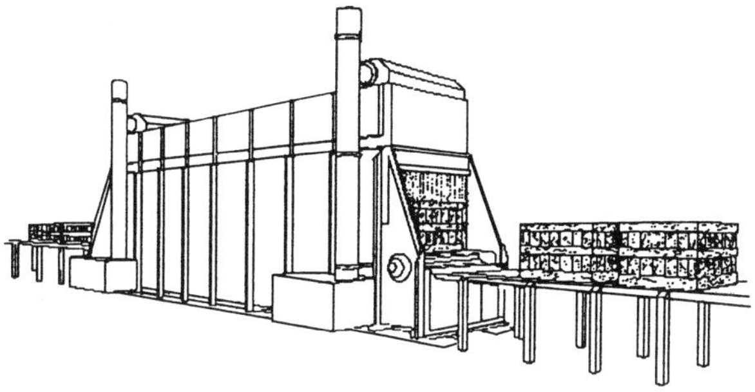

Figure 11-3. This conventional hydrocooler passes fruit cartons or bulk bins along a conveyor under a shower of chilled water. Most tree fruits can withstand hydrocooling, but it would stimulate too much decay in small fruits. Taken from North Carolina State University publication AG414-4, *Hydrocooling*.

Biocontrol treatments of several beneficial microorganisms, when applied to fruit before harvest or to harvested fruit, can prevent or reduce the attacks of fruit-rotting fungi that cause brown rot of stone fruits, blue mold of pears and apples, and other fruit rots. These biocontrol agents include several yeast species, several species of the filamentous fungus *Trichoderma*, the bacterium *Bacillus subtilis*, and several species of *Pseudomonas* bacteria. Many of these beneficial organisms occur naturally on fruit surfaces, but their populations may be insufficient to prevent disease, or they may be reduced by hydrocooling, washing and packing.

Growers can spray these biocontrol agents on fruit a few days before harvest, using a normal orchard sprayer that is not contaminated with any fungicide, bacteriocide or other chemical that could kill the biocontrol organism. Such preharvest application may work well if fruit will not be hydrocooled or subjected to washing or extensive handling after harvest. Alternatively, a packing line spray applicator or drench can coat harvested fruit with the biocontrol agent, prior to any waxing, in fruit packing houses.

Several biocontrol agents are now commercially available and registered for postharvest use on fruit crops, and more are being developed. Aspire™, produced by Ecogen, Inc. of Langhorne, PA (www.ecogeninc.com), contains a yeast that controls blue mold and green mold on stored pears and apples, and storage rots of stone fruits and berries. EcoScience (www.ecosci.com/biosave; 877-866-5773) produces three Bio-Save™ products, each containing a particular strain of the beneficial bacterium *Pseudomonas syringae* that controls blue mold, green mold and mucor rot, and each most effective on particular fruit crops. Use Bio-Save 100 on apples, Bio-Save 110 on pears, and Bio-Save 1000 on cherries. These biocontrol agents are naturally occurring, not genetically modified, and are considered completely safe so require no special labeling.

Combining a preharvest iprodione fungicide treatment with a postharvest yeast treatment reduced postharvest brown rot in sweet cherries by 89 to 97% in an Oregon study led by Dr. Robert Spotts. Following these two treatments with modified atmosphere packaging (see below), which elevates carbon dioxide levels around the fruit, reduced brown rot even further to 1% of fruit, compared with 41% postharvest brown rot in nontreated fruit. Ecogen is now developing this beneficial yeast, currently called EXC 9001, as a new biocontrol product to be marketed in 2001.

The USDA has also signed an agreement with Micro-Flo Company of Memphis, TN, to develop another beneficial yeast discovered by Dr. Charles Wilson into a biofungicide coating. In addition to the yeast, the coating contains a synthetic sugar which pathogenic fungi perceive as food but can't use, causing the pathogen to deplete its resources in attempting to grow. The coating also contains chitosan, a naturally-occurring fiber that forms a film on the fruit, holding in carbon dioxide to inhibit pathogen growth and increase storage life just like modified atmosphere packaging.

Heat treatment and calcium infiltration of harvested fruit can reduce storage rots in long-storing fruits such as apples and pears, especially when combined with biocontrol treatments. Although rapidly cooling harvested fruits is usually the best way to preserve quality, researchers have found that apples harvested in the pre-climacteric stage, and held 4 days after harvest at 100 F and 85% relative humidity before being placed in cold storage at 32 F, developed much less rot 6 months later. Pressure-infiltrating harvested apples with a calcium chloride solution can increase fruit calcium content beyond that achievable through orchard fertilization, and fruit cell walls with higher calcium concentrations better resist breakdown induced by enzymes of fruit-rotting pathogens. Treating apples first with heat treatment, then with calcium infiltration, then with the biocontrol agent Pseudomonas syringaeprevented decay in 91% of fruit later inoculated with the pathogen causing blue mold.

Hot water and ethanol dips can reduce postharvest fruit rots. Ethanol (ethyl alcohol) is very volatile, so it evaporates off the fruit surface almost immediately after treatment, leaving no chemical residues like fungicide treatments. This ensures safety for the consumer, but means that packers and handlers must take care not to expose the treated fruit to new fruit- rotting spores.

A California test of this method on peaches and nectarines, which had been exposed to high spore levels of the brown rot fungus in the orchard, proved it to be as effective as fungicide treatments. Dipping fruit for 2½ minutes in a 10% solution of ethanol in water, heated to 122F, reduced subsequent incidence of decayed fruit (after 4 days at 68F) to 25%, as compared to 83% for untreated fruit. Fruit treated with triforine fungicide exhibited 33% incidence of decay. The ethanol/hot water treatment did not injure the fruit or produce off- flavors, and it beneficially increased fruit firmness.

Adding a hot ethanol treatment tank to a packing house will require increased equipment costs, including an ethanol vapor abatement system. Treating with ethanol before hydrocooling should eliminate disease spores in the hydrocooling water. Whatever cooling system is used, workers should wear gloves and sanitize them with alcohol after they handle any infected fruit. Sanitized packing containers and storage areas can further reduce the likelihood of recontaminating treated fruit.

Cleaning and drying nuts. Nut growers generally run their mechanically harvested nuts through a cleaning machine to remove debris (Photo 11-14, Figure. 11-4). An aspirator blows out leaves and dust, then the nuts pass over a set of chains, which removes large sticks. A rotating or vibrating screen then sifts out sand and soil particles, and a gravity table removes dirt clods, rocks and other heavy contaminants.

After cleaning, nuts that were dry when harvested are hauled directly to a hulling or cracking plant. Nuts such as almonds or walnuts retain the fruit (hull or shuck) around the nut, and must be

Photo 11-14. Nut growers run the harvested nuts, such as these pecans, through a mechanical nut cleaner to remove sticks and debris.

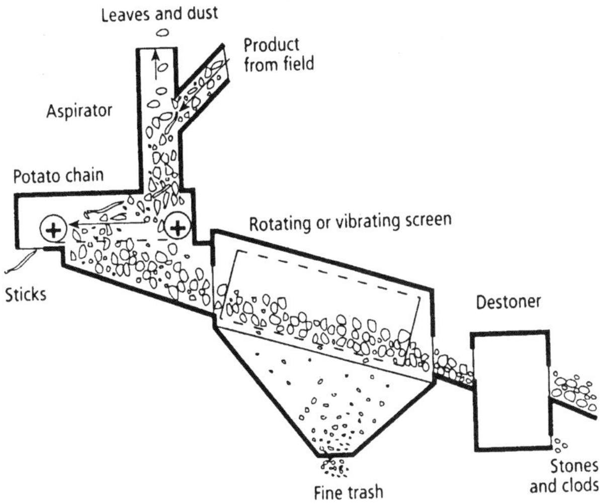

Figure 11-4. This diagram of a mechanical nut cleaner details how the device removes trash from the harvested nut crop. Taken from University of California publication 3364, Almond Production Manual.

Photo 11-15. Hand-harvested fruit, such as these wholesale blueberries, are often transported to a packing house

Photo 11-16. Workers in the packing house sort out cull fruit, grade the fruit by size and color, and pack it for sale.

hulled to remove this shuck before the nuts can be cracked to remove the shells. In nuts such as pecans, hazelnuts or chestnuts, the shuck splits open at maturity to release the nut, so these nuts need only be cracked to remove the shells.

Hull moisture levels must be below 10% for safe storage of nuts to be hulled, and below 8% during hulling to avoid slowing the hulling equipment. Nuts that need no hulling, along with hulled nuts, need moisture levels below 7%. If nuts are shaken from the trees with too much moisture to hull or crack, the nuts can be left to dry on the orchard floor before harvesting, or they be harvested and run through an electric dryer. High humidity during the harvest season can make electric drying economical, while high electricity costs and low humidity favor on-tree or orchard-floor drying.

Sorting and grading fruits. Worker-harvested fruit to be sold fresh, rather than for processing, is often brought to a packing house after harvest for grading and packing (Photos 11-15, 11-16). Workers generally use machines along a conveyor belt to sort out cull fruit and separate the fruit by size, sometimes also by color, into various grades determined by the USDA for each type of fruit. Modern packing houses use multi-lane computer-controlled graders, with optical scanners to sort out culls. The USDA grades limit the number of fruits with defects that may be packed, and specify the extent of the defects. For example, U.S. no. 1 pears can have no scab markings on the fruit surface, while U.S. no. 2 pears can each have a scab marking of ½-inch diameter or less.

The larger-sized grades produce a much greater profit for the grower, so growers aim to maximize that grade size through optimum fruit thinning, irrigation, fertilization, pest control and other cultural practices. Smaller fruit might be specially packed for certain markets (such as lunchbox apples) or sold to fruit processors. Sound fruit with defects that do not hurt internal quality, such as superficial scab markings or insect stings, can also be sold to processors.

Workers cull fruit with any signs of rot or internal breakdown. This fruit should be promptly removed from the packing house and destroyed each day, usually by burning, to avoid spreading diseases. Clean and sanitize all surfaces that contact fruit daily. Avoid bruising the fruit during grading by using cushioned bumper pads along the grading line.

Packing. Wholesale growers utilize waxing machines after grading for some types of large fruit, especially apples and pears, to prevent moisture loss and enable the fruit to store longer. Then workers pack graded fruit into vented corrugated cardboard boxes, with the interior waxed or lined with vented polyethylene, for storage or shipment. High-grade fruit may be individually machine- or hand-wrapped in paper before placing them in boxes, with a foam or molded corrugated cardboard sheet separating each layer of fruit. Smaller fruit is often packed into 3- and 5-pound vented polyethylene bags, and the bags placed in the cartons for storage or shipment. Pack cushioning material around the fruit to prevent bruising from fruits rolling into each other during transit. Many wholesale growers utilize computer programs to record packout tracking, order fulfillment and other fruit contract management.

Retail growers have workers pack fruit directly into the containers that will display the fruit for sale. Many retail growers find that attractive baskets lined with open plastic bags make eye-catching display containers for large fruits (Photo 11-17). When a customer makes a purchase, the sales clerk simply pulls the bag from the basket and ties it, keeping the basket for future display use at the retail market.

Modified-atmosphere packaging. Wholesale growers of small fruits and cherries pack into shallow containers and shrink- wrap them with a layer of polyethylene. This reduces moisture loss of the fruit, and also gradually changes the atmosphere inside the pack. As fruit respire, they use up oxygen (O_2) and give off carbon dioxide (CO_2). As O_2 levels decline and the packaging prevents more from getting in, this limits further fruit respiration and therefore softening and deterioration.

Photo 11-17. Attractive baskets lined with plastic bags make eye-catching retail displays for fruit. When a customer buys a basket of fruit, the sales clerk simply pulls out the bag and ties it, retaining the basket at the market for future displays.

Modified atmosphere packs change the atmosphere immediately upon packing, by vacuuming the existing air (21% O_2, 0.035% CO_2, 79% N_2) out of the container and replacing it with the levels of O_2 and CO_2 prescribed in Table 11-1 for each crop. This requires specialized equipment on the packing line, but keeps the fruit fresher longer, so long as it is held at 29 to 31 F throughout the distribution system. At higher temperatures, fruit with little O_2 can start to ferment.

Shipping. Wholesale fruit growers ship their packed products in refrigerated trucks. Grocery store produce managers keep the fruit cool until it is sold, except for fruits such as pears that need further ripening.

Small retail growers who want to precool fruit and transport it short distances to farmers' markets or restaurants can construct a portable forced-air cooler and mount it on a pickup truck. North Carolina State University's web site http://www.bae.ncsu.edu/programs/extension/publicat/postharv provides instructions for building such a unit in their publication AG414-7, *Cool and Ship*. This portable forced-air cooler costs much less to build and uses less energy to run than a stationary cooling facility, and eliminates the need for a refrigerated truck.

STORAGE

Fruit and nut storage time can range from just a few hours for retail operations, to a few weeks or months for stone fruits or grapes, to up to a year for apples and pears and up to two years for nuts. Keeping the temperature at about 30 to 32 F and the humidity level at 90 to 95% provides optimum storage conditions for most fruits, while nuts store best at sub- freezing temperatures of about 0 F and humidity levels of 55 to 75%. Some fruits store longer under controlled-atmosphere conditions of lower oxygen levels and higher carbon dioxide levels. Controlling ethylene levels helps to retain fruit quality, so store ethylene-producing climacteric fruits separately from ethylene-sensitive fruits.

Refrigerated storage. Most temperate fruit and nut crops can withstand temperatures from a few degrees above freezing to a degree or two below freezing without damage, due to the sugars and other substances dissolved in the cells' sap that act like antifreeze. Set coolers at 30 to 37 F, depending on the crop species (see Table 11-1). Use a humidifier to bring humidity levels up to 90 to 95% for fruits. Stack containers so that they don't block air circulation in the cooler. Potassium permanganate filters to remove excess ethylene may be useful when storing certain climacteric fruits such as apples.

A room cooler using regular refrigeration costs less than controlled-atmosphere (CA) room storage. Regular refrigeration is a necessity for retail growers who must remove some fruit from the storage each day for sale.

Generally, later-ripening cultivars of a particular fruit can store much longer than earlier-ripening cultivars. Fruit held in cold storage longer than indicated in Table 11-1 can develop internal breakdown: the flesh becomes dry and mealy, or wet and mushy, and often turns brown. Sell the fruit fresh or for processing, or transfer it to CA storage, before that happens.

Controlled-atmosphere storage. Controlled-atmosphere (CA) storage works similarly to modified-atmosphere packaging, except that the gas components of the atmosphere are changed in the entire cold storage room, rather than in each individual fruit package. A CA storage room, an airtight room cooler equipped with instruments to monitor and maintain specific gaseous atmospheres, is filled with fruit, then sealed for a few months or up to a year until all the fruit is marketed at once. Temperature is maintained at the usual 30 to 31 F (or the optimum for that crop) and humidity at 90 to 95%, like a regular room cooler.

A room cooler will start with an atmosphere of normal air (21% O_2, 0.035% CO_2, 79% N_2), but as time passes, O_2 levels drop as fruits use it in respiration, and CO_2 levels, a product of respiration, increase. CA storage rooms reduce the O_2 levels immediately, which reduces fruit respiration, thus retarding softening and deterioration. To compensate for the reduction in O2, CA storage rooms increase levels of CO_2 and/or N_2. Table 11-2 lists the optimum concentrations of O_2 and CO_2 for each type of fruit. CO_2 levels higher than those indicated could cause damage, so N_2, which is inert, can make up the rest of the atmosphere.

CA storage reduces the ability of the ethylene produced by ripening fruit to further accelerate ripening, but ethylene levels can accumulate in CA storage and cause fruit fermentation. Therefore, CA storage rooms monitor ethylene levels, as well as those of other gases, and use potassium permanganate filters to remove excess ethylene. A new product called EthylBloc™, containing 1-methylcyclopropene (1-MCP), has blocked ethylene production of CA-stored apples in trials, when fruit were drenched in the compound after harvest. The compound also reduces superficial scald, a storage disorder that commonly arises in CA storage of apples, and shows promise for replacing the potentially hazardous diphenylamine (DPA) drench that most apples currently undergo before CA storage.

CA storage has increased the storage time and quality of long-storing fruits such as apples, pears, and grapes, but it doesn't benefit all fruits, or nuts. Table 11-1 details the increase in storage life that CA storage provides over regular cold storage for each crop. CA storage is most useful for wholesale growers or co-ops of long-storing fruits, or for large retail growers who are open year-round.

PROCESSING

Most growers aim to sell most of their fruit to the fresh market, which generally returns much greater profits than does fruit sold for processing. These growers usually rely on processing, however, to market 20 to 30% of their crop: some fruit from an extra-large crop, or fruit that is smaller or damaged by superficial insects or diseases, early fall frost or hail. Growers may haul their fruit to an independent processing company, belong to a grower cooperative, or invest in on-farm processing equipment themselves.

Some fruit growers raise particular crops specifically for processing. Growers who produce canning peaches plant clingstone cultivars preferred by canneries, rather than the freestone cultivars grown for fresh market. Lowbush blueberries, mayhaws, saskatoons, sandhill plums and other tart native fruits are harvested strictly for processing.

Processing methods include canning, freezing and drying whole fruit or fruit slices; processing fruit into jams, syrups, and pie fillings; roasting nuts and making nut candies; using nuts and fruits in ice cream, breads, cakes, cookies, and bread and cake mixes; and crushing fruit into cider and juice and fermenting juice into wine. Americans consume much greater amounts of processed fruits and nuts than they do fresh fruits and nuts.

Contracts with processors. Fruit processing companies have arisen in the major fruit growing areas (Photo 11-18). Many processing companies enter into contracts with growers at the beginning of each growing season, specifying a price or price range for a certain tonnage of fruit or nuts or the production from a certain number of acres. This guarantees the grower a market for the crop, providing the fruit meets processors' specifications.

Many processors now dictate fruit growers' pest control programs, as they aim to eliminate any pesticide residues in the processed product. For example, many canneries and baby food companies

Photo 11-18. Fruit processors, such as this blueberry processor with its individual quick freeze chamber for freezing blueberries, have arisen in areas of concentrated

now require peach growers to use mating disruption, rather than pesticides, to control Oriental fruit moth. Processors may reject fruit due to pesticide residues, but will also reject fruit if inadequate pest control results in worm damage or decay. Bruised fruit, fruit with internal browning, or fruit that has hit the ground is also rejected.

Fruit for certain uses must be above a specified size. "Peeler" apples, for example, that will be frozen, canned, dehydrated or made into applesauce, must be larger than 2.5 inches in diameter. Fruit to be crushed into juice can be smaller. Juice processors may pay a premium for certain cultivars, such as high-acid apples like 'Granny Smith' or 'Braeburn'.

Fruit growers in some areas have formed bargaining associations to deal with independent fruit processors, to negotiate fair minimum prices that the processors will pay. Otherwise, processors set prices based on the world supply of processing fruit, and heavy or light crops in other countries can cause wild swings in prices paid to U.S. growers from year to year.

Most fruit processors will buy some fruit from non-contracted growers who find that a large crop or damaging weather left them with fruit not suitable for fresh market. A grower in such a situation may find it difficult to sell the fruit at a price that produces any profit, however.

Cooperative processors. Several well-known fruit and nut processing companies are grower-owned cooperatives. This means that all profits from processing and marketing the products are returned to the growers. Successful examples of grower co-ops include:

- *Tree Top*, owned by 2500 apple growers in Idaho, Oregon and Washington, makes apple juice, blended fruit juices, apple cider, and applesauce.
- *Ocean Spray* began in 1930 as a cooperative of three cranberry growers in New Jersey and Massachusetts, and now includes over 800 cranberry growers in the United States and Canada plus 126 Florida grapefruit growers. Ocean Spray processes and markets fresh cranberries, sweetened dried cranberries, sauces and juices.
- *Welch's,* owned by 1461 grape growers cultivating over 43,000 acres in Michigan, New York, Ohio, Pennsylvania, Washington and Ontario, produces grape juices, juice concentrates, jams and jellies.
- California cooperatives include *Sun-Maid* raisins, *Sunsweet* prunes, *Valley* figs, and *Diamond* walnuts. Diamond walnut growers now manage 15,000 acres, and the co-op has beat independent processors' field prices for eight of the past ten years.

Cooperatives require a concentration of growers of a particular crop, or group of crops, within a region. Growers who belong to cooperatives have a voice in deciding how the crops will be processed and marketed. Each grower signs a contract with the cooperative, to deliver all fruit or nuts grown on the contracted acreage, or a certain tonnage of the crop, to the cooperative processor. Successful cooperatives return greater profits to growers than does selling to independent fruit processors.

The National Council of Farmer Cooperatives offers downloadable information at its web site, http://www.ncfc.org/resources. *Co-ops 101* describes the characteristics, structure, people, finance and taxation of co-ops. *How to Start a Co-op* includes samples of a member- user questionnaire, a legal document, a schedule of fixed asset costs and depreciation, a schedule of financing needs and sources, and helpful resources.

On-farm and custom processing. Growers who invest in on-farm processing equipment can reap the most profits from value-added processing, provided they develop an adequate market for their products and adhere to all food safety regulations. Some retail growers have developed on-farm bakeries making pies, muffins and other products from their fruits and nuts. Others process their fruits into cider, jams, or syrups, or their nuts into candies or spiced nuts. Some retail farm markets feature cafes where people can eat as well as buy products to take home. Some fruit and nut growers offer mail-order catalogs and sell their products on Internet web sites.

Smaller growers not ready to invest in expensive processing equipment, and the additional labor required to process fruit, can have their fruit custom processed by cider mills, juice or candy manufacturers, or jam processors. Develop your own label with your farm's logo, and contact your state Board of Agriculture to make sure your label includes all information required.

MARKETING

Every step in fruit and nut production leads up to marketing, the final and most important phase. Profitable growers keep marketing in mind throughout the year, during every phase of operation.

Wholesale marketing. Most wholesale fruit growers contract their crops through a broker, who determines the size and quality of the crop (usually a few weeks before harvest) and then works to get the best bids for it. Brokers work on commission, receiving a percentage of the sales price, so get paid more if they negotiate a better price for the grower. Brokers also coordinate transportation of the crop, and can write contracts for growers who wish to pre-sell their crop to a buyer early in the season. Brokers can represent growers entering a new marketing area, and promote products to increase sales. Some wholesale growers may contract their crop directly with a buyer, such as a supermarket.

Brokers and growers can now trade globally over the Internet. Companies that offer this service include:

- eFruit International (www.efruitinternational.com)
- Foodtrader.com (www.foodtrader.com)
- FreshPlex (www.freshplex.com)
- ProduceOnline.com (www.produceonline.com)
- ProduceWorld (www.produceworld.com)
- Tradingproduce.com (www.tradingproduce.com)

These sites are accessible for trading only by authorized buyers and sellers. The companies provides targeted buying and selling and uses firewalls, certificates and encryption technology to create a secure trading site. Transaction fees averaging about 1%, based on a per-package fee, are charged to the seller. On-line trading can offer lower administrative costs and savings on shipping, storage and handling.

Retail marketing. Retail growers do everything they can to keep their operation in the public eye. They place attractive signs on all highways leading to their farm, and signs indicating where to turn off to get to the farm. They befriend features editors of local newspapers and magazines, and offer interesting events at the beginning of each harvest season to generate free publicity. They also place eye-catching ads in such publications, and may buy local radio ads. They list their farms in a state Extension publication on farm markets and pick-your-own farms, many of which are now available on the Internet.

Some farm marketers or pick-your-own growers ask customers to sign a guestbook with a mailing address or e-mail address, to let customers know when each harvest season is starting. Mailing costs are expensive, but can be reduced by using postcards printed with the farm logo, and obtaining a bulk mail permit. E-mail costs virtually nothing for growers already using the Internet—just the cost of bulk e-mail software—and can reach the one-third to one-half of customers who have e-mail.

Some retail growers also now run Web sites, both to attract new customers and to keep existing customers informed of upcoming harvest seasons, new cultivars or crops, and other developments. A few growers sell fresh fruit directly to customers over the Internet, usually by shipping different cultivars and kinds of fruit to customers as they ripen or are taken from storage, through a fruit-of-the-month club.

Take advantage of a telephone answering machine to inform and attract customers. Regularly update the recording as to hours of operation, fruits and cultivars that are currently available or will soon be ready to harvest, and any processed products offered.

Develop a memorable farm name and an eye-catching farm logo, and use these consistently on all signs, ads, Web sites, e-mails, mailings, and picking containers. This creates a brand that increases customer loyalty to your farm and products. Of course, maintain high fruit quality, keep the farm neat and the harvesting or marketing experience positive for customers, and always treat customers with complete honesty and courtesy. The best retail advertising is word-of-mouth from satisfied customers.

Further Reading

Benbow, J.M. and D. Sugar. 1999. Fruit surface colonization and biological control of postharvest diseases of pear by preharvest yeast applications. *Plant Dis. 83:839-844.*

Conway, W.S., W.J. Janisiewicz, J.D. Klein, and C.E. Sams. 1999. Strategy for combining heat treatment, calcium infiltration, and biological control to reduce postharvest decay of 'Gala' apples. HortScience 34:700-704.

Funt, R.C., T.E. Wall and J.C. Scheerans. 1999. Yield, berry quality, and economics of mechanical berry harvest in Ohio. Ohio State Univ. Res. Circ. 299-99, 22 pages. Downloadable from http://www.af.ohio-state.edu/~ohioline/rc299

Henry, R. 1995. Extending the shelf life of berries. *Amer. Fruit Grower*, May issue: 31, 34.

Hong, C.X., T.J. Michailides, and B.A. Holtz. 1998. Effects of wounding, inoculum density and biological control agents on postharvest brown rot of stone fruits. *Plant Dis.* 82:1210-1216.

Margosan, D.A., J.L. Smilanick, G.F. Simmons, and D.J. Henson. 1997. Combination of hot water and ethanol to control postharvest decay of peaches and nectarines. *Plant Dis.* 81:1405-1409.

Morris, J.R. 1999. Developing mechanized systems for producing, harvesting and handling brambles, strawberries and grapes. *HortTechnology* 9:22-31.

Spotts, R.A., L.A. Cervantes, T.J. Facteau, and T. Chand-Goyal. 1998. Control of brown rot and blue mold of sweet cherry with preharvest iprodione, postharvest Cryptococcus infirmominiatus, and modified atmosphere packaging. *Plant Dis.* 82:1158-1160.

Qing, F. and T. Shiping. 2000. Postharvest biological control of Rhizopus rot of nectarine fruits by Pichia memranefaciens. *Plant Dis. 84:1212-1216.*

North Carolina State University's web site (www.bae.ncsu.edu/programs/extension/publicat/postharv) provides much information on designing and using a cooling room, forced-air cooling and hydrocooling.

Chapter 12
Strawberries

Strawberries are among the most widely grown fruits. The species is widely adaptable and breeders have exploited its genetic variation to breed many cultivars, each suited to a particular region. Strawberries offer fruit growers the advantage of bearing a full crop within 14 months of planting, although plantings must also be replaced more often than other deciduous fruits.

Cultivated strawberries are a human-developed species, created by hybridizing two species native to the Americas. One parent was the meadow strawberry native to the eastern half of North America, *Fragaria virginiana*. The meadow strawberry bears soft but tasty fruit that are quite small. The other parent, the beach strawberry (*Fragaria chiloensis*), native to the west coast of both North and South America, offers large, firmer fruit with a generally milder flavor. The original hybrid was bred in France from specimen plants collected in Virginia and Chile, thus giving rise to the specific names of the two parents. Many additional hybridizations have been made since, and other wild species have now been incorporated into the germplasm of some strawberry cultivars. All cultivated strawberries are classified as *Fragaria* x *ananassa*, with the x indicating that this species arose from hybridizing two naturally occurring species within the genus.

PLANT DESCRIPTION AND BOTANY

Strawberries grow as semi-evergreen, herbaceous perennials. A rosette of leaves arises from each crown. The leaves rise only 6 to 8 inches of the ground, and almost all roots occur in the top 6 inches of soil, making irrigation especially important. Each rosette sends out stolons or runners, horizontal stems that root at their tips, forming additional rosettes or "daughter plants" a few inches to a foot or so from the "mother plant" (Photo 12-1). Long days in late spring and early summer stimulate runner plant production. Nurserymen propagate cultivars from these daughter plants, or rooted runners.

Each mother plant can live for several years, and a planting can be maintained for 8 years or more by encouraging daughter plants to grow and replace mother plants as they die. However, the recent trend in strawberry production has been to keep plantings for fewer years, often just one or two years, and to plant new acreage each year with disease-free plants.

The strawberry, like many fruit crops, is a member of the rose family. Each fruit is an aggregate

Photo 12-1. Strawberry plants grow as semi-evergreen, herbaceous perennials that send out stolons (runners) to spread.

fruit, arising from a single flower containing many pistils. Following pollination and fertilization, the ovary within each pistil ripens into the botanical fruit, commonly called a seed, but really a hard achene that contains a seed. Tissues beneath each achene enlarge and ripen into the horticultural fruit, the strawberry, with many achenes dotting its surface.

Flowers occur in clusters that bloom sequentially over several weeks (Photo 12-2). The first blossom to open, the king blossom, will produce the earliest-ripening and largest fruit. Two secondary blossoms below the king blossom bloom next. Then four tertiary blossoms open, two below each of the secondary blossoms, and finally eight quartenary blossoms open, two below each of the four tertiary blossoms. Each pollinated bloom takes about a month to develop into a ripe fruit, so the later the bloom opens, the later the fruit ripens. Fruit size decreases as the ripening season progresses, because the plants' energy is spread over increasing numbers of fruit. Older cultivars especially exhibit this tendency, whereas newer cultivars have been selected for less reduction in fruit size over the season.

Photo 12-2. Primary flowers develop into the first and largest fruit, which have already set here. Two secondary blooms (which have shed their petals here) open below each primary bloom, and two tertiary flowers (in full bloom here) open below each secondary blossom, each producing slightly smaller fruit. Finally two quartenary flowers (in the bud stage here) bloom below each tertiary blossom, producing the smallest and latest-ripening fruit.

TYPES OF STRAWBERRIES AND ADAPTATION

Spring-bearing (short-day) strawberry cultivars initiate flower buds only during the shortening days of late summer and early fall, then bloom in early spring and ripen fruit in late spring. Where winters are mild, as in California and Florida, some cultivars may continue to initiate flower buds through mid-winter, and the strawberry ripening season may extend for several months in winter or spring. In most areas, spring-bearers produce a concentrated crop, ripening over about a month in May or June.

Day-neutral cultivars flower and fruit any time temperatures remain between 35 and 85 F, regardless of daylength. Day-neutrals can provide three seasons of fruit in some areas: late fall, winter and spring where winters are mild; or late spring, summer and fall where summers are cool. Unfortunately, throughout most of the central United States, hot summers and cold winters limit day-neutral fruiting to late spring, early summer and again in fall. In California strawberry production areas, certain day-neutral cultivars can continue cropping for 9 to 10 months.

California now produces 80% of the U.S. commercial strawberry crop. Florida produces 12%, Oregon produces 2%, and all other states combined produce 6%. Some commercial strawberry production takes place in almost every region of North America, from Alaska and Nova Scotia to Mexico and southern Florida. The key is to choose cultivars adapted to your area and its disease pressures. Strawberries are also very adaptable as to soil type, from sandy loams to silt and clay loams. Sandy soils will require more frequent irrigation, while heavier soils will require the construction of raised beds before planting.

CROP USES AND PRODUCTION SYSTEMS

About 70% of strawberry production goes to the fresh market. Many California and Florida strawberry growers raise large acreages for wholesale fresh market sales, relying on hired workers to harvest the fruit. Fresh-market growers in eastern North America usually raise strawberries in conjunction with other fruit and/or vegetable crops, devoting just a few acres to strawberries. Many such growers use hired workers to harvest about half their strawberries for retail sales, and pick-your-own customers harvest the rest.

About 30% of the strawberry crop is processed, with 20% frozen as whole or sliced berries, and 10% processed into juice or puree for making syrups, jams, pie fillings and other products. About 90% of Oregon's strawberry production goes to processing. Hired workers still harvest most of the processing acreage, while harvesting machines are used in a few major processing areas, such as Oregon and Ontario.

The traditional production system for strawberries has been the perennial matted-row system (Photo 12-3), which keeps a field in production for 3 to 7 years but requires much hand labor. Over the last 20 years, annual hill-system plasticulture production (Photo 12-4) has become more popular, as strawberry farms have become larger and better able to justify labor-saving equipment. Large growers in California, Florida, and several other Southeastern states have adopted annual hill-system plasticulture, and some smaller growers in the Middle Atlantic states and Midwest are trying it also. Plastic tunnel production systems (Photo 12-5) have become very popular in Europe and Japan, and may be adopted by American growers in the near future.

Perennial Matted-Row Production System

This system involves planting in early spring, growing the planting for 13 to 14 months before harvesting in late spring of year 2, then renovating the planting each year after harvest for an additional 2 to 6 years of harvests. Day-neutral cultivars grown under this system will produce a first crop 4 to 5 months after planting. Growers set nursery plants 18 inches to 2 feet apart within rows spaced 4 feet apart, often on raised beds, and rely on the runners produced by these original plants to fill in the planting rows (Photo 12-6).

As runner plants form, many growers set the runners into desired positions, until each row reaches 2 feet wide and plants are spaced evenly about 1 foot apart. Later-forming runners are removed, so the row doesn't become too matted. Late-forming runners act as weeds: they will not produce fruit the following year, and compete for water and nutrients with the earlier-rooted runners that will fruit next year.

Growers protect matted-row plantings from winter damage to the crowns by covering the plants with straw in late fall, then remove the mulch and place it between the plants in spring. Straw mulch between the plants helps to control weeds, provides

Photo 12-3. Perennial matted-row strawberries are planted about 2 feet apart in single rows per bed, and runner plants will grow to fill in the rest of the planting bed. Spring-bearers will fruit about 14 months later, and day-neutrals bear about 3 months after planting.

Photo 12-4. Annual plasticulture strawberries are planted closer, about 1 foot apart in double rows per bed. Growers remove any runners that form, to grow large crowns in separate "hills" that will fruit in just a few months. (Photo from Aaron Creek Farms, Buffalo Junction, VA.)

Photo 12-5. Strawberries grown under plastic tunnels can produce fruit before the normal strawberry season, which may result in higher prices. (Photo from Univ. of Florida.)

Photo 12-6. Most perennial matted-row strawberry growers set the first runners that form and remove later runners, so rows don't become too dense and matted.

a clean surface for harvesters, and reduces fruit rots by keeping ripening berries off the soil.

Well-managed matted-row plantings can remain productive for several years, but typically yields decline each year as weeds invade and diseases build up. Proper renovation after harvest each year can prolong the life of a planting. Renovation involves mowing off the older leaves which carry leaf diseases, fertilizing with nitrogen to encourage rapid decomposition of the older leaves with their disease spores and growth of new leaves, disking up the older plants with root diseases to allow growth of new runners, irrigating the new growth, herbiciding and adding fresh straw mulch to control weeds.

Many small retail strawberry growers still use the perennial matted-row system, as it is the least expensive system in terms of plants, materials, equipment, planting labor, and frost protection costs. Typical costs for establishing matted row strawberries run $2400 to $3200 per acre, while establishing plasticulture strawberries costs $4000 to $6000 per acre, and the matted-row planting should produce 3 to 5 crops while the plasticulture planting produces 1 to 2. Small growers who can't afford transplanting machines, and must plant by hand, like the fact that the matted-row system requires planting only 5% as many plants over a 5-year period as does the plasticulture system. In areas with spring frosts during strawberry bloom season, this system cuts frost protection costs in half as compared to a plasticulture system.

Annual hill-system plasticulture production

This system involves planting in late summer to fall into fumigated raised beds covered with plastic mulch, cropping the plants the following winter to spring, or spring to fall for day-neutral cultivars, then destroying the planting. Runner plants are not desired and few will form during this time period, so growers set nursery plants only 1 foot apart, usually in double rows down each bed (photo 12-4). Each plant develops a large crown and forms a distinct "hill", instead of plants growing together into a matted row.

The color of plastic mulch used depends on the climate, planting time, and weed control methods. Winter plantings in mild-winter areas of California and the Southeast, set in late fall, use clear plastic mulch applied in early November to heat the soil during short winter days, stimulating crown development and causing an early flush of fruit production in winter or spring. Summer plantings, set in late summer or early fall while temperatures are still high, use white plastic mulch to cool the soil, which helps plants become established. White plastic mulch induces larger fruit in some cultivars, and is the least likely mulch color to cause fruit burn during hot weather. Fall plantings in cold-winter areas, such as the mid-Atlantic states, often use black plastic mulch, which warms the soil in late fall and delays fruit production in spring. Black plastic can cause fruit burn when temperatures exceed 90 F, but the harvest season may be over by that time in the areas where it's used.

Clear plastic mulch requires soil fumigation before planting to kill weed seeds. Most white mulch is translucent and does not control weeds, so also requires fumigation. Black plastic mulch does control weeds, better than straw mulch, so organic growers and others who don't fumigate use black or other opaque mulches. Opaque mulches in white and other colors are now available and may gain in popularity as common fumigants are phased out.

For a week or two after planting, water and cool the newly set plants with sprinkler irrigation for several hours each day during the afternoons. This is especially important when using black plastic mulch, which can become very hot. In many areas, sprinkler irrigation will also be needed for frost protection once blooming begins. Trickle irrigation lines are laid under the mulch to provide most of the regular irrigation. Thus, plasticulture strawberry systems often require both trickle and sprinkler irrigation, an added expense.

Spring-bearing cultivars used for plasticulture ripen over a longer season, which lasts for 6 to 8 weeks in most Eastern areas, 3 to 4 months in Florida and the Gulf Coast, and up to 6 months in California. Day-neutral plasticulture strawberries produce a winter-spring crop from December to May in southern California and Florida, and a summer-fall crop from April to December in California's central coast. The widely spaced, well-developed mother plants and excellent weed control produce larger berries throughout the harvest season and higher yields per acre than a matted-row system. California growers produce 30 to 50 tons per acre per year with this system, thanks to extended winter flower bud formation and thus extended ripening, while Eastern yields range from 5 to 14 tons per acre per year. Eastern matted row yields generally range from 4 to 8 tons per acre per year, so when everything is done right, plasticulture can potentially double yields.

By destroying plasticulture strawberry plantings after one harvest season, growers avoid carrying over diseases from year to year. Eastern growers and California central coast growers sometimes keep a planting for a second year, if anthracnose and other diseases are not present in the planting. This saves planting labor, cost of new plants, fumigation costs, and plastic mulch.

Plastic tunnel production

This system modifies annual hill-system plasticulture by installing a plastic tunnel, supported by wire hoops 16 to 18 inches tall, over each row from fall to spring (see Photo 12-5). Some growers erect larger tunnels over several rows, allowing headroom for workers and tractors. Such tunnels allow production from day-neutral cultivars to continue over the entire winter in regions which would not normally be warm enough for such production. Tunnels also allow spring-bearing cultivars to begin fruiting a month early, with the tunnels providing frost protection. Such off-season production often commands the highest prices, which pay for the increased labor of installing the hoops and plastic, and temporarily removing the plastic during unseasonably warm periods over the winter or early spring.

Tunnel culture has rapidly increased in southern Europe and in Japan. More than 90% of the strawberry acreage in southern France is under tunnels, along with 60% of Spain's strawberry acreage. Tunnel production can double the yields per unit of land per year, and spread economic returns out over the entire year. Tunnels also create brighter red fruit and can protect the fruit from certain pests, allowing organic production and freedom from pesticide residues, which is very important in Europe.

Researchers in Maryland and Illinois have demonstrated that tunnel culture can induce spring-bearing cultivars to begin ripening a month earlier and continue through the normal harvest season, doubling yields. Tunnel culture has not caught on much yet with growers in North America, since the major commercial strawberry production areas do not need season extension, and the increased materials and labor expense puts off growers in shorter-season areas.

CHOOSING CULTIVARS

Choose cultivars that are adapted to your growing and marketing systems and your area's climate, and that resist the most important diseases of your area. Tables 12-1 and 12-2 detail the adaptation, fruit traits and disease resistance of many of the most widely-grown short-day (spring-bearing) and day-neutral strawberry cultivars.

See the following key to all cultivar charts in this book.

Pest Resistance Ratings for Cultivar Charts

VR = very resistant: No economic damage occurs even where pest pressure is severe, so no pesticide treatments are needed.

R = resistant: Pesticide treatments are not usually needed to prevent economic damage, but a reduced pesticide schedule may be needed during seasons or in areas where pest pressure is severe.

MR = moderately resistant: If pest pressure is slight or moderate, pesticide treatments are not usually needed to prevent economic damage. A reduced pesticide schedule may be needed if pest pressure is average, and a full pesticide schedule will be needed during seasons or in areas where pest pressure is severe.

MS = moderately susceptible: Reduced pesticide treatments are usually needed under slight or moderate pest pressure to prevent economic damage, and a full pesticide schedule will probably be needed where pest pressure is average or severe.

S = susceptible: A full pesticide schedule is usually needed under moderate to average pest pressure to prevent economic damage, and extra pesticide sprays may be needed during seasons or in areas where pest pressure is severe.

VS = very susceptible: A full pesticide schedule is required if pest pressure is slight or moderate, and extra pesticide sprays will be needed with average pest pressure. Where pest pressure is severe, do not plant this cultivar.

Table 12-1.
POPULAR SHORT-DAY STRAWBERRY CULTIVARS*

Cultivar	Regions Where Adapted [z]	Flavor	Fruit Size	Flesh Firmness	Freezing Quality	Resistance to [y]: Leaf Spot	Leaf Scorch	Red Stele	Verticillium Wilt	Powdery Mildew	Fruit Rots	Anthracnose Crown Rot	Spider Mites	Yield	Ripening Season
Earlibrite	GC, SE	good	large	firm	fair	MS	MS	S	S	MR	MR	S	VS	high	very early
Sweet Charlie	SE, GC	good	medium	medium	fair	MS	MS	S	S	MR	MS	S	S	high	very early
Pelican	GC, SE	good	very large	medium	poor	MR	MR	R	S	MS	MR	R	S	very high	very early
Chandler	SW, GC, SE	good	very large	firm	good	MR	R	S	MR	MS	S	S	MR	very high	v. early-late
Camarosa	SW, GC, SE	good	very large	very firm	good	MR	MR	S	S	MS	VS	S	MR	very high	v. early-late
Earliglow	all but FN, GC, SW	best	medium	firm	best	MR	MR	R	MR	MR	R	S	MR	medium	very early
Gaviota	SW, GC, SE	very good	very large	very firm	good	MS	MS	MS	MS	MR	MS	MR	MR	very high	early-late
Delmarvel	MA, SE	very good	large	firm	very good	MR	MR	R	R	MR	MR	R	MR	medium	early
Hood	NW	best	large	firm	very good	MS	R	VR	R	MR	S	S	MR	medium	early middle
Honeoye	NE, FN, NC	fair	large	medium	good	MR	MR	S	S	R	MR	S	MS	high	middle
Primetime	MA, SC	very good	large	medium	good	MR	MR	R	S	S	S	S	MS	medium	middle
Redchief	NC, SC	good	large	firm	very good	MS	MR	R	MR	R	S	VS	MS	high	middle
Cavendish	FN, NE, NC	very good	large	firm	very good	R	R	R	MR	S	S	S	MR	high	middle
Puget Reliance	NW	best	large	firm	best	S	MR	R	S	R	MR	S	S	high	middle
Allstar	NC, NE, MA, SC	good	large	firm	fair	MR	MR	R	R	MR	S	MR	VR	high	late middle
Totem	NW	very good	medium	medium	best	S	S	VR	S	S	MR	S	MS	medium	late middle
Winona	FN, NE, NC, SC	very good	large	firm	very good	MR	MR	R	MR	MR	S	S	MR	high	late
Latestar	MA, SE, NE, NC	good	large	firm	fair	R	MR	R	MR	R	S	S	MR	high	late
Jewel	NE, FN, NC	very good	med-large	very firm	very good	R	R	S	S	MR	MS	S	S	medium	late
Firecracker	NW	very good	medium	firm	best	MR	MR	S	S	VS	MR	S	MS	high	very late

[z] **Regions**:
SE = Southeast: from southeast Virginia and Georgia to northern parts of Gulf Coast states, USDA zones 7b-8b
GC = Gulf Coast: from eastern Texas to Florida along the coast, zones 9 to 10
SW = Southwest: from western Texas to southern and central California, zones 7 to 10
NW = Northwest: from northern California and Nevada to Washington, Idaho and British Columbia, zones 5 to 9
FN = Far North and Rocky Mountains, zones 2b to 4a
NE = Northeast: New England, northeastern Canada, New York and northern Pennsylvania, zones 4b to 6a
MA = Mid-Atlantic: from New Jersey and southern Pennsylvania to Virginia, zones 6a to 7b
NC = North Central: from Ohio to South Dakota and Nebraska, zones 4b to 6a
SC = South Central: from West Virginia and western North Carolina to Kansas and Oklahoma, zones 6a to 7b

[y] **Pest resistance ratings**: VR = very resistant, R = resistant, MR = moderately resistant, MS = moderately susceptible, S = susceptible, VS = very susceptible

Notes on Cultivars:

Earlibrite, a 2000 Florida release, blooms and ripens very early, and is too frost-prone further north. It's less prone to gray mold than Sweet Charlie.

Sweet Charlie, from Florida, resists anthracnose fruit rot but not anthracnose crown rot, gray mold or Phomopsis fruit rot. Fruit can crack after rains.

Pelican, from Louisiana, resists both anthracnose fruit rot and anthracnose crown rot. Fruit skin may crack after rains.

Chandler is a main plasticulture cultivar in California and the Southeast, due to high yields and long ripening time. Chill plants 1-2 weeks before planting.

Camarosa, a 1993 California release, has greater fruit size and firmness than Chandler and holds up well under shipping, but has greater susceptibility to field diseases.

Earliglow, still a leading matted-row cultivar after 27 years, shows the greatest resistance to gray mold fruit rot. Berry size declines greatly over season.

Gaviota, a 1997 California release, is short-day with extended flowering under cool conditions, like Chandler. It resists anthracnose crown rot and rain cracking of fruit.

Delmarvel, from Maryland, resists anthracnose fruit rot and crown rot, and is less prone to gray mold than many others.

Hood, a 1965 Oregon cultivar, is still widely grown for processing into jams and ice cream. It is very susceptible to viruses, so don't plant near older plantings.

Honeoye, from New York, is very winter hardy. Tart fruit can turn bitter if allowed to turn dark red or ripening temperatures are high.

Primetime, a 1996 Maryland release, is most productive in USDA zones 6 and 7. Its large fruit are susceptible to fruit rots.

Redchief, an older cultivar, is still widely grown in the Midwest. It resists most diseases, but is very susceptible to leaf blight and anthracnose.

Cavendish, from Nova Scotia, is very winter hardy. Fruit have better flavor than Honeoye but may bruise easily and color irregularly.

Puget Reliance from Washington resists viruses and produces high yields of fruit suited to processing.

Allstar is popular in the Midwest due to high yields of large fruit, which are orange-red with a pink interior.

Totem, from British Columbia, is the most widely planted cultivar in the Pacific Northwest. It's dark red internally and externally, desirable for processing.

Winona, a 1996 Minnesota release, is very winter hardy and tolerant of black root rot. Large, flavorful fruit, but skin may be weak in warm, wet seasons.

Latestar, from Maryland, blooms late so needs less frost protection. It has mildly acidic, tough-skinned fruit. It's susceptible to leaf blight and fruit rots.

Jewel, from New York, has large, firm, flavorful fruit with a relatively long shelf-life.

Firecracker is a very late Oregon processing cultivar released in 1998. It yields more and resists fruit rots better than Independence, another newer cultivar.

---

*In approximate order of ripening

Table 12-2.
IN A NUTSHELL: POPULAR DAY-NEUTRAL STRAWBERRY CULTIVARS *

						Resistance to [y]								
Cultivar	Regions Where Adapted [z]	Flavor	Fruit Size	Flesh Firmness	Freezing Quality	Leaf Spot	Leaf Scorch	Red Stele	Verticillium Wilt	Powdery Mildew	Fruit Rots	Anthracnose Crown Rot	Spider Mites	Yield
Tristar	all but FN, GC, SW	very good	med.-small	firm	good	MS	MS	R	R	MR	MS	S	MS	medium
Tribute	all but FN, GC, SW	good	med.-large	very firm	good	MR	R	VR	R	VR	MR	S	MR	medium
Pacific	SW, GC	good	very large	firm	good	MS	MS	S	MS	MR	S	MS	MR	very high
Aromas	SW, GC	very good	large	firm	fair	MS	MS	S	MS	MR	S	MR	MR	very high
Diamante	SW, GC	very good	extra large	medium	poor	MS	MS	VS	MS	MR	VS	MS	MR	very high

[z] **Regions**: SE = Southeast: from southeast Virginia and Georgia to northern parts of Gulf Coast states, USDA zones 7b-8a
GC = Gulf Coast: from eastern Texas to Florida along the coast, zone 9
SW = Southwest: from western Texas to southern and central California, zones 7 to 10
NW = Northwest: from northern California and Nevada to Washington, Idaho and British Columbia, zones 5 to 9
FN = Far North and Rocky Mountains, zones 2b to 4a
NE = Northeast: New England, northeastern Canada, New York and northern Pennsylvania, zones 4b to 6a
MA = Mid-Atlantic: from New Jersey and southern Pennsylvania to Virginia, zones 6a to 7b
NC = North Central: from Ohio to South Dakota and Nebraska, zones 4b to 6a
SC = South Central: from West Virginia and western North Carolina to Kansas and Oklahoma, zones 6a to 7b

[y] **Pest resistance ratings**: VR = very resistant, R = resistant, MR = moderately resistant, MS = moderately susceptible, S = susceptible, VS = very susceptible

Notes on Cultivars:
Tristar, a cold-hardy day-neutral cultivar from Maryland that's adapted to the Northeast, has good flavor but fairly small fruit.
Tribute, another cold-hardy, Maryland day-neutral cultivar, offers better fruit size and firmness, disease and mite resistance than Tristar.
Pacific, a 1997 California cultivar, stores and ships well. Needs 3-4 weeks cold storage before planting, throws few runners, tolerates later planting, needs more nitrogen.
Aromas, a 1997 California release, is the most productive day-neutral cultivar, with high fall yields, and shows greatest pest tolerance.
Diamante, a1997 California cultivar, has huge fruit with very good flavor but internal color too light for processing. Plant early in fall, without chilling, or it runners too much.

*In approximate order of ripening.

Most California cultivars, developed for the high-yielding plasticulture system, depend on soil fumigation and have little or no resistance to soilborne fungal diseases such as red stele, Verticillium wilt, anthracnose, black root rot and charcoal rot. Eastern breeding programs have developed cold-hardy cultivars resistant to several of these diseases, but these cultivars do not produce such high yields over as long a time as the California cultivars.

Many plasticulture strawberry producers in California and the Southeast grow the short-day, but long-ripening and high-yielding, California cultivar 'Chandler', which is winter hardy in much of the East. Another short-day California cultivar widely planted in California and Florida is 'Camarosa' (Photo 12-7). Some Central Coast California growers are replacing the very anthracnose-susceptible 'Pajaro' with the more recently-released 'Gaviota' , which resists anthracnose crown rot, and growers in other areas are trying 'Gaviota'.

Southeastern plasticulture growers have also relied on 'Sweet Charlie', a Florida cultivar that resists anthracnose fruit rot (a devastating disease in the Southeast) but is susceptible to anthracnose crown rot, and ripens early. The new Florida cultivar 'Earlibrite', which ripens from early December to February in Florida, offers better resistance to gray mold than 'Sweet Charlie'. 'Strawberry Festival', another new Florida cultivar, ripens from late December to late March.

These California and Florida cultivars bloom too early for plasticulture growers in the mid-Atlantic states or the Southern mountains, where severe freezes still threaten throughout their bloom season. For those areas, the late-blooming Maryland cultivar 'Latestar' performs best for plasticulture.

Perennial matted-row strawberry growers should generally choose cultivars that resist red stele and Verticillium wilt, two soilborne diseases that can kill plants, especially if not fumigating the soil before planting. Resistance to the foliar diseases, leaf spot and leaf scorch, and to common fruit rots can minimize the need for fungicide sprays or allow organic production. For matted-row growers, the Maryland-developed cultivar 'Earliglow' (Photo 12-8) has shown the greatest overall resistance to the most common fruit rots: gray mold, leather rot and anthracnose fruit rot. 'Earliglow' blooms and ripens early but also continues fruiting until late in the spring-bearing season.

Northern matted-row strawberry growers, who need extra cold hardiness, like the New York cultivar 'Honeoye', the Minnesota cultivar 'Winona', and Canadian cultivars such as 'Cavendish'. 'Honeoye' (Photo 12-9) does poorly in warmer areas, where the fruit turn bitter, and has shown greater susceptibility than other cultivars to black root rot and bacterial angular leaf spot, two diseases which are increasing in some areas, although it resists tarnished plant bug. 'Winona' has performed well in the southern Plains states as well as northern areas. Growers in the Pacific Northwest rely on cultivars suited to processing and the climate and resistant to the strong virus pressure, mainly 'Totem', and also the older 'Hood', 'Puget Reliance' and the new 'Firecracker'.

Day-neutral strawberry cultivars produce winter to spring crops in southern California and the Gulf Coast, and summer to fall crops in cool-

Photo 12-7. 'Camarosa', one of the newer short-day strawberries released from the University of California breeding program, produces extra-large fruit and very high yields, but is susceptible to anthracnose crown rot and fruit rots. (Photo by Toni Ham, Ham's Farm, Plant City, FL.)

Photo 12-8. 'Earliglow', released in 1975 from the USDA breeding program in Maryland, is still widely grown for its excellent flavor, firmness, early ripening and relatively high degree of resistance to gray mold and other fruit rots.

summer areas such as California's central coast, upstate New York and Minnesota. The most recent California day-neutral cultivars include 'Aromas', 'Diamante', and 'Pacific'; all of these are more susceptible to powdery mildew and anthracnose fruit rot than short-day plants, and more susceptible to mites unless kept growing vigorously. Avoid two older California day-neutral cultivars, 'Fern' and 'Selva', as they yield less fruit of lower quality and are very susceptible to mites and anthracnose. 'Tribute' and 'Tristar' are day-neutrals adapted to northern and eastern areas, with 'Tribute' exhibiting greater fruit size and disease resistance than 'Tristar'.

SITE PREPARATION AND PLANTING

Choose a site with excellent air drainage, especially for plasticulture strawberries, which will bloom earlier and be more prone to frost-killed blooms. Sloping sites work well, even for trickle irrigation, as long as beds are laid out on the contour. Create a uniform slope of 0.75 to 1% for trickle-irrigation.

Take soil samples and test for pH, phosphorous and potassium several months before planting. Apply any lime or gypsum needed for pH modification several months before planting and mix it thoroughly with the soil. This allows time for the material to adequately disperse, and for rains or irrigations to leach excess salts from the root zone before planting. If adding gypsum to improve water penetration, however, apply it to the soil surface without mixing it in.

Turning under a cover crop of white clover or another legume, or tilling in compost, several months before planting will add nitrogen as well as loosening a tight soil and improving water drainage, and increasing water retention in a loose, sandy soil. However, do not add organic matter if planning to fumigate or if using herbicides that get tied up by high levels of organic matter. Turning under two successive cover crops of rape (canola), as outlined in chapter 2, can substitute for fumigation in controlling soilborne diseases. Applying biocontrol products containing beneficial soil fungi, also discussed in chapter 2, is another alternative to fumigation for soilborne disease control.

Soil fumigation or solarization

The high-yielding plasticulture system developed in California has relied on soil fumigation to control most weed seeds, soilborne fungal diseases, nematodes and soil-dwelling arthropods. Major strawberry producers in California, Florida and other Southeastern states have fumigated each year with methyl bromide and chloropicrin. Now that these chemicals are being phased out, researchers are developing other fumigants such as metam sodium (Vapam™), but the newer chemicals do not kill as broad a spectrum of soil life, especially weed seeds.

Methyl bromide fumigation was most commonly applied to flat ground, before building raised beds, and the treated ground was immediately sealed with large plastic tarps (Photo 12-10). After the fumigant's waiting period of several days, tarps were removed and beds formed. Two weeks after tarp removal, strawberries could be planted and plastic mulch applied.

Most strawberry growers today strip-fumigate after forming the planting beds and laying out the plastic mulch, which will remain in place. They use herbicides to control weeds between the rows rather than the fumigant, which reduces the fumigant's harmful effects on beneficial soil life. Some fumi-

Photo 12-9. 'Honeoye', developed by the Cornell University breeding program in New York, produces higher yields in cold Northern climates than other early cultivars.

Photo 12-10. The high-yielding annual plasticulture system of growing strawberries, developed in California and now used in the Southeastern U.S. and elsewhere, has depended on soil fumigation to control weeds and soilborne diseases. (Photo from the University of California)

gants can be applied through the trickle irrigation system. After waiting at least 3 weeks after fumigation, workers place transplants through holes in the plastic mulch.

Soil solarization can substitute for fumigation, although it doesn't control pests as completely, or solarization can supplement fumigation when applying the fumigant at lower rates. Irrigate or allow rainfall to bring soil moisture to field capacity before solarizing, since sunlight will heat moist soil more quickly and deeply, and pathogens and weed seeds are more susceptible to heat when moist.

For best control of soilborne pathogens and nematodes, apply two layers of clear plastic tarps separated by about 1 inch to flat ground, before removing the plastic tarps and forming raised beds. Where weed control in the beds is more critical than pathogen control, form beds first and strip-solarize them with rolls of clear plastic mulch. After solarization, keep the plastic mulch in place, paint it white to cool the soil and block light to weeds, and plant into the white-coated plastic. Solarize for at least 6 weeks during the hottest months in warm-summer climates, and for cool-summer climates such as California's central coast, solarize for 8 to 12 weeks before planting.

Forming raised beds, fertilizing and mulching

Raising strawberry planting beds 6 to 10 inches over the aisles ensures good soil drainage, which lessens the threat of many potential soilborne disease problems. Growers use their own or a custom applicator's bedmaking equipment to form beds about 2 feet wide, spaced 3½ to 4 apart from center to center, for most strawberry plantings (Photo 12-11). Plasticulture hill-system growers will set two rows of plants per bed, with plants spaced 12 to 14 inches apart. Perennial matted-row growers will set one row of plants per bed, spacing short-day plants about 24 inches apart and day-neutral plants about 18 inches apart. Winter plantings in southern California and Florida, intended to produce large, early crops over a shorter season, may utilize wider beds with four rows of plants set per bed.

Apply most of the nitrogen fertilizer needed for summer or fall plantings just before or during bed shaping, so the nitrogen ends up only in the planting beds. The amount to apply depends on soil type, climate, projected yields and marketing system. Higher amounts of nitrogen can increase fruit size and yields, but also reduce fruit flavor and increase susceptibility to fruit rots, so wholesale strawberry growers generally use more nitrogen than retail growers who sell locally. California growers who will crop their strawberry fields for many months often apply 120 to 240 pounds of actual nitrogen before planting, while Eastern growers apply about 60 pounds of actual nitrogen before planting in late summer or fall.

Most nitrogen fertilizers release the nutrient quickly, stimulating rapid vegetative growth of the crowns and roots, which is the goal in most planting systems. Winter planting systems aiming for early fruit production utilize slow-release fertilizers at planting, placed about 1 inch below the plants' roots. When spring planting perennial strawberry fields, use only 25 to 40 pounds actual nitrogen before planting, then apply 20 to 35 pounds of actual nitrogen in early fall when fruit buds are forming.

When planting strawberries after a heavily fertilized vegetable crop, nitrogen is often the only fertilizer element needed. In other situations, phosphorous or potassium may be needed, so add fertilizers containing these elements to the nitrogen based on soil tests taken a few months earlier. In California and other Western states, zinc may also be deficient and require a fertilizer containing zinc as a trace element.

When using trickle irrigation, lay out the lines down the center of each bed before applying mulch or planting. Super-bedder machines (Photo 12-12) apply trickle lines while forming beds, fertilizing, laying plastic mulch, covering the mulch edges with soil, and strip-fumigating.

Most plasticulture growers apply rolls of plastic mulch to the beds and strip-fumigate before plant-

Photo 12-11. Forming raised beds before planting strawberries reduces problems with soilborne fungal diseases such as red stele and black root rot. This bedding machine was farmer-built.

ing. Those who fumigate the entire field before forming beds may apply mulch strips to the beds before or after planting. Matted-row growers apply straw mulch after planting when using transplanting machines, but small growers may apply straw mulch before planting by hand to prevent more weed seed germination.

Handling transplants or plugs

Matted-row strawberry fields planted in spring, and plasticulture fields planted in fall, are started with field-grown transplants grown from disease-free runner tips. Plasticulture fields planted in late summer are best started with greenhouse-grown plugs (Photo 12-13), very small plants grown in plastic trays of 50 or so plants per flat, like those used in the bedding plant industry, rather than regular field-nursery plants.

Strawberry nurseries dig transplants for spring or summer plantings from December to February, then hold the plants in cold storage to keep them dormant until they are shipped. Growers should plant such transplants immediately. High-elevation California nurseries or high-latitude Canadian nurseries produce transplants for winter plantings, digging them in October and early November. Growers plant these immediately if they've already received enough winter chilling; if not, they place short-day cultivars in cold storage for 1 to 2 weeks and day-neutral cultivars in cold storage for 2 to 4 weeks. Inadequate chilling causes poor stands and reduced plant vigor that leads to increased susceptibility to mites and other pests, but too much chilling causes excessive runner production and reduces yields.

Plugs are actively growing small plants so should not be chilled, and should be kept well watered until planting. Keep dormant transplants wrapped in the polyethylene box liners they were shipped in, to prevent them from drying out. Prewet the planting beds with a trickle irrigation system (set under the plastic mulch in plasticulture fields with mulch already applied) before planting transplants or plugs.

Most large strawberry growers use a 4-row transplanting machine, pulled by a tractor, to set the plants (Photo 12-14). Such machines feed the transplants or plugs at regular intervals to workers sitting behind the machine, who grab each plant and set it in the hole created by the machine. Plasticulture growers who apply mulch before planting use a water-punch wheel planter aid to pierce the mulch and create small planting holes.

When plastic or straw mulch is applied after planting, growers often use a different type of transplanter that places plants into grooves cut into each bed, while also applying starter fertilizer below or to the side of the planting slot. Where plastic mulch is unrolled over planted beds, workers use heated cylinders to cut holes through the plastic over each plant.

Remove old leaves from transplants (Photo 12-15) before setting them, since old leaves can be a source of foliar and fruit diseases. Placing strawberry plugs or transplants at the correct depth is critical. Place plugs just deep enough to barely cover the plug's potting soil with field soil. If the potting soil is exposed to the air, it rapidly dries out, causing the young plant to desiccate and die, but if the crown is buried the plant may rot. Place transplants with the crown just at soil level and all roots covered (Photo 12-16), as transplants may also rot if the crown is buried and grow poorly if planted too shallow.

Photo 12-12. Super-bedder machines form beds, lay drip irrigation and plastic mulch, fertilize and fumigate the beds all at once. (Photo from Univ. of Florida.)

Photo 12-13. Plugs, tiny greenhouse-grown plants like those used in the bedding plant industry, work best for late-summer plasticulture plantings in warm climates. (Photo from Aaron Creek Farms, Buffalo Junction, VA.)

Photo 12-14. Large strawberry growers use 4-row transplanter machines to set the plants. (Photo from Govin's Strawberries, Menomonie, WI.)

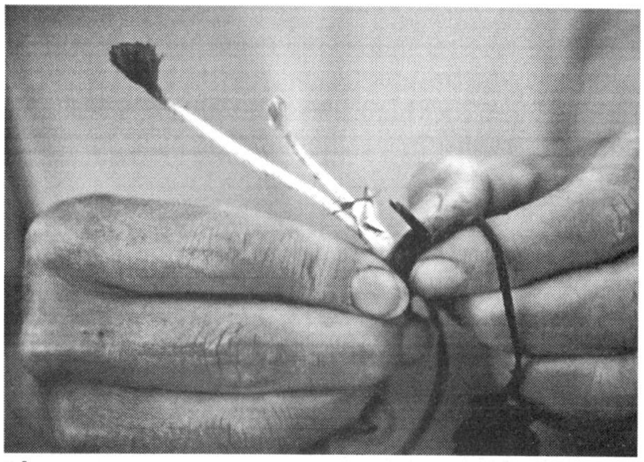

Photo 12-15. Removing old leaves from transplants greatly reduces the main source of foliar and fruit diseases, such as leaf spot and gray mold, in new plantings.

CARE AFTER PLANTING

Run the irrigation system immediately after planting. Irrigate spring-planted matted-row fields about once a week for the next month, and throughout the summer, if no rain occurred during the previous week. Irrigate hill-system plasticulture plantings every day for the first week after planting, and once a week thereafter. A sprinkler system during the first week will help cool the plants during the heat of the day, which can aid establishment when planting in late summer, while a trickle irrigation system provides later irrigations using less water. For growers wanting just one system, microsprinklers can help establish plants and provide later irrigation and frost protection, using less water than larger sprinklers. Fertigation through the trickle or microsprinkler system can provide up to 30 pounds per acre of the planting's nitrogen requirement.

Hill-system plasticulture growers want each strawberry plant to grow a large crown and produce many blooms and fruit in a few months, so they have workers remove runners that form in the fall after planting. Crews go over the fields several times for the first 2 to 3 months after planting, removing runners and hand-pulling weed as necessary. As days begin to lengthen again in winter, runner production ceases and plants begin to bloom and fruit.

Matted-row strawberry growers want to encourage root growth of new plantings and runnering, so the row will fill in and produce a good crop the next year, so they have workers pick off the flower buds that develop in spring during the first few weeks following planting. This eliminates a first-year crop of short-day cultivars, and delays the first-year harvest of day-neutral cultivars until plants are well established. Remove blossoms of day-neutral matted row strawberries for the first six weeks, then allow blooms to develop and harvest fruit for the rest of the season.

Photo 12-16. Setting transplants at the correct depth, with roots just covered and the crown at the soil surface, is critical to plant survival. Plants set too deeply can rot, while if roots are exposed they dry out and die.

Matted-row growers set the first runners that form on both types of strawberries, placing each about 8 to 12 inches from the parent plants and from other runners, and anchoring the runners with mulch. After sufficient runners have filled the planting bed, they pull off later-forming ones (photo 12-17). Late runners act as weeds by producing little or no fruit themselves the next season, and competing with the productive first-set runners.

WINTER AND FROST PROTECTION

Commercial strawberry growers in California, Florida, and other Southeastern and Gulf Coast states do not need to protect plants from winter damage, but growers in colder climates do. When soil temperatures drop below 40 F in early winter, about a month after strawberry plants have entered dormancy and some leaves have turned wine-red, cold-climate growers apply several inches of straw mulch on top of the plants (Photo 12-18). This insulates the crowns and flower buds from extreme cold and keeps the soil from freezing and thawing, which can heave up the crowns and expose the roots, killing the plants. In early spring, after soil temperature has reached 40 F and most danger of a heavy freeze (25 F or below air temperature) has passed, and growers find new strawberry leaves growing when they push aside the mulch, they remove the excess covering. Workers place the mulch in the aisles to inhibit weeds and provide a dry surface for picking.

Some growers use fabric row covers for winter protection instead of straw mulch, where they want to promote fall growth and accelerate growth in early spring. Winter use of row covers works best where the winter is short or uniformly cold and few spring frosts threaten, or with a late-blooming cultivar, since row covers accelerate spring bloom. With an early-blooming cultivar like 'Earliglow', frosts often prove fatal to those extra-early flowers, despite the protection of the row cover put back on before the frost. Most growers use heavier row covers weighing 1.25 to 1.5 ounces per square yard, such as Kimberly Farms' Strawberry Freeze Cover™, that provide 6 to 8 F protection while decreasing light by 50%. Such heavy covers, treated for protection from ultraviolet light, can last at least four years, while lighter covers provide less cold protection and may last only one or two years.

Growers have found it takes 5 to 6 workers to apply each row cover, and that winds of 10 mph or more make this difficult. Workers use metal pins made for landscape fabric to hold covers in place, along with 1000 sandbags per acre on breezy sites. In early spring, workers fold the covers off the strawberry beds and place them on the aisles, securing them with the sandbags. If an overnight frost or freeze threatens during bloom, secure the covers over the beds again early in the day to trap latent ground heat. Remove row covers soon after the freeze, to allow for pollination and drying. Row covers can also be used to extend fall harvests of day-neutral cultivars beyond the first fall frost.

Photo 12-17. Pulling off late-forming runners, which act as weeds, keeps matted-row plantings from getting too crowded and prone to fruit rots.

Photo 12-18. Strawberry growers in cold climates must protect the crowns from winter heaving due to soil freezing and thawing. Most growers apply several inches of clean straw mulch over the crowns after the first hard freeze. (Photo by G.R. Brown, Kentucky State Univ.)

Sprinkler irrigation during bloom (Photo 12-19) can provide greater frost protection than can row covers. The system must be capable of providing 60 to 80 psi water pressure at the nozzle tips, usually using 1/8" or 9/64" nozzle tips to provide 1/10" of water per acre per hour. When temperatures drop into the lower 20's F, the water application rate will double (see chapter 8, table 8-3). Microsprinklers can reduce the water needs of frost protection by 20 to 75%, compared to conventional sprinklers. Using a computer-controlled system, that cycles water applications while maintaining plant temperatures above critical levels, can further reduce water use for frost protection. This is important, both to save water and energy, and since excess water application can cause waterlogged soils that lead to root rots.

Cryoprotectant sprays tested on strawberries have failed to lessen frost damage during bloom. Some of the cryoprotectant sprays have increased the number of fruit with deformities, so using cryoprotectants is not recommended for strawberries.

Plastic tunnel production provides another way to protect strawberry plants from winter cold and spring frosts, while simultaneously promoting growth and earlier fruiting. Tunnel production requires much greater materials cost, including wire hoops and heavy-duty plastic strips to create the tunnels, but allows off-season production during winter, which can command a higher price.

Many large strawberry growers rent beehives during bloom, setting out one hive per acre. Native bees can provide superior pollination of small plantings, if abundant enough. Growers should check the first flowers to determine if pollination is adequate. As petals begin to drop, those flowers should have uniformly dark pistils; if many petalless flowers still have bright yellow or white pistils, you may need to bring in more bees.

MAJOR PESTS OF STRAWBERRIES

Soilborne fungal diseases and viruses that lead to decline and death of plants are among the worst pests of strawberries. Several leaf diseases and fruit rots can greatly reduce yields, and weeds have limited many plantings. Insect and mite pests generally pose less of a threat to strawberries, although some can cause much damage in certain areas. Slugs and snails prefer the fruit and can damage large amounts of it. Deer can destroy plantings where deer populations are high, and birds damage the fruit in some areas.

Photo 12-19. Running sprinkler irrigation during bloom can protect blossoms down to even lower temperatures, so long as the system is turned on before temperatures dip below freezing, and runs all night until the ice melts completely.

Root and crown diseases

Soil fumigation, solarization or cover cropping, combined with raised beds, prevents many root and crown diseases, which are caused by soilborne fungal pathogens and/or nematodes. Always buy certified disease-free transplants, since diseased transplants could reintroduce the pathogen(s) into clean beds. Strawberry breeders have also developed some cultivars resistant to several races of some of these pathogens. Always check with your state's Extension specialist on strawberry diseases to be sure that resistant cultivars you choose are resistant to the pathogen races occurring in your area. When cultivar resistance isn't adequate and the soil hasn't been fumigated, solarized or cover cropped to reduce these diseases, soil fungicides such as metalaxyl (Ridomil™ or Subdue™) or fosetyl-Al (Aliette™) can help, but these also inhibit beneficial mycorrhizae.

Anthracnose crown rot, caused by several *Colletotrichum* species, first appears as dark spots on leaf petioles and runners, then decays crown tissues and roots, causing entire plants to wilt and die. Warm, wet weather favors this pathogen, which can survive in the soil for 9 months without host plants, and also infects many weeds such as chickweed and vetch. It also causes a fruit rot with sunken lesions that develop orange spore masses, to which day-neutral cultivars are especially susceptible. Watch young fruit carefully, and if anthracnose fruit rot develops, apply foliar fungicides to avoid the spread of the disease and subsequent crown rot. Several cultivars resist the crown rot, the fruit rot, or both.

Red stele root rot, caused by the fungus *Phytophthora fragariae*, causes young roots to rot from the tips upward and produces a red discoloration of the stele, the root's central cylinder, above the rotted areas (Photo 12-20). The pathogen forms resistant spores that can survive in the soil for several years. Cool, wet conditions favor this disease. Many Eastern cultivars listed in Tables 12-1 and 12-2 resist several races of this pathogen.

Phytophthora crown rot, incited by several other *Phytophthora* species, is not usually a problem but may cause serious losses during wet years on heavy clay soils. Symptoms first appear as wilted young leaves, then decay spreads around the crown

Photo 12-20. Red stele, a common root-rotting disease that causes root centers to turn red and plants to wilt and die, can be prevented by planting only cultivars resistant to red stele. (Photo by Jay W. Pscheidt, Oregon State Univ.)

until an entire plant is girdled, wilts and dies, although roots may remain healthy. No cultivars are highly resistant, but some are less susceptible to this disease.

Verticillium wilt, caused by the fungus *Verticillium dahliae*, attacks a wide range of broadleaved plants and can survive in the soil for many years without hosts. Avoid planting strawberries within 4 years of growing susceptible crops such as melons or tomatoes, or after heavy infestations of susceptible weeds such as pigweed or lambsquarters. Older leaves of infected strawberry plants turn brown along the margins and between veins and dry up as the disease progresses, until the crown discolors and entire plants wilt and die. Many Eastern cultivars resist this disease, and some California cultivars are less susceptible than others.

Black root rot is a disease complex, in which roots injured by lesion nematodes or frost heaving are attacked by weakly pathogenic fungi such as species of *Pythium* and *Cylindrocarpon*. Roots of affected plants turn black and slowly die, without a red stele or crown discoloration (Photo 12-21). This disease complex has been increasing in recent years, and no cultivars resist it, although many cultivars resistant to red stele show some tolerance to black root rot. 'Honeoye' is especially susceptible, so avoid it if black root rot has been a problem. Cover cropping with oats and raised beds have reduced black root rot. Previous fescue crops reduce lesion nematodes, and treating the soil with the nematode-attacking fungus in DiTerra™ (www.abbott.com) controls lesion, root knot, and other nematodes.

Photo 12-21. Black root rot, a disease complex of several fungi that attacks roots weakended by nematodes or winter damage, has become more prevalent in the Northeast and Midwest in recent years. Cover crops of 'Saia' oats before planting and raised beds help prevent this

Virus and mycoplasma diseases

Using certified virus-free transplants or plugs eliminates almost all virus and mycoplasma problems in annual plantings. Perennial plantings may become infected with these sub-microscopic pathogens that spread throughout the plant's cells, especially if older plantings or wild strawberries are located nearby, since most of these diseases are spread by insects. Five virus diseases and two mycoplasma diseases may be of concern.

Strawberry viruses are spread by the strawberry aphid and several other aphids. **Necrotic shock**, caused by tobacco streak virus, is also spread in pollen. If just one type of virus infects a planting, it usually causes no visible symptoms, although it can substantially reduce yields. When any two of strawberry mottle virus, strawberry crinkle virus, mild yellow edge virus, or vein banding virus are present, **strawberry yellows** occurs, which reduces yields, fruit size and fruit quality. Strawberry yellows stunts and distorts leaves, curling them into a cup shape with yellow edges, and turns older leaves prematurely red.

Mycoplasma diseases are spread by leafhoppers but not aphids. **Lethal decline** is caused by a mycoplasma that also causes X-disease in cherries and peaches, and that has a wide host range among broadleaved weeds. Younger leaves of infected strawberry plants become stunted and yellow with edges curled upward, and plant vigor gradually declines until plants die. **Green petal**, caused by a mycoplasma that also infects various clovers, induces infected strawberry and clover plants to develop flowers with green petals. Strawberry fruit either fail to develop from these flowers, or develop into clusters of large, green achenes.

Immediately remove and destroy any plants with symptoms of mycoplasma or virus infection.

Foliar diseases

Fungal and bacterial pathogens that cause leaf diseases are spread by splashing water or wind. Eastern strawberry breeders have selected most cultivars to have at least moderate resistance to these diseases, since the higher rainfall encourages these diseases. In drier areas, using trickle irrigation instead of sprinkler irrigation, unless the latter is needed for plant establishment or frost protection, can reduce leaf diseases. When using sprinklers, microsprinklers or a close spacing of larger sprinkler heads (30 feet by 30 feet) reduce the need to overirrigate parts of the field, thus minimizing foliar diseases. Destroying old foliage, by disking in annual plantings after harvest finishes or renovating perennial plantings, also reduces these diseases.

Common leaf spot lesions have tan centers with dark brown margins (Photo 12-22). The disease, caused by the fungus *Mycosphaerella fragariae*, reduces plant vigor, yields and fruit quality. Begin weekly monitoring for leaf spot when first leaves expand. Fungicide treatments will reduce but not eliminate pathogen populations, and the pathogen has developed resistance to some fungicides, so use fungicides only when leaf spot appears on many plants, and rotate between fungicides to avoid developing pathogen resistance. Resistant cultivars can avoid the need for fungicides.

Leaf scorch, incited by the fungus *Diplocarpon earliana*, appears as small dark purple spots on upper leaf surfaces, leaf petioles and runners. If spots become numerous, entire leaves dry up and die. Many cultivars are resistant; otherwise apply fungicides when symptoms first appear.

Powdery mildew, caused by the fungus *Sphaerotheca macularis* f.sp. *fragariae*, affects leaves, flowers and fruit, reducing yields. It is worst under high humidity conditions, such as in coastal areas, although rainy weather inhibits the disease. Powdery mildew first appears as an upward curling of leaf edges with purplish patches developing on leaf undersides, then patches of powdery white fungal strands appear on leaf undersides as the disease progresses. Fruit infected while green remain hard and fail to ripen. Choose resistant cultivars or apply a fungicide or biocontrol agent, such as AQ 510™ (www.ecogeninc.com), when disease symptoms first appear.

Photo 12-22. Leaf spot, a common foliage disease of strawberries, can reduce yields and quality of susceptible cultivars. Planting resistant cultivars can avoid the need for fungicide sprays for this disease.

Fruit rots

These destructive diseases can decay 50% or more of the berries during some seasons. Fruit rot symptoms often don't show up until fruit start to ripen, but infections often begin during bloom, as rainfall or irrigation splashes pathogens from decaying leaves or the soil onto the blossoms. Buying disease-free transplants, and removing old leaves at planting time, can prevent a new planting from starting with a huge amount of spores that cause fruit rots. Using plastic or thick straw mulch can greatly reduce these diseases by reducing fruit contact with the soil, and keeping plants well-spaced and dead leaves removed provides good air circulation and reduces spores. Avoid excess nitrogen, which causes overly vegetative plants that dry more slowly and extra-large, watery fruits more susceptible to all fruit rots. Sprinkler irrigation can spread fruit rots, so use trickle irrigation whenever possible.

Gray mold, also called "Botrytis rot" is caused by *Botrytis cinerea*, a widespread pathogen that infects dead or dying plant parts and causes decay on many crops. Prolonged cool, rainy weather can produce severe outbreaks of gray mold. The pathogen overwinters in old leaves, from which spores infect blossoms, either killing them so fruit doesn't develop, or more often remaining dormant until fruit begins to ripen. Diseased fruit areas first turn light brown, then become covered with a velvety gray growth as the fungus produces spores (Photo 12-23). No cultivar is highly resistant to gray mold, but 'Earliglow' is much more resistant than most other strawberry cultivars, and some growers have raised it successfully without using fungicides.

Removing all dead leaves from strawberry plantings before bloom can greatly reduce gray mold infections. When weather forecasts predict cool wet weather during bloom season, apply preventative fungicides starting when new leaves emerge and continuing through early bloom, with no fungicides applied during fruit ripening. Fungicides applied during fruit ripening would be too late to do much good, and could leave pesticide residues on fruit. Safer fungicides to control gray mold are now available, such as potassium bicarbonate (Kaligreen™ from www.montereychemical.com). Sprays of biocontrol agents have not proved effective, but adding the biocontrol agent sold as T-22 Planter Box™ (www.bioworksbiocontrol.com) to beehive inserts has reduced gray mold. Instruct pickers to harvest all ripe fruits from plants, and have other workers wearing gloves remove any fruit with gray mold, placing them in plastic bags to be disposed of so the spores don't spread to healthy fruit.

Anthracnose fruit rot develops most when warm, rainy weather occur during fruit ripening, so this disease is worst in Southern growing areas. Caused by several *Colletotrichum* species, it starts as small, sunken brown spots, which later develop orange spore masses (Photo 12-24). Fungicides provide only partial control. Southern growers should choose cultivars with some resistance to anthracnose fruit rot, such as 'Sweet Charlie' or 'Earliglow', and remove infected fruit as described for gray mold.

Leather rot produces tough, leathery lesions that may turn brownish red or light pink, and the entire fruit is filled with strands of the pathogen, several *Phytophthora* species from the soil, that give it a nasty taste. This rot can be a pernicious pest, because infected fruit may be hard to distinguish visually, and a single infected fruit harvested by a pick-your-own customer can ruin an entire batch of homemade jam. Raised beds and plastic or thick straw mulch prevent most leather rot, and sprays for gray mold also help.

Weeds

Strawberry plants are highly susceptible to weed competition, due to their low plant height and shallow root system. Before planting strawberries, eradicate any perennial weeds, such as nutsedge, bermudagrass or field bindweed, usually by spraying with glyphosate (Roundup™) herbicide.

For the last 30 years or so, most strawberry growers have killed many annual weed seeds by fumigating the soil before planting. (Soil fumigation does not control most perennial weeds.) Methyl bromide has been the most common fumigant, but as previously mentioned it is being withdrawn from use, and metam-sodium is taking its place. Certain

Photo 12-23. Gray mold is the most common fruit rot of strawberries. The pathogen overwinters in the dead leaves, so removing them and promptly harvesting all ripe fruit can greatly reduce the amount of fruit with gray mold.

Photo 12-24. Anthracnose fruit rot is worst in hot, humid areas such as the Southeastern U.S. Some cultivars resist this disease better than others. Day-neutral cultivars are more susceptible than spring-bearing strawberries.

annual weeds, such as burclover, puncturevine and sweetclovers, are resistant to fumigation (Table 12-3), so choose fields that are not infested with these weeds. Soil solarization can substitute for fumigation and controls some fumigation-resistant weeds such as burclover, but doesn't control other annual weeds such as purslane.

Dr. Marvin Pritts of Cornell University has developed a method of planting strawberries into a killed grass cover crop that helps smother out many weeds and reduces parasitic nematodes. Wheat reduced weeds the most without hurting strawberries in his trials, although he did not test oats, which reduced harmful nematodes the most in Massachusetts tests. (Rye greatly reduced weeds but also hurt strawberry plant growth with the allelopathic chemicals exuded by rye roots.) Plant the grains in fall before spring strawberry planting, using either spring cereal crops which will naturally die over winter, or winter cereal crops to be killed with glyphosate (Roundup™) a few weeks before planting strawberries.

Corn gluten meal, a natural byproduct of corn milling, serves as an excellent pre-plant herbicide for strawberries, preventing the germination of many annual weed seeds. The material is much more expensive than traditional pre-plant herbicides, but it could be cost-competitive with fumigation if soilborne diseases don't need to be controlled also. However, corn gluten meal does not control barnyardgrass or foxtails (giant, green, or yellow), so if any of those grassy weeds are a problem, kill them after they emerge with sethoxydim (Poast™), a post-emergent herbicide that kills only grasses.

Few post-plant herbicides are registered for use in strawberries. Where only grassy weeds present a problem, growers can use sethoxydim (Poast™) when strawberries are not flowering or fruiting. The pre-emergent herbicides DCPA (Dacthal™) and napropamide (Devrinol™) can be applied at planting, and later from late summer through early spring but not during flowering or fruiting, before weeds emerge, to control annual grasses and some annual broadleaved weeds. Oxyfluorfen (Goal™) has received special labeling in many states for fall broadleaf weed control in spring-bearing strawberries after they have hardened for winter, and provides excellent post-emergent control of winter annuals such as chickweed, henbit, and oxalis. 2,4-D Amine (Formula 40™) controls dandelions and other broadleaf weeds when applied to strawberry plants that are highly dormant, but can injure strawberries that aren't fully dormant, although this may be desired to thin plantings during renovation. Growers may kill weeds in row middles with a directed spray of glyphosate (Roundup™) or paraquat (Gramoxone™), using shields to protect strawberry rows from these chemicals which would kill strawberry plants too. Table 12-3 details the weed species controlled by each herbicide or other control measure.

Black plastic mulch or a thick straw mulch prevents many weed seeds from germinating. Some regular hand weeding is still necessary with strawberries. Instruct workers to remove from the field and destroy any weed parts that could regrow if left on moist soil in the field, such as nutsedge, purslane, bermudagrass or field bindweed. Some pick-your-own growers encourage some customer weeding by saying customers may eat berries while harvesting, but to pull a weed for each berry they eat.

Foliage-feeding insect and mite pests

Two-spotted spider mite (*Tetranychus urticae*) and **cyclamen mite** (*Phytonemus pallidus*) are serious pests of strawberries in the warm, dry climate of California and sometimes in Florida, but not usually a problem in cooler growing areas where higher rainfall washes mites from leaves and encourages fungal parasites of mites. Spider mites live mainly on the undersides of strawberry leaves, where they suck the sap, which reduces plant vigor. Continued heavy mite feeding eventually decreases yields and fruit size, and can kill plants. California growers monitor pest and predator mites regularly throughout the growing season, encourage biological control by the predators, and apply selective miticides or release predator mites when monitoring indicates these are needed. Sulfur fungicide and some insecticides used to control lygus bugs, aphids or other pests can destroy predator mite populations and lead to spider mite outbreaks, so avoid using such pesticides until late in the season. Day-neutral cultivars are generally more susceptible to mite damage than short-day cultivars, and cultivars within each group differ in susceptibility (See Tables 12-1, 12-2, pages 150-152), so growers might choose the more resistant cultivars where mites are especially damaging.

Aphids, including the **strawberry aphid** (*Chaetosiphon fragaefolii*) and the **melon aphid** (*Aphis gosypii*), also feed on plant sap from the undersides of leaves. These pests can reduce plant vigor if populations get high, but often do little damage unless they transmit viruses as they feed. Using virus-free transplants and destroying old plantings and wild strawberries reduces the risk of virus transmission. Aphids also secrete a sugary honeydew onto leaves and fruit which serves

Table 12-3.
CONTROL MEASURES FOR COMMON WEED SPECIES*

	Preplant Treatments						Postplant Treatments			
	Soil Solarization	Metam-Sodium (Vapam)	Methyl Bromide	Paraquat (Gramoxone)	Glyphosate (Roundup)	Corn Gluten Meal	DPCA (Dacthal)	Napropamide (Devrinol)	Sethoxydim (Poast)	2,4-D Amine
Perennial Broadleaf Weeds										
Clovers	XXX	XXX	XXX	X	XX	XX	0	0	0	XXX
Curly dock	0	0	X	0	XX	0	0	0	0	XXX
Dandelion	X	X	XX	0	XX	0	0	0	0	XXX
Field bindweed regrowth	X	X	0	X	XXX	0	0	0	0	XXX
Field bindweed seedling	XXX	0	0	XXX	XXX	XX	0	0	0	XXX
Thistles	X	X	XX	X	XX	0	0	0	0	XXX
Perennial Grasses & Sedges										
Bermudagrass regrowth	XX	XX	XX	X	XXX	0	0	0	XX	0
Bermudagrass seedling	XXX	XXX	XX	XX	XXX	X	XXX	XXX	XXX	0
Quackgrass	XX	XX	XX	X	XXX	0	0	0	XX	0
Yellow nutsedge	X	XX	XXX	X	XXX	0	0	X	0	X
Annual Broadleaf Weeds										
Burclover	XXX	XX	0	XXX	XX	XX	0	0	0	XXX
Burning nettle	XXX	XXX	XXX	XXX	XXX	XXX	XXX	XX	0	XXX
Chickweed	XXX	XXX	XXX	XX	XXX	XXX	XXX	XXX	0	X
Groundsel	XXX	XXX	XXX	XXX	XXX	XXX	0	XXX	0	XXX
Horseweed	XXX	XXX	XXX	XXX	XXX	XXX	0	0	0	XXX
Lambsquarters	XXX	XXX	XXX	XX	XXX	XXX	XXX	XXX	0	XXX
Little mallow	XXX	XX	0	0	XXX	XXX	XX	XXX	0	XXX
Oxalis	XXX	XXX	XXX	XXX	XXX	XX	X	X	0	XX
Pigweeds	XXX	XXX	XXX	XXX	XXX	XXX	XXX	XXX	0	XXX
Pineappleweed	XXX	XXX	XXX	XXX	XXX	XXX	XX	0	0	XXX
Puncturevine	XX	XX	XX	XXX	XXX	XXX	XX	XX	0	XXX
Purslane	XX	XXX	XXX	XXX	XXX	XXX	XXX	XXX	0	XXX
Sheperd's purse	XXX	XXX	XXX	XXX	XXX	XXX	X	X	0	XX
Sowthistle	XXX	XXX	XXX	XXX	XXX	XXX	XXX	XXX	0	XXX
Sweetclovers	0	XX	0	XX	XXX	XX	0	XX	0	XXX
Annual Grass Weeds										
Annual bluegrass	XXX	XXX	XXX	XXX	XXX	XXX	XXX	XXX	XXX	0
Barnyardgrass	XXX	XXX	XXX	XXX	XXX	XXX	XXX	XXX	XXX	0
Crabgrass	XXX	XXX	XXX	XXX	XXX	XXX	XXX	XXX	XXX	0
Foxtails	XXX	XXX	XXX	XXX	XXX	X	XXX	XXX	XXX	0
Ryegrasses	XXX	XXX	XXX	XXX	XXX	XXX	XXX	XXX	XXX	0
Volunteer grains from mulch	XXX	XXX	XXX	XXX	XXX	XXX	XX	XXX	XXX	0
Wild barley	XXX	XXX	XXX	XXX	XXX	XXX	XXX	XXX	XXX	0

0 = no control, X = fair control or topkill only, XX = partial control, XXX = excellent control

* Adapted and expanded from *Integrated Pest Management for Strawberries*, Univ. of California pub. 3351, and *Midwest Small Fruit Pest Management Handbook*, Ohio State Univ. bull. 861.

as a food substrate for a black fungus called sooty mold, which disfigures fruit and renders them unmarketable. Growers monitor populations of aphids and their predators, including lady beetles and lacewings, each week all season, and apply an insecticide or release predators when aphid populations exceed thresholds. In some areas, such as California's central coast and many East Coast states, aphids rarely reach damaging levels and treatments usually aren't recommended. If aphids are allowed to build up in spring in such areas, they attract general predators that control lygus bugs and spider mites later on.

Fruit-feeding insect pests

Lygus bugs (*Lygus hesperus*) feed on strawberry fruit directly, causing a distorted growth of berries called catfacing (Photo 12-25). If catfacing occurs early in the season, it might be caused instead by cold weather during pollination or early fruit development, whereas lygus bug populations increase during the growing season and are most damaging to day-neutral strawberries that ripen in summer and fall. This pest is most troublesome in the West. Legume crops or weeds can produce large numbers of lygus bugs, so avoid growing clovers as a groundcover near strawberries, disk in legume cover crops before they flower, and control weeds such as chickweed, wild mustards, curly dock and smartweed by early March.

Monitor lygus bug populations and degree-day accumulations to determine if and when treatments are needed. Most insecticides effective against lygus bugs also harm natural enemy populations, and can cause spider mite outbreaks. Some strawberry growers use tractor-mounted vacuum devices to remove about half the lygus bug nymphs; the vacuums also reduce beneficial spiders and bigeyed bugs, but aren't so hard on predator mites, lacewings and other predators. Both vacuum devices and available pesticides are most effective on the youngest lygus bug nymphs and not effective on adults, so if control is necessary, treat as soon as the nymphs appear on strawberries.

Tarnished plant bugs, the closely related *Lygus lineolaris*, also feed on fruit and cause catfacing, and also build up on legume crops and weeds. This pest causes the most damage in the Northeast, especially to day-neutral cultivars ripening later in the season after bug populations have built up. 'Honeoye' has shown some resistance in New York trials, with only 10% early-season damage and 25% late-spring damage in unsprayed, weedy plots in 1997, when 'Earliglow' and day-neutral cultivars had 20% early damage and 40% late damage. During that same 1997 season, tarnished plant bug damage on 'Honeoye' not treated with insecticides reached only 2 to 5% on three Iowa strawberry farms. Vacuum treatments and biological control sprays with *Beauveria bassiana* (Mycotrol™) have not reduced tarnished plant bugs below economic thresholds. Avoiding nearby legumes and weeds, and choosing resistant cultivars or using insecticides, are key control measures where this pest is a problem.

Strawberry sap beetles (*Stelidota germinata*) and **picnic beetles** (*Glischrochilus species*) eat cavities in the fruit, and disseminate fruit-rotting organisms from diseased to healthy fruit. Prompt harvest of ripe fruit, along with removing and destroying all damaged or diseased fruit, can prevent much damage from these beetles. Annual plasticulture reduces these pests, as does annual renovation of perennial beds after harvest and removing dead foliage in early spring. Early-ripening cultivars and those that produce firm fruit tend to escape attack from these beetles.

Strawberry clipper weevil, also known as strawberry bud weevil (*Anthonomus signatus*), was considered a major pest by many Eastern strawberry growers, but recent research has refuted this. Adult female weevils lay eggs in strawberry buds, then clip the bud stalks, so these buds never open or develop into fruit. Strawberry plants compensate for this damage, however, by maturing more of the other buds, and producing larger fruit. Most years, strawberry clipper weevils actually act as beneficial thinning agents that increase fruit size.

Photo 12-25. Lygus bugs and other plant-eating bugs cause "cat-facing" damage to strawberry fruit. Avoid growing legume crops, which harbor such bugs, near strawberries.

Slugs and snails

These mollusk pests also feed directly on the fruit, and can be very destructive. Damp, cool conditions favor them, so they are worse in dense matted rows with thick straw mulch than in plasticulture hill-system plantings. Perennial strawberry producers can reduce slug problems by spacing the first runners and removing excess runners, avoiding excess mulch during harvest, and renovating plantings after harvest.

Matted-row strawberry growers frequently must also apply a granular or pelletized toxic bait to control slugs or snails (Photo 12-26). Metaldehyde, the slug bait usually used, is also toxic to mammals who injest it, posing a potential threat to young children with pick-your-own customers who might pick up pellets and put them in their mouths. Therefore, if applying metaldehyde pellets to pick-your-own plantings, place them under the straw mulch. A newly available substance, iron phosphate (Sluggo™ from www.montereychemical.com) has proven just as toxic to slugs but is not toxic to mammals.

Deer, birds and other vertebrate pests

Where deer populations are high, they can destroy strawberry plantings by eating all the foliage. Growers in such areas have found the only solution to be erecting a slanted electric fence.

Birds may attack ripening fruit, and in some areas cause much damage. Growers have used netting or rotating scare devices, discussed in chapter 9, to reduce bird damage. The bird repellent chemical methyl anthranilite, derived from 'Concord' grapes, is not currently registered for use on strawberries but may receive registration in the next few years.

Ground squirrels sometimes become serious pests in California by feeding on ripening fruit. Control small populations by placing baited traps or Conibear traps over burrow entrances. Larger populations may require placing bait stations every 150 to 200 feet around the strawberry field, using grain treated with a slow-acting rodenticide.

Meadow voles (field mice) reach cyclic population peaks every 4 to 7 years, when their potential for causing damage is greatest. Meadow voles thrive in pastures or grassy roadsides where a thick mat of grassy vegetation provides food and cover, so eliminating such habitats near the strawberry field will reduce damage. Encourage predators, including hawks, owls and cats, by providing habitats for them. If field mice build up and cause problems, use poison baits in late winter, to control populations before spring breeding.

Photo 12-26. Slugs do more damage under cool, moist conditions, so are worse with straw mulch than plastic mulch. Growers often need to apply granular or pelletized slug bait

HARVEST AND POST-HARVEST MANAGEMENT

Harvest each section of a strawberry planting every 2 to 3 days throughout the season, which can last for 3 to 12 weeks for short-day cultivars, and 3 to 9 months for day-neutral cultivars. Pick-your-own growers give customers flags to mark their stopping points, so all areas get picked. Instruct workers or customers to pick all ripe berries regardless of size, to pick fruit with the caps on but without fruit stems, and to remove any cull berries and place them in plastic bags for disposal. It's best to have designated workers remove any rotten berries, wearing gloves, so the disease spores aren't transferred to healthy fruit.

Six hired workers can harvest each acre of strawberries over the season. The average worker harvests 10 to 12 quarts (20 to 24 pints, or 14 to 18 pounds) of berries per hour, and is paid 22% to 33% of the selling price. Pick-your-own growers need 350 to 400 customers per acre of strawberries. Pick-your-own customers usually pay about two-thirds the price of pre-picked fruit.

Both workers and customers harvest strawberries directly into the shallow containers they are sold in (Photo 12-27), to minimize handling and bruised of the fruit. Wholesale strawberry containers are usually covered with plastic to provide modified atmosphere packing, which increases shelf life by a few days. Adding E-2-hexenal, a natural volatile compound that contributes to the aroma of fresh-mown grass, to the storage atmosphere will kill many fungal spores without leaving any residue or off-flavor on the fruit. Rapidly cooling the harvested berries to 32 F with a forced-air cooling system will preserve fruit quality and minimize post-harvest fruit rots.

RENOVATING STRAWBERRY PLANTINGS

Most perennial, matted-row growers fruit short-day cultivars for 3 to 5 years, and day-neutral cultivars for 2 years. Plasticulture growers may fruit short-day cultivars for 2 or occasionally 3 years. Plasticulture day-neutral strawberries are generally not kept past 14 months, since they have produced fruit for 8 to 10 months within that time.

Renovating short-day cultivars in matted rows. After the spring-bearing strawberry harvest ends, growers should renovate perennial plantings to keep them productive the following year. If broadleaf weeds are a problem and the planting is thick, treat with 2,4-D amine after harvest to kill the weeds and reduce runnering. If more runners will be needed to fill gaps and few broadleaf weeds are present, skip this treatment. If grassy weeds are a problem, use sethoxydim (Poast™) also. One week after any herbicide treatment, mow 2 to 3 inches above the crowns to remove old leaves, and rake mown leaves to the aisles where cultivation will shred them to minimize leaf diseases. Then till up plants along the edges of raised beds, leaving a row width of 10 to 14 inches. Where growers use flat beds, they may till up the center of the rows to remove the oldest crowns, allowing new runners along the edges to form new rows between the old rows. Fertilize lightly with nitrogen at 25 to 35 pounds per acre actual N, cultivate between beds to destroy old leaves and throw soil onto bed edges, irrigate, and spread clean straw mulch around the strawberry crowns.

As the plants regrow, they again send out runners. Hand-set the first ones in bare spots, anchoring them with mulch, and remove the later-forming ones so the planting doesn't get too crowded. Some growers have modified springtooth harrows to remove weakly-rooted new runners but leave large plants, to minimize the hand labor needed for runner removal. The fewer runners needed, the less nitrogen should be applied at renovation. Instead, apply more nitrogen in early fall when fruit buds are forming, to increase next spring's yields.

Avoiding crowding and keeping the plants vigorous through renovating are the keys to keeping a spring-bearing strawberry patch productive for several years. Well-managed perennial strawberries produce good yields for about four or five years; after that, it's time to start a new planting.

Options for renovating day-neutral cultivars in matted rows. Like spring-bearers, perennial day-neutral strawberries also grow too crowded after a couple of years, but often can't be renovated in early summer because of their continued production. Many growers therefore keep them for only one or two seasons, and keep planting more in new locations. To keep a planting longer, pull out excess runners to prevent them from getting too dense. Set new runners and pull out old crowns after a couple of years, because as day-neutral plants age they form multiple crowns, which produce very small fruits due to within-plant competition. Fertilize day-neutrals during the lull between the spring and early summer harvests, or between the summer and fall harvests.

Photo 12-27. Customers or workers harvest strawberries directly into the shallow containers they are sold in to minimize handling and bruising.

An alternative production system for perennial day-neutral strawberries could be to renovate in spring, by mowing them off and fertilizing lightly with nitrogen as they begin to flower. This would sacrifice the spring day-neutral crop, but when growers also raise spring-bearing cultivars, the more rot-prone day-neutral spring crop might not be missed. Early spring renovation of day-neutral cultivars should take some stress off the plants and induce them to produce a larger summer and fall harvest. Autumn is generally drier, reducing rot and slug problems, and the cooler days sweeten the fall crop. Eliminating the spring crop to favor the fall crop hasn't been done much on day-neutral strawberries, because many growers treat them as annuals, but is standard practice when growing everbearing raspberries such as 'Heritage', and may prove profitable in some areas.

Where summer heat limits day-neutral strawberry fruiting, partial shade in early afternoon, white plastic or heavy straw mulch, and mid-day sprinkling can cool the plants, coaxing a longer fruiting season as long as the temperature is kept below about 85 F.

CARRYING OVER PLASTICULTURE PLANTINGS FOR A SECOND YEAR

Only carry over healthy plantings for a second year. If anthracnose or any other plant-killing diseases are present in a planting, destroy it. Some growers remove the strawberry plants and plant a melon or vegetable crop directly into the plastic mulch in order to utilize it for double cropping.

After harvest and before July 1, broadcast spray 2,4 D amine over plantings to carry over, to reduce runnering and control broadleaf weeds. Two weeks later, mow berry plants 2 inches above the plastic mulch. Check sample plants for the number of branch crowns, and if more than 5 per plant, thin them out to reduce competition and increase fruit size next year. Crown thinning can be accomplished by spraying paraquat (Gramoxone™) with a shielded sprayer that exposes the outer half of plants on each row on the beds to the chemical.

Trickle irrigate the planting the day after crown thinning, and minimally as needed throughout the summer, to keep the plants alive but slow growing. Do not fertilize during renovation or during the summer, because runner plants are not desired. Starting in early fall, fertigate through the trickle system as leaf analysis indicates is needed, to build large healthy crowns.

Spray aisles between rows with paraquat (Gramoxone™) in mid-fall, using a shielded sprayer to protect the beds, to kill weeds and rooted runners between the beds. Then treat the planting like a first-year planting, applying winter and frost protection and preparing for harvest.

Further Reading

Dale, A. and J.J. Luby, eds. 1991. The Strawberry into the 21st Century. Timber Press, Portland, OR.

O'Dell, C.R. and J. Williams. 2000. Hill system plastic mulched strawberry production guide for colder areas. Virginia Coop. Ext. Pub. no. 438-018. Downloadable from: www.ext.vt.edu/pubs/fruit/438-018/438-018.html

Pritts, M., T. Crocker and C. Chandler, eds. 1996. Proc. IV North American Strawberry Conference. Timber Press, Portland, OR.

Strand, L.K. 1993. Integrated pest management for strawberries. Univ. of Cal. pub. 3351.

Strawberry IPM Update current and past issues are available at: www.exnet.iastate.edu/Pages/plantpath by selecting "newsletters".

Chapter 13
Raspberries and Blackberries

Raspberries and blackberries, also called brambles or cane fruits, are both members of the genus *Rubus* in the rose family. Both are cup-shaped aggregate fruits, composed of many fused drupelets, with each drupelet a tiny fleshy fruit with a stony endocarp surrounding the single seed. Each aggregate fruit, popularly called a berry although not botanically a berry, surrounds a fleshy white cone called a torus or receptacle. When ripe, each raspberry separates freely from the torus and has a hollow center, while with blackberries the torus separates from the plant and remains inside the blackberry (Photo 13-1).

Blackberries turn dull black when ripe, while diverse species of raspberries may be red, purple, black or yellow when ripe. Various bramble species are native and are cultivated throughout North and South America, Europe, and Asia. Various *Rubus* species have ploidy levels (number of sets of chromosomes) ranging from diploid (two sets of chromosomes) to dodecaploid (twelve sets of chromosomes), but most crossing barriers can be overcome. Distinct *Rubus* species often cross naturally in the wild and have been crossed by breeders, producing many hybrid species.

Brambles have perennial roots and crowns that grow biennial stems, called canes, which are usually thorny. Canes grow vegetatively the first year and are called primocanes. The second year, these canes flower and fruit and are called floricanes (Photo 13-2). After fruiting, the floricanes die; new primocanes grow which will bear fruit the next year. The upper halves of primocanes can flower and fruit in the autumn of their first year in some cultivars

Photo 13-1. A harvested raspberry (left) is hollow, with the receptacle or torus left on the plant, while a harvested blackberry (right) retains the receptacle within it.

Photo 13-2. Primocanes, the first-year bramble canes, grow vegetatively (up the middle of this row) and do not fruit except in primocane-bearing red and yellow raspberries. The next season, these canes become floricanes, which flower, fruit (along the edges of this V-shaped trellis) and then die.

of red and yellow raspberries, called fall-bearing or primocane-bearing cultivars.

Blackberry canes are generally stouter and armed with larger thorns than raspberry canes, but some cultivars of blackberries have been selected or bred to be thornless. All raspberries have some spines, though these are minimal in some cultivars. Most bramble canes are upright or semi-upright, but some blackberries have long, trailing canes that are grown on trellises.

Raspberries and blackberries are among the most perishable of fruits, and are highly prized in season, commanding high prices for fresh fruit. Many brambles are also grown for processing into frozen berries, jams, and syrups.

BRAMBLE SPECIES AND ADAPTATIONS

Most raspberries are well adapted to moderate to cold winters and cool summers, while excessive heat seriously threatens plant survival and yields. Blackberries are less cold tolerant, but thrive with warm to hot summers. Blackberries offer much larger fruit size and longer shelf life but also have a stronger flavor and larger "seeds" (true seeds enclosed by stony endocarp) than raspberries. Each bramble species and cultivar offers its own unique flavor, and certain cultivars have developed a reputation for fine fresh fruit and processed products that keeps them in demand.

Red Raspberries

Cultivated red raspberries arose mainly from two diploid subspecies of *Rubus idaeus*, the native red raspberry of North America and eastern Asia (*R. idaeus strigosus*) and the native red raspberry of Europe (*R. idaeus vulgatus*). Older literature lists these as two separate species, *R. strigosus* and *R. vulgatus*, but they intercross with little or no sterility in the resulting seedlings and are now considered subspecies. Some cultivars are tetraploids, with four sets of chromosomes instead of the two in diploids. The American subspecies offers greater winter hardiness, to USDA zone 3 or 4, and most American cultivars have a greater percentage of this species in their ancestry than do most European cultivars. Red raspberries are the most heat-sensitive type of raspberry, requiring relatively cool summers.

Red raspberry breeders have incorporated other species into some cultivars to achieve greater environmental tolerance. The Arctic raspberry (*R. arcticus*) contributes extreme winter hardiness, to zone 2, and early primocane fruiting to some commercial cultivars. Two Asiatic species, *R. kuntzeanus* and *R. parvifolius*, provide tolerance to summer heat and fluctuating winter temperatures in cultivars bred for warmer climates, such as 'Southland' and 'Dormanred', but these cultivars do not have the fruit quality of other red raspberry cultivars. Genes from the black raspberry, *R. occidentalis*, have also been incorporated into red raspberry cultivars through crossing a red with a purple raspberry and then backcrossing to the red. These genes provide greater fruit firmness, later ripening, resistance to fruit rots or resistance to the aphid vector of raspberry mosaic virus.

Red raspberries grow mainly upright, with little branching of canes. Shorter cultivars may be grown without a trellis (Photo 13-3), or with temporary string trellises to hold fruiting canes up during harvest, while cultivars with longer canes require a trellis. Red raspberries increase by suckering, sending new canes up from the roots.

Worldwide, Russia and eastern European countries produce the greatest amount of red raspberries. In North America, the Pacific Northwest region (northern California, Oregon, Washington and British Columbia) raises the greatest amount of commercial red raspberries, much of it processed by freezing or into jam. Small retail growers produce red raspberries in the Northeast U.S. and Eastern Canada, as well as in New Zealand, Australia, Chile and northern European countries.

Yellow Raspberries

Most yellow raspberry cultivars are single-gene variants of red raspberries, with the same growth habit and climate adaptation. Yellow-fruited single-gene variants of black raspberries also occur, but haven't been propagated as cultivars. Some yellow raspberries result from crossing *R. idaeus* with

Photo 13-3. These red raspberries and some blackberry cultivars grow sufficiently upright that no trellis is needed.

Asiatic species. Many of the yellow-fruited cultivars offered by nurseries are primocane-fruiting fall bearers.

Yellow raspberries have novelty value for retail growers, and tend to be very sweet. Birds can't easily distinguish yellow from green, so the fruit often suffer less bird predation than other colors of raspberries.

Black Raspberries

Most black raspberry cultivars are derived almost entirely from *R. occidentalis* L., a diploid species native to eastern North America. A few black raspberries result from crosses of this species with a purple raspberry, with a black-fruited raspberry native to the Pacific Northwest, or with a subtropical black-fruited Asiatic raspberry.

Black raspberries are not as cold hardy as red raspberries, generally only to USDA zone 5, and have only slightly greater heat tolerance, so thrive in an even narrower climate range. Black raspberries suffer from virus diseases and anthracnose more than red raspberries, so often plantings only yield well for 2 to 3 years before they must be destroyed. They produce lower yields than red raspberries, of fruit with a stronger taste that is prized by the food industry as a natural dye. Black raspberries are grown commercially only in North America, with the Pacific Northwest producing 90% of the crop, mainly for processing into jams, yogurts and ice cream.

Black raspberries branch freely and the branches arch down to the ground and root at the tips, producing new canes. Nurserymen propagate cultivars by these "tip layers". Pruned canes remain erect and do not generally need a trellis, although black raspberries grown for processing may be trellised to facilitate mechanical harvest.

Purple Raspberries

Purple raspberries arose as natural or human-created crosses between black raspberries and red raspberries. Some of the best purple raspberries today are hybrids between the original purples and red raspberries, so they are ¾ red and ¼ black, sometimes called red-purple raspberries. Purple raspberries demonstrate hybrid vigor: the plants are more vigorous than either red or black raspberries, and produce higher yields of much larger fruit. The best cultivars offer the cold tolerance of red raspberries (to zone 4) with the greater heat tolerance of black raspberries.

Purple raspberry acreage is very small, mainly in North America, and mostly for local retail sales. Some New York growers raise purple raspberries for processing. Pick-your-own purple raspberries can be very profitable in areas such as the Midwest, where summers are too hot for red raspberries and black raspberries don't yield well.

Purple raspberries grow like black raspberries, with arching canes, but yield best when the longer, more vigorous canes are trellised rather than pruned back severely.

Erect and semi-erect blackberries

Upright and semi-upright blackberries descended from natural and human hybridization of erect species native to North America, such as *R. allegheniensis*, *R. argutus*, *R. setosus* and *R. frondosus*, with some use of European blackberry species and descendents of a cross between a dewberry and a red raspberry. The erect, sometimes arching, canes are almost always heavily armed with large spines, although breeding programs have developed erect thornless cultivars (Photo 13-4). Erect blackberries are tetraploid, and must carry four recessive alleles of the gene for spinelessness to express this trait, so a cross between a thorny cultivar and a thornless cultivar yields only 1 thornless plant for every 16 offspring.

Erect blackberries are very heat tolerant and are widely grown in the Midwestern and Southeastern U.S. Some thorny cultivars are cold tolerant to USDA zone 5 and are grown in the Northeast, mainly for processing. Erect blackberries produce

Photo 13-4. Blackberry breeders at the University of Arkansas have developed thornless erect blackberries.

heavy yields of large, elongated, sweet fruit. Thorny cultivars produce the largest fruit and highest yields, and are machine-harvested for processing. Pick-your-own growers plant thornless erect cultivars in USDA zones 6 and warmer, and thornless semi-erect cultivars in zones 5 and 6.

Erect blackberries produce many primocanes from crowns and root suckers, and also produce arching branches which often root at the tips, forming tip layers. They are therefore easy to propagate, but fruiting plantings require much pruning and control of unwanted plants. The biggest advantage of the fully upright blackberry cultivars is saving the expense of a trellis.

Trailing blackberries

Most of these prostrate-caned blackberries developed from species native to or naturalized in the Pacific Northwest of North America, largely from *R. ursinus*, *R. macropetalus*, and *R. loganobaccus*, and the Pacific Northwest remains the major world production area for trailing blackberries. Individual cultivars have become famous, particularly 'Marion', which accounts for half the acreage; Oregonians call them "Marionberries" rather than blackberries. Two other cultivars grown in the Pacific Northwest were selected from European species, 'Himalaya' from *R. procerus* and 'Thornless Evergreen' from *R. lacinatus*. Trailing blackberries grown in the Southeastern U.S., called dewberries, derive mainly from a heat-tolerant species native to that area, *R. trivialis*.

Trailing blackberries produce very vigorous, flexible canes that require trellising or staking. In nature, they propagate by tip-layering rather than suckering as erect blackberries do. The large, roundish to elongated, wine-colored to black fruit have a distinctive flavor. Most cultivars are hardy only to USDA zone 7 or 8.

Raspberry-blackberry hybrids

Private fruit breeders in the U.S. created some of the most famous bramble cultivars by crossing trailing and erect blackberries with red raspberries. 'Boysenberry', 'Loganberry' and 'Youngberry' all originated this way between 1880 and 1930. Scottish breeders created 'Tayberry' in the same way in 1970. 'Loganberry' and 'Tayberry', hexaploids with six chromosome sets, derived from crosses between an octoploid trailing blackberry (*R. vitifolius*) and a tetraploid red raspberry (*R. idaeus*). 'Boysenberry' and 'Youngberry', septaploids with seven sets of chromosomes, resulted from crossing such hexaploid hybrids with an octoploid erect blackberry, *R. baileyanus*.

Like trailing blackberries, these hybrids produce vigorous, prostrate canes that require trellising. Most are hardy only to USDA zone 7, or zone 6 with winter protection of primocanes. The very large, reddish-black fruit are prized for their individual flavors.

Growers observed thornless "sports" (mutations) of 'Boysenberry', 'Loganberry' and other cultivars. Nurserymen soon found that such sports must be propagated by tip layering, for root suckers or root cuttings of such thornless blackberries produced thorny plants. Breeders who used the thornless sports as parents discovered that they produced only thorny offspring. Anatomical investigations revealed that these thornless blackberry sports were periclinal chimeras, with smooth stems developing from the outer epidermal layers of cells, while the internal tissues that produce the roots and gametes (sperm and egg) retained the genes for thorns. Growers must remove all thorny shoots that come up from root suckers in plantings of these thornless sports, since the thorny shoots are more vigorous and can outgrow the thornless canes, causing the latter to die out.

CROP USES AND PRODUCTION SYSTEMS

Brambles grown for processing are machine harvested, while brambles grown for the fresh market are hand-harvested by hired workers or pick-your-own customers. Mechanically harvested plantings are often mechanically pruned also. Historically, the two greatest expenses of growing brambles have been hand-picking the numerous small fruits, and hand-pruning the canes which die each year, so mechanization has greatly reduced costs. The growing system depends on the harvesting and pruning methods, cultivars and training methods.

Freestanding, biennial cane, hand-pruned system

This system is used for erect blackberries and some shorter-cane raspberries of all types, which are hand-harvested for fresh market. Growers plant transplants 2½ to 4 feet apart in hedgerows 10 to 12 feet apart, in fall or spring, with no trellis (Photo 13-5). The plants grow vegetatively for a season, growers prune back the canes in summer and early spring, then the floricanes flower and fruit in the second season. After harvest, workers hand-prune out the canes that just fruited, since they will die. New primocanes grow to replace the removed canes, workers prune them back, and they bear the next season.

Photo 13-5. Erect brambles such as these 'Heritage' red raspberries are grown in freestanding hedgerows. Workers must hand-prune out dead floricanes, seen here, after they fruit in summer on raspberries and blackberries that bear only on floricanes, or for both a summer and fall crop of cultivars such as 'Heritage' that also bear on primocanes.

This system requires less initial cost per acre than a trellis system. Erect thornless blackberries tend to grow only semi-erect in the first season, and may require tying canes to temporary bamboo poles for the first harvest (Photo 13-6), but the poles can be removed after that. Summer tipping keeps black raspberries more upright without a trellis. Red and yellow raspberries may require temporary twine, stretched on both sides of each row between stakes at the ends of each row, to hold up fruiting branches during harvest. Using this system for primocane-fruiting red and yellow raspberries produces two harvest seasons each year, in early summer and in fall.

This system does require considerable hand pruning, which is expensive. In addition to pruning out the dying floricanes just after harvest, workers must thin out the primocanes and head back their branches to keep plants upright.

Freestanding, biennial cane, machine-hedged system

This system is used for erect blackberries and some erect summer-bearing raspberries of all types, that are hand-harvested or machine harvested. Growers establish plantings as in the system above. Workers top primocanes during summer, by hand pruning or hedging machines, at 3 to 4 feet tall to keep them upright and stimulate branching. The subsequent branches are machine hedged in late winter (Photo 13-7). Old floricanes are not removed after fruiting, but are left in the rows to add support to new primocanes, and eventually the dead canes break off and decompose.

This system greatly reduces hand-pruning expenses, but risks greater cane disease problems by not promptly removing floricanes after fruiting. This system is most popular among erect thornless blackberry growers in the Pacific Northwest, where lower rainfall reduces anthracnose problems. Old floricanes do not bother hand harvesters of this fresh-market crop, so long as they are thornless.

Photo 13-6. Erect blackberries grow in freestanding hedgerows from the second season on, but often need temporary support the first season by tying canes to bamboo stakes.

Photo 13-7. Machine hedging of erect blackberries for processing can eliminate hand pruning, although it risks increased cane diseases. Growers in the Southeast mow plantings near the ground every 5 to 6 years to eliminate old floricanes. (Photo by Justin Morris, Univ. of Arkansas.)

This system is also used for machine-harvested thorny erect blackberries, grown for processing in the Southeast. Every 5 to 6 years, growers in Texas, Arkansas and other Southern states mow these blackberry plantings at 3 to 4 inches off the ground immediately after harvest, and remove all mowed canes from the field, to prevent a buildup of canes diseases. The following year, canes will be shorter and can not be machine-harvested, but the smaller crop can be hand-harvested if sufficient labor can be found. Southern Florida erect blackberry growers mow canes to the ground after harvest each year, since the longer post-harvest growing season allows new primocanes to fully develop and produce a good crop the next year.

Freestanding, annual cane, machine-pruned system

This system enhances fall production of primocane-bearing red and yellow raspberries, and allows mechanization of pruning, by sacrificing the spring crop. The greater fall crop does not yield as much as a summer crop and fall crop combined, but fall crops generally command a higher fresh-market price because of the unusual season, and mechanization of pruning cuts production costs. This system is usually used for hand-harvested, fresh market fall raspberries, but may also be used for fall-bearing red raspberries that are machine-harvested for processing.

Growers set transplants closer, 1½ to 2 feet apart in hedgerows 8 to 10 feet apart, in fall or spring. Plants grow vegetatively through spring and summer, then bear a fall crop. Just before growth begins the following spring, growers mow off all canes a few inches above the ground, and remove the canes from the field to avoid any disease carryover. New primocanes will grow and fruit the next fall. This system generally avoids the need for temporary twine to hold up fruiting branches.

Adding a second mowing to this growing system can enable growers in the upper South and lower Midwest to raise fall-crop raspberries. Without mowing, the "fall" crop begins ripening during the hottest part of summer in these areas, causing limited yields of very soft, highly perishable fruit. Mowing in early spring, and again when canes are about 1 foot tall, will delay fruiting until early fall and improve fruit quality. Portuguese researchers have demonstrated that this double mowing also increases winter production of primocane crops in unheated plastic hoop houses in warm-winter regions. Such off-season raspberries can command high prices in some areas.

Freestanding or trellised, biennial cane, alternate-row machine-pruned system

Mechanized pruning has now been adapted for purple and red raspberries that bear a summer crop on floricanes only. This system may be used for both hand harvesting and machine harvesting. Growers plant as in the erect, biennial cane, hand-pruned system above. After the first crop in the second summer, growers mow every other hedgerow at about 12 inches off the ground. This eliminates the floricanes that will die, and prunes back the primocanes, greatly reducing the crop they will bear next year. The remaining half of the hedgerows are hand-pruned that first bearing season, removing floricanes at ground level and only lightly heading back primocanes, so they will produce a full crop the next year. This system reduces overall pruning costs by almost half in the first bearing season.

The next year after harvest, growers mow the hedgerows that were hand-pruned the previous year. The hedgerows that were mowed the previous year should be hand-pruned to remove the dead floricanes if the crop will be hand-harvested. For a machine-harvested crop where cane diseases aren't a problem, hand pruning can be eliminated, as the dead floricanes will be removed by mowing the following year, and meanwhile they may help support the new canes.

Many alternate-row growers use a primocane-suppressing chemical, Ignite™, on the fruiting rows in early spring, from the time primocanes are 4 to 6 inches tall until fruiting canes bloom. This keeps down primocanes that would otherwise interfere with harvest, especially machine harvest.

This alternate-row mowing system does reduce per-acre yields by 10 to 33%, depending on the cultivar, but the cost savings may offset the yield loss. New York trials showed that 'Titan' red raspberries produced 89% the yield of hand-pruned plots, while 'Royalty' purple raspberries produced 67% as much as hand-pruned plots. These summer-bearing red and purple cultivars have taller canes that are less upright than fall-bearing cultivars, and the lack of tall primocanes to help support the floricanes often necessitates a trellis. When a trellis is needed, posts and wires must be removed before mowing and replaced afterward when using a common mower such as a BushHog™, or a special mower must be purchased that can be operated between trellis posts.

Trellised, biennial cane, hand-pruned system

Trailing blackberries, raspberry-blackberry hybrids and semi-erect thornless blackberries require a trellis, and many purple raspberries and summer-bearing red raspberries yield best when trellised. Numerous trellis systems have been devised, some suited for hand harvesting and some for machine harvesting. These trellis systems require hand pruning of old floricanes for every-year production.

Trailing blackberries, raspberry-blackberry hybrids and red raspberries in the Pacific Northwest are generally trained to a 2- or 3-wire **vertical trellis**, with the top wire at 5 ½ feet off the ground and the lower or middle wire at 4 feet off the ground, and sometimes a third wire about 2 ½ feet off the ground (Photo 13-8). This system works well for machine harvesting for processing, as done on 98% of the acreage, as well as for hand harvesting for fresh market. Primocanes grow very long and sprawl along the ground until they are tied to the trellis in August for the summer-trained system, or they are tied to the trellis after they become floricanes the next February in the winter-trained system. Summer-trained plants produce about 46% greater yield than winter-trained plants through an increased percentage of lateral buds developing into shoots, and greater fruit production per lateral shoot. However, growers in areas with colder winters often delay training of Marionberries and hybrid berries until late winter, to protect these cold-sensitive canes from winter damage. Regardless of the timing, workers tie the canes in bundles and weave them around the wires, bringing half the canes from one direction and half from the other. Immediately after harvest, floricanes are pruned out by hand.

Southeastern growers of dewberries, Boysenberries and other trailers have traditionally practiced winter training to **wooden posts** (Photo 13-9). They let primocanes sprawl on the ground, cover them with straw mulch to protect them over winter if need be, then wrap the floricanes around the posts in early spring. This system must be hand harvested.

The **V-trellis** provides the best light interception for summer red and purple raspberries, which can increase yields in many areas. V-shaped end posts, created by installing fence posts at angles or welding two metal pipe pieces together, anchor two wires on each side of each row (Photo 13-10). Place the wires 5 to 6 feet and 3 to 4 feet off the ground, and 3 to 4 feet apart for the top wires and 2 to 2½ feet apart for the bottom wires.

Photo 13-8. The most popular bramble trellis is a 2- or 3-wire vertical trellis, especially when machine harvesting for processing, since the harvester straddles the row.

Photo 13-9. Dewberries, trailing blackberries native to the Southeast, have traditionally had floricanes wrapped around wooden posts. Primocanes remain on the ground the first season and over winter to protect them from cold.

Photo 13-10. A V-trellis provides maximum light interception, as floricanes are tied to the outside wires and primocanes grow up the middle. It works well for summer-bearing red and purple raspberries that are hand-harvested for fresh market.

With semi-erect thornless blackberries and purple or summer red raspberries, let the primocanes grow up the middle without tying, and tie the floricanes to the wires so they form the V-shape down the row. This positions the fruit on the edges of the rows where they are easier to harvest by hand or machine, and prevents the floricanes from shading out the primocanes, allowing them to grow stronger to produce a good crop next year. This system was designed for hand harvesting, but can be machine harvested if the V-shape is narrowed enough to fit the harvester.

The **single-T trellis** has a T-shaped end post with two wires, spaced 1½ to 3 feet apart, at the ends of the T and 5 to 6 feet off the ground. It produces almost the same effect as the V-trellis with less wire, if the cultivar's canes rapidly grow to a length that will reach the wires. It works well for some less vigorous trailing blackberries and raspberry-blackberry hybrids, which do just that. This system was also originally designed for hand harvesting, but the wider T spacing allows machine harvesting.

The **double-T trellis** has a T-shaped end post with a narrower crossbar 2 feet below the T. Four wires, two spaced 2 to 3 feet apart on the top T and 5 to 6 feet off the ground, and two spaced about 1½ feet apart on the crossbar and 3 to 4 feet off the ground, provide basically the same shape as the V-trellis, but a little narrower. This trellis is often used for purple or summer red raspberries, and can be harvested by hand or machine.

The **shift trellis** moves to increase fruit set and later shield the fruit from excess light that can cause sunscald, a big problem with raspberries and some semi-erect blackberries in warmer climates. This flat-plane trellis is like the vertical trellis but hinged, allowing the trellis to move in an arc from one side of the row to the other. At bloom, the trellis is positioned almost parallel to the ground, which concentrates flower development on the upper part of the canopy. As the fruit starts to mature, growers move the trellis upward to about 120 degrees from horizontal and slanted westward, which shades the fruit.

The **rotatable cross arm trellis** (Figure 13-1) was developed for semi-erect blackberries to be harvested by a newer machine that does little damage to the fruit, allowing sales for fresh market. This trellis has a Y shape with one shorter arm and one longer arm, and the longer crossarm rotates. Growers position the longer crossarm over the shorter one at about 22.5 degrees above horizontal from early fall through early spring, to encourage flower development on the upper part of the canopy. In late spring, they swing the longer crossarm to the other side for harvest, which positions the fruit below the canes for easy machine removal with little damage, and shades fruit to protect it from sunscald. This newer system has not been widely implemented yet, but shows promise for machine harvest of fresh-market blackberries.

Trellised, biennial cane, alternate-year machine-pruned system

About 20% of trailing blackberry acreage is managed for alternate year production, which yields 80% to 85% as much over a two-year period as every-year production, and cuts pruning and training costs in half. Trailing blackberry growers who follow this system usually manage entire blocks for production one year and other blocks for production the next, rather than every other row as practiced for red and purple raspberries.

After harvest, a fruiting block is mowed near the ground, and the mowings are chopped up with a flail mower or removed and burned. This reduces cane diseases and eliminates the need for fungicide sprays after harvest or the next year. This also removes primocanes, so eliminates next year's crop. The following year, primocanes will grow and be trained, and produce fruit two years later. Canes growing after an "off year" have shown greater cold tolerance than those in every-year production. The biggest drawback to the system is the need for specialized mowers or removing the trellis to mow the canes.

CHOOSING CULTIVARS

Bramble cultivars depend greatly on a favorable environment to develop their full quality and yield potential. Researchers in different regions often report completely contradictory rankings of plant survival, yields, fruit characteristics and ripening times for the same cultivars tested. Therefore, choose cultivars adapted to your climate as your first priority. Many bramble growers must choose the old favorite cultivars that their customers prefer, or possibly new cultivars with quality similar to the old but improved disease resistance. Choosing cultivars resistant to the main diseases and insects of your area (see Tables 13-1 to 13-5, pages 178-184) can greatly prolong the life of the planting and reduce the need for pesticides.

Red and yellow raspberries

Red raspberries dominate U.S. and world raspberry production. Yellow raspberries are a very minor crop but may command higher prices as a novelty item for local sales.

Summer red cultivars: The prevailing raspberry cultivar in the Pacific Northwest is 'Meeker,' (Photo 13-11) a summer-bearing red cultivar suitable for machine harvest and for canning, freezing and preserves. 'Meeker' is less susceptible to root rot than 'Willamette', an older summer red cultivar that's still grown, but more susceptible to winter injury. 'Chilliwack' has nearly spine-free canes well-suited for machine harvest, good hardiness for the Northwest, resistance to the aphid vector of raspberry mosaic virus, some root rot resistance, and fine-flavored fruit with considerable resistance to postharvest fruit rots. 'Qualicum' also has firm fruit that resist postharvest decay and resists the aphid vector of raspberry mosaic virus, but is susceptible to root rot so needs sandier soils.

'Titan' from New York has been an important summer red cultivar in the Northeast for hand harvest, due to its high yields of large fruit, smooth canes and resistance to the aphid vector of raspberry mosaic virus, but it requires trellising and is very susceptible to root rot, so needs lighter soils and raised beds. 'Sentry' from Maryland also yields heavily, has nearly smooth canes, is vigorous and needs a trellis. 'Nova' from Nova Scotia is nearly thornless, very hardy, resistant to late yellow leaf rust, and widely grown in eastern Canada where that disease has been troublesome. 'Lauren', from

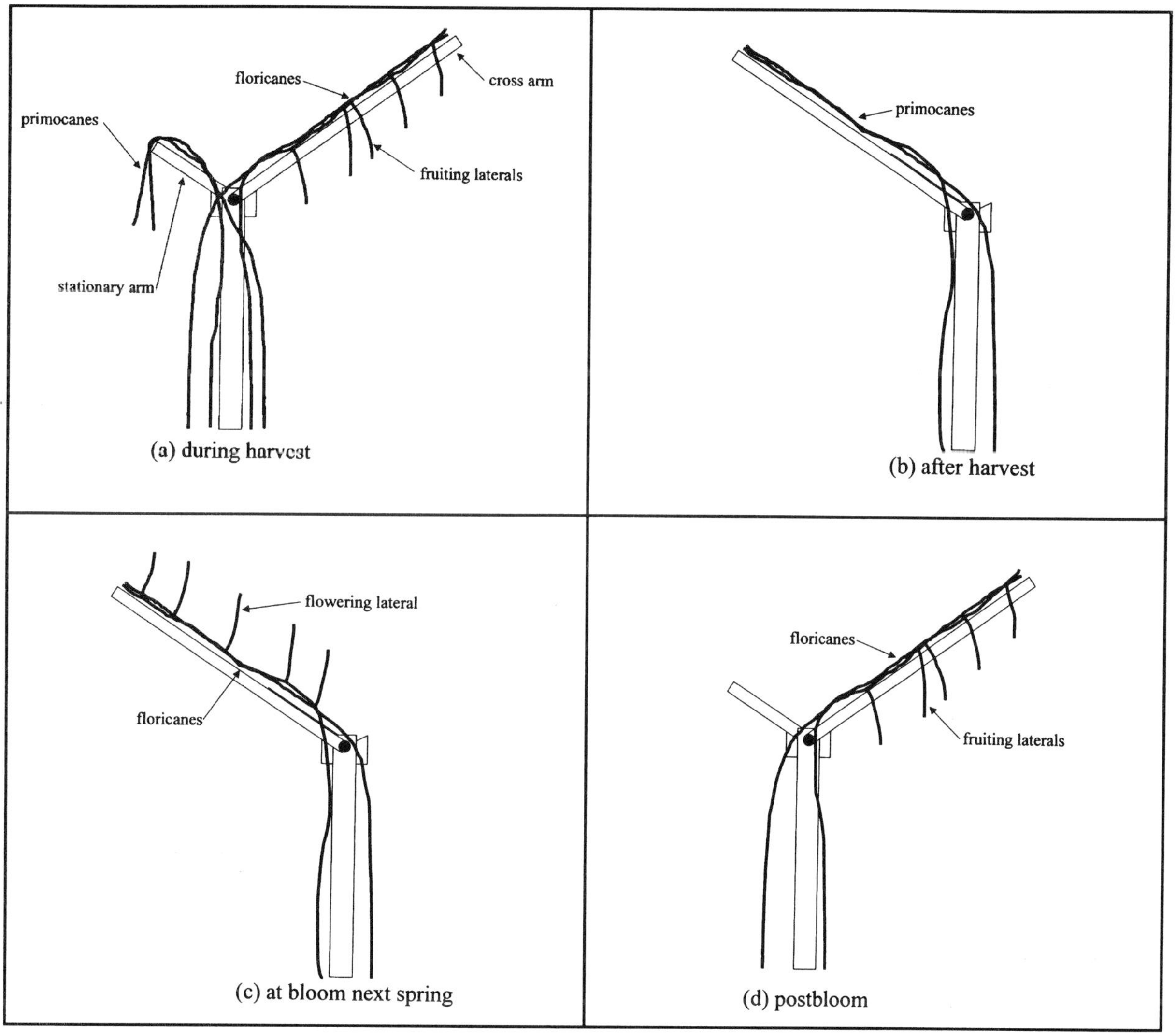

Figure 13-1. End view of a rotatable cross-arm Y trellis (a) during summer harvest of fruiting canes on the long arm, while the short arm supports new primocanes, (b) after harvest and removal of floricanes, with primocanes tied to the long arm, now rotated to the side with the short arm (c) at bloom in late spring, with flowering laterals growing upward, and (d) during fruit ripening, with fruiting laterals hanging down so the mechanical harvester can remove them with less damage. Redrawn from *HortTechnology* 9:16-21.

Table 13-1.
ONE-CROP, SUMMER-BEARING RED RASPBERRIES

Cultivar	Where Adapted*	Hardy to Zone	Fruit				Resistance^z to			Cane Erectness	Spininess	Ripening Season	Machine Harvest
			Flavor	Size	Firmness	Yields	Mosaic	Root Rot	Other Pests, Environment				
Prelude	MA	5b	fair	medium	medium	medium	S	MR		semi-erect	medium	very early	no
Southland	SE, SW	5b	fair	medium	medium	medium	S	S	R to heat, leaf spot, anthracnose, mildew	sprawls	medium	early	no
Reveille	MA, SC, NC	4a	good	large	medium	high	R	MR	R to spur blight; S to mildew, fire blight	erect	very low	early	no
Nova	FN	3a	fair	medium	firm	very high	R	MS	R to yellow rust, cane wilt, fire blight	erect	very low	early	yes
Malahat	NW	7a	good	large	firm	high	R	S	R to postharvest fruit rot	semi-erect	very low	early	yes
Regency	FN	3a	good	medium	firm	medium	S	MR	S to anthracnose, cane wilt	erect	low	early	no
Willamette	NW	7b	good	medium	medium	medium	S	S	R to bushy dwarf, leaf curl, mildew, anthracnose, crown gall; S to cane wilt	sprawls	high	early-mid	yes
Sentry	MA, NC	5b	very good	medium	firm	high	S	S	R to spur blight; VS to leaf spot, anthrac.	erect	low	early-mid	
Lauren	MA, NE, NC	6a	good	very large	medium	very high	S	S	S to varying winter temps.	semi-erect	low	early-mid	no
Emily	MA	7a	good	very large	firm	high	S	MR	S to cold below 0F; ships well	semi-erect	low	early-mid	no
Titan	NE, NC	5b	good	very large	firm	very high	R	VS	VS to crown gall	semi-erect	low	early-mid	no
Canby	NW, NC	7a	very good	medium	medium	medium	VR	VS	S to anthracnose, leaf spot, spur blight	sprawls	low	early-mid	yes
Meeker	NW	6b	good	large	medium	very high	S	MS	R to fruit rots, cane wilt, anthracnose	sprawls	medium	early-mid	yes
Chilliwack	NW	6a	best	large	firm	medium	VR	MR	R to bushy dwarf, spur blight, fruit rots; VS to Botrytis cane wilt, anthracnose	erect	very low	middle	yes
Qualicum	NW	5b	very good	large	firm	very high	R	S	MR to bushy dwarf, spur blight, fruit rots	erect	very low	middle	yes
Liberty	NC, SC	5b	very good	large	medium	high	VR	S		erect	medium	mid-late	no
Sumner	NW	5b	very good	medium	medium	high	S	R	R to yellow rust	erect	medium	mid-late	no
Citadel	MA, SC	5b	good	large	medium	medium	S	S	R to leaf spot, bushy dwarf, mildew	erect	low	late	no
Tulameen	NW	6b	very good	very large	firm	high	R	VS	R to crown gall, MR to bushy dwarf	erect	low	late	yes
Coho	NW	7a	very good	large	very firm	high	R	MR		semi-erect	medium	very late	yes
Encore	NE, MA, NC	5b	very good	very large	firm	medium	S	MR	VS to fire blight, varying winter temps.	semi-erect	medium	very late	no
Dormanred	SE, SW	6a	fair	very large	firm	high	S	S	R to heat, mites; fruit sour till very ripe	sprawls	very high	very late	no

* Regions: SE = Southeast: from southeast Virginia and Georgia to northern parts of Gulf Coast states, USDA zone 8

GC = Gulf Coast: from eastern Texas to Florida along the coast, zone 9

SW = Southwest: from western Texas to southern and central California, zones 7 to 10

NW = Northwest: from northern California and Nevada to Washington, Idaho and British Columbia, zones 5 to 9

FN = Far North and Rocky Mountains, zones 2b to 4a

NE = Northeast: New England, northeastern Canada, New York and northern Pennsylvania, zones 4b to 6a

MA = Mid-Atlantic: from New Jersey and southern Pennsylvania to Virginia, zones 6a to 8a

NC = North Central: from Ohio to South Dakota and Nebraska, zones 4b to 6a

SC = South Central: from West Virginia and western North Carolina to Kansas and Oklahoma, zones 6a to 7b

[z]Resistence Ratings:

VR = very resistant; no other controls usually needed for that disease
R = resistant; minimal other controls needed for that disease
MR = moderately resistant; fewer than average other controls needed
MS = moderately susceptible; average amount of other controls needed
S = susceptible; more than average other controls needed
VS = very susceptible; very difficult to grow where disease is a problem

Average start of ripening, which lasts 3 to 6 weeks

at Beltsville, MD: very early – early June; early – mid-June; middle – late June; late – early July; very late – mid July

at Corvallis, OR and Geneva, NY: very early – late June; early – early July; middle – mid-July; late – late July; very late – early August

One-crop, summer-bearing red raspberries -- in approximate order of ripening (varies with area).

Table 13-2.

TWO-CROP OR FALL-BEARING RED AND YELLOW RASPBERRIES

Cultivar	Where Adapted*	Hardy to Zone	Fruit					Resistance^z to			Cane Erectness	Spininess	Fall Ripening Season
			Color	Flavor	Size	Firmness	Yields	Mosaic	Root Rot	Other Pests, Environment			
Autumn Bliss	MA, NC	4b	red	very good	large	firm	v. high	R	R	R to powdery mildew, S to bushy dwarf	semi-erect	low	very early
Polana	FN, NE	4b	red	good	medium	medium	high	S	MR	may form double fruit from same flower	erect	medium	very early
Summit	NW	5b	red	very good	medium	firm	high	S	R		erect	low	very early
Golden Summit	NW	5b	yellow	very good	medium	firm	high	S	R		erect	low	very early
Caroline	MA, NC	4b	red	best	large	firm	high	S	MR	MR to yellow rust	erect	medium	very early
Redwing	FN, NC	3b	red	good	medium	medium	high	S	S	S to anthracnose	semi-erect	medium	very early
Amity	NW	5a	red	good	medium	firm	medium	MR	MR	MR to spur blight, S to bushy dwarf	semi-erect	very low	early
Dinkum	MA, NE	5b	red	best	medium	firm	medium	S	S	S to bushy dwarf, R to fruit rot	erect	medium	early
Autumn Briton	MA,NC	4b	red	good	large	firm	medium	R	MR	S to bushy dwarf	erect	medium	early
Golden Harvest	NE, MA, NC	4b	yellow	fair	small	firm	medium	S	S	R to bushy dwarf	sprawls	medium	early
Double Delight	FN	3a	red	good	small	firm	medium	S	S	may form double fruit from same flower	erect	high	early
Ruby	NE, MA, NC	5a	red	fair	very large	medium	low	S	VS	VS to crown gall & fruit rot	semi-erect	medium	middle
Heritage	NE, MA, NC	4a	red	very good	medium	firm	medium	MR	S	MR to leaf curl, mildew, leaf spot, R to anthracnose, S to yellow rust	erect	high	middle
Goldie	NE, MA, NC	4a	yellow	very good	medium	medium	medium	MR	S	same as Heritage	erect	high	middle
Kiwi Gold	MA, NE	4a	yellow	good	medium	firm	medium	MR	S	same as Heritage	erect	high	middle
Anne	MA, NC	4b	yellow	best	very large	firm	low	S	MR	S to fruit rot & birds	semi-erect	medium	mid-late
Josephine	MA, SC	4b	red	very good	very large	firm	medium	S	MR	R to fruit rot	erect	medium	late
Cherokee	SC, SE	5a	red	best	medium	firm	high	S	R	R to heat	erect	medium	late
Nordic	FN	3a	red	good	medium	firm	medium	MR	R	MS to anthracnose & spur blight	erect	medium	late
Sceptor	MA, SC	5b	red	very good	large	medium	medium	S	S	R to bushy dwarf, anthracnose, leaf spot	semi-erect	medium	very late

* Regions:

SE = Southeast: from southeast Virginia and Georgia to northern parts of Gulf Coast states, USDA zone 8
GC = Gulf Coast: from eastern Texas to Florida along the coast, zone 9
SW = Southwest: from western Texas to southern and central California, zones 7 to 10
NW = Northwest: from northern California and Nevada to Washington, Idaho and British Columbia, zones 5 to 9
FN = Far North and Rocky Mountains, zones 2b to 4a
NE = Northeast: New England, northeastern Canada, New York and northern Pennsylvania, zones 4b to 6a
MA = Mid-Atlantic: from New Jersey and southern Pennsylvania to Virginia, zones 6a to 8a
NC = North Central: from Ohio to South Dakota and Nebraska, zones 4b to 6a
SC = South Central: from West Virginia and western North Carolina to Kansas and Oklahoma, zones 6a to 7b

[z]Resistence Ratings:

VR = very resistant; no other controls usually needed for that disease
R = resistant; minimal other controls needed for that disease
MR = moderately resistant; fewer than average other controls needed
MS = moderately susceptible; average amount of other controls needed
S = susceptible; more than average other controls needed
VS = very susceptible; very difficult to grow where disease is a problem

Average start of ripening, which lasts 3 to 6 weeks
at Beltsville, MD: very early – mid Aug.; early – late Aug.; middle – early Sept.; late – mid Sept.; very late – late Sept.
at Corvallis, OR and Geneva, NY: very early – very early Sept.; early – early Sept.; middle – mid-Sept.; late – late Sept.; very late – early Oct.

Two-crop or fall-bearing red and yellow raspberries -- in approximate order of ripening (varies with area)

Table 13-3.
BLACK AND PURPLE RASPBERRIES

Cultivar	Where Adapted*	Hardy to Zone	Fruit					Resistance[z] to				Cane Erectness	Spininess	Ripening Season	Suited to Machine Harvest
			Color	Flavor	Size	Firmness	Yields	Mosaic	Anthrac-nose	Powdery Mildew	Root Rot				
Earlysweet	NW, MA	5b	black	very good	medium	firm	high	S	MR	MR	R	semi-erect	high	early	yes
Bristol	NE, NC, MA	5b	black	best	small	firm	high	MR	MS	MR	R	semi-erect	high	early	yes
Haut	MA, SC	5a	black	very good	small	firm	medium	S	MR	MS	R	semi-erect	high	early	yes
Allen	NE, MA	6a	black	good	med-large	firm	medium	S	VS	MR	R	semi-erect	high	early-mid	yes
Lowden	NC, NE	6a	black	very good	medium	medium	low	S	R	R	R	semi-erect	medium	early-mid	yes
Munger	NW, NC	6a	black	best	large	firm	high	S	MR	MR	S	semi-erect	high	middle	yes
Jewel	NE, NC, SC	5a	black	best	medium	firm	high	S	MR	MS	R	erect	medium	middle	yes
Allegheny	MA, SC	5a	black	good	large	firm	high	S	MR	MR	R	semi-erect	medium	middle	yes
Dundee	NE, NC	5a	black	good	medium	firm	medium	MR	S	S	R	semi-erect	medium	middle	yes
New Logan	NC,SC	5b	black	good	large	firm	medium	MR	MR	MR	R	semi-erect	medium	middle	yes
Black Hawk	NC	4b	black	good	medium	medium	medium	MR	R	S	R	sprawls	high	mid-late	no
Success	FN	4a	purple	best	small	firm	high	S	S	R	R	sprawls	high	mid-late	no
Brandywine	NE, NC	4a	purple	good	medium	firm	high	S	S	R	R	semi-erect	very high	mid-late	no
Lowden Sweet Purple	NE, NC	4b	purple	very good	medium	firm	medium	S	VR	R	R	erect	medium	late	no
Royalty	MA, SC, NC	4b	purple	very good	very large	medium	very high	R	S	R	MR	semi-erect	high	late	no
Estate	FN	4a	purple	good	large	firm	high	S	S	R	MS	semi-erect	high	very late	no

* Regions:

SE = Southeast: from southeast Virginia and Georgia to northern parts of Gulf Coast states, USDA zone 8

GC = Gulf Coast: from eastern Texas to Florida along the coast, zone 9

SW = Southwest: from western Texas to southern and central California, zones 7 to 10

NW = Northwest: from northern California and Nevada to Washington, Idaho and British Columbia, zones 5 to 9

FN = Far North and Rocky Mountains, zones 2b to 4a

NE = Northeast: New England, northeastern Canada, New York and northern Pennsylvania, zones 4b to 6a

MA = Mid-Atlantic: from New Jersey and southern Pennsylvania to Virginia, zones 6a to 8a

NC = North Central: from Ohio to South Dakota and Nebraska, zones 4b to 6a

SC = South Central: from West Virginia and western North Carolina to Kansas and Oklahoma, zones 6a to 7b

[z]Resistence Ratings:
VR = very resistant; no other controls usually needed for that disease
R = resistant; minimal other controls needed for that disease
MR = moderately resistant; fewer than average other controls needed
MS = moderately susceptible; average amount of other controls needed
S = susceptible; more than average other controls needed
VS = very susceptible; very difficult to grow where disease is a problem

Average start of ripening, which lasts 3 to 6 weeks

at Beltsville, MD: very early – early June.; early – mid June; middle – late June; late – early July; very late – mid July

at Corvallis, OR and Geneva, NY: very early – late June; early – early July; middle – mid-July; late – late July; very late – early Aug.

Black and purple raspberries -- in approximate order of ripening (varies with area)

Table 13-4.
ERECT AND SEMI-ERECT BLACKBERRIES

Cultivar	Hardy to Zone	Fruit					Resistance* to				Cane Erectness	Thorniness	Ripening Season
		Flavor	Size	Firmness	Seed Size	Yields	Orange Rust	Anthrac-nose	Double Blossom	Other Pests, Environment			
Choctaw	6b	very good	large	medium	small	high	R	MR	S	blooms early; frosts may hurt	erect	very thorny	very early
Brazos	7b	good	large	soft	large	very high	MS	R	VS	R to heat & drought	erect	thorny	early
Cheyenne	6b	good	large	firm	medium	high	R	MR	S	MR to fruit sunscald	erect	thorny	early
Cherokee	6a	very good	medium	firm	small	medium	R	R	VS	S to root rot	erect	mod. thorny	early-mid
Rosborough	7b	very good	large	firm	large	high	MS	R	VS	R to heat & drought	erect	very thorny	early-mid
Darrow	5a	good	medium	firm	medium	high	S	MS	S	productive in Northeast	erect	thorny	early-mid
Arapaho	6b	good	medium	firm	small	medium	MR	MR	R	VS to stink bugs	erect	thornless	middle
Shawnee	6b	good	large	medium	medium	very high	R	MR	VS	S to cane blight if high N added	erect	thorny	middle
Illini Hardy	4b	fair	medium	medium	medium	high	S	MS	S	most cold-hardy	erect	very thorny	middle
Ouachita	6a	best	large	firm	small	high	R	MR	R	full crops may need trellis	erect	thronless	middle
Kiowa	6a	very good	very large	firm	medium	high	R	MR	MS		erect	very thorny	middle
Chickasaw	6a	very good	very large	firm	large	very high	R	MR	MR	MS to powdery mildew	erect	thorny	middle
Navaho	6a	best	medium	very firm	small	medium	MS	MR	MR	hardens late, S to late fall freezes	erect	thornless	mid-late
Dirksen	5b	good	large	medium	medium	medium	MS	R	S	MR to leaf spot	semi-erect	thornless	mid-late
Apache	6a	very good	large	firm	small	high	R	MR	S		erect	thornless	mid-late
Triple Crown	5b	best	very large	firm	large	high	MR	MS	MR	needs much pruning of laterals	semi-erect	thornless	mid-late
Chesapeake	6a	best	very large	medium	medium	high	R	MS	MR		erect	thorny	mid-late
Hull Thornless	5b	very good	large	medium	very large	high	MR	MS	R		semi-erect	thornless	late

* Resistence Ratings:
VR = very resistant; no other controls usually needed for that disease
R = resistant; minimal other controls needed for that disease
MR = moderately resistant; fewer than average other controls needed
MS = moderately susceptible; average amount of other controls needed
S = susceptible; more than average other controls needed
VS = very susceptible; very difficult to grow where disease is a problem

Note: all erect and semi-erect blackberries can be machine harvested for processing.

Average start of ripening, which lasts 3 to 6 weeks
at Clarksville, AR: very early – late May.; early – early June; middle – mid June; late – latr June; very late – early July
at Beltsville, MD: very early – late June; early – early July; middle – mid-July; late – late July; very late – early Aug.

Erect and semi-erect blackberries -- in approximate order of ripening (varies with area).

Table 13-5.
TRAILING BLACKBERRIES AND RESPBERRY-BLACKBERRY HYBRIDS

Cultivar	Hardy to Zone	Fruit						Resistance* to		Cane Thorniness	Ripening Season
		Color	Flavor	Size	Firmness	Seed Size	Yields	Anthracnose	Other Pests, Environment		
Flordagrand	9a	shiny black	good	large	firm	medium	high	MR	plant with Oklawaha for pollination	thorny	very early
Oklawaha	9a	shiny black	good	large	firm	medium	high	MR	plant with Flordagrand for pollination	thorny	very early
Thornless Boysen	8a	maroon	best	very large	soft	medium	medium	MR	VS to Verticillium wilt	thornless	early
Lucretia Dewberry	7a	shiny black	best	very large	medium	medium	high	MS	adapted to Southeast	thorny	early
Gem Dewberry	7b	shiny black	best	very large	medium	medium	high	MS	R to rosette, adapted to Southeast	thorny	early
Siskiyou	7b	dull black	very good	large	firm	small	high	R	R to leaf spot, cane & leaf rust	thorny	early-mid
Black Butte	8a	dull black	good	large	firm	medium	medium	S	S to leaf spot, fruit rot	thorny	early-mid
Tayberry	7b	red-purple	very good	very large	soft	medium	high	VS	S to bushy dwarf, leaf & cane spot	very thorny	early-mid
Boysenberry	7a	maroon	best	very large	soft	medium	medium	MR	VS to Verticillium wilt	thorny	early-mid
Lavaca	6b	maroon	best	very large	medium	medium	high	MR	Boysenberry seedling; hardier	thorny	early-mid
Youngberry	8a	burgundy	very good	very large	medium	medium	medium	MS	grown in Northwest & Southeast	thornless	middle
Kotata	6b	shiny black	very good	medium	firm	small	very high	S	R to Botrytis; VS to budmoth	very thorny	middle
Marion	8b	shiny black	best	medium	medium	small	medium	S	S to leaf spot, MR to fruit rot	thorny	middle
Olallie	9a	shiny black	good	medium	firm	medium	high	S	R to Verticillium wilt	thorny	mid-late
Waldo	8a	shiny black	very good	medium	firm	small	very high	S	VS to Botrytis, budmoth	thornless	mid-late
Thornless Evergreen	8a	shiny black	fair	medium	firm	large	very high	S	S to leaf & cane spot	thornless	late
Everthornless	8a	shiny black	good	medium	firm	large	very high	S	S to leaf & cane spot	thornless	late

* Resistence Ratings:
VR = very resistant; no other controls usually needed for that disease
R = resistant; minimal other controls needed for that disease
MR = moderately resistant; fewer than average other controls needed
MS = moderately susceptible; average amount of other controls needed
S = susceptible; more than average other controls needed
VS = very susceptible; very difficult to grow where disease is a problem

Note: all trailing blackberries and hybrid brambles can be machine harvested for processing.

Average start of ripening, which lasts 3 to 6 weeks

at Raleigh, NC: very early – early June.; early – mid June; middle – late June; late – early July; very late – mid July

at Corvallis, OR: very early – early June; early – mid July; middle – late July; late – late Aug.; very late – mid Aug.

Trailing blackberries and raspberry-blackberry hybrids — in approximate order of ripening (varies with area).

Maryland in 1996, ripens its large fruit early and is less susceptible to root rot than 'Titan'. 'Encore', a 1998 New York cultivar, has some resistance to root rot and very large, sweet fruit. It ripens its summer crop late and also produces a smaller fall crop, and is best grown for both crops rather than just the fall crop.

Summer yellow cultivar: 'Honey Queen' from Alberta, Canada is very hardy, productive and tolerant of root rot. Shorter plants don't need a trellis.

Fall red cultivars: These are mostly erect so don't need a trellis. 'Heritage' (Photo 13-12) from New York became the first important fall red cultivar in 1969 and remains the most widely planted one. It produces a moderate summer crop and a much larger fall crop, so is well suited to annual mowing to produce only the fall crop. 'Heritage' is winter-hardy, has some resistance to root rot, resists powdery mildew and some viruses, and produces high yields of fruit with excellent processing qualities. 'Heritage' can be machine harvested, while many other fall red raspberries are not suited to machine harvest. Most other fall red cultivars, including 'Amity', 'Autumn Bliss', 'Caroline', 'Polana', 'Redwing', 'Ruby', and 'Summit', were selected in short-season areas to start ripening earlier than 'Heritage', so 'Heritage' is still the best cultivar for most longer-season areas. In California's central coast area, cultivars such as 'Autumn Bliss' can produce successive crops from late April to December. Some growers in northern areas are now producing primocane-bearing red raspberries from December to April in greenhouses.

Fall yellow cultivars: 'Fallgold,' (Photo 13-13) the first fall-bearing yellow raspberry from New Hampshire in 1967, is erect and very hardy, but produces low yields of soft fruit that is good fresh but not for processing. Newer cultivars are more productive. 'Goldie', a yellow-fruited sport of 'Heritage', ripens with and yields about as much as 'Heritage', 2 to 3 times the yield of 'Fallgold'. 'Goldie' fruit are moderately firm and turn pink when ripe. 'Kiwigold', another Heritage sport from New Zealand, also ripens its firm, high-quality, good-sized fruit with 'Heritage'. 'Golden Harvest', yet another 'Heritage' sport, has smaller fruit with less flavor and lower yields. 'Anne' from Maryland in 1998 has very sweet fruit that's unfortunately more attractive to birds than other yellow cultivars.

Black and purple raspberries

Most black raspberries are grown for processing, with very small amounts grown for local fresh sales. Purple raspberry acreage is quite low but is increasing for pick-your-own sales especially.

Black cultivars: 'Munger' is the most-planted cultivar worldwide, with most of its acreage in Oregon. Plantings often fruit for just 2 or 3 seasons due to high susceptibility to viruses and other diseases. Growers in Eastern North America are planting more black raspberries, especially Eastern cultivars with resistance to some troublesome diseases. 'Earlysweet' from Maryland in 1998 has been disease-free there. 'Allegheny', a 1970 Maryland cultivar, has moderate resistance to anthracnose and powdery mildew and does well in mid-Atlantic states. 'Jewel' from New York in 1973 also has some anthracnose and mildew resistance, bears large fruit, and is hardy enough for the Northeast. 'Black Hawk' from Iowa in 1955 offers mosaic and anthracnose resistance and hardiness to zone 4b and is suited to the Midwest, where powdery mildew is generally not serious, as it is susceptible to that disease. 'Lowden' (Photo 13-14) from Ontario in 1961 has some red raspberry in its ancestry and also resists anthracnose and mildew.

CAPTIONS FOR PHOTOS 13-11 THROUGH 13-18

Photo 13-11. 'Meeker' is the predominant red raspberry cultivar grown in the Pacific Northwest. (Photo from Washington State Univ.)

Photo 13-12. 'Heritage', the first primocane-bearing red raspberry with a large fall crop, is still the most widely planted.

Photo 13-13. 'Fallgold', one of the oldest yellow-fruited raspberries, is still grown but newer yellow cultivars are much more productive.

Photo 13-14. 'Lowden' black raspberry resists anthracnose.

Photo 13-15. 'Royalty' purple raspberry bears very large fruit and resists the aphid vector of raspberry mosaic virus.

Photo 13-16. 'Navaho' was the first erect thornless blackberry. It has excellent flavor and small seeds.

Photo 13-17. 'Marion' trailing blackberry has become the standard processing cultivar in the Pacific Northwest. (Photo by Dalton Hobbs, Oregon Dept. of Agric.)

Photo 13-18. 'Thornless Boysen' a sport of 'Boysen' raspberry-blackberry hybrid, has a unique flavor prized for jam processing. (Photo from Oregon State Univ.)

Photo 13-11.

Photo 13-12.

Photo 13-13.

Photo 13-14.

Photo 13-15.

Photo 13-16.

Photo 13-17.

Photo 13-18

Purple cultivars: 'Royalty' (Photo 13-15) is the premier purple raspberry, producing high yields of large, sweet, red-purple fruit and resisting the aphid vector of raspberry mosaic virus. 'Royalty', from New York in 1982, does well for pick-your-own and local sales. 'Brandywine' from New York in 1976, produces medium-sized, tart, dark purple fruit that is excellent for processing, but is susceptible to raspberry aphid. 'Lowden Sweet Purple' from Ontario is very resistant to anthracnose. 'Royalty' and 'Lowden Sweet Purple' are hardy to USDA zone 4b, while 'Brandywine', 'Success' (from New Hampshire) and 'Estate' (from Minnesota) are a bit hardier, to zone 4a.

Erect and semi-erect blackberries

The acreage of erect and semi-erect blackberries is much less than that of trailing blackberries, but is growing while that of trailing blackberries is declining. Thornless plantings are especially growing for pick-your-own sales, while thorny types are planted for processing.

Thorny erect cultivars: Americans began selecting superior wild erect blackberries over a century ago, and several states developed breeding programs. 'Darrow' from New York in 1958 is very winter hardy and productive, and is still grown in the Northeast for processing. 'Brazos', a high-yielding Texas cultivar from 1959, is still the standard for processing in the Southwest. The University of Arkansas breeding program utilized these two cultivars and other genotypes to produce a series of high-yielding and widely-adapted erect blackberries, all given the names of American Indian tribes, that are suited to mechanical harvest and now dominate production. 'Cherokee' resists orange rust and anthracnose, while 'Cheyenne', 'Choctaw', 'Chickasaw', 'Kiowa' and 'Shawnee' resist orange rust and are moderately resistant to anthracnose. 'Chickasaw' and 'Kiowa' produce huge fruit.

Thornless semi-erect cultivars: The first thornless erect blackberries, bred in England and by the USDA in Maryland, were really only semi-erect and required trellising. They derived their thornlessness from the southern European species *R. rusticanus* var. *inermis*, which unfortunately came with linked genes for an arching growth habit, susceptibility to winter injury, acid fruit, large seed size, and late ripening season. This group includes the British cultivar 'Merton Thornless', and the early USDA cultivars that resulted from crosses of this cultivar with American erect blackberries, including 'Thornfree', 'Smoothstem', 'Black Satin', and 'Dirksen Thornless'.

The newer USDA cultivars 'Hull Thornless', 'Chester Thornless', 'Chesapeake' and 'Triple Crown' were selected for winter hardiness to zone 5 in Illinois and have sweeter fruit, but still need a trellis. Pick-your-own growers in zones 5 and 6a of the upper Midwest, where erect thornless cultivars aren't fully hardy, plant these four semi-erect thornless cultivars.

Thornless erect cultivars: The Arkansas breeding program has released four fully-upright thornless cultivars so far, which do not require a trellis and have small seed size. 'Arapaho' ripens earliest, followed by 'Ouachita.' 'Navaho' (Photo 13-16) ripens its very sweet fruit in mid-season, and 'Apache' ripens latest with very large fruit. 'Arapaho' and 'Ouachita' resist double blossom (rosette) and are moderately resistant to orange rust and anthracnose. 'Navaho' offers moderate resistance to anthracnose and rosette but is moderately susceptible to orange rust, while 'Apache' resists orange rust and is moderately resistant to anthracnose but susceptible to rosette. These cultivars are hardy to zone 6a, and are widely grown in the Southeast and lower Midwest for pick-your-own and retail sales.

Trailing blackberries and raspberry-blackberry hybrids

About 98% of trailing brambles are grown for processing, mainly in the Pacific Northwest.

Trailing blackberries: 'Marion' (Photo 13-17), an Oregon cultivar from 1956, accounts for over half of trailing blackberry production in the Pacific Northwest. The processing industry prizes 'Marion' for its excellent flavor and very small seeds. A vigorous grower with thorny canes up to 20 feet long, it can produce 5 to 6 tons of high-quality fruit per acre, almost double the yield of erect blackberries. However, 'Marion' is relatively cold sensitive, and cold injury reduces the crop some years.

Other trailing blackberry cultivars offer thornless canes and better cold hardiness, but aren't planted as much in the Pacific Northwest because they can be sold to processors only as a generic "blackberry" product, which returns only about two-thirds the price to growers as "Marionberries". 'Thornless Evergreen', a periclinal chimera selected from naturalized ("wild") stands of an introduced English trailing blackberry species, offers greater cold tolerance and produces higher yields than 'Marion' and is the second most-planted cultivar. A genetically thornless version of this cultivar, called 'Everthornless', is now produced by tissue culture. 'Waldo', an Oregon cultivar from 1989, also produces heavy yields of highly flavored, firm berries with small seeds, and canes are thornless. This was the first genetically thornless trailing blackberry, not a chimera, so root suckers are thornless also. However, fruit lack the unique 'Marion' flavor and plants are less vigorous than 'Marion', making them harder to manage on traditional trellises, so 'Waldo' is not widely planted. 'Black Butte' and 'Siskiyou', 1998 and 1999 Oregon cultivars, are early and very large-fruited trailing blackberries for fresh market, with thorny canes. 'Flordagrand' and 'Okinawa' are large-fruited trailing blackberries for fresh markets in Florida, and must be planted together to cross-pollinate each other. 'Gem' and several other dewberries are still grown for local fresh markets in the Southeast.

Raspberry-blackberry hybrids: 'Boysenberry' (Photo 13-18), from California in 1935, is the most widely planted raspberry-blackberry hybrid. The large reddish-purple fruit have a distinctive, rich, tangy flavor that's valued for processing as a cultivar and local fresh sales, but the fruit is too soft to ship. The thornless sport is most widely planted and is hardy to USDA 8a, but the original thorny cultivar is hardier, to USDA zone 7a. 'Tayberry', from Scotland in 1970, produces huge fruit and is grown on a limited scale for local markets in the Mid-Atlantic region, but is only hardy to zone 7b.

SITE PREPARATION, FERTILIZATION AND PLANTING

Choose a site for brambles that hasn't had any problems with Phytophthora root rot, Verticillium wilt or crown gall on other crops, and hasn't been planted to crops or had high populations of weeds susceptible to Verticillium wilt. This means avoiding fields devoted to tomatoes, peppers, potatoes, melons, strawberries, or stone fruits within the last 5 years, or fields infested with pigweeds, lambsquarters, or nightshades. Also choose a site that's at least 1000 feet from any wild brambles, or remove all wild brambles within that distance, to prevent virus diseases and cane borers from invading the planting from the wild plants.

Like all fruit crops, brambles need a site with good air drainage, but since they bloom late and have few troubles with frosts during bloom, they don't need the least frost-prone site on a farm. A winter windbreak, usually to the northwest of the site, to reduce winter wind-chill is very advantageous. The site should receive full sun in most regions, although afternoon shade benefits bramble plantings in hot-summer areas.

Increasing organic matter, fertilizing and preventing soil pests

Brambles require a well-drained soil and benefit from high amounts of soil organic matter, about 3 to 4%. Select the site a year or two before planting, and grow a cover crop or add compost or manure, to build up organic matter and increase beneficial microorganisms that fight pathogens causing soilborne diseases.

Microbial breakdown of these organic materials requires additional nitrogen (N). Legume cover crops add enough nitrogen to the soil to fuel microbial decomposition plus supply the bramble crop with about 30 pounds N per acre. If using a small grain or grass cover crop to reduce nematodes (a good idea on sandier soils), add 30 pounds of actual N per acre when working the cover crop into the soil, to aid decomposition and avoid tying up all the existing soil N. If the site has been in pasture sod, turn it under the season before planting and grow corn or wheat to reduce white grubs and wireworms, soil insects that build up under sods. Do not use persistent triazine herbicides, such as atrazine (Atrazine™ or Aatrex™) or simazine (Princep™ or Aquazine™) in corn the year before planting brambles.

Test manure or compost for N and salt content before applying them. Electrical conductivity should test less than 4; if the material exceeds that, it is too salty and could damage brambles, so do not use it. If not too salty, apply 5 to 6 tons per acre poultry manure, 8 to 12 tons per acre cattle manure, or 10 to 15 tons per acre compost. Add about 15 pounds per acre of actual N for each ton of cattle manure or compost added but no added N for poultry manure; adjust this for the nitrogen content of the material. Ideally, incorporate the cover crop, manure or compost into the top 4 to 6 inches of soil between 18 months and 6 months before planting brambles, to give the ammendment time to decompose and improve the soil.

Test the soil to a depth of 12 inches for pH, phosphorous (P) and potassium (K) before incorporating the ammendments, and apply any needed lime or sulfur, P and K when turning under the organic materials. The pH should fall within the range of 5.8 to 7.5 for raspberries and 6.0 to 7.8 for blackberries, and the soil should provide 25 to 40 pounds of available P per acre and 280 to 320 pounds of K. A pH of 5.5 or less will reduce availability of P, K, calcium and magnesium, stunt growth and reduce yields unless the soil is limed. A pH of 8.0 or more can cause iron chlorosis in brambles, which will require foliar iron sprays every 3 to 4 weeks unless the pH is brought down with sulfur.

Test the soil for damaging nematodes too, especially on sandier soils. If nematode populations are high or pathogenic fungi or persistent weeds are known to exist, some growers fumigate the soil in fall before planting brambles, although fumigation is expensive and not as widespread as for strawberries. Small grain cover crops or biocontrol treatments, described in chapter 2, can greatly reduce nematodes and certain fungi. Avoiding planting after Verticillium-susceptible plants and building raised beds reduces soil fungal problems. Control any perennial weeds before planting with glyphosate (Roundup™), and control many annual weeds by cultivating several times after turning under organic ammendments.

Forming planting beds, planting, applying nitrogen and mulching

Build raised beds, 8 to 10 inches high and 3 to 4 feet wide, spaced 8 to 12 feet between bed centers depending on equipment needs, for heavy soils to prevent root rot (Photo 13-19). Researchers have found this prevents root rot more than installing drainage tiles over an entire field. Convert any areas where water stands after a rain into grassed waterways, preferably, or install tile drains in those areas.

Pacific Northwest growers often grow winter cover crops between the beds and cultivate in summer to minimize water competition with the brambles (Photo 13-20), but growers in other areas seed a permanent sod to control erosion and provide a clean surface for harvesters (Photo 13-21). Seed row middles to creeping red fescue or another fine-leaf fescue at 25 pounds per acre, mixed with 5 pounds per acre dwarf perennial ryegrass. Creeping red fescue is long-lived, drought-tolerant and needs less mowing than other grasses, but is slow to establish, so the quick-growing ryegrass fills in for it during the first few years until the ryegrass dies out from drought and competition. Apply 40 pounds actual N per acre to the row middles only and seed the grass in early fall before planting brambles.

Plant brambles in late fall or early spring in most areas. In USDA zone 5 and colder areas, plant in spring to avoid winter damage to newly set plants. Always obtain bramble transplants from a reputable nursery that certifies their plants to be virus-free. Most growers plant field-grown transplants started from layers that have been inspected for viruses, but some like to start with tissue-cultured plantlets which are almost always

Photo 13-19. Building raised beds on heavy soils can greatly reduce problems with Phytophthora root rot on brambles.

Photo 13-20. Bramble growers in arid areas practice clean cultivation between the bramble rows, such as in these red raspberries, to minimize competition for water and nutrients. (Photo by Larry Tremaine, Tremaine Farms, Lynden, WA.)

Photo 13-21. Bramble growers in most areas plant sod, such as this creeping red fescue mixed with annual rye, between the bramble rows to provide a clean picking surface and control erosion.

virus-free. Tissue-cultured plants are smaller, so need more frequent irrigation and stricter weed control to get them established.

Plant spacing depends on the growing, pruning, trellising and harvesting system. Generally, space fall-bearing red and yellow raspberries 1½ to 2 feet apart, summer-bearing red raspberries and black raspberries 2 ½ to 3 feet apart, purple raspberries and erect blackberries 3 to 4 feet apart, and semi-erect and trailing blackberries and hybrid brambles 5 to 6 feet apart.

Prevent crown gall by dipping or spraying the roots of transplants with a solution of the biocontrol agent *Agrobacterium radiobacter* (Galltrol™ or Norbac 84™). Prevent anthracnose and other cane diseases by cutting the tops of dormant transplants back to 3 or 4 inches, just above a bud, and destroying the old cane "handles" removed. Irrigate immediately after planting.

Apply about 20 pounds of actual N per acre of planting, to the planting beds only, a week or two after planting, unless following a legume cover crop. Six to eight weeks later, apply 10 to 15 pounds additional actual N per acre, so that the total amount applied during the first growing season equals about 3 the rates in table 13-6. Use a low-nitrogen fertilizer such as calcium nitrate (if pH is on the low side) or ammonium sulfate (if pH is on the high side), to avoid damaging brambles' sensitive, newly planted root systems. If pH, P and K were adjusted before planting, N should be the only fertilizer nutrient needed for the first 3 years after planting.

Applying a 3- to 4-inch thick straw mulch to the planting rows will prevent many weed seeds from germinating, and will keep bramble roots cooler in summer, helping to reduce heat stress . If mulch is not used, apply pre-emergent herbicides to control annual weeds, and irrigate immediately after applying the herbicide to get it into the soil.

TRAINING AND PRUNING

Semi-erect and trailing blackberries, hybrid brambles, and black and purple raspberries generally grow as distinct plants (the "hill system") where they were planted, with only a few additional plants growing from tip layers or root suckers in between the originally-set plants. Trellised red raspberries are generally grown by the hill system also in the Pacific Northwest, by removing root suckers that arise between the set plants, but elsewhere red and yellow raspberries and erect blackberries form hedgerows, as root suckers fill in the spaces between the original plants.

Higher cane densities per hill or foot of row increase yields, but can decrease fruit size as canes compete with each other for water and nutrients. Workers remove excess spindly primocanes from summer-bearing brambles, keeping the more robust primocanes, when they remove dying floricanes after harvest, and further thin out any canes less than ½ inch in diameter early next spring. Leave 4 to 8 strong canes per hill for hill-system brambles, and canes spaced 6 to 8 inches apart in rows 12 to 24 inches wide for hedgerow brambles, providing 3 to 5 fruiting canes per linear foot of row (Photo 13-22). If primocane-bearing raspberries get overly dense, spindly primocanes are also removed in early summer.

Erect blackberries and black and purple raspberries benefit from tipping the primocanes back in early summer. This shortens the canes, keeping them more upright, and induces lateral branches to grow below the cut, increasing the bearing surface. Tip black raspberry shoots when they reach 24 to 30 inches tall, purple raspberry shoots when they reach 32 to 36 inches tall, and erect blackberry shoots when they reach 36 to 42 inches tall.

Do not summer-tip primocanes of trailing or semi-erect blackberries or hybrid brambles that must be trellised. Also do not summer-tip primocanes of fall-bearing raspberries, as this would remove the fall crop. Tipping summer-bearing red raspberries is not generally recommended, especially in colder regions, since the lateral branches induced are more prone to winter injury than unbranched canes, and most growers use a trellis anyway to support the long canes. However, growers in mild-winter climates who want to produce summer red raspberries without a trellis might try summer tipping of cultivars with shorter, more erect canes.

Trellised summer red raspberries and semi-erect blackberries are generally cut back early next spring instead, to the height of the trellis, often using a mechanical hedger. Some growers allow these canes to grow longer and weave them between the wires instead, as do growers of trailing blackberries and hybrid brambles. Weaving longer canes increase yields over cutting canes back, but reduces fruit size slightly.

The lateral branches which result from summer tipping of erect blackberries and black and purple raspberries often grow 3 feet or more during the summer and arch to the ground, and must be headed back early next spring to keep fruit off the ground and increase fruit size. Cut back black

Photo 13-22. Prune out excess fruiting canes of brambles in early spring. Leave 3 to 5 strong fruiting canes per linear foot of hedgerow, or 6 to 8 strong canes per hill for hill-system brambles.

Photo 13-23. Cut back fruiting laterals of erect blackberries to 15 to 18 buds in early spring, to increase fruit size and keep fruit off the ground. Cut black raspberry laterals to 8 to 10 buds, and purple raspberry laterals to 12 to 14 buds.

raspberries to 8 to 10 buds per lateral, purple raspberries to 12 to 14 buds per lateral, and erect blackberries to 15 to 18 buds per lateral (Photo 13-23).

Remove dead floricanes from all brambles, either right after harvest or early the following spring, by hand or machine pruning as already discussed.

IRRIGATION AND EVAPORATIVE COOLING

Raspberries have shallow root systems and are very sensitive to drought. They require irrigation any time the available moisture in the top 2 to 3 feet of soil drops below 50%, and irrigation during bloom and while berries are sizing is especially critical. Raspberries grown in the Pacific Northwest generally require weekly irrigation over the entire summer from bloom to post-harvest periods, while those grown in rainier areas need irrigation any time during the growing season when rainfall is less than one inch per week. Blackberries have extensive root systems and are quite drought-tolerant, but yield best when regularly irrigated during fruit sizing whenever rainfall is less than one inch per 7 to 10 days.

Most bramble growers use trickle irrigation, especially drip tape, because brambles bloom late and do not usually need sprinkler irrigation during bloom for frost protection. Trickle irrigation reduces water needs and also reduces diseases as compared to sprinkler irrigation, which disperses disease spores and wets the fruit, foliage and canes, encouraging pathogen growth. Some growers use furrow irrigation, which also avoids wetting the above ground parts and costs less to install than a trickle system, but uses much more water. Both trickle and furrow irrigation require level fields or setting the irrigation system along contours.

Sprinkler irrigation provides benefits for raspberries in some situations. Where summers are hot and raspberries don't normally do well, midday sprinkling during the hottest, driest weeks of summer can provide evaporative cooling. Use micromist nozzles for 3 to 4 hours per day to avoid saturating the soil, and turn the system off in late afternoon so water will evaporate off the plants before nightfall and won't aggravate diseases. Some growers of fall-bearing raspberries also use sprinkler irrigation to protect against fall freezes.

FERTILIZING ESTABLISHED PLANTINGS

Take leaf samples in August during the first fruiting season, to determine if N levels are within the desired range of 2.75 to 3.0% N. If leaf N levels are adequate, leaves are dark green and canes grew to a length expected for each cultivar, apply N at $^2/_3$ to $^3/_4$ the rate in table 13-6 during the second growing season, and follow the guidelines in Table 13-6 for the third year and on. Spindly growth, yellowish leaves and low leaf N levels call for higher rates of N fertilization, while excessive growth or high amounts of fruit rots call for reduced nitrogen. Test soil every 3 years or so, to determine if P or K need to be added or if the pH needs adjustment.

Apply $^2/_3$ the nitrogen needed each year in early spring just before growth starts, and the other $^1/_3$ about 6 weeks later, at least a month before fruit begin to ripen. Delay the second application until early summer for fall-bearing raspberries. Avoid applying nitrogen, or organic materials that release nitrogen, during fruit sizing, since that makes fruit much more susceptible to fruit rots. Sandy soils and those with less than 1.3% organic matter need the most nitrogen, while soils high in clay or with more than 3.5% organic matter need the least, as Table 13-6 shows.

Fall-bearing raspberries require the most nitrogen, followed by trailing blackberries, hybrid berries and summer-bearing red raspberries. Erect and semi-erect blackberries and purple raspberries need less nitrogen than red raspberries, and black raspberries need less than purple raspberries. Almost all bramble growers irrigate their plantings, but those who don't irrigate should keep first-year N rates the same, reduce second-year N rates by $^{1}/_{8}$ to $^{1}/_{4}$, and reduce N rates for years 3 and on by $^{1}/_{4}$ to $^{1}/_{3}$.

Bramble plantings in high-pH soils, such as those of the Southwest, may develop symptoms of iron or zinc chlorosis. Treat such plantings with foliar applications of iron or zinc sulfates, or with soil or foliar applications of iron or zinc chelates.

POLLINATION

Almost all brambles are self-fruitful, although a few trailing blackberry cultivars, noted in Table 13-5, require cross-pollination by a different cultivar that's also adapted to the region. Bramble flowers are highly attractive to bees, and in most cases bumblebees and other wild bees provide plenty of natural pollination. Where wild bees are lacking, some growers provide 2 beehives per acre of brambles.

Incomplete pollination in raspberries can cause crumbly berry, in which fewer drupelets than normal develop and berries often crumble apart when picked. Crumbly berry may also be caused by viruses, drought, deep cultivation, nematodes, or winter damage, so rule out all these factors before concluding you need to improve pollination.

PEST MANAGEMENT

Virus and fungal diseases have caused many bramble plantings to decline, so choosing cultivars that resist the prevalent diseases, and the aphids that transmit the virus diseases, of your area is the best way to prevent this. Cane borers must be controlled in bramble plantings, and in some areas various insects that attack the fruit and foliage require treatments. Weed control is also important.

Virus Diseases. Raspberries suffer greater infection and more serious damage from virus diseases than any other fruit crop. Blackberries do not have as many virus problems but still have some. Symptoms similar to those of viruses can be caused by late spring frosts, pesticide injury, powdery mildew or insect feeding, so send leaf samples to a University testing lab for a positive diagnosis. Prevent many virus problems in brambles by starting with certified virus-free transplants, choosing cultivars resistant to the aphid vectors of these viruses if possible, and removing any wild brambles within 1000 feet of a planting. Do not plant black or purple raspberries near old red raspberry plantings, even though the red raspberries appear healthy, since they may have latent infections. Plant black raspberries upwind from red raspberries, as far away as possible, since virus-spreading aphids are carried further with the wind.

Inspect plantings in early and late summer of each year, test tissues of any plants you suspect are infected, and promptly remove and destroy plants

Table 13-6.
NITROGEN NEEDS PER ACRE PER YEAR

	Soil type & percent organic matter		
Bramble Crop	Sandy or < 1.3% O.M.	Loamy or 1.3-3.5%	Clayey or > 3.5% O.M.
Fall red & yellow raspberries	100 - 120	80 - 100	60 - 80
Summer red raspberries	60 - 80	50 - 70	40 - 60
Purple raspberries	60 - 70	50 - 60	40 - 50
Black raspberries	55 - 60	45 - 50	40 - 45
Erect & semi-erect blackberries	60 - 80	50 - 70	40 - 60
Trailing blackberries & hybrids	80 - 100	70 - 80	60 - 70

Table 13-6. Approximate pounds of actual nitrogen/acre needed per year, for brambles 3 years old and older.

that test positive for a virus disease. Spray infected plants with an insecticide a day before removing them, to prevent aphids on these plants from dispersing to healthy plants and infecting them. Maintain strict aphid control in late spring and early summer, when populations are highest.

Raspberry mosaic is caused by a complex of up to four different raspberry viruses, all spread by the large raspberry aphid (*Amophorophora agathonica*), which can infect all types of raspberries but not blackberries. This disease generally damages black and purple raspberries more than red or yellow raspberries, although 'Royalty' purple raspberry and 'Chilliwack', 'Titan' and several other red raspberries escape mosaic because they are non-attractive to large raspberry aphids. Symptoms of raspberry mosaic vary considerably with the cultivar, which of the four viruses are present, and time of year. The disease is easiest to recognize on new canes during the cooler weather of spring and fall, as symptoms may disappear during the heat of summer, although plants remain infected for life and become progressively more stunted each year. Infected canes are usually less vigorous than healthy canes, and leaves have mottled, light green spots on a darker green background. Leaves of highly susceptible cultivars pucker with large, dark green blisters surrounded by yellow-green areas, and tops of newly-infected black or purple raspberry canes often curl downward, blacken, and die. Raspberry mosaic reduces yield of all raspberries, and causes fruit to be dry, seedy, crumbly and lack flavor.

Raspberry leaf curl virus, spread only by the small raspberry aphid (*Aphis rubicola*), causes leaf curl on all types of raspberries but not blackberries. This disease does not often occur west of the Rockies because this aphid is not common there. From the Rockies east, it is less common than the raspberry mosaic complex, but even more destructive, and no cultivars are resistant. Leaves on infected plants are all small, dark green, crinkled and tightly curled downward and inward. Remove any plants with these symptoms as soon as you see them, as the virus reduces yields from infected plants by up to 70%, and it spreads easily to other plants. Berries from infected plants are small, dry, seedy and crumbly.

Raspberry streak, caused by tobacco streak virus, infects black and purple raspberries, but is presently limited to northern Ohio, western New York, and western Pennsylvania. This virus spreads through pollen and possibly by thrips. Numerous purple streaks appear on lower parts of infected canes, and terminal leaves of these canes are often darker green than normal and hooked, twisted or rolled. Fruits from infected canes are smaller than usual, dull, seedy, crumbly and lack flavor, and drupelets often ripen unevenly, giving the fruit a blotched appearance.

Tomato ringspot virus, a soilborne virus spread by the dagger nematode (*Xiphinema americanum*), attacks many species of plants, but only red raspberries among the brambles. This disease is widespread in the Pacific Northwest and the Northeast of North America. Infected plants may appear normal, although somewhat less vigorous, but the most obvious symptom is crumbly berries that fall apart when touched. Where this disease occurs, test the soil for dagger nematodes before planting, and if present fumigate or turn under cover crops that reduce nematodes.

Blackberry sterility, a virus-like disease of presently unknown cause that is genetically transmitted, occurs in all blackberry growing regions of North America, and can infect black raspberries as well as blackberries. Infected plants usually grow more vigorously than healthy plants and bloom heavily, but either fail to set fruit or produce only small, misshapen berries. Remove and burn such plants, including the roots, and do not replant in the spot for several years.

Root and Crown Diseases

Prevent these diseases before planting, as they are much more difficult to control once present in a planting.

Phytophthora root rot (caused by several species of fungi in the genus *Phytophthora*) can decimate red, yellow and purple raspberry plantings, but doesn't usually trouble black raspberries or blackberries. This fungal disease is worst on heavy clay soils, where raised beds should be built before planting. Planting a cover crop of 'Saia' oats, then planting raspberries into raised beds, reduced the disease most in a Massachusetts test. Pre-plant soil treatments of metalaxyl (Ridomil™) and the biocontrol fungus *Trichoderma virens* did not significantly reduce root rot in New York trials. Raspberry cultivars vary greatly in their susceptibility to Phytophthora root rot, and growers with heavy soils should choose adapted cultivars that are most resistant to this disease.

Verticillium wilt, caused by the soilborne fungus *Verticillium dahliae*, seriously reduces raspberry yields by wilting, stunting and eventually killing the fruiting canes or entire plants. The disease hurts black and purple raspberries more than red raspberries, and seldom causes severe losses on blackberries. Verticillium wilt is worst on poorly drained soils following cold, wet springs, and leaves yellow, wilt and drop in summer as the weather turns hot and dry. Avoid planting raspberries on land planted within the past 5 years to Verticillium-susceptible crops, such as tomatoes, potatoes, peppers, melons or strawberries. Cover crops and raised beds used to prevent Phytophthora root rot will also reduce Verticillium wilt.

Crown gall, a bacterial disease caused by *Agrobacterium tumefaciens*, causes swollen galls (Photo 13-24) to form on the roots and crowns of both raspberries and blackberries, which reduce productivity and eventually kill infected plants. The disease is most common on heavy clay soils. Where it occurs, prevent crown gall by dipping or spraying transplant roots with a highly effective biocontrol agent in the same genus, *Agrobacterium radiobacter* (Galltrol™ or Norbac 84™), which colonizes the roots and crown of treated plants and blocks the crown gall pathogen from infecting the plants. If crown gall occurs in a planting, dig out and burn infected plants to prevent its spread.

Photo 13-24. Many brambles are very susceptible to crown gall, which causes these swollen areas at the crown that reduce yields and eventually kill plants. Prevent this by dipping transplants in a biocontrol solution at planting time.

Cane and Leaf Diseases

These diseases are caused by fungi spread by wind or rain, and can cause significant damage. Prevent them with a combination of resistant cultivars, prompt pruning, and fungicide applications.

Anthracnose, caused by *Elsinoe veneta*, attacks all brambles, but is especially serious on black and purple raspberries and some blackberry cultivars, and occurs almost everywhere brambles are planted. Infected leaves and canes develop purplish spots that enlarge, develop gray centers that may drop out of leaves and become sunken on canes over the season (Photo 13-25). Anthracnose stunts the canes, often girdling them before the crop matures, which causes the berries to shrivel and dry, and many canes eventually die. Choosing resistant cultivars can prevent much cane dieback. Even resistant cultivars benefit from one application of lime-sulfur or copper fungicide in spring just after bud break, when new leaves are about 1/8 inch long. Prune out and burn any infected canes as soon as you see them.

Cane blight (caused by *Leptosphaeria coniothyrium*) and **spur blight** (caused by *Didymella applanata*) can cause extensive damage to raspberry canes, especially during wet springs and summers. Cane blight does the most damage to black raspberries, but also occurs on other raspberries and occasionally on blackberries, especially those whose primocanes have been damaged by machine harvesting. Spur blight occurs only on red and purple raspberries. Cane blight causes lateral shoots to wilt and die, weakens the canes, and reduces yields. Both diseases cause purple-brown cankers on the canes, which can continue to produce spores that spread the disease for several years after the canes die. Removing floricanes right after harvest helps prevent these diseases, and the single delayed-dormant lime-sulfur spray for anthracnose provides further control of spur blight.

Raspberry leaf spot, also called Septoria leaf and cane spot because it's caused by *Septoria rubi*,

Photo 13-25. Anthracnose causes purplish spots on leaves and canes that develop gray centers. The disease can greatly reduce yields and kill many canes. Where it's a problem, choose cultivars resistant to anthracnose and apply lime-sulfur or a copper fungicide once each spring, just after bud break.

afflicts both raspberries (Photo 13-26) and trailing blackberries, while most erect and semi-erect blackberry cultivars resist it. Raspberry leaf spot causes small, tan to brown spots on the leaves and canes but usually does not reduce yield or fruit quality, except in the lower Midwest and Southeast, where prolonged wet weather can aggravate the disease enough to cause defoliation. The anthracnose spray will generally provide enough control for raspberry leaf spot also. 'Royalty' purple raspberry is very susceptible to raspberry leaf spot, and may require additional fungicide sprays in humid areas during wet seasons.

Powdery mildew, caused by *Sphaerotheca macularis*, causes a powdery white growth of fungal mycelium on the undersides of leaves, and sometimes on flower buds and fruits, of raspberries. Powdery mildew usually does not attack blackberries or blackberry-raspberry hybrids, and several raspberry cultivars possess moderate to strong resistance. Warm, dry weather with high humidity favors this disease, and in regions where this weather is common in late spring to early summer, choosing species or cultivars resistant to powdery mildew can avoid the need for several fungicide sprays. When powdery mildew requires fungicides beyond the spray for anthracnose, the safest materials include potassium bicarbonate (Kaligreen™) and a biocontrol for powdery mildew (AQ10™).

Yellow rust, also called late leaf rust or autumn rust because symptoms appear in late summer, attacks only red and purple raspberries. Caused by *Pucciniastrum americanum*, this disease is not serious in many areas, but has become increasingly important in California, the upper Midwest and the Northeast, especially Nova Scotia. Small yellow spots appear on mature leaves during harvest, which turn brown and develop spores that spread the disease onto fruits of late-ripening summer cultivars and primocane-fruiting cultivars. Late yellow rust rendered 30% of 'Heritage' fruits unsalable during a bad year in Ohio. The Nova Scotia breeding program has selected cultivars that resist yellow rust, such as 'Nova'. The delayed-dormant spray used for anthracnose and spur blight provides some control, and a Bordeaux mixture spray after harvest can provide additional control. Where yellow rust is severe and growers cultivate between rows rather than using a permanent sod, they can reduce yellow rust by waiting to tie up primocanes until the leaves fall, then disking the leaves under.

Orange rust (Photo 13-27) can severely attack some thorny upright blackberry cultivars, and may also occur on thornless blackberries and black and purple raspberries to a lesser extent, but doesn't attack red raspberries. *Arthuriomyces peckianus* or *Gymnoconia nitens*, a related fungus with a shortened life cycle, cause orange rust. Infected leaves turn yellowish in spring and are soon covered with bright orange pustules full of

Photo 13-26. Raspberry leaf spot, also called Septoria leaf spot, is generally controlled by the early-spring spray for anthracnose. 'Royalty' purple raspberry, shown here, is very susceptible to raspberry leaf spot and may require additional fungicide sprays.

Photo 13-30. Orange rust infects susceptible blackberries systemically, so dig up and burn any infected plants, and protect the planting with spring and fall sprays of a fungicide such as myclobutanil (Nova™). 'Apache', 'Arapaho' and many thorny blackberry cultivars resist orange rust.

spores, which spread the disease as they scatter. Infected canes are spindly and do not bear fruit, and the disease spreads throughout all new canes the plant produces, so infected plants should be dug out and burned. Remove all nearby wild brambles, especially wild blackberries, which can be severely infected. Choose blackberry cultivars resistant to orange rust whenever possible, but susceptible cultivars can be protected with spring and fall sprays of a fungicide such as myclobutanil (Nova™), a sterol inhibitor.

Blackberry rosette or **double blossom** is caused by a fungus (*Cercosporella rubi*) which may chronically infect blackberry plants in the southeastern U.S. Infected plants develop abnormal, broom-like growths of shoots with many thin leaves, and enlarged flower buds with reddish, twisted petals that appear like doubled flowers and do not produce fruit. Many thornless blackberry cultivars have more resistance to this disease than thorny cultivars, and choosing the most resistant cultivars makes sense in the Southeast. If infections occur, cut all canes of infected plants close to the ground and destroy them, and apply a fungicide before and after harvest.

Fruit Rots

The worst fruit rot of raspberries, like strawberries, is **gray mold**, caused by the fungus *Botrytis cinerea*. It also attacks raspberry canes, causing cane Botrytis. This disease may also attack blackberries, but usually isn't a problem on them. Raspberry cultivars differ in their susceptibility to gray mold and cane Botrytis, so avoid highly susceptible cultivars, especially in regions where weather during harvest is often cool and wet, favoring the disease. This is especially important now that strains of the fungus have developed that are no longer controlled by most fungicides, although newer fungicides such as potassium hydroxide (Kaligreen™) or a strobiluron fungicide can help. Harvest raspberries every other day and cool them promptly after harvest to further minimize gray mold. Modified atmosphere packaging helps protect raspberries to be shipped from post-harvest fruit rots.

Leaf-Feeding Insect and Mite Pests

Arthropods that feed on bramble leaves reduce their ability to photosynthesize, thus reducing yields and fruit sugars. Several of these pests also transmit diseases that can kill the plants or greatly reduce their yields.

Raspberry aphids (*Amphorophora agathonica* and *Aphis rubicola*) feed on the undersides of bramble leaves, causing leaves to curl and to photosynthesize less, but the worst damage these aphids do is to spread several viruses. Avoid virus infections by choosing cultivars that are non-attractive to the aphid vectors of these diseases, such as 'Chilliwack' or 'Titan' red raspberries or 'Royalty' purple raspberry, and by starting with certified virus-free plants and removing nearby wild brambles. Natural predators, such as lady beetles and lacewings, and rains usually keep aphids under control. Scout for aphids periodically in spring and summer, and apply control measures if aphids infest more than 10% of the shoots. Sprinkler irrigation, releasing aphid predators, or a selective insecticide application can provide control.

Spider mites (*Tetranychus urticae* and *Eotetranychus carpini borealis*) are tiny but can build up to huge numbers under hot, dry conditions in dusty fields. Leaves appear bronzed after spider mites feed on their undersides, then leaves turn brown and are covered by fine webbing under heavy infestations. Such heavy damage reduces yields and fruit quality, and the adult mites may move onto the fruit, contaminating it. Prevent spider mite outbreaks by growing a groundcover between bramble rows rather than cultivating the soil, by utilizing a sprinkler system both for irrigation and for evaporative cooling, and by using only pesticides that won't hurt the natural enemies of spider mites that usually keep them in check. Monitor about 50 leaves weekly during hot, dry weather, and release predator mites or apply a selective miticide if spider mite populations average more than 15 to 25 mites per leaflet.

Japanese beetles (*Popillia japonica*), where they occur in Eastern North America, can skeletonize bramble leaves during mid- to late-summer. Japanese beetles do the worst damage to brambles during the first two or three years of a planting. Commercial traps using a feeding attractant or sex attractant, placed near but not within the planting, can reduce damage but may not provide adequate control if populations are high. An insecticide treatment after harvest may be needed.

Clay colored weevils (*Otiorhynchus singularis*) are becoming more widespread and damaging in the Pacific Northwest. They feed on developing buds in early spring, and can cause 15 to 20% yield loss. Monitor early-season bud damage and use a beating tray in the evening to monitor weevil populations, and apply an insecticide if needed.

Cane-Feeding Insect Pests

Cane borers can attack all types of brambles, killing the canes and thus greatly reducing yields. These pests include a clearwing moth larvae called the raspberry crown borer, a fly larvae called the raspberry cane maggot, and three beetle larvae: the raspberry cane borer with a round head, and two flat-headed borers, red-necked cane borer and bronze cane borer. Eliminating all wild brambles within 1000 feet of commercial plantings will reduce these pests. Control depends on timely, thorough pruning and one or two well-timed insecticide applications per year.

Wilted tips of young shoots in early summer (Photo 13-28) usually indicate **raspberry cane borer** (*Oberea bimaculata*). The adult female, a ½-inch long black beetle with a yellow thorax and antennae nearly as long as her body, kills each cane tip by making two girdling rows of punctures about ½ inch apart, and laying her eggs between the girdles. Regular scouting for wilted tips in early summer, pruning them off below the girdles and destroying the prunings, can eliminate most eggs before the borers hatch, thus preventing most damage, since brambles benefit from early summer tip pruning. Larvae that hatch will bore down through the canes by fall and into the crown by the next summer, causing extensive damage. Insecticide sprays applied when beetles are present on lower foliage, usually just before bloom, can also prevent some damage. Another insect that also causes wilted stem tips, **raspberry cane maggot** (*Pegomya rubivora*), is also controlled by pruning off the wilted tips during summer, but doesn't require any insecticide sprays.

The flat-headed borers, **red-necked cane borer** (*Agrilus ruficollis*) and **bronze cane borer** (*Agrilus rubicola*), burrow into bramble canes, which develop swellings at those sites on the canes (Photo 13-29), usually within 1 foot of the ground but up to 4 feet off the ground. Infested canes die or are severely weakened. Prune out infested canes as soon as you see them in early summer, and at dormant pruning, and burn or otherwise destroy them. An insecticide spray aimed at adult beetles just before bloom, as for raspberry cane borer, can prevent some damage.

Newly hatched **raspberry crown borer** (*Pennisetia marginata*) larvae crawl down the canes in fall and form blister-like hibernation cavities at the bases of canes. In spring, larvae tunnel into new canes and crowns, girdling them and causing canes to become weak and spindly. All canes from infested crowns can die, and the entire crown may eventually die. The larvae pass the second winter in bramble roots, then pupate inside the plant in the second summer, with moths emerging in late summer to lay eggs on the leaves. Fall insecticide sprays to the lower canes and crown can kill some newly-hatched larvae, but the long life cycle means it takes 2 to 3 years of treatments to reduce populations. Again, promptly remove and destroy infested canes and crowns.

Photo 13-28. Wilted cane tips in early summer usually indicate raspberry crown borer, or sometimes raspberry cane maggot. Prune off and burn wilted tips.

Photo 13-29. Cane borers such as red-necked cane borer tunnel into bramble canes, creating swollen areas where they overwinter. Infested canes die or are severely weakened. Prune out infested canes as soon as you see them in early summer, and at dormant pruning, and burn or otherwise destroy them. Reduce damage with an insecticide spray aimed at adult beetles just before bloom.

Fruit-Feeding Insect Pests

Raspberry fruitworm and **Western raspberry fruitworm**, both in the genus *Byturus*, are beetle larvae that may feed in fruit receptacles and end up contaminating harvested berries, especially blackberries with their retained receptacles, and machine-harvested raspberries. Fall-fruiting raspberries escape injury, because the larvae drop to the ground in early July to pupate in the soil. Adult beetles also cause damage by chewing holes in unfolding leaves and flower buds, which skeletonizes the leaves and causes reduced fruit set, distorted berries, or premature fruit drop. Late-ripening fruit is less prone to damage than early-ripening fruit. Growers should monitor flower buds for small holes and new leaves for long, ragged holes, and if damage occurs should apply an insecticide as blossom buds first appear and again just before flowers open. Good weed control can also minimize this pest.

Tarnished plant bug (*Lygus lineolaris*), **southern green stinkbug** (*Nezara viridula*) and other plant bugs can cause malformed berries with some drupelets failing to develop, similar to the damage to strawberries. Plant bug feeding on mature fruit drains the color from damaged drupelets, so they turn white or lavender (Photo 13-30). Minimize these bugs by controlling weeds and avoiding nearby forage legume crops or clover in the aisles between brambles, or keeping clover closely mowed. Do not mow clover or weeds when brambles are flowering or setting fruit, however, as that can cause the bugs to move from the groundcover or weeds onto the bramble blossoms and fruit. Monitor tarnished plant bug by sampling 50 plants in mid-morning, tapping one flower or fruit cluster per plant over a small white tray to dislodge the bugs. If you find more than an average of 0.5 bugs per cluster, treat with an insecticide just before flowers begin to open, and again when fruits begin to color.

Photo 13-30. Tarnished plant bugs and other plant bugs cause stunted, misshapen berries when they feed on developing bramble fruits, and turn individual drupelets white when they feed on ripe fruit. Reduce such problems by controlling weeds and nearby legume crops, and if more than 0.5 bugs occur per flower cluster, treat with an insecticide just before flowers begin to open and again when fruits begin to color.

Picnic beetles and **sap beetles** (*Glischrochilus* species) and other beetles bore into fully ripe and overripe fruit to lay eggs, and the beetles become contaminants in harvested fruits and infect fruits with fruit-rotting fungi. **Yellowjacket wasps** (species in the genera *Paravespula*, *Vespa* and *Vespula*) are also attracted to fully ripe and injured fruit, and pose a danger to harvesters. Populations of these beetles and yellowjackets build to highest levels in late summer and fall, so late-ripening raspberries are most prone to damage. Prevent most damage by keeping ripening berries off the ground, by trellising or temporary support of canes if necessary, and regularly picking all ripe fruit. Have one worker pick damaged or overripe fruit into plastic bags, place these away from the field to attract and trap picnic and sap beetles, then tie and burn or otherwise destroy the bags to eliminate some beetles. Do not allow workers or customers to bring sweet drinks or other foods that attract yellow jackets into the field. Where yellowjackets pose a problem despite good harvest practices, put traps up around the perimeters of the planting before berries begin to ripen. Commercial traps work for only certain species, but fish traps made with a fish suspended over soapy water work for all yellowjacket species.

Adult **beetles** of three species of **root weevils**, all in the genus *Otiorhynchus*, are the most problematic fruit contaminant in the Pacific Northwest, but thankfully do not occur east of the Rockies. The larvae attack bramble roots, reducing plant vigor and yields. Monitor these weevils with a beating tray from mid-May to late June, and apply an insecticide in late June before harvest to control them.

Various **caterpillars** can also infest harvested fruit, especially machine-harvested fruit. These include **raspberry looper** (*Autographa ampla*), **speckled green fruitworm** (*Orthosia hibisci*), **orange tortrix** (*Argyrotaenia citrana*) and **obliquebanded leafroller** (*Choristoneura rosaceana*). Two to three pre-bloom and bloom sprays of the biocontrol agent *Bacillus thuringiensis* (Bt; many trade names include Dipel, Agree, Javelin, MVP) can control these, and since Bt poses no hazard to humans it can be applied up to the day of harvest if needed.

Weed Management

Control perennial weeds before planting brambles, by treating with glyphosate (Roundup™). Mulching in the row will prevent many annual weeds, while post-emergent sprays of sethoxydim (Poast™) will control grassy weeds, but should not be applied within 45 days of harvest. Early spring or late fall pre-emergent sprays of napropamide (Devrinol™) or oryzalin (Surflan™) prevent many annual broadleaf weeds.

Pacific Northwest bramble growers often till between the rows to control summer annual weeds, using a rotary-type cultivator at a depth of just 3 to 4 inches to avoid excessive damage to brambles' shallow roots. These growers usually seed a winter cover crop after harvest and turn it under in early spring, to provide organic matter and control winter annual weeds. Most Eastern growers plant a permanent sod between bramble rows that chokes out most weeds, and keep the sod mowed during the growing season to minimize competition with the brambles.

HARVEST AND HANDLING

Machines harvest over 75% of all brambles in North America, and the percentage of the crop that is machine harvested is increasing as labor costs rise and laborers are harder to find. Essentially all berries for processing are machine harvested, and newer machines can harvest blackberries, although not yet raspberries, for fresh markets. Fruit sold to individually quick-frozen berry processors can fetch prices almost as high as for the wholesale fresh market, although fruit sold to juice processors generally returns only 30 to 40% of fresh market prices. Growers machine harvest raspberries once every 2 to 3 days, and blackberries once every 5 to 7 days. Manufacturers have improved the efficiency of machine harvesters over the last 20 years, but the machines still damage, fail to remove, or drop to the ground 20 to 25% of the crop.

Growers use various types of self-propelled, over-the-row harvesting machines. The rotary shaker (Photo 13-31) is replacing older machines with horizontal slappers. Vibrating rods pass along both sides of the row, shaking the entire plant, and the fruit fall onto spring-loaded, overlapping catch plates. Conveyor belts then carry the fruit through a pneumatic cleaner that blows off leaves and debris, onto an inspection belt that removes unsound fruit, and into containers. The machine operator controls machine speed and beater speed and centers the machine on the row, so a skilled operator can reduce harvest losses.

Machine harvesting shakes many insects off of raspberry plants that would normally remain on plants that are hand harvested. These insects become objectionable contaminants in harvested fruit, and processing companies dock prices for contaminated lots. Growers who machine harvest must therefore apply an extra insecticide treatment shortly before harvest, to kill insects which wouldn't be considered pests when hand harvesting. Growers must also fine-tune the pneumatic cleaner to blow off as many insects as possible without blowing off any fruit.

The expense of harvesting machines requires at least 25 acres of brambles, or brambles plus other fruits that can be harvested with the same machine (such as blueberries). Many small bramble growers near population centers still rely on hired harvesters and pick-your-own customers (Photo 13-32) to hand-harvest their crops. A grower will need 8 to 10 trained workers to harvest each acre of brambles for sale, about 250 to 300 pick-your-own customers for each acre of raspberries, and about 350 to 400 pick-your-own customers for each acre of blackberries. Train workers or customers to gently harvest all ripe fruit when they are fully colored but still firm and separate easily from the plant. Most raspberry growers try to hand harvest each row every 2 days, while blackberries can be hand harvested every 3 to 4 days.

Whether harvesting by hand or machine, harvest brambles early in the day while temperatures are cooler, directly into the shallow containers

Photo 13-31. Self-propelled, over-the-row harvesting machines, such as this one in red raspberries, harvest 75% of all brambles in North America. (Photo by Larry Tremaine, Tremaine Farms, Lynden, WA.)

Photo 13-32. Fresh bramble growers need 250 to 400 pick-your-own customers and/or 8 to 10 hired harvesters per acre.

they'll be sold in. Place machine-filled flats in the shade until they can be loaded onto trucks to transport them to the processor, which should be within 2 hours. Precool berries to be shipped within 1 hour of harvest with a forced-air cooler, then load them into refrigerated trucks or planes and keep them at 32 to 34 F. Shipping berries are often enclosed in polyethylene film, and the CO_2 level is raised to 30% to reduce decay during transit.

Further Reading

Brambles - production, management and marketing, 1999, Bull. 782-99 from Ohio State Univ., is downloadable from: www.ag.ohio-state.edu/~ohioline/b782/b782

Ellis, M.A., R.H. Converse, R.N. Williams, and B. Williamson. 1991. Compendium of raspberry and blackberry diseases and insects. Amer. Phytopathol. Soc., APS Press, 3340 Pilot Knob Rd., St. Paul, MN 55121-2097. 100 pp.

Pritts, M. and P. Handley. 1994. Bramble production guide. NRAES-55, Northeast Regional Agricultural Engineering Service, Ithaca, NY 14853. 200 pp.

Strik, B.C. and H.K. Cahn. 1999. Pruning and training affect yield but not machine harvest efficiency of 'Meeker' red raspberry. *HortScience* 34:611-614.

Takeda, F. and D.L. Peterson. 1999. Considerations for machine harvesting fresh-market eastern thornless blackberries: trellis design, cane training systems, and mechanical harvester developments. *HortTechnology* 9:16-21.

Vanden Heuvel, J.E., J.A. Sullivan, and J.T.A. Proctor. 2000. Trellising system and cane density affect yield and fruit quality of red raspberry. *HortScience* 35:1215-1219.

Wilcox, W.F., M.P. Pritts, and M.J. Kelly. 1999. Integrated control of Phytophthora root rot of red

Chapter 14
Grapes

Grapes produce the world's largest fruit crop, both in acreage and dollar value. Vineyardists raise different species and cultivars of grapes for specific purposes: wine, juice, table grapes and raisins. Wine production accounts for 54% of U.S. grape acreage and an even greater percentage of European acreage. In the U.S., 24% of grapes are grown for raisins, 14% for table grapes, and 8% for unfermented grape juice and jelly.

All grapes belong to the genus *Vitis*, in the family Vitaceae. These names mean "vine" and "vine family", reflecting the fact that all members are perennial, woody, deciduous, climbing vines. Grape plants have clasping tendrils opposite many leaves, which wrap around tree branches in the wild or support trellises under cultivation, allowing the vines to climb toward the light they seek. Grape leaves are simple and palmately lobed. Flowers are inconspicuous and occur in clusters, with flower parts in fives or multiples of five. The fruit are botanical berries, with two to five seeds, or seed traces in seedless cultivars, per berry.

TYPES OF GRAPES

The genus *Vitis* is divided into two subgenera, *Euvitis* and *Muscadinia*, both diploids, with two sets of chromosomes. *Euvitis* includes the bunch grapes native to Asia, Europe and North America, which have 38 chromosomes. **Bunch grapes** (Photo 14-1) ripen over a concentrated period, so that entire clusters are harvested together, and the individual berries are elongated.

Muscadine grapes (Photo 14-2), in the subgenus *Muscadinia* with 40 chromosomes, have round berries with thick skins that ripen over an extended period, so are often harvested as individual berries. Various bunch grape species are adapted to a wide range of climates in temperate areas, and breeding efforts have produced some cultivars that are cold-tolerant to USDA zone 4 and others that have low chilling needs and grow in USDA zone 10 and subtropical climates. Muscadine grapes have a more limited range of adaptation, from USDA zones 7 to 10. Both types of grapes are long-season crops that ripen relatively late in the growing season.

The bunch grapes account for almost all grape acreage, and include many species. ***Vitis vinifera***, native to Asia Minor in the region of the Black and Caspian Seas, has been cultivated the longest, for over 10,000 years. Widely disseminated in the

Photo 14-1. Bunch grapes, such as this 'Monticello' American grape, produce somewhat elongated berries that ripen over a concentrated period, allowing entire clusters to be harvested. Bunch grapes account for almost all grape acreage.

Middle East and Europe, cultivars of this species are often called "European grapes" or "*vinifera* grapes". Over 5000 cultivars of wine grapes and seedless table and raisin grapes derived from this species. However, it is susceptible to cold and several important grape diseases and insects, so pure *vinifera* cultivars are difficult to grow outside of warm, dry climates with low humidity and little pest pressure. The Mediterranean climates of southern Europe, California, and parts of Australia and Argentina create ideal conditions for *vinifera* grapes.

"American bunch grape" cultivars derive from several bunch grape species native to North America, which have been crossed with each other and with *V. vinifera*. *Vitis labrusca*, the fox grape, contributes its strong "foxy" flavor to cultivars such as 'Concord' and 'Niagara', grown mainly for juice and in a limited local way for seeded table grapes. *Vitis riparia*, the riverbank grape, has contributed to several grape cultivars. Several other native American species, including the rock or mountain grape (*V. rupestris*) and the summer or winter grape (*V. aestivalis*), have been used to breed various grape cultivars and rootstocks. Breeding programs in New York, Arkansas and other states have developed several hardy, seedless table grape cultivars by crossing seedless *vinifera* grapes with cold-hardy and disease-resistant American cultivars. American grapes generally have "slipskin" berries, in which the thick skin separates from the flesh and is not eaten, but several of the seedless hybrids have adherent, thin, edible skin like those of *vinifera* grapes.

"French hybrid" wine grapes resulted when breeders and nurserymen in France crossed French *vinifera* cultivars with American species and cultivars with resistance to downy and powdery mildew and phylloxera, a soil insect pest. Many French hybrids combine the excellent wine quality of *vinifera* grapes with the resistance to pests and cold of American bunch grapes, and grow well in areas with greater winter cold, rainfall and pest pressure than do *vinifera* grapes. French hybrids also produce higher yields than American bunch grapes where spring frosts are common, because if the primary buds or developing shoots are destroyed by frosts, French hybrids can produce a commercial crop from secondary or even tertiary buds. When frosts do not hurt the primary shoots, however, French hybrids require much pruning or cluster thinning or they will overbear, resulting in much smaller fruit of lower quality.

Muscadine grape cultivars derive mainly from the muscadine grape species, *Vitis rotundifolia*, with some genes from a related *Muscadinia* species, the Munson or bird grape, *V. munsoniana*. Muscadine grapes show considerable resistance to many grape diseases and are well adapted to the hot and humid Southeastern U.S., where they are native. Berries are very large and round, generally with thick skins that are not eaten, although some newer cultivars have edible skins. The muscadine grape was the first American grape species to be cultivated, and the industry is currently expanding in the Southeast for both wine production and pick-your-own fresh fruit.

Photo 14-2. Muscadine grapes produce large, round berries that often ripen unevenly, so are usually harvested individually. Muscadines are native to and adapted to the Southeastern U.S.

SELECTING AND PREPARING A VINEYARD SITE

Many grape cultivars ripen late in the growing season and are prone to damage from both winter cold and summer heat, so select a site that provides a sufficiently long growing season for the cultivars you plan to grow, as well as protection from temperature extremes. Planting near a body of water, if possible, will provide a longer growing season and moderate both winter and summer temperatures. Sloping sites generally offer excellent air drainage, which prevents much winter or spring frost damage, if the bottom 50 feet of the slope are left unplanted. Do not plant closer than 75 feet from thick woods, which harbor colder air in winter and spring. Avoid slopes greater than 15%, as they make equipment operation difficult, unless terraced on the contour.

Where growing seasons are short and summers are cool, as in the Northern U.S. and Canada, a slope facing southeast, south or southwest can make it possible to grow later-ripening cultivars, by advancing spring growth and ripening times and providing more needed summer heat. However, where growing seasons are long and summer heat stress can reduce grape quality, as in the Southern U.S., avoid such slopes, as they can aggravate both winter and summer injury. Slopes facing east, northeast or north are best in the South, as vines will be less subject to winter sunscald, start growth later in spring to avoid much frost damage, and stay cooler in summer. Consider the type of grape and cultivars you'll grow when choosing the site. European, French hybrid and muscadine grapes generally need 2500 to 3500 hours of temperatures above 50 F (heat units) during the growing season, while American bunch grapes do best with 1800 to 2500 heat units, and seedless American hybrid grapes need 2500 to 3000 heat units.

The wide variety of grape cultivars and rootstocks ensures that some will grow well on almost any soil type, from clay to sandy to gravelly soil, from acid to alkaline soils, and from well drained to somewhat poorly-drained. Vineyard soils should generally be well drained, and must have a rooting depth of at least 30 to 40 inches. Grapes generally need irrigation, so the site must have a water source.

Consider your neighbors' farming operations when choosing a site for grapes. Many grain growers use the broadleaf herbicide 2,4-D, and drift from this herbicide can cause extreme damage to many grape cultivars. Avoid sites downwind from grain fields, or select only grape cultivars with resistance to 2,4-D.

Site preparation.

Test soil for pH, phosphorous (P), potassium (K), and any other nutrients that may be deficient in that soil type, at least 6 months before spring planting, to allow time for any needed ammendments to disburse through the soil. Grapes can thrive under a wide pH range of 5.5 to 8.5 by choosing an adapted rootstock; add lime well before planting if the pH is less than 5.5, or add sulfur if the pH is greater than 8.5. Add P fertilizer if soil P levels are less than 10 ppm, and add K fertilizer if K soil levels are less than 100 ppm.

Also test the soil for harmful nematodes, including root knot, dagger and ring nematodes. If high levels of such nematodes exist, turn under canola and/or grass cover crops that reduce such nematodes (see chapter 2), or fumigate the soil before planting. If the site was previously in cotton or alfalfa, which can each cause a large buildup in a particular root knot nematode species, plant the land to nematode–reducing crops for 18 months before planting grapes. When replanting an old vineyard site, fumigate the soil before planting to kill old roots and nematodes in the top 5 to 6 feet of soil.

Where nematodes aren't a problem and soil pH doesn't need adjusting, and the site is in pasture grass or hay that would make a good groundcover between the grape rows, retain the sod and prepare only the planting rows. Apply needed P and K to a 2-foot-wide strip along each row and till them in, or if no ammendments are needed, kill the sod in a 2-foot strip with glyphosate (Roundup™) and plant directly into the killed sod.

If the site must be cultivated before planting but nematodes aren't a problem, grow a winter legume cover crop such as winter peas, bell beans, or purple or white vetch, to turn under before spring planting. This can eliminate the need to add nitrogen (N) at planting time and add organic matter, but most legume cover crops cause nematodes to increase, so avoid them where nematodes are a problem.

If the soil is somewhat poorly drained, construct raised beds and choose a rootstock that can withstand somewhat poor soil drainage. Most grape growers do not need to construct raised beds.

CHOOSING CULTIVARS

Choose grape cultivars that are adapted to your climate, suited to your market, and resist as many pests of your area as possible. Tables 14-1 to 14-8, pages 204-218, detail the zones of adaptation, disease resistance and other qualities of many popular grape cultivars.

***Vinifera* wine grapes** (Table 14-1) are most prized for wine production, especially for "varietal" wine made from mainly one cultivar, but these grapes are the most cold-tender and disease-susceptible so can only be grown in favorable areas. Famous cultivars such as 'Chardonnay' and 'Cabernet Sauvignon' (Photo 14-3, page 220) predominate in California, and are also grown in a few frost-protected sites in the East, Midwest and Northwest of North America.

French hybrid wine grapes (Table 14-2) are generally more disease-resistant and frost-tolerant than pure *vinifera* cultivars, so predominate among wine grapes in Eastern North America, and are also grown in California and other Western areas. Some of the most widely grown French hybrids include 'Seyval' (Photo 14-4), 'Vidal Blanc', 'Baco Noir' and 'Chancellor'.

Table 14-1.

VINIFERA WINE GRAPES

in approximate order of ripening (varies with area).

Cultivar	Winter Hardy to Zone	Summer Climate Needs	Berry Color	Wine Style	Susceptibility to Powdery Mildew	Susceptibility to Bunch Rot	Susceptibility to Black Rot	Susceptibility to Downy Mildew	Vine Vigor	Type of Training	Type of Pruning	Yields	Harvest Season
Pinot noir	6b	cool	black	burgundy	VS	S	VS	VS	medium	head	cane	low	early
Chardonnay	6b	cool	white	semi-sweet	S	VS	VS	VS	high	cordon	cane	low	early
Gewurztraminer	7a	cool	bronze	spicy	S	VS	VS	VS	medium	head	cane	medium	early
Semillon	6a	warm	white	sauterne	MS	S	S	VS	high	head	cane	medium	early-mid
Sauvignon blanc	7a	warm	white	sauterne	VS	S	S	S	very high	head	cane	medium	early-mid
Chenin blanc	7a	warm	white	sweet	S	VS	S	S	very high	cordon	cane	high	early-mid
Thompson Seedless	7b	hot	white	bulk blends	S	MR	S	MR	very high	head	cane	very high	middle
Tokay	7b	warm	red	brandy	VS	S	S	S	high	cordon	spur	high	middle
French Colombard	7a	warm	white	dry, acid	MR	MS	VS	S	high	cordon	spur or cane	high	middle
Zinfandel	**7a**	cool	red	fruity, spicy	MR	VS	S	S	medium	cordon	spur or cane	high	middle
Grenache	**7b**	**warm-hot**	red	rose & port	MR	S	S	MR	very high	cordon	spur	high	middle
Rubired	**7b**	**hot**	red	port, blends	MR	MR	S	S	high	cordon	spur	high	middle
White Riesling	**6b**	**cool**	white	fruity, spicy	MS	VS	VS	VS	high	cordon	cane	medium	mid-late
Merlot	**6b**	**warm**	black	soft, full	VS	MR	S	VS	medium	head	cane	medium	mid-late
Barbera	6b	hot	black	acid blends	MR	S	S	S	medium	head	spur or cane	medium	mid-late
Muscat of Alexandria	7b	warm	white	muscatel	VS	MR	S	S	medium	cordon	spur	high	mid-late
Cabernet Sauvignon	6b	warm	black	dry claret	S	MR	VS	VS	high	cordon	cane	low	late
Burger	7a	hot	white	bulk blends	S	VS	S	S	high	cordon	spur or cane	high	late
Ruby Cabernet	7a	warm-hot	red	blends	VS	MR	S	S	high	cordon	spur	high	very late
Carignane	**7a**	**warm**	black	bulk blends	VS	VS	S	S	very high	cordon	spur	very high	very late

Disease susceptibility ratings: VS = very susceptible, S = susceptible, MS = moderately susceptible, MR = moderately resistant

Average ripening times

at Fresno, CA: early: late July, early-mid: early Aug., middle: mid-Aug., mid-late: late Aug. to early Sept., late: mid-Sept. very late: late Sept.

at Medford, OR & Geneva, NY: early: mid-late Sept., early-mid: late Sept. to early Oct., middle: mid-Oct., mid-late: mid- to late Oct.; late: late Oct., very late: doesn't ripen

Notes on Cultivars (from France unless specified otherwise):

Pinot noir needs cool, dry summers; fruit sunburn in hot areas and crack in rain. "Mariafeld" is a clone with greater bunch rot resistance.

Gewurztraminer, a Geman cultivar, needs cool, dry summers, as warm regions produce overly acid wines and fruit cracks in wet weather.

Chardonnay, the most-planted white wine grape, breaks bud early, so needs the least frost-prone site. Summer hedge to reduce excess vigor.

Semillon makes excellent sweet wines or sauternes, where "noble rot" (Botrytis) is allowed to grow on berries to create unique flavor.

Sauvignon blanc is often blended with Semillon to make sweet wines or sauternes.

Chenin blanc is used to make naturally sweet, dry and sparkling wines.

Thompson Seedless, an old table/raisin grape from Iran, is harvested in large amounts in CA for bulk white wine blends, to increase sugars.

Tokay, from Algeria, was once the leading CA table grape, but has seeds, so is now used for wine blends, especially brandy.

French Colombard, the second most-planted white wine grape in CA after Chardonnay, is used for dry and sparkling wines.

Zinfandel ripens irregularly, is very susceptible to cluster rot, and tends to raisin, but produces a uniquely fruity, spicy wine.

Grenache, a Spanish cultivar, makes rose wines and is blended for darker port wines.

Rubired has very intense red juice, used for blends, port wines and concentrates.

White Riesling, from Germany, needs cool, dry summers; fruit crack and rot if rains before harvest. Late bud break prevents frost damage.

Merlot ages faster than Cabernet Sauvignon, with which it is often blended.

Barbera, from Italy, has very high acidity, useful for blending into red table wines in warm regions.

Muscat of Alexandria, a very old North African raisin/table grape, is used for musacatel, a dessert wine, in CA. Fruit sunburn in hot areas.

Cabernet Sauvignon is the most-planted red wine grape in CA Summer hedge or use a divided canopy to control excess vigor.

Burger produces very high yields, used to blend in bulk white wine.

Ruby Cabernet is a hybrid of Carignane and Cabernet Sauvignon, used for blends.

Carignane, from **Spain, is used for** bulk red wine blends.

Table 14.2

FRENCH HYBRID AND OTHER HYBRID WINE GRAPES

in approximate order of ripening (varies with area).

All are cordon-trained and spur-pruned.

Cultivar	Winter Hardy to Zone	Summer Climate Needs	Berry Color	Wine Style	Susceptibility to Powdery Mildew	Bunch Rot	Black Rot	Downy Mildew	Anthrac-nose	Pho-mopsis	Vine Vigor	Yields	Harvest Season
Aurore	5b	warm-cool	white	delicate	MS	VS	S	MR	MS	MR	med-high	med-high	very early
Leon Millot	4b	warm-cool	black	burgundy	S	VS	MS	MS	S	S	med-high	med-high	very early
Foch	4b	cool-warm	black	burgundy	MR	MS	MS	MR	S	S	medium	medium	early
St. Croix	3a	cool	blue	claret	MS	MS	MS	MS	S	S	high	medium	early
Cayuga White	5b	warm-cool	white	sweet or dry	MR	MR	MR	MS	VS	MR	high	very high	early
Chancellor	4b	cool-warm	black	Bordeaux	S	MR	MR	VS	S	VS	medium	medium	middle
Seyval	5b	warm-hot	white	fruity, dry	MS	VS	MS	MR	MS	S	medium	med-high	middle
Vignoles	5a	warm-cool	white	dessert	S	S	MR	MR	VS	S	medium	low-med	middle
Baco Noir	6a	warm-hot	black	burgundy	S	MS	VS	MR	MS	MR	very high	medium	middle
Melody	5b	cool-warm	white	light, floral	MR	MR	VS	MS	S	S	high	high	middle
Chardonel	6a	cool-warm	white	semi-sweet	MS	MR	S	MS	S	S	medium	high	middle
Vidal Blanc	6a	warm	white	semi-dry	VS	MR	S	MS	VS	MR	high	very high	mid-late
Traminette	5b	cool	white	spicy, fruity	MR	MR	MR	S	S	S	medium	medium	mid-late
Chambourcin	6b	warm-cool	blue	dry claret	MS	MS	MS	MR	MS	S	low	high	late
Villard Blanc	6a	warm-hot	white	dry, lemony	VS	MR	MS	MS	VS	S	medium	high	late
MO Riesling	5a	warm	white	spicy, fruity	S	MS	MR	VS	S	S	high	high	late
Villard Noir	4b	warm-hot	blue	dry blends	VS	MR	S	MR	S	S	medium	high	late
Couderc Noir	6a	warm-hot	black	rose, blends	S	MS	MR	MR	S	S	low	low	late

Disease susceptibility ratings:
VS = very susceptible, S = susceptible, MS = moderately susceptible, MR = moderately resistant, R = resistant

Average ripening times

at Fresno, CA: very early: mid-July, early: late July to early Aug., middle: mid- to late Aug., mid-late: late Aug. to early Sept., late: mid- to late Sept.
at Mountain Grove, MO: very early: early Aug., early: mid- to late Aug., middle: early Sept., mid-late: mid-Sept., late: late Sept.
at Geneva, NY: very early: late Aug., early: early to mid-Sept., middle: late Sept., mid-late: early Oct., late: mid-Oct. (may not ripen)

Notes on Cultivars:

Aurore is a dependable producer that breaks bud late and ripens early, so is widely planted though wine quality is fair. Berries crack in wet weather, so harvest promptly. It's VS to angular leaf spot.

Leon Millot wine has distinct berry aromas, like Foch, but vines are more vigorous, productive. Protect from birds. It's MR to angular leaf spot.

Foch needs a fertile site and a vigorous, phylloxera-resistant rootstock as it's susceptible. The early, small black berries attract birds, so protect. Don't use sulfur. It's MR to angular leaf spot.

St. Croix is very hardy, very precocious in bearing, and may need cluster thinning to prevent overbearing.

Cayuga White, from NY in 1972, makes semi-sweet, fruity wines or dry wines with oak aging.

Chancellor breaks bud early so is prone to frosts, needs the best sites. It's prone to crown gall if winter-damaged after overcopping, so cluster-thin heavy crops. Don't use sulfur or copper. It's VS to angular leaf spot.

Seyval does best on fertile soils, or grafted on a vigorous rootstock. It breaks bud early so needs the least frost-prone sites. Thin clusters to prevent overbearing.

Vignoles breaks bud late, so resists spring frosts. Fruit develops very high sugar content while retaining acidity if allowed to hang longer before harvest.

Baco Noir breaks bud very early so needs frost protection, and is virus-sensitive so grow on a virus-resistant, less vigorous rootstock.

Melody, from NY in 1985, yields heavily with no cluster thinning, if well-exposed canes are chosen when pruning.

Chardonel is a 1990 NY cross between Chardonnay, the leading white wine grape, and Seyval. It makes wine like Chardonnay, without cluster thinning.

Vidal Blanc breaks bud very late, resisting frosts. Use a virus-resistant, invigorating rootstock. Prune heavily, cluster thin to avoid overcropping.

Traminette, a Gewurztraminer hybrid from NY in 1996, produces similar wines, but is hardier and more disease-resistant.

Chambourcin produces fine wine but needs a site less prone to cold, an invigorating rootstock and extra nitrogen to stimulate growth, and cluster thinning to prevent overbearing. Don't use sulfur.

Villard Blanc needs a long season, moderate winters. Use an invigorating rootstock. Cluster thin to prevent overcropping.

Missouri Riesling needs a long season. It makes a spicy wine like 'Catawba' American grape, not like a true 'Riesling'.

Villard Noir needs a long season and an invigorating rootstock. Cluster thin to prevent overcropping.

Couderc Noir needs a long season and a fertile site or invigorating rootstock.

Table 14-3.

SEEDLESS VINIFERA TABLE AND RAISIN GRAPES

for USDA zones 7b-10,

In approximate order of ripening (varies with area).

Cultivar	Summer Climate Needs	Berry				Cluster		Susceptibility to		Vine Vigor	Type of Pruning	Yields	Harvest Season
		Color	Flavor	Sugars	Size	Size	Density	Powdery Mildew	Bunch Rot				
Perlette	hot-cool	white	good	high	medium	large	very compact	S	VS	high	spur	medium	very early
Beauty Seedless	hot-warm	black	good	medium	medium	small-med.	medium	S	S	high	spur	medium	very early
Black Emerald	hot-cool	black	good	medium	medium	medium	compact	S	S	med.-high	spur	high	very early
DOVine	hot	white	good	high	medium	large	medium	VS	S	very high	cane	very high	early
Flame Seedless	hot-warm	red	best	high	medium	medium	loose	S	MS	medium	spur	very high	early
Fiesta	hot	white	very good	high	medium	large	medium	VS	VS	high	cane	very high	early
Thompson Seedless	hot	white	good	very high	medium	large	medium	S	MR	very high	cane	very high	middle
Sugraone	warm	white	good	very high	large	large	compact	S	S	medium	cane	medium	middle
Fantasy Seedless	hot-warm	black	good	high	med-large	large	compact	S	VS	very high	cane	medium	middle
Ruby Seedless	hot	red	very good	high	medium	very large	compact	VS	VS	very high	spur	very high	middle
Black Monukka	hot-cool	black	very good	high	medium	large	loose	S	MS	very high	cane	very high	middle
Princess (Melissa)	hot	white	good	high	large	med.-large	medium	S	S	high	cane	very high	mid-late
Emerald Seedless	hot-cool	white	good	high	med-large	large	loose	S	S	high	spur	high	mid-late
Summer Royal	hot-warm	blue	good	high	medium	medium	loose	S	S	medium	spur	high	mid-late
Blush Seedless	hot-warm	red	very good	medium	large	medium	medium	S	S	very high	spur	very high	late
Autumn Seedless	hot-warm	white	good	high	large	large	medium	S	S	medium	cane	very high	late
Redglobe	hot-warm	red	fair	low	very large	very large	medium	S	S	low	spur	high	late
Crimson Seedless	hot-warm	red	best	high	medium	medium	medium	S	MS	very high	spur	high	very late
Autumn Royal	hot-cool	black	very good	high	large	very large	compact	S	VS	med.-high	spur	very high	very late

Disease susceptibility ratings: VS = very susceptible, S = susceptible, MS = moderately susceptible, MR = moderately resistant

Average ripening times at Fresno, CA: very early: early to mid-July, early: late July to early Aug., middle: early to mid-Aug., mid-late: mid- to late Aug., late: early to mid-Sept., very late = late Sept. to mid-Oct.

Notes on Cultivars:

Perlette fruit has mild flavor, rarely sunburns. Very compact clusters will rot severely unless heavily thinned with brushes; GA thinning sprays don't work. Apply GA or girdle at berry set to increase fruit size.

Beauty Seedless needs thinning to 2 shoots per spur at 6-8 inches, cluster thinning to 1 cluster per shoot, a GA thinning spray and 1-2 GA sizing sprays or girdles.

Black Emerald, a 1997 CA release, needs one 40 ppm GA spray or girdling at berry set to increase fruit size, and may need cluster thinning to reduce overbearing.

DOVine is a raisin grape suited for drying on the vine, that ripens 2-3 weeks before Thompson Seedless. It needs an open gable or overhead trellis.

Flame Seedless berries are crisp, with fine flavor. Thin shoots when 6-8 inches long, also thin clusters to 1 per shoot for larger berries and use GA. Birds love this cultivar, so must protect.

Fiesta ripens 10 to 14 days earlier than Thompson Seedless so does better for DOV raisins, with a distinct muscat flavor. Plantings are increasing.

Thompson Seedless needs heat and both GA thinning and sizing sprays for best fruit quality. Do not thin shoots. After berry set and cluster elongation, remove poor-quality clusters, leaving 8 per cane.

Sugraone responds to summer evaporative cooling irrigation to increase fruit quality, postharvest N fertilizer and irrigation to improve canes for next year, and GA at bloom to reduce bunch compactness.

Fantasy Seedless produces large berries without any GA sprays (phytotoxic to it), girdling or cluster thinning. Berries have thin skin subject to cracking and bunch rot, so remove basal leaves to reduce this.

Ruby Seedless needs frequent powdery mildew sprays to prevent fruit cracking and severe bunch rot. Thin shoots when 8-14 inches long, and after berry set, remove alternate branches on each cluster to reduce compactness and bunch rot. Do not apply GA.

Black Monukka has mild flavor, develops seeds in some years. It starts growth early in spring, so needs best sites and frost protection. Do not thin shoots or apply GA. It needs bird protection.

Pricess (Melissa) produces large berries without GA if girdled at berry set, or very large berries with lower rates of GA at berry set.

Emerald Seedless

Summer Royal, released by CA in 1999, needs just one 40 ppm GA spray at berry set to increase fruit size.

Blush Seedless needs shoot thinning when 6-8 inches long. Overcropped vines are more susceptible to mildew and bunch rot, so thin flower clusters, leaving 25-30 per vine.

Autumn Seedless needs no shoot thinning or GA sprays, which can cause poor fruit set; it has naturally large berries.

Redglobe is seeded, but still accounts for 13% of table grape production as it needs no shoot thinning, cluster thinning, girdling, GA or ethephon sprays to produce huge, well-colored berries. Use leaf removal 7-10 days before harvest to increase color.

Crimson Seedless needs a GA bloom thinning spray and girdling at berry set to increase fruit size. Do not use a GA sizing spray; that reduces fruit color and causes berry shatter at harvest.

Autumn Royal produces naturally-large berries with no GA sizing sprays or girdling. Use a GA bloom thinning spray to reduce cluster compactness, and tip clusters to reduce excess cluster size and crop load.

Table 14-4.

SEEDLESS AMERICAN HYBRID TABLE GRAPES

in approximate order of ripening (varies with area).

All are cordon-trained, and pruned to short or long canes.

Cultivar	Winter Hardy to Zone	Summer Climate Needs	Berry				Cluster		Susceptibility to						Vine Vigor	Yields	Harvest Season
			Color	Flavor	Sugars	Size	Size	Density	Powdery Mildew	Bunch Rot	Black Rot	Downy Mildew	Anthrac-nose	Angular Leaf Spot			
Venus	6b	hot	blue	fair	medium	large	large	tight	MR	VS	MR	MS	S	MR	high	medium	very early
Interlaken	6a	cool	white	good	medium	medium	medium	compact	VS	S	VS	MS	S	MR	medium	medium	very early
Himrod	6b	warm-cool	white	best	high	medium	medium	very loose	MS	MR	MS	MR	VS	MR	high	low	early
Jupiter	5b	warm-hot	blue	very good	high	large	large	compact	MR	MR	MR	MS	MR	MR	high	high	early
Reliance	5a	hot-cool	red	best	very high	medium	medium	loose	MR	MR	MR	S	VS	VS	high	medium	early
Canadice	6b	cool	red	very good	very high	medium	medium	tight	MS	S	VS	MS	S	S	medium	high	early
Einset Seedless	6a	cool-warm	red	best	high	medium	medium	compact	VS	MS	VS	MS	S	S	medium	high	early
Neptune	6a	warm-hot	white	very good	high	medium	very large	tight	MS	MS	MR	MS	MR	MR	medium	medium	early-mid
Mars	5b	warm-hot	blue	good	medium	large	small	compact	R	MR	MR	MS	MR	MR	med-high	high	early-mid
Saturn	5b	warm-hot	red	very good	high	large	medium	compact	VS	S	S	S	MR	MS	medium	high	early-mid
Challenger	6b	warm-hot	red	very good	very high	large	medium	loose	VS	MS	VS	S	VS	S	medium	medium	early-mid
Vanessa	5b	cool-warm	red	good	high	medium	medium	compact	MS	S	VS	MR	S	S	medium	medium	middle
Glenora	5b	cool	blue	good	high	small	small	compact	VS	S	VS	VS	VS	S	medium	low	middle
Lakemont	6b	cool-warm	white	good	high	medium	large	compact	MS	VS	S	VS	MS	S	high	high	middle
Marquis	6b	cool-warm	white	best	high	large	very large	loose	S	MR	MR	S	S	S	medium	high	middle
Suffolk Red	7a	warm	red	best	high	large	medium	very loose	S	S	VS	MS	S	S	high	low	middle
Remaily Seedless	6b	cool	white	fair	high	large	large	compact	S	MS	S	MS	S	MR	medium	high	mid-late
Concord Seedless	5a	cool	blue	fair	medium	small	small	loose	MS	MR	S	MR	MS	S	medium	low	mid-late
Romulus	5b	cool	white	good	medium	small	large	compact	MS	VS	MS	S	S	S	medium	medium	late

Disease susceptibility ratings: VS = very susceptible, S = susceptible, MS = moderately susceptible, MR = moderately resistant, R = resistant

Average ripening times

at Mountain Grove, MO: very early: early Aug., early: mid-Aug., early-mid: late Aug., middle: early Sept., mid-late: mid-Sept., late: late

at Geneva, NY: very early: late Aug., early: early Sept, early-mid: mid-Sept., middle: late Sept., mid-late: early Oct., late: mid-Oct. (may not ripen)

Notes on Cultivars:

Venus needs hot summers, as its seed traces become seeds in Northern areas. Skin is tough. Harvest as soon as ripe to avoid bunch rot.

Interlaken needs a less frost-prone site. It often overbears, so thin clusters to avoid excess winter damage. It responds to GA.

Himrod responds well to GA sprays to increase berry size and fill clusters better. Harvest promptly, as clusters shatter when ripe.

Jupiter has a mild muscat flavor and adherent skin that resists cracking. Thin clusters to reduce overcropping.

Reliance has fine spicy flavor, but thin skin may crack during rainy seasons. Do not use GA or girdle; those increase cracking. Cluster thin after bloom. Use leaf removal to increase fruit color.

Canadice has flavor like Delaware. Do not use GA: it causes overly tight clusters and fruit cracking. Use plastic brushes before bloom to reduce cluster tightness. Thin to 24-40 clusters per vine after fruit set. Harvest promptly to avoid bunch rot.

Einset Seedless skin resists fruit cracking and bunch rot. It doesn't respond well to GA or girdling to enlarge berries, and needs cluster thinning to avoid overcropping.

Neptune has a fruity flavor, crisp texture, and adherent skin that resists cracking. Do not thin clusters, as that can cause overly tight clusters prone to rot.

Mars is slipskin with a labrusca flavor somewhat like 'Concord'. It resists most diseases. Thin to 24-50 clusters per vine after bloom.

Saturn has crisp texture and flavor like vinifera grapes, but skin may crack and seed traces get woody in Northern areas. Thin clusters to prevent uneven ripening.

Challenger has crisp fruit like vinifera seedless grapes, with thick, adherent skin but 1 or 2 soft seeds. It responds to GA and girdling.

Vanessa has crisp, high-quality fruit with thin skin, but skin may crack in wet weather, and seed traces are noticeable some years. Don't girdle or use a GA thinning spray, but apply a 25 ppm GA sizing spray when berries reach 5-7 mm. Thin to 25-35 clusters per vine.

Glenora needs cool summers; fruit develop little flavor in heat. It responds to GA and girdling to increase berry and cluster size.

Lakemont needs cluster thinning to prevent overbearing. Harvest as soon as ripe to avoid fruit cracking and bunch rot.

Marquis has the largest fruit of hardy white seedless grapes, and thick skin that resists rain cracking. Do not use GA; it damages the plant. Thin clusters to prevent overbearing.

Suffolk Red needs fertile soils to fruit well. Overly loose clusters respond well to GA thinning. Remove basal leaves to increase fruit color. Don't use sulfur.

Remaily needs cluster thinning and GA to produce clusters the size of 'Thompson Seedless', with similar neutral flavor. Don't remove leaves; fruit russet and sunburn.

Concord Seedless is a bud sport of 'Concord', with similar flavor. Low yields limit plantings. Don't use sulfur.

Romulus ripens too late for some Northern areas. Thick skin can crack severely in humid climates, leading to fruit rots. Cluster thin and harvest promptly.

Table 14-5.

SEEDED AMERICAN GRAPES FOR JUICE, JELLY, WINES AND SEEDED TABLE GRAPES

in approximate order of ripening (varies with area).

All are cordon-trained (or trained to multiple trunks in far northern areas) and cane-pruned.

	Winter Hardy to	Summer Climate	Berry				Cluster	Susceptibility to						Vine		Harvest
Cultivar	Zone	Needs	Color	Flavor	Sugars	Size	Size	Powdery Mildew	Bunch Rot	Black Rot	Downy Mildew	Anthrac-nose	Angular Leaf Spot	Vigor	Yields	Season
Valiant	4a	cool	blue	good	high	small	small	MS	S	S	VS	S	S	high	medium	very early
Edelweiss	4a	cool	white	very good	medium	medium	medium	MS	MS	MS	R	S	S	high	high	very early
Seneca	5b	cool	white	best	very high	medium	medium	VS	S	VS	MS	S	VS	medium	high	early
Alwood	5b	warm	blue	good	very high	medium	medium	MS	MS	MR	MR	S	S	high	high	early
Price	6a	warm	blue	very good	medium	medium	medium	MS	S	MR	MS	S	S	high	high	early
Monticello	6a	warm	red	best	high	large	large	MS	MS	S	MS	S	S	high	medium	early
Delaware	5b	cool	red	best	high	small	small	MS	MS	MS	VS	S	MR	medium	medium	middle
Niagara	5b	cool	white	very good	high	large	large	S	MS	VS	VS	S	MR	high	very high	middle
NY Muscat	5b	cool-warm	red	very good	high	medium	medium	VS	S	S	MS	S	MR	medium	high	middle
Alden	5b	cool	red	very good	medium	large	large	S	S	MS	MS	S	S	medium	high	middle
Sunbelt	6a	warm-hot	blue	fair	medium	large	small	R	MR	MR	R	MR	S	high	medium	middle
Vinered	5a	cool	red	best	high	large	large	MS	MS	S	MS	VS	S	high	high	mid-late
Steuben	5b	warm-cool	blue	very good	high	large	large	MR	MS	MS	MR	S	S	high	high	mid-late
Yates	6a	warm	red	good	high	large	large	S	MS	S	S	S	S	high	high	mid-late
Concord	4b	cool	blue	fair	medium	medium	medium	S	MS	VS	MR	MS	S	very high	very high	mid-late
Cynthiana	6a	warm	blue	good	high	small	small	MS	MS	MS	S	MS	S	medium	medium	late
Catawba	5a	cool-warm	red	very good	high	medium	medium	S	MS	S	VS	S	MR	high	high	late
Golden Muscat	5b	warm-cool	gold	best	high	very large	very large	S	VS	S	S	S	S	medium	medium	late

Disease susceptibility ratings: VS = very susceptible, S = susceptible, MS = moderately susceptible, MR = moderately resistant, R = resistant

Average ripening times

at Mountain Grove, MO: very early: early Aug., early: mid-Aug., middle: late Aug. to early Sept., mid-late: mid-Sept., late: late Sept.

at Geneva, NY: very early: late Aug., early: early Sept, middle: mid- to late Sept., mid-late: early Oct., late: mid-Oct. (may not ripen)

Notes on Cultivars:

Valiant, from South Dakota, is a superior cultivar for very cold climates with a tangy flavor.
Edelweiss, also for very cold climates, has fruit like 'Ontario' and makes good semi-sweet wine if picked at the early mature stage to avoid foxiness.
Seneca has excellent quality, adherent skin and clear juice.
Alwood has strong, sweet flavor like 'Concord' and tough skin. It's very susceptible to 2,4-D injury.
Price has a non-foxy flavor, good as a table or juice grape.
Monticello has a spicy flavor, a mix of labrusca and muscat.
Delaware is a high-quality labrusca grape, but tender skins often crack if rains fall near harvest, and fruit won't color where summers are hot.
Niagara is the most widely-planted white labrusca grape, for juice, wine, and some fresh sales.
New York Muscat has a distinct muscat flavor, used to make muscatel wine. Prune severely or thin clusters to prevent overbearing, winter injury.
Alden has meaty berries with a muscat flavor and adherent skin. Prune to short canes or cluster thin to prevent overbearing and winter damage.
Sunbelt, from Arkansas, was developed as a juice grape with flavor like 'Concord' that colors well in hot-summer areas and resists diseases.
Vinered has thin skin and seeds that separate easily from the flesh, but it starts growth early, so needs the best sites to avoid frost damage. It's very susceptible to anthracnose (bird's eye rot), so needs extra sprays.
Steuben is spicy and high quality, in strong demand by wineries for a rose wine. Prune severely or cluster thin to prevent overcropping.
Yates needs cluster thinning after bloom to ensure fruit coloration and ripening. It has an exceptional cold storage life as a table grape.
Concord is the most widely-planted labrusca grape, but fewer new plantings go in as more consumers prefer grape products with less "foxy" flavor.
Cynthiana, also called 'Norton', is relatively disease-resistant and is often grown in the Midwest to make a spicy, dry red wine. Don't use sulfur.
Catawba has a sweetly spicy flavor with a slight muscat tang, often used for wine. Plant on a south slope and cluster thin to induce earlier ripening.
Golden Muscat ripens too late for many Northern areas, but has fine labrusca-muscat flavor where it ripens. Prune severely or cluster thin.

Table 14-6.

LOW-CHILL AMERICAN BUNCH GRAPES RESISTANT TO PIERCES'S DISEASE

For USDA zones 8 to 10, in approximate order of ripening (varies with area).

		Berry				Cluster		Susceptibility to									
Cultivar	Use	Color	Flavor	Sugars	Size	Size	Density	Powdery Mildew	Bunch Rot	Black Rot	Ripe Rot	Anthrac-nose	Downy Mildew	Vine Vigor	Type of Pruning	Yields	Harvest Season
Orlando Seedless	table	white	very good	very high	small	large	medium	R	MR	S	S	S	R	medium	cane	medium	very early
Blanc du Bois	wine	white	best	medium	medium	medium	medium	MR	MR	S	S	S	R	high	spur	medium	very early
Suwanne	wine, juice	white	good	medium	large	medium	compact	R	MS	MR	MR	S	S	low	cane	high	early
Stover	wine	white	good	medium	medium	medium	medium	S	MR	MR	R	S	R	medium	short cane	medium	early
Conquistador	juice, jelly	blue	very good	medium	medium	medium	medium	S	MR	R	R	MR	R	medium	short cane	medium	middle
Miss Blanc	wine	white	good	medium	medium	medium	medium	R	MS	S	S	MS	R	medium	short cane	medium	middle
Lake Emerald	wine, juice	white	good	high	small	very large	medium	S	MR	S	S	S	R	medium	short cane	high	late
Daytona	table	red	best	medium	very large	very large	loose	MR	MR	MR	MS	MS	R	high	spur	low	late
Champanel	jelly	black	very good	medium	large	small	medium	R	MR	R	MR	R	R	very high	spur	high	late
Black Spanish	wine, juice	black	good	low	small	large	compact	S	S	VS	S	S	VS	medium	short cane	very high	late

Disease susceptibility ratings: VS = very susceptible, S = susceptible, MS = moderately susceptible, MR = moderately resistant, R = resistant

Average ripening times at Leesburg, FL: very early: late June to early July, early: early to mid-July, middle: late July, late: early to mid-Aug.

Notes on Cultivars: All except 'Blanc du Bois', 'Lake Emerald' and 'Miss Blanc' require grafting on Florilush, Tampa, Dog Ridge or SO4 rootstock.
Orlando Seedless needs GA to increase berry size and eliminate gritty seed traces, and may also need girdling or cluster thinning to give berries adequate size.
Blanc du Bois produces a spicy premium table wine. On its own roots it resists root knot nematodes, but not alkaline or poorly drained soils.
Suwanne has a slight muscat flavor and good tolerance to ripe rot. It starts growth late, avoiding spring frosts.
Stover starts growth very early, so prune late to delay growth. Harvest when berries soften and skin slips off easily.
Conquistador has 'Concord' as 2 of its grandparents, so is valued for juice and jelly. It starts growth late, avoiding frosts.
Miss Blanc, from Mississippi, is being increasingly planted for semi-dry wine production.
Lake Emerald fruit does not keep well, but high sugar content is useful in wine or juice production.
Daytona, though seeded, has vinifera-like fruit similar to 'Tokay', with adherent skins. It resists grape leaffolder and tolerates grape leafhoppers.
Champanel needs no fungicides due to outstanding disease resistance, and resists most insects except grape leaffolder. Fruit are very acidic until fully ripe.
Black Spanish, also called **'Le Noir'**, produces very dark red, acidic juice high in tannins, which makes outstanding port wines or juice blends. It needs frequent fungicides.

Table 14-7.

MUSCADINE GRAPES FOR JUICE, JELLY AND WINE

In approximate order of ripening (varies with area).

All are perfect-flowered, resist Pierce's disease, and need a long, hot growing season.

Cultivar	Winter Hardy to Zone	Berry		Use	Susceptibility to							Vine Vigor	Yields	Harvest Season
		Color	Sugars		Powdery Mildew	Bitter Rot	Macro-phoma Rot	Ripe Rot	Bird's Eye Rot	Black Rot	Angular leaf Spot			
Carlos	6b	Bronze	low	white wine, juice	MR	MS	MR	MS	MR	MS	MR	very high	very high	early-mid
Noble	7a	Black	low	red wine, juice	MR	VR	R	MR	MS	S	R	high	high	early-mid
Dixie	8a	Bronze	high	juice blends, wine	VS	VS	S	S	R	S	MS	very high	med-high	middle
Golden Isles	7a	Bronze	low	non-musky juice	S	MS	MS	MS	MS	S	MS	medium	med-high	middle
Welder	9a	Bronze	high	juice blends	VS	R	VR	MS	MS	S	MS	high	medium	mid-late
Doreen	7b	Bronze	very high	juice	MR	MS	MR	MS	MS	VS	R	medium	high	late

Disease susceptibility ratings:
VS = very susceptible, S = susceptible, MS = moderately susceptible, MR = moderately resistant, R = resistant

Average ripening times at Leesburg, FL: early-mid: mid-July, middle: late July, mid-late: early Aug., late: mid-Aug.

Notes on Cultivars:
Carlos is the main bronze muscadine for wine and juice, but may develop Pierce's disease if stressed by drought or overcropping.
Noble is the most popular muscadine for red wine and juice, with a less musky flavor than most muscadines.
Dixie produces very sweet berries good for blending in juices and wines, but berries may crack from powdery mildew.
Golden Isles has a non-musky flavor desirable for juice. It did well in GA trials but poorly in AR.
Welder is less cold hardy and requires more fungicide sprays than most other muscadines.
Doreen produces elongated berries high in sugar that make excellent juice. It needs extra magnesium.

Table 14-8.
MUSCADINE TABLE GRAPES

will not grow here

In approximate order of ripening (varies with area).

All need a long, hot growing season, resist Pierce's disease.

Cultivar	Winter Hardy to Zone	Flower Type	Berry					Susceptibility to							Vine Vigor	Yields	Harvest Season
			Color	Flavor	Sugars	Size	Scar	Powdery Mildew	Bitter Rot	Macro-phoma Rot	Ripe Rot	Bird's Eye Rot	Black Rot	Angular Leaf Spot			
Early Fry	7b	female	bronze	good	high	large	wet	MR	VS	VS	S	S	VS	S	high	very high	very early
Triumph	9a	perfect	red	good	high	medium	very dry	MR	MR	MR	MR	MR	MR	R	med-high	low-med.	early
Sugargate	9a	female	black	best	very high	very large	wet	MR	VR	R	MR	MR	MR	MS	high	low	early
Darlene	7b	female	pink	very good	very high	very large	wet	S	VS	S	S	S	S	S	high	low-med.	early
Summit	8a	female	pink	very good	high	large	dry	MR	MR	S	VS	S	S	R	high	high	early-mid
Tara	8a	perfect	bronze	very good	high	very large	very dry	MS	R	MR	MR	MR	S	R	high	high	early-mid
Florida Fry	7b	perfect	bronze	fair	med-high	large	very dry	VS	VS	VS	MR	MR	MR	S	medium	medium	early-mid
Sweet Jenny	7b	female	bronze	very good	very high	very large	wet	S	S	R	S	S	S	S	high	low-med.	early-mid
Pam	7b	female	bronze	very good	high	very large	very wet	MR	MR	MR	MR	MR	MR	S	high	very high	early-mid
Fry	8b	female	bronze	best	high	very large	very wet	MR	VS	VS	VS	S	VS	S	medium	medium	middle
Seedless Fry	7b	perfect	dark red	good	high	very small	very dry	MR	VS	VS	S	S	VS	S	high	low-med.	middle
Nesbitt	8a	perfect	black	very good	medium	large	dry	MS	R	MS	MR	MR	S	S	med-high	medium	middle
Black Beauty	8b	female	black	best	very high	very large	wet	MR	MR	S	MR	S	MR	MS	medium	high	middle
Scarlett	7b	female	red	very good	high	large	dry	MR	MR	MR	MR	MR	MR	MS	high	high	middle
Polyanna	8a	perfect	purple	good	medium	large	dry	R	R	R	R	VR	R	R	medium	high	middle
Supreme	8a	female	purple	best	very high	very large	wet	MR	R	S	MR	MR	MR	MS	medium	very high	mid-late
Pineapple	7b	perfect	bronze	very good	medium	large	very wet	MR	VS	VS	MR	MR	MR	MS	high	very high	mid-late
Loomis	9a	female	black	best	med-high	medium	dry	MR	MR	MR	MR	MR	MR	MS	med-high	medium	mid-late
Granny Val	8b	perfect	bronze	good	low	large	dry	S	S	S	S	S	S	S	low-med.	very high	late

Disease susceptibility ratings:
VS = very susceptible, S = susceptible, MS = moderately susceptible, MR = moderately resistant, R = resistant, VR = very resistant

Average ripening times at Leesburg, FL: very early: late June, early: early July, early-mid: mid-July, middle: late July, mid-late: early Aug., late: mid-Aug.

Notes on Cultivars: Female-flowered cultivars need a perfect-flowered cultivar nearby to provide pollination.
Early Fry resembles the popular Fry, but is 2 weeks earlier. Clusters ripen uniformly so can be clipped for shipping.
Triumph berries must be individually picked, as they shatter if entire clusters are cut. It's not hardy enough for some areas.
Sugargate has very sweet, fine flavored, very large berries, but skins tear when picking and it isn't hardy enough for some areas.
Darlene berries may tear when picked individually, but ripen evenly, so clip entire clusters for shipping.
Summit is not quite as attractive as Fry, but has better disease resistance and yields.
Tara has softer fruit than most muscadines, which may shrivel if water is lacking.
Florida Fry berries may develop skin splitting from powdery mildew if this is not controlled.
Sweet Jenny berries may appear rough, but soft skin is very edible and berries are very sweet.
Pam has very large clusters that can be clipped for shipping, since it ripens evenly but stem scars are wet.
Fry is the standard for pick-your-own and local fresh markets. Uneven ripening and wet stem scar limit shipping.
Seedless Fry berries are very small but offer no seeds.
Nesbitt may show symptoms of Pierce's disease if allowed to overcrop.
Black Beauty has edible skin, fine flavor and very sweet fruit, but isn't hardy enough for some areas.
Scarlett, from Georgia, has a beautiful red color, unusual for a muscadine.
Polyanna has multiple pest resistance and has not required any fungicide or insecticide sprays in FL, where it was bred.
Supreme has very sweet, very large fruit with edible skin, generally the best black muscadine.
Pineapple has a pineapple-like flavor and large clusters.
Loomis has a great taste but a tough skin with numerous lenticels, and isn't hardy enough for some areas.
Granny Val extends the season but needs extra fungicides, and isn't hardy enough for some areas.

Photo 14-3. 'Cabernet Sauvignon', a popular vinifera wine grape from France, is the most-planted red wine grape in California, and is also grown in a few favorable Eastern U.S. areas. (Photo by Bernadine Strik, Oregon State Univ., Corvallis, OR.)

Photo 14.4. 'Seyval', also called 'Seyval Blanc', is one of the most widely planted French hybrid grapes.

Photo 14-5. 'Canadice' seedless grape, from the New York breeding program, is one of the hardy seedless grapes that withstand colder winters than European seedless grapes.

Photo 14-6. 'Niagara' grape, an American grape with "foxy" flavor, is the main white American grape for juice and jelly processing.

Photo 14-7. 'Noble' is the most widely planted black-fruited muscadine cultivar for wine production, and may also be sold fresh. (Photo from TyTy Nursery, TyTy, GA.)

Photo 14-8. 'Fry' has long been the standard bronze-fruited muscadine cultivar for pick-you-own and local fresh markets. (Photo from www.smallfruits.org)

Seedless *vinifera* table and raisin grapes (Table 14-3) are produced for wholesale markets on large acreages, mainly in California in the U.S., and also in countries such as Chile, Mexico and Mediterranean nations with mild winters and dry summers. These cultivars are only hardy to about USDA zone 8 and are susceptible to diseases. 'Thompson Seedless,' a white (green) cultivar, dominates production of table grapes and accounts for almost all raisin grapes. Raisin grapes naturally darken as they dry, unless treated with sulfur to make golden raisins. 'Flame Seedless' is the leading red table grape cultivar.

Seedless American hybrid table grapes (Table 14-4) were developed for areas with cold winters and humid summers, and are grown for pick-your-own and local retail sales. These cultivars produce smaller fruit than the seedless *vinifera* table grapes, but several offer outstanding flavor as well as resistance to cold and diseases. The red Arkansas cultivar 'Reliance' has proven hardy to zone 5a, and has a delicious spicy flavor. Other seedless cultivars from Arkansas and from New York, such as 'Canadice' (Photo 14-5) and 'Marquis', are gaining favor with growers in the Northeast, Mid-Atlantic, Midwest and Northwest of North America.

American grapes (Table 14-5) for processing into juice and jelly are produced under contract with Welch's, a grower cooperative, in Washington, New York, Pennsylvania, Michigan and Ohio. The blue (purple) cultivar 'Concord' accounts for two thirds of production but plantings are declining, while the white (green) cultivar 'Niagara' (Photo 14-6) produces the remaining tonnage and plantings are increasing. Many other high-quality American grape cultivars are grown in very small plantings for local sales as seeded table grapes.

Low-chill American bunch grapes resistant to Pierce's disease (Table 14-6) were developed for the Southeastern U.S., where that disease kills susceptible bunch grape cultivars within a few years. Grape growers in Florida, the eastern half of Texas, and the southern halves of Louisiana, Mississippi, Alabama and Georgia must plant either these bunch grapes resistant to Pierce's disease, or muscadines. 'Blanc du Bois' and several other bunch grapes resistant to Pierce's disease are gaining popularity for wine production, several cultivars are grown for juice or jelly production, and 'Daytona' and 'Orlando Seedless' are being planted as table grapes.

Muscadine grapes for juice, jelly and wine (Table 14-7) produce more juice than other muscadine cultivars. Again, these are adapted to the Southeastern U.S. 'Carlos' and 'Noble' (Photo 14-7) have been planted the most for processing.

Muscadine table grapes (Table 14-8), also adapted to the Southeastern U.S., are sold pick-your-own or for local retail sales, traditionally by picking individual berries as they ripen unevenly, rather than entire clusters. Some newer cultivars ripen more evenly, so entire clusters may be cut for retail shipping, to reduce harvest labor and postharvest spoilage. Customers prefer cultivars with larger berries, and growers have found that a ratio of 80% bronze-fruited cultivars to 20% black- or red-fruited cultivars sells the best. 'Fry' (Photo 14-8) has long been the most popular cultivar, but many others, increasingly with greater disease resistance, are planted.

GRAPE ROOTSTOCKS

American bunch grapes and muscadine grapes are generally propagated from cuttings, so grow on their own roots, which resist most soil pests. European and French hybrid grapes may also be grown on their own roots in areas without soil pests, but are generally propagated by grafting or budding the desired cultivar onto a rootstock with resistance to soils pests, since the cultivar is susceptible to these. Loams and clay loams often require rootstocks with resistance to phylloxera, a soil insect pest, while sandy soils frequently need rootstocks resistant to root knot nematodes and/or dagger nematodes, which spread fanleaf virus. Various rootstocks also provide adaptation to different soil conditions, and increase or decrease the vigor of the cultivars grafted to them. Choose well-adapted rootstocks that provide the vine vigor needed for your cultivars, soil type and climate.

Phylloxera-resistant rootstocks include Coudrec 3309, Coudrec 1616, 5C Teleki, St. George, SO4, 5BB Kober, 140 Ruggeri, and 110 Richter. **Coudrec 3309** is recommended for cultivars with poor fruit set because it increases this, but young vines on this rootstock require crop removal. This rootstock advances fruit maturity and is less susceptible to lesion and dagger nematodes and crown gall than many rootstocks, but is susceptible to root knot nematodes and viruses. It requires deep, well-drained, irrigated soils, and does not tolerate drought, saline or shallow soils, or heavy clay soils.

Coudrec 1616 also increases cultivar fruit set and advances fruit maturity, and offers high resistance to root knot nematodes, poorly drained and saline soil, but is sensitive to drought and high pH soils and requires fertile soils. **5C Teleki** also increases cultivar fruit set and advances fruit maturity, has good resistance to root knot and dagger nematodes, and does well on well-drained fertile soils, but it tolerates high soil pH, moderate drought and heavy clay soils.

St. George, an older phylloxera-resistant rootstock, is very susceptible to root knot nematodes, susceptible to dagger nematodes and very susceptible to the fan leaf virus that they spread, and moderately resistant to lesion nematodes. It induces very vigorous growth that can reduce fruit set and delays fruit maturity, so is now being replaced by SO4, which is less vigorous. **SO4** resists root knot nematodes and tolerates both acid and alkaline soils, but does best on light, well drained, irrigated soils of low fertility. On fertile soils SO4 can cause excess vigor that leads to reduced fruit set and Botrytis fruit rot. SO4 assimilates magnesium poorly, so do not use it for cultivars with high magnesium demand, such as 'Merlot' or 'Cabernet Sauvignon'.

5BB Kober is another vigorous rootstock best suited to less fertile soils, as it can reduce fruit set and delay maturity on fertile soils, and is hardy only to zone 8b. It has high tolerance for alkaline soils but no tolerance for saline or poorly drained soils, and its poor potassium uptake makes it unsuitable for high-yielding cultivars such as 'Merlot' or 'Cabernet Sauvignon'. It has moderate resistance to root knot and dagger nematodes. **140 Ruggeri**, another vigorous rootstock, also does best on less fertile soils and assimilates potassium poorly, but shows great tolerance for drought, salinity, and both acid and alkaline soils. It has moderate resistance to root knot nematodes. **110 Richter** is adapted to warm, arid areas, tolerating both drought and poorly drained, shallow clay soils. It tolerates alkaline soils and moderately resists root knot nematodes, but assimilates potassium and magnesium poorly.

Rootstocks resistant to root knot nematodes include Harmony, Freedom, Ramsey, Dog Ridge, Coudrec 1613, Tampa and Florilush. Three phylloxera-resistant rootstocks described above—Coudrec 1616, 5C Teleki and SO4—also offer good resistance to root knot nematodes. Root knot nematodes often develop populations able to feed on **Harmony**, **Freedom** and **Ramsey** rootstocks after 10 to 15 years; after that, these rootstocks show tolerance to nematode feeding, with less yield loss than susceptible rootstocks, but no longer true resistance. If replanting vineyards that formerly used any of these rootstocks, fumigate or use cover crops first to reduce nematode populations, then switch to a different rootstock, outside this group, that resists root knot nematodes.

Dog Ridge rootstock also resists cotton root rot, frequently a problem on alkaline soils in the Southwest. **Harmony** and **Freedom**, which descended from Dog Ridge, are also adapted to alkaline soils and have moderate resistance to cotton root rot.

Florilush, a cross between Dog Ridge and Tampa, was developed in Florida for bunch grapes resistant to Pierce's disease that do poorly on their own roots. Besides resisting root knot nematodes, it also resists Pierce's disease and ring, sting and stubby root nematodes, and tolerates grape root borer, which often kills Tampa. Now that Pierce's disease is becoming more of a problem in California, Florilush could be useful there too.

Rootstocks resistant to dagger nematodes, which spread fanleaf virus, include O39-16 and O43-43. These both resulted from crosses between vinifera grapes and muscadine grapes, and both also resist root knot nematodes. **O39-16** is very resistant to dagger nematodes and also tolerates fanleaf virus, but is moderately susceptible to phylloxera, while **O43-43** has only moderate resistance to dagger nematodes but good resistance to phylloxera. **Coudrec 3309, 5C Teleki, Harmony** and **Freedom** also are less susceptible to dagger nematodes than most rootstocks.

VINEYARD LAYOUT, PLANTING AND TRELLIS INSTALLATION

A planting of well-adapted grapes cultivars can produce for 50 years or more, so take time to carefully plan the vineyard layout and trellis system before planting.

Dividing the vineyard into blocks

Plant cultivars with different needs, such as those with different levels of disease or frost resistance, in different blocks so that each block can be managed separately. This can greatly reduce pesticide inputs by enabling growers to spray only the cultivars that need each pesticide application. Plant cultivars that are most sensitive to winter cold or that break bud earlier in spring on the highest sites to protect them if possible, or install sprinkler irrigation in blocks that will need frost protection.

The maximum length of rows within each block depends on the soil type and irrigation system. Trickle-irrigated vineyards can have rows up to 300 feet long on sandy soils and up to 600 feet long on heavier clay soils, while furrow-irrigated vineyards can have somewhat longer rows, up to 360 feet on sandy soils and up to 720 feet on clay soils. Sprinkler-irrigated vineyards can have the longest rows, up to 1000 feet long, depending on the system's output.

Row orientation

On nearly level sites, orienting rows north to south will maximize light exposure, leading to higher yields and fruit quality. Vineyardists may adjust row orientation to the southeast or southwest to maximize row length and minimize row numbers, in order to reduce the number of expensive end-post assemblies needed. Table grape growers in California and other high-sun areas run rows northeast to southwest, to minimize sunburning of clusters. Raisin growers orient rows east to west, so that trays of harvested grapes placed between the rows to dry will receive maximum sunlight and dry quickly so they do not mold. The newer dried-on-the-vine (DOV) system for raisin also needs east-west rows, so fruiting canes can be trained to the south side of each row to dry faster, and renewal canes trained to the north side. Where strong winds during the growing season create problems, plant rows parallel to the wind direction to minimize damage to fruit and vines.

Plant in straight rows on slopes less than 10%, generally across the slope to minimize soil erosion. However, vineyards to be furrow irrigated, which should be on slight slopes of 2 to 5%, need rows running down the slope. On slopes greater than 10%, plant on a modified contour, keeping rows straight as long as possible to make trellising easier, and creating breaks where the rows bend. (If using the head system of training, with no trellis, plant on the contour.)

Row and vine spacing

Space bunch grape rows 8 to 10 feet apart for single curtain or vertical training systems, and 10 to 12 feet apart for double curtain or T- trellis systems, depending on equipment widths. Muscadines require 10 to 12 feet between single-curtain rows and 12 to 14 feet between double-curtain rows. Leave 20 to 30 feet at the ends of rows, beyond the end posts, to provide space to turn equipment around.

Vine spacing in rows depends on cultivar and rootstock vigor as well as soil fertility. Space most European and French hybrid wine grapes 6 to 8 feet apart, although a few weaker-growing cultivars such as 'Gewurztraminer' may be planted just 4 to 5 feet apart. Planting most other wine grapes that close might increase yields for the first 2 or 3 years of cropping, but after that would reduce yields due to increased competition and shading, as well as increasing pruning costs to control vine size. Plant table grapes, raisin grapes and American bunch grapes 7 to 8 feet apart, and muscadines 10 to 12 feet apart. Use the closer spacing for less fertile soil, an intermediate spacing for soil of average fertility, and the wider spacing for more fertile soil.

Lay out the vineyard on a windless day, marking the vineyard corners and using triangulation to ensure correct alignment and that each row remains straight. Mark planting rows by creating a deep furrow with a tractor. Mark the future location of all trellis posts, at a spacing between 22 and 30 feet apart within rows, that allows even planting of vines (such as 28 feet apart when vines are planted 7 feet apart, or 24 feet apart when vines are planted 6, 8 or 12 feet apart). Then use different types of markers or different-colored flags to show where each vine should be planted.

Planting

Plant in early spring using dormant, 1-year-old plants. Dig holes about 12 to 18 inches deep and 8 to 12 inches in diameter. Some large growers use mechanical tree transplanters to dig holes and set plants. If using a tractor-driven auger to dig holes, use a narrow shovel to break up the sides of the hole in soils with high clay content. Before or during planting, prune off overly long or broken roots from each plant, and cut off all branches and suckers, leaving just one main stem with 2 to 4 buds. Keep plants' roots moist until planting, set plants an inch or so deeper than they grew in the nursery, and tamp soil firmly around the roots. Keep the graft union of grafted grapevines 2 to 3 inches above the soil so the scion doesn't root, since it is susceptible to phylloxera and/or nematodes. Do not add any fertilizer to planting holes. Irrigate immediately after finishing planting a block.

Nitrogen fertilizer and groundcover management after planting

Apply 20 to 30 pounds per acre of actual nitrogen (N) to the planting rows only, soon after planting, unless a legume cover crop was turned under before planting. If a grass cover crop was turned under before planting, the row middles will require an additional 15 to 20 pounds of N per acre to break down the straw into organic matter.

Grape growers in California, and other irrigated areas with little to no summer rainfall, generally cultivate between the rows during the growing season, to eliminate groundcover competition for water and nutrients. In fall, they plant winter cover crops to till under in spring, to increase soil organic matter and water-holding capacity. However, vineyards on sloping ground in non-arid areas should be planted to a permanent groundcover to prevent soil erosion. Most growers plant cool-season grasses such as creeping red fescue or tall fescue, which reduce nematodes and compete less than warm-season grasses with the grapes. Cool-season legumes might be planted instead of or with the grass, if nematodes aren't a problem on that soil, to provide nitrogen. Such cool-season groundcovers may be planted in spring after planting the grapes, but seed generally germinates better in fall, so a summer annual cover crop might be sown and turned under in fall before planting the permanent groundcover. Choosing canola as the summer cover crop can reduce any harmful nematodes present.

Trellis installation

Install trellis posts, row-end braces, and at least one wire within the first growing season after planting grapes. Eight-foot-long softwood posts, pressure-treated with wood preservative, are standard for line posts within the row. When driven 2 feet deep into the ground, they support wires up to 6 feet off the ground. Line posts should be at least 3 inches in diameter, and end posts should be at least 5 inches in diameter and 9 or 10 feet long, so they can be driven 3 to 4 feet deep.

Most growers use an external row-end brace, which is adequate for vertically trained trellises up to 600 feet long and double curtain or T-trellises up to 400 feet long. End posts are driven into the ground slanted away from the row, 15 to 30 degrees off vertical, to keep wires taut. Then install a steel screw-type anchor, or "deadman", 4 feet from the base of the end post, and attach them with a double loop of 9-gauge bracing wire. Twist the bracing wire clockwise, the same direction used to screw the deadman into the ground. Longer rows, especially those of double curtain or T- trellises, require a stronger but more expensive internal brace assembly that uses more posts rather than an anchor.

Once all end posts and line posts are installed, attach the lowest wire, so it may be used in vine training. Use high tensile, galvanized steel wire, of 12 or 12.5 gauge. Mark wire heights on the posts, then use a wire jenny or spool to dispense the coiled wire. Position the jenny at one end of the row, then pull the loose end of the wire to the opposite end of the row, on the windward side of the posts. (Do this on as calm a day as possible, and wear gloves and eye protection.) Attach the loose end of wire to the end post, at the correct height, with 2 compressible wire sleeves, or extend the wire beyond the end post to an earth anchor. Then cut the opposite end of wire from the jenny, fit the wire with an in-line strainer, and attach it to the other end post with wire sleeves or extend it to an earth anchor. The in-line wire strainer should be mounted in the middle of the row for rows that are longer than 500 feet, but may remain near the end for shorter rows. Staple the wire loosely to all posts at the correct height, so the wire can move through the staples, then tighten the wire. Later, attach any additional wires as needed.

TRAINING SYSTEMS

Growers and researchers have developed dozens of different training systems for grapes. The following systems currently dominate production.

Freestanding head system

This traditional system for wine grapes avoids the need for a trellis by retaining a strong, upright trunk and severely pruning back all canes that arise from it, so the vines are eventually self-supporting. This reduces yields the most of any training system, so is only used for high-quality wine grape cultivars in which higher yields cause lower quality. Head-trained vines must be harvested by hand, so few new plantings are trained to this system. However, many head-trained vineyards 30 years old and older are still maintained and produce much fine wine.

A mature head-trained vine (Photo 14-9) has 4 or 5 arms (short branches) radiating out like spokes of a wheel from the top, or head, of a strong trunk. The fruiting units, either canes or spurs, arise from the ends of the arms. Each head-trained vines requires a 2-inch-square wooden stake or fence post, 4 to 6 feet long, to provide support for the first 3 to 5 years, until the trunk reaches a diameter of 3 to 4 inches. Stakes are driven in beside each plant during the first winter after the vineyard is planted, and a single shoot of each vine is tied to the stake to become the trunk. Each spring, lateral growth from the trunk is cut back to short spurs to encourage the main trunk to thicken. When the trunk reaches the top of the stake it is cut back, forming the "head" from which the arms arise. Once bearing begins, growers prune back canes each spring to lengths providing the right number of buds for each particular cultivar. Spur-pruned vines retain almost all canes, but each one is severely pruned back to just 2 to 4 buds. Cane-pruned vines retain just two to four canes per arm, with each cane pruned back to 8 to 20 buds.

Single or bilateral cordon on a single-wire trellis

The single-wire trellis system, with one or two

VINE TERMINOLOGY

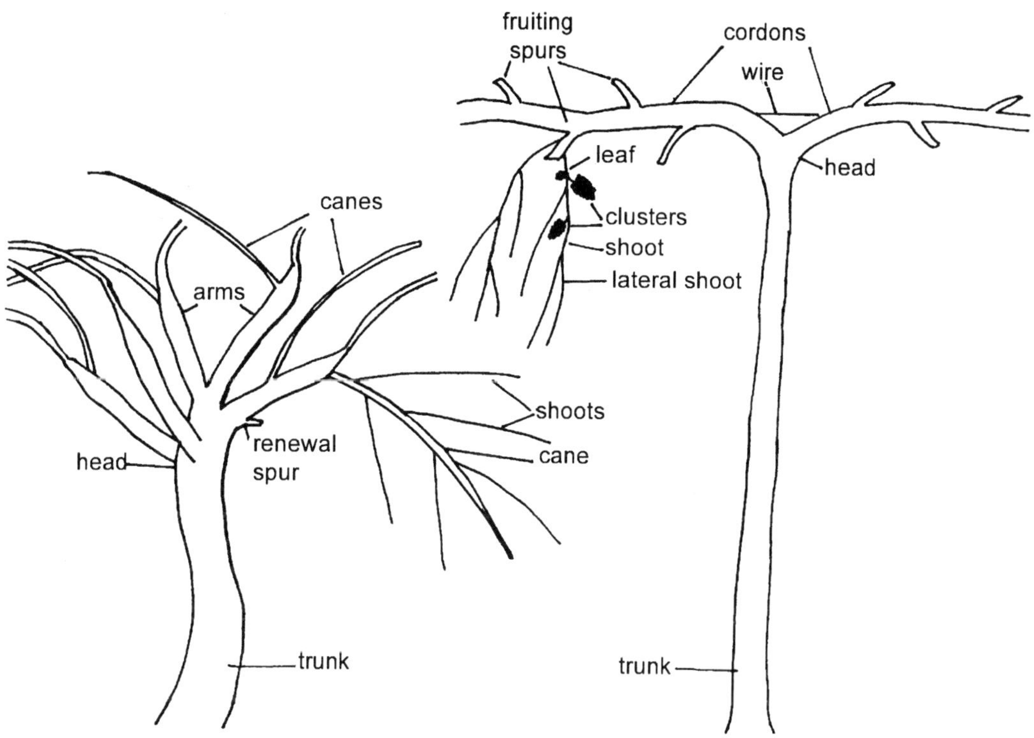

Trunk: the main unbranched stem of the grapevine
Head: the top of the trunk, where arms or cordons arise
Arms: short, thick main branches arising from the trunk
Cordon(s): long extension(s) of a trunk, oriented horizontally and trained along a wire
Shoot: current-season stem with leaves that developed from a bud
Cane: woody stem that matured from a shoot that has overwinteredLong cane: a cane that's been pruned back to 8 to 20 buds
Short cane: a cane that's been pruned back to 4 to 7 buds
Spur: a cane that's been pruned back to 1 to 4 buds
Fruiting spur: a spur that bears buds that will grow into fruiting shoots
Renewal spur: a spur with 1 or 2 non-fruitful buds that will grow vegetatively into shoots to be fruiting canes in later years
Laterals: side branches of a shoot or cane

cordons, is used for many wine grapes in California, especially the more vigorous French hybrids, and is the main system used for wine grapes in other areas of North America. The bilateral cordon system is also widely used for muscadines and for seedless American hybrid table grapes. Both single and bilateral cordon systems are well adapted to over-the-row mechanical harvesters. A trellis of a single wire stretching down each row, about 64 to 72 inches off the ground, is installed before or soon after planting. Workers attach a length of twine from the wire to an anchoring metal loop beside each plant, and tie each vine to the twine to train it up to the wire.

Photo 14-9. Mature grapevines trained to the freestanding head system don't need a trellis, but the severe pruning used for this system limits yields and doesn't allow mechanization. This system works best with high-quality wine grapes where heavier yields reduce quality

For a single cordon, after the vine reaches the wire in the second summer, it is bent horizontally and trained along the wire as a permanent cordon until it reaches the next vine, where it is cut off. For a bilateral cordon (Photo 14-10), the vine is cut back when it reaches the wire, and two side shoots are trained out along the wire in opposite directions, forming two cordons. Workers prune off side shoots to direct the vine's energy into growing the cordon(s), until they have filled their allotted space. Each spring thereafter, canes arising from the cordons are cut back to spurs of 2 to 3 buds or short canes of 4 to 7 buds, leaving the desired number of buds for each particular cultivar. Shoots grow downward from the buds left on the canes or spurs, forming a vertical curtain of foliage. Many growers use equipment to "comb" the shoots into downward positions, to minimize shoots crossing each other and make mechanical harvest easier.

Vertically shoot positioned (VSP) trellis

This system offers higher yields for spur-pruned wine grapes and allows mechanical harvest, but requires shoot training by hand. This system has a single fruiting wire, 52 to 60 inches off the ground, along which the cordons are trained, but also has 1 or 2 "foliage wires" or "catch wires." above the cordon wire. Workers hand-train shoots upward by tucking them into position against the foliage wire(s), each placed 8 to 12 inches above the wire below, which catch the shoots and hold them up. Workers may tape or tie shoots into position if needed. Some cultivars, such as Vignoles and Chancellor, tend to produce very short shoots which stand up well on their own, so they work well with this system and require little hand training.

California researchers have found that shoots trained upward tend to produce higher yields than shoots trained downward, up to 7% higher in some years, and berry sugar content increases. This can justify the increased costs of manual vs. machine training of shoots. Manual shoot training also is more economical for smaller growers without enough acreage to justify the cost of a shoot-combing machine. Adding a second catch wire does not further increase yields over using just 1 catch wire.

Smart-Dyson trellis

This single-cordon trellis for wine grapes, pruned to either spurs or short canes, divides the canopy by training half the shoots upward and half downward. Cordons are trained along a single fruiting wire 42 inches off the ground, and 2 movable foliage wires are each placed from 12 to 30 inches above the fruiting wire, with their height raised as the growing season progresses. Workers train about half the shoots upward between these 2 catch wires, and the rest of the shoots downward.

Dividing the canopy increases light penetration, which increases yields by 7 to 10% over the VSP trellis, improves fruit quality and reduces diseases.

Photo 14-10. The double cordon system involves training each vine into 2 cordons along a single, high wire, and mechanically combing shoots downward to create a single curtain of foliage.

This system requires some hand training of shoots but less than the VSP trellis, and allows mechanical harvest. This system works for cultivars pruned to spurs and those pruned to short canes.

Scott Henry trellis

This system, named after the Oregon vineyardist who developed it, is another divided-canopy trellis for wine grapes but has quadrilateral cordons and more wires. It has 2 fruiting wires positioned vertically, one above the other, plus 2 catch wires above the upper fruiting wire. Growers develop 4 cordons from each trunk, 2 for each fruiting wire, growing out from the trunk in both directions along the wire. Canes arising from the cordons can be either pruned to either spurs or short canes, depending on which works best for each cultivar.

Workers train shoots from the upper fruiting wire upward between the two catch wires, and train shoots from the lower wire downward. This system increases light exposure up to 60% over a single-wire cordon, which can increase yields by 30 to 50%. Separating the canopy also reduces diseases and improves wine grape quality. This system is most effective on fertile soils, where vine growth can become overly vigorous. The Scott Henry trellis is particularly suited to areas which often experience cold damage, since the 4 cordons per vine limit the loss of bearing surface due to death of any 1 cordon, and renewal cordons can more quickly replace the smaller lost cordon.

Geneva double curtain (GDC) trellis

This system was developed in New York for vigorous American juice grapes such as 'Concord' and 'Niagara', and is also used for vigorous muscadines and some vigorous French hybrids. It produces high yields and is well suited to mechanical shoot training and mechanical harvest. T-shaped or y-shaped end posts support two parallel wires about 30 inches apart in a horizontal plane about 6 feet off the ground (Photo 14-11). Vines are planted in a staggered double row under each wire, are trained along each wire as for the bilateral cordon system, and generally pruned to canes or short canes. This forms a "double curtain" of foliage along each row, that can double yields over the single-wire cordon system for these vigorous American grapes, which can take the competition for nutrients and light.

Muscadine and French hybrid growers may plant one vine between the two wires, train it to produce 4 cordons to grow along the 2 wires, and prune to spurs or short canes. Growers mechanically comb shoots downward with this system. Avoid this system with French hybrids that are more susceptible to bunch rot, since the double curtain produces greater humidity, which can increase bunch rot by 40 to 50% in susceptible cultivars.

T-trellis

This standard system for California table and raisin grapes since the 1930's consists of T-shaped end posts similar to those of the Geneva double curtain, with crossarms 2 to 4 feet wide containing 2 to 5 wires spaced evenly apart horizontally and about 5 feet off the ground. Growers use 2 to 3 wires for cordon-trained, spur-pruned cultivars, and develop cordons 18 to 24 inches below the wires, so that shoots from spurs grow upward where the wires support them. Growers use 4 to 5 wires for cultivars that produce better from canes, head the trunk slightly below the trellis, and tie the canes which arise from the head to the center wires, so shoots grow horizontally out and are supported by the outer wires. This creates a horizontal plane of canopy like that of a table, which intercepts 50 to 75% of light falling on the vineyard.

This system produces high yields of high-quality table grapes. It does not lend itself to mechanical harvest, but table grapes must be hand-harvested for fresh market anyway. A T-trellis can increase problems with Botrytis bunch rot over that in a vertical trellis, but attention to crop thinning and leaf removal in table grapes can offset that.

Researchers are experimenting with expanding the width of the T-trellis while keeping row spacing the same in order to capture more light, like an arbor does. Wider crossarms in a T-trellis, a wider Y-shaped trellis, or an overhead gable trellis can all achieve this, and many raisin growers are switching to such systems. These systems work best on

Photo 14-11. The Geneva double curtain system produces high yields by planting staggered double rows, and training each row along 2 horizontally parallel wires. Each vine is trained to a double cordon, and shoots are combed downward, creating a double curtain of foliage.

cultivars with high vigor. Less vigorous cultivars, especially those prone to sunburn and heat damage, respond better to reducing spacing between rows from 12 feet to 9 or 10 feet, and retaining the standard T-trellis.

Overhead trellis

This system was widely used for muscadines, which are generally pruned to short canes or spurs, and its modifications are currently being investigated for use on table and raisin grapes. The original system used an overhead arbor (Photo 14-12) about 6.5 feet off the ground over the entire vineyard, so the canopy captured 100% of sunlight falling on the vineyard. Connecting wires or wood supports radiated out from each upright post in eight directions, and eight cordons developed from the trunk of each vine to follow these supports. This worked well with vigorous muscadine cultivars and allowed planting a vineyard with far fewer plants per acre, but the vineyard took longer to come into full production and required hand pruning, shoot placement and harvesting.

A modified overhead system with 4 cordons per vine, trained on double wires similar to the Geneva double curtain system but set higher, is still used for some muscadines and shows promise for California table grapes. Some new raisin grape plantings designed for dried-on-the-vine (DOV) raisins use an overhead trellis, which can double or even triple yields.

Multi-trunk systems

Several systems that develop multiple trunks per vine, with low renewal wood just a foot or so off the ground, are used to grow cold-sensitive vinifera cultivars in eastern areas where they often suffer winter damage. This helps ensure that some fruitful wood remains even during springs when some trunks must be cut back severely to remove dead wood. All these systems use trellises with 2 to 4 wires arranged in a vertical plane, and require cane pruning so long canes may be tied to the wires.

The **fan system** creates 3 or 4 trunks, renewed about 1 foot off the ground. Canes from these trunks may be covered with soil, straw or snow to protect them over the winter, then in spring they are spread out in a fan shape and tied to 3 or 4 wires in a vertical plane. The **modified Chautauqua system** develops a low cordon along a wire a foot off the ground, from each of 2 trunks per vine, and canes from these cordons are tied to 2 higher wires. The **Keuka high renewal system** uses 2 to 3 wires, heading 2 or 3 trunks below the lowest wire about a foot off the ground, and tying canes to the upper wire(s).

Photo 14-12. An overhead trellis system, or arbor, produces the highest yields by allowing grape foliage to intercept all light falling on the vineyard. This system was traditionally used for muscadines and is now being adapted for use in dried-on-the-vine raisin grapes.

VINE TRAINING AND DORMANT PRUNING

Grapevines require rigorous training during the first few years after planting a vineyard, to develop according to the desired system. Once trained, grapes produce an abundance of vegetative growth, requiring canopy management during the growing season and extensive dormant pruning. Without pruning, a grapevine's fruiting zone will move farther and farther from the trunk each year, and the vines will become unproductive.

Pruning helps to regulate crop load. Overcropping delays fruit maturity, reduces vine growth and hence return cropping potential for the following year, and delays wood hardening for winter thus increasing winter injury. Undercropping reduces current season yield and causes excessive vegetative growth, which increases shading, reducing fruit quality and bearing potential of canes to produce next year's crop.

Training new vines

Almost all grape training systems aim to grow a single strong, straight trunk per vine up to bearing height as rapidly as possible, so training for the first couple of years is very similar among the various systems. At planting time, each vine was pruned to a single stem with 2 to 4 buds, so 2 to 4 shoots should grow. Some growers install a temporary stake or bamboo pole next to each vine (Photo 14-13), or tie a string from the trellis wire to an anchor in the soil next to each vine, during the first growing season. Workers then choose the strongest shoot from each vine during the summer, and loosely tie it to the stake or string, sometimes stubbing back the other shoots to encourage greater height gain by the chosen shoot. (For a multi-trunk system, choose and retain shoots for the desired number of trunks.) Other growers allow vines to grow unhindered during the first growing season to produce maximum leaf area and root system development. They wait until the dormant season to choose the strongest cane per plant that developed from the first season's shoots, install a post or twine by each vine, tie the chosen cane to the support, and cut it back to 2 to 4 buds. Whether the cane that will become the vine's trunk was chosen in summer or winter, during that first dormant pruning, remove all other canes.

During the second and third springs, numerous shoots will emerge from each vine, requiring "suckering" or removal of extra shoots. Wait until the longest shoots have reached at least 8 to 12 inches long to begin suckering, since shoots are very tender before this stage and it's easy to accidentally remove desired shoots. Workers may break off extra shoots by hand until they reach a length of 18 inches, when they become woody at their bases and must be cut off with hand pruners. Leave 2 or 3 strong shoots per vine during the first suckering operation, and tie the best shoot per vine to the twine or pole if it's at least 12 inches long. During the second suckering operation, tie up the best shoot of each vine that didn't get tied before, and retie shoots that have been tied further up. If any chosen shoots broke off, replace them with one of the other shoots, but otherwise remove all other shoots during the second suckering operation, leaving just 1 shoot tied to each support.

Every 10 days or so during the rest of the growing season, tie this main shoot further up to prevent breakage and assure a straight trunk. Remove lateral shoots arising from the bottom 2 to 3 feet of the main shoot as they arise.

Once the main shoot reaches the desired height, usually during the second summer, train it for the specific system chosen. For a single cordon (including the VSP and Smart-Dyson trellis systems), bend it horizontally along the wire and tie it to the wire. For a bilateral cordon (including the Scott Henry and GDC trellis systems), top the main shoot when it gets to within 2 inches of the wire, select 2 lateral shoots that grow from buds below this cut, and train each along the wire in opposite directions. For a T-trellis or overhead trellis, also head the main shoot when it almost reaches the trellis wires, and train as many lateral shoots as are needed, for the desired number of next year's fruiting canes, along the wires. For head training, head the main shoot when it reaches the desired working height. Multiple trunk systems, as described above, require low heading of the multiple stems at 1 to 2 feet off the ground, just below the lowest wire.

Most vineyards produce some fruit during the third growing season, but crops must be rigorously thinned (see below) to prevent overbearing at such a young age, which could hurt vine establishment. From the third season on, practice dormant pruning as outlined below, and from the fourth season on, practice canopy management.

Cane pruning vs. spur pruning

Growers prune canes, spurs and other woody growth during the dormant season. Growers also continue to remove suckers (unwanted shoots from the trunk) each summer, and many growers also prune leaves and shoots during the growing season, as discussed below under Canopy Management. Dormant-season pruning (Photo 14-14) involves

Photo 14-13. Each young grapevine must be trained up a stake, or a string tied to the trellis wire and anchored in the ground, to develop a straight trunk as quickly as possible.

Photo 14-14. Prune grape canes, the 1-year-old growth that will produce shoots this season, during the dormant season. This double-cordon vine, before pruning, has long canes that must be thinned out, and the canes to remain must be shortened.

selecting which canes (last year's shoots that have become woody) to retain, and pruning selected canes back to either **long canes** of at least 8 buds, **short canes** of 4 to 7 buds (Photo 14-15), or **spurs** with just 2 to 4 buds (Photo 14-16).

Many European grapes and most French hybrids and muscadines are pruned to spurs, because the basal buds of each cane are fully fruitful and this spread the fruiting zone evenly over the entire cordon or head. Some of these grapes are pruned to short canes, and some European table and raisin grapes are pruned to long canes. American grapes are pruned to short or long canes, because the first two buds on each cane generally produce only leaves and bear no fruit.

Hand pruning remains the method of choice for table and raisin grapes, and for wine grape cultivars with large clusters or low to medium vine vigor. Hand pruning allows workers to selectively retain the most fruitful canes, from the outside of the canopy on the top and sides of the trellis or head-trained vine. Such canes received the most light during the previous growing season so have the greatest fruiting potential, and matured earlier than shaded canes so developed greater winter hardiness. Workers also select canes for darker color and a diameter of about ½-inch, as these are the most fruitful. Spindly canes produce less growth and fruit, while overly thick "bull canes" grow excessively vegetatively with little fruit, so workers remove such canes along with interior canes which were shaded as they grew.

After removing the less fruitful canes, workers keep many canes and prune each back to just 2 to 4 buds for the spur pruning system, or keep just 2 to 6 canes and prune each back to 8 to 20 buds for the cane pruning system. A typical grapevine may have 200 to 300 buds on mature canes capable of producing fruit, and a total of 35 to 90 buds may be left on the vine after pruning.

When spur pruning, space spurs 9 to 12 inches apart on the cordon, or about equally spaced around the head. Leave a total of 35 to 75 buds, not counting unfruitful buds within ½ inch of the base of each spur, depending on the cultivar, vine vigor, cropping level, and light intensity of the area. Leave 1 fruitful bud per spur on spurs ¼ inch in diameter (about pencil diameter), and 2 to 4 buds each on spurs of larger diameter.

When cane pruning, select 2 to 6 of the best canes, depending on the training system, that are in good positions and can be easily tied to wires, or well-spaced along the cordon or around the head. Leave a total of 40 to 90 buds, including the basal unfruitful buds, with the number often determined by balanced pruning (below). For every fruiting cane retained, also retain a renewal spur with 1 or 2 vegetative buds, to produce a cane for next year's fruiting.

Balanced pruning, a form of hand pruning used for cane-pruned American grapes and some French hybrids, involves leaving a number of buds on a vine proportional to the total vine growth produced the previous year, as measured by the amount of 1-year-old pruning wood removed. After removing all less fruitful canes, these are weighed with a small portable scale, and the weight determines how many buds to leave on the remaining canes. Do not include any older wood removed in the weight.

In Western areas with high light intensity, leave 50 buds for the first pound (or less) of prunings,

Photo 14-15. Many growers still practice hand pruning. This vine is being pruned to short canes of 4 to 7 buds each.

Photo 14-16. This vine was pruned to spurs of 2 to 4 buds each.

then 10 buds for each additional pound of prunings. For example, if the undesirable 1-year-old prunings weigh 2.5 pounds, leave 65 buds; if prunings weigh 3.75 pounds, leave 77 or 78 buds. In Eastern areas with less light intensity, leave 30 buds for the first pound (or less) of prunings, then 10 buds for each additional pound of prunings. For example, if the undesirable 1-year-old prunings weigh 2.5 pounds, leave 45 buds; if prunings weigh 3.75 pounds, leave 57 or 58 buds.

Machine hedging increases yields of some wine and juice grape cultivars, but often delays harvest by 1 to 2 weeks. Vigorous cultivars with small clusters respond best. Cultivars such as 'French Colombard' and 'Sauvignon Blanc' respond very favorably the first year, with little delay in ripening. Other cultivars, such as 'Ruby Cabernet' and 'Carignane', may take 1 or 2 years to adjust to the higher bud numbers left by machine pruning, and will show delayed fruit maturation for the first few years. Some commercial vineyards have been successfully machine-hedged for over 16 years. Large vineyards that can support the cost of hedging equipment can greatly reduce labor costs for pruning.

Relying solely on mechanical hedging for pruning, however, can increase incidence of Phomopsis cane and leaf spot in humid areas where this disease occurs. Where this disease is a problem, workers should remove dead and infected canes by hand before machine hedging occurs, to avoid spreading the disease.

Minimal pruning of wine grapes also increases yields and delays fruit maturity by 1 to 3 weeks, and offers important advantages of reduced labor costs and reduced diseases. Minimal pruning almost eliminates bunch rot, and greatly reduces incidence of Eutypa dieback, a fungal disease that enters pruning wounds and can eventually kill vines. Reduced pruning works best on vigorous vineyards with a high bilateral cordon system. It involves only bottom trimming or "skirting" of canes, removing only the ends of canes within 20 inches above the ground, by hand or usually by machine.

Severe pruning creates an unproductive cycle in which mineral nutrients are tied up in stems that are later removed by pruning. Minimal pruning allows increased nutrient flow to leaves and fruit, which increases the number of shoots and buds per vine, clusters per vine, and yields. Some commercial wine vineyards have used minimal pruning successfully for over 10 years.

IRRIGATION, FERTILIZATION AND GROUNDCOVER MANAGEMENT

Most vineyards require irrigation, although some wine grapes can be grown without it in favorable regions. Growers use the amounts and timing of irrigation water, fertilizer additions, and organic matter from cover crops or permanent groundcovers to help regulate vine growth.

Irrigation needs

A mature vineyard requires between 19 and 30 inches of water during the growing season, depending on the trellis system and vine vigor. Raisin grapes and less vigorous European wine grapes may need just 19 to 24 inches of water, while more vigorous French hybrids, American grapes and muscadines usually need 24 to 30 inches of water. Growers may apply additional irrigation water for frost protection or leaching salts from the soil. Increasing numbers of grape growers are starting to adopt **regulated deficit irrigation**, which eliminates early-season irrigation, both to keep vines from growing too vigorously and to conserve water resources and energy.

Grapevines' water needs can be divided into 4 stages. **Stage 1** lasts from shortly after budbreak until bloom, a period of about 6 weeks in spring. Grapes need only 2.5 to 3 inches of water during this period for growth, but some cultivars may need frost protection irrigation during the first 2 or 3 weeks of this stage. If not irrigating for frost protection, growers who follow regulated deficit irrigation can completely eliminate irrigation during stage 1, except on very sandy or shallow soils with very little soil water storage, or when cover crops have not yet been turned under. Withholding irrigation may help improve berry set.

Stage 2 lasts about 6 to 8 weeks, from bloom to veraison, when grape berries begin to change color and soften. Vines use 5 to 9 inches of water during this period, when cell division and elongation are occurring in the fruit. Water stress during stage 2 will reduce berry size and yield.

Stage 3 lasts from veraison until harvest, a period of about 7 to 9 weeks, depending on cultivar and region. Vines require 7 to 12 inches of water during this stage when fruit is gaining in size, but most growers stop irrigating anywhere from 6 weeks to 2 weeks before harvest, depending on soil type and type of grapes grown. Excessive irrigation late in this period can encourage fruit rots, delay fruit maturity, and delay wood maturity as vines should be hardening for winter.

Stage 4 lasts from harvest until leaf abscission in late fall, a period of 2 to 9 weeks, depending on cultivar and region. Vines need 2 to 6 inches of water during this period. Irrigate only enough to maintain the canopy during this stage and prevent drought-induced early defoliation. Mild to moderate water stress is beneficial in stopping shoot growth and promoting wood maturity in preparing for winter. Arid soils that need leaching of excess salts can be irrigated heavily just before leaf fall, when temperatures are too low to encourage new shoot growth, and this will also replenish the soil water reservoir for next spring's growth.

Irrigation methods

Many grape growers use **drip irrigation** (Photo 14-17), especially for new vineyards and where water costs are increasing. Trickle irrigation tubing may be tied along a lower trellis wire, but increasingly growers bury the tubing to provide subsurface irrigation. Burying the tubing can reduce animal damage and water loss due to evaporation, and can cut the seasonal number of herbicide applications needed to control weeds in the vineyard row from 3-4 to 1-2.

Water needs for drip irrigation can be cut even further, up to 50%, with **partial root drying** irrigation. This involves installing 2 tubes per row, one on each side about a foot out from the trellis line, and alternating the water supply so each side is irrigated every other time, on a 5- to 14-day interval. This allow roots on the non-irrigated side to dry down, causing leaf stomata to close more over the entire vine, which reduces total vine water use even while roots on the irrigated side take up enough water for good fruit growth. Partial root drying can reduce vine leaf area by 25%, which reduces excess growth of overly vigorous cultivars and exposes clusters to more light, increasing berry color and flavor compounds by up to 30% and reducing leaf and fruit diseases.

Many older vineyards are still **furrow irrigated**, which requires 20-45% more water than trickle irrigation due to evaporation losses from the irrigation ditches and furrows. However, practicing **alternate-furrow irrigation**, which also causes partial root drying, can also cut up water use for furrow irrigation to half. This involves closing off the sluice gates to every other furrow during one irrigation, and the next irrigation opening those gates and closing the ones that were open last time, so that half the root system dries down and reduces vine water use. Alternate-furrow irrigation can be up to 85% efficient on clay soils but only up to 65% efficient on sandy soils, so sandier soils are probably better converted to trickle irrigation, with its 90 to 98% efficiency.

Photo 14-17. Many vineyards utilize trickle irrigation, which reduces water use over sprinkler or furrow irrigation. This Missouri vineyard has a permanent grass groundcover between the grape rows, and mulch to conserve water and reduce weeds within the grape rows.

Some grape cultivars break bud early, and growers of such cultivars may need **overhead sprinkler irrigation** to provide frost protection unless an elevated site provides excellent air drainage. A sprinkler system is much more expensive to install than trickle or furrow irrigation, so in many cases growers would be better off choosing other cultivars that break bud later unless particular early-leafing cultivars command higher prices. Where high temperatures often occur in the month after bloom, reducing fruit set, a sprinkler system can also provide evaporative cooling irrigation and may be justified.

Fertigation or fertilization

Grape growers with trickle irrigation often apply soluble fertilizers through the drip lines about 4 or 5 times during the growing season, a process termed **fertigation**. Most soils will need nitrogen (N) and potassium (K), while some also require phosphorous (P), zinc (Zn), magnesium (Mg) or sulfur (S). Applying N through fertigation reduces the amount of N lost to leaching and thus reduces pollution of surface or groundwaters, and allows the N to be taken up more evenly over the growing season. Fertigation with K, Zn or other needed nutrients can similarly improve uptake over one-time soil applications, and reduce deficiency symptoms. Fertigation has brought young vines into production sooner, increased yields, and improved berry color and size over those of conventionally fertilized vineyards.

Take petiole samples at bloom to determine which nutrients to add and how much of each. A 2-

10-15 NPK solution is common in Sonoma County, CA, with Zn, Mg or S added as needed. This solution is applied 4 or 5 times at 15 gallons per acre per application for non-bearing vineyards and 25 gallons per acre for bearing vineyards.

Fertigation can not be practiced on furrow- or sprinkler-irrigated vineyards. On these, **split nitrogen additions** into at least 2 applications to minimize losses to leaching. The best times to apply N in warm-winter areas are in late spring between fruit set and veraison (start of berry color development), and in fall between harvest and leaf fall. In cold-winter areas, N applied in fall could delay vine hardening and increase winter damage, so apply N in mid-spring and early summer instead.

In addition to petiole sampling at bloom to determine N levels in plant tissues, adjust N rates based on vine vigor and production. Growers want to reduce vegetative growth on many vigorous cultivars by keeping N additions low, while some less vigorous cultivars may need greater N additions. In general, grapes need 25 to 50 pounds of N per year.

Soil tests every 3 or 4 years may be adequate to determine P and K needs, although growers with a history of K deficiencies, common on some cultivars and rootstocks, may need to test plant tissues annually for K. Zinc deficiencies are common on grapes on alkaline Western soils, and often require 2 to 4 zinc oxide or neutral zinc foliar sprays each spring, from 3 weeks before bloom until bloom.

Vineyard floor management. Turning under a winter legume cover crop each spring can provide the equivalent of 50 pounds of N, and eliminate the need for additional N fertilizer. Purple vetch, bell beans, yellow sweet clover and field peas are among the most widely grown winter legume cover crops. However, continued use of legume cover crops can increase nematodes on sandier soils, so on those soils use legume cover crops only every other year, and use grass cover crops and add N fertilizer in alternate years. Turn under the cover crop several weeks before bloom, so the soil has a chance to settle before spring frosts threaten, since a firm, bare soil releases the most heat to help vines avoid frost damage.

Vineyards need permanent groundcovers where soil erosion is a problem. Growers can reduce or eliminate the need for N fertilizers by using cool-season legumes such as white or red clover on heavier soils. Again, such legumes can cause nematodes to build up on sandier soils, where it's best to use a grass groundcover and add N fertilizer. Vineyards with permanent sods may also require fertilization with calcium, sulfur, iron or boron, since the sod competes with vines for these nutrients.

Control all vegetation in the grape rows with herbicide treatments and/or mulch, to reduce competition for water and nutrients.

CANOPY MANAGEMENT

Grape growers use many techniques to increase light penetration into vineyard canopies. These include removing excess shoots or leaves and setting shoots into desired positions.

Shoot thinning

Growers of spur-pruned table grape cultivars often remove excess shoots, to reduce shoot congestion in the fruiting zone and let in more light, and to reduce the number of grape clusters so the remaining clusters become bigger and sweeter. When shoots reach 6 to 12 inches long, such growers remove all latent (non-fruiting) shoots from older wood, and remove all but two shoots per spur during a heavy crop year. If the crop is light, only non-fruiting shoots are removed.

Cane-pruned table grapes generally require thinning only the latent shoots, except when several canes are wrapped together on a single wire and competing fruitful shoots must be thinned. Some table grape cultivars, such as 'Redglobe' produce very few latent shoots and seldom require crop adjustment, so rarely need shoot thinning. Do not thin shoots of table grape cultivars that are very susceptible to heat damage or sunburn, such as 'Thompson Seedless' or 'Redglobe', since shading from extra shoots helps protect the fruit from damage.

Shoot positioning

Many grape cultivars produce long shoots that tend to grow horizontally. This habit fits in well with a head system or overhead trellis, but with many standard trellises the horizontal shoots run along the top of the wires, significantly shading the fruit and renewal zone and reducing yields or fruit quality. Growers of all types of grapes therefore train such shoots to grow either up or down. A few wine grape cultivars tend to produce very short shoots that stand up well on their own, and so need no positioning.

Growers who use high-cordon systems generally use machines to comb shoots to grow downward twice in early spring. The first shoot combing takes place one to two weeks before bloom; if significant shoot breakage occurs when starting the combing, wait a week before resuming this first combing. Perform the second combing about two weeks later, but before tendrils start to attach to wires, since many shoots with tendrils already attached will break.

Lower-cordon systems may use catch wires above the cordons or canes to support shoots that are manually trained upward. Set the first catch wires 8 to 12 inches above the cordon, and additional wires every 8 to 12 inches as needed. Tuck the shoots in upright positions, taping or tying them if needed, shortly before and after bloom as with the mechanically combed system.

California researchers found that canopies oriented upward tended to produce more leaf area and higher yields than canopies oriented downward, along with higher berry weights, greater sugar accumulation and lower acidity in the fruit, although fruit sugars and acidity did not differ significantly at harvest. The high prices of some fine wine cultivars may justify the added cost of hand-training shoots upward, while producers of lower-priced wine or juice grapes usually use machines to train shoots downward to reduce costs.

Early basal leaf removal

Many table grape and wine grape growers remove some basal leaves and lateral shoots between bloom and berry set. This increases light to the berries, resulting in better color development, and reduces humidity in the fruit zone, which lessens the threat of fruit rots. It can improve the character of some wines, such as 'Sauvignon Blanc', which produces a wine with a more floral and less vegetative flavor when workers remove basal leaves. Early leaf removal also facilitates manual operations, especially cluster thinning and harvest. However, leaf removal can render the more exposed fruit more susceptible to bird predation, heat damage and sunburn. Therefore, do not remove any leaves before veraison (the beginning of berry color development) on cultivars susceptible to heat damage or sunburn, such as 'Thompson Seedless' or 'Redglobe' table grapes or 'French Columbard' wine grape.

High-vigor table grape cultivars on T-trellises, which tend to produce many overlapping shoots, and high-vigor trellised wine grape cultivars benefit most from early leaf removal. Moderately vigorous vineyards may require removing only a few basal leaves and side shoots, while vineyards with very dense canopies may benefit from removing all primary leaves and lateral shoots from the base of each shoot to the leaf opposite the top cluster of grapes. Low-vigor vineyards do not require early leaf removal.

Leaf removal costs less the earlier it is performed, but removing leaves before bloom reduces berry weight and increases shot berries (tiny, unmarketable berries) in seeded table grape cultivars such as 'Cardinal' and 'Black Ribier', so wait until berry set to remove leaves on these cultivars.

Leaves can be removed from seedless table grape cultivars, such as 'Flame Seedless', before bloom without ill effect, at one-third the cost of waiting until several weeks after berry set. Probably the application of gibberellic acid to seedless grape cultivars to increase berry size overcomes any negative effects of early leaf removal.

Leaf removal after veraison

Some red and black table grape cultivars require greater light exposure for two to three weeks before harvest in order to develop full color. Growers provide more light by removing canes around fruit clusters on the north side of the row, to provide indirect light that won't sunburn the fruit. They may also have to remove leaves around the clusters.

High-vigor vineyards require much more shoot and leaf removal than medium-vigor vineyards, and low-vigor vineyards may not require any. Some cultivars, such as 'Crimson Seedless' and 'Emperor', require considerable light to color properly, while most other red and black table grape cultivars need some additional light but not as much as these two. Ethephon treatments also increase coloring, and can substitute for shoot and leaf removal in a few cultivars, such as 'Flame Seedless'.

POLLINATION, FROST AND HEAT PROTECTION, GROWTH REGULATORS AND FRUIT THINNING

Bunch grapes mainly self-pollinate but can also be cross-pollinated by wind. Some muscadine grape cultivars are self-fertile, but some muscadine cultivars have only female flowers and require a self-fertile cultivar nearby for cross-pollination. Under most conditions, berry set is quite adequate, and growers may need to thin some cultivars. Table grape growers commonly use chemical growth regulators to thin fruit and increase berry size, or to increase fruit color, and table and wine grape growers may practice hand thinning of flower or fruit clusters or individual berries to increase fruit size or quality or reduce fruit rots. In some areas, high temperatures during and after bloom can reduce berry set, and evaporative cooling irrigation can ameliorate this and increase yields.

Gibberellic acid (GA) sprays

Seedless table grape growers commonly apply GA as a bloom thinning spray, to reduce the number of flowers which set, and/or as one or two sizing sprays about a week and two weeks after bloom, to increase berry size. 'Thompson Seedless' generally receives a thinning spray of 10 to 15 ppm GA when 50% of clusters show 50 to 70% of flowers in bloom (Photo 14-18). Then it receives a sizing spray of 40 ppm GA when berries reach an average diameter of 3/16 inch, and possibly a second sizing spray 5 to 7 days later. Some seedless table grape cultivars do not respond well to GA thinning sprays, so receive only a sizing spray or two. Eastern seedless table grape cultivars often respond well to a 20 ppm GA bloom thinning spray and a 50 ppm sizing spray about 2 weeks after bloom.

GA sprays can damage the foliage of some cultivars; this is called phytotoxicity. Workers may use directed manual GA sprays to treat clusters while minimizing contact with the foliage, or even dip clusters by hand into GA solutions.

Photo 14-18. Seedless table grapes often receive gibberellic acid (GA) sprays when 50 to 70% of the flowers have opened, as here, to prevent excess berry set. This "thinning" spray allows remaining berries to grow larger, and reduces bunch rot in the clusters. Later GA "sizing" sprays can induce berries to grow even larger.

Cluster thinning by hand

In addition to shoot thinning, which reduces an excess crop and thus improves berry size, wine and table grape growers practice cluster thinning of many cultivars. Removing excess clusters increase the size of remaining clusters and berries and prevents overcropping, which can reduce sugar levels in both berries and wood, lead to winter damage in cold areas. Soon after bloom, workers remove all but one flower cluster per shoot for most shoots. Two clusters may be left on very vigorous shoots, while all clusters are removed from weak shoots.

Mechanical thinning

Certain wine grape cultivars that tend to overbear, such as 'Cabernet Sauvignon', may benefit from cluster thinning, and growers who use mechanical harvesters can also use these machines to thin the clusters. Harvesters equipped with bo-peep rods that turn at 220 or 250 rpm will remove an adequate number of clusters soon after bloom. Larger growers may purchase a machine specially designed for thinning clusters, the Morris-Oldridge thinner, with bo-peep rods that turn at 400 or 600 rpm for greater thinning.

Frost protection and evaporative cooling irrigation

Frosts during bloom do not present a problem in most areas where grapes are grown, although a few growers in northern regions use sprinklers for frost protection. If frosts kill primary buds, many grape cultivars will set about two-thirds of a crop from secondary buds, or even one-third of a crop from tertiary buds if secondary buds are also killed. Late-ripening grape cultivars in short-season areas may require sprinklers for protection against early fall frosts just before harvest.

Temperatures above 90 F during the month following bloom can greatly reduce berry set in some cultivars, particularly American grape cultivars and certain wine grape cultivars adapted to climates with cooler summers. Moisture stress during this period can also reduce berry set in many cultivars, especially seedless table grapes. Where frosts may occur during bloom, and high temperatures often present a threat after bloom, installing a sprinkler irrigation system may prove economical, both for frost protection during bloom and evaporative cooling after bloom.

Other methods of avoiding frost damage include delaying pruning until after budbreak, and leaving more buds if some have been damaged. Applying an alginate gel to grapevines in spring can also delay budbreak of cultivars that tend to start growth early.

Vine girdling

Seedless table grape growers often girdle vines, in addition to applying GA and/or ethephon, to increase berry size, sugars or color. Girdling involves removing a thin strip of bark and underlying phloem tissue entirely around the trunk or cane, without hurting the wood (xylem) beneath. This stops the flow of sugars and natural plant growth regulators downward through the phloem to the roots, enabling the berries to get bigger and sweeter, but doesn't stop the upward flow of water and mineral nutrients from the roots through the xylem. Girdles take about a month to heal, and during this time roots will be stressed, so may need more frequent irrigation.

Girdling at berry set increases berry weight of 'Thompson Seedless' from 10 to 30% more than applying GA alone, and increases berry size of 'Flame Seedless' about 10 to 15% over GA alone. Start girdling when berry diameter averages 3 mm and complete the operation before berries reach 6 to 8 mm in diameter, about a week after cap shatter.

Occasionally, growers may girdle 'Thompson Seedless' again later, when berries start to soften or lighten in color, to hasten ripening when pursuing a more lucrative early market. This can advance harvest a week but response diminishes when the crop is heavy. Girdling 'Flame Seedless' at about the same time (veraison) enhances berry color and sugar development and reduces acidity, but again the response lessens when the crop is heavy. Such double girdling is tough on vines, so practice it only on very vigorous vineyards, and not every year.

Cytokinen sprays

A synthetic cytokinen called CPPU or forchlorfenuron (trade name: KT-30) received an experimental use permit for grapes in California in 2001. When applied just before bloom, CPPU increases berry set, but this is usually undesirable except for some cultivars with very loose clusters. When applied near fruit set, this compound stimulates cell division and elongation, resulting in increased berry size, but also delays fruit maturity, fruit coloring and the accumulation of sugars. It will be most useful for white cultivars in which GA sprays cause a phytotoxic reaction, such as 'Princess' ('Melissa'), or when delayed maturity is desirable.

Ethephon sprays

Many growers of red and black table grapes, both seedless and seeded, apply ethephon at veraison to increase color development. The combination of a heavy crop, high rates of gibberellin and a berry set girdle can reduce coloring enough that most of the crop fails to meet grade, so such conditions especially call for an ethephon spray. Ethephon does not increase berry sugar content.

PEST CONTROL

A large number of diseases and arthropod and vertebrate pests can attack grapes, but in most areas only a few of these present a regular threat. Organic grape production is easiest in the arid West, where powdery mildew is often the only serious disease and grape berry moth the only serious insect pest. Grape growers in more humid areas such as the Pacific Northwest, the Midwest, and the Mid-Atlantic states must also deal with black rot, downy mildew, Eutypa dieback and other diseases as well as more bird predation. Southeastern growers face all these pests plus additional fruit rots and Pierce's disease. Fortunately, grape breeders have developed cultivars for each region with at least moderate resistance to the prevalent diseases (see Tables 14-1 to 14-8), and new pest management materials provide ecological ways to control insects, birds and other pests. Where demand for certain wine, juice or table grape cultivars dictates the planting of cultivars that are highly susceptible to prevalent diseases, at least plant these susceptible cultivars in separate blocks, so that they may be sprayed more frequently without spraying the more resistant cultivars more often.

Diseases of roots, canes and shoots

Crown gall, caused by the soil-borne bacterium *Agrobacterium tumefaciens*, induces swollen galls on grape roots and trunks, which hurt vine growth and yields. The pathogen enters plants through wounds, especially those from freeze damage, so crown gall outbreaks often appear following severe winters on cold-sensitive cultivars such as vinifera grapes or certain French hybrids. Prevent much crown gall by choosing cultivars hardy enough for your region, and by pruning and adjusting crop loads to minimize winter injury. On heavy soils where crown gall is known to occur, dip transplants in a biocontrol solution (Galltrol™), as described in chapter 3, just before planting.

Virus diseases can also kill grapevines, so always buy transplants certified to be free of these diseases from a reputable nursery. Leafroll, corky bark, and fan leaf viruses are the most common grape viruses. **Fan leaf virus** is spread by dagger nematodes through the soil, so soil fumigation has been used to kill the nematode vectors where they are present in high numbers (mainly on sandy soils), and preplant canola cover crops and biocontrols for nematodes (described in chapter 3) can reduce the disease. Use O39-16 rootstock where dagger nematodes are a problem but phylloxera is not, or O43-43 where phylloxera is also a problem.

Root knot nematodes also attack grapes, causing problems mainly on sandy soils to vinifera and French hybrid cultivars. Some vinifera cultivars, including 'Thompson Seedless', 'Emperor' and 'Tokay', show field tolerance by reducing root knot nematode penetration or development, so can be planted on their own roots if phylloxera-resistant rootstocks aren't needed. When planting cultivars susceptible to root knot nematodes, use cultivars grafted onto rootstocks resistant to the nematode biotypes present. No single rootstock provides resistance to all biotypes of root knot nematodes, so have soil tested before planting to determine the predominant biotypes present. Cotton and alfalfa cause large buildups in different species of root knot nematodes, so plant such land to nematode-reducing cover crops for 18 months before plantings grapes. When replanting an old vineyard site, fumigate the soil before planting to kill old roots and nematodes in the top 5 to 6 feet of soil. Plant to different rootstocks or cultivars than were previously planted on the site, to prevent the buildup of nematode biotypes able to overcome a particular type of nematode resistance.

Pierce's disease occurs in parts of North America with mild winters, especially the Southeast, California and Mexico, and kills grapevines. Southeastern growers should plant only muscadine or bunch grape cultivars resistant to this bacterial disease, caused by *Xylella fastidiosa* and spread by several species of leafhoppers called sharpshooters. The glassy-winged sharpshooter, native to Florida and imported into California about 10 years ago on ornamental nursery plants, spreads the disease much faster than other leafhoppers, especially in California where it has no natural enemies. The state of California and the USDA have spent over $100 million over the last 6 years trying to eradicate this leafhopper, including rearing and releasing a tiny wasp native to Mexico that parasitizes sharpshooter larvae. Southern California growers have already lost about $14 million of susceptible vinifera vineyards, and the glassy-winged sharpshooter is moving north. Growers limit damage by controlling Bermudagrass and other grassy weeds that harbor leafhoppers, trapping sharpshooters, and controlling them with a systemic insecticide, imidacloprid (Provado™). Recent research has also demonstrated that removing plants that serve as breeding hosts for sharpshooters (wild grapes, blackberries, elderberries, periwinkle, California mugwort, Mexican tea, mulefat, stinging nettle and cocklebur) from vineyards and nearby riparian woodlands reduces abundance of these vectors of Pierce's disease by 70 to 99%. Where the disease is epidemic, do not plant the most susceptible cultivars, such as 'Chardonnay' and 'Pinot Noir' wine grapes or 'Fiesta' raisin grape. No cultivars commonly grown in California are highly resistant, but 'Ruby Cabernet' wine grape and 'Thompson Seedless' table and raisin grape are relatively resistant.

Eutypa dieback, caused by the fungus *Eutypa lata*, produces cankers that generally form around pruning wounds in older wood of main trunks. Cankers appear as flattened areas that are difficult to spot since they are covered with bark, but removal of the bark reveals darkened, dead wood in the flat area bordered by lighter-colored, healthy wood. Shoots from above the cankers will eventually be stunted, with small, cupped, yellow leaves, but these symptoms may not develop until 2 to 4 years after the vine was first infected.

Recent research has shown that machine pruning or minimal pruning reduces incidence of Eutypa dieback compared to traditional hand pruning. Coating pruning wounds with a fungicide (1% benomyl) greatly reduces the disease, and a natural substance, trans-cinnamic acid from cinnamon, has completely inhibited growth of this fungus on pruning wood in the lab, so may receive registration for grapes soon. Also inspect vineyards each spring and cut off any trunks with cankers, sterilizing tools between cuts, and burn or bury the prunings. Where this disease is serious, some growers use the double trunk training system. That way, if a diseased trunk must be removed, workers can prune the remaining trunk to leave the full number of buds until a new second trunk can be established.

Phytophthora crown and root rot, sometimes called collar rot and caused by several species of the water mold fungus *Phytophthora*, may occur on poorly drained soils, especially under drip irrigation. Such soils may require building raised beds, and place trickle irrigation emitters a foot or so away from the vine trunks.

Foliage and fruit diseases

Powdery mildew caused by the fungus *Uncinula necator* (Photo 14-19), induces a powdery white mycelial growth on leaves and a net-like russet on fruit of susceptible grape cultivars, that can severely reduce yields and fruit quality. This disease is responsible for most fungicide sprays to grapes grown in California and other drier climates. Most *vinifera* cultivars and French hybrids are much

Photo 14-19. Powdery mildew, a white powdery growth on leaves and fruit, is responsible for most fungicides applied to grapes in arid areas. Some cultivars are less susceptible to powdery mildew, and need fewer fungicide sprays. (Photo by Edward Hellman, Oregon State Univ., Corvallis, OR.)

more susceptible to powdery mildew than most American cultivars, and American cultivars may not require fungicide sprays for powdery mildew except under weather conditions very conducive to the disease. Temperatures between 68 and 77 F, with relative humidity between 40 and 100% but little rainfall, favor powdery mildew development. When fungicides are needed for cultivars susceptible to powdery mildew and no other diseases are a problem, sulfur has long been used and is considered "organic". However, it can damage some cultivars, so apply it only on cultivars known to be sulfur-tolerant and only when temperatures are less than 80 F. A dormant lime sulfur spray reduces powdery mildew without harming any cultivars.

California growers currently use JMS Stylet-Oil, a highly refined petroleum distillate, extensively for powdery mildew control during the growing season on wine grapes. It also works well on juice grapes, but shouldn't be used on table grapes because it removes the waxy "bloom" covering the fruit. Do not apply stylet oil within 2 weeks before or after applying sulfur or captan fungicides, or severe damage to vines can occur. The biocontrol agent AQ-510 also controls powdery mildew. So do sterol-inhibiting fungicides (trade names include Bayleton, Nova, Procure and Rubigan; these also control black rot) and strobiluron fungicides (trade names include Abound, Flint and Sovran; these also control black rot and downy mildew). Wine and raisin grape growers can stop applying fungicides when the berries achieve an average sugar content of 12% and are no longer susceptible to powdery mildew. However, table grape growers must continue fungicide applications as long as temperatures remain favorable for infection, because the stem caps remain susceptible and reduce table grape storage life if infected.

Black rot (Photo 14-20), caused by the fungus *Guignardia bidwellii*, can destroy an entire crop if not controlled, and preventing this disease is the goal of most fungicide applications to grapes in the East, Midwest and Pacific Northwest of North America. Black rot does not occur in the Southwest. Black rot causes a leaf spot that reduces yields and produces more spores that spread to the fruit, cane lesions that are especially abundant in mechanically pruned or hedged vineyards, and a fruit rot that starts with light brown spots and eventually shrivels the berry into a hard, black mummy. The fungus overwinters in mummified fruit from last year, on the vine and on the ground. These mummies produce spores that induce primary infections on new leaves and fruit, from the time of 1-inch shoot growth until 10 to 14 days after bloom. If timely fungicide applications during this primary infection stage can prevent infection of young fruit and leaves, no further black rot controls are needed once the overwintering spores are depleted. However, if primary infections aren't adequately controlled, additional fungicide applications will be needed until berries turn color and reach 6-8% sugar, to protect against secondary infections. Removing from the vineyard and burning as many mummies and infected shoots as possible, including all mummies remaining on the vines or trellises, will also greatly reduce primary infections.

Downy mildew, caused by the water mold fungus *Plasmopara viticola*, can rot flower and fruit clusters and shoots, and causes premature defoliation of infected leaves. Downy mildew requires free moisture to spread its spores, and is a major disease in Eastern North America and the Pacific Northwest. Some cultivars are quite resistant, but susceptible cultivars require applications of fungicides with good activity against downy mildew, such as azoxystrobin (Abound™; it also controls powdery mildew and black rot), mancozeb (also effective against black rot and phomopsis) or ridomil (controls only downy mildew).

Angular leaf spot, caused by the fungus *Mycosphaerella angulata*, can defoliate both bunch grapes and muscadines in humid regions where it occurs, especially in the Southeast. Choose cultivars with greater resistance to the disease where it's a problem, and apply fungicides.

Phomopsis cane and leaf spot, caused by the fungus *Phomopsis viticola*, can infect shoots

Photo 14-20. Black rot is the worst grape disease in the East, Midwest and Northwest of North America. It requires 4 to 8 fungicide sprays from budbreak until berries start to color (veraison) to prevent it.

and fruit in humid areas, and during some years in arid areas such as California. Black lesions on shoots can appear very severe, but do not usually cause any direct crop loss. Shoot lesions do provide inoculum for fruit infections that remain dormant until near harvest. Berry infections first appear light brown, then produce black spores, and berries shrivel, resembling the mummies caused by black rot. A dormant application of lime sulfur for powdery mildew control also reduces phomopsis. Where phomopsis is a problem, growers include mancozeb or the strobilurin fungicide azoxystrobin (AboundTM) in the early-season spray program. Cultivars differ in susceptibility to this disease, and those with moderate resistance need fewer sprays. Avoiding overhead sprinkler irrigation also minimizes this disease.

Botrytis bunch rot or gray mold, caused by the fungus *Botrytis cinerea*, often rots between 1 and 10% of the fruit of all types of grapes, and can cause severe losses up to 40% during bad years. Botrytis bunch rot is worst on tight-clustered French hybrids, such as 'Seyval' and 'Vignoles', and many vinifera cultivars, especially 'Chardonnay', 'Chenin Blanc', 'Gewurztraminer', 'White Riesling' and 'Zinfandel'. Growth regulator sprays to loosen such clusters, or leaf removal around the clusters before bunch closing to improve air circulation and spray penetration, reduces losses. Grapes grown on cross-arm trellises, such as T-trellises or Geneva Double Curtain trellises, developed 47% more bunch rot than the same susceptible cultivars grown on 2-wire vertical trellises in California trials, so growers of susceptible cultivars may want to use vertical trellises.

Botrytis infects female flower parts during bloom, then the fungus remains dormant until close to harvest, when individual berries turn brown and rot, then produce gray spore masses that spread to infect other fruit. Rotting berries may attract fruit flies carrying vinegar bacteria that produce a sour-smelling rot, or other secondary rotting organisms may infect the fruit. A fungicide spray at full bloom can protect the flower parts from infection, and two more sprays two weeks apart protect young berries from infection via adhering senescent flower parts. Two more fungicide applications, when the berries reach 5% sugar or the disease is first seen, and two weeks later, protect ripening berries. The registered fungicides that provide the best control of Botrytis rot — potassium bicarbonate (Kaligreen™), cyprodinil (Vangard™), iprodione (RovralTM) and fenhexamid (Elevate™) — do not generally control other diseases, so must be applied in addition to any other fungicides needed, although the early potassium bicarbonate sprays also reduce powdery mildew. Stobiluron fungicides applied early in the season for other diseases also reduce bunch rot, but not as well as the fungicides above.

Several other fruits rots caused by fungi can be devastating in warm, humid regions such as the Southeastern U.S. **Anthracnose** or **bird's eye rot** (Photo 14-21), caused by *Elsinoe ampelina*, produces orange spore masses on fruits and also attacks young shoots and mature leaves. A dormant lime sulfur spray reduces overwintering inoculum for anthracnose, helping to reduce the disease. **Bitter rot**, caused by *Greeneria uvicola*, attacks only ripening fruit, especially muscadines, and produces a bitter flavor in them. **Ripe rot**, caused by *Colletotrichum gloeosporiodes*, and **macrophoma rot**, caused by *Botryosphaeria dothidea*, also attack ripening fruit and are most damaging on muscadine grapes. Choose cultivars with some resistance to these diseases in the Southeast, thoroughly apply the pre-bloom and first post-bloom fungicide sprays because infections actually start then and remain latent until near harvest, and remove and destroy all infected berries. Pre-harvest captan sprays may help if these rots pose a large problem.

Insects that feed on roots, trunks, canes and leaves

Grape phylloxera (*Daktulosphaira vitifoliae*), a soft-bodied insect related to aphids, causes swollen galls on smaller roots and may rot larger roots, and does the most damage to vinifera cultivars, which it often kills. Prevent root damage from this pest by

Photo 14-21. Anthracnose is another common fruit rot of grapes. A dormant lime-sulfur spray helps prevent this, as do early-spring fungicides. Less susceptible cultivars may not need additional sprays, but more susceptible cultivars often need captan sprays before harvest.

planting vinifera cultivars grafted onto American or hybrid grape rootstocks that resist phylloxera. A foliar form of phylloxera (Photo 14-22) present in eastern North America also causes galls on leaves of susceptible vinifera and French hybrid cultivars, which require insecticide sprays at bloom and again 10 to 14 days later to prevent this.

Western grape rootworm (*Diabrotica virgifera*) on the West Coast and grape rootworm (Fidia viticida) on the East Coast are beetle larvae that consume small roots and tunnels into larger roots and vine crowns. These brown-headed, C-shaped white grubs overwinter deep in the soil but approach the surface to pupate in early spring, when disking under a winter cover crop usually controls them. Sprays for grape berry moth also control grape rootworm. Growers who use mating disruption to control grape berry moth, and have a permanent sod groundcover, may need to apply one insecticide spray, such as carbaryl, when adult beetles make chains of small holes on leaves.

Grape root borer (*Vitacea polistiformis*) larvae tunnel into larger roots and crowns, weakening and eventually killing grapevines. The adult is a clearwing moth that mimics a wasp, with a brown body with yellow markings. Control this pest with mating disruption, applying 100 pheromone twist ties per acre (one every 6 or 7 vines) in late spring, or by planting into polyethylene mulch and reapplying the mulch every 3 to 4 years as it starts to break down. Some rootstocks, such as Florilush, tolerate some grape root borer feeding without showing yield reductions or vine death.

Grape trunk borer (*Clytolepus albofasiatus*), longhorn beetle larvae, feed in dead or dying wood that is usually 2 or more years old. Drought stress, sunscald or mechanical injuries predispose grape trunks to borer injuries, so avoiding these with irrigation, white paint or white plastic tree guards, and careful cultivation, can reduce borer problems. Promptly remove dead or dying vines.

Photo 14-22. Grape phylloxera attacks grape roots, and many cultivars require grafting onto phylloxera-resistant rootstocks to prevent this, especially on heavier soils. A foliar form of phylloxera also causes leaf galls in Eastern North America and may require an insecticide spray.

FOLIAGE INSECTS AND MITES

Leafhoppers include several *Erythroneura* species that suck sap from grape foliage and overwinter as adults about 1/8 inch long. Heavy infestations cause leaves to turn yellow or brown, and many leaves fall off, reducing fruit yields and quality. Grapevines can generally tolerate populations of up to 10 to15 leafhopper nymphs per leaf without damage, so monitor leafhopper populations and treat with a contact insecticide if populations exceed this threshold. 'Thompson Seedless' grown for raisins or wine tolerates up to 20 nymphs per leaf without damage, while late-ripening table grapes may require treatment at thresholds of 5 to 10 nymphs per leaf. Monitor leafhoppers every 7 to 10 days starting in early to late June (depending on region), based on degree-day accumulations. Beneficial wasp parasites (*Anagrus* species) of grape leafhopper eggs can usually control these pests if plants harboring other leafhoppers that overwinter in the egg stage — such as blackberries, prunes, or roses — are planted or grow naturally within ½ mile of the vineyard. Parasitized eggs turn red, so are easily recognized. Mowing an orchard groundcover with a flail mower before spring budbreak can greatly reduce overwintering adult grape leafhoppers, if done in early morning before temperatures warm up enough to permit leafhopper flight. Basal leaf removal during berry set and the next 2 weeks can also reduce leafhopper populations by 30 to 50%.

Grapeleaf skeletonizer (*Harrisina* species) larvae feed on leaves, completely consuming them between the veins, leaving only the "skeleton" of the main veins. One of the two skeletonizer species, easily recognized by their yellow larvae with bright blue and black stripes, occurs almost everywhere in North America where grapes are grown. Western grapeleaf skeletonizer can be a severe pest especially in California, while Eastern grapeleaf skeletonizer occurs throughout eastern North America and is worst in Florida. Second and third generations of this blue-black moth pest have defoliated entire vineyards, reducing yields and fruit quality and causing fruit sunburn. In addition, hairs on the larvae cause skin welts on workers whose hands brush against them.

Two parasitic biocontrol agents of this pest have been released in California's San Joaquin Valley, and one of the parasites (*Amedoria misella*) is now common in many vineyards, providing natural control. A granulosis virus disease of this pest has also been introduced in southern California with excellent success. Skeletonizer larvae infected with the virus shrink instead of growing, consume tiny patches of tissue rather than an entire leaf, and most eventually die, while the few that live pass the virus disease to the next generation. If parasites or the granulosis virus are not present in your area, control grapeleaf skeletonizer with sprays of the biocontrol agent *Bacillus thuringiensis* (Bt, Dipel, Thuricide or other trade names) while larvae are still small. Timing these sprays for leafrollers (see below) will also control skeletonizers.

Grape leaffolder (*Desmia funeralis*) caterpillars fold or roll up leaves and feed from inside the protected area, almost everywhere in North America that grapes are grown, but several tiny wasp and fly parasites often keep this pest in check. Wine and raisin grapes seldom require pesticide sprays for this pest, but table grapes may. Scout vineyards regularly and if leaffolders infest many vines, treat with Bt while larvae are still small.

Japanese beetle (*Popillia japonica*) adults feed on more than 250 plant species, but grapes are among the most preferred hosts. This pest occurs only in the eastern third of North America. It prefers French hybrid cultivars with thin, smooth leaves, so apply an insecticide to such cultivars as soon as adult beetles appear and about 15% of leaves are damaged. Commercial Japanese beetle traps may be used to monitor the population, but these often aren't needed since the beetles are easily detected when walking through the vineyard. Japanese beetles avoid American cultivars with thick, hairy leaves, so these often don't require special treatments for this pest.

Spider mites (*Tetranychus* and *Eotetranychus* species) may be serious pests in California vineyards, especially those heavily treated with insecticides that disrupt the predator mites that normally feed on these tiny, eight-legged pests. Excessive spider mite populations suck the sap from leaves, causing foliage bronzing and defoliation. Dry, clean-cultivated vineyards suffer the most damage, while well-irrigated vineyards with grass strips between the rows suffer less from mites. Overhead sprinkler irrigation is also effective in washing pest mites off the vines, without harming predator mites. Spraying dirt roads in and adjacent to the vineyard with road oil reduces dust, which keeps spider mite populations down. Where mite problems occur, growers monitor populations of both pest mites and predators, and treat with a selective miticide or release predator mites when pest mites get out of hand.

Insects that feed on flowers or fruit

Grape berry moth (*Endopiza viteana*), the worst insect pest of grape berries in eastern North America, can destroy up to 90% of the berries in untreated vineyards (Photo 14-23). Mottled brown adult female moths lay eggs on or near flowers or berry clusters, and the caterpillars that hatch form webs that tie buds or berries together. First-generation larvae feed externally on the berries in early summer, while larvae from 2nd and 3rd generations tunnel into the fruit, destroying them and attracting bunch rot fungi. Growers with vineyards at least 5 acres in size control this pest by mating disruption, applying 400 pheromone twist ties per acre. Vineyards bordered by woods containing wild grapes may also require 2 to 3 treatments of border rows with an insecticide such as phosmet (ImidanTM), just before and 7 to 10 days after bloom or timed by pheromone trap catches, in addition to mating disruption.

Leafrollers have become more important pests of grapes since 1960, especially **omnivorous leafroller** (*Platynota stultana*) in California and red banded leafroller (*Argyrotaenia velutinana*) in Eastern North America. This occurred due to

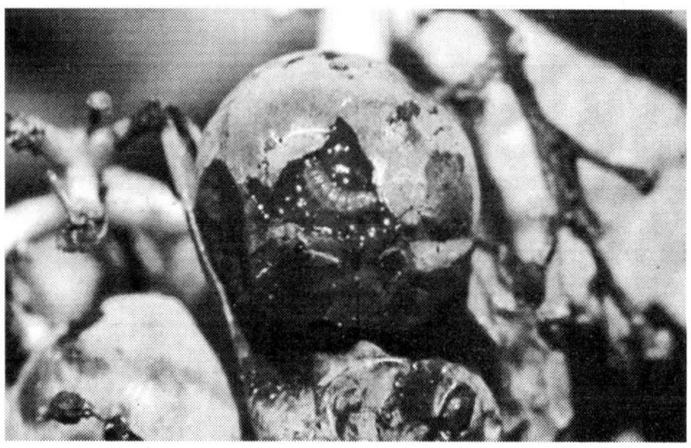

Photo 14-23. Grape berry moth, the worst insect pest to attack grape berries in Eastern North America, can now be controlled by mating disruption in vineyards of at least 5 acres. Applying 400 pheromone-impregnated twist ties per acre prevents adult moths from mating and laying eggs in the fruit without insecticide treatment.

increased use of insecticides for other pests that killed leafrollers' natural enemies, and increased control of weeds in vine rows with herbicides rather than grape hoes, which formerly buried mummified grape clusters where leafrollers overwintered. Moths lay egg masses on leaves or fruit, and larvae invade flower clusters, spinning protective webbing among the flower parts that limit ability of pesticides to reach and kill the larvae. Feeding larvae kill portions of clusters, and as berries ripen fruit-rotting organisms attack the feeding injuries. The second generation generally does the most damage. Where leafrollers are a problem, use pheromone mating disruption or sprays of the biocontrol agent *Bacillus thuringiensis* (Bt, Dipel, Thuricide, and other trade names), timed by pheromone traps (generally before bud closing, in late June and late July), to control them. Basal leaf removal will improve leafroller spray coverage.

Grape orange tortrix (*Argyrotaenia citrana*) larvae, which occur west of the Rocky Mountains, cause the same type of damage as leafrollers, by feeding on buds and webbed leaves in spring, then entering bunches and making webbed nests to feed on the berries. This moth pest does have an effective parasitic wasp, the tiny *Exochus nigripalpus subobscurus*, which has provided season-long control in California's Salinas Valley, especially when coyote brush is allowed to grow near the vineyard. The parasite overwinters on orange tortrix and other host caterpillars on the coyote brush, when the pest is more scarce in vineyards, allowing the parasite population to maintain itself at a high enough level to control orange tortrix in vineyards starting early in the growing season. If parasites do not adequately control this pest, Bt sprays applied for leafrollers will reduce it.

Rose chafer (*Macrodactylus subspinosus*) adult beetles attack grapes at bloom, destroying the blossoms and often skeletonizing the leaves as well. This pest emerges from the soil and is worst on sandy soils in northeastern North America, attacking many different fruits, vegetables and ornamental plants. Where rose chafer occurs, huge populations can appear suddenly and cause severe damage. Intensive mass trapping, using 35 Japanese beetle traps per acre and a feeding attractant for rose chafer, over a 4-year period can reduce populations below the treatment threshold level of two beetles per vine. Where populations exceed this threshold, treat with an insecticide such as carbaryl (Sevin™) just after bloom and again 10 days later if needed.

Tiny, metallic blue grape flea beetle (*Altica chalybea*) adults appear in early spring, feeding on buds and unfolding leaves and laying eggs. Dark brown larvae then attack flower buds, hollowing them out and severely reducing yields if beetle populations are high. This pest is particularly serious on bunch grapes in the Southeast, which may require an early-season application of carbaryl (Sevin™) to control it.

Vertebrate pests

Birds can consume or render unmarketable large amounts of grapes, especially early-ripening cultivars with dark, small berries that fit easily in birds' mouths. Grape growers manage pest birds by choosing cultivars less prone to bird damage, reducing bird habitats on the farm, using scare devices, or applying bird repellent sprays. An effective bird repellant, methyl anthranilate, is a component of 'Concord' grapes, which birds don't like, and some other American cultivars repel birds so don't need protection from them. Some smaller growers find bird netting (photo 14-24) cost-effective.

Deer feeding and rubbing on young grapevines can kill many vines, so deer need control where their populations are high. An electric fence (see chapter 9) provides the best control but is expensive, so where populations are moderate growers may control deer with hunting and repellents.

HARVEST, HANDLING, PROCESSING AND MARKETING

Harvest grapes at their peak ripeness level for their intended use, since grapes do not ripen further after harvest. Growers sample each cultivar at least twice a week as harvest time approaches, monitoring fruit color, sugar content, acidity and

Photo 14-24. Birds can be serious pests in vineyards. Small pick-your-own or local market growers may use plastic netting to prevent bird damage.

pH. As berries ripen, flavor develops fully, fruit color and sugar concentration increase, titratable acidity decreases, and pH increases.

Winegrowers harvest certain cultivars a bit early to produce light, fruity styles of wine. For heavier, full-bodied wines, they allow berries to fully ripen and develop full flavor, color, and tannins. Sugar content and juice pH are the most important measurements used in deciding when to harvest wine grapes, and growers take random samples of 250 berries from 250 vines for lab analysis several times a week as harvest approaches. Most winemakers prefer white wine grapes with 19.5 to 23% sugars, with an optimum pH of 3.1 to 3.2 and no more than 3.3. Red wine grapes should have 20.5 to 23.5% sugars, with an optimum pH of 3.3 to 3.4 and no more than 3.5. American and muscadine grapes harvested for wine should have similar pH levels, but often a lower sugar content of 15 to 19% at those pH levels. Higher pH can adversely affect wine color, protein and tartrate stability, ability to clarify, and other attributes.

Growers usually harvest strong-flavored American cultivars, such as 'Concord' and 'Niagara', just before they are fully ripe when they will be used for juice or jelly, to prevent too intense a flavor in the processed products, although this may require more sugar during processing. The grapes must have a minimum sugar content of 15%, however. The few American grapes harvested for the fresh market are picked near peak ripeness, as determined by skin color and soluble solids (sugars) measurements with a refractometer.

Table grape growers also determine harvest time primarily by skin color and soluble solids measurements with a refractometer, for both bunch grapes and muscadines. Most commercial growers start harvesting shipping grapes at 15 to 16% sugar, but local farm markets and pick-your-own operations offer grapes at 17 to 22% sugars. Raisin growers also harvest based on sugar content, which must be 21 to 22% to prevent spoilage during drying.

Machine vs. hand harvest

Machines harvest the majority of wine grapes and raisin grapes, and almost all juice grapes. Mechanical harvesters have horizontal pivoting arms that smack the grape canopy, or vibrating fingers that dislodge clusters and berries from clusters. Machines can harvest 5.5 tons of grapes per hour, the equivalent of 100 hand harvesters. Some small growers of muscadine or bunch grapes for wine or juice processing use catch-frame harvesters, by stretching canvas sheets over sloping frames beneath the vines. In this system, 4 or 5 workers strike the trellis with padded clubs, and the fruit fall onto the frame and are funneled into a harvest container.

Some wine grapes, some raisin grapes and all table grapes are hand-harvested. Wine or raisin cultivars which ripen all at once may need just one hand harvest, but most table grapes and some wine grapes ripen unevenly and require 2 or 3 pickings over several weeks. Workers judge the maturity of individual clusters as they harvest, and clip entire clusters of bunch grapes. Some muscadine cultivars ripen evenly within each cluster, so entire clusters may be cut as with bunch grapes, especially for shipping. Many other muscadine cultivars ripen unevenly within clusters, so are harvested as individual berries by pick-your-own customers or by workers for local retail sale.

Harvest all types of grapes early in the day when temperatures are cooler, and handle fruit carefully to avoid cracked berries and juice leakage. Table grape harvesters trim clusters of small or immature berries, clip the ends off of scraggly clusters, and bag and box clusters for shipping in the field. Move containers of harvested grapes into the shade immediately (except for raisin grapes), and into a cooling facility at 30 to 32 F within 30 minutes of harvest. Ship or sell table grapes as soon as possible, and keep them cool during transport and sales display. Crush and process wine and juice grapes as soon as possible after harvest, especially those that were machine harvested, as they incurred more damage during harvest.

Raisin drying options

Raisin grapes traditionally have been laid on paper sheets between the rows to dry in the sun

Photo 14-25. Raisin grapes dry on paper trays between rows of a traditional T-trellis. (Photo by Bill Peacock, Univ. of Cal. Coop. Ext. Service, Tulare Co.)

Photo 14-25). Such grapes drying in the vineyard need 2 to 3 weeks of clear, sunny days with no rain to become raisins. Any rain during the drying period can destroy the crop, so growers purchase raisin rain insurance. Workers turn the clusters several times during this process, so all berries dry evenly. Four pounds of fresh grapes yields one pound of raisins. Once the moisture content is reduced to 14-15%, the ideal level for raisins, workers roll the paper sheets into bundles, which remain in the field for several more days. This helps equalize moisture levels among all the berries in each bundle.

Workers then load the bundles onto a wagon and bring them to the grower's sorting machine, which uses a vibrating conveyor belt to separate raisins from the clusters and remove larger stems. The sorted raisins go into bins, which are trucked to a packing plant. The packing plant removes remaining stems and chaff by vibration and forced air, washes and dries the raisins, inspects them and packs them.

The total seasonal labor requirement for dried-on-the-ground raisins is about 40 hours per ton of raisins. Many growers are trying an alternative system that requires about ¾ the labor, and spreads it out over the season so a few local workers can be hired rather than many temporary harvest laborers. The dried-on-the-vine (DOV) raisin system utilizes either overhead trellises, or east-west trellises and separation of the canopy into fruiting zones on the south side and renewal zones on the north side. Workers sever fruiting canes from the vines in late summer but leave them on the trellises, so grapes still attached to vines dry into raisins, which are mechanically harvested about 6 weeks later when moisture content has dropped to 14% (Photo 14-26). Normal rains during this drying period do not destroy the crop, since berries are still attached to canes and have more time to redry, so many DOV raisin growers do not purchase raisin rain insurance. However, during some years when unseasonable weather shortens the drying season, DOV raisins must be placed in electric dehydrators to complete the drying process, and insurance policies are being developed to cover added dehydrator costs.

DOV raisins can be planted at higher vine densities to increase yields, and can be irrigated later in the season to improve yields and reduce vine stress. Vineyard floor cultivation can be reduced, or eliminated with permanent groundcovers, which reduces dust and spider mite problems. These advantages are contributing to an increase of DOV raisin plantings, and conversion of existing raisin plantings to DOV management.

Table grape postharvest handling and storage

Pre-cool table grapes with forced air, and ship or sell as soon as possible, keeping the fruit at 31 to 32 F. Many table grapes to be shipped are fumigated with sulfur dioxide at 5000 ppm for 30 minutes, to retard postharvest spoilage. Late-ripening cultivars are especially susceptible to postharvest gray mold (*Botrytis cinerea*) or other pathogens.

Treated vinifera table grapes, kept cool, can last 6 to 12 weeks in cold storage. American grapes last 2 to 8 weeks, and muscadines last only up to 3 weeks in cold storage. Modified-atmosphere storage is rarely used for grapes, because cultivars ripening at various times are shipped in from all over the world to provide a continuous supply of fresh grapes. However, when modified atmosphere storage of 2-5% oxygen and 1-5% carbon dioxide is used, vinifera table grapes may be stored for 4 to 6 months.

WINEMAKING

Winemaking is an elaborate art and science based on a few basic processes. Harvested grapes are dumped into bulk bins and transported to a crushing machine, which extracts the juice. Usually the juice alone is used to make white wines, while red wines are made from "must", which includes juice, skins to add color, seeds and stems. Winemakers evaluate juice quality at this stage, and decide on the type of wine to produce. They may adjust sugar and acid levels by blending various grape harvest lots or cultivars, or by adding refined sugar. Sulfur dioxide is then added to kill bacteria and fungi, which selectively activates desirable

Photo 14-26. These dried-on-the-vine (DOV) raisins are mechanically harvested about 6 weeks after fruiting canes were severed but left on the vines. (Photo by Bill Peacock, Univ. of Cal. Coop. Ext. Service, Tulare Co.)

yeasts on the berry surfaces that cause fermentation, and to prevent enzymatic oxidation and color deterioration.

Juice or must is then transferred to large fermentation vats (Photo 14-27), where yeasts convert sugars into alcohol up to a level of 12 to 14%, which kill the yeasts and stops fermentation. Sometimes sulfur dioxide is added to stop fermentation and act as a preservative. Red wine is fermented at 75 to 82 degrees F, while white wine is fermented at 58 to 64 degrees F. Rosé wine is fermented at intermediate temperatures, and skins are left in the must for less time than with red wines.

White wine is "racked" or filtered at least once to remove sediments, and red wine is racked or filtered several times, first to remove skins, seeds and stems, and later to remove sediments. Many wines, especially white wines, are aged solely in stainless steel containers, for less than a year before they are bottled. Other wines, especially red wines, may be placed in oak barrels to age up to 10 years, so the wine absorbs some of the oak's tannins, before bottling. Sparkling wines, or champagnes, involve a second fermentation in the bottle, with added sugar to produce carbon dioxide.

MARKETING

Grape growing is one of the most competitive fruit production enterprises. Before planting a vineyard, decide on your market and investigate the economics of growing particular types of grapes in your area. If planning to grow for a winery, juice processor or raisin cooperative, contact them to determine the cultivars of grapes they want to buy, the prices they pay, the growing systems they recommend, and whether they offer guaranteed contracts. If planning to grow table grapes for shipping, contact buyers to find out the same things. If planning a pick-your-own operation or roadside market, determine your market size, any existing competition, and the preferences of consumers in your area.

Southern Illinois University at Carbondale has put together a workbook for prospective grape growers to develop a business and marketing plan. This "Vineyard Planning Guide" is available at www.siu.edu/%7Ereadi/viticulture/businessplan.htm and prospective growers would be wise to complete the workbook before planting. The Extension Services of California, Michigan, New York and many other states also offer vineyard planning and marketing assistance.

Photo 14-27. Juice or must from wine grapes is loaded into fermentation tanks, shown here, where yeasts convert sugars into alcohol. (Photo by Nicholas de Wet, Prospect 1870 Wines, Ashton, South Africa.)

Further Reading

Cornell Univ. in NY offers on-line bulletins on grape cultivars, site selection and more at: http://www.nysaes.cornell.edu/hort/faculty/reisch/bulletin

Grape pest management. 1992. Coop. Extension Univ. of California pub. 4105. 412 pages.

Himelrick, D.G. and W.A. Dozier, Jr. 1996. *Commercial muscadine and bunch grape production guide*. Alabama Coop. Extension ANR-774. Available online at: http://www.aces.edu/department/extcomm/publications/anr/anr-774/anr-774.html

Oregon Winegrape Growers Guide, 5th edition. 2001. Ted Casteel, editor. 258 pages. Oregon Winegrowers Assoc., 1200 NW Front Ave., Suite 400 Portland, OR 97209.

Raisin production manual. 2000. Coop. Extension Univ. of California pub. 280 pages.

Univ. of California Cooperative Extension in Tulare County offers many useful grape-growing publications online at: http://www.ucce.tulare.ca.us/pubgrape

Weigle, T.H. and A. Muza. 2000. *New York and Pennsylvania pest management recommendations for grapes*. Available from Extension offices in NY and PA.

Wolf, T. and E.B. Poling. 1996. *Mid-Atlantic winegrape grower's guide*. Pub. AG-535. Available online at: http://www.ces.ncsu.edu/resources/winegrape/

Zabadal, T. and J. Andersen. 1997. *Vineyard establishment I & II*. Michigan State Univ. bulletins E-2642 and E-2643. Fax 517-353-7168 for copies.

Chapter 15
Cranberries and Blueberries

Cranberries and blueberries, native to North America, have been developed as commercial crops that are now sold throughout the world. Cranberries have been commercially cultivated for over 200 years, while blueberries developed into a crop only during the last 100 years. North America produces over 700 million pounds of cranberries per year and over 400 million pounds of blueberries. North America continues to dominate worldwide cranberry and blueberry production, although plantings have now been made in Poland, New Zealand, Australia, Chile, Norway, Taiwan and Mexico.

Cranberries and blueberries, along with their northern European cousin lingonberries, belong to the genus *Vaccinium* in the heath family, the *Ericaceae*. Members of this family, sometimes called ericaceous plants, are all adapted to sandy or peaty, acid soils, usually with a pH of 4.0 to 5.5, high organic matter, and a high water table. Such adaptation allows ericaceous plants to grow on soils not well suited for other fruit crops, but also limits production on other soil types. However, breeders are continually introducing improved cultivars, particularly combining the genes from several native blueberry species, and extending the range of climates and soil types where blueberries can be grown.

Cranberry, blueberry and lingonberry plants are long-lived woody shrubs or vines, and commercial plantings can last 30 years or more when planted in suitable soils. Constructing a cranberry bog has one of the highest start-up costs of any fruit crop, at $20,000 to $30,000 per acre. Bogs can yield 15 to 25 thousand pounds of cranberries per year, for a gross return of $8000 to $25,000 per acre per year. Blueberry establishment costs and returns vary greatly depending on the type of blueberry and area, but the crop can be very profitable on suitable soils and small growers can much more easily enter the market with blueberries than with cranberries.

All these crops begin to bear fruit about 3 years after planting, and reach full production at 5 to 6 years after planting. All these fruits are true berries (Photo 15-1), with each berry developing from a single ovary and containing many small seeds.

CRANBERRY, LINGONBERRY, AND BLUEBERRY SPECIES

Cranberries belong to the species *Vaccinium macrocarpon*, which indicates they bear large fruit for the genus. This is a diploid species, with two sets of chromosomes. Cranberry plants are trailing, woody evergreen vines that bear fruit on short, 3- to 6-inch-tall vertical branches called uprights

Photo 15-1. Cranberries, blueberries and lingonberries are true botanical berries, as each berry develops from one ovary and contains many seeds. This highbush blueberry cultivar, 'Bluecrop,' is widely produced.

(Photo 15-2). Each upright grows vertically for one or two seasons before weight of fruit bends it down, and new uprights grow from its tip or from buds along horizontal stems. Cranberries are native to acid peat bogs in the northern U.S. and southern Canada, but few wild cranberries are gathered. Improved cultivars of commercial cranberries are planted in acid soil that must be excavated and prepared before planting, in USDA hardiness zones 3 to 7. Cranberries are very acidic so are not eaten fresh. About 5% of cranberries are sold as fresh berries for home cooking, and 95% are frozen or processed into jelly, drinks, snack products or sweetened dried cranberries. Almost all commercial cranberries are machine harvested.

Photo 15-2. Cranberries, native to acid peat bogs in North America, grow as trailing, woody evergreen vines that bear fruit on short uprights. (Photo by T.R. Roper, Univ. of Wisconsin.)

Lingonberries, *V. vitis-idaea*, are also diploid and grow similarly to cranberries except vines are much shorter. Lingonberries are native across northern parts of North America and Europe, in forested regions with cold winters and warm summers. Superior plants have selected from the wild in northern Europe, and a few North American growers are now planting these cultivars. Lingonberries are sweet yet have a pleasantly tangy flavor, so can be eaten fresh and make much-prized jams, tarts and other products. At present, only a few acres of lingonberries have been planted for trial, and most growers must hand-harvest lingonberries with berry rakes similar to those once widely used for cranberries and lowbush blueberries. One larger lingonberry grower has been testing a harvesting machine. As the industry develops and larger plantings are made, harvesting machines developed for lowbush blueberries or for dry harvest of cranberries will probably be adapted for use in lingonberries.

Several blueberry species and hybrids, with heights ranging from 1 foot tall to 12 or more feet tall, are commercially harvested. Various blueberry species have ploidy levels (numbers of sets of chromosomes) ranging from diploid to hexaploid (six sets of chromosomes), but most species that have been commercialized are tetraploids (four sets of chromosomes). Species of the same ploidy level interbreed freely in nature where they occur together, and breeders have now overcome crossing barriers between species of different ploidy levels. The small seed size and thinner skin distinguish blueberries from huckleberries, in a different genus in the heath family and native to North America. Huckleberries, with their large seeds and insipid flavor, are occasionally gathered from the wild but have not been developed as a crop.

Highbush blueberries, derived largely from the tetraploid *V. corymbosum*, provided the first named blueberry cultivars. A commercial cranberry grower in New Jersey, Elizabeth White of Whitesbog, instructed her workers to select superior wild blueberry plants from the surrounding piney woods, and Frederick Coville and other USDA plant breeders started crossing these selections in 1911, developing named cultivars which were first introduced in 1920. Later highbush cultivars also include some germplasm from other blueberry species native to the northern U.S. and southern Canada. Highbush blueberries grow as upright bushes, usually 5 to 8 feet tall, and produce large, high-quality fruit with thin skins. They are best adapted to USDA cold hardiness zones 5 to 7, and still comprise the bulk of planted blueberry acreage (Photo 15-3) and over half the total blueberry crop.

Photo 15-3. The highbush blueberry, derived mainly from the species *Vaccinium corymbosum*, was the first type of blueberry developed into a planted crop, and still accounts for the majority of commercial blueberry production.

About half the highbush crop is hand-harvested by workers or customers for fresh use, and about half is machine-harvested for processing into frozen blueberries, jams and pies.

Lowbush blueberries (*V. angustifolium*) are tetraploids native to Canada, northern New England, northern Michigan, and other northern states in USDA zones 2 to 4. There they form dense mats of scrubby bushes just 1 to 2 feet tall, on large tracts of land called "blueberry barrens." They produce small berries with thick skins, mostly used for processing in blueberry muffin mixes, jams, and other products.

A few superior clones of lowbush blueberries have now been selected, named, and planted on small acreages. However, the vast majority of lowbush blueberry acreage is still harvested from wild stands, for processing. Growers with large tracts use machines to harvest, while smaller growers hire workers who use hand-held rakes or non-motorized push-type mechanical harvesters to harvest the berries. Lowbush blueberries comprise about 42% of the North American blueberry crop.

Highbush-lowbush blueberries, or "half-high" blueberries, are increasingly planted in USDA zones 3 and 4. These tetraploid hybrids between the two species just mentioned combine the fruit quality and upright plant habit of highbush blueberries with the cold hardiness of lowbush blueberries. Plants grow to an intermediate height of about 3 to 4 feet tall, small enough to be covered with snow over the winter, which insulates them and protects them from cold damage. Upright plants make hand harvest for fresh market, mostly local sales and pick-your-own, much easier.

Rabbiteye blueberries derive mainly from *V. ashei*, a hexaploid species native to southeastern U.S. that's adapted to USDA zones 7 to 9. Rabbiteye plants can grow 9 to 18 feet tall, and produce large fruit with a tougher skin and stronger flavor than those of highbush blueberries. Rabbiteye bushes need less organic matter in the soil, and are less susceptible to both drought and excess water than are highbush plants. Collections of wild rabbiteye plants from Florida and Georgia began in 1925, and a USDA rabbiteye breeding program began in 1940. Where highbush and rabbiteye blueberries can both be grown, highbush berries ripen first, about a month before the first rabbiteye berries, and then rabbiteyes start to ripen as the last of the highbush finish ripening. This can extend the blueberry ripening season to 2 ½ or 3 months. Most rabbiteye blueberries are hand-harvested by customers or workers for fresh sales. Some larger growers harvest rabbiteyes with machines for processing, and some rabbiteye cultivars can be machine-harvested for the fresh market.

Southern highbush blueberries, the most recent development, resulted from crossing Northern highbush cultivars with several blueberry species native to the Southeast, including *V. darrowi*, *V. myrsynites*, *V. elliotti*, *V. constablaei*, and a little bit of the rabbiteye blueberry. The resulting tetraploid cultivars, usually 5 to 8 feet tall, are adapted to USDA zones 7 to 10a. Southern highbush blueberries ripen very early in the season, and are grown mainly for shipping to Northern supermarkets for more lucrative early-season sales, so are mostly hand-harvested.

CRANBERRY AND BLUEBERRY PRODUCTION AREAS AND SOIL NEEDS

Cranberry production is concentrated in a few northern states and southern Canadian provinces, which provide the proper soils and cool temperatures under which cranberries thrive. These same conditions favor lingonberries, and some cranberry growers are starting to grow lingonberries also. Wisconsin produces over half the North American cranberry crop, while Massachusetts grows about 24%. British Columbia produces about 9% of North American cranberries, New Jersey provides about 7%, Oregon grows about 6%, Quebec produces about 4% and Washington raises about 2%.

Despite blueberries' unique soil requirements, blueberry production is spread more diversely across the U.S. and Canada than is the production of many fruit crops. Maine produces about 20% of the total North American blueberry crop, Michigan raises about 18%, British Columbia and New Jersey each grow about 10%, Nova Scotia raises about 8%, Quebec produces about 7%, New Brunswick harvests about 5%, and North Carolina and Georgia each produce about 3%. Other states and provinces with significant blueberry production include Florida, Oregon, Washington, New York, Arkansas, Missouri, Ontario, Alabama, Indiana, Illinois, Wisconsin, Minnesota and Virginia, and other states and provinces have smaller areas of blueberry plantings.

The most important factor in determining where cranberries and blueberries can grow is the soil. Most good cranberry and blueberry sites have well-drained sandy loam soil with 3 to 20% organic matter and a pH of 4.5 to 5.5, and a perched water table due to a change in subsoil texture (Photo 15-4). The pH can be lowered with sulfur if it's a bit too high, but calcium saturation of the soil cation-

Photo 15-4. Most cranberry, blueberry and lingonberry sites have well-drained sandy loam topsoil with 3 to 20% organic matter and a pH of 4.5 to 5.5. Usually a different-textured subsoil, as here, provides a perched water table that ensures adequate water when needed.

exchange complex should not exceed 20%. Acid loams or clay loams may also make good blueberry soils, if they are well drained and high in organic matter. Peat moss can be added to planting holes to increase organic matter in mineral soils. Muck soils, which formed in swamps and have organic matter contents of 20 to 50%, make excellent cranberry bogs and good blueberry soils if they can be drained. This was often done in the past but is now mostly prohibited by the Wetlands Protection Act. Where marshland can be used, permits must be obtained from the state's or province's Natural Resources Department and from the U.S. Army or Canadian Corps of Engineers before any development can commence.

CHOOSING CULTIVARS

Cranberry growers rely on just a dozen cultivars, some selected from the wild and most others first-generation hybrids of those selections. Table 15-1 details the traits of the most commonly-grown cultivars. Growers select cultivars mainly for high yields, good red skin color and time of ripening. Resistance to false blossom is an important trait in most cranberry-growing regions, and moderate resistance to fruit rots at harvest can avoid losing much of the crop.

Lingonberry cultivars are still being tested, and Table 15-2 describes some under test. Most lingonberry cultivars bloom twice so can produce two crops per year, but in most parts of North America the first bloom is prone to frost damage and the first crop is small, so only the second crop is harvested.

Blueberry growers produce a more diverse range of cultivars than do growers of many other fruits. Unlike other fruits, a few cultivars do not account for the bulk of production for any type of blueberry. Rather, growers choose several cultivars based on adaptation to the climate, ripening times, and resistance to prevalent diseases.

The USDA winter hardiness zone is the main factor determining which type(s) of blueberries to grow in a particular area. Choose cultivars hardy enough for your area, and in Southern areas choose cultivars that need about as many winter chilling hours as your area normally gets. Cultivars that need less chilling than your area provides will bloom too early and often lose crops to frosts, while those that need more chilling than your area gets will start growth erratically and not produce well. Local disease occurrence determines whether a grower must choose cultivars resistant to a certain disease. Tables 15-3 to 15-6 detail the winter hardiness, chilling hours, disease resistance, ripening times and other traits of many important cultivars of each type of blueberry.

The type of market determines selection of cultivars based on ripening times. Growers who machine-harvest for the processing market plant large acreages of just a few cultivars, each ripening during a concentrated period, with fruit that shakes off easily with little bruising. Growers who produce fresh fruit to ship to supermarkets often plant mainly cultivars that ripen early in the season, when prices are highest.

Growers for fresh local sales, whether worker-harvested or pick-your-own, plant enough cultivars to provide a continuous harvest season over as long a period as possible in that area. The highbush blueberry season in USDA zones 5 and 6 can last 2 months or more, and the highbush-lowbush season in zones 3 and 4 can last 1 ½ months. Growers in zones 7 to 9, in the Southeast and the Pacific Northwest, can raise both highbush or Southern highbush cultivars and rabbiteye cultivars, providing a total blueberry harvest season of 3 months or more.

PREPARING THE SITE

Select a site for cranberries or blueberries at least a year before planting. Consider only sites with acid, at least moderately well-drained soil with at least 3% organic matter. The site must also have good air circulation and access to large volumes of irrigation water that has a low pH and is low in calcium. Cranberries and blueberries need full sunlight in most areas, although in some marginal areas with hot summers, such as the Southeast or southern Midwest, a few high-pruned pine trees west of the site could provide needed afternoon shade for blueberries.

Completely eliminate any perennial weeds, such as Johnsongrass, Bermuda grass, or rushes, with glyphosate (Roundup™) well before planting. Test the soil for pH, phosphorous (P), potassium (K) and organic matter, and add sulfur if needed to lower the pH, along with any P or K needed, at least 6 months before planting time. A pH above 5.5 will cause iron deficiency, calcium or potassium toxicity, and poor growth of cranberries or blueberries. Table 2-2 in chapter 2 provides the approximate amounts of sulfur to add for different soil types and pH levels. Increase soil organic matter, if needed, by growing and incorporating a summer cover crop, then a winter cover crop if it can be turned under at least 6 weeks before spring planting time.

Creating cranberry bogs. Cranberry growers create new bogs by scraping away the topsoil, stockpiling it for use later, and excavating the subsoil to about 18 inches above the final water table depth. Rectangular beds, about 150 feet wide and 600 or more feet long, are then constructed and leveled with a laser level so water will flow through the beds from the inlet point to the outlet point (Photo 15-5). Stockpiled soil is used to build berms for trucks and equipment between the bogs. Growers construct a reservoir for irrigation water, irrigation ditches, dikes and other water control structures. They also install a sprinkler system including pumps, mainlines and lateral lines, to protect against frost and provide additional irrigation water.

Photo 15-5. Rectangular cranberry beds are constructed with a laser level so water will flow through the beds from the inlet point to the outlet point. Stockpiled soil is used to build berms for trucks and equipment between the bogs. (Photo by T.R. Roper, Univ. of Wisconsin.)

Once bogs are constructed and sprinkler systems installed, growers apply about 2 inches of coarse sand evenly over bog surfaces. The sand reduces weed seed germination and provides a better rooting medium for cranberry plants, and prevents the soil below, which is high in organic matter, from becoming hard and dry.

Photo 15-6. Where *Phytophthora* root rot is a problem and soils have less than perfect drainage, growers build raised beds on which to plant blueberries.

Constructing blueberry and lingonberry beds. Where soil is only moderately well-drained and *Phytophthora* root rot is a problem, many blueberry growers construct raised beds, 8 to 12 inches high and 4 feet wide, to create excellent water drainage around the plants' roots (Photo 15-6). Some growers install tile drains on Coastal Plain sites where the water table is less than 20 inches below the soil surface. Where soil drainage is already good and root rot is not a problem, however, planting on flat ground will reduce irrigation needs for the planting.

Photo 15-7. Lingonberry growers often prepare narrow, low raised beds similar to those made for strawberry plantings, with rows spaced 4 feet apart. (Photo from Lingonberry Corp. of America.)

Lingonberries are often grown on narrow, low raised beds (Photo 15-7) similar to those made for

Table 15-1.

COMMON COMMERCIAL CRANBERRY CULTIVARS

in approximate order of ripening (varies with area).

Cultivar	Fruit			Susceptibility to:				Harvest time
				False	Fruit Rots			
	Quality	Size	Yields	blossom	Harvest	Storage	Other pests, environment	
Early Black	very good	small-med	medium	R	MS	MS	MS to tipworm	early
Hyred	very good	large	high	S	MR	S	T to tipworm	early
Wilcox	best	med-large	high	R	MR	S	R to leafhopper	early
Ben Lear	good	med-large	high	S	MR	S	VS to red leaf spot, fruitworm	early
Franklin	very good	med-large	med	R	MR	S	R to leafhopper, upright dieback	early
Crowley	good	small-med	high	MR	MS	VS		early-mid
Stevens	very good	very large	high	MR	MR	S	VS to red leaf spot, weevil; T to tipworm, R to bitter rot	middle
Bergman	good	med-large	medium	R	S	S	R to leafhopper, VS to tip blight	middle
Searles	fair-good	medium	med-high	S	S	VS	VS to upright dieback, ringspot	middle
Howes	very good	small-med	med-high	S	MS	R	R to fall frost, S to tipworm	late
McFarlin	best	large	med-high	R	MS	MS	blooms late	late
Pilgrim	very good	very large	medium	R	MS	R	R to leafhopper, VS to tip blight	late

Susceptibility/Resistance ratings: VR= very resistant, R = resistant, MR = moderately resistant, MS = moderately susceptible, S = susceptible, VS = very susceptible, T = tolerant

Average ripening times in **WI, MA and BC**: early: mid- to late Sept., middle: early Oct., late: mid- to late Oct. Harvest is about 2 weeks earlier in **WA** and about 1 week earlier in **southern NJ**.

Notes on Cultivars:
Early Black, selected from the wild in Cape Cod, MA in 1852, is still widely grown in MA and NJ.
Hyred, a new hybrid of 'Stevens' x 'Ben Lear' released in 2003, was selected for early, intense red color.
Wilcox is a hybrid of 'Howes' x 'Searles'.
Ben Lear, selected from the wild in WI, is widely grown there, but tends to biennial bear in the East.
Franklin resulted from a cross of 'Early Black' x 'Howes'.
Crowley, a hybrid of 'McFarlin' x 'Prolific Washington', is little planted due to its small size.
Stevens resulted from a cross of wild selection 'Potter' x 'McFarlin'. It's the most-planted cultivar in recent years.
Bergman resulted from a cross of 'Early Black' x 'Searles'.
Searles, selected from the wild in WI in 1893, is widely grown there.
Howes, selected from the wild in Cape Cod, MA in 1843, is still grown but no longer planted due to small fruit.
McFarlin, selected from the wild in Cape Cod, MA in 1874, has been widely planted in all cranberry regions.
Pilgrim is a hybrid of 'Prolific Washington' x 'McFarlin'.

Table 15-2.

LINGONBERRY CULTIVARS FOR TRIAL IN NORTH AMERICA

in approximate order of ripening (varies with area).

Planting at least 2 cultivars for cross-pollination improves yields.

	Plant	Bloom		Fruit				Harvest
Cultivar	height	First	Second	Size	Sugars	Acidity	Yields	time
Splendor	6-8"	mid-late	profuse	med-large	medium	medium	high	early
Ida	4-8"	early	profuse	large	medium	medium	med-high	early
Regal	7-9"	late	profuse	med-large	med-high	medium	high	early
Red Pearl	10-14"	early-mid	profuse	large	medium	low	high	early-mid
Koralle	10-12"	early-mid	profuse	small	medium	medium	high	middle
Sanna	6-10"	middle	none	medium	high	low	medium	mid-late
Sussi	6-10"	middle	profuse	large	medium	medium	high	mid-late
Erntesegen	10-12"	middle	profuse	large	medium	low-med	high	late
Linnea	6-10"	early-mid	sparse	medium	med-high	high	high	late

Average ripening times in **WI, MA and BC**: early: early Aug. and early Oct., middle: mid-Aug. ar mid-Oct., late: late Aug. and late Oct. Harvest is about 1 week earlier in **OR** and **southern NJ**.

Notes on Cultivars:

Splendor plants, selected in WI, are precocious. Blossoms have some frost tolerance.
Ida plants, from Sweden, are precocious, set fruit in small clusters that may have less fruit rot.
Regal plants, selected in WI, are precocious. Blossoms have some frost tolerance.
Red Pearl, a Dutch cultivar, is R to Phytophtora root rot. Berries have a mild flavor.
Koralle, a Dutch cultivar, is strongly branched so has high yield potential. Fruit are tart.
Sanna plants, from Sweden, are not for areas with spring frosts since no second crop.
Sussi is slow to establish but produces high yields of large fruit once established.
Erntesegen, a German cultivar, has fruit with mild flavor.
Linnea plants, from Sweden, are precocious but not for areas with spring frosts.

Table 15-3.

HIGHBUSH BLUEBERRY CULTIVARS

In approximate order of ripening (varies with area).

All grow 5-7' tall, do best with cool to warm summers within USDA hardiness zones 4b to 7b, and all need about 1000 hours of winter chilling between 32 and 45 F.

Cultivar	Best in USDA Zones	Berry				Resistance to								Yields	Suited to Machine Harvest	Harvest Season
		Flavor	Size	Firmness	Scar	Spring Frosts	Mummy Berry Rot	Mummy Berry Blight	Anthrac-nose	Phomopsis Canker	Red Ringspot	Shoe-string	Other Pests, Environment			
Chanticleer	5b-7a	good	large	firm	medium	S	MR	R	S	S	S	S	MR to stem blight	medium	yes	very early
Weymouth	5a-7b	poor	small	soft	medium	S	R	MR	MR	VS	MR	S	VS to stem canker, R to root rot	medium	no	very early
Duke	5b-7b	good	large	very firm	small, dry	R	MR	VR	MS	MR	R	S	S to stem blight, stem canker	high	no	very early
Spartan	4b-7a	good	large	firm	small, dry	VR	MR	S	S	VS		S	VS to stem canker	med-high	yes	early
Reka	5b-7b	best	medium	firm	small, dry	MS	R	R	MR	S	S	S	R to stem blight, root rot	very high	yes	early
Bluejay	5a-7a	good	medium	very firm	small, dry	MS	R	VR	MS	VS		R		med-high	yes	early-mid
Blueray	4b-7b	very good	large	firm	medium	R	S	S	VS	S	S	VS	MR to fusicoccum, stem canker	med-high	no	middle
Bluecrop	5a-7b	very good	medium	firm	small, dry	MR	MR	MS	VS	S	MR	VR	R to stem blight & stem canker, MR to mildew, VS to fusicoccum	high	yes	middle
Toro	5a-7b	very good	large	firm	medium	R	S	MR	S	MR	S	S	S to stem canker	very high	no	middle
Puru	5b-7b	best	large	firm	small, dry	MR	R	S	MS	S	S	S	R to stem blight, root rot	medium	no	middle
Berkeley	6a-7b	fair	large	firm	large, dry	S	VS	MR	VS	S	S	S	R to mildew, MS to fusicoccum	medium	no	middle
Rubel	5a-7a	fair	small	firm	medium	MR	MS	R	S	MR	S	VS	MR to fusicoccum, S to mildew	med-high	no	middle
Brigitta	5b-7b	best	large	very firm	small, dry	R	MR	R	MR	S	S	S	S to scale insects, VS to shock	high	yes	mid-late
Legacy	6b-7b	best	medium	very firm	small, dry	MS	MR	MR	MS	S	S	S	R to stem blight			
Chandler	6b-7b	best	very large	firm	medium	R	MR	R	S	S	S	S	S to sunscald	high	no	mid-late
Nelson	4b-7b	best	large	firm	small, dry	R	MR	MR	S	S	VR	S	S to fruit cracking, VS to shock	high	yes	late
Bluegold	5b-7b	good	medium	firm	small, dry	R	R	VS	S	S	VR	S	VS to blueberry shock virus	high	yes	late
Jersey	5a-7a	fair	medium	firm	medium	R	MS	R	MS	S	VR	VS	MR to stem canker; VS to leaf spot, mildew, fusicoccum, stunt	med-high	yes	late
Elliott	5a-7b	good	medium	very firm	small, dry	VR	S	VR	R	R	S	VS		very high	yes	very late

Resistance ratings: VR= very resistant, R = resistant, MR = moderately resistant, MS = moderately susceptible, S = susceptible, VS = very susceptible

Mummy berry occurs wherever highbush blueberries are grown, causing both a fruit rot and a twig blight; cultivar resistance to the two phases varies. **Anthracnose** is a problem mainly where summers are hot and humid, while **phomopsis canker** is a concern where winters are cold. Prevent **shoestring and red ringspot viruses** with virus-free planting stock; resistance is needed only in MI and southern NJ.

Average start of ripening

at Mountain Grove, MO and **eastern North Carolina**: very early: mid- to late May, early: late May to early June, early-mid: early to mid-June, middle: early to late June, mid-late: mid- to late June, late: late June to early July, very late: early to mid-July. Ripening in **southern New Jersey** and **central Illinois** is about 3 weeks later.

at Southaven, MI: very early: late June to early July, early: early to mid-July, early-mid: mid-July, middle: mid- to late July, mid-late: late July to early Aug., late: early to mid-Aug., very late: mid- to late Aug. Ripening in **Geneva, NY** is about 2 weeks later than in MI. Harvest of each cultivar lasts 2-5 weeks.

Notes on Cultivars:

Chanticleer blooms and ripens very early, so needs frost protection. It does best in the Northeast, poorly in the Northwest.
Weymouth, an older cultivar, is no longer planted but still yields 30% of the NJ crop.
Duke flowers late, escaping frost damage, yet ripens very early. Harvest can be done in 2 to 3 pickings.
Spartan also blooms late but ripens relatively early. Plants perform poorly on ammended upland soils.
Reka, from New Zealand, is heavily productive from a young age; prune to reduce overcropping. It tolerates wetter soils.
Bluejay berries hang on bush without losing quality until most are ripe, good for machine harvest.
Blueray may overproduce if not pruned heavily. Berries may drop in hot weather, so harvest often.
Bluecrop, the main cultivar in MI, NJ, OR and WA, ripens over a prolonged season. Berries turn blue before reaching full sweetness, so delay picking 5 to 7 days.
Toro ripens in a short period, so only 2 harvests are needed. It tolerates fluctuating winter temperatures well.
Puru, from New Zealand, is popular for local sales due to excellent flavor. It may crop again in fall.
Berkeley, an older cultivar, is not planted much now but still yields 25% of the OR crop.
Rubel, an older cultivar selected from the wild, is still produced on large acreages in MI for processing, as it's well-suited to machine harvest.
Brigitta, from Austrailia, stores well for a month, and longer with controlled atmosphere storage.
Legacy is 1/4 V. darrowi, a low-chill species native to the Southeast, so not suited to Northern areas.
Chandler ripens over 4-6 weeks for pick-your-own or retail markets. It does well in the Northwest's moderate winters with high chilling, and cooler summers.
Nelson has fine flavor, but berries may split during hot, dry weather with irrigation.
Bluegold has medium-size fruit which are good for processing.
Jersey, released in 1928, is still widely grown and is very good for machine harvest.
Elliot extends the highbush blueberry season into late summer.

Table 15-4.
SOUTHERN HIGHBUSH BLUEBERRY CULTIVARS
in approximate order of ripening (varies with area).
All do best with warm to hot summers within USDA hardiness zones 7a to 10a.

Cultivar	Best in USDA Zones	Chill Hours Needed	Berry				Resistance to							Bush Height	Yields	Suited to Machine Harvest	Harvest Season
							Spring Frosts	Stem Blight	Stem Canker	Mummy Berry		Anthrac-nose	Other Pests, Environment				
			Flavor	Size	Firmness	Scar				Rot	Blight						
Jewel	9b-10b	150-250	good	med-large	firm	small,dry	VS	MS	MR	S	VS	S		5-6'		no	very early
Sapphire	9b-10b	100-150	good	med-large	firm	small,dry	VS	MS	MR	S	MS	S	S to fruit cracking	5-6'		no	very early
Star	8b-9b	400-500	very good	large	medium	medium	MS	R	R	S	S	S	S to fruit cracking	6-7'	medium	no	very early
Windsor	8b-9b	350-400	good	very large	firm	medium	VS	MS	MR	S	S	S	S to fruit cracking	6-7'	medium	no	very early
Millennia	9a-10a	300-350	good	large	firm	small, dry	VS	MS	MR	S	S	S	MR to leaf spot	6-7'	high	no	very early
Emerald	9a-10a	300-350	very good	large	firm	small,dry	VS	MR	MR	S	S	S	MR to root rot	6-8'	medium	no	very early
O'Neal	8b-9b	400-500	best	very large	very firm	small, dry	S	MR	MR	MS	S	S	MS to shoestring	6-8'	med-high	no	very early
Sharpblue	9a-10a	200-300	good	medium	medium	large, wet	VS	MS	R	MR	S	R	MR to root rot	5-6'	medium	no	very early
Biloxi	8b-9b	350-400	very good	medium	very firm	small, dry	VS	MR	MR	MR	R	S		6-8'	high	yes	early
Misty	9a-10a	300-400	good	large	firm	small, dry	S	VS	R	S	S	S	VR to shoestring	5-6'	medium	no	early
Bladen	7a-8b	500-600	very good	med-large	med-firm	small, dry	S	MR	R	R	VR	R	MR to fruit cracking	6-8'	high	yes	early
Reveille	7a-8b	600-800	very good	med-large	very firm	small, dry	S	MR	R	R	MR	MS	MS to fruit cracking	6-8'	high	yes	early
Southmoon	8b-9b	400-500	very good	large	firm	small,dry	MR	R	S	S	MS	S	R to root rot	6-8'	med-high	yes	early
Avonblue	8b-9b	400-450	very good	med-large	firm	small, dry	S	MS	MR	S	S	S	MR to root rot	4-5'	medium	yes	early
Gulfcoast	8b-9b	200-300	good	medium	firm	small, dry	VS	MS	VS	MR	S	S	R to root rot	5-6'	medium	no	early-mid
Georgiagem	8b-9b	350-400	good	medium	firm	medium	VS	S	S	MS	MS	S	VR to shoestring	5-6'	medium	yes	early-mid
Magnolia	7b-8b	500-600	good	med-large	firm	small, dry	MR	MS	MR	S	S	S		5-6'	medium	yes	early-mid
Jubilee	7b-8b	500-600	good	med-large	firm	small, dry	MS	MS	MR	S	VS	S		6-8'	medium	yes	early-mid
Cooper	8b-9b	400-500	good	med-large	firm	small, dry	MS	S	VS	MS	VS	VS		5-6'	medium	yes	early-mid
Cape Fear	7a-8b	500-600	good	large	medium	small, dry	MR	R	S	R	VS	VS		6-8'	high	no	middle
Croatan	7a-8b	600-700	good	medium	firm	small,dry	MR	S	R	S	S	S	R to bud mite	6-8'	very high	no	middle
Blue Ridge	7a-8b	500-600	very good	large	firm	large, wet	MR	MR	S	MS	S	S		6-8'	med-high	no	middle
Pearl River	7b-8b	500-600	very good	large	firm	small, dry	R	MR	MR	R	S	MS		6-8'	medium	yes	middle
Bluechip	7a-8a	700-800	best	very large	very firm	small, dry	R	VS	MR	S	MS	VS		5-6'	med-high	yes	middle
Summit	7a-8b	600-650	best	large	firm	small,dry	R	S	S	MS	VS	S		6-7'	high	yes	mid-late
Ozarkblue	6b-8a	800-900	very good	med-large	firm	small, dry	R	MS	S	MS	S	S	MR to mildew	7-8'	very high	yes	mid-late

Resistance ratings: VR= very resistant, R = resistant, MR = moderately resistant, MS = moderately susceptible, S = susceptible, VS = very susceptible

Average start of ripening in south Georgia: very early: late April to early May, early: early to mid-May, early-mid: mid- May, middle: mid- to late May, mid-late: late May to early June. Harvest of each cultivar lasts 2-4 weeks.

Notes on Cultivars:
Jewel, suited to Florida south of Gainesville, blooms extremely early so needs sprinkler irrigation for frost control.
Sapphire sets flower buds very heavily so needs heavy pruning to remove many, but also needs frost protection.
Star blooms later than other early cultivars (10 days after Sharpblue), but fruit develop rapidly.
Windsor blooms very early so needs frost protection irrigation.
Millenia sets flower buds heavily so needs heavy pruning to ensure good leaf growth and keep berry size large.
Emerald blooms very early so needs frost protection irrigation. Berries maintain good size through the harvest season.
O'Neal starts blooming very early but continues over a long period, so later blooms may escape frosts.
Sharpblue blooms very early, so needs frost protection. Berry size goes down during the last half of harvest season.
Biloxi needs frost protection.
Misty overcrops from a young age, leading to stem blight, so needs crop removal for the first 2 years and thinning-out pruning to reduce crops after that.
Bladen may be machine harvested for fresh market as well as processing. Machine hedging after bloom increase fruit size next year.
Reveille may also be machine harvested for fresh market as well as processing. It yields well from a young age, but blooms early.
Southmoon can crop well in FL even in low-chilling years, but blooms later, avoiding many frosts. Plant survival has been very poor in some GA fields.
Avonblue produces many flower buds, so needs heavy annual pruning to ensure good fruit size.
Gulf Coast is not suited to machine harvest, because many stems remain attached to berries.
Georgiagem has low vigor and a poor root system, so plant closer.
Magnolia blooms late, helping to avoid frosts. It does well on ammended upland soils, but is very difficult to establish.
Jubilee also blooms late and does well on ammended clay loam soils. 95% of berries can be harvested in 2 pickings, and berries store very well.
Cooper has serious stem canker problems in FL but is healthy in GA.
Cape Fear is good for local sales but fruit are too soft for shipping.
Croatan, an older cultivar resistant to the main pathogen races causing stem canker, still accounts for 60% of NC highbush acreage.
Blue Ridge has very large, light blue fruit great for pick-your-own but not for retail or wholesale due to wet picking scar.
Pearl River is a highbush-rabbiteye hybrid, with fruit skin tougher than most highbush and very dark, almost black fruit. It blooms very late, so escapes frosts.
Blue Chip has very large fruit desired for pick-your-own, but plant only in areas where stem blight isn't a problem.
Summit extends the southern highbush blueberry ripening season in the mid-South.
Ozarkblue needs more chilling than other Southern highbush, so is adapted to the upper South, where it blooms later and avoids frosts.

Table 15-5.

RABBITEYE BLUEBERRY CULTIVARS

In approximate order of ripening (varies with area).

All do best with warm to hot summers within USDA hardiness zones 7a to 9b. Rabbiteyes tolerate both drought and damp soils better than highbush. Rabbiteye harvest begins as the Southern highbush harvest season ends, and can last until early August.

Cultivar	Best in USDA Zones	Chill Hours Needed	Berry				Resistance to							Bush Height	Yields	Suited to Machine Harvest	Harvest Season
			Flavor	Size	Firmness	Scar	Spring Frosts	Stem Blight	Stem Canker	Mummy Berry Rot	Mummy Berry Blight	Anthracnose	Other Pests, Environment				
Windy	9a-9b	300-350	very good	medium	med-firm	small,dry	VS	MS	MS	VS	VS	S	S to root rot, gall midge	8-12'	med-high	yes	very early
Snowflake	7b-9a	450-500	very good	medium	medium	medium	S	R	MR	S	VS	S	S to root rot	7-9'	med-high	no	very early
Beckyblue	9a-9b	300-400	good	large	firm	small, dry	VS	MS	MS	R	VS	MS		8-12'	med-high	yes	early
Montgomery	7b-8b	500-550	very good	med-large	firm	small, dry	S	MR	MR	R	S	S	R to stunt, fruit cracking	8-12'	med-high	yes	early
Premier	7a-9a	600-650	best	large	medium	small, dry	S	R	R	R	S	S	R to stunt, S to gall midge	12-15'	high	yes	early
Climax	7a-9a	450-500	very good	med-large	very firm	small, dry	S	MS	S	R	VS	MR	VS to shoestring	8-12'	high	yes	early
Austin	7a-9a	450-500	good	med-large	firm	small, dry	R	MS	MS	MR	VS	S		8-12'	very high	yes	early
Bluegem	8b-9b	350-400	good	medium	very firm	small,dry	S	S	S	R	S	S	S to root rot	8-12'	medium	yes	early
Bonita	9a-9b	350-400	very good	very large	firm	small, dry	VS	S	S	MR	VS	S	MR to root rot, mildew	10-14'	medium	yes	early
Ira	7a-8a	750-800	good	med-large	firm	small,dry	VR	MR	MR	R	S	S		8-12'	high	yes	middle
Brightwell	8a-9b	550-600	good	med-large	firm	small, dry	R	S	S	MR	VS	S	R to stunt, shoestring	10-14'	high	yes	middle
Woodard	8a-9b	350-400	fair	med-large	soft	medium	MR	MR	MR	R	S	MS	R to stunt	7-9'	med-high	no	middle
Chaucer	8a-9b	350-400	good	medium	soft	wet	MR	S	S	R	S	S		10-14'	high	no	middle
Bluebelle	7b-9b	450-500	very good	large	soft	wet	S	S	S	R	S	MR		8-12'	high	no	middle
Briteblue	7a-8b	500-600	good	large	very firm	small, dry	MR	S	S	MS	VS	R	R to stunt	8-12'	high	yes	middle
Powderblue	7a-9a	550-600	very good	med-large	firm	small, dry	R	S	S	MR	VS	MR	MR to fruit cracking	10-14'	high	yes	mid-late
Tifblue	7a-8b	550-650	good	med-large	firm	small, dry	R	VS	R	R	VS	MR	S to fruit cracking	10-14'	high	yes	mid-late
Choice	7b-9b	450-500	very good	medium	medium	medium	R	S	S	R	S	S	S to root rot	10-14'	high	yes	late
Centurion	7a-8b	550-650	very good	medium	medium	small, dry	R	R	R	R	MS	MR	MS to fruit cracking	10-14'	high	no	late
Baldwin	8a-9b	450-500	good	medium	firm	small, dry	R	S	S	R	S	S		10-14'	med-high	no	late

Resistance ratings: VR= very resistant, R = resistant, MR = moderately resistant, MS = moderately susceptible, S = susceptible, VS = very susceptible

Average start of ripening in south Georgia: very early: mid- to late May, early: late May to early June, middle: early to mid-June, mid-late: mid- to late June, late: late June to early July. Harvest of each cultivar lasts 2-4 weeks.

Notes on Cultivars:
Windy needs well-drained sites to prevent root rot. Post-harvest top-hedging controls height and delays bloom next year, reducing frost risk.
Snowflake also needs well-drained sites to prevent root rot. It produces many basal sprouts so needs more pruning.
Beckyblue needs a frost-free site or sprinkler irrigation for frost protection.
Montgomery resists blueberry leafhoppers, which transmit blueberry stunt.
Premier needs heading-back pruning of limber canes on young plants to support fruit loads. It blooms late for an early-ripening cultivar.
Climax has a short ripening period, good for machine harvest. It blooms very early so is frost-prone.
Austin is starting to replace 'Climax' as it blooms late and is even better for machine harvest, with less berry loss and up to 3 times the yield.
Bluegem needs very well-drained soil.
Bonita needs very well-drained soil.
Ira blooms late, avoiding frosts, but ripens in midseason. Prune heavily to prevent overcropping and stems sticking to fruit.
Brightwell resists frosts and is a major cultivar. It can overcrop, leading to biennial bearing, so prune heavily to reduce a heavy crop and stimulate return bloom next year.
Woodard works for fresh local sales, but is too soft for shipping and not for processing because skins toughen when frozen.
Chaucer is good for pick-your-own, but wet stem scar is poor for shipping and mechanical harvest.
Bluebelle ripens over a long period, good for pick-your-own, but not for shipping or processing because skin may tear.
Briteblue berries ripen in clusters and hang well on plants, so it's a top pick-your-own choice.
Powderblue is replacing Tifblue as the main mid-late season cultivar due to resistance to fruit cracking.
Tifblue produces many canes, increasing pruning costs for machine harvest, and fruit may crack in some years.
Choice needs well-drained sites to prevent root rot. It's good for local fresh sales, shipping and processing.
Centurion is a high-quality late cultivar for pick-your own, but fruit may crack after very heavy rains.
Baldwin is good for pick-your-own, but too dark-colored for fresh pre-picked sales and not machine harvestable for processing.

Table 15-6.
HIGHBUSH-LOWBUSH (HALF-HIGH) AND EXTRA-HARDY HIGHBUSH BLUEBERRY CULTIVARS in approximate order of ripening (varies with area).

All do best with cool summers within USDA hardiness zones 3a to 5a. Bushes must be covered with snow in zone 3 for adequate protection below -30 F.

Cultivar	Best in USDA Zones	Berry				Resistance to					Bush Height	Yields	Suited to Machine Harvest	Harvest Season
		Flavor	Size	Firmness	Scar	Spring Frosts	Mummy Berry Rot	Mummy Berry Blight	Anthracnose	Other Pests, Environment				
Bluetta	4a-5a	good	medium	medium	medium	VS	MR	R	VS	VS to root rot	4.5'	med-high	yes	very early
Polaris	3a-4b	very good	medium	firm	small, dry	S	MR	S	VS		3.5'	medium	yes	very early
St. Cloud	3a-4b	very good	medium	firm	small, dry	S	VR	S	VS		4'	med-high	yes	early
Patriot	3b-5a	good	large	firm	small, dry	VS	R	MS	S	MR to root rot	4.5'	high	yes	early
Northcountry	3a-3b	fair	small	soft	small, dry	VS	VR	MS	S		2.5'	low	no	early
Chippewa	3a-4b	very good	large	firm	medium	S	MR	MS	S		3.5'	high	yes	early-mid
Northblue	3a-4a	good	medium	medium	medium	S	R	VS	S		3'	med-high	no	early-mid
Northsky	3a-3b	fair	small	soft	small, dry	VS	VR	S	MS		2'	low	no	early-mid
Friendship	3a-3b	good	small	soft	small, dry	VS	R	MS	MR		2.5'	high	no	middle
Northland	3b-5a	fair	medium	soft	medium	S	MR	MS	S	R to shoestring	4'	high	yes	middle

Resistance ratings: VR= very resistant, R = resistant, MR = moderately resistant, MS = moderately susceptible, S = susceptible, VS = very susceptible

Average start of ripening in central Minnesota: very early: late June to early July, early: early to mid-July, early-mid: mid to late July, middle: late July to early Aug. Harvest of each cultivar lasts 2-4 weeks.

Notes on Cultivars:
Bluetta often winter-kills at the tops of bushes so needs removal of dead tips; below 4' survives.
Polaris has very good flavor and stores very well. It's self-unfruitful, so plant with other cultivars.
St. Cloud often winter-kills at the tops of bushes so needs removal of dead tips; below 4' survives.
Patriot berries are quite tart until fully blue.
Northcountry bushes are short, good for snow protection, and have fruit like lowbush blueberries.
Chippewa berries are sweet and light blue, like highbush.
Northblue berries are quite tart until fully blue.
Northsky bushes are short, good for snow protection, and have fruit like lowbush blueberries.
Friendship bushes are short, good for snow protection, and produce high yields of fruit like lowbush blueberry.
Northland, one of the oldest half-high cultivars, was widely planted in northern MI and still producing.

strawberry plantings, with rows spaced 4 feet apart. Make beds that are 2 feet wide and 6 to 8 inches high. Like blueberries, raised beds aren't needed where soil drainage is good.

PROPAGATING OR PURCHASING PLANTS AND PLANTING

Many cranberry and blueberry growers propagate their own plants for use in new plantings. Such growers must take special care to prevent spreading diseases into new fields.

Propagating and planting cranberries and lingonberries. Cranberries and lingonberries are usually grown from cuttings planted directly in place without previous rooting. Cuttings are obtained by mowing or disking existing plantings and raking up the uprights or rhizome (underground stems) pieces, or if new cultivars are being planted, by ordering rhizome cuttings from a nursery. Growers should test plants from which they'll take cuttings for viruses and phytoplasmas (mycoplasma-like organisms that attack plants), to be sure they're disease-free, before taking cuttings.

High-yielding cranberry fields are mown in early spring to produce cuttings, which are baled and transported to the new bog (Photo 15-8). Each bog is planted to just one cultivar, since cranberries self-pollinate. Workers spread cuttings on the sand of a newly-prepared bed at the rate of 2 tons of rhizomes per acre, and use dull disks to push plants into the sand (Photo 15-9). After disking in one direction, disk again at a right angle to ensure that vines become properly anchored. Sprinkler irrigation must be applied immediately after disking, and at least every other day for the next 3 weeks after planting (Photo 15-10). Many rhizomes do not receive adequate sand coverage and thus dry out and die, but the very heavy rate of cutting application ensures that enough cuttings form roots within a few weeks and then begin top growth.

Lingonberry cuttings are taken in early spring by using a single-disc coulter to cut off a 4- to 5-inch-wide strip from the outer edges of existing rows. These rhizome pieces are then planted in furrows spaced 4 feet apart (made in raised beds or on flat ground), using a vegetable transplanter. Plant at least 2 cultivars of lingonberries to provide cross-pollination. Each cultivar can be planted in a separate row for ease of harvest, since lingonberries are partially self-fruitful.

New cranberry and lingonberry plantings often benefit from applying P fertilizer at planting time

Photo 15-8. New cranberry bogs are usually started by spreading cuttings, obtained by mowing high-yielding fields in spring. (Photo by T.R. Roper, Univ. of Wisconsin.)

Photo 15-9. Cranberry cuttings must be disked into the sand in 2 directions. (Photo by T.R. Roper, Univ. of Wisconsin.)

Photo 15-10. Immediately after disking, irrigate new cranberry beds with sprinklers. Run sprinklers every other day for the next 3 weeks to ensure rooting. (Photo by T.R. Roper, Univ. of Wisconsin.)

if soil P tests low, plus a slow-release nitrogen (N) fertilizer. This enables the vines to grow faster and cover more ground within the first 2 growing seasons, often inducing a harvestable yield by the third year after planting rather than the fourth. Massachusetts researchers recommend applying 100 pounds per acre of triple superphosphate (0-46-0) to new cranberry bogs before disking in the cuttings, so the fertilizer will be incorporated when vines are disked. They advise broadcasting a slow-release N material, that provides 50 to 200 pounds per acre of actual N, just after planting. These materials could be banded down rows of newly-planted lingonberries.

Propagating blueberry plants. Many established blueberry growers propagate some of their own plants from hardwood or softwood cuttings. They construct outdoor propagation beds with mist irrigation adjacent to their fields, about 4 feet wide and 16 to 24 feet long. Cinderblocks or treated lumber form the walls, and the hardware cloth or screen bottom is raised a few inches off the ground to provide excellent drainage. The propagation bed contains an 8-inch-deep rooting media of half sand and half peat moss (Photo 15-11).

Photo 15-11. Many blueberry growers propagate some of their own plants from hardwood cuttings, in rooting beds filled with a mixture of half peat and half sand or perlite, and equipped with a misting system.

Where diseases caused by viruses and phytoplasmas are a problem, as in the long-established blueberry-growing areas of Michigan and New Jersey, growers create special propagating blocks of bushes where they take extra steps to prevent viruses and phytoplasmas from encroaching. They fumigate the soil before planting this propagation area to control dagger nematodes, and spray regularly for blueberry aphids and leafhoppers, because all these pests transmit viruses or phytoplasmas. They take leaf samples from these plants each year and have them tested to be sure they are still free of viruses and phytoplasmas. In other parts of the country where virus and phytoplasma diseases do not generally occur, growers may propagate from bushes in the commercial planting, but should still send leaf samples from these bushes to a testing lab to be sure they are disease-free.

Growers usually take hardwood cuttings from highbush blueberries during pruning operations in late winter, just before plants break dormancy. Rabbiteye blueberries are more difficult to root from hardwood cuttings, so many growers take softwood cuttings in late spring.

Workers select hardwood cuttings from long, excess branches removed from some bushes. They remove the top section from each pruned branch, which contains flower buds, and cut the remaining branch into 5-inch sections. Each section is stuck into the rooting media without any added rooting hormone, because highbush blueberries root readily without it. Softwood cuttings are taken from young shoots and are dipped in rooting hormone for rabbiteye blueberries.

Periodic misting keeps the rooting bed moist, and misting frequency is increased as the cutting develop leaves and temperatures rise during the season. Growers apply soluble fertilizer through the misting system every 2 weeks, from about a month after leaves appear until late summer. The rooted cuttings then overwinter in the propagation bed, and are generally planted the next spring. Growers may transplant the cuttings into pots or a nurse bed in spring and grow them for another year before planting into the field, or they may transplant the 1-year-old rooted cuttings directly into their field. Fields planted to 1-year-old rooted cuttings will need more frequent irrigation during the first couple of years than fields planted to 2-year-old plants.

Purchasing blueberry plants. New blueberry growers must purchase plants, as must established growers who are planting new cultivars. Growers who may have viruses or phytoplasma in their existing plantings, such as blueberry stunt or red ring virus, should also purchase certified virus-free and phytoplasma-free plants.

Most growers purchase 2-year-old plants that have grown in pots (Photo 15-12), then been removed from the pots for shipping. These have larger root systems that survive shipping and transplanting to distant fields better than 1-year-old plants.

Blueberries have less-developed root systems than many other fruit plants, so bare-root, field-grown transplants do poorly, and pot-grown transplants are needed to preserve the roots they have. Always buy from a reputable nursery that guarantees its plants to be disease-free when shipped.

Some growers buy rooted liners or 1-year-old rooted cuttings and grow them for a year in a nurse bed before planting them into the field. This practice reduces start-up costs but delays field establishment by a year. The nurse bed consists of a raised bed of acid peat moss 8 to 10 inches deep, with a support frame around it. Place rooted cuttings in the bed at a spacing of 8 to 12 inches apart, and keep the bed well-watered with drip irrigation or a soaker hose. Provide wind protection and shadecloth to protect young plants from excess heat, and fertilize every 2 weeks until late summer with an acid fertilizer. Place mulch or a winter freeze cover over plants after they go dormant, to protect them over the winter before planting into the field in early spring.

Planting blueberries. Plant blueberries in early spring. Avoid fall planting except in the Deep South, because blueberries establish slowly and frost heaving can kill new plants.

Where soil organic matter is low, as in much of the Southeast, most growers incorporate peat moss into each planting hole. Peat moss soaks up more water during rains and irrigation, which blueberries need in great quantities, and retains positively-charged ions such as ammonium, which is the only form of nitrogen that blueberries can take up. One study showed that mixing peat into the planting holes of an upland soil increased blueberry plant survival by 30%, and doubled plant growth over the next 2 years.

Photo 15-12. When purchasing blueberry plants to start a field, growers buy 2-year-old plants that were grown in pots, then removed from the pots for shipping. Blueberries have a less-developed root system than many fruit crops, and do poorly as bare-root transplants.

Thoroughly water container-grown plants before planting, and roughen up their root balls while planting so roots will grow out into the soil. Soak the roots of bareroot plants for several hours before planting. Blueberry roots do not have root hairs like those of most plants, which take up most of the water and nutrients for the plant. Instead, blueberries depend on mycorrhizae, beneficial soil fungi that grow outward from the plants' roots, to perform the same function of sucking water and dissolved nutrients into the roots. Most soils naturally contain such beneficial mycorrhizae if they have not been fumigated in the past 5 years or so, but previously fumigated soils will benefit from dipping transplants' roots in a solution of commercially produced mycorrhizae.

Be careful not to plant blueberries too deep, as that is a common reason for poor plant performance. Set container-grown plants so the surface of the potting soil is level with or slightly above the soil surface, and cover with mulch rather than soil. Spread the roots of bareroot plants outward in the planting hole, rather than down, and cover with no more than ½ inch of soil with mulch over that.

Highbush blueberry growers usually space rows 10 feet apart to accommodate equipment, and space plants about 4 feet apart in the row. Newer Southern highbush cultivars are often planted at the same spacing, but some older Southern highbush cultivars are less vigorous and are planted 3 feet apart and perhaps 9 feet between rows. Rabbiteye blueberries grow more vigorously, so growers plant them 5 to 7 feet apart in rows 12 to 14 feet apart. Half-high blueberries are planted 2 to 3 feet apart in rows 8 to 9 feet apart.

Alternating rows of 2 or more cultivars that bloom at the same time increases cross-pollination, which increases fruit set and size in many highbush and highbush-lowbush cultivars, and is essential for fruit set of most rabbiteye blueberries. Planting cultivars that bloom early in one block, middle bloomers in another, and late bloomers in another not only improves cross- pollination, but is also more efficient for frost protection of early-blooming cultivars.

Irrigate new blueberry fields immediately after planting, and keep them well-irrigated during the first growing season. Prune off all flower buds soon after planting, so plants' energies go into growing new roots and leaves and becoming established.

MANAGING NATIVE LOWBUSH BLUEBERRIES.

Almost all lowbush blueberries are produced by managing native stands, rather than by planting a field. Many of these stands have been managed for over 100 years, and before that Native Americans often burned areas with lowbush blueberries to increase berry production. The major management tools are mowing or fire, plus herbicides. Fire, the traditional method, controls other woody plants but simply prunes lowbush blueberry plants, causing them to produce vigorous shoots from their rhizomes during the season following the fire. Mowing has gained favor in recent years as an alternative way to prune lowbush blueberries, since it costs much less, produces similar yields and preserves soil organic matter which burning destroys. Mowing does not kill other woody plants, however, so may require more herbicide use than burning. Continued mowing allows a buildup of diseases or insects in the blueberry leaf litter, so burning may be needed every 4 to 8 years to reduce diseases or insect problems.

Growers mow or burn each tract every other year, as studies have shown this produces greater yields than mowing or burning every third year. Lowbush plants flower and fruit during the second growing season after each mow or burn. Growers divide their land in 2 sections and prune half each year, so the other half produces the crop that year. When burning, build a fireguard around the section to be burned by plowing 3 to 4 furrows around it, so fire doesn't spread to the section burned the previous year.

Mow or burn lowbush blueberries while dormant, in late fall or early spring. Many growers who burn mow the tract before burning. They then use tractor-pulled burners, which consume 400 to 500 gallons of furnace oil or 200 to 275 pounds of propane gas per acre, to burn the mowed area. If these materials become expensive, some growers spread 1300 to 2200 pounds of straw or marsh hay per acre over the blueberry areas, and light them to burn the field.

Mower pruning requires fields that aren't too uneven or rocky. Flail mowers are better adapted to lowbush blueberry fields than rotary mowers, and many fields require narrow mowers, 3 feet wide or less, to follow the contours of the field. Mow as close to ground level as possible to achieve good pruning, and drive slowly, usually just 1 to 2 miles per hour. Most fields will have some areas in dips or around rocks that can't be mowed, so these areas will need spot burning.

Growers who mow rely on herbicides to kill grasses and woody plants other than blueberries. Burning eliminates many species that would act as weeds, but stimulates some other species, such as bunchberry and sweet fern, to produce new shoots. Growers apply herbicides, either over the entire field while blueberries are dormant or as spot treatments, to control such weeds.

Low rates of nitrogen (N) fertilizer, about 35 pounds per acre applied soon after mowing or burning, improve lowbush blueberry production in the following year. Do not apply N in summer, or it can prolong new growth into fall, rendering it more susceptible to winter killing. Fertilizers other than N have shown no benefit to lowbush blueberries, and complete fertilizers, containing phosphorous (P) and potassium (K) as well as N, have stimulated the growth of grassy weeds.

Lowbush blueberry fields usually have some spots bare of blueberries, which reduce yield potentials and are subject to soil erosion where herbicides are used. Growers can apply mulch to these areas to increase spread of blueberry rhizomes into them and reduce soil erosion. Transplanting sections dug from neighboring plants, or cuttings previously rooted in propagation beds, into the bare spots in early spring will speed blueberry coverage of the field. Spacing plants about 2 feet from each other in all directions results in complete coverage in about 6 years. Allow these transplanted blueberry plants to grow without pruning for 4 years.

SOIL MANAGEMENT, WEED CONTROL, MULCHES AND FERTILIZERS

Many cranberry growers prevent many weeds by applying a fresh layer of sand to bogs every few years (Photo 15-13). The sand must be free of weed

Photo 15-13. Cranberry growers prevent many weeds by applying a fresh layer of sand to bogs every few years. (Photo by T.R. Roper, Univ. of Wisconsin.)

seeds. Cranberry growers also use pre-emergent herbicides to prevent weed seed germination, and selective post-emergent herbicides to control grasses. Existing broadleaf weeds can be killed with glyphosate using a "wiper" applicator bar with a roller, brushed against the broadleaf weeds when they are 6 to 12 inches taller than the cranberry plants, so no chemical gets on the crop. Lingonberry growers can use any herbicides registered for cranberries, and also use an acid-forming organic mulch such as pine wood chips or pine sawdust, applied to a depth of about 2 inches.

Highbush and rabbiteye blueberry growers on level, sandy Coastal Plain sites, where soil erosion is not a problem, generally cultivate between blueberry rows to provide maximum water for the blueberry plants. On sloping upland sites with heavier soils more prone to erosion, growers plant a sod of bluegrass, creeping red fescue, white clover or some other less competitive groundcover between the blueberry rows.

Growers control weeds within blueberry rows with herbicides, and on upland soils, also with organic mulches. Larger blueberry growers on Coastal Plain sites do not usually use mulch due to the large expense of applying it, and because the perched water table in the subsoil provides consistent moisture for the plants throughout the growing season. On upland soils that dry out much more during the summer, mulch has great value in preserving moisture around blueberry roots, as well as adding organic matter that helps to maintain soil acidity, and reducing weeds.

Where using mulch, apply it soon after planting. Choose an organic material for mulch that acidifies the soil as it disintegrates slowly, needing replenishment less often. Pine sawdust (Photo 15-14), pine bark chips, and other acidic materials are good, but avoid hardwood sawdust or chips that have a pH closer to neutral. Maintain a mulch thickness of 4 to 6 inches on blueberries, which usually requires adding more mulch each spring.

Fertilizing new plantings. Ericaceous plants have shallow, rudimentary roots with no root hairs, and depend on mycorrhizae, beneficial fungi that live on and in their roots, to absorb most water and nutrients. These plants need little fertilizer, and excess commercial fertilizer can hurt the mycorrhizae so should be avoided. Ericaceous plants are adapted to low soil pH and thus have a high need for iron (Fe), which is more available at low pH, and can become deficient if the pH rises.

Use ammonium sulfate or urea, which breaks down into ammonium, as an N source for all ericaceous plants, since their roots take up ammonium much more readily than nitrate forms of N. Both materials help to maintain soil acidity, but ammonium sulfate has a greater effect, so choose it where the pH is 5.0 or more, and use urea where the pH is less than 5.0. Use potassium sulfate as a K source if K is needed, avoiding potassium chloride because chlorides can harm blueberry roots. Avoid adding more K than needed, because excess K can cause magnesium (Mg) deficiencies. If the soil contains less than 100 ppm Mg, it is often included in the complete fertilizer mix as 3% magnesium oxide, or as magnesium sulfate (Epsom salts).

Organic fertilizer sources can also be used for cranberries, blueberries, and lingonberries, if the materials are acidifying, such as cottonseed meal or fish emulsion. Avoid materials that increase pH, such as bone meal, or materials high in phosphorous and chlorides, such as fresh turkey manure, as these can cause iron deficiencies. Well-composted manure has been used successfully on blueberries at the rate of 10 tons per acre. Organic fertilizers must break down before they begin to release nutrients, so must be applied 1 to 3 months before the plants need N.

Wait until a few weeks after planting cranberries, blueberries or lingonberries, when newly planted roots have become established and tops have started to grow, to apply fertilizer. Fertilize

Photo 15-14. Acidifying organic mulches such as pine sawdust, shown here, conserve moisture and provide needed organic matter and acidity on upland sites, as well as helping to control weeds in the blueberry rows.

first-year cranberry bogs every 2 to 3 weeks until mid-August, with 5 to 10 pounds of actual N per acre each time, for a total of 25 to 40 pounds actual N per acre. Urea is the best N source during this first year, since ammonium sulfate may burn new roots. If no P was added at planting time, apply 50 pounds per acre of triple superphosphate with the N, but do not add additional P if it was supplied at planting.

Use a complete fertilizer for blueberries and lingonberries if the soil has less than 20 ppm P or 100 ppm K. Highbush and Southern highbush blueberry growers on sandy soils apply about 10 pounds of actual N per acre when growth starts, along with P and K as needed. They repeat this application every 4 to 6 weeks until mid-August, for a total of 40 to 50 pounds actual N per acre applied in 4 to 5 installments. On heavier soils, growers apply 50 to 60 pounds actual N per acre, but split this over just 2 or 3 applications. The heavier N application to upland soils compensates for some N being tied up by soil microorganisms that break down the mulches used. Lingonberries need about the same amount of N as highbush blueberries on heavier soils. Rabbiteye blueberry growers apply N at only about ½ the above rates because rabbiteyes are naturally vigorous, and half-high blueberry growers use about ¾ the above rates. Do not over-fertilize, as that is the most common cause of plant loss in new fields.

During the second growing season, fertilize cranberries like the first year, from late April until mid-August, with a total of 30 to 40 pounds actual N per acre. Increase highbush blueberry N rates to 50 to 60 pounds per acre per season for sandy soils, and 70 to 80 pounds per acre per season for heavier upland soils. Rabbiteye blueberries again need only ½ these N rates, and half-high blueberries need about ¾ these N rates. Fertilizers can be broadcast or applied through a trickle irrigation system. Fertigation works especially well on sandy soils, when N must be added in 4 to 6 installments per season.

Fertilizing bearing plantings. Cranberries and lingonberries usually bear their first good crop in their fourth growing season, and generally need only N fertilizer from then on. Cranberries need only 15 to 40 pounds of actual N per acre per year. Lingonberries are less vigorous and need more N fertilizer, usually 90 to 120 pounds of actual N per year. For both cranberries and lingonberries, provide N in 2 or 3 smaller applications rather than all at once. Apply N just before bloom (early to mid-May in most cranberry-growing areas), after fruit set (late June), and in late July or early August. Do not apply N later than mid-August as it could delay plant hardening for winter, and could delay ripening and cause softer fruit that is prone to fruit rots.

Blueberry growers begin harvesting in the third growing season. Increase N rates slightly each year as harvests increase until the fifth or sixth harvest year, when berry yields start to level off. Mature highbush plants need about 60 to 80 pounds of N per acre per season on sandy soils, and between 90 and 120 pounds of N per acre per season on heavier upland soils. Mature rabbiteye blueberries need just 30 to 40 pounds of N per acre per season on sandy soils, and 45 to 60 pounds of N per acre per season on loams or clay loams. Highbush-lowbush blueberries need 45 to 60 pounds of N per acre per season on sandy soils, and 60 to 75 pounds of N per acre per season on loams or clay loams.

Certain cultivars, or certain parts of a field, may exhibit low vigor or N deficiency symptoms, and need higher N rates than other cultivars or areas of a field. Take leaf samples in mid-summer to determine N content of such plants and of normal-looking plants, and adjust N rates accordingly for each part of the field. The optimum N level for mid-shoot leaves of fruiting shoots is about 2.0 % for all ericaceous plants.

Check soil pH at least every other year for blueberries and lingonberries, and at least every 3 to 4 years for cranberries. Upland plantings may need sulfur applications each winter or every other winter, even when using acidifying fertilizers, to maintain soil pH below 5.2. Too high a pH can cause iron deficiency (Photo 15-15), which causes interveinal

Photo 15-15. A pH greater than 5.5 can cause iron deficiency, indicated by the interveinal yellowing of new leaves, which reduces blueberry plant growth and yields.

chlorosis, stunting of plants, and reduced berry set, berry size and yields. Chelated iron or iron sulfate sprays can alleviate such symptoms while waiting for sulfur applications to reduce pH.

TRAINING AND PRUNING

Cranberries and lingonberries spread from rhizomes. Cranberries are allowed to cover the entire bog surface so need no training if they are to be water-harvested, but need training so vines grow in one direction if they are to be dry-harvested. This training is accomplished by pulling runners all one way with a rake, and the field must be harvested in that same direction. Dry harvesters also prune excess vegetative growth while harvesting cranberries.

Wet-harvested cranberry bogs need special pruning to rejuvenate them every few years after harvest begins. Growers usually accomplish this by mowing these cranberry fields after harvest, to stimulate growth of uprights and prevent runners from rooting. Eastern growers also apply a new, 1-inch-thick layer of sand to cranberry bogs every 3 to 4 years, to cover the old stems and stimulate new rooting of stems closer to plants' fruiting zones. Sanding also reduces weeds and some insect pests and is widely practiced from Wisconsin eastward, but is not recommended in the Pacific Northwest, because with little rainfall there, sand would eventually build up to 5 or 6 inches thick and roots would grow only in the sand layer. Cranberry bogs with such a thick sand layer would require heavy fertilizer applications to keep them productive.

Lingonberries need no training for the first few years since they grow slowly. Once lingonberry rhizomes spread so much that rows start to touch, rows must be narrowed with disk coulters, and the rhizomes cut off can be used to plant new fields.

Blueberries grow as upright bushes. New shoots emerge from the crowns, or from the rhizomes of lowbush blueberries, and grow vertically. After overwintering, they become canes, and produce flowers and fruit from this 1-year-old wood. The canes continue to live for several years, producing lateral branches that also bear as 1-year-old wood, but fruitfulness declines as canes age past about 4 years old, or past 2 years old for lowbush blueberries.

Training young blueberry plants and crop control. Prune blueberry bushes in early spring, just after they have broken dormancy. Young blueberry bushes need very little training. During the first couple of years, retain all canes and as many branches as possible, removing only those that droop to the ground. Head back the branches that produce flower buds during the first 2 growing seasons, to prevent plants from fruiting too soon, which can cause young plants to die. Pruning off all flower buds, which are produced toward the ends of branches, during the first 2 growing seasons ensures that plant energy goes toward root production and becoming well-established to produce larger crops in future years. Some flowers may be left on irrigated, vigorous rabbiteye blueberries during the second season in the field, so a small crop can be harvested that year.

During the third growing season, lightly head back branches to remove some flower buds, but allow about 75% of the flower buds to remain to produce the first crop. From the fourth growing season (second fruiting season) on, adjust the level of pruning to the density of live flower buds on each bush. When many flower buds are produced and survive cold weather, some must be removed or the fruit will be small and less sweet, and plants will lose vigor and produce much fewer flower buds the next year. When fewer flower buds are produced, prune branches back only lightly.

Waiting to prune until the flower buds have started to swell allows workers to judge the extent of winter cold damage, and to prune off damaged branch tips and leave an appropriate amount of flower buds to produce a good crop. Where spring frosts threaten during bloom, waiting until late bloom when most threat of frosts has passed can enable workers to adjust the number of flowers left to the extent of frost damage.

Renewal pruning and height control of mature blueberry plantings. From the fifth growing season on, pruning also involves removing a few canes per bush that are 5 years old or older. This renewal pruning encourages the growth of new, more vigorous canes that will produce better in a couple of years. First remove diseased or winter-damaged wood and any drooping or weak canes, with a diameter less than 0.2 inches. Then thin the remaining canes. Aim to keep an even number of canes in each of 3 ages classes: less productive 5- to 7-year-old canes, highly productive 2- to 4-year-old canes, and 1-year-old renewal canes. A well-pruned blueberry bush should have 12 to 24 total canes, depending on the cultivar's growth habit and productivity.

Rabbiteye blueberries and some more vigorous highbush cultivars also need pruning to control excess height. Cut back tall first-year shoots in summer or 1-year-old canes in winter by 1/3 to ½, and selectively remove the taller canes older than 4 years old. Cut the remaining tall canes back to a strong lateral or within 1 foot of the ground, to induce formation of lateral branches. Larger rabbiteye growers mechanically top bushes after harvest at 5 to 6 feet tall, hedging a couple inches higher every year to prevent excessive branching at the same height. Follow-up hand pruning is still needed to remove jagged stubs left by the mower, and to adjust crop load. After 6 or 7 years of hedging, bushes must be rejuvenated in spring to bring the height down to 3 ½ to 4 feet tall. This is normally done during a spring when all flower buds were lost to a freeze, or else heights must be gradually lowered over several years, to avoid pruning off all flower buds.

Highbush-lowbush blueberries may need some height control also, to keep bushes under the height normally covered by snow during the coldest part of winter where they are grown. Burning every other winter accomplishes pruning for lowbush blueberries.

IRRIGATION AND TEMPERATURE PROTECTION

All ericaceous plants have rudimentary roots so need lots of water, and irrigation is essential. All also need protection from temperature extremes.

Water management in cranberry bogs. Cranberries need large amounts of water for irrigation, spring frost protection, summer heat protection, winter protection, and usually for harvest. Most growers use both flood and sprinkler irrigation. Cranberry bogs do not need to be constantly flooded, although historically they were often flooded to provide some weed control, since cranberry plants are adapted to standing water while most other plants are not. Cranberry bogs in most areas need to be flooded over the winter to protect plants from cold and frost heaving (Photo 15-16), but some parts of the Pacific Northwest which do not experience sub-freezing temperatures can grow upland cranberries using only sprinkler irrigation and no flooding. Most cranberry growers use water-harvesting machines, and must flood the bogs before harvest, but some growers use dry-harvesting machines. Altogether, cranberry growers who water harvest and use winter flooding need 6 to 8 feet (72 to 96 inches) of water per year.

Bogs that are water harvested can remain flooded after harvest if freezing temperatures will occur soon, and dry-harvested bogs that need winter protection will need flooding before temperatures drop below about 20 F. Wisconsin growers start applying the winter flood in late November or early December, but growers in Massachusetts and New Jersey must wait until late December or early January before temperatures are low enough to freeze the flood. Water needs to cover the plants, so it will protect them as it freezes. Growers commonly build up the winter flood by applying 2 to 3 inches of water at a time, allowing that layer to freeze, then applying another layer of water until vines are covered. Once ice is frozen solid, a 1-inch layer of new sand may easily be applied by driving dump trucks over the ice, once every 3 to 4 years.

The winter flood is drained off in early spring, usually in mid- to late March in New Jersey and Massachusetts and in early to mid-April in Wisconsin. When plant buds begin to grow a few weeks later, automated temperature sensors turn on sprinkler irrigation to protect the new growth whenever temperatures drop below 34 F, and remain on until temperatures rise above 36F. Sprinklers apply water at the rate of 0.1 to 0.2 inches per hour for frost protection. Growers who lack sprinkler irrigation may hold the flood for a month longer to delay growth until after most chance for frost has passed, but this reduces yields due to a shorter growing season. Cranberries do not bloom until late spring or early summer, so open blossoms are not threatened by frosts. Sprinklers may be needed in late fall, however, to protect ripening fruit from fall frosts just before harvest.

Another management option, called "late water," involves draining bogs in early spring but then re-flooding them to a depth of at least 12 inches for 30 days just before bloom. Late water stimulates cranberry plant growth and aids in pest control as well as providing frost protection, but should only

Photo 15-16. Cranberry bogs in most areas need to be flooded over the winter to protect plants from cold and frost heaving. (Photo by T.R. Roper, Univ. of Wisconsin.)

be practiced once every 2 or 3 years, as crops may be reduced if done every year. Late water increases upright height, allowing growers to reduce N fertilizer by 20 to 40%, and suppresses invasive perennial weeds, particularly dewberries. Late water suppresses early-season cutworms, cranberry fruitworm and southern red mite enough that no sprays for these pests are generally needed during the following growing season, unless moths of the first two pests fly in later in the season. Late water also reduces fruit rots at harvest and during storage, so fungicides for fruit rots can often be eliminated in that growing season. The reduction in pesticide and fertilizer costs more than pays for increased water pumping costs, but late water requires 12 inches of water rather than just 2 to 4 inches for 2 to 4 nights of sprinkling for frost protection. Late-flooded bogs may need an algaecide, generally applied through the sprinkler system, and if algal growth gets severe or water temperature rise above 65 F, the flood must be removed early. Late water is recommended when the previous summer had many sunny days and winter temperatures were about average, and should be avoided if the previous summer had many cloudy days or a prolonged drought or the winter was very cold or unusually warm.

During the growing season, cranberries need frequent irrigation since sandy soils in the beds dry out quickly, and most growers use sprinklers to irrigate. Sunscald can injure immature cranberry fruit when temperatures exceed 85 to 90 F, and sprinkling during summer provides evaporative cooling that prevents sunscald. Sprinklers should switch on when temperature sensors reach 80F to provide evaporative cooling.

Growers who use water harvesters flood the bogs with 8 to 10 inches of water shortly before harvest. Many growers who use dry harvesters also flood bogs with a few inches of water, to cushion berries during harvest and reduce bruising, but some cranberry fields are dry-harvested without flooding.

Irrigation and cold protection for blueberries and lingonberries. Blueberries and lingonberries need between 1 and 1¾ inches of water per week during the growing season, or 45 to 60 inches of water per year. Periods of highest water needs occur when fruit are gaining size, from late April to early August depending on region and cultivars, and when fruit buds form in late August and September for next season's crop. Droughts between these periods necessitate irrigation to keep plants alive and healthy. Irrigate at least twice per week during hot, dry conditions.

Many blueberry cultivars bloom early in the season when spring frosts still threaten. Many growers therefore choose sprinkler irrigation (Photo15-17), which can also provide frost protection. The relatively low height of blueberry bushes allows the use of lower-cost, movable low sprinklers with above ground aluminum lateral lines. Permanent overhead irrigation systems with underground laterals and supply lines are much more expensive to install, but require less labor to operate and may be justified on large acreages of blueberries with no later-blooming crops on which to use frost protection.

Some growers in less frost-prone locations choose later-blooming blueberry cultivars and use trickle irrigation to reduce water and pumping costs. Lingonberry growers often use trickle irrigation and harvest only the second lingonberry crop, which is not threatened by frosts. Growers operate the trickle system daily on sandy soils or every 2 to 3 days on upland soils. Typically, each bush has a single emitter providing a 1-gallon per hour flow rate. Trickle irrigation enables fertigation, which can deliver N and other fertilizer nutrients more evenly over the growing season.

Half-high blueberry growers often use snow management to protect plants from winter damage. They may install snow fences parallel to blueberry rows to catch drifting snow and enable it to build up over the blueberry bushes enough to cover them, or they may plow snow onto blueberry rows to cover the plants. Some growers use polypropylene row covers over half-high blueberries in winter, as these help catch snow and protect branches above the snow line from desiccation by winds.

Photo 15-17. Sprinkler irrigation can provide frost protection for early-blooming blueberry cultivars.

POLLINATION AND GROWTH REGULATORS

Cranberries, blueberries and lingonberries are pollinated by several species of bees (Photo 15-18), and a profitable commercial yield requires at least 70% of the flowers that remain after pruning to set fruit. Cranberry cultivars are self-fruitful. Individual highbush blueberry and lingonberry cultivars are usually self-fruitful, but more fruit set and berries grow larger if cross-pollination between cultivars occurs. Feral honeybees have historically provided over half the pollination needed by cranberries and blueberries, but mite parasites are decimating wild and managed honeybee colonies.

Wild bumblebees or blueberry bees may still provide sufficient pollination in many areas, but where they are scarce, growers contract with beekeepers to put out 1 or 2 strong honeybee colonies per acre starting at about 25% bloom. Later-blooming cranberries (such as 'McFarlin') and blueberry cultivars that are moderately attractive to honeybees (such as 'Bluecrop' and 'Blueray') need just 1 hive per acre. Earlier-blooming cranberries (such as 'Stevens') and blueberry cultivars that are less attractive to honeybees (such as 'Elliot' and 'Jersey') need 2 hives per acre.

Some blueberry growers have started raising other bee species, including mason bees, which do a better job of pollinating blueberries than do honeybees. Many cranberry growers leave idle land undisturbed to provide nesting sites for bumble bees, which pollinate cranberries much more efficiently than do honeybees. Some cranberry growers provide artificial bumble bee nesting boxes along bog margins, and plant bumble bee forage plants that bloom later in the season, such as borage, bee plant and catmint, to help increase bumble bee populations.

Lowbush blueberries are self-sterile, and grow as large colonies spread by rhizomes from a single clone, so require cross-pollination by bees that forage to other clones. Bringing in at least 1 honeybee hive per acre, to supplement the native bees, helps achieve this. Producers must protect hives by screening from bears in native lowbush blueberries.

Most rabbiteye blueberry cultivars are also self-sterile, so plant alternating rows of 2 or more cultivars that bloom at the same time to provide cross-pollination. Rabbiteye growers may apply gibberellic acid to avoid poor fruit set due to pollination problems or early spring freeze damage. Gibberellic acid sprays applied just before most flowers shed pollen, or after a freeze, can induce parthenocarpic fruit to form without complete pollination, so these fruit are less seedy. Parthenocarpic fruit are usually smaller and ripen later than normal fruit. A new cytokinen compound, CPPU, has increased both fruit set and fruit weight of some rabbiteye cultivars when applied at 5 ppm at petal fall. This compound also delays ripening slightly.

Photo 15-18. Cranberries and blueberries require pollination by various species of bees in order to set fruit. Where native bees are scarce, bring in hives of honeybees or raise blueberry mason bees.

PEST MANAGEMENT

Many diseases and insects can attack cranberries and blueberries in the main growing regions, but blueberries grown outside their native regions and lingonberries generally suffer from few pests. Birds pose a serious threat to blueberries almost everywhere.

Diseases of roots and stems. These diseases can kill plants, and generally should be prevented before planting by modifying the site or choosing cultivars resistant to the disease, or minimized after planting by reducing plant stress.

Phytophthora root rot, caused by several soil fungi in the genus *Phytophthora*, is the most common disease of cranberries, blueberries and lingonberries on soils that are somewhat poorly drained. On such sites, install drainage tiles in cranberry bogs and construct raised beds for blueberries before planting, and avoid susceptible blueberry cultivars. When this is not done, the disease can kill over half the plants in a field. Rabbiteye blueberries generally show greater resistance to Phytophthora root rot than do highbush blueberries.

Upright dieback of cranberries and lingonberries, caused by the fungus *Phomopsis vaccinii* or another fungus, *Synchronoblastia crypta*, causes uprights to die back (Photo15-19) during times of high heat or drought stress, or after winter damage or oxygen deficiency during winter. Minimize upright dieback by using sprinkler irrigation to reduce heat and drought stress, and by choosing cultivars with some resistance to the disease such as 'Franklin'. If needed, apply fungicides from late April to mid-May.

Phomopsis twig blight of blueberries, also caused by the fungus *Phomopsis vaccinii*, attacks mainly canes that have been mechanically damaged and stressed by cold. Prevent much damage by choosing cultivars hardy enough for your area, and by stopping fertilizer additions by mid-summer to avoid stimulating late growth more susceptible to cold damage. A dormant lime-sulfur spray, and fungicide applications every 2 weeks from bud break until petal fall, can reduce the disease.

Stem canker of blueberries, caused by the fungus *Botyosphaeria corticus*, weakens and kills canes of susceptible highbush cultivars (Photo 15-20) over several seasons where it occurs in the Southeast, especially in eastern North Carolina. Most rabbiteye cultivars show moderate resistance to stem canker, whereas most highbush cultivars are susceptible unless bred for resistance. Where stem canker occurs, plant only blueberry cultivars resistant to the main pathogen races in the area. Propagate only from canker-free plants, and prune out and destroy all cankers in a planting each year. Applications of captan or another fungicide from green tip until the interval before harvest can reduce stem canker.

Stem blight, also prevalent among highbush and rabbiteye blueberries in the Southeast, is caused by the fungus *Botyosphaeria dothideae*. It can cause shoots to wilt and plants to die within a single season, so choose cultivars resistant to the prevalent pathogen races where the disease occurs. Stem blight usually infects plants through wounds caused by storms, pruning, mechanical harvesters or paraquat herbicide. Prevent much disease by follow-up hand pruning of branches damaged by storms or machine hedging, by calibrating mechanical harvesters to minimize damage, and shielding plants from paraquat herbicide.

Fungicide applications for stem canker also reduce stem blight. The stem blight pathogen has been recovered from hardwood cuttings of all cultivars tested, so where stem blight is a problem, propagate from softwood cuttings or buy certified disease-free plants. North Carolina growers routinely plant double the number of bushes needed in every fifth row, to use as replacements when other plants die from stem blight.

Photo 15-19. Upright dieback of cranberries and blueberries causes uprights to die back during times of stress from drought or winter damage. Minimize heat and drought stress by using sprinkler irrigation. (Photo by S. Vicen, Univ. of Wisconsin.)

Photo 15-20. Stem canker often kills blueberry canes in the humid Southeast. Where stem canker occurs, choose resistant cultivars.

Fusicoccum canker of blueberries, caused by the fungus *Fusicoccum putrefaciens*, attacks winter-injured canes and is most troublesome in Northern areas. Choosing cultivars hardy enough for your area can prevent many infections, as can pruning out diseased canes containing cankers. Spores from cankers are released from early spring until mid-summer, and infect new tissues following rains. Infected leaves turn red-brown and hang on late until fall, and red-brown lesions with a bulls-eye pattern form around leaf scars. Infected shoots wilt suddenly during hot, dry weather, and infected young stems may be killed to the ground. After pruning, apply lime sulfur at late dormancy to kill remaining spores that have overwintered, and where fusicoccum canker is serious, apply captan or other fungicides from green tip until petal-fall.

Bacterial canker, caused by the bacterium *Pseudomonas syringae*, may attack blueberries in the Pacific Northwest. Infected 1-year-old stems first appear water-soaked in late winter, and these areas swiftly develop into red-brown to black cankers. Cankers usually surround the stems, killing buds within each canker, and sometimes girdling stems and killing buds and shoots above the canker. Wind, rain, insects, mechanical harvesters and pruning tools can spread the bacteria to other plants, so prune out diseased branches as soon as they're seen and again before fall rains start. Dip pruning loppers in a 10% bleach solution between bushes, to avoid spreading bacteria on the pruners. Where bacterial canker is a problem, choose cultivars resistant to it, such as 'Elliot', and avoid susceptible cultivars such as 'Jersey' and 'Patriot'. Avoid late-summer N applications, which would prolong succulent growth susceptible to bacterial canker. Apply Bordeaux mixture or fixed copper fungicide twice in fall, in early October and early November, to reduce the number of infected stems next spring.

Diseases caused by phytoplasmas and viruses. Several virus and phytoplasma diseases attack cranberries and blueberries. Always start plantings with certified virus-free and phytoplasma-free plants from a reputable nursery, or test plants you will propagate from to be sure they are disease-free.

False blossom of cranberries, caused by a phytoplasma spread by the blunt-nosed leafhopper, nearly wiped out the cranberry industry in the early 1900's, when it spread in diseased plants from Wisconsin to most cranberry-growing areas. Flowers on infected plants become enlarged and are sterile, and many normally latent axillary buds grow, creating a "witch's broom" of uprights. Selection of cultivars with resistance to false blossom saved the industry, and cranberry growers should choose resistant cultivars if the disease occurs in their area. Insecticide sprays to control the leafhopper vector, applied before bloom, 10 to 14 days after mid-bloom, and 28 days after bloom, will also reduce disease spread.

Blueberry stunt disease, caused by a phytoplasma spread by the sharp-nosed leafhopper, is a problem in Michigan, New Jersey, New York, Massachusetts and North Carolina. Infected plants are stunted and excessively branched, with shortened internodes and small leaves that curl downward (Photo 15-20). Leaves turn yellow between the veins and along the margins, then turn prematurely red in late summer while midribs remain green. Fruit from infected plants remain small and hard, ripen late or fail to ripen, and hang much longer on bushes than do fruit on healthy plants. If the disease occurs, spray affected plants with an insecticide to kill leafhoppers on them, so leafhoppers do not spread the disease to adjacent healthy plants, then remove the infected bushes.

Necrotic ringspot of blueberries, caused by tobacco ringspot virus, is a soil-borne virus spread by dagger nematodes. These nematodes are common on sandy soils across North America, especially land previously planted to fruit trees, so test such soils for nematodes before planting. If high nematode numbers occur, plant cover crops to reduce nematodes (see chapter 2), or fumigate the soil before planting and then reinoculate with mycorrhizae. Necrotic ringspot infections in a blueberry planting occur in concentrated areas where nematode numbers are high, and slowly spread outward each year as nematodes transmit the virus to more bushes. Leaves of infected plants become puckered, with small necrotic spots whose centers fall out, creating holes with necrotic or red rings in the leaves. Confirm the presence of tobacco ringspot virus by sending leaves to a lab which tests for viruses, and remove and burn all infected plants.

Blueberry shoestring virus is spread by blueberry aphids and occurs mainly in Michigan, where it has infected 0.5% of blueberry plants, and occasionally in New Jersey. Infected plants develop elongated red streaks along new stems, and leaves and flowers may also show a red banding. Diseased leaves are narrow, wavy and somewhat sickle-shaped, and berries turn purple prematurely. Production drops dramatically within a few years of infection. Where blueberry shoestring virus occurs, regularly inspect fields for these symptoms, and immediately remove any plants with such symptoms. Send leaves of surrounding plants to a testing lab, and remove these plants also if they test positive for the virus. Removing any wild blueberries near the field, and controlling aphids with insecticides, can also reduce infections. Some highbush blueberry cultivars have more resistance to blueberry shoestring virus than do others, and planting more resistant cultivars may be wise in areas where the disease is severe.

Photo 15-21. Blueberry stunt, caused by a phytoplasma, causes shorter branches with smaller leaves (left) than normal (right). Stunt is most serious in Michigan, where growers should choose resistant cultivars and control leafhoppers that spread the pathogen.

Photo 15-22. Red ringspot virus causes red rings on leaves and stems, and reduces yields by about 25%. Prevent it by starting with virus-free plants, and choose resistant cultivars in New Jersey and other areas where the disease often spreads.

Red ringspot virus is a serious problem in some Wisconsin cranberry bogs, and on blueberries in New Jersey, parts of Michigan, and Arkansas. The disease has also been found in New England, New York, North Carolina and Oregon, but only at a very low incidence. Infected planting material is the main source of the disease. Mealybugs apparently spread red ringspot virus, which has spread naturally only in 'Searles' cranberry bogs in Wisconsin and in New Jersey blueberry plantings. Older leaves of infected plants develop distinct red spots, rings and oak-leaf patterns on their upper surfaces, and these markings may appear on stems (Photo 15-22) and berries also in some cultivars. Blueberries are small, often deformed, and ripen late, resulting in a crop loss of 25% or more. Cranberry yields don't appear to suffer, but keeping quality of 'Searles' fruit is markedly reduced. Starting with virus-free plants is enough to avoid red ringspot in most areas, but Wisconsin cranberry growers should probably avoid 'Searles', and New Jersey blueberry growers should also choose resistant cultivars. The old highbush cultivar 'Jersey' and several newer cultivars appear immune to red ringspot, while other cultivars are very susceptible.

Blueberry shock virus and **blueberry scorch virus**, spread by blueberry aphids, are the worst blueberry viruses in the Pacific Northwest. Both diseases cause young leaves to wilt and develop blackened veins and black streaks down the petiole, or blight to an orange color. Infected flowers drop, causing significant yield loss. Some cultivars, such as 'Bluecrop' and 'Jersey', tolerate blueberry scorch with little yield loss, but serve as reservoirs of the virus when planted in fields with susceptible cultivars. Do not establish new plantings next to infected fields, and always start with virus-free plants.

Diseases of shoots, flowers and fruits. Minimize these diseases, where they are serious, by choosing resistant cultivars when possible. Susceptible cultivars must receive fungicides to prevent them.

Cottonball disease of cranberries, caused by the fungus *Monolinia oxycocci*, causes tip blight and fills berries with a cotton-like mat of fungal mycelia (Photo15-23). This is the most important cranberry disease in Wisconsin and often requires control in British Columbia, but generally does not occur or needs no treatment in other areas. Cottonball-infected fruit and mummies float, so re-flooding beds after harvest, and using harvest machines to remove them, can reduce the disease for next year. Sanitize harvesting equipment after removing infected berries. Apply a fungicide once or twice during shoot elongation in early spring, and twice during bloom, to bogs that suffered high amounts

Photo 15-23. Cottonball disease of cranberries causes tip blight and fills berries with a cotton-like mat of fungal mycelia. Avoid planting cultivars very susceptible to tip blight where it's a problem. (Photo by S. Vicen, Univ. of Wisconsin.)

of cottonball the previous year. Scout other fields for tip blight just before bloom, especially in high-risk wetter areas along ditches. If symptoms of tip blight are found, apply a fungicide twice during bloom. Avoid planting cultivars very susceptible to tip blight, such as 'Bergman' and 'Pilgrim', in areas where it's a problem.

Mummy berry blight of blueberries, caused by the related fungus *Monolinia vaccinii-corymbosi*, infects young shoots, causing rapid wilting. Also called leaf and twig blight or bud and twig blight, this disease can also kill flower buds. Highbush blueberry cultivars vary greatly in susceptibility to mummy berry blight (see Table 15-1), so choosing less susceptible cultivars can greatly minimize this disease. Burning greatly reduces this disease in lowbush blueberries, but growers should carefully monitor mowed fields for weather conditions during bloom, and apply a fungicide if a frost occurred followed by wetness of 6 hours or more and temperatures above 40 F. Cultural practices and fungicides to reduce mummy berry fruit infections also reduce mummy berry blight.

Mummy berry also attacks fruit, turning infected fruit into hard, pink mummies that don't ripen. This disease can be serious in most blueberry-growing areas. The fungus overwinters on mummified fruit that fall to the ground, so destroying mummies on the ground in early spring can greatly reduce the disease. Do this by cultivating or disking between the rows or by covering mummies with at least 2 inches of new mulch, and applying urea to speed decomposition of the mummies.

Prevent mummy berry in isolated areas, where the disease is not present in cultivated or wild blueberries nearby, by planting only bare-root plants. This avoids introducing infected leaves or sclerotia, the overwintering form of the fungus. If planting potted plants, first scrape off and destroy all plant debris on the surface of the potting soil, to avoid planting sclerotia.

Blueberry cultivars vary greatly in susceptibility to mummy berry fruit infection, and resistance to the bud and twig blight form of mummy berry has little correlation to resistance to the fruit infection. Where the disease is a problem, choosing cultivars with greater resistance can minimize the need to spray for mummy berry. Where growers use cultural practices to reduce mummies and choose cultivars with some resistance, just 2 fungicide sprays are needed to prevent primary shoot infections and reduce secondary infections of blossoms and fruits. Apply a fungicide such as azoxystrobin (Abound™) or triforine (Funginex™) at budswell and again 2 weeks later.

Bitter rot or anthracnose fruit rot, caused by the fungus *Colletotrichum gloeosporioides* or *C. acutatum*, may attack cranberries, blueberries and lingonberries, causing berries to rot and produce orange-colored spore masses (Photo 15-24). Hot, humid weather encourages this disease, so it is worst on cranberries in Massachusetts and New Jersey, and on blueberries in the Southeast. Fungal spores overwinter in dead or diseased twigs, fruit spurs and cankers, so raking away all cranberry prunings and removing all dead and diseased blueberry canes and twigs during dormant pruning helps reduce anthracnose. Pruning out old canes and branches with many dense twigs also reduces the disease by increasing air circulation.

No cultivars are highly resistant to anthracnose, but cultivars such as 'Stevens' cranberry, 'Elliot' Northern highbush blueberry, 'Bladen' and Reveille' Southern highbush, and 'Briteblue' rabbiteye blueberry have some resistance and need fewer sprays. 'Blueray' and 'Jersey' highbush blueberries are more susceptible than most cultivars because their ripe fruit can hang on bushes for a prolonged time before picking, so harvest such cultivars more frequently to minimize anthracnose. During hot, humid seasons, growers must also apply fungicides. Cranberry growers spray from early bloom until 2 to 3 weeks after petal fall, while blueberry growers apply cover sprays from 7-10 days after petal fall until the interval required before harvest.

Photo 15-24. Anthracnose often infects susceptible blueberry cultivars such as 'Blueray', producing orange spore masses, during hot, humid summers. Harvest such cultivars frequently and apply fungicides, or choose more resistant cultivars.

Several other **fruit rot** diseases caused by fungi may attack cranberries before or after harvest. **Blotch rot**, incited by *Physalospora vaccinii*, may seriously reduce yields in New Jersey and Wisconsin. **Early rot**, caused by *Phyllosticta vacinii*, can cause large field losses in Eastern cranberry regions, and **ripe rot**, caused by *Coleophoma empetri*, may reduce yields in all cranberry-growing areas. Several other fungi may also cause pre-harvest fruit rots. Sprays to control cottonball disease or bitter rot generally prevent these other pre-harvest fruit rots also. Storage rots include **end rot**, caused by *Fusicoccum putrefaciens* and **black rot**, caused by any of three fungi: *Allantophomopsis lycopodina*, *A. cytisporea*, or *Strasseria geniculata*. When cranberries will be processed soon after harvest, as 95% of them are, storage rots rarely arise. When harvesting fruit for fresh market, growers usually use dry harvesters to minimize fruit exposure to water and bruising. Avoiding excess N and prolonged periods of flooding will reduce both pre-harvest and post-harvest fruit rots.

Botrytis blight or **gray mold**, caused by the fungus *Botrytis cinerea*, may be a problem on ericaceous plants where cool, wet weather prevails during bloom. It can be serious on cranberries, lingonberries and highbush blueberries in the Pacific Northwest, lowbush blueberries in Maine and the Maritime Provinces of Canada, and rabbiteye blueberries (but usually not highbush blueberries) in the Southeast. Blighted blossoms develop a fuzzy gray mold, and may drop without fruit developing. Fruit that do develop from infected blossoms retain the dormant fungal spores, which can cause serious postharvest decay. Pruning to keep blueberry bushes open, and reducing nitrogen fertilizer applications, can reduce Botrytis blight. When cool, humid weather occurs during bloom, apply fungicides. Where gray mold is serious, avoid highly susceptible cultivars such as 'Blueray' and 'Weymouth' highbush blueberries.

Foliage Diseases. These leaf diseases can cause considerable damage during warm, humid spring conditions. Susceptible cranberry cultivars may need fungicide treatments in the Pacific Northwest and susceptible blueberry cultivars often need fungicides in the Southeast, but treatments are rarely needed elsewhere. The southern highbush cultivars 'Reveille' and 'Bladen' are very resistant to blueberry leaf diseases even when not sprayed.

Twig blight of cranberries, a fungal disease incited by either *Lophodermium oxycocci* or *L. hypophyllum*, sometimes causes serious losses in Washington and Oregon. Uprights become infected in summer, but do not show symptoms until early the next spring, when foliage on one-year-old wood turns brown. The disease progresses into one-year-old wood and kills it, but does not spread further down uprights. Most cultivars grown in the Pacific Northwest appear susceptible, including 'Stevens', 'McFarlin', 'Bergman' and 'Crowley'. Growers must protect new growth with three fungicide sprays spaced 2 weeks apart during the infection period, based on pest management advisors' scouting reports.

Rose bloom and **red leaf spot** of cranberries, caused by the fungi *Exobasidium oxycocci* and *E. rostrupii*, respectively, may occur in all cranberry regions but usually cause damage only in the Pacific Northwest. Both diseases occur mainly in bogs where available N is high and vines are making excessive growth. Rose bloom attacks axillary buds, causing them to produce short, thickened, rose-colored leaves. Red leaf spot produces circular red spots on upper leaf surfaces and occasionally on young green berries, and the disease may enter vines, causing new growth to wither and die. Where red leaf spot is a problem, keep N levels low, and avoid very susceptible cultivars, such as 'Ben Lear' and 'Stevens'. Spraying bogs with a copper or manganese fungicide soon after budbreak protects new leaves from infection.

Gloeosporium leaf spot of blueberries, caused by the fungus *Gloeosporium minus*, is sometimes also called anthracnose leaf spot. The disease causes small, reddish flecks on young leaves and shoot stems, then brown lesions ½ to 1 inch wide on the leaves. This can cause severe defoliation and yield reductions. Highly susceptible cultivars, such as 'Jersey', exhibit dieback of stems up to 20" long.

Septoria leaf spot of blueberries, caused by the fungus *Septoria albopunctata*, produces smaller, ⅛- to ½-inch-wide white or tan leaf spots with purple borders (Photo 15-25), as well as stem lesions. These are most severe on young plants and on lower parts of older plants.

Powdery mildew of blueberries, caused by the fungus *Microsphaera alni*, appears as white fungal growth on the upper or lower leaf surfaces, or sometimes just chlorotic spots with reddish borders that may be mistaken for red ringspot virus. Powdery mildew often does not develop on blueberry leaves until after the crop is harvested in the Southeast.

Summer mowing after harvest to reduce bush height reduces all these blueberry leaf diseases by removing older, infected leaves. New leaves that

Photo 15-25. Septoria leaf spot commonly infects blueberry leaves where hot, humid summers prevail, as in the Southeast U.S. Top-hedging plants after harvest reduces this and other leaf spot diseases.

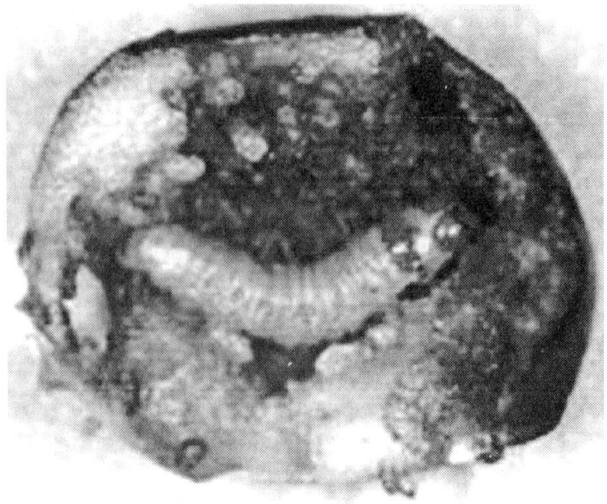

Photo 15-26. Cranberry fruitworm caterpillars attack developing and ripening cranberries and blueberries. Control it with a molting inhibitor insecticide. (Photo by J.F. Landry, Agric. Canada, Agassiz, BC)

grow from midsummer on usually avoid foliage diseases. Mowing can eliminate the need for 3 or 4 fungicide sprays from after harvest until mid-August. Southeastern growers should still make at least 1 fungicide application to highbush blueberries, and at least 2 to rabbiteye blueberries, before harvest to control leaf spots, except on the most resistant cultivars.

Fruit and flower bud insects and mites. Several insect pests can attack cranberry and blueberry fruit, especially in the major commercial growing districts where they require control. Isolated blueberry and lingonberry plantings in other regions may not need any insecticide treatments.

Cranberry fruitworm (*Acrobasis vaccinii*) caterpillars attack developing and ripening cranberries and blueberries (Photo 15-26). This is often the most important insect pest of cranberries, and can cause serious damage to blueberries where cranberries are also grown, especially in New Jersey. **Cherry fruitworm** (*Grapholitha packardi*) is a similar moth pest that attacks blueberries in regions where cherries are grown commercially, especially in Michigan. Cranberry fruitworms are mostly green with some brownish-red color on the top surface and up to ½ inch long, while cherry fruitworms are orange-pink with brown heads and up to 1/3 inch long. Both have just one generation per year, and can be controlled with or a molting inhibitor such as tebufenozide (Confirm™).

If cherry fruitworm is a problem in the area, apply an insecticide at petal fall and again 10 days later. If cranberry fruitworm is a problem, apply an insecticide 10 days and 20 days after petal fall. Using late water enables a cranberry grower to skip this treatment that year.

Blackheaded fireworm (*Rhopobota naevana*) is another main cranberry insect pest in Wisconsin and in the Pacific Northwest, and a sporadic pest in Eastern cranberry areas. **Spotted fireworm** (*Choristoneura parallela*) is the most important insect pest of cranberries in New Jersey. First-generation larvae of both these moths feed between cranberry uprights that they have webbed together. Second-generation larvae feed on fruit surfaces, blemishing them. Late water or spring flooding can kill overwintering larvae of both these pests, and selective insecticide sprays for moths, such as Confirm™ or the biological insecticide *Bacillus thuringiensis* (Bt), also reduce them.

Sparganothis fruitworm (*Sparganothis sulfureana*), another potential moth pest of cranberries, also has two generations per year. First-generation larvae feed on new foliage and flowers in spring, often webbing several terminals together. Second-generation larvae also feed on foliage and bore into fruit, with each larva consuming 3 to 5 berries during development.

This pest increases where broad-spectrum organophosphate insecticides have been used routinely, as these kill the natural enemies that normally help control it. Choose the most selective insecticides possible for other pests, such as Confirm™ or Bt for cranberry fruitworm or fireworms. Sparganothis fruitworm has become serious in New Jersey, where mating disruption using a sprayable pheromone has proved effective in controlling it.

Brown spanworm (*Ematurga amitaria*) and **green spanworm** (*Itame sulphurea*) are sporadic moth pests of cranberry, but a severe infestation can eliminate a crop. These moth pests have one generation per year, and larvae hatch just before and during bloom to feed on foliage, buds and blossoms. Large spanworm populations can destroy all flowers and cause the bog to turn brown, but infestation is usually patchy. Scout for spanworms just before bloom and treat with Confirm™ or Bt as soon as any are seen.

Cranberry blossomworm (*Epiglaea apiata*) is a cutworm that feeds on leaves and bores into buds of cranberries in mid-spring. Late water or spring flooding greatly reduces this pest, which is mainly a problem in Eastern cranberry-growing areas. Again, selective insecticide sprays for moths also reduce them. Other potential cranberry cutworms controlled in the same way include **false armyworm** (*Xylena nupera*) and **spotted cutworm** (*Xestia* species).

Blueberry maggot (*Rhagoletis mendax*), a fly larva, attacks blueberries in Northern regions, especially in the main blueberry-growing areas of Michigan and New Jersey. This pest does not oversummer well in the lower Midwest or the South, so usually needs no control there. Adult flies emerge over a prolonged 6-week period during summer, and females lay eggs under blueberry skins just as they begin to turn blue. Maggots feed for about 3 weeks inside ripening and harvested fruit, and can contaminate packed berries if undetected (Photo 15-27).

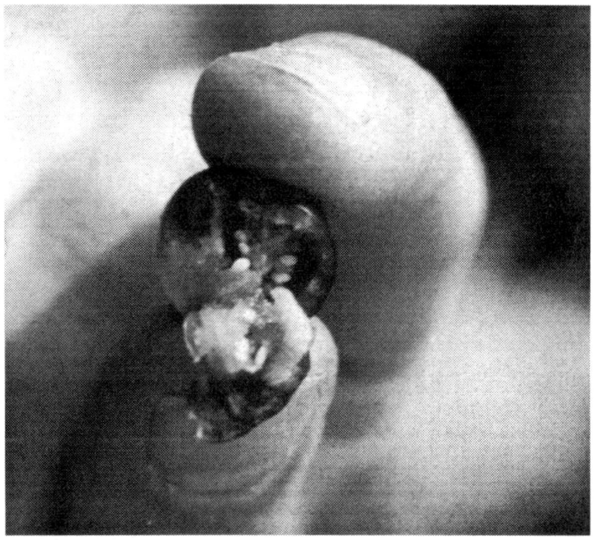

Photo 15-27. In regions where it occurs, blueberry maggot can infest harvested berries if not controlled with insecticides .

Insecticides to control blueberry maggot, such as carbaryl (Sevin™) or phosmet (Imidan™), must be applied close to harvest, so timing is very important to control the pest without violating the preharvest interval for the pesticide treatment. Growers use yellow sticky traps, baited with ammonium acetate crystals that attract the flies, along field borders to monitor blueberry maggot adults. Check traps twice per week and replace them every 2 to 3 weeks, and apply an insecticide when traps catch several flies per week.

Cranberry weevil or **blueberry blossom weevil** (*Anthonomus musculus*) is a problem in the main cranberry- and blueberry-growing areas, where they grow wild. This is a snout beetle whose adults emerge in early spring, and females lay eggs in expanding flower buds and leaf buds. An insecticide treatment made immediately after punctures are first noticed can control this pest. Removing wild cranberries and blueberries near plantings also reduces this pest.

Plum curculio (*Conotrachelus nenuphar*) is a common pest of blueberries east of the Rocky Mountains, especially where they are planted near peaches, plums, apples, wild plums or wild cherries. Adults are weevils that overwinter in wooded areas and fly into fruit plantings in early spring. Females cut a crescent-shaped egg-laying scar in each berry in which they deposit eggs, which hatch into legless worms that feed inside the developing berries for 2 weeks before dropping to the ground to pupate. Apply an insecticide such as phosmet (Imidan™) at petal fall and 10 days later, as for cherry fruitworm, to control plum curculio.

Blueberry bud mite (*Acalitus vaccinii*) occurs along the East Coast from Massachusetts to Florida, and has become an increasing problem in the Southeast and New Jersey especially. These tiny mites suck sap from the buds, causing distorted flowers that may fail to set fruit, or berries may only partially develop and become red, blistered and misshapen. Early-ripening cultivars generally show the most damage. Reduce bud mite populations by pruning out and destroying old canes, and by applying a miticide after harvest and again 3 to 4 weeks later to kill mites in the developing fruit buds for next year.

Insects and mites that feed on shoots and roots. Most of these insects are problems only in certain areas, so do not need control everywhere.

Cranberry tipworm (*Dasineura oxycoccana*) larvae feed on leaves of developing uprights in spring or summer, killing the terminals (Photo 15-28). This moth pest is serious in Wisconsin and Massachusetts cranberry bogs. Where growing seasons are longer and plants have more time to recover after tipworm feeding, little damage is sustained, but when feeding occurs later in the season, crop losses are much higher. Some cultivars, such as 'Stevens' and 'Hyred', produce more regrowth after tipworm feeding and so resist this pest, while cultivars such as 'Early Black' and 'Howes' suffer more damage. Adding sand to bogs every 3 to 4 years can reduce early-season tipworm populations, and again selective insecticide sprays for moths reduce them.

Blueberry tip borer (*Hendecaneura shawiana*) can build to damaging levels in the major blueberry-growing districts. These borers, pink when newly hatched and white when fully grown to 3/8 inch long, tunnel through shoots, causing shoots to wilt while stems turn purple and leaves become yellow with red veins. Control small infestations by pruning out damaged tips as soon as they're seen, and burning them to kill larvae inside. This is a moth pest, so a Bt spray or Confirm™, applied when eggs are hatching in early summer and before larvae tunnel into shoots, can prevent damage. The second cranberry fruitworm spray should also help control tip borer. Watch for the emergence of adult moths, brown with orange markings and a gray spot, to aid in determining when eggs are being laid and sprays are needed.

Cranberry girdler (*Chrysoteuchia topiaria*) moth larvae feed on bark and wood of cranberry vines from late summer until after harvest, girdling and killing vines. Thin or dead spots in bogs develop, but these often aren't seen until the next spring. Bogs that receive new sand every 3 to 4 years usually don't have a problem with cranberry girdler. Holding the flood water for 4 to 5 weeks after harvest also suppresses this pest. Dry-harvested cranberry beds can be treated twice with beneficial nematodes in early to mid-August.

Black vine weevil (*Brachyrinus sulcatus*) grubs attack cranberry roots in the Pacific Northwest. Weevils may completely girdle crowns or roots, causing plants to wilt or turn brown. Other beetle larvae that similarly attack cranberry roots in Eastern cranberry-growing regions include **cranberry white grub** (*Phyllophaga anxia*) and **cranberry root grub** (*Lichnanthe vulpina*). Root weevils or grubs usually are not problems in water-harvested cranberries, and if they do occur in such bogs, root weevils can be controlled by holding the flood water for 4 to 7 days after harvest. Dry-harvested cranberry beds can be treated once with beneficial nematodes in May or September to control root weevils or grubs.

Photo 15-28. Cranberry tipworm larvae may kill upright terminals. Adding sand to bogs every 3 to 4 years reduces this pest, and some cultivars produce more regrowth after tipworm feeding, so suffer less damage. (Photo by R. Kachadoorian, Univ. of Wisconsin.)

Gypsy moth (*Lymantria dispari*) can devour cranberry foliage and terminal buds in spring where this pest occurs, especially in New Jersey and Massachusetts. Attacked bogs yield poorly for the next 2 years unless the infestation is controlled with a selective insecticide such as Confirm™.

Japanese beetles (*Popillia japonica*) can skeletonize blueberry leaves by midsummer in eastern North America where they occur. Use Japanese beetle traps near but not in the planting, to determine when to apply an insecticide without attracting beetles to the blueberry plants.

Scale insects of cranberries and blueberries include several species of soft-bodied insects covered with a waxy or hard shell-like structure that protects them from most insecticides. **Oystershell scale** (*Lepidosaphes ulmi*), **lecanium scale** (*Lecanium corni*) and several other species may attack cranberries, while **Putnam scale** (*Ospidiotus ancylus*), **terrapin scale** (*Mesolecanium nigrofasciatum*), and others may attack blueberries. Scale infestations on cranberry uprights or blueberry canes, where they feed on plant sap, can reduce plant vigor and yields.

Some scales also feed on or near fruit and secrete a sugary waste product called honeydew, which serves as a growth medium for sooty mold, a black fungus that disfigures the berries. Prevent scale problems by applying superior oil during the dormant or delayed dormant period. (If leaf spots are also a problem, apply oil while plants are still dormant and temperatures are above 50 F, as a delayed dormant oil application reduces the leaves' waxy coating, rendering them more susceptible to leaf spots.) Some cultivars, such as 'Brigitta' highbush blueberry, are more susceptible to scale insects, so avoid these where scale is a recurring problem.

Leafhoppers commonly feed on cranberries and blueberries, and do the most damage by spreading phytoplasma diseases. **Blunt-nosed leafhopper** (*Euscelis striatulus*) spreads false blossom of cranberries, while **sharp-nosed leafhopper** (*Scaphytopius magdelensis*) spreads blueberry stunt disease. Reduce leafhopper populations by applying insecticides before bloom, 10 to 14 days after bloom and 28 days after bloom.

Southern red mites (*Oligonychus ilicis*) may build up on cranberry bogs in Massachusetts. Severe infestations cause foliage bronzing and reduce yields. Late water can control them, as can a delayed-dormant oil spray.

Mammal pests. Several mammals may become pests in cranberry bogs, but usually don't bother blueberries or lingonberries.

Muskrats (*Ondatra zibethicus*) can be serious pests in cranberry bogs, if they tunnel through dikes and around water control structures, causing washouts and dike breakage. Muskrats may also feed on cranberry plants under ice in winter, and can cause much damage to small areas. **Beavers** (*Castor canadensis*) may occasionally disrupt water management in cranberries by building their own dams. Control both these mammals by trapping and relocating them, or if populations are high, by obtaining necessary permits and shooting them or using poison baits.

White-tailed deer (*Odocoileus virginianus*) can cause serious problems in cranberries and sometimes blueberries or lingonberries where populations are high, especially in Wisconsin and New Jersey. Deer can make paths across cranberry bogs, trampling plants and eating some berries, but the worst damage is leaving their droppings in cranberry bogs. Deer droppings are about the same size and shape as cranberries and float, so are picked up by harvesting machines as a contaminant, and bounce like a cranberry through the bounce-board separator, so must be removed by hand. Deer damage to blueberries and lingonberries consists mainly of eating stems and buds over winter. Where deer populations are high, encourage hunting and if necessary put up electric fencing.

Controlling birds on blueberries. Birds can be the worst pests of blueberries, especially in small fields with woods nearby. With no control, birds can consume or peck at enough fruit to make a blueberry planting unprofitable. Cranberries are not bothered by birds because of high fruit acidity, and birds have not yet been serious pests of lingonberries due to other preferred food sources.

Small growers often rely on bird netting, which pays for itself by the second harvest and can last for many years, although it is labor-intensive to install. Growers first build a support structure around the planting, usually by setting posts 9 feet above the ground around the edges and running wire between the poles to form a grid over the blueberry plants (Photo 15-29). When berries begin to turn reddish-blue, workers hang netting over the grid and lace the sections of netting together to form a continuous barrier. Anchor the netting edges to the ground or bury them. After harvest is over, remove the netting and store it in a cool, dry place.

The bird repellent chemical methyl anthranilate (Bird Shield™), with the odor and taste of 'Concord' grapes, provides some protection for blueberries, but since most blueberries ripen over a 3- to 4-week period, it does not always work well. It must be applied just before berries first begin to turn color for birds to learn to stay away and do little damage, and reapplied a week later. Yet it also must be applied more than a week before harvest, or be washed off by rain or postharvest rinsing, so customers do not taste the pungent, slightly bitter chemical. Methyl anthranilate is most useful for processing cultivars with a concentrated ripening period.

Photo 15-29. Many smaller blueberry growers construct frames for bird netting over their plantings, to prevent severe losses to birds which can destroy all profits.

Noise deterrents and visual bird-scaring devices alone often do little to deter birds in blueberries, but a combination of noise and visual effects may keep birds away. Some growers hire workers to regularly drive motorcycles or all-terrain vehicles through blueberry plantings while berries are ripe. This has proved effective, but drivers must stay away from hired pickers to avoid accidents, and shouldn't drive when customers are present at pick-your-own farms. A combination of recorded bird distress calls and helium-filled balloons or kites above the planting, with hawk silhouettes hanging from them, repels birds well in some areas. Providing habitat for real hawks and other predators also helps.

Weed control. Control perennial weeds before planting with glyphosate (Roundup™). This material may be needed as a spot treatment after planting also, as woody perennials such as dewberries may overrun cranberry bogs, and greenbriar (*Smilax* species) and sumac have invaded some blueberry fields and become major pests. Cut these woody perennial weeds back to a few inches from the ground after harvest or during winter, then treat the new shoots that sprout with glyphosate (at the higher rate for woody perennials).

Several pre-emergence herbicides can prevent many weeds from sprouting in cranberry bogs or blueberry rows. Careful timing of pre-emergence herbicide treatments, in early fall and early spring, is important, along with annual mulch applications, because few materials are registered for post-emergence weed control in bearing cranberries or blueberries.

Where grasses have germinated in cranberry bogs or within blueberry rows, control them with sethoxydim (Poast™) up to 30 days before harvest. Broadleaf weeds can be controlled after emergence only by hoeing, hand-pulling, glyphosate spot treatment if far from plants, or paraquat directed spray with blueberry bushes shielded. Paraquat drift onto blueberry plants can render them much more susceptible to stem canker, so many growers are using it less.

Organic growers may use concentrated vinegar or soap-based herbicides as directed sprays against germinated weeds, being careful that these materials do not contact cranberry or blueberry foliage. Other alternatives to herbicides in blueberries include using black woven polyethylene mulch (landscape fabric weed barriers) rather than organic mulch, or weeder geese if they are kept out of the planting when berries are ripening.

HARVEST AND HANDLING

Essentially all cranberries are machine-harvested. Most growers use water harvesters, which harvest cranberries more efficiently than dry harvesters although subjecting berries to more bruising, since 95% of cranberries are processed. Wet-harvested cranberries are prone to post-harvest fruit rots, so must be frozen or otherwise processed immediately. Growers who sell berries fresh use dry harvesters.

Blueberries for fresh market are generally hand-harvested by workers for local retail sales and for wholesale shipping, or by customers at pick-your-own farms. Machines harvest blueberries marketed to processors, and some fresh-market rabbiteye blueberries. About 35% of the total blueberry crop is sold fresh, and about 65% is processed. The highbush and rabbiteye crops go about half to fresh sales and half to processing, while almost all lowbush blueberries go to processing, and almost all Southern highbush and half-high blueberries go to fresh markets.

Harvest time for both cranberries and blueberries is determined by skin color development. Growers should allow berries to develop as much color as possible to enhance sugar development and marketability.

Mechanical harvest and handling of cranberries. Each cranberry bog is harvested just once per season, so growers should wait until most berries have turned red to begin harvest. Cultivars that develop early and intense red color, such as the old wild selection 'Early Black' and the newly-released 'Hyred', are often preferred so harvest can begin sooner. Ethephon, the growth regulator which releases the ripening hormone ethylene, is not widely used in cranberries because it is not effective at temperatures below 60F, which often occur during cranberry harvest season.

The night before a wet cranberry harvest begins, a grower will flood the bog so it has 8 to 10 inches of standing water. The next day, workers use water reels, nicknamed "egg beaters", to knock berries from vines, so they float to the water surface (Photo 15-30). Other workers corral the floating berries with wooden booms into a section of the bog, where they are pushed onto conveyor belts that carry them up elevators and dump them in trucks (Photo15-31). Berries are quickly delivered to a central receiving station, sorted through a bounce-board separator to remove unsound fruit that don't bounce, and immediately frozen or otherwise processed.

Many growers who dry harvest also apply a few inches of water the night before, to cushion fruit and reduce bruising during harvest. Dry harvesters, which look like giant lawn mowers, comb berries from vines with moving metal teeth. A conveyor belt on the harvester then carries berries to a plastic receptacle at the back of the machine, or a wooden large box towed behind it on a small barge. Receptacles are hand-dumped into wooden bins, and bins or boxes are moved by fork-lifts or cranes onto trucks and transported to packing facilities. There, fruit is sorted, cleaned, dried in boxes with slatted bottoms, and kept in refrigerated storage until it is packaged into plastic bags and shipped to grocery stores.

Hand harvesting blueberries. Most blueberry cultivars ripen their fruit over a period of 3 to 4 weeks. Harvest each cultivar 3 to 4 times at weekly intervals, as blueberries hang well on the bush as compared to other small fruits, and more frequent harvesting is inefficient. Blueberries are not fully ripe until about 5 days after they start to turn blue, so do not begin harvest until about a week after the first berries start to appear blue. Harvesting too early produces smaller, acidic fruit with lower sugar content. Instruct workers or customers to examine the bases of berries, near the stems, and to pick berries only if those areas have turned from red to blue.

Pick-your-own operations need about 450 customers per acre of blueberries, and about 40 person-hours of supervisory labor per acre over the average 6-week harvest season. Harvests can last 4 to 10 weeks, depending on the number of cultivars chosen and the ripening period of each cultivar. Growers generally provide picking buckets (Photo 15-32), flats or quarts, and either weigh the harvested fruit or sell by container volume.

Growers who sell harvested blueberries at retail markets, or who ship wholesale berries to supermarkets, need to hire 10 to 15 pickers per acre during the season's peak. Because blueberries ripen during schools' summer vacations in most regions, many growers hire students, and many operate buses to transport pickers to the farm. Pickers harvest berries directly into the pint or quart containers in which they will be sold. This minimizes bruising from excessive handling, and removes as little as possible of the waxy "bloom" covering the dark blue berries, that gives them their desirable light blue appearance. A good picker can harvest up to 20 flats of 12 pints each per day if the crop is heavy, although a light crop can cut this rate in half. Some growers pay pickers by the hour, while others

Photo 15-30. Cranberries are often water-harvested with water reels, nicknamed "egg beaters," that knock berries from vines so they float to the surface. (Photo by T.R. Roper, Univ. of Wisconsin.)

Photo 15-31. Other workers corral the floating berries into a section of the bog, where they are pushed onto conveyor belts that carry them up elevators and dump them in trucks (Photo by T.R. Roper, Univ. of Wisconsin.)

Photo 15-32. Pick-your-own growers generally supply customers with small buckets or other containers to hold the blueberries they pick.

find that paying by the amount picked is worth the extra supervision needed, as it creates incentives for workers to earn more by picking faster.

Rabbiteye blueberry growers without enough acreage to justify a mechanical harvester may use hand-held, battery-powered mechanical harvesters. These small machines with picking fingers gently shake berries from individual canes onto tarps spread below the bushes. Some growers have workers use plastic baseball bats or PVC pipes to jar the canes and shake berries onto tarps. Since rabbiteye blueberries have tough skins, these methods work well even for fresh-market berries.

Small lowbush blueberry or lingonberry growers may use non-motorized push-type mechanical harvesters, similar to old-fashioned, non-motorized push-type lawnmowers. These harvest berries faster than the traditional hand-held rakes used by lowbush blueberry growers.

Growers should cool harvested fruit, or tell pick-your-own customers to refrigerate their berries, as quickly as possible after harvest. Keep blueberries below 40 F, but not below 32 F. Most highbush blueberries will keep at 32 F for 2 weeks and at 40 F for 1 week. Rabbiteye blueberries and a few highbush cultivars keep longer, up to 4 weeks at 32 F. Growers or buyers who ship blueberries run the harvested flats through packing lines where machines cap individual pints in plastic wrap. The plastic wrap eventually reduces oxygen in the pints, which extends storage life by a few days. Additional modified-atmosphere storage does not provide much benefit to blueberries, extending storage life by only 1 week.

Machine harvesting blueberries. A grower must produce at least 40 acres of blueberries to justify purchasing a mechanical harvester. Small growers may hire custom mechanical harvesting, or several small growers may jointly purchase a harvester. A single mechanical harvester can replace up to 100 hand pickers, but the fruit is usually suitable for sale only to processors, who often pay only about half the price of the fresh market. Mechanical harvesters miss some fruit so yields are generally lower, especially with certain cultivars, and some hand harvest and inspection labor is still needed for quality control. Mechanically harvesting and grading blueberries for processing costs about 30 cents per pound, while hand harvesting and packing fruit for shipping costs about 60 cents per pound. The cost of mechanical harvesting represents about 50% of total production costs for processing blueberries, while hand harvesting costs contribute about 75% of total production costs of harvested fresh market blueberries.

Mechanical harvesters detach highbush and rabbiteye blueberries from the bushes by vibrating the bushes with rubber fingers or nylon wands (Photo 15-33). Lowbush blueberry harvesters use canoe-shaped steel teeth on a rotating head or large scoop similar to the traditional hand rake used for harvest. Careful machine adjustment allows harvesters to remove a high percentage of ripe fruit, which are attached to bushes less securely, while removing few of the green fruit that are still strongly attached. However, machine removal of green fruit can be a serious problem early in the harvest season and on cultivars with tight fruit clusters. Growers often need to hire workers to hand harvest the earliest-ripening berries of each cultivar, to avoid significant fruit losses. Many growers employ hand harvesters to pick fruit for the first 2 or 3 harvests of each cultivar, then use mechanical harvesters for the later-ripening fruit which will go to processors. Lowbush blueberry growers wait until 90% of the berries are blue before they harvest.

Machines also harvest overripe or shriveled fruit and berries damaged by birds, insects or diseases, which hand pickers would normally drop on the ground or leave on the bushes. Machine-harvested fruit must therefore go through an air-blast sorter conveyor, either as part of a newer machine harvester or as a separate operation after an older machine harvester. This extra handling increases bruising and reduces storage life of the fruit, making it important to get the fruit to the processor as soon as possible.

Minimize bruising of machine-harvested fruit by adjusting machine vibration to the minimum

Photo 15-33. Mechanical blueberry harvesters detach fruit with rubber fingers or nylon wands.

level needed to remove ripe fruit, which can vary with cultivar. Growers who plan to machine-harvest should choose cultivars that are better adapted to this, and that ripen their fruit over a more concentrated period so fewer passes with the harvester will be needed. Cooler temperatures reduce fruit damage, so harvest during the early morning or evening; some growers run mechanical harvesters at night. Fill field containers to a depth of no more than 4 to 5 inches, and limit fruit depth to 2 inches on the feed conveyor of the air-blast cleaner. Adjust the air volume of the air-blast cleaner so berries are not lifted or dropped more than 1 inch. Handle and dump all field containers gently. Clean harvesters at least twice a day, and sorting equipment every 2 hours, with a 10% bleach solution, to kill disease spores that could otherwise spread to healthy berries.

Marketing and Processing. The vast majority of cranberries are marketed either through Ocean Spray, a cooperative owned by over 800 U.S. and Canadian cranberry growers and 126 U.S. grapefruit growers and headquartered in Massachusetts, or through Northland Cranberries, a vertically-integrated and publicly traded Wisconsin cranberry company. Growers deliver harvested fruit to processing and packing facilities owned by Ocean Spray or Northland Cranberries, and accept the price offered them. Most berries are frozen for later use in production of juices, jellies, sauces and other products, while berries for fresh market are refrigerated until they are packaged for sale.

Some large wholesale blueberry growers pack, market and ship their berries directly to supermarkets. Many wholesale growers deliver harvested fruit to local receiving stations owned by wholesale buyers or grower cooperatives, or these buyers may send refrigerated trucks to pick up harvested berries at the farm. The Michigan Blueberry Growers Association handles about half the Michigan crop and also markets blueberries for growers in Indiana, Florida, Georgia, Mississippi, Louisiana, and North Carolina. Tru-Blue Cooperative, based in New Jersey, markets about half the berries in that state and also many from North Carolina. Cooperatives generally pool the crops from a number of growers to meet the needs of large buyers. Grower cooperatives often own refrigerated trucks, cooling areas and packing houses, and provide inputs such as harvest containers and pesticides to growers at a lower cost. Blueberries intended for the wholesale fresh market are cooled, transported to packing houses, graded, packaged, inspected to verify that they meet all federal and state standards, and shipped to supermarkets. Any lots that do not meet grades are diverted to processing.

Retail and pick-your-own blueberry operations sell fruit directly to customers, so providing a good farm experience to customers is the most important goal of the operation. Blueberries are easy to pick, especially when growers choose cultivars with larger fruit. Choosing cultivars resistant to the worst diseases of the area can minimize sprays needed, which many customers appreciate. Hire enough courteous employees to provide quick customer assistance and minimize customers waiting in line, and provide a water cooler, bathroom, and a shaded checkout area.

Blueberries intended for processing are sold directly to the processor. These berries pass through a blower to eliminate leaves and other light debris, then a water separator that allows green berries to float and be skimmed off. Berries are dried, then frozen in bulk or individually. Most processed blueberries are individually quick-frozen in giant IQF cylinders (Photo 15-34). Frozen berries pass through a squirrel-cage stemmer and are graded again before being packaged in 30-pound polyethylene-lined cartons. These are later used to make pie fillings, jams, and other products. Some lowbush blueberries are dehydrated for packaged blueberry muffin mixes.

Photo 15-34. This individual quick freeze cylinder at a blueberry processing plant allows rapid freezing of berries that stay separate and are easily utilized for many products.

Further Reading

Caruso, F.L. and D.C. Ramsdell. 1995. *Compendium of blueberry and cranberry diseases.* Amer. Phytopathol. Soc., APS Press, 3340 Pilot Knob Rd., St. Paul, MN 55121-2097. 87 pp. Available by calling 1-800-323-7560.

Eck, P. 1988. *Blueberry science.* Rutgers Univ. Press. 284 pp.

Eck, P. 1990. *The American cranberry.* Rutgers Univ. Press. 420 pp.

Foulk, D., E. Hoover, J. Luby, T. Roper, C. Rosen, W. Stienstra, D.Wildung and J. Wright. 2001. *Commercial blueberry production in Minnesota and Wisconsin.* Univ. of Minnesota Extension Service pub. WW-02241. Available online at http://www.extension.umn.edu/distribution/horticulture/DG2241.html

Hancock, J. and E. Hanson. 2001. *Blueberry varieties for Michigan.* Michigan State Univ. Extension. Available online at http://www.msue.msu.edu/fruit/bbvarbul.htm

Himelrick, D.G., A.A. Powell, and W.A. Dozier, Jr. 1995. *Commercial blueberry production guide for Alabama.* Alabama Coop. Extension System pub. ANR-904. Available online at http://www.aces.edu/department/extcomm/publications/anr/anr-904/anr-904.html

Kuepper, G.L., G.K. Ames, and S. Diver. 1997. *Organic blueberry production. Appropriate Technology Transfer for Rural Areas (ATTRA).* Available by calling 1-800-346-9140 or online at http://www.attra.org/attra-pub/blueberry.html

Lingonberry Corporation of America, LLC has a website with useful cultural information, http://www.wislingonberry.com/farming.html

Lowbush blueberry production. Agriculture and Agri-Food Canada Research Branch pub. 1477/E. Available online at http://wysiwyg://5/http://res2.agr.ca/kentville/pubs/pub1477/1477_e.htm

Mahr, D.L., T.R. Roper, D.S. McManus and R.A. Flashinski. 2004. *Cranberry pest management in Wisconsin.* Univ. of Wisconsin pub. A3276, available online at http://www.uwex.edu/ces/pubs/pdfs/A3276.PDF

Roper, T.R. and T.D. Planer. 1996. *Cranberry production in Wisconsin.* Univ. of Wisconsin pub., available online at http://www.hort.wisc.edu/cran/Publications/productn.html

University of Maine Cooperative Extension offers many fact sheets on producing lowbush blueberries, available online at http://www.umext.maine.edu/pubs/blubrry.htm

Chapter 16
Apples and Crabapples

Apples produce by far the largest tonnage of deciduous fruit trees, over 5 million tons (10 billion pounds) per year in the United States alone. Although planted on only about half the acreage of that devoted to grapes in the U.S., they produce about 85% of the tonnage of grapes. Apples are widely planted throughout temperate areas of the world, and recent breeding efforts have produced cultivars that are grown commercially in subtropical regions.

Like grapes, apples have a very long history of cultivation, for at least 8000 years, and over 7000 named cultivars have been selected. Cultivated apples (Photo 16-1) derive mainly from *Malus pumila*, a species native to eastern Europe and Western Asia. In recent years, breeders have incorporated many other *Malus* species into apple cultivars, mainly as a source of disease resistance or cold hardiness. *Malus* species other than *M. pumila* are called crabapples, and generally have much smaller fruit (Photo 16-2) that is quite tart and often coarse-textured.

Photo 16-1. Apple cultivars, such as this 'Gala' derive mainly from the species *Malus pumila*, while breeders have intergressed genes for disease resistance and cold hardiness from several other *Malus* species.

Photo 16-2. *Malus* species other than *M. pumila* are called crabapples, and generally have smaller fruit with a coarse texture and more acidity.

Some crabapple species are native almost everywhere in the temperate zone, including several species native to North America. Several hundred named cultivars of various crabapple species and hybrids have been selected, and some of these are grown commercially, both as pollenizers in apple orchards and for the fruit, which is processed into jellies and spiced crabapples.

Apples and crabapples have long proven popular because the fruit stores well over the winter, and can be utilized in many processed products including cider, juice, canned pie fillings, jelly, pectin for making other fruit jellies, and dried apples. Apples used to be stored in root cellars over winter for up to 6 months, but refrigeration and then controlled atmosphere storage have lengthened that storage life to over 12 months. The rise in apple production in Australia, New Zealand, Chile and other Southern Hemisphere countries has contributed to the current year-round availability of apples. Apple production has also greatly increased in China, which now produces 4 times as many apples as the U.S., which has hurt prices for North American apple growers in recent years.

The state of Washington leads apple production in North America, producing about 47% of the total U.S. apple crop. New York , California and Michigan each produce about 9 to 12% of the U.S. crop. Oregon, Pennsylvania, Maryland, Virginia, North Carolina, Ohio, Illinois, New England, and many Canadian provinces also produce many apples. About 57% of the U.S. apple crop is sold fresh, and about 43% is processed. Of processed apples, 57% go to juice and cider, 32% are canned, 6% are dried and 5% are frozen.

Apples and crabapples, like all the main temperate-zone tree fruits, belong to the rose family, the Rosaceae. They are long-lived small trees, growing about 25 to 30 feet tall and wide in nature and living 30 to 50 years. The flowers have 5 petals, 5 sepals, stamens in multiples of 5, and an inferior ovary with 5 carpels. The ovary develops into the core of the fruit, a pome, and the edible flesh develops mainly from accessory tissues. Fruit develop mainly on spurs, thick twigs with very short internodes that continue to bear fruit for many years.

TREE SIZE AND ORCHARD DENSITY

Commercial apple trees, like other fruit trees, are not grown from seed because seedlings are extremely variable in fruit quality, disease resistance and other important characteristics. Instead, nurserymen propagate particular cultivars by grafting scions of the cultivar (Photo 16-3) onto rootstocks, young trees grown from seed or produced by cuttings or mound layering. Apple rootstocks grown from seed, called standard rootstocks, produce trees that grow 25 to 35 feet tall and wide. Planting cultivars on standard rootstocks at a spacing of 20 to 25 feet apart in the row and 24 to 30 feet between rows produces a low-density orchard of 58 to 90 trees per acre.

The greatest change in apple production over the last 40 years has been the development of dwarfing rootstocks, propagated clonally by cuttings or layering, that allow planting trees closer, at higher tree densities per acre. Orchard systems now range from semi-vigorous trees in moderately low densities (100 to 194 trees/acre), to semi-dwarf trees in moderately high densities (226 to 444 trees/acre), to dwarf trees in high densities (518 to 968 trees/acre), to super-dwarf trees in ultra-high density systems (1000 or more trees/acre).

Planting apple orchards at higher densities greatly increases yields in the early years of the planting, when only a small portion of a low-density orchard's total area is covered by apple trees. Most dwarfing rootstocks also induce precocity, or early bearing, in scions, further increasing yields in early years.

High-density orchards continue to out-yield low-density orchards in later years also, because smaller trees allow light penetration and bear fruit throughout the entire tree canopy, whereas larger trees receive light and bear fruit mainly around the tree edges.

The smallest, most densely planted trees put most of their energy into fruit production instead of growing a supporting framework of a strong trunk and limbs, so they need a permanent support structure of a trellis or stakes.

Different regions reach the limit to yield increases with increasing tree densities at different points, depending on light intensity, summer heat, soil fertility and other factors. Some growers in Washington and British Columbia, where light intensity and soil fertility are high yet summer temperatures remain cool, have put in orchards with 2300 to 2500 trees/acre.

Researchers in the Netherlands, which has cool summers and fertile soil but medium light, have calculated the optimum apple orchard density to be about 2000 trees/acre. New York researchers found that orchard profitability in that cool-sum-

Photo 16-3. Nurserymen propagate apples by grafting a bud (as here) or stem piece onto a rootstock.

mer but lower-light region increased up to about 1000 trees/acre, then leveled off. North Carolina research showed that densities of 450 to 600 trees/acre produced the most profit for apple growers in the Southeast, with its lower light and soil fertility and higher summer heat.

The higher the tree density, the higher the establishment costs per care, as each apple tree costs between $5 and $10, depending on the number of trees ordered, the cultivars and the rootstocks. Ultra-dwarf trees and many dwarf trees are poorly anchored so require a trellis or supporting stakes, adding to establishment expenses, and will require an irrigation system in areas where semi-dwarf or semi-vigorous trees might not. On the plus side, higher density orchards allow use of smaller, less expensive sprayers, and greatly reduce labor costs to prune and harvest.

Dwarf cultivars, dwarfing rootstocks and interstems. A few **genetic dwarf** cultivars exist, such as 'Cascade Delicious' and 'Starkspur Compact Mac'. These naturally grow into dwarf-size trees, even when grafted onto standard or semi-vigorous rootstocks. Where such cultivars fit the climate and market, they offer the advantage of being able to use a standard or semi-vigorous rootstock with good resistance to drought and pests, and enough anchorage to avoid the need for trellising or staking.

Spur-type strains of a cultivar are genetic semi-dwarfs, growing to about ¾ the size of most cultivars. Such strains produce more fruit-bearing spurs and less vegetative shoots, which reduces vegetative growth and increases yields. Examples include 'Spur Red Delicious', 'Spur Red Rome', and 'Spur Criterion'. Such cultivars produce a semi-dwarf tree when grafted onto a standard or semi-vigorous rootstock, and a dwarf tree when grafted onto a semi-dwarfing rootstock.

Most cultivars, however, will need to be dwarfed by the rootstock or an **interstem**. An interstem is a stem piece 6 to 12 inches long, grafted between the rootstock and the cultivar. A dwarfing interstem can keep the tree smaller, while a standard or semi-vigorous rootstock provides good anchorage and resistance to drought and soil pests (Figure 16-1). Interstem trees cost more for the nursery to produce since they involve making two grafts per tree, so growers must pay extra for them and often order them in advance to get the cultivar/interstem/rootstock combinations they want. Interstem trees may be worth the extra cost, however, in savings on tree support, irrigation and soil pest control.

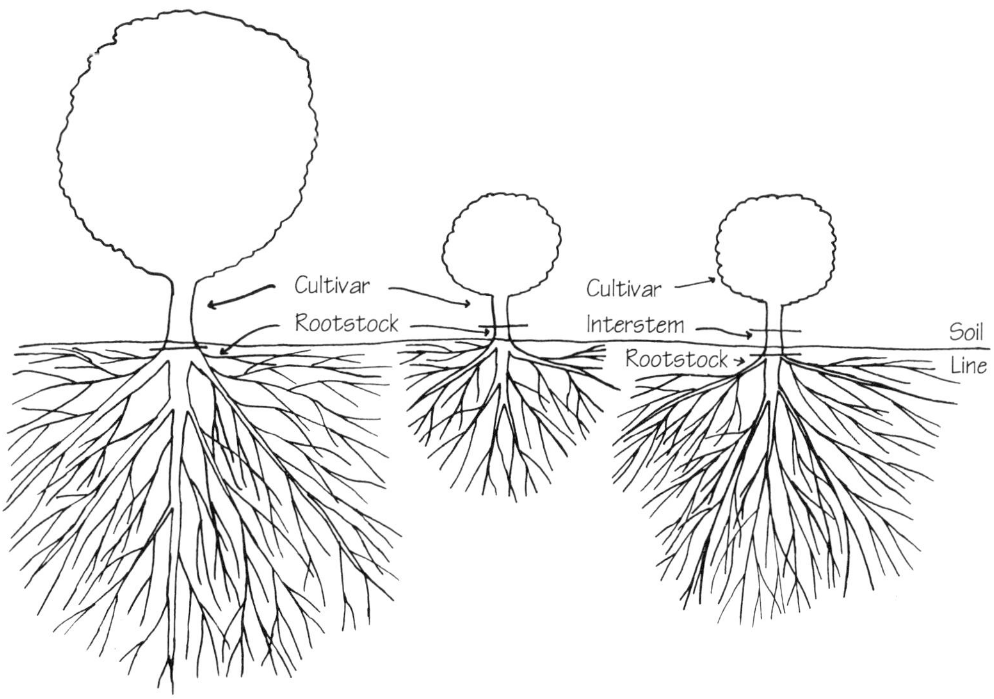

Figure 16-1. Root systems of (a) standard-sized, (b) rootstock-dwarfed and (c) interstem-dwarfed apple trees. In (c), a dwarfing interstem keeps the tree smaller, while a semi-vigorous rootstock provides good anchorage and resistance to drought and soil pests.

Many dwarfing rootstocks were classified or developed at a research station in East Malling, England, and these are designated by the letter "M" followed by a number, such as M9 or M26. (The sequence of numbers bears no relation to the rootstock's degree of dwarfing.) Later cooperative breeding work between the Malling station and the John Innes Institute at Merton, England resulted in the Malling-Merton series of semi-dwarfing rootstocks, designated by "MM" followed by a number, such as MM 106 and MM111.

These M and MM rootstocks, like most rootstocks clonally propagated over many years, often contain 4 or more viruses that generally cause no problems, but some cultivars are sensitive to certain viruses. For example, M7 and M9 commonly carry rubbery wood virus, which reduces yields of 'Golden Delicious', and tomato ringspot virus can kill cells at the graft union of 'Delicious'/MM106 trees, causing trees to die.

The East Malling and Long Ashton research stations in England have teamed up to produce virus-free propagating material of the M and MM rootstocks, and these are designated as "EMLA" or just "E" followed by the number. For example, virus-free M9 is called EMLA 9 or E9, and virus-free MM111 is called EMLA 111 or E 111.

Building on England's pioneering work, breeding by other nations has produced additional apple rootstocks with varying degrees of dwarfing, cold hardiness and resistance to several serious pests. Very cold-hardy dwarfing rootstocks include the "P" series from Poland and the Budagovski or "Bud." series, developed by Dr. Budagovski in Russia.

Other dwarfing apple rootstocks include the Geneva or "G" series developed in Geneva, NY; "Mark" and the Michigan Apple Clone ("MAC") series developed in Michigan; and Ottawa 3 or O3 developed in Ottawa, Canada.

All these rootstocks have strong points and weaknesses, and different climates, soil types, pest problems, and orchard management systems call for different rootstocks or rootstock/interstem combinations. Figure 16-2 illustrates the relative tree size produced by some common apple rootstocks and interstem/rootstock combinations.

Table 16-1 details the dwarfing, soil adaptation, hardiness, pest resistance and other traits of commonly used apple rootstocks and interstem/rootstock combinations. Using a spur-type cultivar can reduce tree size of a particular rootstock or combination by another 25% beyond the size listed in the table.

Low-density orchards. Many older, low-density apple orchards are still in production (Photo 16-4), though few are being planted. Some of these are on **seedling rootstocks**, grown from seeds of commercial cultivars obtained from juice factories and canneries. In Northern apple-growing areas, growers often chose **Antonovka** standard rootstock for its greater cold hardiness, to USDA zone 3 rather than zone 5 for standard seedling rootstocks.

Photo 16-4. Few apple growers plant low-density orchards anymore, but many older ones are still producing.

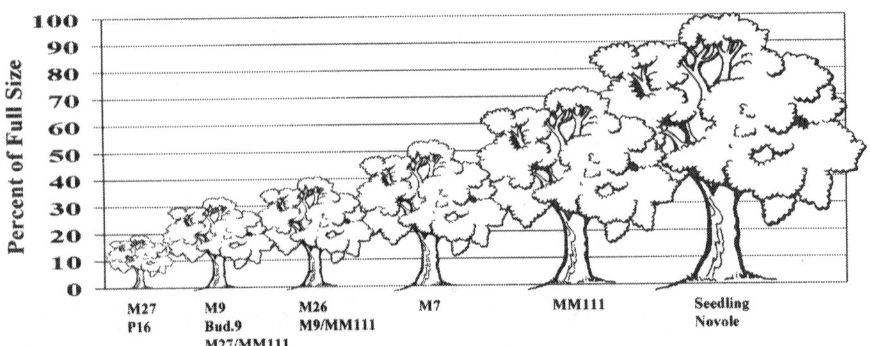

Figure 16-2. Relative tree size produced by some common apple rootstocks and interstem/rootstock combinations.

Table 16-1.

COMMONLY USED AND PROMISING APPLE ROOTSTOCKS AND INTERSTEM/ROOTSTOCK COMBINATIONS

In approximate order of tree height.

Rootstock or Interstem/ rootstock	Height in feet	Support needed	Hardy to zone	Susceptibility to Drought	Susceptibility to Crown Rot	Susceptibility to Wooly Aphids	Susceptibility to Fire Blight	Tendency to Grow burr-knots	Tendency to Grow root suckers
M27	4.5-6	yes	6	VS	R	VS	MS	medium	low
P16	4.5-6	yes	3	VS	VR	S	S	none	low
Bud. 491	5-7	yes	3	S	MS	S	MS	medium	low
P2	6-8	yes	3	S	R	S	S	none	very low
P22	7-9	yes	3	S	R	S	S	none	low
M9	8-10	yes	5	S	R	S	VS	high	high
Bud. 9	8-10	yes	3	S	VR	S	S	medium	low
M27/MM111	8-10	no	6	R	R	VR	MR	medium	medium
G16	9-11	maybe	3	MS	MR	S	VR	none	very low
Mark	9-11	maybe	5	S	R	MS	MR	medium	low
Bud. 491/A	9-11	no	3	R	R	MS	MR	none	very low
Ottawa 3	9-11	maybe	3	MS	R	VS	S	low	high
P22/MM111	9-11	no	4	R	R	VR	MS	low	medium
M26	10-12	yes	4	S	MS	VS	VS	high	low
M9/MM111	10-12	no	5	R	R	VR	MS	medium	medium
G30	12-14	maybe	4	MR	R	S	R	low	low
M7	14-16	maybe	4	MS	MR	S	R	medium	high
MM106	16-18	no	5	MR	VS	VR	MS	medium	low
MM111	18-22	no	4	R	R	VR	MR	medium	medium
Antonovka	24-28	no	3	R	R	MS	MR	low	medium
Novole	25-30	no	4	R	R	MS	MR	none	very low

Pest resistance ratings: VR = very resistant, R = resistant, MR = moderately resistant, MS = moderately susceptible, S = susceptible, VS = very susceptible

Another standard rootstock developed more recently, **Novole**, resists damage from voles, which are serious pests in some regions. All these rootstocks produce full-size trees, so trees were planted at very low densities of 58 to 90 trees per acre. These orchards often took 4 to 7 years after planting to begin bearing, and 14 to 16 years to come into full production.

Medium-low density orchards Over the last 30 years, many orchards were planted at slightly higher tree densities (100 to 136 trees/acre) on **MM111** rootstock. This rootstock produces a ¾-size tree but does not induce earlier bearing; most orchards began bearing 4 to 5 years after planting and reach full production after 12 to 13 years. MM111 provides good anchorage, hardiness to zone 4 and resistance to drought, crown rot, fire blight, and wooly apple aphid. Its one problem is production of burrknots, bumps along the trunk where root initials developed that create openings for borer insects to attack.

Many medium-density orchards, with 140 to 194 trees/acre, have been planted on **MM106**, a rootstock hardy to zone 5 that produces a tree 70% the size of standard rootstocks and does induce earlier bearing, 3 to 4 years after planting. However, MM106 is very susceptible to both crown rot and fire blight, resulting in early death of many trees, so should only be planted on soils with excellent drainage where fire blight isn't a problem. **Bud. 490** produces a tree about the same size as MM106, and offers moderate resistance to crown rot and to fire blight.

Medium-high density orchards. M7 produces a tree about 60% of standard size, and has been the most widely planted semi-dwarfing rootstock, planted at densities of 226 to 300 trees/acre. M7 resists fire blight and is moderately resistant to crown rot, but it produces burrknots and the rather shallow root system is very susceptible to drought, so is not suited to very sandy soils. M7 is hardy to zone 4 and usually does not require support, but if cropped heavily 3 or 4 years after planting it may need staking on some soil types.

Another way to achieve orchards of the same density as M7 is to plant **spur-type cultivars on MM111**.

Geneva 30, or G30, is a newer rootstock hardy to zone 3 that produces trees slightly smaller than M7, that bears well at an even younger age of 2 to 3 years after planting. G30 has very high yield efficiency, and allowing trees to carry large crops from a young age can dwarf them almost to the size of M26, which can allow a higher planting density of 272 to 345 trees/acre. G30 resists fire blight and crown rot, has some tolerance to apple replant disease, has wider soil adaptation than M7. In some soil types it does not need support, but in most soils it does. It produces very few burrknots but many spines, which must be rubbed or cut off in the nursery.

M26, hardy to zone 4, produces trees about 45% the size of standard, planted at 340 to 388 trees/acre. It was fairly widely planted 25 to 35 years ago, but better alternatives to M26 are available today, such as the just-mentioned G30. M26 is poorly anchored so requires support, is very susceptible to fire blight and wooly apple aphids, is moderately susceptible to crown rot, produces many burrknots, and is very susceptible to drought.

Several dwarfing rootstocks may be used as **interstems on MM111** rootstock to produce trees about the size of M26. MM111's good anchorage and drought resistance ensure that such interstem trees will not need staking, and they usually don't need irrigation in areas with 40 inches or more of rain per year. The MM111 root also provides resistance to wooly apple aphids and crown rot.

M9/MM111, the most commonly offered interstem/rootstock combination, produces a tree slightly larger than M26 that's hardy to zone 5 and has moderate susceptibility to fire blight. **P22/MM111** produces a tree slightly smaller than M26, also with moderate susceptibility to fire blight, that's hardy to zone 4, while **P22/Antonovka** keeps trees the size of M26 and is hardy to zone 3. These interstem trees can be planted at densities of 340 to 444 trees/acre.

High-density orchards. These are usually trellised (Photo 16-5) or staked (Photo 16-6), although orchards of interstem trees are freestanding. Most growers have planted in single rows, although some plant in beds of 2 rows each to cover more of the land with apple trees. Trellised bed plantings 2 rows wide can be easily harvested by hand from both sides (Photo 16-7).

M9 has been the main dwarfing rootstock used for the last 30 years, producing trees 30% of standard size that are planted at densities of 518 to 792 trees/acre. It requires support and irrigation, is susceptible to wooly apple aphids and fire blight, and produces many burrknots, but resists crown rot and is hardy to zone 5. **EMLA 9**, the virus-free version, is more vigorous than M9, producing trees 35% of standard size, so shouldn't be planted quite so closely. **M9-337** is a newer virus-free clone that produces trees the size of the original M9.

Several newer rootstocks produce trees about the size of M9. Trees on **Mark** rootstock grow

Photo 16-5. High-density apple orchards generally need support, such as this trellis.

Photo 16-6. High-density apple orchards may be staked instead, as with these stakes of electrical conduit pipe.
Photo 16-7. Most high-density apple growers plant in single rows, but some use double rows, as here, to cover more ground surface with apple trees. Each narrow tree is easily hand-harvested from one side.

slightly larger and may not need support on some soil types, but on tighter soils they do need support, especially if allowed to crop heavily from a young age. Mark resists fire blight and crown rot but is just as susceptible as M9 to drought and wooly apple aphids, with similar hardiness to zone 5. **P2, P22** and **Bud. 9** are all similar to M9 in pest susceptibility and need staking and irrigation, but all are cold hardy to zone 3. **Geneva 16** produces trees the size of EMLA 9, but with few burrknots or root suckers, and is hardy to zone 3, very resistant to fire blight and tolerant of apple replant disease.

Spur-type cultivars on G30 or on M9/MM111 or P2/Antonovka will also produce trees the size of M9. The interstem trees will not need staking and need much less irrigation. Non-spur type cultivars on **M27/MM111 or Bud. 491/MM111 interstem trees** also grow to the size of M9 trees, and need no support and less irrigation.

Ultra-high-density orchards. Bush-like super-dwarf trees are supported by trellises (Photo 16-8). In the last 10 years, growers have established plantings with 1000 to 2500 trees per acre in Washington, British Columbia and other major apple-growing areas. Spacing in single-row plantings ranges from 8 to 10 feet between rows and from 2 to 4 feet apart within rows. This ultra-high tree density requires very high costs for establishment, irrigation and fertilization, but begins yielding in the second growing season (18 months after planting) and produces very high yields.

In Europe and other areas the trellises are sometimes grouped into beds, with 3 to 6 rows of trees spaced close together in each bed and an aisleway between beds to allow harvesting, pruning and other operations. This bed system has not been used much in North America but is often used now in parts of Europe, where specialized equipment has been developed to hedge-prune, spray and sometimes harvest over the tops of the beds.

M27, the first and most widely used super-dwarfing rootstock, produces trees just 15% the size of standard, or 5 to 6 feet tall. It is hardy to zone 6, very susceptible to drought and wooly apple aphids, moderately susceptible to fire blight and resistant to crown rot. **P16** produces a similar-size tree, also resistant to crown rot, and is hardy to zone 4 but is more susceptible to fire blight. **Bud. 491** produces a slightly larger tree and is susceptible to crown rot but is extremely hardy, to zone 3.

Selecting Cultivars. Most apple growers choose cultivars based on their current popularity with wholesale buyers for supermarkets. Over the last 50 years this has caused more concentrated

Photo 16-8. Ultra-high density apple orchards have trees planted as close as 2 feet apart. (Photo by Dave Mather, www.meaford.com)

planting of fewer cultivars, and has sometimes led to gluts of certain older cultivars as newer ones eclipse them in popularity. Cultivars with larger fruit size, strong skin coloration and smooth finish have proven most popular, but every few years a new cultivar gains popularity due to its unique flavor, despite lack of size or coloration. Over the next few years, nurserymen tend to find and propagate sports of that cultivar that produce larger or better-colored fruit, then growers increasingly plant those sports, called strains, instead of the original cultivar even if the strains are less flavorful. Current U.S. grading standards pay growers more for apples with greater red color, even on cultivars which are naturally bicolored or yellow with a red blush, and these grading standards aren't likely to change in the next few years.

Table 16-2 details the adaptation, disease susceptibility and other attributes of some of the currently most popular apple cultivars, and lists some popular strains of each cultivar. 'Delicious' has been the main cultivar for the last 70 years or so, but plantings are declining as it has fallen in favor with customers. 'Delicious' is an average producer, yielding about 22 tons per acre in good years and about 17 tons per acre in off-years. High yielders like 'Golden Delicious' and 'Jonagold' produce about 26 tons per acre in good years and about 20 tons per acre in off-years in the Pacific Northwest region. Some of the newer cultivars that have been widely planted in recent years, such as 'Gala' and 'Fuji', produce lower yields of about 18 tons per acre in good years and 13 tons per acre in off-years.

Growers should make sure that a cultivar will ripen well in their area before planting it, and can save greatly in pest control costs by choosing cultivars that aren't highly susceptible to serious

Table 16-2.

COMMON COMMERICAL APPLE CULTIVARS

In approximate order of ripening (varies with area).

	Best in	Fruit				Resistance to									Cold
Cultivar	Zones	Color	Flavor	Texture	Size	Apple Scab	Cedar Rust	Fire Blight	Powdery Mildew	Black Rot	White Rot	Bitter Rot	Bloom Time	Ripening Season	Storage Life
Lodi	4-7	green	tart	crisp	large	MS	VS	VS	R	S	S	S	early	very early	1 mon.
Zestar	3-6	red	sweet/tart	crisp	medium	MS	MR	MR	MR	S	S	S	mid	very early	2 mons.
Sansa	5-7	red/yellow	sweet/tart	crisp	small	MR	R	S	MS	S	S	S	mid	very early	2 mons.
Gala	5-8	red/yellow	sweet	crisp	medium	MS	S	S	MS	MS	MS	S	mid	early	1 mon.
McIntosh	4-6	red/green	tart	tender	medium	VS	R	S	S	VS	MR	MR	early-mid	early-mid	4 mons.
Honeycrisp	3-6	red/yellow	sweet, rich	crisp	medium	MS	MR	MR	MR	S	MR	VS	mid	early-mid	6 mons.
Cortland	4-7	red	sweet/tart	tender	large	VS	S	S	S	VS	MR	MR	mid	early-mid	5 mons.
Jonathan	5-8	red	spicy tart	crisp	medium	S	VS	VS	VS	S	VS	S	mid	early-mid	5 mons.
Spartan	4-7	red	sweet/tart	coarse	small	S	R	S	R	MS	MR	MR	mid	middle	6 mons.
Golden Delicious	5-8	yellow	sweet	coarse	medium	S	VS	MS	S	MS	S	S	mid-late	middle	5 mons.
Delicious	5-8	red	bland	coarse	med/large	MS	VR	R	R	MR	MR	MR	mid	middle	6 mons.
Empire	4-7	dark red	sweet/tart	crisp	small	VS	R	MR	MS	S	VS	VS	early-mid	middle	6 mons.
Jonagold	5-8	red/yellow	sweet, rich	crisp	large	S	S	MS	VS	S	VS	S	mid	middle	7 mons.
Melrose	5-8	red	sweet/tart	crisp	large	S	S	MS	S	S	VS	S	late	middle	7 mons.
Idared	4-7	red	bland	firm	med/large	S	S	VS	S	VS	S	VS	early-mid	middle	7 mons.
Northern Spy	4-7	red/green	tart	hard	medium	MS	S	MS	S	VS	S	S	late	mid-late	7 mons.
Cameo	5-8	red/yellow	sweet/tart	crisp	large	S	MS	MS	MR	S	VS	S	mid-late	mid-late	8 mons.
Mutsu	5-9a	yellow	bland	coarse	large	VS	VS	S	VS	S	S	VS	mid	mid-late	6 mons.
Suncrisp	4-7	red/yellow	sweet	crisp	large	S	MS	S	MS	S	MR	VS	mid	mid-late	7 mons.
Stayman	5-8	red	tart	firm	med/large	VS	MR	MR	MS	VS	VS	S	mid	late	7 mons.
Winesap	5-8	red	tart	firm	medium	VS	MS	MR	MR	S	S	MS	late	late	5 mons.
York Imperial	5-8	red/green	bland	hard	large	VS	S	VS	MS	S	S	MR	late	late	6 mons.
Rome Beauty	5-8	red	bland	very firm	large	VS	VS	VS	S	S	VS	MS	late	late	7 mons.
Braeburn	6-8	red/yellow	sweet/tart	hard	medium	S	VS	MS	VS	S	VS	S	early	late	7 mons.
Fuji	6b-9a	red/yellow	sweet	crisp	large	VS	MR	VS	R	VS	MS	S	mid-late	very late	8 mons.
Granny Smith	6b-9a	green	tart	crisp	large	MS	R	VS	S	S	S	VS	mid	very late	8 mons.
Pink Lady	6b-9a	pink	sweet/tart	crisp	med/large	S	S	VS	MS	S	S	S	early-mid	very late	7 mons.

Disease resistance ratings: VR = very resistant, R = resistant, MR = moderately resistant, MS = moderately susceptible, S = susceptible, VS = very susceptible

Average start of ripening in northern CA, southern IL and western VA: very early: early to late July, early: early to late Aug., early-mid: late Aug. to early Sept., middle: early to late Sept., mid-late: late Sept. to early Oct., late: early to late Oct., very late: late Oct. to early Nov. Ripening times will be about 2 weeks earlier in central AR and northern GA, about 1 week earlier in central CA and western NC, about 1 week later in southern MI and south-central PA, about 2 weeks later in western OR and central WI, about 3 weeks later in western WA and central NY, and about 4 weeks later in central MN and central VT, where late-ripening cultivars can't be grown.

Notes on Cultivars:

Lodi, an early summer cooking apple, produces heavily and needs dwarfing rootstocks to control excess vigor.
Zestar, from MN, is very cold-hardy but does poorly where summers are hot. Crops annually.
Sansa, from Japan, has fine fruit but trees have a genetic mottle that weakens trees if viruses are present, so order virus-free trees. VS to mites.
Gala, a high-quality early apple for fresh eating, is yellow with red stripes or blush, though more red strains are now grown.
McIntosh is still widely grown in New England and NY despite pre-harvest drop, very disease-prone. 'Pioneer Mac' strain drops less.
Honeycrisp, from MN, is very cold-hardy but does poorly where summers are hot. Don't delay harvest, or off-flavors develop; store at 35 F.
Cortland is a McIntosh-type apple with very large fruit that hangs very well on the tree.
Jonathan, popular in the Upper Midwest, has spicy flavor but is very disease-prone.
Spartan is a McIntosh-type apple but more firm, grown in the Northeast and Pacific Northwest. It suffers severe pre-harvest drop.
Golden Delicious, the standard yellow apple, can russet badly if cool, wet weather prevails after bloom. 'Smoothee', 'Gibson' strains russet less.
Delicious, still the most widely-grown red apple despite bland flavor, has many strains that color early.
Empire, a McIntosh x Delicious cross from NY, has excellent eating quality. Fruit hang well on tree.
Jonagold, a Jonathan x Golden Delicious cross, is popular in Europe and the Pacific Northwest. A triploid, it has sterile pollen and large fruit.
Melrose, a Jonathan x Delicious cross from OH, has high eating quality with the sugars of Delicious and the spice of Jonathan.
Idared, from ID, bears very early and heavily. Grown for processesing and fresh market despite bland flavor, VS to sooty blotch.
Northern Spy, an older cultivar with fine flavor, starts bearing very late so needs dwarfing rootstocks to induce earlier bearing.
Cameo, from WA, is a semi-spur tree with good vigor and precocity, needs heavy thinning. High-quality fruit resist scald, watercore.
Mutsu, a Japanese triploid with sterile pollen, has fruit like Golden Delicious but much larger, more bland, with little russet but S to blister spot.
Suncrisp, a Golden Delicious x Cox Orange Pippin cross from NJ, needs a warm season for best flavor. Thin heavily to avoid biennial bearing.
Stayman, or Stayman Winesap, has larger, firmer and less tart fruit than Winesap but fruit can crack badly. A triploid, it has sterile pollen.
Winesap a popular cultivar in Mid-Atlantic states, has tart fruit with a vinous flavor, less cracking than Stayman. Also a triploid with sterile pollen.
York Imperial is grown for processing in Mid-Atlantic states.
Rome Beauty is still widely grown in the East for processing. 'Spur Red Rome' strain is compact, needs much less pruning.
Braeburn, a late New Zealand apple, stores well. Spur-type tree has low vigor. It attracts mites, is prone to frost damage, watercore.
Fuji, a high-quality, very late Japanese apple that stores well, bears biennally unless heavily thinned. Prone to watercore (preferred in Japan).
Granny Smith, a very late, tart green Austrailian apple that stores well, is popular in CA and other areas with a long growing season.
Pink Lady, another very late, New Zealand apple with a unique color that stores well, does best in warm areas.

Table 16-3.

DISEASE-RESISTANT APPLE CULTIVARS

In approximate order of ripening (varies with area).

Cultivar	Best in Zones	Fruit				Resistance to							Bloom Time	Ripening Season	Cold Storage Life
		Color	Flavor	Texture	Size	Apple Scab	Cedar Rust	Fire Blight	Powdery Mildew	Black Rot	White Rot	Bitter Rot			
Pristine	5-7	yellow	mild, sweet	crisp	medium	VR	S	MS	VR	MS	MS	VS	mid	very early	4 weeks
Williams' Pride	3-7	dark red	sweet/tart	crisp	large	VR	MR	R	R	MS	MS	VS	mid	very early	6 weeks
Redfree	4-6	red	sweet	coarse	medium	VR	VR	MS	MR	VS	VS	VS	mid	very early	5 weeks
Wynooche	5-7	red	sweet/tart	tender	medium	VR	R	MS	MR	S	MR	S	early	very early	3 weeks
Dayton	5-8	red	sweet/tart	crisp	medium	VR	MS	MR	MR	MR	MR	MR	mid	early	2 mons.
Blairmont	6-9a	red	mild, sweet	coarse	medium	R	S	VR	R	R	MR	MR	late	early	2 mons.
Prima	5-7	red	mildly tart	crisp	medium	VR	VS	MR	R	R	S	S	early	early	1 mon.
Galarina	5-8	red blush	sweet	crisp	medium	VR	S	R	R	MS	MS	S	mid	early-mid	4 mons.
Red Baron	3-6	red	tart	crisp	large	MR	MS	MR	S	MR	MR	S	mid	early-mid	3 mons
Chehalis	5-8	yellow	sweet	crisp	large	R	S	MR	MR	MR	S	S	early	early-mid	2 mons.
Jonafree	5-8	red	sweet/tart	crisp	medium	VR	VS	R	S	MR	R	MR	mid	early-mid	3 mons
Belmac	4-7	red	sweet/tart	crisp	medium	VR	MR	S	MR	MR	S	S	mid	early-mid	3 mons.
Liberty	4-7	red	sweet/tart	crisp	medium	VR	VR	R	R	MS	MR	MS	early	middle	3 mons
Freedom	4-7	red	spicy tart	coarse	medium	VR	S	MR	R	MS	MS	S	late	middle	2 mons.
Priscilla	5-8	red	mild, sweet	coarse	medium	VR	VR	MR	R	R	S	S	mid	middle	1 mon.
Moira	4-7	dark red	sweet/tart	coarse	small	VR	R	S	S	MS	S	S	mid	middle	3 mons.
Keepsake	3b-6	red	tart	hard	medium	MR	MR	MR	MS	MR	MR	S	mid	mid-late	6 mons.
Novaspy	4-7	yellow	tart	hard	large	VR	S	R	MR	MR	S	S	late	mid-late	5 mons.
Sundance	5-8	yellow	sweet/tart	crisp	large	VR	VR	R	MR	S	S	MS	late	mid-late	6 mons.
Juliet	5-8	red	mild, sweet	crisp	large	VR	MS	R	R	MS	MS	MS	early	mid-late	9 mons.
Florina Querina	4-7	red	mild, sweet	crisp	small	VR	S	MR	MS	MS	MS	S	mid	late	6 mons.
Enterprise	5-8	red	spicy sweet	crisp	medium	VR	R	R	MR	MR	MR	MS	mid	late	6 mons.
GoldRush	5-8	yellow	sweet/tart	crisp	medium	VR	S	MR	MS	MR	MS	MS	late	very late	7 mons.

Disease resistance ratings: VR = very resistant, R = resistant, MR = moderately resistant, MS = moderately susceptible, S = susceptible, VS = very susceptible

Average start of ripening in northern CA, southern IL and western VA: very early: early to late July, early: early to late Aug., early-mid: late Aug. to early Sept., middle: early to late Sept., mid-late: late Sept. to early Oct., late: early to late Oct., very late: late Oct. to early Nov. Ripening times will be about 2 weeks earlier in central AR and northern GA, about 1 week earlier in central CA and western NC, about 1 week later in southern MI and south-central PA, about 2 weeks later in western OR and central WI, about 3 weeks later in western WA and central NY, and about 4 weeks later in central MN and central VT, where late-ripening cultivars can't be grown.

Notes on Cultivars:

Pristine is a high-quality summer apple for cool-summer areas, little sooty blotch even under low-spray programs. Biennial bearer.

Williams' Pride has full flavor, bears young & heavily annually, little sooty blotch. Not for hot regions: softens, skin & flesh brown.

Redfree has very good flavor in Northern areas but gets soft quickly where summers are hot. Prone to biennial bearing.

Wynooche, adapted to the Pacific Northwest, resists anthracnose but suffers pre-harvest drop.

Dayton has a spicy flavor like 'Jonathan' but less tart, is a dependable annual producer.

Blairmont, with fruit like 'Delicious', does well in the hot, humid Southeast, with good resistance to all fruit rots.

Prima can't tolerate cold winters or hot summers.

Galarina has fruit like 'Gala' but resists scab, ripens a week later. It needs heavy fruit thinning to maintain fruit size.

Red Baron, from MN, does well in Northern areas but not where summers are hot.

Chehalis, adapted to the Pacific Northwest, has fruit like 'Golden Delicious' but disease-resistant.

Jonafree has spicy fruit similar to 'Jonathan'. Tip-bearer, may suffer from pre-harvest drop, biennial bearing.

Belmac, from Quebec, has fruit like 'Spartan' but larger disease-resistant. Suited to Northern areas.

Liberty is the highest-quality scab-immune cultivar with fruit like 'McIntosh'. Needs heavy thinning. Attracts Japanese beetle.

Freedom has spicy flavor that some love, some hate. Needs summer black rot control. Severe storage scald if picked too soon.

Priscilla has lively flavor that intensifies in storage, but fruit soften and drop fast. Thin heavily to avoid biennial bearing.

Moira, from Ontario, is a McIntosh-type apple for the Northeast.

Keepsake does well in Northern areas, stores very well, tastes best after stored 3 mons.

Novaspy has fruit like 'Northern Spy' but resists scab and starts bearing sooner.

Sundance fruit have a rich, spicy flavor but a stem-end russet, hang well on tree, can harvest over a 6-week season.

Juliet is a semi-spur type tree, very productive, crops annually on sturdy branches, little blind wood, no pre-harvest drop.

Florina Querina, from France, has 'Jonathan' as one parent but very crisp fruit has less acid.

Enterprise has large fruit during 'Rome Beauty' season but juicier, better flavor in warm areas, hangs well.

GoldRush may russet and crack severely in humid areas; avoid by harvesting early and storing a few weeks to ripen like a pear.

Table 16-4.

COMMON CRABAPPLE CULTIVARS

In approximate order of bloom for pollenizers and ripening for fruiting cultivars (varies with area).

				Resistance to							
Cultivar	Best in Zones	Use	Fruit Diameter	Apple Scab	Cedar Rust	Fire Blight	Powdery Mildew	Leaf Spot	Bloom Time	Ripening Season	Comments
Manchurian crab	3-7	pollenizer	small	MR	R	R	MR	MR	early-mid	no harvest	heavy annual blooms
White Angel	5-8	pollenizer	small	VR	R	VR	VR	R	middle	no harvest	heavy annual blooms
Snowdrift	5-8	pollenizer	small	MR	R	S	VR	MS	mid-late	no harvest	heavy annual blooms
Whitney	3-8	spiced, sauce	2"	MR	MR	MS	R	MR	early	early	bears young, sweet-tart fruit
Dolgo	2-7	jelly	1.5"	MS	VR	R	VR	MR	very early	early	biennial bears, drops fruit
Centennial	3-8	sauce, spiced	2"	MR	R	MR	R	R	early	early-mid	compact tree, sweet-tart fruit
Ralph Shay	4-8	jelly	1.25"	MR	S	R	R	MS	early	early-mid	no pre-harvest drop
Chestnut	3-8	sauce, spiced	2"	MR	R	R	R	R	early	middle	nutlike flavor, drops fruit
Evereste	5-8	spiced, sauce	1.75"	VR	R	MR	R	MS	middle	middle	compact tree, no pre-harvest drop

Disease resistance ratings: VR = very resistant, R = resistant, MR = moderately resistant, MS = moderately susceptible, S = susceptible, VS = very susceptible

Average start of ripening in northern CA, southern IL and western VA: very early: early to late July, early: early to late Aug., early-mid: late Aug. to early Sept., middle: early to late Sept., mid-late: late Sept. to early Oct., late: early to late Oct., very late: late Oct. to early Nov. Ripening times will be about 2 weeks earlier in central AR and northern GA, about 1 week earlier in central CA and western NC, about 1 week later in southern MI and south-central PA, about 2 weeks later in western OR and central WI, about 3 weeks later in western WA and central NY, and about 4 weeks later in central MN and central VT.

diseases in that region. Some diseases, such as apple scab and fire blight, do not occur in many arid Western regions, so apple growers in those areas may grow cultivars susceptible to these diseases without additional pesticides. Where such diseases occur, such as along the humid Pacific Northwest coast, growers can greatly reduce costs by choosing cultivars resistant to the prevalent diseases. Where spring frosts often threaten, choosing cultivars that bloom later can avoid the need for frost protection equipment.

Direct marketers often find customers eager to try different apple cultivars than they can find in supermarkets (Photo 16-9). Such growers have the opportunity to grow many different apple cultivars, including some excellent ones released in the last 40 years that resist diseases important in many areas.

Breeding for resistance to apple scab, a serious threat to apples in the northeastern and north-central parts of North America, began in 1942 in Illinois by then-graduate student C. Frederick Hough. He cooperated with other researchers at University of Illinois and at Purdue University in Indiana, then moved to Rutgers University in New Jersey where he continued this cooperative work. The program expanded to select for resistance to additional diseases, and other programs selecting for disease resistance in apples began at Cornell University in New York, in Nova Scotia, Quebec, France and later Australia. Today, apple growers may choose from over 2 dozen high-quality cultivars with good resistance to several important diseases (Table 16-3), which can cut the need for pesticide sprays, where these diseases occur, from 15 or more per season to just 3 or 4.

In past centuries many apple cultivars were grown specifically for processing into hard (alcoholic) or sweet (non-fermented) cider, but today most processing apples are simply those which didn't make the grade as fresh-market apples. Some growers plant "antique" apples for nostalgia, but most of the older cultivars do not ship or store as well as modern cultivars, and the modern disease-resistant apples have much greater resistance to most diseases than do most older apples. Retail growers who want to try older apples should test just a few trees of each cultivar to see if customers accept them, and how well they grow in that location, before planting more. Retail growers may also raise a few fruiting crabapples (Table 16-4) for farm processing, or for fresh sales to customers who like to make jellies, sauce or spiced crabapples.

Photo 16-9. Retail growers find customers like to try apple cultivars they can't find in grocery stores, including disease-resistant cultivars that need far fewer pesticide sprays.

SITE SELECTION, PREPARATION AND PLANTING

Apples bloom later than most other fruit trees so don't need the least frost-prone sites, but do need good air drainage and soils with good water drainage. Apples do well on a range of soil textures and soil fertility levels, so long as growers choose rootstocks and cultivars suited to that soil texture and fertility. Apples thrive in a fairly wide soil pH range, from 6.0 to 8.0. If pH levels are less than 6.0, work in lime at least 6 months before planting time, and if pH levels are above 8.0 work in sulfur.

Ideally a site for a new apple planting was previously planted in a grass pasture or hay crop, or a small grain crop. Unfortunately, many apple growers are now planting high-density orchards on land from which old low-density orchards were removed. This has led to an increase in apple replant disease, caused by the buildup of several root-rotting soil fungi and sometimes lesion nematodes. Removing as much of the root system as possible when old orchards are taken out, then cover cropping over the winter with three quick crops of 'Eltan' soft white winter wheat, reduces these soil pathogens and increases growth and survival of trees planted the next spring. Such cover cropping can avoid the need to fumigate the soil when replanting an old apple site.

Take soil samples for pH and soil nutrients at least 6 months before planting, with separate samples for topsoil (to 8 inches deep) and subsoil (8 to 16 inches deep). Get soil analyzed for levels of phosphorous (P), potassium (K), calcium (Ca), magnesium (Mg), and any micronutrients that are commonly lacking in the area, such as zinc in alkaline soils. The soil test should also provide the soil's cation exchange capacity (CEC), which is its ability to store and release to plants positively charged ions of Ca, Mg, K and others, and the proportion of the CEC taken up by each nutrient.

Amend topsoil before planting, if needed, so that Ca makes up 67% of the CEC and Mg makes up 13% of the CEC. If lime is added to raise pH it will provide Ca, and if Mg is needed use dolomitic lime. Add K if soil tests show less than 300 ppm K in topsoil with a low CEC and less than 450 ppm K on topsoil with a high CEC, adding at least 120 pounds of K_2O per acre. Add P if soil tests indicate a low to moderate soil P level.

Orchard layout and tree support. Most apple growers today still plant in single rows. Rows running north-south intercept the most light. Plant freestanding trees on erodable slopes on the contour to reduce soil erosion. Most new apple orchards will be trellised, so need straight rows.

Most new orchards are high- or ultra-high-density, with rows 8 to 14 feet apart and trees 2 to 7 feet apart within the rows. Some plantings use individual posts to support each tree, but most use a single-wire trellis similar to that used for single-curtain grapes. Install 10-foot-tall posts 2 feet into the ground and about 50 feet apart down each row, to support the wire 7 to 8 feet off the ground. Growers install a pressure-treated wood stake of 2-inch diameter, or ¾-inch electrical conduit pipe, next to each tree and tie it to the wire, to provide support for each tree. Some growers use a trellis with 4 or 5 wires instead of the individual stake or pipe at each tree, but this limits access.

If the soil is somewhat poorly drained, build raised beds before planting, otherwise plant on flat ground. Raised beds increase root exposure to winter injury, so avoid them in colder regions unless necessary to improve drainage.

Providing for cross-pollination. Most apple cultivars require cross-pollination from a different cultivar of apple or crabapple. The two cultivars' bloom times must overlap, so plant early-blooming cultivars near others that bloom early and late bloomers in blocks with others that bloom late. Some apple cultivars are triploids, with three sets of chromosomes instead of the usual two; these produce desirably larger fruit but have sterile pollen. When growing triploids with other apple cultivars, interplant the triploid with two different diploid cultivars, and make sure all three cultivars bloom at similar times. Triploid cultivars can be grown in solid blocks if crabapples are also planted to provide pollen for them.

Retail growers usually plant several to many cultivars that provide abundant cross-pollination. Bees tend to fly straight down a row, so it's better to change cultivars every 50 to 100 feet within each row than to plant solid rows of each cultivar. Plant cultivars in blocks across the rows to enable harvesting the same cultivar from both sides of most aisleways.

Wholesale growers often plant large blocks of a single cultivar, and plant crabapples as **pollenizers** (trees that provide pollen) which are usually not harvested. Growers usually choose two different crabapple cultivars, one blooming during the early part of the apple cultivar's bloom period and the other blooming during the later apple bloom period, and alternate these every 50 to 100 feet down each row. Crabapple pollinizers thus make up 5 to 10% of each apple block (Photo 16-10).

Choose apple or crabapple pollenizer cultivars that have about the same degree of resistance to the area's major apple diseases as does the main apple cultivar in the block. This will prevent the need for extra sprays for the pollenizer trees. Crabapple pollenizer cultivars should also bloom heavily each year and have white petals like apples, rather than pink, since foraging bees don't readily switch to different-colored blossoms. All crabapple pollenizer cultivars listed in Table 16-4 bloom heavily each year and have white flowers. The fruiting crabapples listed also have white blossoms, but some tend to biennial bear.

Planting. Plant in fall wherever possible, since apples are cold hardy and the roots of fall-planted trees will grow over the winter and become well established by spring. In USDA zone 5 and colder regions, plant in early spring. Use a tree planter or

Photo 16-10. Wholesale apple growers often plant a single cultivar in a large block to make harvest easier, and interplant crabapples (usually not harvested) to provide cross-pollination.

dig holes with a tractor-mounted auger of at least 18-inch diameter, and break up sides of each hole with a shovel before planting.

Plant trees on dwarfing or semi-dwarfing rootstocks with the graft union a few inches above the soil surface, so the scion doesn't root and overcome the rootstock's dwarfing effect. Plant interstem trees with the rootstock fully buried and the bottom 2 inches of the interstem section underground also. This minimizes problems with burrknots on the rootstock, especially when MM111 is used as the rootstock as it usually is with interstem trees.

Water in each new tree, or irrigate a new planting immediately after finishing planting for the day. When planting in fall, place white plastic tree guards around each tree to protect from sunscald and rodent damage over winter. Wait until the next spring to prune back fall-planted trees, to avoid cold damage to new pruning cuts, but prune spring-planted trees right after planting. Follow specific pruning guidelines for the training system chosen, outlined below and in chapter 4.

Generally, head unbranched trees (whips) back to about 30 inches above the soil to induce branching. With branched (feathered) trees, head the central leader back to 10 to 16 inches above the uppermost branch, and remove any narrow-angled branches or those arising less than 24 inches above the ground. Select 4 or 5 well-spaced, wide-angled branches for the first tier of scaffolds and remove others. Cut selected branches back by 1/4 to 1/3 if they exceed 24 to 30 inches in length, cutting to an outward bud where the branch begins to turn upward.

TREE TRAINING SYSTEMS

Researchers and growers have developed many different training systems for apples. All of these systems aim to increase light penetration within the tree canopy, so that more fruit buds develop and trees produce greater yields per acre. Each training system works best for a certain size of tree and orchard density. Choosing a compatible training system and rootstock are the two keys to making higher-density orchard systems work.

Older **low-density orchards** have usually been trained according to the **modified leader** system, described in chapter 4, which creates a vase-shaped tree to allow more light penetration within the wide tree canopy. Growers of medium- and high-density orchards aim for a narrow pyramidal or Christmas tree shape, to minimize shading from upper branches in these shorter, narrower trees. **Medium-low and medium-high density orchards** have been trained mainly to the **central leader system**, also outlined in chapter 4.

Many trellised or staked **high-density orchards** have been trained to the **vertical axis** system, also described in chapter 4, and this system is the most popular one for new plantings today (Photo 16-11). Growers also use slender spindle training, described below, for staked or trellised high-density orchards. Trellised high-density orchards may also be trained to a flat vertical plane with rows planted closer together, or to a flat horizontal plane such as the Lincoln Canopy, described below.

Ultra-high density systems aim for a narrow columnar tree, with no main side limbs, but just a straight trunk with many fruiting branches and spurs arising from it. The slender spindle system and super spindle system, both outlined below, are widely used.

Slender Spindle training. This system produces a pyramidal tree like a miniature central leader, maintained at a maximum size of 8 to 10 feet tall and 6 to 7 feet wide (Photo 16-12). Trees need a permanent stake or trellis for support, and workers tie the leader to this support every 15 to 18 inches. Develop one whorl of 4 or 5 permanent lateral limbs from side branches arising between 2 feet and 3 feet off the ground. Above that, only smaller branches are permitted.

Training at planting is the same as described above. In addition, whether trees planted are whips or feathered, slender spindle training requires increasing the development of lateral fruiting branches from the leader by bagging or snaking the leader. Bag or snake the leader each year until the

Photo 16-11. The vertical axis training system is the most popular one for new apple plantings today.

Photo 16-12. The slender spindle training system produces a pyramidal tree like a miniature central leader, maintained at a maximum size of 8 to 10 feet tall and 6 to 7 feet wide.

tree reaches its desired height after 2 to 4 growing seasons.

Bagging (Figure 16-3) is accomplished by placing a 3-mil polyethylene sleeve over the previous year's unbranched growth in mid-winter, 4 to 6 weeks before expected bud break. Bagging is only effective on 30 to 32 inches of leader, so cut back longer leaders to 32 inches above the highest branch. Tightly close the ends of the tube with bag ties or tape, and keep the bags on until the new lateral branches induced reach 1 to 2 inches long, about bloom time. Immediately after removing bags, apply a 250 ppm solution of Promalin™, a growth regulator that promotes branching, and an anti-desiccant to the previously bagged portion.

Snaking (Figure 16-4) is done several times during the growing season, starting when the leader's new growth reaches 18 inches long. Bend this new growth to a 45-degree angle and tie or tape it to the support stake. When the leader's extension growth reaches 18 inches, bend this new section to a 45-degree angle in the opposite direction, and tie or tape it to the support stake. This technique produces a zig-zag leader shape, and induces more lateral buds to grow. It's most useful for very vigorous cultivars that tend to grow 36 inches or more of leader per season, and allows the grower to use that leader growth rather than cut it back.

Once the desired tree height of 8 to 10 feet is reached, maintain that height by **weak-leader renewal**. Each year during dormant-season pruning, cut the leader back to a side branch that can be tied to the post to form a new, weaker leader. Maintain the bottom whorl of limbs at the desired length by heading each one back to a lateral branch each year, also. Remove any upper branches that grow larger than 1/3 the leader's diameter at the point just below the weak leader extension.

Super spindle training. This system produces a miniature spindle, or spindle bush, no more than 6 to 7 feet tall and 2 to 4 feet wide and trained to a trellis. Do not develop any side limbs, but develop many small, fruitful branches all along the leader from 2 feet off the ground to the top of the tree. If planting feathered trees, cut the side branches back to 12 to 18 inches to keep trees within the spacing allotted. Practice snaking or bagging on whips or on the leaders above branched trees.

These trees should fill their allotted space by the end of the 2 or 3 growing seasons. Maintain them in the same way as for slender spindle trees, by weak leader renewal and heading side branches.

Lincoln canopy training. This system allows mechanization of pruning and harvest by creating a flat, horizontal fruiting area like a table top about 6 feet off the ground (Photo 16-13). Equipment can operate partially under the trellis, so the canopy spreads out to intercept more light falling on the land, increasing yields. The Lincoln canopy is well adapted to mechanical harvest, because fruit fall only a very short distance to the catching system of the Lincoln harvester designed for this system.

This system requires a T-shaped trellis with 4 to 5 wires in a plane parallel to the ground, similar

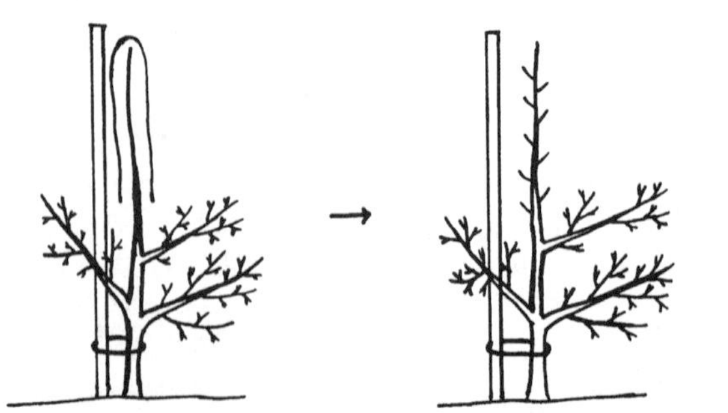

Figure 16-3. Bagging an unbranched leader with a 3-mil polyethylene sleeve, 4 to 6 weeks before expected bud break, can induce lateral branching.

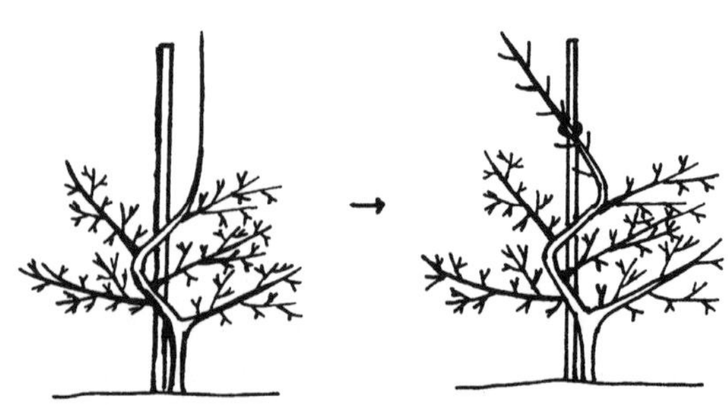

Fig. 16-4. Snaking a leader, by repeatedly bending it in opposite directions, is another technique to induce lateral branching.

Photo 16-13. The Lincoln canopy training system creates a flat plane of branches parallel to the ground, allowing mechanization of pruning and harvest.

to the traditional T-trellis used for table grapes. When training to this system, plant whips, since lower branches must be cut off. Develop 4 to 6 limbs from branches arising near the top of the trellis, and tie them to the trellis to train them horizontally. Cut off the central leader above the point where the limbs arise.

Side shoots growing up from the horizontal limbs grow excessively vigorously, so this system requires repeated mechanical mowing of these shoots during the summer. This system's increased summer pruning requirement, plus high capital costs of the harvesting equipment and trellis, has limited its use in North America. Growers in Australia and New Zealand use the Lincoln canopy more.

TREE TRAINING AND PRUNING.

All training systems involve selecting certain branches to remain as scaffolds or fruiting branches, and removing other branches. Chosen branches must often be headed back to induce lateral branching or spur formation, and must often be spread to develop at a desirable wide angle. Conduct this pruning and training both in the dormant season and during summer.

Each year, remove any branches arising below the lowest desired branches, and remove all suckers arising from the roots. Retain vigorous branches arising at a wide angle and in desired positions for the training system chosen. Head back the chosen branches when they reach the desired length, to induce growth of lateral branches or fruiting spurs. Many apple cultivars grow vigorously and require summer pruning each year of current season shoots when they reach 15 to 18 inches long, to keep the tree from growing too vegetatively and shading the fruit.

Spreading young branches. Even branches that arise from the trunk at a wide angle often turn up to grow vertically a foot or two from the trunk. Spreading these branches so that they grow at an angle between 30 and 45 degrees from horizontal will strengthen them and enable them to bear heavier crops sooner without breaking.

Apple growers use several methods to spread branches. When training to a trellis, tie branches to the trellis wires as they develop. New branches of freestanding trees may be spread with clothespins, and later in the season these branches can be spread with wooden strips notched at both ends (Photo 16-14). Weights can also be hung on the branches of freestanding trees to pull them down (like fruit would do), or branches can be tied to stakes set in the ground under the trees. Once a branch has hardened, remove the ties, spreaders or weights.

Growth regulators for tree training and pruning. Several chemical growth regulators can reduce the need for physical training and pruning. Promalin™, a combination of **6-BA and gibberellic acids 4 and 7**, promotes side branching of young trees, greatly reducing the need to head back the main branches by hand. However, do not apply it to trees on M9 or more dwarfing rootstocks, or to trees under stress from drought, winter injury or other problems.

Promalin™ may be applied as a high-concentration latex paint mixture with a paintbrush or sponge in early spring, or as a low-concentration foliar spray when new terminal growth reaches 1 to

Photo 16-14. Branches of many apple cultivars tend to grow too upright, and benefit from spreading while young in order to form a stronger, wider branch angle that can withstand heavy crop loads without breaking.

3 inches long. Apply it as a paint mixture at 5000 to 7500 ppm (0.2 to 0.33 pint Promalin™ per pint of white latex paint) to one-year-old wood only, when terminal buds begin to swell but before green tissues emerge, and thoroughly cover the bark surface where branches are desired.

For a foliar spray, use 1 to 2 pints Promalin™ per 100 gallons water to make a concentration of 125 to 500 ppm.

An **auxin**, 1-napthaleneacetic acid (**NAA**; trade name Tre-Hold Sprout Inhibitor), can control regrowth of root suckers and water sprouts. Many root suckers can develop around the bases of trees, and water sprouts are fast-growing vertical shoots that arise from scaffold branches and next to pruning cuts. Both types of growth are succulent so harbor fire blight, aphids and other pests, and both can shade fruitful branches. After pruning off old root suckers and making pruning cuts on scaffold branches, apply a 0.5 to 1.0% NAA solution to the pruning cuts to prevent regrowth.

NAA can be mixed with white latex paint and applied with a paintbrush to the pruning cuts, or can be applied as a spray. Once trees start bearing, wait until 4 weeks after petal fall to apply NAA to pruning cuts, so it won't also cause excessive fruit thinning.

Some young apple trees are slow to begin bearing, and **ethephon** (Ethrel™), which produces the hormone gas ethylene, can induce them to bear sooner. Apply Ethephon 4 to 5 weeks after budbreak to young trees that have not yet cropped, to induce or increase bloom the following year. Do not use ethephon on young trees that are weak or on M9 or smaller rootstocks, as it can cause too many blossoms to set and stunt tree growth.

Many apples growers have started using a new growth retardant, **prohexadione-calcium** (Apogee™) to greatly reduce pruning and reduce fire blight, a disease which thrives on lush new growth. Two sprays of this material per year have reduced dormant pruning time by 25 to 50% and practically eliminated summer pruning in commercial orchards, when applied to trees that were already well pruned.

Make the first application when terminal shoots are just 1 to 2 inches long (usually at late bloom or petal fall), because it takes 10 to 14 days after spraying for Apogee™ to begin to slow growth. Use 6 to 12 ounces of material per 100 gallons water for the first spray, and half that rate for the second spray, which can be made 3 weeks later or when the first signs of regrowth appear in the tops of the trees. Where excessive growth is located mainly in the tops of trees, a directed spray to just that area, with lower nozzles of the sprayer closed off, works well and saves money.

Use a water conditioner such as ammonium sulfate in the spray if hard water (high in calcium) is used, because hard water reduces Apogee™ effectiveness since the material contains calcium. Similarly, do not tank-mix this material with calcium or boron nutrient sprays, although it is compatible with pesticides. Under some but not all conditions, Apogee™ may increase fruit set, which may call for a higher rate of a chemical thinner or two applications of fruit thinners instead of one (see Fruit Thinning section below). Do not use Apogee™ on the cultivar 'Empire', as it can cause fruit cracking and cork formation on that cultivar.

GROUNDCOVER MANAGEMENT AND FERTILIZING YOUNG TREES

Growers in high-rainfall areas usually maintain a grass sod between trees, and use herbicides to control all vegetation within a 2- or 3- foot-wide strip along tree rows. Growers in arid areas often grow winter cover crops between trees, and either cultivate in summer or grow summer cover crops, depending on soil moisture available and the need for soil organic matter.

Mulching within tree rows increases levels of soil nutrients, leaf nitrogen (N) and potassium (K) levels, and tree growth, but growers avoid mulches in areas where voles cause problems, because mulch harbors voles. Growers usually maintain a herbicide strip 2 or 3 feet wide across each row of trees, controlling all vegetation in this zone to minimize competition with apple roots for water and nutrients and reduce vole populations.

Apple growers are applying higher amounts of NPK fertilizers to young trees now than in the past, so trees will grow rapidly during the first and second growing seasons, and produce a commercial crop by the second or third year. Terminals of young trees should grow about 3 feet per year, and lateral branches should grow about 2 feet per year, during the first and second seasons.

Determine the amount of fertilizer to add by preplant soil tests, then take leaf samples in midsummer to guide fertilizer applications for next year. Once trees fill their allotted space and begin to bear, reduce fertilizer amounts so terminals grow just 15 to 18 inches per year and lateral branches grow just 12 to 15 inches per year.

POLLINATION, FRUIT THINNING AND GROWTH REGULATORS FOR CROP CONTROL

Apple flowers occur mainly on spurs, and to a lesser extent on lateral buds of one-year-old branches. Fruiting spurs continue to bloom and bear fruit for many years, but each particular spur tends to produce flowers every other year, a process called biennial bearing. Ideally, about half the spurs will bloom each year, and the other half will bloom the next year, creating fairly consistent yields from year to year. However, crop loss one year to a severe freeze or other problems can push an entire block of trees into a biennial bearing cycle, in which it yields excess crops of small fruit during the "on" year and very little during the "off" year. Apple growers must manage pollination, frost protection, crop thinning, tree nutrition and pest control to avoid such biennial bearing.

Many commercial apple growers place beehives in the orchard during bloom, to ensure adequate pollination. Where native bees are plentiful, this may not be necessary. Cool, cloudy weather during bloom reduces bee activity, especially that of honeybees, so under those conditions more hives are needed.

When favorable weather occurs during pollination, far too many fruit will set, so fruit thinning is needed. Apple growers generally rely on chemical thinning sprays, as several effective thinning compounds are available for apples. Selecting the proper rate and timing for chemical thinning is critical to profitable apple production. Excess thinning reduces yields, while inadequate chemical thinning will require follow-up hand thinning, which is much more expensive and will not increase return bloom the following year.

Climate and weather conditions greatly influence the amount of thinning that chemical thinners will cause, so follow recommendations for your area and for expected weather forecasts. Generally, growers in cool maritime areas need to apply higher rates than do inland growers where warmer weather prevails during bloom, but reduce rates if high humidity or frosty nights are expected. Also reduce chemical thinner rates or eliminate them if poor pollination conditions occur or if the previous year's crop was heavy, which often causes a lighter crop the next year. A computer program, Thin-it (www.tfrec.wsu.edu/Horticulture/applethin.html) can help growers determine concentrations for chemical thinners of apples.

Thinning chemicals. A few apple growers use **bloom-thinning chemicals**, salts or acids that burn the flowers' stigmas and styles, thus preventing fertilization and fruit growth. Bloom thinners include endothall (ThinRite™), Wilthin™, and Thinex™. However, bloom thinners do not work on some apple cultivars and in most cooler areas, and on other cultivars and in warmer areas they can cause excessive thinning. Therefore, most growers do not use bloom thinners except on certain compatible cultivars, and only when they anticipate a very heavy crop with no frosts during bloom. Apply bloom thinners at 65 to 80% bloom (Photo 16-15), or when 80 to 90% of the king blooms have already been pollinated, since these chemicals will remove lateral blooms that have not yet been pollinated but leave the king blooms that have already been fertilized.

Post-bloom thinners include several growth regulators and several pesticides, and many growers use both a growth regulator and an early-season pesticide to provide thinning. The most common growth regulator used for thinning is an **auxin**, 1-napthaleneacetic acid (**NAA**; trade names include Fruitone N, Fruit Fix 200 or 800, and Alphaspra 200) or its amide form, 1-napthaleneacetamide (**NAD** or AmidThin™). NAD is applied earlier, at 7 to 14 days past full bloom, when the largest fruit are 2-5 mm in diameter, but temperatures after spraying should rise above 50 F and preferably above 65F.

Do not use NAD on 'Delicious', as it can cause many abnormally small (pygmy) fruit. Apply NAA 15 to 25 days after full bloom, when the largest fruits are 10-15 mm in diameter and temperatures are expected to rise above 50 F for the next 5 days. Both NAD and NAA produce their effect from absorption of the sprayed solution, and fruit no longer absorb it once it dries, so choose a calm, humid day to apply either thinner so they will dry slowly and

Photo 16-15. Apply bloom thinners, where they can be safely used, at 65 to 80% bloom.

allow greater absorption. These auxins, applied at a rate of 2-25 ppm depending on cultivar, stimulate production of ethylene, a hormone gas that causes fruit abscission.

Another way to produce ethylene is to apply **ethephon** (Ethrel™), which breaks down into ethylene. Growers may apply both NAD and ethephon to certain hard-to-thin cultivars, and ethephon will also increase return bloom the next year when applied 6 weeks after bloom during an "on" year to biennial-bearing trees.

A third type of growth regulator thinner, the synthetic **cytokinen** 6-benzyladenine (**6-BA**; trade name Accel), increases fruit size as well as providing some thinning. Do not apply both NAA and 6-BA in the same year, as that can cause many pygmy fruit to form, especially on the cultivar 'Delicious'. Many states recommend using 6-BA in conjunction with the insecticidal thinner carbaryl, especially on hard-to-thin cultivars such as spur-type 'Delicious' and 'Fuji', since the thinning effect of 6-BA is weak.

Carbaryl (Sevin™), an insecticide often used to control early-season codling moth or other insect pests, is the most common pesticide used that also provides fruit thinning. Lime sulfur fungicide or oxamyl (Vydate™) insecticide, though less commonly needed, also provide some thinning if used between 5 and 30 days after full bloom.

Carbaryl may be combined with NAD or NAA in a single spray to increase thinning, but usually growers apply carbaryl as a separate spray later, after determining how much thinning the NAD or NAA spray did and how much additional thinning is needed, if any. Temperatures should rise above 70 F for the 5 days after applying carbaryl, as cooler temperatures cause poor thinning and small seedless fruit, but temperatures over 90 F cause excess thinning.

Carbaryl is highly toxic to bees, so apply it only 10 to 25 days after full bloom (Photo 16-16) when bees are not foraging in the orchard, and use the less toxic formulations Sevin XLR Plus or Sevin 4F rather than the more toxic Sevin WP. Carbaryl is also highly toxic to predator mites and can disrupt a biological mite control program. Minimize this problem by applying carbaryl only early in the season when additional thinning is needed (using a different insecticide for pest control when no more thinning is needed), and by directing sprays towards tops of the trees and avoiding spraying trunks and lower limbs. Oxamyl (Vydate™) insecticide, a carbamate like carbaryl, has similar thinning properties but less detrimental effects on predator mites, but can increase russeting on 'Golden Delicious', 'Stayman Winesap' and other cultivars prone to russeting.

Photo 16-16. Apply carbaryl as a thinner 10 to 25 days after full boom, when it also helps to control several fruit-feeding insects.

Cultivar differences in thinning response. 'Golden Delicious', 'Fuji', 'Gala' and several other cultivars with 'Golden Delicious' parentage often need more thinning than other cultivars to ensure adequate fruit size, especially when a "snowball bloom" occurs, meaning that a high percentage of spurs and lateral buds bloom in a single season Photo 16-17). A snowball bloom usually develops the year after the crop was greatly reduced by a severe frost or other problem, and if not thinned aggressively, an overly heavy crop will result and set

Photo 16-17. Apple growers must apply higher rates of chemical thinners when a "snowball bloom" occurs, as here.

up a cycle of biennial bearing. Use 2 to 3 thinning sprays to maintain annual production.

First use a bloom thinner, if one works for these cultivars in your area, then apply NAD and carbaryl when 80% of the flowers have reached the petal fall stage and fruit are 3-5 mm in diameter. A third thinning spray of NAA and carbaryl when fruit reach 10-15 mm in diameter provides more thinning, if needed after evaluating trees. To increase thinning even further and also increase return bloom the following spring (to avoid an overly light crop during the "off" year), add ethephon to the third thinning spray.

Spur-type strains of 'Delicious' are also more difficult to thin than non-spur 'Delicious'. When spur-type 'Delicious' bloom heavily, use both a bloom thinner and a spray of NAA and carbaryl at 10-15 mm fruit diameter. Do not use NAD on 'Delicious', and do not apply 6-BA during heavy-bloom years when NAA was needed along with carbaryl for thinning, as such applications could cause many pygmy fruit on 'Delicious'. Do not apply ethephon to 'Delicious' until at least 3 weeks after full bloom, as it can cause fruit with pinched calyx ends on this cultivar if applied earlier.

Other spur-type strains of cultivars, such as spur-type 'McIntosh', are harder to thin than their non-spur counterparts. Non-spur 'McIntosh', 'Empire', 'Cortland' and 'Northern Spy' are moderate in thinning response. Cultivars that thin easily include non-spur 'Delicious', 'Jonagold', 'Mutsu', 'Idared' and 'Jerseymac'. Apply just NAD or NAA to easily thinned cultivars, wait 14 days to determine how many fruit will drop, and then apply carbaryl or oxamyl only if needed, which is not probable.

With all cultivars, direct most of the thinning spray to the top two-thirds of the tree canopy, where higher light levels cause more flower buds to form. Avoid spraying lower and inside limbs, where fewer flowers form and those that do are more easily thinned. Use lower rates of chemical thinners for young trees, and for mature trees that are low in vigor or experienced stress the previous season from pests, nutrient deficiencies or overly wet or dry weather.

Growth regulators to increase fruit size or quality. The synthetic **cytokinen** 6-benzyladenine (**6-BA** or Accel™), mentioned above as a weak thinner, is more often applied to increase fruit size. Make 2 or 3 applications, 5 to 10 days apart, between petal fall and fruit diameter of 10-12 mm. Increasing fruit size requires only 10-15 grams of active ingredient per acre per spray, but double that rate is needed to provide some thinning along with fruit size enhancement, and 6-BA should be tank-mixed with carbaryl or oxamyl to provide adequate thinning under most conditions. Optimum size enhancement occurs when temperatures between 70 and 80 F and slow drying conditions prevail for several days after a 6-BA application.

Promalin™, a combination of **6-BA and gibberellic acids 4 and 7**, increases fruit length, and is especially used for 'Delicious' (Photo 16-18). Apply it when the king blossom is just opening and secondary flowers are at full ballon stage.

ProVide™, **gibberellic acids 4 and 7**, reduces russeting or cracking in apples. To reduce russeting of cultivars such as 'Golden Delicious' that are susceptible to russet, apply at petal fall and then 1 to 3 more sprays at 7- to 10-day intervals; 2 sprays are often sufficient. When applying more than 2 sprays, use an aggressive fruit thinning program to avoid reducing return bloom the next year. To reduce cracking of cultivars such as 'Stayman Winesap' that are prone to crack, apply 3 to 6 times at 2- to 3-week intervals later in the season, starting at least 3 weeks before fruit cracking is likely. If ProVide™ was already used to reduce russeting, it may not be used to reduce fruit cracking.

Ethephon may be applied 2 to 3 weeks before the normal harvest date and 1 to 2 weeks before the desired harvest date, to induce earlier skin coloration and ripening and allow earlier harvesting. Ethephon will also cause fruit to loosen and be more prone to preharvest drop, so be ready to quickly mechanically harvest the entire block sprayed with ethephon, or apply a stop-drop spray 4 or 5 days before applying ethephon.

Photo 16-18. Promalin™, a combination of cytokinen and gibberellic acid, was applied to these 'Delicious' apples to increase fruit length.

Stop-drop sprays. Some apple cultivars tend to drop ripening fruit before it is fully mature, due to internal production of ethylene. Lower concentrations of the same **auxins** that cause fruit thinning early in the season will, near harvest, counteract ethylene and act as stop-drop sprays to hold fruit on the trees until they can be harvested. Apply NAA 7 to 14 days before harvest, in 1 or 2 applications, and harvest at least 3 days after the last NAA spray.

Aminoethoxyvinylglycine (**AVG** or ReTain™) inhibits ethylene synthesis by apples, so also acts as a stop-drop, but must be applied several weeks earlier. It delays harvest and reduces watercore and internal browning, physiological disorders that can arise in cold storage, in addition to reducing preharvest drop and increasing fruit firmness. Apply it 4 weeks before anticipated normal harvest. Growers with large blocks of a few cultivars ripening all at once may use ReTain™ to spread out harvest, and designate apples treated with it for the longest cold storage periods. AVG is a natural product of fermentation, so is considered organic.

Neither NAA nor AVG stop-drop sprays work well if trees have been under drought stress or leaves have severe mite damage, so irrigate or control mites, if needed, before applying stop-drop sprays.

IRRIGATION AND CROP NUTRITION

Most older apple orchards in Eastern and Midwestern North America were not irrigated, because apples on standard or MM111 rootstocks are fairly drought-resistant. Orchards in arid Western areas have always relied on irrigation, and newer orchards on dwarfing rootstocks require irrigation even in high-rainfall areas.

Most new orchards use trickle irrigation to reduce water and energy costs. Most apple cultivars bloom late enough to avoid most spring frosts, so few growers install sprinkler irrigation for frost protection. However, in hot-summer climates, apples may benefit from evaporative cooling irrigation, so growers in the Southeast and Southwest may install sprinkler irrigation (Photo 16-19) to provide both spring frost protection and evaporative cooling in summer.

Evaporative cooling irrigation. High summer and early fall temperatures often reduce red color development, sugar accumulation and firmness of apples, and can cause sunburn of fruit, particularly with newer training systems that open tree canopies to higher light intensities. Some apple growers in hot-summer regions, such as the arid parts of the Pacific Northwest, California's Central Valley, the Southwest and the Southeast, install over-tree sprinkler irrigation to provide evaporative cooling to reduce these heat effects. The system must apply water at a rate of at least 30 gallons/minute per acre. Evaporation of the water applied cools the fruit, which reduces sunburn and hastens development of red color, sugars, and flavors. This permits earlier harvest, allowing growers to take advantage of higher early-season prices.

Thermocouples inserted in at least 4 fruit, one at each compass point of a tree, generally activate the irrigation system whenever any thermocouple exceeds 92 to 95 F and switch it off when all thermocouples fall below 90 F, in order to minimize the effects of excess midday heat. However, growers who don't have a problem with sunburn and are mainly trying to speed up red color development, to start harvesting as early as possible, may run evaporative cooling irrigation from ½ hour before to ½ hour after sundown. This rapidly cools fruit to near 70 F, the ideal temperature for red color development.

Evaporative cooling irrigation may add 12 to 18 inches of water per season to the irrigation needs for mature apple trees, compared to trickle irrigation, so an adequate water supply is crucial. High water quality is also very important, since the high evaporation rates can leave deposits of calcium carbonates, silicates and other salts on the fruit. Avoid calcium carbonate deposits by lowering water pH, if needed, to 6.5 to 6.6.

Photo 16-19. Most apple growers use trickle irrigation (foreground) to conserve water and energy costs, but overtree sprinkler irrigation (rear) can be useful for early-blooming, frost-prone cultivars or those that benefit from evaporative cooling during summer heat.

Where both frost protection and evaporative cooling irrigation are often needed, installing an over-tree sprinkler system can pay for itself. Evaporative cooling also can reduce certain pest control costs, as studies have shown that it washes European red mites off of leaves and reduces mite damage. However, evaporative cooling can also wash off pesticide residues and may increase incidence of bitter rot, where that fruit rot occurs.

Fertilizers and nutrient sprays. Apples benefit from application of specific nutrients at certain times during the growing season. Growers with trickle irrigation may apply these nutrients through fertigation. Some nutrients are best applied as foliar sprays, so the leaves take up the nutrients directly.

Many commercial apple orchards need phosphorous (P) at the beginning of each growing season, and if so growers can start fertigation or apply dry P fertilizers to the soil under the trees at full bloom. Nitrogen (N) is often not needed in bearing orchards on fertile soil, but if needed, add it with the P.

Many Western apple growers need to apply micronutrients, usually as foliar sprays for best absorption of the material into apple tissues. Alkaline soils often require zinc (Zn) and copper (Cu), and these nutrients are most effectively absorbed at the silver tip stage. Growers who need to apply fungicides at that time for disease control may choose those that contain Zn (ziram) or Cu (copper sulfate). Boron (B) is also often needed to combat cork spot, applied as 1 to 4 sprays which may be added to pesticide sprays applied from pink through petal fall. If manganese (Mn) is needed, apply it with pesticide sprays 7 to 10 days after petal fall.

Magnesium (Mg) demand in apples increaes after petal fall, and applying Mg through fertigation from petal fall until 40 days later can increase yields and fruit quality. A common fertigation rate is 3.5 to 4 pounds of actual Mg per acre per week over the 6-week period. Some cultivars with high Mg needs, on soils low in Mg, may also benefit from foliar Mg applications, such as adding up to 15 pounds of Epsom salts per 100 gallons to pesticides sprayed during this period. However, apples are also most susceptible to russeting during this period, so avoid mixing Epsom salts with any material in which the combination increases russeting, and avoid spraying during humid weather when sprays dry slowly.

Demand for potassium (K) and calcium (Ca) in apples rises sharply about 45 to 60 days after petal fall, as fruit start to rapidly increase in size. Some cultivars, such as 'Jonagold' require especially high amounts of K as well as of Mg. Adding 6 to 7 pounds of K_2O per acre per week through fertigation is helpful during this period. Growers who don't fertigate can apply dry K fertilizer under the trees before a rainfall or irrigation. Take leaf samples to determine if K rates are within the recommended range, and adjust K fertilization accordingly.

Apples have a high calcium (Ca) requirement, and getting enough Ca into the fruit from the soil can be difficult. Lack of Ca causes several fruit disorders, including cork spot (sunken brown spots that develop in the orchard) and bitter pit (small dried spots just beneath the fruit skin that develop in cold storage). Most growers who will store their crop longer than a couple weeks apply 2 to 4 foliar sprays of calcium chloride, adding this material to pesticides applied from 45 to 60 days after petal fall. Growers also apply calcium sprays to cultivars that are quite prone to cork spot, such as most 'Delicious' strains.

PEST MANAGEMENT

Many different diseases, insects, mites and mammalian pests can attack apples, but many of these can be avoided by choosing a location where the pests don't occur, or cultivars or rootstocks that resist the pests where they do occur. Apple growers in drier regions, where the majority of North American apples are produced, do not have to worry about many diseases or insects such as plum curculio.

Apple breeders have produced many quality cultivars that resist several important diseases of humid areas, and several serious insect pests of apples can now be controlled non-chemically with mating disruption. When pesticides are needed, a wide range of newer chemicals is now available that pose less risk to people and the environment than most older chemicals.

Diseases, insects and mammals that can kill trees. Growers must prevent these pests in new plantings because once damage occurs, it can't be remedied and trees often die. Careful choice of cultivars and rootstocks, modifying the soil before planting, and excluding tree-killing insects and mammals can prevent much damage.

Crown rot or **collar rot**, caused by the soil-borne fungus *Phytophthora cactorum* or other species in this genus, has killed many apple trees, especially on wetter soils. This fungus rots roots and trunks up to the graft union in crown rot, and trunks above the graft union in collar rot. Prevent this by preparing raised beds before planting in less well-drained soils, as described above under Site Preparation.

Also avoid rootstocks that are very susceptible to crown rot, such as MM106 or M26, on all but very well drained soils. During unusually wet seasons, early spring applications of fungicides effective against *Phytophthora*, such as metalaxyl (Ridomil™) or fosetyl-Al (Aliette™), can help prevent but not cure crown rot.

Fire blight, caused by the bacterium *Erwinia amylovora*, can also kill apple trees when susceptible cultivars or rootstocks are grown in areas where it occurs, which include many North American apple-growing districts except the driest ones. Avoiding highly susceptible cultivars (such as 'Jonathan', 'Gala', 'Rome Beauty', or 'Idared') and rootstocks or interstems (such as M26, M9 or Mark) is the most effective means of preventing fire blight in new plantings. Where susceptible rootstocks or interstems are used, prune off root suckers promptly at least once a week during early growth and bloom, or these suckers can become infected and blight can then spread into the trunk, killing the tree.

Delay nitrogen (N) fertilizer applications to susceptible trees until late spring, after the worst period for fire blight has passed, and keep N levels lower than usual to avoid stimulating lush growth that is more prone to fire blight. Prune off any blight-infected shoots, easily identified by the wilting tips that resemble a shepherd's crook (Photo 16-20), as soon as they appear and at least 12 inches below the visible disease margin, and sterilize pruning shears in a 10% bleach solution between cuts to avoid spreading the bacteria. Sprinkler irrigation can also spread the disease, so use trickle irrigation whenever possible.

Flower blossoms are especially susceptible to fire blight, so spray susceptible cultivars with a bacteriocide, such as streptomycin or copper sulfate, if daily mean temperatures during bloom exceed 60 F and humidity is high. Begin treatments when blossoms first start to open, and treat every 3 to 4 days as long as warm, humid weather persists. Many growers monitor daily temperature and humidity and type the data into a computer software program, MARYBLYT, which determines when infection events have occurred and predicts the risk of future infection events.

Apple replant disease, caused by several soil-borne fungi and sometimes exacerbated by lesion nematodes, has become more common in recent years as growers remove older low-density orchards to plant high-density apples. Minimize this disease by removing as much of the root system as possible when taking out an old apple orchard, then cover cropping the land to a grass crop such as 'Eltan' winter wheat before replanting apples. Soil fumigation may help in some cases, but if old apple roots are still present, the fungi causing this disease build up to damaging levels again within a few years after fumigation. Cover cropping stimulates beneficial soil microorganisms that compete against the harmful fungi, while fumigation kills such beneficial organisms.

Borer insects have also become more of a problem in recent years, as many dwarfing rootstocks are more susceptible to borers than seedling rootstocks. **Dogwood borer** (*Synanthedon scitula*), a clearwing moth larvae common in the eastern half of North America, feeds mainly on burrknot tissues of clonal rootstocks, so choosing a rootstock with lower tendency to form burrknots can prevent this. Planting susceptible rootstocks so the graft union is just 1 inch above the soil line inhibits development of burr knots. Mounding sand or soil over dwarfing rootstocks that were set higher, before dogwood borers begin laying eggs in late spring, can prevent much egg-laying and avoid the need for a trunk insecticide. Painting exposed burrknots with white interior latex paint also helps.

Photo 16-20. Fire blight, a bacterial disease kills flowers and young shoots of susceptible apple cultivars, causing dead terminals to bend like a shepherd's crook. Prune out blighted shoots promptly to prevent blight spread, which can kill trees. Blight-resistant cultivars are best wherever this disease occurs.

Several beetle larvae also bore into apple trunks. **Flatheaded apple tree borers** include *Chrysobothris femorata* (Photo 16-21) in the eastern half of North America, and *C. mali* on the Pacific coast. These often lay eggs near the graft union, and may be excluded by wrapping trunks with metal, fiberglass or nylon screen material so that it covers trunks from the soil line to above the graft union. Close the top of the screen with a wire twist-tie, and loosen this and retie it at least once a year to avoid girdling the tree.

Roundheaded apple tree borer (*Saperda candida*) attacks trunks near ground level so is harder to exclude with screening, although burying the material several inches into the soil around the trunk may work. Removing nearby Juneberries (*Amelanchier* species), the preferred host, may reduce this pest. Keeping young apple tree well irrigated minimizes problems with all types of borers. Trunk sprays of insecticides such as chlorpyrofos (Lorsban™) applied in late spring can also prevent borers.

Voles, including the woodland or pine vole (*Microtus pinetorium*) and the meadow vole (*M. pennsylvanicus*), attack apple roots and trunks just below the soil line, and can kill trees when populations are high. Keeping all vegetation controlled with herbicides in a 2-foot-wide strip down tree rows will greatly minimize vole damage, and do not use mulch around trees where voles are a problem. White plastic tree guards or screen cages used for borers can also prevent much vole damage if buried several inches into the soil. Where vole populations are high, many growers use poison baits such as zinc phosphide pellets, and some choose rootstocks, such as Novole, that resist vole damage.

Deer can strip the bark from young apple trees during winter feeding, which can girdle and kill the trees. Bucks rubbing on trees in fall can also remove bark (Photo 16-22), killing branches or entire trees. Problematic species include whitetail deer (*Odocoileus virginianus*) throughout most of North America, and mule deer (*O. hemionus*) in the western half of the continent. Where deer populations are high, an electric fence is necessary to prevent damage. Where populations are low to moderate, repellent chemicals can reduce damage by 34 to 46%, but must be reapplied monthly during fall and winter.

Diseases attacking foliage, fruit and branches. These diseases pose problems every year where they occur, so growers must choose cultivars resistant to them, practice strict orchard sanitation, and/or apply fungicides regularly.

Apple scab, caused by the fungus *Venturia inaequalis*, is the most important apple disease in northeastern North America, and is also serious in the humid Pacific Northwest and the southern Appalachian Mountain apple areas. Cool, humid spring weather favors the disease, which infects leaves and fruit with scabby brown lesions (Photo 16-23), and can cause defoliation and crop losses of 70% or more when it is severe. Apple scab is usually not a problem in drier Western apple districts, and the pathogen does not oversummer in most of the Southeast and lower Midwest where summers are hot. Where apple scab does occur, growers can completely avoid it by choosing cultivars with strong resistance, described in Table 16-3. Susceptible cultivars will need 12 to 14 fungicide applications per season, starting at green tip and continuing until a few weeks before harvest.

Time these fungicide applications based on temperature and leaf wetness data, and weather

Photo 16-21. Several borer insects, such as this flatheaded apple tree borer, can kill trees if not prevented with insecticide trunk sprays or tree wraps.

Photo 16-22. Winter deer feeding on bark can kill apple trees. Where deer populations are high, an electric fence around the orchard offers the best solution.

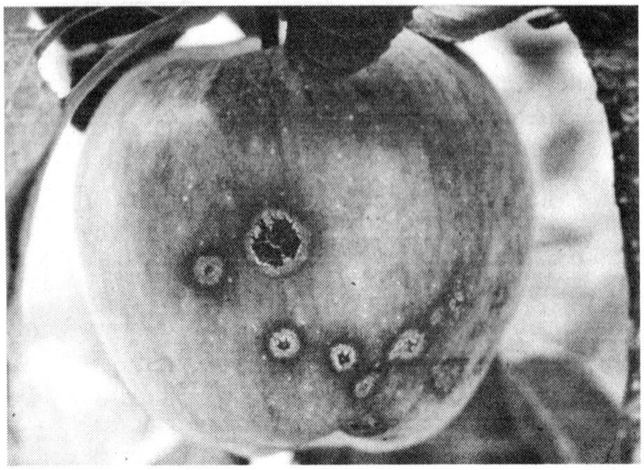

Photo 16-23. Apple scab can disfigure fruit or keep it from developing, and leaf infections can defoliate trees. Where this disease is serious, choose scab-immune cultivars or apply a diligent spray program of 10 to 15 fungicide sprays per season.

forecasts, so sprays are applied during infection periods for the fungus. Rotate using several fungicides with different modes of action, to avoid inducing the pathogen to develop resistance to a fungicide. David Rosenberger, plant pathologist at Cornell University's Highland, NY station, recommends a copper spray between silver tip and ¼-inch green to control fire blight as well as scab and the leafspot phase of black rot.

About 7 days later apply a contact fungicide such as mancozeb or captan, which prevent fungal infection if applied before spores germinate but won't stop infections that have already started, so spray just before a predicted rain and before tight cluster. During drier springs, contact fungicides alone may be adequate for all sprays up to petal fall, but years with heavy rains and extended wetting periods call for systemic fungicides with some curative action, that can kill infections that have already started when sprayers couldn't get into the orchard.

The sterol-inhibiting fungicides such as myclobutanil (Nova™) or fenarimol (Rubigan™) provide at least 3 days of curative activity, and Rosenberger has achieved good scab control when he applied them 10 days after an infection period. From tight cluster to petal fall, apply a sterol-inhibiting fungicide every 10 days or so, for a total of 2 or 3 applications. Include a contact fungicide with the last spray of a sterol inhibitor at petal fall or first cover (2 weeks later), to protect against fruit scab and summer fruit diseases.

Cover sprays continue every 2 weeks or so during the summer to protect against scab and summer fruit diseases such as black rot, sooty blotch and flyspeck. Paul Pecknold, plant pathologist at Purdue University in Indiana, suggests using strobilurin fungicides such as trifloxystrobin (Flint™) or kresoxim-methyl (Sovran™) in the 1st, 3rd and 7th cover sprays and contact fungicides in the other cover sprays. Limiting the use of strobilurin and sterol-inhibiting fungicides to the times when they're most needed should avoid development of fungal resistance to these compounds, which is likely if they are used too often.

Cedar-apple rust, caused by the fungus *Gymnosporangium juniperi-virginianae*, causes rusty orange lesions on leaves (Photo 16-24) and fruit of susceptible apple cultivars and is a problem in most parts of North America. This disease requires an alternate host, Eastern red cedar, so if none of these grow within 2 miles of the orchard, it can be avoided, but eliminating cedars is rarely practical. Many apple cultivars resist rust and need no sprays for it, but others are very susceptible. The sterol-inhibiting fungicides are very effective against rust, when applied twice in spring when orange fruiting structures appear on cedar galls, usually at about pink bloom and first cover.

Powdery mildew, caused by the fungus *Podosphaera leucotricha*, is a problem mainly in cool, humid areas such as the Northeast and Pacific Northwest. Many cultivars resist powdery mildew even under those conditions, but susceptible cultivars develop a powdery white growth on leaves that can defoliate them if severe, and a netlike russet on fruit that can render them unmarketable except for processing. Powdery mildew does not need a wetting period to develop, so susceptible cultivars (such

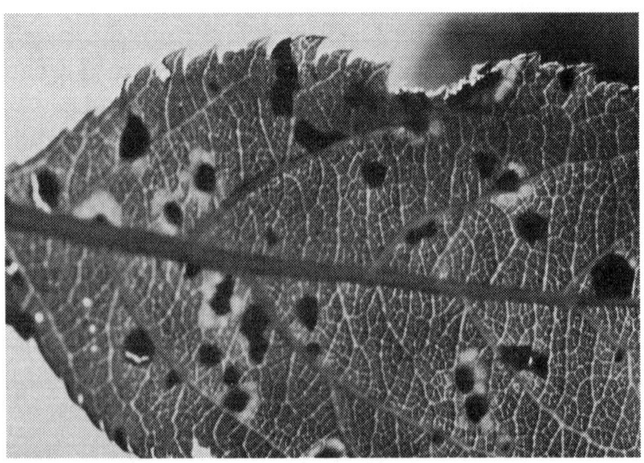

Photo 16-24. Cedar apple rust leaf infections can also defoliate susceptible cultivars, and may disfigure fruit. Rust-resistant cultivars or a few early-season fungicide sprays prevent this.

as 'Jonathan', 'Rome Beauty', 'Idared' and 'Cortland') need sprays even during dry periods, from tight cluster or pink until first or second cover. The sterol-inhibiting fungicides are effective against mildew, when also used to control scab and/or rust. When only mildew needs control, wettable sulfur works well but can harm plants at temperatures greater than 80 F, and does not control scab or rust. The biocontrol agent *Ampelomyces quisqualis* (AQ10™) is effective against powdery mildew if no other diseases need control.

Three **fruit rot fungi** commonly attack apples in humid areas, and can cause branch cankers and limb diebacks as well as rotting fruit. **Black rot**, caused by *Botryosphaeria obtusa*, also causes frogeye leaf spot and is the worst rot disease in cooler areas such as the Northeast. **White rot**, caused by *Botryosphaeria dothidea*, presents a severe problem in warmer areas such as the Southeast and Midwest, as does **bitter rot**, caused by *Glomerella cingulata*. Black rot fruit symptoms generally start at the blossom end and spread upward in alternating black and brown rings, while white rot start as small tan spots and eventually turn the entire fruit tan, soft and watery (Photo 16-25). Bitter rot lesions are sunken and produce bright orange spores (Photo 16-26).

All of these fruit rots can destroy up to an entire crop under warm, humid conditions. Prevent them by regularly pruning trees to keep them open, and promptly removing dead and weak wood, including the current year's fire blight strikes, to prevent these fungi from colonizing them.

A good fruit thinning program that breaks up clusters minimizes fruit rot spread from fruit to fruit, and prompt removal of any infected fruit will prevent them from becoming mummies that infect other fruit. Alternating strobilurin or thiophanate-methyl (Topsin-M™) and contact fungicides, applied as later summer cover sprays during the 2 months before harvest, can prevent much fruit rot damage.

University of Georgia plant pathologist Floyd Hendrix advises growers to spray for white rot when fruit sugar content reaches 10.5% and the weather gets wet, although cultivars with some resistance to white rot may need no sprays in orchards practicing strict sanitation of prunings and mummified fruit. Scout for bitter rot on every fruit on a tree in the lowest, most humid part of the orchard, and spray as soon as a single fruit with bitter rot appears.

Sooty blotch (caused by several different fungi) and **flyspeck** (caused by *Zygophiala jamaicensis*) form a disease complex that infects just the waxy surface of apple fruits, disfiguring them (Photo 16-27). Sooty blotch appears as dark, cloudy smudges, while flyspeck forms clusters of black dots. These diseases are worst in orchards subject to heavy dews, fogs, or rainy periods when nighttime temperatures remain between 65 and 75 F. Summer pruning reduced incidence of these diseases by 50% in Massachusetts trials, but complete control relies on later fungicide cover sprays or postharvest chlorine dips.

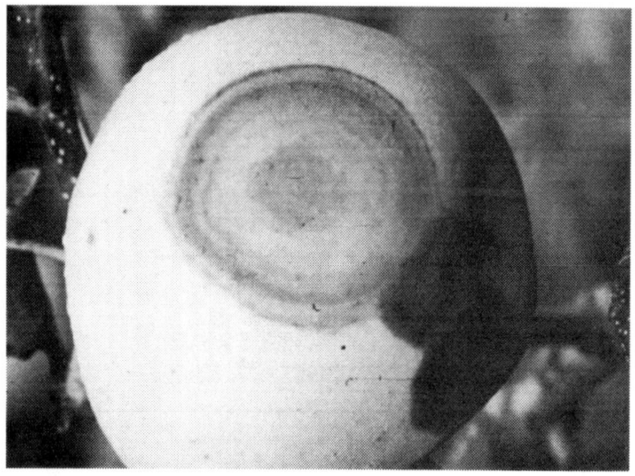

Photo 16-25. White rot forms concentric rings on this apple's side, while two sunken bitter rot lesions mar its base.

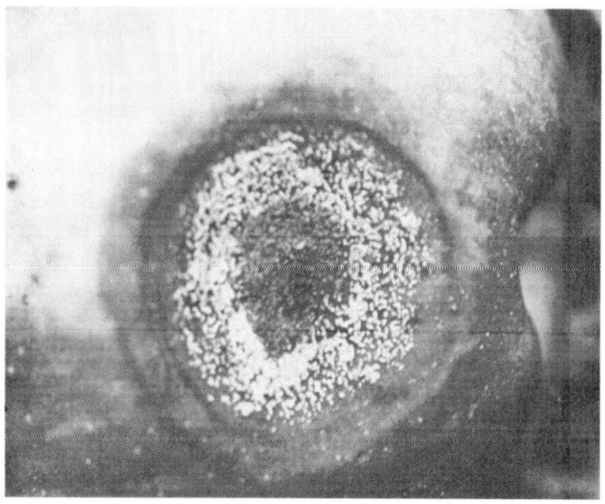

Photo 16-26. Bitter rot lesions produce masses of orange spores that cause secondary infections, and can rot every apple in an orchard within 3 days during hot, humid, late summer weather if not controlled.

Strobilurin fungicides provide excellent control, so include them in cover sprays when favorable disease conditions occur. Some growers use a 15-minute post-harvest fruit dip into a 500 ppm solu-

Photo 16-27. Sooty blotch (smoky gray areas) and flyspeck (larger black spots), two superficial apple diseases, infect only the wax on the outside of fruit, in late summer and early fall. Summer pruning reduces these diseases, as do certain late-summer fungicide sprays.

Photo 16-28. Codling moth tunnels into fruit, leaving damage or worms at harvest. Control this pest with mating disruption on orchards of 5 acres or more.

tion of swimming pool chlorine, which is labeled for use to remove sooty blotch and flyspeck from infected fruit. Any residues volatilize away within minutes of treatment.

Fruit-infesting insects. These insects are major pests because they directly attack the fruit, and must be controlled each year wherever they occur.

Codling moth (*Cydia pomonella*) is the most widespread and often the most serious fruit pest of apples, occurring almost everywhere they are grown. This moth pest lays eggs on the fruit, which hatch into larvae that tunnel into the fruit, creating fruit with worms or worm damage at harvest (Photo 16-28). Wormy apples cannot be sold on the fresh market, and codling moth damage leaves fruit susceptible to rots, which keep many culled fruit from being used for juice processing.

Codling moth can now be controlled without pesticides by using mating disruption, on orchards of 5 acres or more. Pheromone dispensers can be tied to each large tree or every third dwarf tree in the orchard during bloom, then replaced in midsummer before the second generation moth flight, or a sprayable codling moth pheromone can be applied at those two times.

Where codling moth is the only major insect pest, mating disruption costs as little as pesticide applications. However, where several other insect pests are serious, pesticides can control them all at less expense. Where only moth pests such as codling moth, Oriental fruit moth and leafrollers are problems (and plum curculio and apple maggot aren't), growers can use non-toxic insecticides that inhibit larval molting such as azadirachtin derived from neem seed (Aza-Direct™, Azatin™ or Neemix™), tebufenozide (Confirm™), or methoxyfenozide (Intrepid™).

Oriental fruit moth (*Grapholita molesta*) causes similar damage to apples as codling moth, and is an even worst pest in the South because it can have 5 to 6 generations per year instead of codling moth's 2 to 3 generations. Mating disruption works very effectively against this pest, but growers must use a pheromone specific to each pest, applying 2 sets of twist ties or 2 pheromone sprays if also treating for codling moth. Sprays for plum curculio and codling moth (below) also control the first 2 generations of this pest. Where later-generation Oriental fruit moth does more damage than codling moth, as in southern Missouri and Arkansas, mating disruption for just the later generations of this pest may be economical following early-season sprays for all fruit pests.

Plum curculio (*Conotrachelus nenuphar*), a snout beetle that lays eggs in fruit, is a major pest of apples east of the Rocky Mountains. Egg-laying scars and surface feeding by adult beetles can cause misshapen, poorly formed fruit, and feeding by larvae can cause many young fruit to drop and internal damage to remaining fruit. In Northern areas where plum curculio has only one generation per year, this pest can be controlled with just two insecticide sprays, at petal fall and again 10 days later, with a material such as phosmet (Imidan™) or indoxacarb (Avaunt™) that has minimal effect on beneficial insects. Those sprays will also control first generations of codling moth and Oriental fruit moth.

In Southern regions where plum curculio has a second generation, a third spray is needed in mid-summer when monitoring shows that adult beetles have emerged or when moth traps show second generations of codling moth and Oriental fruit moth have emerged, as this spray controls them also. Trials using the kaolin-clay material Surround™, the only "organic" material with any activity against plum curculio, have produced disappointing results, with 25% of fruit culled due to plum curculio damage plus 20% damage from internal moth pests at harvest.

Apple maggot (*Rhagoletis pomonella*), a fly pest, causes severe damage in Northeastern North America but thankfully doesn't occur in the Southeast, South Central or Western parts of North America. Adult flies emerge in June and July, and growers use yellow sticky traps (Photo 16-29) or red spheres coated with fruit aromas to monitor them. Place traps along orchard edges closest to a woods or abandoned orchard if these are nearby, or along the southern edge of the orchard. Spray when each trap catches an average of five apple maggot flies, with a material such as kaolin (Surround™), phosmet (Imidan™) or a synthetic pyrethroid such as fenpropathrin (Danitol™), permethrin (Ambush™ or Pounce™), or esfenvalerate (Asana™). However, the pyrethroids can hurt mite predators and thus bring on an outbreak of pest mites

Several **leafrollers** and related caterpillars in the Tortricid moth family feed on fruit surfaces (Photo 16-30), and have caused increasing damage to apples in recent years, as more specific chemicals or mating disruption are used to control the pests that feed internally in fruit. These pests of apples include **tufted apple bud moth** (*Platynota idaeusalis*), **red-banded leafroller** (*Argyrotaenia velutinana*), **oblique-banded leafroller** (*Choristoneura rosaceana*), **fruittree leafroller** (*Archips argyrospila*), **omnivorous leafroller** (*Platynota stultana*), **apple pandemis** (*Pandemis pyrusana*), and **orange tortrix** (Argyrotaenia citrana). Leafroller larvae feed on leaves and buds early in the season, webbing or rolling leaves to protect themselves from predators, which also protects them from pesticide sprays. As fruit grow, these insects feed on fruit surfaces, which can allow entry to fruit rots. These moth pests can be controlled with bloom and petal fall sprays of the biological insecticide *Bacillus thuringiensis* (Bt, Javelin™, Dipel™, etc.) or non-toxic insecticides that inhibit larval molting such as azadirachtin derived from neem seed (Aza-Direct™, Azatin™ or Neemix™), tebufenozide (Confirm™), spinosad (SpinTor™), or methoxyfenozide (Intrepid™). Mating disruption with pheromones for the damaging species also controls leafrollers.

Indirect insect and mite pests. These pests generally reduce yields indirectly by sapping tree

Photo 16-29. Apple maggot is a severe pest in Northern apple-growing regions. This apple maggot trap attracts the adult flies with its yellow color and a pheromone, and enables growers to apply insecticides at the right times and minimize sprays.

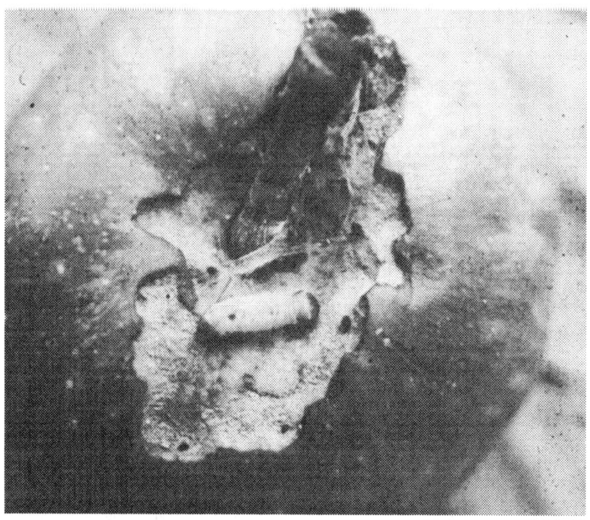

Photo 16-30. Several leafroller species feed shallowly on apple surfaces, causing fruit to be culled at harvest. Selective pesticides that target moth pests control leafrollers.

energy from leaves, rather than directly attacking fruit, although some of these pests may affect fruits secondarily. Often predators and parasites will naturally control these indirect pests, although outbreaks can occur due to unusual weather that slows natural enemy development but favors pest population growth, or disruption of biocontrols by broad-spectrum pesticides. Always consider the effects of any pesticide on beneficial insects and mites before choosing to apply it.

Aphid pests include **rosy apple aphid** (*Dysaphis plantaginea*), **green apple aphid** (*Aphis pomi*) and **spirea aphid** (*Aphis spiraecola*). Rosy apple aphid is usually the most serious, because it feeds on developing fruit, stunting and deforming them, as well as causing leaves to curl (Photo 16-31), protecting the aphids from pesticide sprays. Green apple aphid and spirea aphid feed only on new leaves, and general predators such as lacewing larvae and lady beetles usually keep populations low enough to avoid damage, unless disrupted by broad-spectrum insecticide applications.

A dormant oil spray reduces populations of all aphids by smothering overwintering aphid eggs on apple twigs, without hurting beneficial insects that overwinter in the groundcover or elsewhere. Rosy apple aphid may also need a pyrethroid insecticide such as fenpropathrin (Danitol™), permethrin (Ambush™ or Pounce™) or esfenvalerate (Asana™) applied at pink before leaves start to curl and before beneficial mite predators are active. Cinnamaldehyde (Valero™), a natural component of cinnamon, is effective against aphids and mites but the spray must directly contact the pests and has residual activity of 1 day or less.

Mite pests of apples include **European red mite** (*Panonychus ulmi*), **two-spotted spider mite** (*Tetranychus urticae*), and **Pacific mite** (*Tetranychus pacificus*). Predator mites that feed on these pests usually keep their populations in check, but heavy mite pest infestations (Photo 16-32) can cause foliage bronzing and premature defoliation. A dormant oil spray is very important, because it controls pest mites without hurting predators. Keep trees adequately irrigated because pest mites cause more damage to drought-stressed trees, and avoid using insecticides that disrupt mite predators. If pest mite outbreaks occur anyway, use a selective miticide that doesn't hurt predator mites much, such as hexythiazox (Savey™), pyridaben (Pyramite™), or cinnamaldehyde (Valero™).

San Jose scale (*Quadraspidiotus perniciosus*) has a soft body but is covered by a hard shell that protects it from most insecticides. Scales feed on

Photo 16-31. Rosy apple aphid, generally the worst aphid pest of apples, causes leaf curling in early spring and can distort developing fruit. A dormant oil spray and one or two early-season insecticide sprays control it. (Photo from British Columbia Ministry of Agriculture.

branches, which can cause dieback, and when populations are high can infest fruit (Photo 16-33). Prevent this with a dormant oil spray, and prune out heavily infested branches. If summer control is needed, use a safer material such as pyriproxyfen (Esteem™), an insect growth regulator that stops pest development.

Leafhopper pests include **white apple leafhopper** (*Typhlocyba pomaria*) and **potato leafhopper** (*Empoasca fabae*). These leafhoppers feed on leaves, doing little damage to them but excreting a sugary honeydew that can accumulate on fruit, fostering the growth of sooty mold fungi and posing a nuisance to harvesters.

Summer pruning of watersprouts, rapidly growing shoots that leafhoppers prefer to colonize, can reduce pest populations. Usually just the first generation needs an insecticide treatment. Many

Photo 16-32. High populations of mites, like these European red mites, can cause leaf bronzing and defoliation, reducing yields. Use a dormant oil spray and selective mitcides as needed in summer.

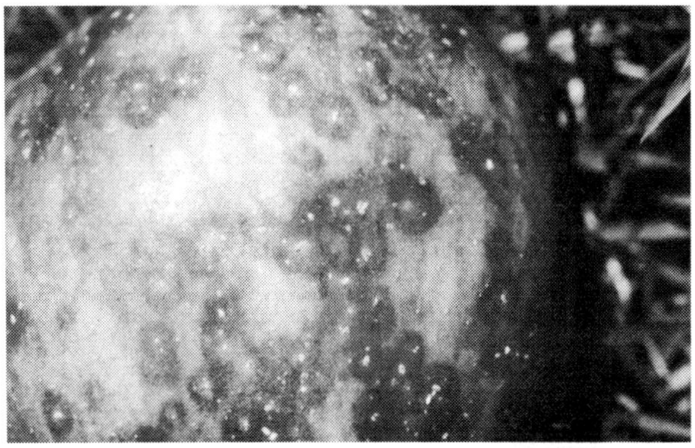

Photo 16-33. San Jose scale can infest fruit surfaces as well as limbs, and can kill trees over several seasons if populations get high. Prevent this with a dormant oil spray.

insecticides and miticides control leafhoppers, including pyribaden (Pyramite™), which is often used to treat mites at about the time leafhopper control is needed.

HARVEST AND STORAGE

Most apples continue to be hand-harvested for the fresh market, and many are quickly placed into long-term storage, to provide fresh-market apples all year long. Apples culled during the grading operation, but still free of decay and worm damage, are usually sold to processors who produce juice, cider, applesauce and canned pie slices. About 25 to 35% of the North American apple crop is diverted to processing each year.

Determining when to harvest. Apples for storage are harvested 1 to 2 weeks earlier than apples to be sold immediately, because they will store better if picked at a firmer stage, and starches will gradually break down into sugars during storage. Fruit that have developed to the stage where they can ripen to good quality during storage are considered *mature*, though they are not yet fully ripe.

If fruit are harvested too early, while still immature, they will be smaller, lack flavor, more prone to water loss and fruit shrivel during storage, and more likely to develop storage disorders such as bitter pit or scald. Fruit allowed to become overmature before harvesting are softer, may develop off flavors, have a shorter storage life and are more prone to decay and internal breakdown during storage.

Harvest apples to be stored before they enter the final stage of rapid ripening called the climacteric, during which fruit respiration rates and ethylene production increase. Apples sold for immediate eating should be allowed to enter the climacteric before harvest, so they are sweeter and less starchy as the climacteric begins to break down starches into sugars. Apples for immediate use should be ripe, but not overripe, or they will become too soft and acidity levels drop to the point that flavor is impaired.

Growers use several **maturity indices** to determine when to harvest. **Fruit firmness** should generally measure 16 to 18 pounds force for storage, and 14 to 15 pounds force for immediate market. **Fruit ground color** should still be light green for storage apples, and turn yellow-green or light yellow for immediate market. **Starch scores** from a starch-iodine test, available at drugstores, should generally read 2 to 3 for storage apples, and 4 to 5 for immediate market. Individual cultivars have their own guidelines. For example, 'Gala' apples can have starch scores of 1 to 2 for storage and 3 to 4 for immediate sales, while 'Fuji' starch scores can be 4 to 5 for storage and 5 to 6 for immediate sales. Ground color is a more reliable harvest indicator than starch for both these cultivars. When sampling fruit for harvest maturity, test 20 to 30 apples from at least 5 different trees that are typical of the entire block.

Fruit handling during harvest, packing and storage. Most large apple growers have workers harvest into canvas picking bags, and empty these into bulk pallet bins holding 14 to 25 bushels. Workers use forklifts or tractor lifts to load these pallet bins onto flat-bed trucks or trailers, which transport the bins to a cooler and then to a packing house. Train workers to harvest and move apples carefully, as they are easily bruised. Bruised fruit ripen faster than undamaged fruit, and cuts and bruises reduce marketability and allow decay organisms to enter the fruit.

Cool apples rapidly after harvest to reduce respiration and transpiration, which cause losses of fruit sugars and water. Many growers submerge bins of harvested fruit in a hydrocooler, then run fruit along a conveyor belt where they are sprayed with water to further clean them, and sprayed with a wax coating to retard moisture loss.

Keep chlorine levels in water systems at 50 to 100 ppm, to kill spores that could cause storage diseases such as blue mold (caused by several *Penicillium* species), gray mold (caused by *Botrytis cinerea*), mucor rot (caused by *Mucor piriformis*) or bull's-eye rot (caused by *Pezicula malicorticis*).

Mummified fruit from last year, the source of these diseases, remain in bins and a single contaminated plastic bin can carry ½ billion spores, while a contaminated wooden bin may hold up to

2 billion spores. Therefore, before harvest season, inspect all bins and remove any mummies by hand, and sanitize bins plus all packinghouse and storage walls with quarternary ammonium. Without adequate sanitation, grocery store surveys in the Northeast from February to April found 21% of 'McIntosh' bags had some blue mold, and 37% of 'Empire' bags contained blue mold.

Packinghouses generally treat wholesale apples with the fungicide thiabendazole (TBZ) to reduce postharvest fruit rots. Captan may be used also to reduce rots further, but some export markets prohibit its use. Some packinghouses also use 2 biocontrol agents registered for postharvest treatment of apples and other fruits, usually in addition to but sometimes instead of TBZ. Aspire™ contains the yeast *Candida oleophila*, while Bio-Save 100™ contains the non-pathogenic stain ESC-10 of the bacterium *Pseudomonas syringae*. When these biocontrol agents coat the fruit soon after harvest, they colonize the fruit surface and inhibit any pathogens from later establishing there and causing fruit rots.

Growers or packinghouses generally run harvested fruit through sorting machines to remove any blemished fruit, then through sizing and color-sorting lines to separate fruit into various grades based on fruit size and skin color. Smaller apples go into bags holding 3 pounds, 5 pounds or 10 pounds of fruit, while larger apples are packed into cartons with molded trays holding each fruit, to be sold individually for higher prices.

Early-ripening summer apples, intended mainly for immediate sale, usually last only a few weeks in cold storage. Late-summer or early-fall apples such as 'Gala' may be stored in refrigerated storage for 6 to 8 weeks, and in controlled-atmosphere (CA) storage for 6 to 7 months. Most apple cultivars that ripen later in fall store well for 3 to 6 months in refrigerated storage and for 10 to 12 months in CA storage.

During storage, apples produce ethylene, which triggers further ripening and continued respiration. Ethylene scrubbers in storage systems can reduce buildup of ethylene in the storage atmosphere, and thus reduce respiration loss of fruit sugars and crispness, but scrubbers do not reduce internal fruit ethylene production. A formulation of 1-methyl cyclopropene (1-MCP) labeled as Smart Fresh™ does reduce internal fruit ethylene production and respiration, in both CA and refrigerated storage, when applied once when fruit are placed in storage. This compound improves flavor, juiciness and firmness of apples after storage, and reduces storage scald.

Storage disorders. Several physiological disorders may develop on apples during storage, especially extended storage. Prevent these with careful timing of harvest and attention to storage conditions.

Scald, the most widespread physiological disorder of apples, causes a diffuse browning of fruit just under the skin (Photo 16-34). Scald is worst on apples harvested early in the season with poor color development, and when fruit are held at above 32 F. Certain cultivars, such as 'Rome Beauty' and 'Stayman Winesap', are especially susceptible to scald, so allow them to reach full maturity before harvesting. Scald usually doesn't develop until after apples have been stored for several months and then been placed at room temperature for several days. Reduce scald by using CA storage for apples to be stored 2 months or longer, or by dipping newly harvested apples in a diphenylamine (DPA) solution before placing in regular refrigerated storage, or by using 1-MCP in either refrigerated or CA storage. Apples to be stored less than 2 months generally need no treatment for scald if harvested when mature and stored at 32 F.

Bitter pit causes small sunken brown spots to appear on apple surfaces. Prevent this during the growing season by avoiding high nitrogen levels that could produce extra-large fruit low in calcium, and by applying calcium sprays. Also avoid harvesting too early. Large fruit from vigorous young trees, or fruit harvested early, may need postharvest calcium dips to avoid bitter pit.

Internal browning of apple flesh develops if carbon dioxide levels in storage get too high. Keep CO2 levels in storage below 0.5%, and do not place late-harvested apples in CA storage, as they are more prone to internal browning.

Water core causes glossy, water-soaked regions high in sugars to develop around the core of apples (Photo 16-35). This disorder develops just before harvest, especially when intense heat and sunlight prevail as apples approach maturity, but is often not noticed until fruit are taken out of storage. Water core is worst on trees with a very light crop and on over-mature apples, so harvest before extensive water core develops. Apples with water core are actually preferred in Japan and a few other areas, since the water core regions are very sweet, so Japanese cultivars such as 'Fuji' tend to be more prone to this disorder, and need prompt harvesting in North America.

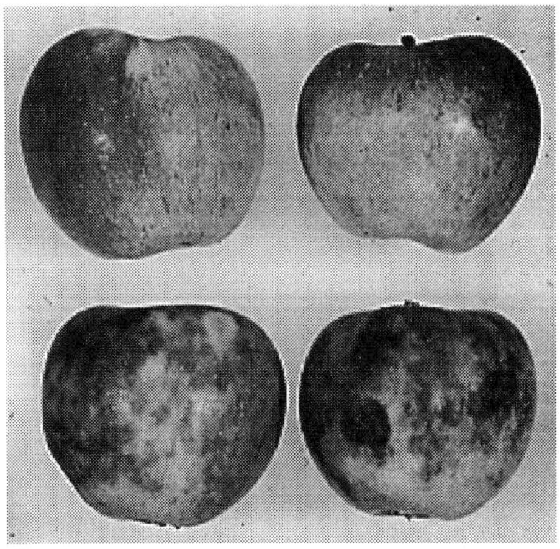

Photo 16-34. Scald is the most common storage disorder of apples. Prevent it by not harvesting too early, using CA storage for apples to be stored more than 2 months, and treating with 1-MCP or DPA. (Photo from Agric. Canada.)

Photo 16-35. Water core causes glassy areas of high sugar content to develop in pockets, often around the core. Prevent it by harvesting before fruit become over-mature.

Further Reading

Barritt, B. 1992. *Intensive orchard management.* 211 pp. Good Fruit Grower, 1005 Tieton Dr., Yakima, WA 98902.

Dunn, J. S. and M. Stolp. 1987. Apples on the Lincoln Canopy—mechanized management. *HortScience* 22-568-572.

Evans, R.G. 2000. Overtree evaporative cooling system design and operation for apples in the Pacific Northwest. Washington State Univ. report. Available online at: http://www.bsyse.prosser.wsu.edu/report/cool.html

High density apple orchard management. 2000. North Carolina Cooperative Extension Service pub. 581. Available online at: http://www.ces.ncsu.edu/depts/hort/hil/pdf/ag-581.pdf

Integrated pest management for apples and pears, 2nd ed. 1999. Univ. of California Agric. and Natural Resources pub. 3340.

Jones, A.L. and H.S. Aldwinckle, eds. 1990. *Compendium of apple and pear diseases.* APS Press, St. Paul, MN.

Olcott-Reid, B. 1990. Low-spray apple program. Amer. *Fruit Grower*, vol. 110, no. 12 (Nov. 1990): 15-18.

Tustin, S.D., P.M. Hirst, W.M. Cashmore, I.J. Warrington and C. J. Stanley. 1990. The principles and practices of training slender pyramid trees for high density production. *Compact Fruit Tree* 23:83-92.

Chapter 17
European and Asian Pears

Pears are deciduous fruit trees in the genus *Pyrus* of the rose family (Rosaceae) that produce pome fruits and are closely related to apples. Many apple-growing regions also produce a much smaller amount of pears, although some areas of the Pacific Northwest, where pears grow especially well, specialize in pears. The West Coast region (Washington, California, Oregon and British Columbia) produces 97% of the pears grown in North America.

Most North American pear-growing districts raise European pears, derived mainly from the species *Pyrus communis*, which has been cultivated on this continent for over 300 years. Over the last 20 years, Asian pears derived from *P. serotina* (sand pear) and *P. ussuriensis* (Siberian pear) have been widely planted in California and Southern states, especially those with large Asian-American populations. Both Asian pear species have also been crossed with European pears to produce hybrid pears, generally selected for European-type fruit and the disease resistance plus cold hardiness or heat tolerance of the Asian species.

Photo 17-1. European pears have a pyriform shape, are harvested before they ripen, and are allowed to ripen and soften at room temperature before eating. 'Bartlett', shown here, accounts for almost half of North American production. About 75% of the 'Bartlett' crop is canned, with the rest sold fresh.

European pears have a pyriform shape that is rounded on the calyx end and narrows to a slender neck at the stem end (Photo 17-1). Of the Asian pears, Japanese cultivars are derived mainly from sand pear and have a rounded, apple-like shape (Photo 17-2), while Chinese cultivars are derived from hybrids of sand pear and Siberian pear and have a blocky pyriform shape, as do European-Asian hybrids. Japanese pears, sometimes called "apple pears" due to their shape, include many cultivars that ripen over a long season like that of apples, from late June to mid-October. Chinese pears and European pears ripen mainly from late August to late October, although some European pears and European–Asian hybrids ripen earlier as "summer pears".

The biggest difference between European and Asian pears is fruit texture. European pears ripen to a soft, smooth, buttery texture, while Asian pears remain crisp like an apple. Both types of pears are very juicy and sweet, with a distinctive pear taste.

European pears do not tree-ripen well, as mature fruit develops internal browning if left on trees to soften. Consequently, they are harvested while still very firm, can be kept in cold storage for up to 8 months, and then are allowed to ripen

Photo 17-2. Asian pears are crisp and juicy, eaten like apples, and can be tree-ripened. Japanese cultivars, such as this 'Shinseiki', have a round shape like apples, while many Chinese cultivars and Chinese-Japanese hybrids have a short pyriform shape.

Photo 17-3. Fire blight can devastate most commercial cultivars of European and Asian pears, killing trees in a single season. The semi-arid West Coast region produces almost all North American pears, partly because of lower fire blight pressure. (photo from British Columbia Ministry of Agric.)

at room temperature before or during marketing. Asian pears are allowed to ripen on trees and can be marketed immediately or kept in cold storage for several months after harvest. About 53% of European pears are sold fresh, while 47% are processed, mainly into canned pears and baby food. Almost all Asian pears are sold fresh.

China produces almost half of the world's pears, over 7000 metric tons per year. Italy and the U.S., the second- and third- largest pear producing countries, each grow less than 1000 metric tons of pears per year.

LIMITING FACTORS IN PEAR PRODUCTION

Pear consumption in North America is much less than that of apples, limiting demand. Consumers generally prefer fruit they can eat immediately, so few buy supermarket European pears that must be ripened at room temperature for several days before eating. Asian pears are becoming more popular, but many North Americans are still unfamiliar with them and hesitate to buy these strange-looking fruit, especially those with russet skin. However, direct marketers have readily stimulated demand by pre-ripening European pears and cutting samples of both Asian and European pears for customers to sample.

Pears suffer from fewer diseases and insects than do apples, but are generally much more susceptible than apples to fire blight, a disease that often kills pear trees in most parts of North America (Photo 17-3). Fire blight causes damage in semi-arid West Coast pear orchards during some years, but is not often severe there and can be more easily managed than it can in the East, a prime reason for pear production's concentration in the West. However, breeders have developed cultivars of both European and Asian pears with strong resistance to fire blight and other diseases, and some pear cultivars can be grown almost anywhere in North America, from USDA hardiness zones 2 to 10. Where not threatened by diseases, pear trees can be very long-lived, to 100 years or more.

Pear flowers are similar to those of apples, with flower parts in multiples of 5, but open earlier in spring (Photo 17-4). This earlier bloom subjects pears to frost problems in many areas, another reason they are grown mainly in West Coast regions that seldom suffer spring frosts. Some cultivars bloom later than others, and growers can select late-blooming cultivars and better sites to minimize frost problems. Pear flowers and fruit develop mainly on short spurs that arise from 2-year-old or older branches.

Pear trees generally grow much more upright than do apple trees, reaching a height of 30 to 35 feet in nature and a spread about half as wide (Photo 17-5). Many pear cultivars also come into bearing late, making them less economical to produce than apples. Newer dwarfing rootstocks, earlier-bearing cultivars, and tree-spreading techniques are enabling pear trees to produce sooner and stay smaller, increasing their profitability. However, overall costs of producing pears keep rising while prices received for wholesale pears do not, and many pear growers have operated at a loss for the last few years.

ORCHARD DENSITY AND ROOTSTOCKS

Traditional low-density pear orchards have been planted in rows 20 to 24 feet apart, with trees 16 to 20 feet apart within rows, for a density of 90 to 136 trees per acre. Growers generally train such trees on standard rootstocks to the central leader or modified leader system. Almost all pear orchards in the Pacific Northwest of North America are 35 to 100 years old, on 'Bartlett' seedling rootstocks, with an average density of 111 trees/acre and maintained at a height of about 15 feet by pruning.

Newer pear orchards often use dwarfing or semi-dwarfing rootstocks and are planted more closely. The size-controlling rootstocks most often used are well anchored so require no support, but trellises may be used to aid in tree training. Free-standing trees often utilize the central axe training system, which has enabled tree densities as high as 388 trees per acre, at a spacing of 14 feet between rows and 8 feet within rows. Trellised pear orchards usually follow some form of the palmette training system, which involves spreading branches to a 45-degree angle and tying them to wires to hold them in place. Table 17-1 details tree height, pest resistance and other traits of commonly used pear rootstocks.

Standard rootstocks. Seedlings of several European pear cultivars are widely used as standard rootstocks for European pears, semi-dwarfing rootstocks for Chinese pears and dwarfing rootstocks for Japanese pears. 'Bartlett' seedlings are most commonly used due to a large supply of seeds from canneries, but are very susceptible to fire blight, so growers should avoid this rootstock in most areas. Some nurseries offer 'Winter Nelis' seedling rootstocks, which has moderate resistance to fire blight and increases yields and fruit size of cultivars over those induced by 'Bartlett' seedling rootstocks. Both 'Bartlett' and 'Winter Nelis' seedlings are susceptible to crown gall, root lesion nematodes and pear root aphids.

Pyrus betulaefolia, a vigorous Asiatic pear species, is the most common standard rootstock for Asian pears and is also used for European pears. It offers good resistance to fire blight and crown gall and strong resistance to root lesion nematodes, and has the most tolerance to very heavy or wet soils. This rootstock induces larger fruit and is often preferred for Japanese pears, especially the smaller-growing cultivars that may be overly dwarfed by European pear rootstocks. However, it can induce overly vigorous growth of Chinese, Chinese-Japanese, and European pears, which is especially a problem where fire blight is serious, since vigorous growth is more prone to the disease. Most strains of this rootstock are hardy only to 0 F (USDA zone 7a), and the more cold-hardy 'Reimer' strain is hardy only to –10 F (zone 6a), so it's best avoided in zones 5 or colder.

Pyrus calleryana, another Asiatic pear species which produced the ubiquitous 'Bradford'

Photo 17-4. Pears bloom early, so are prone to frost damage. The lower blossoms opened before a frost and were killed, while the upper flowers opened after the frost so are still alive.

Photo 17-5. If not trained and pruned, pear trees grow very upright and can become about 30 feet tall.

Table 17.1

COMMONLY USED AND PROMISING PEAR ROOTSTOCKS

In approximate order of tree height.

	Height in feet			Hardy	Susceptibility to			
Rootstock	European cultivars	Japanese cultivars	Chinese cultivars	to zone	Fire Blight	Root Lesion Nematodes	Pear Root Aphids	Comments
Provence Quince	9	not used	not used	6b	VS	R	R	Old Home interstem used for compatibility
OH x F 51	10	6	10	4b	R	MS	MR	
Pyrodwarf	12	6	12	5a	R	MS	MR	new from Germany, induces fruiting in 2-3 years
Pyro 2-33	14	7	13	5a	R	MS	MR	new from Germany, induces fruiting in 2-3 years
OH x F 40	14	7	13	4b	R	MS	MR	
OH x F 87	16	7	14	4b	R	MS	MR	
OH x F 69	16	7	14	4b	R	MS	MR	
OH x F 333	18	7	15	4b	R	MS	MR	
Bartlett seedling	28	8	18	5a	VS	MS	VS	
OH x F 97	30	9	20	4b	R	MS	MR	
Winter Nelis seedling	30	9	20	4b	MS	MS	VS	induces larger fruit
P. calleryana	30	15	28	6b	R	R	R	best for Southeast; R to crown gall, crown rot
P. betulaefolia 'Reimer'	32	18	30	6a	R	MR	R	induces larger fruit

Table 17-2.

COMMON COMMERCIAL EUROPEAN PEAR CULTIVARS

In approximate order of ripening (varies with area).

Cultivar	Best in Zones	Fruit					Susceptibility to			Tree		Bloom Time	Harvest Time	Cold Storage Life
		Quality	Flavor	Size	Skin Color	Yields	Fire Blight	Pear Scab	Leaf Spot	Size	Growth			
Clapp's Favorite	4-7	good	bland	very large	yellow/red	high	VS	R	MS	very large	upright	middle	early	2 weeks
B.P. Morettini	5-8	good	tart	large	yellow/red	high	MS	S	S	large	upright	very early	early	2 mons.
Bartlett	5-9	very good	mild	large	yellow	high	VS	MR	MR	large	upright	middle	early-mid	2 mons.
d'Anjou	5-8	good	bland	large	green	high	S	S	S	very large	spreads	early	mid-late	8 mons.
Bosc	5b-8	best	spicy	large	russet	high	VS	S	VS	large	upright	mid-late	late	4 mons.
Concorde	5-8	best	rich	large	green	high	VS	S	S	large	upright	middle	late	4 mons.
Comice	5-8	best	rich	large	green/red	low	MS	R	MS	medium	upright	mid-late	late	4 mons.

Disease susceptibility ratings: VS = very susceptible, S = susceptible, MS = moderately susceptible, MR = moderately resistant, R = resistant.

Average harvest time in Sacramento, CA and central NC: early: late July, early-mid: early to mid-Aug., mid-late: late Aug. to early Sept., late: mid- to late Sept. Ripening times will be about 1 week earlier in southern CA and northern GA, about 1 week later in northern CA and western VA, about 2 weeks later in southern MI and southern PA, about 3 weeks later in central OR and central WI, and about 4 weeks later in western WA and central NY.

Notes on Cultivars:
Clapp's Favorite, also called 'Bennett', crops reliably. 'Kalle', or 'Red Clapp's Favorite', is a strain with red skin.
B.P. Morettini, also called **Juno**, is shaped like 'Bartlett' but ripens 2 weeks earlier, so grown as an "early Bartlett".
Bartlett, called **Williams** in Europe, is a top canning cultivar, also leads fresh sales and provides 47% of North American pear production. 'Sensation' and 'Max Red Bartlett' are Bartlett sports with red skin.
Anjou, or **Beurre d'Anjou**, needs dwarfing rootstocks to bear before 7 years. 'Red Anjou' is a strain with red skin that ripens later.
Bosc, or **Buerre Bosc**, fruit have a rich, spicy flavor. Tree grows slowly and is slow to bear, but yields well when mature.
Concorde, a Conference x Comice cross from England, is being planted more as a high-quality, high-yielding alternative to Comice.
Comice is grown less due to slow bearing, poor crops after frosts. 'Red Comice' has red skin; 'Taylor's Gold' has russet skin.

Table 17-3.

DISEASE-RESISTANT EUROPEAN PEAR CULTIVARS

In approximate order of ripening (varies with area).

Cultivar	Best in Zones	Fruit					Susceptibility to			Tree		Bloom Time	Harvest Time	Cold Storage Life
		Quality	Flavor	Size	Skin Color	Yields	Fire Blight	Pear Scab	Leaf Spot	Size	Growth			
Tyson	4-8	best	spicy	medium	yellow	medium	R	MR	S	very large	upright	middle	very early	1 mon.
Harrow Delight	5-8	very good	mild	med-large	yellow/red	high	VR	MR	S	medium	upright	late	very early	1 mon.
Harvest Queen	5-8	very good	mild	medium	yellow	medium	R	MR	S	medium	spreads	middle	early	2 mons.
Flordahome	9-10a	good	sweet/tart	med-small	green	high	VR	S	MR	medium	upright	early	early	3 weeks
Moonglow	5-8	good	mild	large	green	high	VR	MR	R	large	upright	middle	early	4 mons.
Honeysweet	5-8	best	spicy	med-small	russet	medium	MR	S	MS	large	upright	mid-late	early-mid	3 mons.
Rescue	5-8	best	mild	large	yellow/red	medium	R	R	S	medium	upright	middle	early-mid	2 mons.
Maxine	5-8	good	sweet	large	green	high	MR	MR	MR	large	upright	middle	early-mid	2 mons.
Magness	5-8	best	mild	medium	green	low	R	R	MS	large	spreads	late	middle	4 mons.
Warren	5-8	best	mild	medium	green	low	R	R	MS	large	spreads	late	middle	4 mons.
Seckel	5-8	best	spicy	small	russet	medium	MR	VS	S	medium	dense	mid-late	middle	3 mons.
Blake's Pride	5-8	very good	rich	med-large	russet	high	VR	VR	MS	large	upright	mid-late	middle	2 mons.
Doyenne Gris	5-8	best	spicy	medium	russet	medium	R	R	MS	medium	upright	middle	middle	4 mons.
Mericourt	5-9	good	sweet/tart	medium	green	high	VR	R	MS	medium	upright	early-mid	mid-late	2 mons.
Luscious	3-7	very good	sweet	medium	green	medium	MR	MR	MS	medium	upright	early-mid	mid-late	3 weeks
Potomac	5-8	very good	mild	med-large	yellow	high	R	R	MS	medium	upright	early-mid	mid-late	2 mons.
Harrow Sweet	5-8	very good	rich	large	yellow/red	high	VR	MR	S	medium	spreads	middle	late	2 mons.
Duchess	6-8	very good	rich	large	yellow	high	MR	S	MR	medium	dense	mid-late	late	2 mons.
Tenn	8-9	good	sweet/tart	medium	yellow	high	VR	S	R	large	upright	early	late	1 mon.

Disease resistance ratings: VR = very resistant, R = resistant, MR = moderately resistant, MS = moderately susceptible, S = susceptible

Average harvest time in Sacramento, CA and central NC: very early: mid-July, early: late July, early-mid: early Aug., middle: mid-Aug., mid-late: late Aug. to early Sept., late: mid- to late Sept., very late: early Oct. Ripening times will be about 1 week earlier in southern CA and northern GA, about 1 week later in northern CA and southern VA, about 2 weeks later in southernl MI and southern PA, about 3 weeks later in central OR and central WI, and about 4 weeks later in western WA and central NY.

Notes on Cultivars:
Tyson has high quality for an early pear, but is slow to begin bearing.
Harrow Delight, from Ontario, has fruit like 'Bartlett' but resists blight, ripens earlier. It cans well.
Harvest Queen, from Ontario, has fruit very similar to 'Bartlett' but a bit smaller, ripens 5 days earlier and resists blight.
Flordahome, a hybrid of European and sand pears from FL, provides good-quality pears for zones 9 and 10a.
Moonglow bears young, has mild flavor but astringent skin if trees lack water.
Honeysweet has fruit like Seckel but a bit larger and earlier.
Rescue, a pear seedling from BC, has flavor like 'Clapp's Favorite', bears annually and rarely needs thinning.
Maxine, also called **Starking Delicious**, bears young and heavily, cans well.
Magness, a USDA hybrid of Seckel seedling x Comice, is slow to bear, blooms late, has sterile pollen.
Warren, a sister seedling to Magness, is similar but has fertile pollen, fruit close to 'Comice'.
Seckel has small fruit with a rich, spicy flavor. It's slow-growing and slow to begin bearing.
Blake's Pride, a 2002 USDA release from MD, has an aromatic flavor.
Doyenne Gris or **Golden Doyenne**, an old French pear, has fruit like Bosc but smaller.
Mericourt, a hybrid of European and sand pears from TN, does well in hot summers.
Luscious, a hybrid of European and Siberian pears from SD, is very cold-hardy.
Potomac, a USDA cross of Moonglow x d'Anjou, has fruit like d'Anjou, with a patchy russet. It may scald in storage.
Harrow Sweet, from Ontario, bears young, needs thinning to maintain fruit size and annual bearing.
Duchess or **Duchesse d'Angouleme**, an old French pear, bears young and reliably.
Tenn, or **Tennessee**, a hybrid of European and sand pears, is a bit slow to bear but one of the better cultivars for the Deep South.

Table 17-4.

COMMON ASIAN PEAR CULTIVARS

In approximate order of ripening (varies with area).

All do best in USDA hardiness zones 5 to 9.
Cultivars are apple-shaped and originate from Japan unless otherwise noted.

Cultivar	Fruit				Skin	Yields	Susceptibility to		Tree		Bloom	Harvest	Cold
	Quality	Flavor	Texture	Size	Color		Fire Blight	Pseudo-monas	Size	Growth	Time	Time	Storage Life
Ichiban Nashi	good	very sweet	fine, soft	medium	russet	high	MR	MR	medium	upright	mid-late	very early	1 mon.
Shinsui	very good	very sweet	bit gritty	small-med	russet	high	MS	S	large	upright	early	very early	1 mon.
Yakumo	good	melon-like	fine	small-med	yellow	low	MS	MS	medium	upright	middle	very early	2 mon.
Kosui	best	very sweet	fine	small-med	russet	medium	S	VS	medium	upright	mid-late	very early	2 mon.
Shinseiki	good	mild	coarse	medium	yellow	very high	S	VS	medium	spreads	early	early	3 mon.
Hosui	best	very sweet	fine, tender	med-large	russet	high	VS	MS	large	spreads	early	early	1 mon.
Seigyoku	good	sweet/tart	medium	med-large	yellow	very high	VS	MR	medium	spreads	early-mid	early	3 mon.
Yongi	very good	spicy	medium	large	russet	high	S	S	medium	upright	late	early-mid	5 mon.
Kikusui	very good	sweet/tart	fine	medium	green	high	VS	MS	medium	spreads	middle	early-mid	6 mon.
20th Century	very good	sweet/tart	medium	medium	yellow	high	VS	S	small-med	upright	early-mid	middle	5 mon.
Yoinashi	best	sweet	fine	large	russet	medium	S	S	medium	upright	middle	middle	5 mon.
Chojuro	fair-good	bland, dry	gritty	med-large	russet	very high	MS	S	medium	spreads	early	middle	5 mon.
Atago	very good	sweet	medium	large	russet	high	S	S	medium	spreads	middle	middle	5 mon.
Shinko	very good	rich, sweet	fine, tender	med-large	russet	very high	VR	R	small-med	upright	late	mid-late	2 mon.
Seuri	best	spicy	fine	med-large	russet	very high	MR	S	very large	upright	very early	mid-late	1 mon.
Niitaka	fair-good	bland	coarse	large	russet	high	MR	MS	large	upright	early	mid-late	3 mon.
Ya Li	good	mild	medium	med-large	green	very high	MS	MS	large	spreads	very early	mid-late	6 mon.
Tsu Li	good	mild	medium	large	green	medium	S	MS	large	upright	very early	late	8 mon.
Dasui Li	very good	sweet	medium	very large	green	medium	R	S	very large	upright	early-mid	late	6 mon.
Shin Li	best	sweet	fine	very large	green	medium	VS	S	very large	upright	early-mid	late	6 mon.
Olympic	fair	bland	coarse	very large	russet	high	MR	S	very large	upright	very early	very late	5 mon.
Okusankichi	fair-good	mild, tart	gritty	very large	russet	very high	VS	MR	very large	upright	late	very late	8 mon.

Disease resistance ratings: VR = very resistant, R = resistant, MR = moderately resistant, MS = moderately susceptible, S = susceptible, VS = very susceptible

Average harvest time in Sacramento, CA and central NC: very early:early to mid-July, early: late July, early-mid: early Aug., middle: mid-Aug., mid-late: late Aug. to early Sept., late: mid- to late Sept., very late: early Oct. Ripening times will be about 1 week earlier in southern CA and northern GA, 1 week later in northern CA and western VA, about 2 weeks later in southern MI and southern PA, about 3 weeks later in central OR and central WI, and about 4 weeks later in western WA and central NY.

Notes on Cultivars:
Ichiban Nashi has larger fruit than other very early cultivars, and the earliest ripening period.
Shinsui is a top cultivar in Japan due to high sugars, but a bit gritty for U.S. markets.
Yakumo is little-grown due to small fruit, but has a pleasing melon-like taste.
Kosui trees are S to mites and many pesticide sprays, but MR to pear scab. Pollen cross-incompatible with Shinsui.
Shinseiki is self-fruitful, needs heavy thinning for good fruit size. Fruit hold well on tree, often color-picked 4 times.
Hosui is overly vigorous on *P. betulaefolia* rootstock and very blight-prone, so plant on OH x F stocks. R to pear psylla.
Seigyoku and Shinseiki are pollen cross-incompatible, so need a third cultivar for pollination of Seigyoku.
Yongi sets fruit heavily so needs heavy thinning.
Kikusui fruit size better than 20th Century if heavily thinned, but suffer preharvest drop.
20th Century, also known by its Japanese name **Nijiseiki**, is one of few cultivars with an English translation. It bruises easily.
Yoinashi fruit size much better than 20th Century if thinned well.
Chojuro, an older cultivar, is losing popularity as it's not as juicy as others. Harvest early or bruises severely.
Atago needs very little thinning to achieve large fruit size.
Shinko, the most blight-resistant cultivar, has excellent flavor in hot climates.
Seuri is a round Chinese pear that blooms very early so needs frost protection, pollination by other very early-blooming Chinese types.
Niitaka has sterile pollen so can't be used as a pollenizer. European pear rootstocks dwarf it severely.
Ya Li is a pear-shaped Chinese pear that's slow to bear, needs frost protection due to very early bloom.
Tsu Li, also a pear-shaped Chinese pear, bears slowly, blooms very early. Fruit quality is fair at harvest, improves in storage up to 8 months.
Dasui Li is a pear-shaped Chinese-Japanese hybrid from CA. Very vigorous trees benefit from dwarfing OH x F rootstocks.
Shin Li, another pear-shaped Chinese-Japanese hybrid from CA, also benefits from dwarfing OH x F rootstocks.
Olympic, also called **Ar-ri-rang**, **Korean Giant**, and **Den Bae**, is a large, round Chinese-type pear from Korea.
Okusankichi has only fair flavor at harvest, but flavor improves in storage of up to 8 months.

ornamental pear, is commonly used as a standard pear rootstock in the Southeast due to its strong resistance to fire blight, a severe problem in that region. It also offers resistance to nematodes, crown rot, crown gall and oak root fungus, and is a semi-dwarfing rootstock for Japanese pears. Most strains are hardy only to 0 F (zone 7a), but a hardier strain does well in zone 6b, especially if the graft union is planted below soil level. This rootstock induces earlier bearing than other standard pear rootstocks, but is prone to iron chlorosis on alkaline soils. It is not as tolerant of pear decline, a disease that is sometimes serious in Western pear districts, as the other pear rootstocks used today.

Breeders have developed a series of clonal rootstocks selected from crosses between two European pears, 'Old Home' and 'Farmingdale', designated as OH x F followed by a number. Most of these are size-controlling rootstocks, but one clone, OH x F 97, produces a standard-size tree for European pears, a semi-dwarf tree for Chinese pears, and a dwarf tree for Japanese pears. Like other OH x F rootstocks, it offers hardiness to –25 F (zone 4b), good resistance to fire blight and moderate resistance to pear root aphids, but is moderately susceptible to root lesion nematodes.

Dwarfing and Semi-dwarfing Rootstocks. Several other OH x F clones are commonly used as semi-dwarfing rootstocks for European and Chinese pears, and sometimes as dwarfing rootstocks for Japanese pears. These include OH x F 40, 87, 69 and 333. Growers have chosen OH x F 87 for most European pear plantings of the last decade or so, with some plantings on OH x F 333.

OH x F 51 produces a fully dwarfed European or Chinese pear, but trees aren't as cold-hardy or productive as those on semi-dwarfing OH x F rootstocks. OH x F 51 causes extreme dwarfing of Japanese pears and also reduces their fruit size, so is not used with them.

The oldest dwarfing rootstock for European pears is quince, which was used in Europe over 400 years ago. Though related to pear, quince is a different genus, and many pear cultivars are graft-incompatible with it. Nurserymen overcome this by using an 'Old Home' pear interstem, which is graft-compatible with quince and all other pear cultivars, between the quince rootstock and the pear cultivar. Quince-dwarfed European pears are propagated mostly for home garden sales, and quince is not used as a rootstock for Asian pears. Quince is very susceptible to fire blight, susceptible to iron chlorosis on alkaline soils, and only cold hardy to –5 F (zone 6b), although it does induce earlier fruiting.

Two newer dwarfing pear rootstocks from Germany also induce earlier fruiting, within 2 to 3 years after planting. 'Pyrodwarf' produces a tree about 40% of standard size, and 'Pyro 2-33' produces a tree about 50% of standard size on European and Chinese pears. Neither has been tested much with Japanese pears but should produce fully dwarfed trees. Like the OH x F series, these rootstocks offer good resistance to fire blight and moderate resistance to pear root aphids, but are moderately susceptible to root lesion nematodes.

PEAR CULTIVARS

Commercial European pear production in North America has one of the narrowest cultivar bases of any fruit industry, concentrating on just a handful of cultivars. Three cultivars—'Bartlett', 'Anjou' and 'Bosc'—make up 90% of production. Red-skinned strains of 'Bartlett', 'Anjou' and cultivars resulting from crosses of 'Comice' with a red-skinned 'Bartlett' were widely planted 10 to 15 years ago, but planting of these has declined since pear markets did not offer the expected high premium for red-skinned fruit. Commercial cultivars, described in Table 17-2, are all susceptible to fire blight so are grown mainly in semi-arid Western regions.

Other regions can grow many fine cultivars with good resistance to fire blight and other common diseases. Table 17-3 describes some of the best of these disease-resistant European pear cultivars. Some are older cultivars selected in France or colonial America, and some resulted from breeding programs by the USDA in Maryland/West Virginia or Agriculture Canada at its Harrow, Ontario research station.

Asian pears include a wider range of cultivars, but most unfortunately are quite susceptible to fire blight, with the exceptions of 'Shinko' Japanese pear and several Chinese pears. Main commercial cultivars in North America include 20th Century, Shinseiki, Hosui and Shinsui, but many other cultivars are grown and described in Table 17-4.

SITE SELECTION, PREPARATION AND PLANTING

Pears' early bloom period requires sites with excellent air drainage, preferably near a temperature-moderating water body or on a north-facing slope, unless located in a region where spring frosts are rare. Soil should be deep, at least moderately well drained and of moderate fertility. Highly fertile soils encourage excessive vegetative growth, which is more susceptible to fire blight and can make trees grow too tall and upright.

The ideal site for a new pear planting, as for other fruit trees, was previously planted in a small grain crop or grass pasture or hay crop. When such a site is available, pears can be planted without tilling up the ground by marking planting rows, treating the rows with glyphosate (Roundup™) herbicide, and planting into the killed sod (Photo 17-6).

A few growers are now replanting old European pear orchards to high-density plantings, though this is much less common for pears than it is for apples, because the same pear cultivars are still in demand 35 to 100 years after orchards were planted. With a replant site, remove as much of the root system of old trees as possible, then treat for nematodes with a year or two of intensive cover cropping (see chapter 2) or fumigation before planting new trees. *Pyrus calleryana* or *P. betulaefolia* are the best rootstocks for replant sites due to their strong resistance to nematodes.

Take soil samples for pH, phosphorous (P), potassium (K), calcium (Ca) and magnesium (Mg) at least 6 months before planting, as with other fruit crops. Pears do well on soils with a pH between 6.0 and 8.0, with P levels above 30 ppm, and K levels above 150 ppm. 'Bosc' and 'Comice' pears have a high Mg requirement so often need Mg applications before planting. If soil pH is below 6.0 and Mg levels are less than 1.0 meq/100 g soil, apply 1.5 tons per acre of dolomitic lime, or the amount required by the pH. If pH levels are fine and lime isn't needed but magnesium is, apply Mg sulfate. 'Anjou' pears need more sulphur S than other pears, which can be provided by magnesium sulfate (if needed) or calcium sulfate (gypsum) at a rate of 50 to 100 pounds actual S per acre.

Pears withstand both excess soil water and drought better than any other fruit trees, so few pear growers build raised beds before planting. Semi-arid pear orchards require irrigation, but those in Eastern North America may not need irrigation, so long as the site is not frost-prone and later-blooming cultivars are chosen.

Orchard layout. Pears are most often grown in hedgerows, with newer plantings usually spaced 14 to 20 feet between rows and 6 to 10 feet apart within rows (Photo 17-7). Rows often run north-south to intercept the most light, although rows may instead run parallel to the direction of prevailing winds to enable dews to dry off quickest and minimize fire blight and other diseases. If pears are planted on a slope, rows should run along the contour, and trees must be freestanding.

Most pear orchards are freestanding due to well-anchored pear rootstocks and the high expense of a trellis, but some growers prefer to establish a trellis because it facilitates tree training and limb spreading. Trellised pears require straight rows, and rows can be spaced closer together because trees will be trained to a flat vertical plane. The most usual trellis is vertical with 4 or 5 wires. Branches are spread to a 45-degree angle and tied to the wires so they grow diagonally and parallel to each other, forming a palmate pattern. Some pear orchardists have trained main limbs horizontally (Photo 17-8), but this produces vertical side branches that grow too vigorously and are more prone to fire blight.

Photo 17-6. Plant pears after a grass crop, if possible. When planting into a perennial grass crop such as this fescue, kill the sod with glyphosate herbicide in the planting strips only.

Photo 17-7. Most pears are planted in freestanding hedgerows, spaced 14 to 20 feet between rows and 6 to 10 feet apart within rows. These are trained to the central leader system.

Providing for cross-pollination. Most pear cultivars require cross-pollination by a different cultivar in order to set viable fruit, although 'Bartlett' will set parthenocarpic crops in the Sacramento River delta region. Commercial pear orchardists usually grow 2 or more cultivars, most commonly 'Bartlett' and 'Anjou' European pears, whose bloom periods overlap to provide for cross-pollination. Sports of a cultivar will not pollinate the parent cultivar; for example, 'Sensation' or 'Max Red Bartlett' will not pollinate 'Bartlett', and 'Taylor's Gold' will not pollinate 'Comice'.

Certain pairs of cultivars are also cross-incompatible, such as 'Seckel' and 'Bartlett', 'Seckel' and 'Potomac', 'Shinsui' and 'Kosui', or 'Shinseiki' and 'Seigyoku'. A few cultivars, such as 'Magness' European pear and 'Niitaka' Japanese pear, have sterile pollen so must be grown with 2 other cultivars that can cross-pollinate each other and the pollen-sterile cultivar.

The bloom periods of various European pear cultivars generally overlap, but in some areas an early bloomer like 'Clapp's Favorite' cannot be counted on to provide pollen for a mid-late bloomer such as 'Bosc', and vice versa. Japanese and European pears can provide cross-pollination for each other where their bloom periods overlap, but Chinese and Chinese-Japanese pears bloom too early for European pears and should only be planted with other Chinese cultivars where frosts are seldom a problem.

Planting two or more pollen-compatible cultivars in alternating rows can work well for widely spaced freestanding pears, but bees tend to fly straight down the rows of trellised pears or tightly spaced hedgerows of freestanding pears, so cultivars need to be mixed in the rows. Changing cultivars every 50 to 100 feet down each row, with the same cultivar planted in blocks across the rows, enables the same cultivar to be harvested from both sides of most aisleways.

Many growers raise one cultivar as their main crop and plant much less of a second cultivar to act as a pollinizer for the main cultivar. 'Bartlett' and 'Bosc' as the main cultivar require at least 10% of the orchard planted to a pollinizer cultivar, which is every third tree in every third row of widely-spaced trees, or every 10th tree in each row of hedgerows or trellises. 'Anjou' and 'Comice' as the main cultivar require 17% of the orchard devoted to pollinizer trees, which could be every third tree in every other row of widely-spaced trees, or every 6th tree in each row of hedgerows or trellises.

Planting. Fall planting works well for pears in most regions, since they're quite cold hardy, but those in USDA zone 5 or colder should plant in early spring. Dig holes with a tree planter or a tractor-mounted auger of at least 15-inch diameter, and break up sides of holes with a shovel before setting trees. Set the graft union of trees on *Pyrus betulaefolia* or *P. calleryana* rootstocks a couple inches below the soil line, to protect these cold-sensitive full-size rootstocks from winter damage. Set pear trees on dwarfing or semi-dwarfing rootstocks so the graft union remains at least several inches above the soil line, to prevent the cultivar from rooting and overcoming the rootstock's dwarfing effect.

Photo 17-8. Some pear growers use trellises to make tree training and branch spreading easier. The palmette training system, with branches trained at a 45-degree angle, is the most popular. A 3-wire trellis system, with branches trained horizontally, as here, is used less often because it induces many vigorous, vertical shoots that need frequent pruning back.

Water in trees immediately after planting, and use tree guards on fall-planted trees to protect against rodents and sunscald. Wait until spring to prune fall-planted trees, but prune spring-planted trees right after planting. Generally, prune pears like apples during the first spring, except not quite as hard so as not to stimulate excess vegetative growth that is more prone to fire blight. Head unbranched trees (whips) back to about 36 inches above the soil line (rather than about 30 inches for apples) to induce branching. Branched (feathered) trees need heading to 16 to 18 inches above the uppermost branch, and branches longer than 30 inches should be cut back by about ¼ of their length.

PEAR TRAINING AND PRUNING

Most pear growers train freestanding trees to central leaders or the vertical axis system, both described in Chapter 4. Both systems have a dominant central trunk and limbs radiating out in all directions. Asian pears on vigorous rootstocks are often trained to the modified leader system due to their fire blight susceptibility, especially in California and the Southeast, so blight-infected limbs may be removed without taking out the whole central part of the tree. Trellised pears are most often trained to the palmette system, a vertical plane of branches growing at a 45-degree angle. Training for a palmette system involves choosing branches that grow parallel to the trellis plane and in desired positions, tying them to the wires so they grow at the desired 45-degree angle, and removing all other branches.

Most pear cultivars produce branches that grow too upright, and spreading these branches during the first few growing seasons (Photo 17-9) is the key to both earlier fruit production and creating strong scaffolds that will remain productive over many years, in both freestanding and trellised trees. Use limb spreaders for freestanding trees, or tie branches to stakes in the ground. With trellised orchards, bend and tie branches to the wires in the desired positions. Workers must spread or tie branches several times during spring and early summer for the first 4 or 5 years after planting to produce the desired tree shape and promote thickening of limbs to support fruit load when trees start to bear. If branches are not spread, they continue to grow too upright and long rather than thick, and may crack when under the weight of fruit loads when trees start to bear in the 3rd or 4th growing season (Photo 17-10).

Summer pruning. Prune pears as little as possible during the dormant season, because dormant pruning stimulates vigorous new growth that is more prone to fire blight. Summer pruning, which includes all pruning done during the growing season, causes less regrowth and so avoids fire blight better while also keeping high-density trees in bounds.

Delayed heading involves waiting to head back the central leader or main scaffold branches until new growth is 2 to 3 inches long. At that time, cut the terminal shoot below the top whorl of growth. This will slow the growth of the leader and induce more lateral buds to grow into side shoots. Heading of scaffold branches can be delayed even more until they reach their desired length in summer; then cut them back to induce branch thickening and growth of side branches and fruiting spurs.

Removal of interior shoots during the growing season reduces shading in the center of high-

Photo 17-9. Pear branches tend to grow very upright, so benefit greatly from branch spreading, as done here on 3-year-old Asian pears trained to the central leader system. (Photo by A.A. Powell, Auburn Univ., AL.)

density pear orchards, especially on cultivars with thick vegetative growth. This increases fruit bud formation for next year's crop, increases fruit size of the current season's crop, and reduces the prime feeding sites for mites and certain insects. Begin interior shoot removal when trees become overly dense, usually 5 to 7 years after planting. Remove shoots about a month after bloom, when they can easily be pulled off by hand, which is faster than using hand pruners and eliminates the need to sterilize pruners between cuts to avoid spreading fire blight. Remove the most vigorous shoots until light penetration is good, leaving enough slower-growing shoots to produce future fruiting spurs.

Remove root suckers and watersprouts in late summer, to reduce regrowth of these fast-grow-

Photo 17-10. Vigorous branches should be spread and headed back while young, or they may split under heavy fruit loads, as here, once trees start to bear.

ing shoots that easily contract blight. Break these off by hand rather than cutting them if possible. If this growth must be cut, dip the pruners in a 10% bleach solution between each cut, to prevent the tools from spreading fire blight or other diseases.

Pruning to increase fruit set. Some cultivars of European pear, notably 'Anjou' and 'Comice', are notoriously shy bearers and need special pruning to increase fruit set. Old, weak or shaded fruiting spurs usually set few fruit, especially when cool weather occurs during bloom. Therefore, each spring remove excess one-year-old branches at the base to reduce shading, and leave remaining branches unheaded so they will form flower buds. Head 2-year-old and 3-year-old branches back to fruiting spurs, again to reduce shading from vegetative growth and to increase fruit set on the spurs next to the cuts. Remove entirely a few older branches with only weak fruiting spurs. This renewal pruning encourages formation of spurs that are more fruitful.

Growth regulators for training and pruning. Many growth regulators described in the apple chapter are also labeled for pears and can be applied the same way. Promalin ™ (cytokinen 6-BA and gibberellins 4 and 7), applied as a latex paint mixture in early spring or a foliar spray in mid-spring, induces side branches of young trees, reducing the need to head back main branches by hand. Tre-Hold Sprout Inhibitor ™ (the auxin NAA) controls regrowth of root suckers and water sprouts when applied to the pruning cuts after removing these. Apogee™ (prohexadione-calcium) reduces vegetative growth of mature pear orchards, which greatly reduces the need for pruning and fire blight shoot infections, although it does not reduce blossom blight. Researchers have tested growth regulators including Promalin ™, Provide ™ (gibberellins 4 and 7), Accel™ (cytokinen 6-BA), and Retain™ (aminoethoxyvinylglycine or AVG) to see if any would increase fruit set in 'Comice' pear, but none did.

GROUNDCOVER AND FERTILIZER MANAGEMENT

Most pear growers maintain a grass sod between pear rows, and use herbicides to control all vegetation within a 3-foot wide strip along tree rows. This permanent groundcover keeps pear trees from growing too vigorously, as well as controlling soil erosion and providing a firm surface for orchard operations.

Nitrogen fertilization. Keep nitrogen (N) levels lower for pears than for apples, again to control tree vigor and reduce fire blight. Excess N also reduces pear fruit sugars, flavor and skin color. Young pear trees should produce 18 to 30 inches of terminal growth per year for the first 3 or 4 years until they begin to bear. Young bearing trees should produce 12 to 15 inches of terminal growth per year from years 5 to 14, and trees 15 years or older should make 8 to 10 inches of terminal growth per year.

Generally, apply 15 to 25 pounds of actual N per acre per year in years 1 to 5, and 25 to 35 pounds of actual N per acre per year in years 6 and 7. Increase N rates to 35 to 50 pounds per acre per year in years 8 to 10, and to 60 to 100 pounds per acre for mature trees. Adjust N rates by measuring terminal growth each summer, and by taking leaf samples for N levels in August. If terminals grow in the desired ranges just mentioned, and leaf N levels fall between 1.9 and 2.5%, continue to add N at the same rate used previously. Increase N fertilizer rates if terminals don't grow to the desired length and leaf N levels fall below 1.9%, and reduce or eliminate N fertilization for a while if terminals grow too long and leaf N rises above 2.5%. Apply N fertilizer to the herbicide-treated strip, generally as a dry form such as calcium nitrate or ammonium nitrate.

The best time to apply dry N fertilizer is in late July to early August, when fruit buds for next year's crop are forming. This timing enables the N to increase the number and quality of fruit buds that form, to increase N reserves for leaf production next year without stimulating excessive shoot growth, and to increase fruit size of the current season's crop once trees are bearing. Older recommendations to apply N in early spring often caused new shoots to grow too long and succulent, leaving them more prone to fire blight and increasing the need for pruning.

By late summer most shoots have stopped lengthening so N applications don't cause a problem unless a large amount is applied at once, which can stimulate late growth and delay tree hardening, leading to winter injury. Therefore, if leaf analysis indicates N levels are below normal, spread only part of the N needed as dry fertilizer in late summer, and apply the rest as a foliar urea spray after harvest time. Growers who fertilize with manure, compost or other organic forms of N should apply them in late spring, since they take time to break down and release N.

Other fertilizer nutrients. Continue to sample soil for pH, P, K, Ca and Mg every 3 to 4 years. Many soils with adequate pH levels do not need fertilizer applications of these nutrients, but if the soil needed lime or any of these nutrients before planting, further additions will probably be needed every few years. Use August leaf analysis to confirm a K deficiency if leaf K levels fall below 0.9%, and to detect a S deficiency if leaf S levels fall below 0.15%.

If K is needed, potassium nitrate is preferred, since salt injury can occur if potassium chloride is applied without enough subsequent irrigation or rainfall to leach salts from the surface root zone. Potassium nitrate can be applied in late summer as part of the N fertilizer. Apply magnesium sulfate if both Mg and S are needed, or other sulfate fertilizers if just S is deficient. Ammonium sulfate provides S and N but also lowers pH, so avoid it unless the pH is high.

Some pear orchards benefit from foliar Ca sprays as for apples. 'Anjou' pears commonly need Ca sprays to reduce cork spot, especially when trees have strong vegetative growth, a light crop or larger than usual fruit, all of which can cause localized Ca deficiencies in fruit. Ca sprays slightly delay fruit maturity, which increases fruit firmness and reduces storage rots, especially blue mold and side rot. These benefits may justify Ca sprays even for cultivars that do not often develop cork spot, such as 'Bosc'. Most growers apply Ca as 2 or 3 calcium chloride sprays added to pesticides applied from 45 to 60 days after petal fall, as for apples.

Boron (B) deficiency may occur on bearing pears in Western regions, and can reduce or prevent fruit set. Once trees start to bear, check leaf samples taken in August for B levels as well as N and S. If leaves contain less than 30 ppm B, apply sodium pentaborate either broadcast on the soil at 15 to 20 pounds per acre, or as 2 foliar sprays per year at a rate of 8 pounds per acre per spray. Spray B in fall before leaves drop and again 3-4 days before bloom in spring, or add B to the prebloom spray and the first cover spray. Where B deficiency commonly occurs, applying sodium pentaborate to the soil every 3 years at a rate of 15 pounds per acre can prevent most problems. If a deficiency does occur, foliar sprays offer the quickest remedy.

Zinc (Zn) deficiency occurs rarely in pears in some alkaline soils. Most growers do not need to routinely test leaf samples for Zn deficiency, but should recognize deficiency symptoms if they occur. Shoot tips develop rosettes of tightly compressed, narrow leaves, while lower leaves on the shoots may become chlorotic and drop early, leaving "blind" shoots with few leaves below the tight whorls at the tips. Trees with these symptoms, or with August leaf Zn levels below 17 ppm, need several sprays of zinc sulfate or other materials containing zinc, such as zineb fungicide if a disease treated by this material also occurs. Apply these materials after harvest and again late in the dormant season.

IRRIGATION AND FROST PROTECTION

Almost all pear orchards are irrigated, because they are located mainly in semi-arid regions. Most pear growers use over-tree sprinkler irrigation so they can also provide frost protection, since pears bloom early. Sprinkler irrigation can also provide evaporative cooling in areas with hot summers, although pears don't need this as much as apples do in these areas, since red-skinned pears are grown mainly in cool-summer regions and heat doesn't impair skin color development of yellow-skinned or russeted pears. Some growers use trickle irrigation (Photo 17-11) and provide frost protection with orchard heaters or wind machines, which reduces problems of irrigation spreading blight bacteria during bloom.

Regular irrigation, especially in the last 8 to 12 weeks before harvest, is critical to development of good fruit size and quality, especially in Asian pears. New Zealand researchers found that water stress in early summer, while Asian pears were still in the slow growth stage, decreased fruit calcium and increased flesh spot decay, a disorder that appears after 2 to 6 weeks of cold storage. Water stress during the 3 weeks before fruit harvest significantly decreased Asian pear fruit size and weight.

Photo 17-11. About 90% of pear growers in the Pacific Northwest region use over-tree sprinkler irrigation which can provide spring frost protection as well as summer irrigation. These Asian pears in the Southeast are drip irrigated. (Photo by A.A. Powell, Auburn Univ., AL.)

POLLINATION, FRUIT THINNING AND GROWTH REGULATORS FOR CROP CONTROL

Pear flowers occur mainly on long-lived spurs like those of apples, and most pear cultivars produce far fewer flowers on one-year-old branches than do most apple cultivars. Biennial bearing poses even more of a threat to pears than to apples, since poor pollination or a frost during pears' early bloom period can cause a crop loss, starting a cycle of alternating years of light and heavy production. Pear growers must provide for adequate pollination, protect against frost, and in most years practice rigorous fruit thinning to ensure annual crops.

Commercial pear growers rent honeybee colonies to provide adequate pollination, because wild bees are not plentiful enough in most pear orchards. Pear flowers are not as attractive to bees as those of many other fruit trees, because they have less nectar with a low sugar content, so pears need higher bee colony rates per acre. Oregon State University researchers recommend at least 2 strong colonies (2 ½ pounds of bees per colony) per acre for the Medford area, and at least 3 strong colonies per acre for the Hood River region.

Wait until 10 to 20% of pear flowers have opened before introducing the bees; if brought in too early bees will forage among other available flowers and ignore pear blossoms once they open. Mow the orchard groundcover just before introducing the bees, to eliminate flowers of weeds such as dandelions, which are more attractive to bees. Mow again every few days during the 10- to 20-day bloom period, to prevent weeds from flowering, and control weeds within tree rows with herbicides. Place beehives in spots sheltered form the wind, in groups of 3 to 5 hives per group distributed throughout the orchard, with hive entrances facing south so the bees will warm up sooner and work longer. Provide drinking water for the bees, and do not spray any insecticides that would harm them during the bloom period.

Fruit thinning. With good pollination and frost protection, European pear cultivars may set 3 to 5 times as many fruit as needed for good fruit size and quality, and Asian pears may set 8 to 10 times the ideal number of fruit for a good crop. This happens because pear flowers occur in clusters, and European pears have usually 5 blossoms per cluster (Photo 17-12), while Asian pears generally have 10 to 12 flowers per cluster. Like apples, some immature fruit will fall naturally during "June drop", but many clusters with 3 to 5 fruit per European pear cluster, or 8 to 10 fruit per Asian pear cluster, remain. Thinning fruit to 1 or sometimes 2 fruit per cluster, for all types of pears, is essential to producing pears with adequate size and sugar content for market. Excess fruit set, if not thinned, also frequently causes limb breakage in pear trees, with their naturally upright growth (Photo 17-13).

Pear growers use post-bloom chemical fruit thinners like those used for apples, but usually at lower rates, especially for Asian pears which are more sensitive to these chemicals. Bloom thinners are not used for pears, since they bloom early and are subject to more frost damage than apples, and are less preferred by bees. A few European pears with low fruit set, notably 'Anjou' and 'Comice', do not require thinning sprays in many seasons. Many growers apply the auxin NAA (naphthalene

Photo 17-12. European pears, shown here, have 5 blossoms per flower cluster, while Asian pears have 10 to 12 blossoms per cluster. If frosts don't reduce the crop and pollination is good, far too many fruit will set, so pears need rigorous fruit thinning to produce good-sized fruit.

Photo 17-13. Inadequate fruit thinning can also cause pear branches to break under the weight of excess fruit.

acetic acid) at 10 to 25 ppm when fruit diameter reaches 5 to 12 mm, usually 7 to 14 days after petal fall. This allows time to check how much fruit set occurred and set the dosage accordingly. Apply NAA only when temperatures are between 60 and 80 F. The auxin NAD (naphthylacetamide) is not used as much because it must be applied earlier, at petal fall to 7 days later, at a rate of 10 to 40 ppm. Ethephon is not used for pears, although it does cause thinning, because it reduces fruit weight and flesh firmness at harvest, and increases flesh spot decay in cold storage.

The insecticide carbaryl (Sevin™), often used to control early-season codling moth, also provides fruit thinning when applied between petal fall and 15 days later (10 to 25 days after full bloom). Carbaryl is highly toxic to bees, so move all beehives out of the orchard before applying it, and use the less toxic formulations Sevin XLR Plus or Sevin 4F rather than the more toxic Sevin WP. Lime sulfur sprays before bloom, to control diseases such as pear scab, can reduce fruit set, as can the fungicide dodine applied during bloom to control scab.

Asian pears require follow-up hand thinning even after applying thinning chemicals, because they are more susceptible to overthinning by chemicals than European pears, so lower rates (5 to 20 ppm NAA) must be used to avoid this. Ten to 14 days after applying chemical thinners, affected fruit should have fallen and workers remove additional fruit, including all fruit from some clusters, so that only 1 fruit remains every 6 to 10 inches apart along each branch. Some Asian pear growers do all fruit thinning by hand, with workers going over the orchard twice to ensure adequate thinning. Some growers have workers cut off excess fruit with clippers, but most have found that twisting them off by hand is just as fast or faster.

Stop-drop sprays. NAA is also registered to prevent pre-harvest fruit drop in pears, as for apples. Apply it when sound fruit first start to drop, 5 to 7 days before expected harvest, and do not harvest within 5 days of applying it.

AVG (aminoethoxyvinylglycine or ReTain™) inhibits ethylene synthesis in pears, as it does in apples, so also acts as a stop-drop if applied 4 weeks before expected harvest. It delays harvest and reduces internal breakdown of pears in cold storage.

PEST MANAGEMENT

Fewer pests generally attack pears than most other fruit trees, but the pests that are problems can be devastating and require very diligent control.

Diseases, insects and mammals that can kill trees. Fire blight is generally the biggest threat to pears in most areas, but several other pests can kill pear trees and often require control also.

Fire blight, caused by the bacterium *Erwinia amylovora*, influences almost every choice that pear growers make. This disease can kill trees of susceptible cultivars in many areas of North America, and even where it's less severe in the semi-arid West, it can greatly reduce crops with blossom blight and hurt development of future fruiting wood with shoot blight. Choosing cultivars and rootstocks with high blight resistance can greatly prolong the life of a pear planting and reduce the need for expensive bacteriocides, especially where the disease is most serious, such as the southeastern U.S. Avoiding excessive vegetative growth through reduced N fertilizer, summer rather than winter pruning, and use of Apogee™ vegetative growth retardant will further minimize blight problems. Sprinkler irrigation can spread fire blight, so avoiding the need for frost protection irrigation by choosing late-blooming cultivars and a site with excellent air drainage also helps.

Most pear producers raise blight-susceptible cultivars, so rely on bacteriocide sprays during bloom to prevent blossom blight, as well as for the next 1 to 2 months of spring and after harvest to prevent shoot blight (Photo 17-14). Apply a copper compound before ½-inch green, and streptomycin (Agri-mycin™) or oxytetracycline (terramycin or Mycoshield™) at first bloom and every 3 to 7 days thereafter when temperatures rise above 60 F and humidity is above 65% or dew occurs. A copper compound can substitute for some antibiotic sprays

Photo 17-14. Control the shoot blight phase of fire blight by limiting N fertilizer, limiting shoot growth by applying prohexadione-calcium (ApogeeTM), or by applying antibiotic or copper sprays during late spring and early summer. Prune out infected shoots, like these, at least 6 inches below the

for russeted pears such as 'Bosc' or 'Hosui', but copper sprays during the growing season cause severe russeting on pears that should have a smooth finish, such as 'Anjou', 'Comice' or 'Shinseiki'. Streptomycin may cause bleaching or burning of fruit and leaves in some Asian pear cultivars, so test it on a small block of each cultivar before using it on the entire orchard. Monitoring daily temperatures, humidity and leaf wetness, and using a computer program such as MARYBLYT™ (available by subscription at http://afrsweb.usda.gov/fireblight/) to predict infection periods, can somewhat reduce the number of sprays applied.

Now that streptomycin has been used to prevent fire blight for almost 50 years, blight bacteria have developed resistance to this material in most pear-growing areas of the Pacific Northwest, and growers there must use oxytetracycline or copper compounds. Growers in areas where streptomycin is still effective should rotate among several antibiotics and/or copper sprays to avoid selecting *E. amylovora* strains that resist streptomycin, or it will no longer be effective in preventing blight. Another approach is to spray a beneficial bacterium, *Pseudomonas fluorescens*, which competes with *E. amylovora* rather than killing it, along with streptomycin. Strain A506 of *P. fluorescens* (BlightBan™) was selected for resistance to streptomycin so it can be applied with that antibiotic at 20% bloom and 50% bloom, and as long as it's sprayed on blossoms before *E. amylovera* bacteria land there, it colonizes the blossoms and prevents fire blight establishment. Do not apply oxytetracycline within 7 days after applying BlightBan™, since this antibiotic will kill the beneficial *P. fluorescens*. If no antibiotics will be applied, another beneficial bacterium, *Bacillus subtilis* (Serenade™), could be applied instead.

Fire blight infections often occur even with regular spraying, and should be promptly pruned out. Cut at least 6 inches below the visible infection line during the growing season, and at least 12 inches below the infection line during the dormant season, and dip pruning tools in a 10% bleach or non-corrosive Lysol™ solution between each cut, to avoid spreading the disease.

Pear decline, caused by a phytoplasma, is transmitted by pear psylla, a tiny aphid-like insect, which injects it into leaf phloem tubes as it feeds. The phytoplasma then spreads throughout the phloem of the branches, trunk and rootstock, causing necrosis and death of phloem cells in susceptible rootstocks. Some pear rootstocks of Asiatic species, notably *Pyrus serotina* and *P. ussuriensis*, are extremely sensitive to pear decline, and if trees on these rootstocks are infected by psylla carrying the disease, they rapidly lose their leaves and die. Many pear trees on these rootstocks died between 1950 and 1975, so few nurseries propagate these rootstocks anymore, and they are illegal to use as rootstocks in New York and some other pear districts. All pear rootstocks commonly used today are tolerant to pear decline, except that *P. calleryana* is only moderately tolerant. Decline-infected pears on commonly used pear rootstocks often develop reddish, curled leaves that drop prematurely. Infected trees produce fewer and smaller spurs and fruits, reducing yields. During the 1960's and 1970's, many pear growers treated decline-infected trees with trunk injections of oxytetracyline, but this practice is expensive and rarely done today. Most growers try to prevent pear decline by controlling pear psylla and keeping trees healthy with timely irrigation and fertilization.

Crown rot or **collar rot**, caused by *Phytophthora cactorum* or other species of *Phytophthora*, a soil-borne fungus, also kills some pear trees on less well-drained soils. Pears tolerate wet soil better than any other fruit tree, but such sites might need raised beds and rootstocks with the most tolerance to this disease, *Pyrus calleryana* or *P. betulaefolia*. During unusually wet springs, early spring applications of fungicides effective against *Phytophthora*, such as metalaxyl (Ridomil™) or fosetyl-Al (Aliette™) can help prevent but not cure crown rot.

Pear replant disease, caused by a combination of soil fungi and root lesion nematodes, kills some trees in new plantings on ground where old pear orchards were removed. Prevent this by removing as much of the old pear root system as possible, cover cropping for a year with grass crops such as sorghum over summer and 'Eltan' winter wheat before planting, and choosing rootstocks with good resistance to lesion nematodes, especially *Pyrus calleryana*.

Borer insects do not pose as great a threat to pears as to apples. They mainly attack trees weakened or damaged by other factors such as sunscald, excess weed competition, mower damage, deer or voles, so controlling these pests usually avoids borer damage. Tree wraps as described for apples can exclude **flatheaded apple tree borer** beetle larvae such as *Chrysobothris mali* in Western North America or *C. femorata* in eastern North America.

Voles and **deer** pose the same threat to pears as they do to apples. Using herbicides to control weeds in the tree rows reduces both vole damage and mower damage. White plastic tree guards, when buried several inches into the soil, can prevent vole damage as well as sunscald. Where vole populations are high, growers may need to apply poison baits such as zinc phosphide pellets. High deer populations require an electric fence, while repellent chemicals reduce damage where populations are low to moderate.

Diseases attacking flowers, foliage and fruit. Each of these diseases poses a threat only in certain regions and not in others, but can be serious every year where they occur. Growers must either choose cultivars resistant to the disease(s) of the area or apply fungicides or bacteriocides regularly, and must also practice strict orchard sanitation in either case.

Pseudomonas blossom blight, caused by the bacterium *Pseudomonas syringae*, kills pear blossoms when cool, wet weather prevails during bloom. This disease is mainly a problem in the humid parts of the Pacific Northwest, where choosing cultivars with some resistance to it, and pruning only in the summer, can minimize damage. A fixed copper spray after leaves drop in fall and again before ½-inch green, and at least 2 antibiotic sprays during bloom, reduce this disease as well as fire blight. Spraying the beneficial bacterium *Pseudomonas fluorescens* (BlightBan™) during bloom can also prevent some infection by this harmful species in the same genus. *Pseudomonas syringae* bacteria provide nuclei for ice formation so cause frost damage as well as blossom blight, which mainly attacks frost-damaged flowers, so the best prevention is good frost protection. If a frost is expected, turn orchard sprinklers on when temperatures drop to 34 F, well before blossoms would be damaged at 28F, to avoid causing evaporative cooling during the frost, and keep water running until all ice melts the next morning. Or use orchard heaters or wind machines. If frost protection is not available, spray with copper or antibiotics before each anticipated frosty night, at least weekly from green tip through full bloom.

Pear scab, caused by the fungus *Venturia pyrina*, is in the same genus as the apple scab fungus but is not as widespread, and causes similar damage of defoliation and corky spots on fruit skin (Photo 17-15). It requires control in humid cool-summer regions including the Pacific Northwest, the Upper Midwest and the Northeast of North America. It does not over-summer in areas with hot summers, such as most of the southern half of the U.S., so usually needs no control there. Where it occurs, choosing cultivars with scab resistance will greatly minimize the need for sprays. Applying urea (at least 50 pounds/acre) and/or dolomitic lime to fallen pear leaves on the orchard floor soon after leaf drop, or spraying urea and lime sulfur on trees just before leaf fall, has reduced spring spore counts of the scab fungus by 50 to 90%. Computer models based on temperature and leaf wetness can predict when to spray fungicides for greatest control and least use of pesticides.

If scab developed during the previous year, fungicide sprays should begin at green cluster and continue every 7 to 10 days until the second cover spray, and beyond that if the season remains wet. Rotate a metal-based fungicide such as mancozeb (which is more effective than ferbam or ziram) with the anilinopyrimidine fungicide cyprodinil (Vanguard™) or a strobilurin fungicide to avoid developing resistance to the anilinopyrimidine or the strobilurin. The strobilurins registered for pear scab include kresoxim-methyl (Sovran™) and trifloxystrobin (Flint™). A sterol inhibitor, fenarimol (Rubigan™) or triflumizole (Procure™), can be substituted for the strobilurin from the petal fall spray on; they may hurt pear fruit shape if used earlier.

Fabraea leaf spot, caused by the fungus *Fabraea maculata*, is worst in hot, humid climates, especially the Southeastern U.S., where cultivars resistant to this disease are advisable. Western pear growers usually don't need to control this disease, and it isn't often a problem during the same season as pear scab in the East, although some growers in mid-Atlantic states must frequently control both diseases. The fungus causes spots on leaves,

Photo 17-15. Pear scab can seriously disfigure fruit in regions with cool summers and humid springs, such as the Pacific Northwest and Northeast of North America. Prevent scab by choosing resistant cultivars or applying fungicides. (Photo by Jay W. Pscheidt, Oregon State Univ., Corvallis, OR.)

fruits and twigs and can defoliate trees, severely reducing yields and tree vigor if this happens for several years. Fortunately, mancozeb rotated with strobilurin sprays from petal fall to the second cover spray, which may be applied anyway for pear scab, also reduce Fabraea leaf spot. It requires additional sprays every 2 to 3 weeks all summer, using ziram or ferbam rotated with lime sulfur (if temperatures remain below 80F and not within 3 weeks of an oil application) or a strobilurin.

Powdery mildew, caused by the fungus *Podosphaeria leucotricha*, is not often a problem on pears, but may occur in the West on cultivars with high susceptibility, such as 'Anjou' and 'Comice'. This fungus produces a white, powdery growth on leaves that may defoliate them, and a net-like russet on fruit. Stobilurin or sterol inhibitor fungicides applied to control scab from the pink stage until second cover will also control powdery mildew, but the metal-based fungicides and cyprodinil will not. Lime sulfur can be added to mancozeb sprays to control mildew on most pear cultivars, once 3 weeks has past since the dormant oil spray, but sulfur fungicides applied after budbreak may reduce fruit set, so don't use them for 'Anjou', 'Comice' or other cultivars with poor fruit set.

Sulfur sprays after fruit thinning will not cause further fruit loss, but apply sulfur only when temperatures will stay below 80 F and not within 3 weeks of a summer oil application, or it can cause leaf damage and severe fruit russeting. The biological fungicide *Ampelomyces quisualis* (AQ10™) works well if mildew is the only disease problem during certain periods.

Fruit rots (caused by several fungi described under apples) pose much less problem for European pears than for apples, since they are picked while still quite firm and ripened off the tree. The thick skin of russeted pears, both European and Asian, also provides good resistance to fruit rots. Smooth-skinned Asian pears such as 20th Century or Shinseiki may need a few summer sprays to prevent fruit rots. Alternate strobilurin fungicides with contact fungicides as for apples. Prompt removal of any mummified fruit or branch cankers caused by fruit rots can reduce or avoid the need for sprays.

Sooty blotch, caused by several fungi, can infect the surface wax of pear fruits with dark smudges as it does on apples. Summer pruning to keep trees open also greatly reduces sooty blotch. Late-summer sprays of strobilurin fungicides can control sooty blotch, or a 15-minute post-harvest dip into a 500 ppm solution of swimming pool chlorine can remove it from fruit.

Sooty mold (Photo 17-16), caused by several other species of black fungi that grow on the surface of fruit, especially thrives on honeydew, the sugary excrement of pear psylla, mealybugs or aphids, so controlling these insects can prevent most problems with the disease. Postharvest chlorine treatment or scrubbing will also remove sooty mold.

Indirect insect and mite pests. Pears are highly susceptible to several indirect insect and mite pests, which spread diseases or secrete honeydew, the substrate for sooty mold. Commercial pear growers therefore make greater use of summer-weight oil sprays plus specialized miticides/insecticides to control these pests than do most other fruit growers.

Pear psylla (*Cacopsylla pyricola*), a soft-bodied insect similar to aphids (Photo 17-17), attacks only pears and is generally the worst insect pest of pears. Psyllas suck sap from the leaves and can greatly weaken trees, in addition to transmitting pear decline and secreting large amounts of honeydew, which drips on fruit and allows growth of sooty mold fungi.

Pear growers control psylla with a dormant oil spray and at least 2 sprays of summer-weight oil during the growing season based on scouting for psyllas in the top shoots of pear trees, where they accumulate. The action threshold is 0.5 nymphs per leaf in spring and 1.5 nymphs per leaf in summer. Summer pruning of new shoots, root suckers and watersprouts in about mid-July can greatly reduce late-season psylla populations, because adults prefer to lay eggs on new growth and have difficulty feeding on older, tougher leaves. Growers may add a pyrethroid insecticide such as esfenvalerate (Asana™) or permethrin (Ambush™ or Pounce™) to the dormant oil spray to kill adult psyllas, since oil alone kills only eggs although it repels adults, but this risks psyllas developing resistance to the pyrethroid.

Most pear growers wait to apply a pyrethroid such as esfenvalerate (Asana™) or the insect growth regulator pyriproxyfen (Esteem™) with 1 to 2% oil at delayed dormant to white bud for psylla control. They also use avermectin (Agri-Mek™) with summer-weight oil at petal fall to 14 days later, to control psylla, mites and aphids. Three pre-bloom applications of kaolin clay (Surround™) at dormant, green cluster and white bud have also provided good control of overwintering psylla adults, but don't control mites, aphids or other pests as well. If psyllas need further control during late spring and summer, avoid pyrethroids, which can hurt beneficial predators of aphids and mites during that

Photo 17-16. Sooty mold grows on the surface of pears, on honeydew excreted by pear psylla or aphids. Control these insects with oil sprays and selective insecticides to prevent the disease. (Photo by Alan Biggs, West Virginia Univ., Kearnysville, WV.)

Photo 17-17. Pear psylla is usually the worst insect pest of pears, generally triggering 5 to 8 sprays of oil or insecticide per season in the Pacific Northwest region. (Photo from Cornell Univ., NY.)

season, and use summer oil instead. Western growers generally spray 5 to 8 times per growing season for psylla. Growers who use mating disruption for codling moth may reduce psylla sprays to 3 to 4 per season, by eliminating use of broad-spectrum organophosphate insecticides that kill psylla predators, which can now provide biocontrol of psylla.

Pear rust mite (*Epitrimerus pyri*) and **pearleaf blister mite** (*Phytoptus pyri*) feed only on pear trees, causing leaf bronzing, stunted growth of young trees, and deformed or russeted fruit if populations are heavy. Russeted pear cultivars such as 'Bosc' or 'Shinko' do not show fruit damage. Mites overwinter on trees as adult females, may damage buds by feeding on them over winter, and begin laying eggs as soon as buds begin to swell. A dormant oil plus lime sulfur spray and a miticide such as pyridaben (Pyramite™) or avermectin (Agri-Mek™) at tight cluster and petal fall are essential to control these pests. Liquid lime sulfur applied in late fall, when high temperatures range between 50 and 75 F but before mites have burrowed under bud scales to hibernate, can greatly reduce next spring's pest population.

Spider mites, including two-spotted spider mite (*Tetranychus urtecae*), Pacific spider mite (*T. pacificus*) and European red mite (*Panonychus ulmi*) may attack pears along with many other fruit plants. These mites also cause leaf bronzing and may stunt growth of young trees, but they form webs and do not cause fruit russeting. Pears have much lower thresholds for economic damage than do apples: just 2 mites per leaf instead of about 10. A dormant oil spray and at least 2 summer oil sprays or miticides as needed, or sprinkler irrigation in summer, can control them. Rotate miticides used, since these mites can have 5 to 9 generations per season, depending on the region and summer temperatures, so populations can rapidly build up resistance to a particular miticide used repeatedly. Using mating disruption to control codling moth reduces the need for spider mite control by conserving predator mites that provide biological control.

San Jose scale (*Quadraspidiotus perniciosus*), a soft-bodied insect covered by a hard shell, causes the same damage to pear branches and fruit as it does to apples. The dormant oil spray and several early-spring sprays used for psylla help to control it, but severe infestations also need pyriproxyfen (Esteem™) plus summer oil when crawlers first emerge in May or June.

Grape mealybug, *Pseudococcus maritimus*, is a common pest of pears as well as grapes in Western North America, and an occasional pest in Eastern North America. These soft-bodied insects, protected by a fuzzy white coating, feed on leaves and shoots, secreting a honeydew that drips onto fruit and serves as a substrate for sooty mold. A delayed-dormant oil spray helps to control grape mealybug. Green lacewings and other predators contribute heavily to grape mealybug control, so avoid using insecticides that disrupt natural predators.

Pear cankerworms include the **spring cankerworm** (*Paleacrita vernata*) and the **fall cankerworm** (*Alsophil pometaria*). Cankerworms, a type of inchworm, may skeletonize leaves, which can greatly reduce yields if populations are high, and occasionally feed shallowly on fruit. Bird predators or insecticides used to control other caterpillar pests usually control cankerworms, but they may reach damaging levels if the area around a pear orchard offers little habitat for birds and if mating disruption is used to control codling moth. The biological insecticide *Bacillus thuringiensis* (Bt), or molting inhibitors specific for moth pests, can control cankerworms.

Fruit-damaging insects. Developing pear fruits are harder than apples, and European pears are harvested before fruit soften, so internal fruit-feeding insects generally cause less damage to pears, although they still must be controlled. Many surface feeders can cause sunken spots that make fruit unmarketable.

Codling moth (*Cydia pomonella*) attacks pears as well as apples, and is controlled as described for apples. Many Western pear growers, where codling moth is the only serious fruit-attacking insect, use mating disruption to control this pest with pheromones instead of pesticides. California studies documented a $40/acre increase in pest control costs in year 1 of using mating disruption, but by year 5 this had changed to a $677/acre decline in pest control costs, due to reduced need for sprays for pear psylla and mites. If leafrollers or green fruitworms are problems as well as codling moth, non-toxic insecticides that inhibit larval molting can control all these moth pests, without hurting predators of psylla, aphids or mites. Molting inhibitors include fenoxycarb (Comply™), methoxyfenozide (Intrepid™), tebufenozide (Confirm™) or azadirachtin derived from neem seed (Aza-Direct™, Azatin™ or Neemix™). If a pyrethroid insecticide such as pyriproxyfen (Esteem™) or fenpropathrin (Danitol™) is used at petal fall to control pear psylla, it will also control first-generation codling moth.

Leafroller pests of pear include **fruittree leafroller** (*Archips argyrospila*), **omnivorous leafroller** (*Platynota stultana*), **oblique-banded leafroller** (*Choristoneura rosaceana*), **apple pandemis** (*Pandemis pyrusana*) and **orange tortrix** (*Argyrotaenia citrana*). Besides feeding on leaves, which they roll up to protect themselves from predators and pesticide sprays, these caterpillars feed on flower parts and fruit early in the season. Fruit feeding leaves depressions that become tough and russeted by harvest.

Both consumers and canneries reject fruit with more than a small amount of leafroller damage. Parasitic wasps, insect and spider predators, and birds frequently keep leafrollers under damaging levels, so scout orchards frequently for leafrollers in spring and early summer, and treat only if needed. Use the biological insecticide *Bacillus thuringiensis* (Bt) or a molting inhibitor. Mating disruption can be used for some species if they are a recurring problem.

Green fruitworms of several species, including *Orthosia hibisci* and *Amphipyra pyramidoides*, feed on leaves and young fruit of pears, gouging out large cavities that become toughened and russeted by harvest, rendering the fruit unmarketable. Green fruitworms often occur only in spots in a pear orchard, usually along the edges as moths have flown in from nearby woods or windbreaks. Scout for the worms, which become quite large and lime-green with a bright yellow stripe so are easy to spot, beginning at petal fall, and spot treat the outer rows or individual trees as needed with the biological insecticide *Bacillus thuringiensis* (Bt).

Stink bugs and other **plant bugs** of many species may feed on young pear fruit early in the season, and inject a toxin that stops future cell growth at the feeding site. This causes dimples or irregularly depressed areas, underlain by pithy tissue, in mature fruit. These pithy areas are too deep to be removed by normal peeling processes, so fruit with such damage is unsuitable for either fresh market or canneries. Damaging species include **consperse stink bug** (*Euschistus conspersus*), **Southern green stink bug** (*Nezera viridula*), **lygus bug** (*Lygus hesperus*), **boxelder bug** (*Leptocoris triviatus*), **leaffooted plant bug** (*Leptoglossus clypealis*) and others. Monitor pear orchards for any of these bugs, damage, or weeds that can harbor these bugs. Control such weeds—including mulleins, docks, mustards, plantains, milkweeds, mallows, morningglories, thistles, and vetches—within and around the orchard. If such weeds are present, check around their bases in early spring for adult bugs of any of these species which have caused damage in past years, and apply an insecticide in you find five or more bugs in about a 1-acre block. Many pear growers do treat for these bugs during the delayed dormant to pink stages, and more sprays may be needed in summer based on sampling trees for bugs and/or damage.

Plum curculio (*Conotrachelus nenuphar*) may also caused misshapen or scarred pears east of the Rocky Mountains in North America. Control it as for apples, by spraying an insecticide such as indoxacarb (Avaunt™), fenpropathrin (Danitol™) or phosmet (Imidan™) at petal fall and again 10 days later, at first cover.

Bird and mammal pests of fruit. Birds do not generally peck at European pears, since they are harvested while still firm and ripened after harvest, but Asian pears that are tree-ripened for local sales may suffer bird attacks. **Crows** can cause huge damage if large flocks descend on the orchard, and pecked fruit attract wasps, which are very hazardous to workers or pick-your-own customers. Asian pear growers often must use several bird-scaring devices, some visual and some noise-making, and rotate these often around the orchard to birds don't get used to them.

Squirrels of various species may reduce yields of smaller-fruited pears in some areas, as they sometimes carry off many immature pears a few weeks before harvest. This is mainly a problem for cultivars with small fruit, such as 'Seckel' (Photo 17-18), 'Honeysweet' or 'Forelle', that didn't receive enough thinning or irrigation to enlarge the fruit. Do not plant such cultivars near woods that can harbor squirrels.

Photo 17-18. Squirrels may greatly reduce yields of smaller-fruited "gourmet" pears such as this 'Seckel'. Don't plant such cultivars near a wooded area.

HARVEST, STORAGE AND RIPENING

Harvest European pears when physiologically mature—when fruit have gained their maximum size and sugar content—but before they begin to ripen, or many fruit will develop internal breakdown on the tree and fruit that are still sound will become gritty and won't store well. Asian pears can be tree-ripened, and should be to develop maximum flavor and sweetness, but may be picked a few days before fully ripe if they are to be shipped long distances or stored for several months.

Fruit maturity indices. Growers determine the time to harvest pears using a combination of several factors, and should not rely on any single maturity index alone. Sample at least 10 fruit from different positions on at least 5 trees for each test.

The number of days from bloom to harvest varies from region to region and somewhat from year to year, but usually falls within a characteristic range for each cultivar within each area. Pear growers in Washington start checking maturity of 'Bartlett' from 100 to 115 days after bloom, of 'Bosc' 130 to 135 days after bloom, and of 'D'Anjou' 145 to 150 days after bloom. This gives growers an idea of when to expect the harvest period, but other methods must determine it more precisely.

Flesh firmness provides the backbone for pear harvest decisions. Test each fruit with a pressure tester, or penetrometer, on two opposite sides after peeling away the skin where the metal rod will be placed (Photo 17-19). Pressure readings indicate the pounds of force needed to push the metal plunger into the fruit flesh, and numbers decline as fruit gets less hard. Harvest European pears when firmness readings fall between 20 and 15 pounds pressure, and Asian pears when firmness lies between 10 and 8 pounds pressure. The optimum pressure varies with cultivars, especially of European pears. For example, the optimum pressure at harvest for 'Bartlett' is 17 pounds, while 15 pounds pressure is optimum for 'Bosc' and 'Anjou'.

Soluble solids are the soluble sugars in the juice of the fruit, so indicates sweetness. European pears should reach at least 10% soluble solids, as measured with a refractometer, before harvesting. Asian pears should have at least 15 to 16% soluble solids for shipping, and at least 18 to 20% soluble solids for local sales. When selling Asian pears locally and allowing them to tree-ripen, taste a fruit after testing with a refractometer to develop a taste correlation to the readings.

Groundcolor changes can be detected on most pears, except those that are heavily russeted. Growers may compare the fruit to standard color

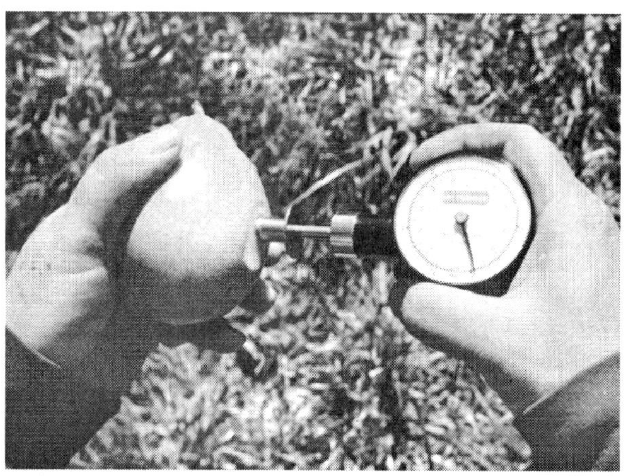

Photo 17-19. Testing flesh firmness is a main method of judging maturity for both European pears, such as this 'Anjou', and Asian pears.

photos for each cultivar that show colors at each maturity stage. Harvest European pears when the skin's groundcolor changes from green to light green, but before it turns yellow. Harvest Asian pears when the groundcolor changes from light green to greenish-yellow. Smooth-skinned Asian pears are often color harvested, with workers going over each cultivar 3 to 4 times over a 10-day period, picking only the fruit that have started to turn yellow each time. The groundcolor is much harder to see on russeted Asian pears, but russeted cultivars such as 'Hosui', 'Yoinashi' and 'Shinko' develop a slight translucence to their skin when fully ripe, that is easy to monitor once seen.

Fruit lenticels, the small dots over the entire fruit surface which act as respiration pores, also change color at maturity. Lenticels are white on immature pears, and turn brown, corky and slightly indented when pears are ready to harvest.

The iodine-starch test indicates how much starch in the fruit remains, and how much has been converted to sugars. European pear growers often use this test, which should show that 60% or less starch remains and at least 40% has been converted to sugars. This test is not often used for Asian pears.

Ease of fruit separation confirms other maturity indices. Mature pears should snap off the tree when twisted upward. If fruit need to be tugged off, they are not mature and should be left on the tree.

Harvest and packing operations. Pears must be harvested by hand. Even those destined for processing would bruise too badly if harvested by machines. Pickers must clip their fingernails short to avoid puncturing fruit, and handle fruit very gently. Asian pears that are tree-ripened are even more subject to bruising than European pears, and any bruises quickly turn black so can not be tolerated. Growers who allow Asian pears to tree-ripen must demonstrate to harvest labor how to pick fruit with a gentle lifting and twisting motion, rather than pulling fruit away from the branch. Workers then place fruit in padded picking buckets, placing fruit so the stems lie between other fruit rather than contacting them, to avoid stem punctures which render fruit unfit for packing.

Such tree-ripened Asian pears must be packed in the field, as transport to a packing house and the effects of a packing line would damage the fruit too much. Growers set up mobile packing stations on wagons, which move through the orchard as harvesting progresses. Experienced labor packs the fruit by size into trays for wholesale to grocery stores, or containers for local retail sales. Workers clip stems from the fruit as they pack them, and boxes are lined with "bubble pads" or paper-covered excelsior pads to prevent rolling in transit. Cool the fruit as soon as it is packed, in a retail stand or in a truck on the way to a grocery store. Asian pears that are harvested a few days before fully ripe can placed into cold storage for 2 to 8 months, depending on the cultivar, and may be graded and packed in a packing house like European pears.

Since European pears are harvested when still quite firm, they can withstand more handling, and they need to be cooled and then ripened after harvest. Workers generally pick into canvas harvesting bags like those used for apples, then unfasten the bags from the bottom to gently pour the fruit into padded bulk bins, which are transported to a packing house. Bulk bins are immediately placed into cold storage for 3 to 5 weeks, until core fruit temperatures fall to 30 to 31F. Bulk bins are then gently dumped into a water transport system, which cleans fruit and minimizes bruising as it carries fruit through sizing and grading lines. Sizing lines separate fruit into up to 8 size categories based on the number of fruit that will fit in a standard box holding 44 pounds of pears, from 70 extra-large pears to 100 medium-sized pears to 150 small pears. In recent years, packinghouses have demanded mostly larger-sized pears with sizes of 70, 80 or 90 fruit per box. Grading lines remove any fruit with punctures, cracks, scab, or rots, and allow no more than one cork spot or hail scar of no more than 3/8-inch diameter.

Sized pears may be placed back into bins and put in cold storage, then packed to order when certain sizes are needed, but many are pre-packed into 44-pound cardboard boxes. Larger-sized pears are individually hand-wrapped in soft paper, then carefully placed into plastic-lined cartons on molded cardboard trays to prevent bruising during transport. Smaller pears may be packed into 3- or 5-pound polyethylene bags, with these placed into cartons. Workers stamp the cartons with the cultivar, size, grade, packing house code and other information. Some cartons go into controlled-atmosphere (CA) storage for long-term storage of up to 12 months, while other go back into regular refrigerated storage and will be sold within 3 to 8 months, depending on the cultivar.

Storage disorders and diseases. European pears may develop **core breakdown** in storage, which causes fruit cores to turn brown and flaccid. **Watery breakdown**, a soft liquid browning of any part of the fruit, especially the blossom end, may also occur in European pears. Chinese or Korean pears may develop **internal breakdown**, brown areas in the core or flesh. Prevent all these types of breakdown by harvesting promptly at optimum maturity, by cooling fruit immediately after harvest, keeping storage temperature at 28 to 30 F, and not storing longer than the cultivar can withstand. High fruit CA levels also reduce these breakdowns, so calcium sprays may be justified.

European pears may also develop **senescent scald**, a brown to black discoloration of the skin (Photo 17-20), if stored too long or at too high a temperature. Affected fruit develop a very disagreeable odor and taste even before the skin discoloration occurs. Watch for skin yellowing in storage—this indicates the end of storage life, so fruit must be sold quickly.

Flesh spot decay may affect Japanese pears. This partial browning of spots or cavities in the fruit may sometimes appear at harvest, but more often after several weeks' cold storage. Prevent this with regular irrigation, prompt harvesting and rapid cooling of harvested fruit.

Many of the same storage rots that may develop on apples can also damage stored pears, often more severely once European pears are allowed to soften before eating. **Blue mold** (caused by *Penicillium expansum*), **mucor rot** (caused by *Mucor piriformis*), **bull's eye rot** (caused by *Pezicula malicorticis*) and **side rot** (caused by *Phialophora malorum*) do the most damage. 'Bosc' is especially susceptible to the last two rots. Minimizing use of N fertilizer and applying Ca sprays helps to prevent all these rots. Getting higher levels of P into 'Anjou' pear fruit, by applying a foliar P spray (nutra-phos-24 at 15 pounds/acre) 3 weeks before harvest, reduced blue mold in storage by 50% in Oregon trials. Applying a fungicide such as ziram 2 to 3 weeks before harvest also reduces pear storage rots. Also sanitize all harvest bins and packinghouse and storage walls with quartenary ammonia, and keep chlorine levels in water systems at 50 to 100 ppm to minimize rots.

Packinghouses generally apply the fungicide thiabendazole (TBZ) to pears. Captan may be applied with TBZ to increase protection, but captan is not permitted in some export markets. Several packinghouses now treat pears with 2 biocontrol agents as well as TBZ: the yeast *Candida oleophila* (Aspire™) and the non-pathogenic bacterium *Pseudomonas syringae* strain ESC-11 (Bio-Save 110™), though both of these work by displacing pathogens and are only partly effective in preventing rot fungi from attacking fruit. A newly-developed yeast that

Photo 17-20. Senescent scald attacks pears that become overripe in storage. Note when skin groundcolor starts to turn yellow, sell pears immediately. Once fruit turn fully yellow, as here, they often scald.

is highly effective, *Cryptococcus infirmo-miniatus* (EXC 9001) will soon be available from Ecogen for postharvest treatment of pears to prevent rots, and could make TBZ fungicide treatment unnecessary.

Conditioning and ripening European pears. Once removed from cold storage, European pears generally take 5 to 10 days to ripen and become soft and sweet enough to eat. Customers prefer to buy ripe or nearly-ripe pears, so grocery stores or other retailers often ripen pears by placing them at room temperature (60 to 70 F) with a relative humidity of 90%. Back rooms used to ripen tomatoes and bananas work well for pears also.

Pears ripen from the inside out, and the stem end will yield to gentle pressure when the inside is ripe, but do not let pears sit so long that they become soft around the middle before offering them for sale, as by then they are overripe. To check on ripening progress, test a few fruit with a pressure tester. A reading of 8 to 9 pounds pressure indicates that pears will be ripe in 3 to 5 days, while ripe pears have a reading of 3 to 4 pounds pressure, and ready-to-eat pears have a pressure reading of 1 to 2 pounds.

'Bartlett' pears turn yellow as they ripen, and red-skinned strains of 'Bartlett' turn darker red, making it easier for produce managers and customers to tell when they are ready to sell and ready to eat. 'Anjou' and a few other cultivars remain green as they ripen, and russeted pears such as 'Bosc' make it hard to see the groundcolor, so these cultivars are more difficult to judge for ripeness. Furthermore, earlier-ripening pears such as 'Bartlett' need only a few days chilling before fruit can be ripened, whereas later-ripening "winter pears" must be chilled for several weeks or months before they can be ripened. These winter pears, especially 'Anjou', benefit from conditioning the fruit before shipping or at the retail store, to bring it to the stage of just starting to ripen. Conditioned winter pears take only 2 to 4 days to ripen, rather than 7 to 10 days.

Conditioning involves treating the fruit with ethylene gas, the natural hormone given off by ripening fruit that induces fruit softening and conversion of starches into sugars. Winter pears that have been held in cold storage for at least 8 to 10 weeks after harvest will accumulate some ethylene naturally, and will ripen without ethylene treatment, but ripen faster if conditioned. Winter pears held in cold storage for less than 8 weeks need an ethylene treatment to ripen properly. Packers condition pears held in cold storage for 3 to 8 weeks by exposing them to warm temperatures and ethylene for 2 days before shipping. Pears held in cold storage for 3 weeks or less need 3 days' exposure to warm temperatures and ethylene before shipping. Conditioned pears should still have a pressure reading of 8 to 12 pounds when shipped. Conditioned winter pears can also be placed back into cold storage, where they will still have about as much storage life as non-conditioned pears.

Further Reading

Beutel, J. A. 1990. Asian pears. p. 304-309. In: J. Janick and J.E. Simon (eds.), *Advances in new crops*, Timber Press, Portland, OR. Available online at: http://www.hort.purdue.edu/newcrop/proceedings1990/v1-304.html

Hart, J., T. Righetti, B. Stevens, B. Stebbins, P. Lombard, D. Burkhart, and P. Van Buskirk. 1997. *Pear fertilizer guide*. Oregon State Univ. Extension Service FG 59. Available online at: http://eesc.orst.edu/agcomwebfile/EdMat/FG59.pdf

Integrated pest management for apples and pears, 2nd ed. 1999. Univ. of California Agric. and Natural Resources pub. 3340.

Jones, A.L. and H.S. Aldwinckle, eds. 1990. *Compendium of apple and pear diseases*. APS Press, St. Paul, MN.

Seavert, C., F. Niederholzer, and D. Burkhart. 1997. Enterprise budget, pears, high density, north central region. Oregon State Univ. Extension Service EM 8486. Available online at: http://eesc.orst.edu/agcomwebfile/EdMat/EM8486.pdf

Stebbins, R.L., W.M. Mellenthin and P.B. Lombard. 1999. Pollination and commercial varieties of pears in Oregon. Oregon State Univ. Extension Service EC 986. Available online at: http://eesc.orst.edu/agcomwebfile/EdMat/html/EC/EC986/EC986.html

Sugar, D. and R.A. Spotts. 1999. Control of postharvest decay in pear by four laboratory-grown yeasts and two registered biocontrol products. *Plant Dis*. 83:155-158.

Chapter 18
Peaches and Nectarines

Peaches and nectarines belong to the same species, *Prunus persica* in the rose family or Rosaceae. In fact, a single gene determines whether the fruit skin will bear many fine hairs or "fuzz" and be a peach (Photo 18-1), or whether it will be a smooth "fuzzless" peach, that is a nectarine. Both peaches and nectarines may have yellow flesh or white flesh, again determined by a single gene. Yellow peaches are the easiest to grow, because white-fleshed peaches bruise very easily and nectarines, without any protective fuzz on the fruit surface, are much more susceptible to brown rot, a common disease. Most customers also think that tree-ripened yellow peaches have the most flavor, with their blend of sugars and acids, while both nectarines and white peaches tend to have lower acidity levels and thus blander flavor. However, white peaches can reach much higher sugar levels of 16 to 17%, compared to 12 to 14% sugars for most yellow peaches, and some customers prefer the strong sweetness with low acidity of white peaches. Some newer cultivars of yellow-fleshed peaches have been bred for high sugars and low acidity, also. Currently, yellow peaches account for about 82% of *Prunus persica* production, nectarines contribute 17%, and white peaches make up only 1% of production, but planting of white peaches is increasing.

Peaches originated in China, which is the world's largest peach producer, followed by the U.S. and Italy. Peaches are less cold-hardy and more heat-tolerant than most other temperate-zone fruit trees, so the main commercial production areas are in warmer temperate regions such as southern California and the Southeastern U.S. The U.S. produces over 2.5 billion pounds of peaches and nectarines per year, with about 77% of that production coming from California. Outside of California, many states produce peaches, with South Carolina and Georgia contributing about 6% and 4% of U.S. production, New Jersey producing about 3% of the total, and Pennsylvania and Washington each producing about 2%. Production of low-chill, early-season peaches and nectarines is increasing in Florida, Texas and Mexico, and several Canadian provinces produce significant amounts of hardier peach cultivars.

Photo 18-1. Many fine hairs or "fuzz" cover the surface of peach fruit, such as these 'Redhaven' peaches, while nectarines have no fuzz but are otherwise the same as peaches. Flesh of both peaches and nectarines may be yellow or white.

CROP USES AND LIMITING FACTORS

Peaches are stone fruits or drupes, with the flesh surrounding a large stony pit that encloses a single seed in the center of the fruit. Some cultivars are called "clingstone" or "cling" peaches, because the flesh clings to the pit at maturity. Other cultivars are called "freestone" peaches because the

Photo 18-2. The flesh of "clingstone" or "cling" peaches and nectarines clings to the pit at maturity. The flesh of "freestone" cultivars separates cleanly from the pit, as with these freestone peaches at left, while the flesh of "semi-freestone" cultivars partially separates from the pit, as in the semi-freestone nectarines at right. Freestone peaches and nectarines are preferred for fresh market, while cling peaches are preferred for canning.

Photo 18-3. Peaches are small, spreading trees, growing about 15 feet tall and 20 feet wide on their own, but are easily kept lower by pruning. They bear young but are short-lived.

flesh separates cleanly from the pit, and others are called "semi-freestone" peaches because the flesh partially separates from the pit (Photo 18-2). Very early-ripening cultivars are generally clingstone, early ripeners are generally semi-freestone, and cultivars that ripen from mid-season on are generally freestone, although some late ripeners are clingstone.

About 45% of North American peaches are sold fresh, and freestone peaches are preferred for fresh markets. About 55% of the peach crop is processed, and cling peaches are generally used for this because they can produce double the yields of most freestone peaches, and much peach processing equipment is geared toward clingstone peaches. About 75% of processed peaches are canned, about 15% are frozen, about 2% are dried, and about 8% go into baby food, wine, brandy or pickled peaches. Processing peaches have "non-melting" flesh that remains firm when ripe, while most fresh-market peaches have "melting flesh" that softens as it ripens. Some newer cultivars of fresh-market peaches have been selected for non-melting flesh so that they can be tree-ripened before shipping.

Peaches are small, spreading trees, growing about 15 feet tall and 20 feet wide if left to their own devices (Photo 18-3). They are easily kept lower by pruning, which is fortunate, because no good dwarfing rootstocks have been developed for peach. Peaches are short-lived trees, lasting 15 to 20 years at most, and often die sooner under cultivation due to many disease problems. However, peaches are precocious trees, often bearing some fruit in their second growing season and a good crop in their third. Nurserymen produce peaches cheaply by budding, and sell the trees at rates $^1/_3$ to $^1/_2$ that of apple trees, so replacing peach orchards rather frequently costs less than replacing apples or other fruit trees.

Peach flowers are borne singly on short peduncles on 1-year-old branches (Photo 18-4), not on spurs on older wood as with most fruit trees, so

Photo 18-4. Peach flowers are borne singly on short peduncles on 1-year-old branches, so good shoot growth each year is important to produce a good fruit crop the next year. Peaches bloom early, so need frost protection in many areas.

good shoot growth each year is important to produce a good fruit crop the next year. Peaches bloom early, so need frost protection in many areas where spring frosts still threaten during bloom. Peaches also need a consistent supply of soil moisture to produce good-sized fruit, so all peach orchards in arid regions and most peach orchards in eastern North America receive irrigation. Even with frost protection, spring frosts eliminate peach crops about 1 year out of 5 in many peach-growing areas of eastern North America. During years when frosts aren't severe, peaches need heavy fruit thinning to produce fruit large enough to market, and most thinning is still done by hand, driving up costs of production.

All fresh-market peaches and nectarines and most processing peaches must be harvested by hand to prevent bruising, and much fruit thinning is still done by hand, so peaches remain one of the most labor-intensive tree fruit crops. Despite the high costs and risks of growing peaches, strong consumer demand for the fruit has kept the total North American peach acreage fairly stable. Yields for processing peaches in California average about 18 tons per acre. Yields for fresh-market peaches average 10 tons per acre in California, and 4 to 4.5 tons per acre in the rest of North America.

SITE SELECTION AND PREPARATION

Most peach cultivars need a region where winter temperatures don't normally drop below 10 F, although a few hardier cultivars can withstand winter temperatures down to -20 F. Choose only sites with the least possible frost risk for peaches, so pick elevated sites where air drains freely and avoid lower slopes where air drains more slowly and spots next to woods, where cold air collects. Growers in the more northern parts of the peach's range can delay bloom by planting on north-facing slopes, but such slopes often do not delay bloom much in southern areas.

Peach roots need a regular moisture supply but are quickly damaged by excess soil moisture, so soil must be well-drained and at least 3 feet deep. Sandy loams are generally best for peaches. Loamy sands will require more irrigation and usually more nematode control. Loams and silt loams may serve well if cover cropped to build up organic matter, and growers may need to build raised beds on such soils to ensure good water drainage. The site must have a source of irrigation water that is clean and low in salts, with total soluble salts below 1000 ppm and a sodium absorption ratio below 7.0.

The ideal peach orchard site was planted to grass crops that reduce nematodes, such as fescue or sorghum, for the past few years. Avoid sites cleared of oak trees in the last few years, because peach roots are very susceptible to oak root rot fungus (*Armillaria mellea*) and this fungus persists in the soil for up to 10 years after susceptible trees have been removed. Also avoid planting after cotton, tomatoes, melons or other crops that are susceptible to Verticillium wilt. Wait at least 3 years after removing a previous peach orchard before replanting to peaches, to avoid these and other soilborne diseases. During that 3-year interval, sow nematode-reducing cover crops as described in chapter 2. Turning under at least one cover crop is a good idea before planting peaches on any site, to increase organic matter that will improve both summer water retention and excess water drainage.

Test sandy soils for nematodes a year before planting, and treat if necessary. Georgia and South Carolina researchers demonstrated that soil solarization followed by introduction of a beneficial bacterium, *Pseudomonas* spp. isolate BG33R, kept ring nematode populations below treatment thresholds for the next 18 months, avoiding the need to fumigate.

Test soil pH, phosphorous (P), potassium (K), calcium (Ca) and magnesium (Mg) at least 6 months before planting, and test for soil salinity if that could be a problem. Peaches need a soil pH between 6.5 and 7.8, so Eastern soils often need lime applications to raise the pH and Western soils occasionally need sulfur to lower the pH. Constructing raised beds, 12 to 18 inches higher than the row middles (Photo 18-5), improves tree performance

Photo 18-5. Forming raised beds, 12-18 inches high, before planting improves peach tree performance on many soils, especially those with less than perfect drainage.

and longevity even on some of the best soils and is standard practice in many areas.

SELECTING ROOTSTOCKS

Selecting a rootstock that's adapted to your climate, soil type and pest problems can greatly increase the life of a peach planting. Rootstocks that promote better tree health and production in one region may reduce these in another region, so obtain data from research trials in your area before trying new rootstocks.

Most peach rootstocks are grown from seed, either from certain cultivars of processing peaches, or from rootstock trees selected for resistance to nematodes or other pests. All commercial peach rootstocks produce standard-sized trees, although some rootstocks induce slightly more or less vigorous tree growth. 'Citation' rootstock, a peach-plum hybrid propagated by cuttings, produces a ½- to ⅔-sized tree but exhibits delayed incompatibility with most peach cultivars, which greatly reduces tree growth and yields. 'Citation' is occasionally used for commercial peach orchards in California, but is used more often for plums and for peach trees sold to homeowners. Other dwarfing rootstocks used for peach trees sold to homeowners come from other *Prunus* species, such as western sand cherry (*P. besseyi*) or Nanking cherry (*P. tomentosa*). These cherry species exhibit delayed incompatibility with peach, so peach trees grafted to them often live only a few years, and are not very productive.

Nematode-resistant rootstocks. Resistance to nematodes is important to most peach growers, especially on sandier soils in warmer regions, but the worst types of nematodes vary. Two species of root knot nematodes, northern and southern, offer the greatest threat in most of California, Texas and parts of the Southeast. 'Nemaguard' seedling rootstock resists both root knot nematodes and is the most widely-used peach rootstock in the Western U.S., but it's susceptible to ring nematodes and bacterial canker, which also threaten peaches in the Southeast. 'Nemared' is a red-leaved strain of Nemaguard that enables nurserymen to easily spot grafts that didn't take, and it has even greater resistance to root knot nematodes but even greater susceptibility to ring nematodes and bacterial canker, so again it does well in the West but very poorly in the Southeast.

'Guardian' seedling rootstock, developed by Georgia and South Carolina researchers and released in 1997, is the best rootstock for the Southeast. It resists ring nematodes, both species of root knot nematodes, and bacterial canker, and enables trees to live longer than they do on other rootstocks such as 'Lovell'. Most older peach orchards in the Eastern U.S. and some in California were planted on 'Lovell' rootstock, which was grown from seed of the formerly common (now hard to obtain) drying peach cultivar. 'Lovell' seedlings are susceptible to root knot nematodes but only moderately susceptible to ring and lesion nematodes and bacterial canker, and moderately resistant to peach tree borer, all of which often pose greater threats in the Southeast. 'Lovell' also offered greater winter hardiness and longer tree life than 'Nemaguard'. 'Halford', a canning peach that is still common today, has traits similar to 'Lovell' and is now the most widely-used rootstock in the Eastern U.S., although it's not quite as cold hardy.

Cold-hardy rootstocks. 'Siberian C' was developed as a very cold-hardy seedling rootstock for parts of Canada, and is still the favored peach rootstock in British Columbia. Where winters are steadily cold without a lot of fluctuating temperatures, 'Siberian C' increases scion flower bud hardiness, delays bloom, and increases yields. Where winter temperatures fluctuate more, as in the lower Midwest and Southeast of the U.S., 'Siberian C' is actually more prone to winter damage and does not delay bloom. It's also more susceptible than other peach rootstocks to peach tree borer, lesion nematodes, bacterial canker and alkaline soil, so does very poorly where any of these are problems.

'Tzim Pee Tao' shows great promise as a very cold-hardy peach rootstock (to USDA hardiness zone 4b) but hasn't been widely propagated by nurseries yet. 'Bailey' is almost as hardy to absolute cold (to zone 5a) and takes fluctuating temperatures well, and is currently the rootstock of choice in Ontario and the upper Midwest of the U.S. 'Rutgers Red Leaf' is relatively cold-hardy (to zone 5b) and has red leaves which are easy to spot in the nursery if grafts don't take. All 3 of these seedling rootstocks offer moderate resistance to lesion nematodes, more than other peach rootstocks, so are especially useful in Northern areas where that is the main nematode pest. They are susceptible to root knot and ring nematodes so aren't well adapted to the Southeast or Southwest.

Rootstocks for low-chill Southern regions. 'Flordaguard' seedling rootstock was developed for low-chill areas of the Southeastern U.S. and resists root knot nematodes, but is very susceptible to alkaline soils so does poorly in the Southwest. Several lower-chill peach-almond hybrid rootstocks, propagated by cuttings, have moderate resistance to alkaline soils and at least moderate resistance to root knot nematodes, so are adapted to the Southwest. These include drought-resistant 'Hansen 536' with low chill needs and 'Nickels' with medium chill needs, both very resistant to root knot nematodes. Other peach-almond hybrids include 'Viking' and 'Atlas', which both increase fruit size and offer moderate resistance to bacterial canker and root knot nematodes, although they're even more susceptible to poor soil drainage than other peach rootstocks.

Table 18-1 summarizes the hardiness, soil adaptation and pest resistance of these peach rootstocks.

Table 18-1.

COMMON ROOTSTOCKS FOR PEACHES AND NECTARINES.

Rootstock	Hardy to USDA zone	Resistance/susceptibility to:						
		Alkaline/ saline soil	Poor soil drainage	Bacterial canker	Nematodes			Peach Tree Borer
					Root knot	Ring	Lesion	
Nemaguard	7b	S	S	S	R	S	S	S
Nemared	7a	S	S	VS	VR	VS	S	S
Lovell	6a	S	S	MS	S	MS	MS	MR
Halford	6b	S	S	MS	S	MS	MS	MS
Guardian	6b	S	S	R	R	R	S	MS
Flordaguard	8a	VS	S	S	R	S	S	S
Siberian C	4b	S	S	VS	S	S	VS	VS
Bailey	5a	S	S	MS	S	S	MR	S
Rutgers Redleaf	5b	S	S	MS	S	S	MR	MS
Tzim Pee Tao	4b	S	S	MS	S	S	MR	S
Citation	6b	S	MR	S	R	R	S	S
Hansen 536	7b	MR	VS	S	VR	S	S	S
Nickels	7a	MR	S	S	VR	S	S	S
Viking	7a	MR	VS	MR	MR	S	S	S
Atlas	7a	MR	VS	MR	MR	S	S	S

Disease resistance ratings: VR = very resistant, R = resistant, MR = moderately resistant, MS = moderately susceptible, S = susceptible, VS = very susceptible

Comments:

Nemaguard is the most common peach rootstock in the West; may delay bloom 1-3 days.
Nemared, red-leaf strain of Nemaguard, is easy to spot if grafts didn't take.
Lovell is the most common rootstock in older Eastern orchards.
Halford is most widely-used in new Eastern plantings.
Guardian is recommended for longer-lived trees in Southeast U.S.
Flordaguard needs 300 chill hours, causes 3-5 days' earlier bloom some years.
Siberian C is adapted only where winters stay very cold, where it delays bloom, increases yields.
Bailey from Iowa increases cold tolerance of cultivar's buds, takes fluctuating winter cold well.
Rutgers Red Leaf is easy to spot if grafts don't take.
Tzim Pee Tao seedlings are R to powdery mildew in the nursery, very cold hardy.
Citation, a peach x plum hybrid, produces 1/2- to 2/3-size tree, delays fruit maturity, increases fruit size and sugars. It has delayed incompatability with many cultivars, so reduces growth and yields.
Hansen 536, a peach x almond hybrid, is MR to drought, makes larger trees, has low chill needs.
Nickels, a peach-almond hybrid, makes larger trees, has higher chill needs than Hansen 536.
Viking, a peach x almond hybrid, increases fruit size, induces early bearing.
Atlas, a peach x almond hybrid, delays fruit maturity, increases fruit size.

Table 18-2.

COMMON COMMERCIAL FRESH-MARKET PEACH CULTIVARS IN ARID WESTERN U.S.

In approximate order of ripening (varies with area).

These cultivars are susceptible to bacterial spot and have little flower bud hardiness, so aren't generally suited to the East.

Cultivar	Chill Hours Needed	Fruit						Avg.Days After Rich Lady
		Stone Freeness	Flesh Color	Flavor	Size	Firmness	Yields	
Super Rich	750	cling	yellow	good, sweet-tart	medium	medium	high	-34
Queencrest	650	cling	yellow	good, sweet	medium	medium	medium	-29
Spring Flame	800	cling	yellow	fair-good, sweet-tart	medium	firm	high	-28
Spring Snow	800	cling	white	very good, low acid	medium	firm	medium	-24
Spring Lady	750	cling	yellow	fair-good, sweet-tart	medium	firm	medium	-19
Springcrest	700	cling	yellow	fair-good, sweet-tart	medium	firm	medium	-17
Sweet Dream	800	cling	yellow	low acid, sweet	very large	firm	high	-17
Crimson Lady	800	cling	yellow	good, sweet-tart	large	very firm	very high	-16
Crown Princess	800	semi-free	yellow	low acid, sweet	large	very firm	high	-16
Sweet Scarlet	850	semi-free	yellow	good, sweet-tart	med-large	firm	high	-6
Snowbrite	850	semi-free	white	low acid, sweet	med-large	soft	medium	-5
Ivory Princess	750	cling	white	low acid, sweet	very large	very firm	high	-5
June Lady	750	semi-free	yellow	fair, sweet-tart	medium	firm	high	0
Rich Lady	800	semi-free	yellow	good, sweet-tart	large	firm	very high	0
Stark Saturn	700	free	white	low acid, sweet	medium	medium	low	0
Flavorcrest	750	free	yellow	good, sweet-tart	large	very firm	medium	2
Country Sweet	850	semi-free	yellow	low acid, very sweet	large	firm	very high	2
Vista	850	semi-free	yellow	good, sweet-tart	large	firm	high	5
Fancy Lady	800	semi-free	yellow	good, sweet-tart	large	firm	high	5
White Lady	800	semi-free	white	low acid, sweet	med-large	very firm	high	7
Klondike	750	free	white	good, low-acid, sweet	very large	firm	medium	9
June Flame	800	free	yellow	good, sweet-tart	large	firm	high	12
Diamond Princess	800	free	yellow	best, sweet-tart	large	very firm	very high	15
Elegant Lady	750	free	yellow	good, sweet-tart	large	very firm	high	18
Summer Sweet	550	free	white	low acid, very sweet	very large	very firm	high	21
July Flame	800	free	yellow	best, sweet-tart	large	very firm	high	27
Sugar Giant	850	free	white	low acid, sweet	very large	firm	high	29
Zee Lady	950	free	yellow	very good, sweet-tart	large	firm	high	34
Summer Lady	750	free	yellow	very good, sweet-tart	large	very firm	high	40
Glacier	750	free	white	low-acid, sweet	very large	firm	high	40
Fay Elberta	750	free	yellow	good, sweet-tart	large	medium	high	44
Summer Flame	800	free	yellow	very good, sweet-tart	large	firm	high	44
Snow King	950	free	white	low acid, very sweet	very large	firm	high	44
Henry II	750	free	yellow	very good, sweet-tart	large	very firm	high	44
O'Henry	750	free	yellow	best, sweet-tart	large	very firm	high	48
Snow Giant	850	free	white	good, low acid, sweet	very large	firm	very high	49
Autumn Snow	800	free	white	best, low-acid, sweet	very large	very firm	high	52
Ryan Sun	500	free	yellow	good, sweet-tart	large	firm	high	59
August Lady	750	free	yellow	very good, sweet-tart	large	firm	high	59
Tra-Zee	850	free	yellow	very good, sweet-tart	large	firm	high	63
August Flame	800	free	yellow	very good, sweet-tart	large	very firm	very high	64
September Snow	950	free	white	good, low-acid, sweet	very large	firm	high	69
September Sun	550	free	yellow	good, sweet-tart	large	firm	very high	73
Fairtime	750	free	yellow	good, sweet-tart	large	firm	medium	75
Autumn Flame	800	free	yellow	very good, sweet-tart	large	firm	high	81
Sweet September	900	free	yellow	low acid, sweet	large	firm	high	86
Full Moon	800	free	white	good,low-acid, sweet	large	firm	high	90
Autumn Rose	800	free	yellow	good, sweet-tart	large	firm	high	96

Day of first harvest, in average days before (-) or after 'Rich Lady', which starts ripening about June 8 in Fresno, CA, June 22 in Modesto, CA and southern GA, July 2 in Sacramento, CA and southern NC, July 22 in western OR, and Aug. 5 in western WA. Days between cultivars increases where nights are cool, decreases where nights are hot.

Notes on Cultivars:
Super Rich has been widely planted recently as a very early peach with better flavor, 75% red blush.
Queencrest is the main very early peach in CA. It ships well.
Spring Flame has been the most-planted cultivar recently due to earliness and high quality.
Spring Snow has the largest fruit of early white peaches, 80% red blush.
Spring Lady is a major early cultivar, though no longer planted due to fair flavor.
Springcrest is widely grown but no longer planted. It may have few fruit buds on 1-year-old branches.
Sweet Dream has full red blush and low-acid, sweet taste.
Crimson Lady has little fuzz, dark red blush, non-melting flesh. Fruit hang on tree for 10 days after ripe.
Crown Princess has 50% dark red blush on pink groundcolor, non-melting flesh that hangs well.
Sweet Scarlet has 80% red blush, crops lightly after low-chill years due to higher chill needs.
Snowbrite has nearly full red blush.
Ivory Princess is one of the more heavily planted white peaches recently.
June Lady, a standard early cultivar, has light crops and button fruit some years, VS to brown rot.
Rich Lady is the leading early cultivar and #3 peach overall. Full red blush may stain fruit flesh.
Stark Saturn, a very sweet donut-shaped or "saucer" peach, has been widely planted in recent years.
Flavorcrest is a major early-midseason peach with nearly full red blush, MR to powdery mildew.
Country Sweet has 85% red blush and newly popular low-acid, sweet taste; widely planted recently.
Vista has full red blush that doesn't stain flesh.
Fancy Lady has nearly fuzzless fruit with 50-75% red blush, ships well.
White Lady, one of the oldest white peach cultivars,ships and stores very well, has 80% red blush.
Klondike has 85% red blush, was fairly widely planted despite light to medium yields.
June Flame has very good size for its early season, full red blush, and has been heavily planted recently.
Diamond Princess has 85% red blush, little fuzz, ships well. Tree bears regularly.
Elegant Lady is a mid-season shipping standard and #2 peach with 85% red blush, VS to powdery mildew.
Summer Sweet blooms early due to low chill needs, has high quality fruit with 85% red blush.
July Flame has superior flavor and size for its season, full red blush, has been heavily planted recently.
Sugar Giant fruit ship well, have 80% red blush.
Zee Lady has high-quality fruit that freezes well.
Summer Lady, a major cultivar, is a mutation of O'Henry that ripens about a week earlier.
Glacier has some problems with stems pulling bark off branches during harvest, so pick carefully.
Fay Elberta, named in the 1870's, was once the most widely planted U.S. peach and is still grown.
Summer Flame has full red blush.
Snow King is widely grown but fruit are prone to flesh-corking.
Henry II is a mutation of O'Henry that ripens a few days earlier.
O'Henry, an older cultivar, still the #1 peach, 75% red blush. MR to powdery mildew, VS to fungal gummosis.
Snow Giant ships well and is the most widely-grown white peach. Pick carefully; some stem-pulling occurs.
Autumn Snow, also called **Yukon King**, has exceptional quality but skin may stain flesh.
Ryan Sun blooms early due to low chill needs. Fruit have nearly full dark red blush, tree is very vigorous, upright.
August Lady has a rich flavor.
Tra-Zee has 75% red blush, freezes well.
August Flame has 90% red blush, very good flavor and has been heavily planted recently.
September Snow is widely-grown but prone to heat damage: flesh reddens when ripe.
September Sun blooms early due to low chill needs, has nearly full red blush.
Fairtime has tender buds so yields less. Fruit crack some years; VS to bacterial spot, powdery mildew.
Autumn Flame has full red blush, very good flavor and has been heavily planted recently.
Sweet September has 70% red blush, low-acid taste and good quality for a late peach.
Full Moon has been widely planted as a very late-ripening white peach.
Autumn Rose has 50-75% red blush, less pre-harvest drop than other extremely late cultivars.

Table 18-3.

COMMON COMMERCIAL PROCESSING PEACH CULTIVARS IN ARID WESTERN U.S.

In approximate order of ripening (varies with area).

All are yellow-fleshed clingstones used mainly for canning, also for freezing. These cultivars are susceptible to bacterial spot and have little flower bud hardiness, so aren't generally suited to the East. All these cultivars need 750 to 850 chill hours.

	Fruit				Powdery		Avg. Days	
		Canned			Mildew		After	
Cultivar	Firmness	Flavor	Size	Yields	Reaction	Tree size	Carson	Comments
Ceres Carson	firm	good	med-large	very high	MR	large	-12	less fruit set, better size than Loadel
Thiara	firm	good	med-large	very high	S	large	-12	no split pits, flesh colors evenly
Loadel	firm	very good	medium	high	MS	large	-7	good size for extra-early cultivar
Stanislaus	very firm	very good	large	very high	S	large	0	better quality, returns than Carson
Carson	firm	good	medium	high	MR	medium	0	S to invisible split pits
Goodwin	medium	good	medium	high	S	large	6	fruit has slight tip
Andross	firm	very good	med-large	high	MS	medium	10	ripens evenly, needs few pickings
Bowen	medium	very good	med-large	high	MS	large	12	fruit may have green pockets
Tuolumne	very firm	best	medium	high	S	large	13	early bloom, precocious yields
Kingsburg	very firm	best	large	very high	VS	large	15	mutation of Dr. Davis, ripens earlier
Klampt	medium	best	med-large	very high	VS	large	16	ripens evenly, needs few pickings
Ross	firm	best	large	very high	VS	medium	18	consistent producer
Rizzi	firm	very good	medium	very high	MR	large	23	stores very well
Dr. Davis	very firm	best	large	very high	VS	large	25	stores very well
Late Ross	firm	best	large	very high	VS	large	26	mutation of Ross, ripens later
Monaco	firm	good	large	high	MS	large	28	consistent producer
Halford	firm	very good	large	high	R	large	31	seed saved for rootstocks
Everts	firm	best	med-large	medium	MR	large	32	somewhat low fruit bud set
Starn	firm	good	large	high	VS	large	34	very popular, blooms late
Hesse	firm	very good	large	very high	MS	large	37	few split pits, needs few pickings
Sullivan #4	firm	good	medium	high	VS	med-large	37	ripens evenly, needs few pickings

Day of first harvest, in average days before (-) or after Carson, which ripens about July 22 in Fresno County, CA.

Disease resistance ratings: VR = very resistant, R = resistant, MR = moderately resistant, MS = moderately susceptible, S = susceptible, VS = very susceptible

Table 18-4.

MEDIUM- TO HIGH-CHILL PEACHES WITH DISEASE RESISTANCE, FOR HUMID AREAS, ZONES 5 OR 6 TO 8

In approximate order of ripening.

Cultivar	Chill Hours Needed	Flower Bud Hardiness		Fruit						Resistance to			Avg.Days After Redhaven
		Winter	Spring	Stone Freeness	Flesh Color	Flavor	Size	Firmness	Yields	Bact. Spot	Leaf Curl	Flesh Browning	
Regal	700	medium	medium	cling	yellow	good	medium	med-firm	medium	VR	S	MR	-28
Springprince	650	medium	medium	cling	yellow	good	small	very firm	medium	MR	S	MR	-24
Goldprince	650	medium	medium	cling	yellow	good	medium	med-firm	med-high	MR	S	MR	-20
Harrow Diamond	950	very hardy	hardy	semi-free	yellow	good	small	medium	medium	MR	S	VR	-20
Candor	950	hardy	hardy	semi-free	yellow	good	small	medium	med-high	VR	R	VR	-18
Surecrop	1000	hardy	hardy	semi-free	yellow	good	medium	medium	med-high	R	MR	MR	-15
Sentry	850	tender	tender	semi-free	yellow	good	med-large	med-firm	medium	VR	S	MR	-13
Juneprince	650	tender	tender	semi-free	yellow	good	medium	med-firm	medium	MR	S	MS	-10
Early Redhaven	950	hardy	medium	semi-free	yellow	good	medium	firm	high	MR	VR	R	-10
GaLa	750	medium	medium	semi-free	yellow	good	medium	firm	high	MR	S	MR	-7
Risingstar	900	hardy	hardy	semi-free	yellow	good	medium	medium	med-high	R	S	R	-7
Southern Pearl	650	medium	medium	semi-free	white	good	med-large	med-firm	high	MR	S	MR	-7
Sureprince	900	hardy	hardy	semi-free	yellow	good	medium	firm	high	MR	S	MS	-6
Redhaven	950	hardy	medium	semi-free	yellow	good	medium	firm	high	MR	MR	R	0
Harvester	750	medium	medium	free	yellow	good	medium	firm	high	MR	S	MR	3
Norman	850	medium	tender	free	yellow	very good	large	firm	med-high	MR	VS	VR	3
Newhaven	950	hardy	medium	free	yellow	good	medium	medium	high	R	R	R	4
Challenger	950	very hardy	very hardy	free	yellow	best	large	med-firm	high	VR	S	R	4
Reliance	950	very hardy	very hardy	free	yellow	fair	medium	soft	high	MR	MR	R	5
Starfire	850	very hardy	very hardy	free	yellow	very good	med-large	medium	high	MR	S	R	5
Nectar	1050	medium	hardy	free	white	mild	large	soft	medium	MR	S	VS	6
Fireprince	850	medium	medium	free	yellow	good	large	firm	med-high	MR	S	MS	8
Topaz	700	tender	tender	free	yellow	best	large	firm	low	MR	S	R	9
Intrepid	1000	very hardy	very hardy	free	yellow	best	large	med-firm	high	R	S	R	11
Majestic	800	medium	medium	free	yellow	very good	large	firm	high	MR	S	VR	12
Jayhaven	850	very hardy	very hardy	free	yellow	good	medium	firm	med-high	R	S	S	12
Winblo	850	medium	medium	free	yellow	best	large	medium	med-high	MR	S	VR	13
Harrow Beauty	950	hardy	medium	free	yellow	very good	medium	firm	high	R	S	VR	13

Continued on next page

Cultivar	Chill Hours Needed	Flower Bud Hardiness Winter	Flower Bud Hardiness Spring	Fruit: Stone Freeness	Fruit: Flesh Color	Fruit: Flavor	Fruit: Size	Fruit: Firmness	Fruit: Yields	Resistance to Bact. Spot	Resistance to Leaf Curl	Resistance to Flesh Browning	Avg.Days After Redhave
Redglobe	850	medium	medium	free	yellow	good	med-large	firm	high	MR	S	S	14
Loring	750	tender	tender	free	yellow	very good	large	firm	high	R	VS	VS	16
Bounty	800	hardy	medium	free	yellow	best	large	firm	high	MR	S	VR	16
Champion	900	very hardy	hardy	free	white	mild	medium	soft	low	R	S	MR	18
Contender	1050	very hardy	very hardy	free	yellow	best	large	firm	high	MR	S	R	19
Madison	850	hardy	very hardy	free	yellow	very good	medium	medium	high	R	MR	MR	20
Allstar	750	hardy	hardy	free	yellow	good	large	med-firm	med-high	MR	S	R	21
White River	850	medium	medium	free	white	best	large	medium	very high	R	S	S	21
Cresthaven	850	hardy	hardy	free	yellow	good	large	firm	high	MR	S	MS	24
Blushingstar	750	very hardy	very hardy	free	white	best	med-large	firm	high	MR	S	R	25
Glowingstar	850	hardy	hardy	free	yellow	very good	med-large	firm	high	R	S	R	25
Sunprince	800	medium	medium	free	yellow	good	very large	firm	high	MR	S	MS	26
Jefferson	850	hardy	hardy	free	yellow	very good	large	med-firm	high	MR	S	MR	31
Jerseyglo	850	medium	tender	free	yellow	very good	large	med-firm	low	R	S	MR	32
China Pearl	1100	very hardy	very hardy	free	white	mild	large	medium	high	R	S	MR	32
White Star	850	medium	medium	free	white	mild	med-large	medium	high	MR	S	S	35
Ouachita Gold	700	medium	hardy	free	yellow	very good	large	firm	high	MR	S	R	40
Stark Encore	900	very hardy	medium	free	yellow	very good	large	firm	high	MR	S	R	40
Flameprince	850	hardy	medium	free	yellow	very good	large	very firm	high	MR	S	MS	42
LaJewel	850	hardy	very hardy	free	yellow	very good	large	medium	high	MR	S	MR	43
Autumnprince	800	hardy	medium	free	yellow	good	med-large	firm	high	MR	S	MS	54

Disease resistance ratings: VR = very resistant, R = resistant, MR = moderately resistant, MS = moderately susceptible, S = susceptible, VS = very susceptible

Day of first harvest, in average days before (-) or after 'Redhaven', which ripens about June 20 in southern GA and central CA, July 2 in southern MO and central NC, July 20 in central MI and central NJ, and Aug. 10 in western WA and northern NY.

Notes on Cultivars:
Regal from LA has large size for a very early peach and few split pits; does well in medium-chill zones.
Springprince from GA has been widely planted in the Southeast in recent years as an extra-early cultivar with 85% red
Goldprince from GA has been widely planted in the Southeast in recent years, replacing 'June Gold'. It has few split pits.
Harrow Diamond from Ontario needs early, heavy thinning to get good size. It has a red blush and few split pits.

Candor from NC requires heavy thinning to size fruit, has many split pits or soft tips some years.
Surecrop blooms late due to high chill need, so avoids much frost damage.
Sentry from WV sets few flower buds but may crop well on best sites in Northeast, has very good fruit size, few split pits.
Juneprince from GA is replacing 'Coronet' in medium-chill areas of Southeast.
Early Redhaven is a sport of Redhaven that ripens 10 days earlier.
GaLa, a joint release from GA and LA, crops consistently in the mid-chilling zones where it's adapted.
Risingstar, a new cultivar, needs heavy thinning most years due to hardy buds, resists peach canker.
Southern Pearl is an early white peach for medium-chill areas, with 70% red blush.
Sureprince produces dependable crops due to hardy buds, late bloom; has 85% red blush.
Redhaven from MI is the industry standard, still the most widely-grown peach in eastern North America.
Harvester is a major older cultivar in GA and SC due to consistent production in mid-chilling zones.
Norman from NC has very dark red skin that is prone to discoloration during washing, so best for local sales, unwashed.
Newhaven from MI is grown in Northern states due to hardy buds, but is less planted now to little red skin color.
Challenger from NC has high chill needs and very hardy buds, so crops well. Fruit have top quality but don't ship well.
Reliance is one of the most cold-hardy peaches, so is grown in Northern areas despite fair flavor, soft fruit.
Starfire crops very consistently due to hardy buds, needs heavy thinning most years, resists peach canker.
Nectar blooms late due to very high chill needs. Sell locally due to soft fruit, some split pits, S to brown rot.
Fireprince, from GA, has almost full red blush.
Topaz sets few flower buds and rarely needs thinning; erratic cropping limits its use to best sites.
Intrepid from NC has high chill needs and very hardy buds, so crops well. Fruit have top quality but don't ship well.
Majestic from LA is starting to replace 'Redglobe' in mid-chilling areas.
Jayhaven from MI is grown in Northern areas due to hardy buds.
Winblo has excellent fruit quality and size, is grown in Carolinas.
Harrow Beauty, from Ontario, is important in the North due to bud hardiness. It needs heavy thinning.
Redglobe, an older cultivar, is still widely grown in the Southeast due to very reliable cropping, though less planted now.
Loring, an older MO cultivar, is still important in lower Midwest though often sets a light crop due to bud tenderness.

Table 18-5.

LOW-CHILL PEACHES AND NECTARINES FOR USDA ZONES 8B-10

In approximate order of ripening.

Cultivar	Chill Hours Needed	Fruit: Peach or Nectarine	Stone Freeness	Flesh Color	Flavor	Size	Firmness	Yields	Resistance to: Bact. Spot	Flesh Browning	Avg.Days After Flordaking
Flordadawn	300	peach	cling	yellow	good	small-med	very firm	medium	MS	VR	-10
UFGold	200	peach	cling	yellow	good	small	very firm	high	R	R	-7
Earlitreat	400	peach	cling	yellow	fair	small	firm	high	S	MS	-7
Flordaglobe	450	peach	cling	yellow	good	small-med	very firm	medium	R	R	-5
UFBeauty	200	peach	cling	yellow	good	small-med	very firm	high	R	R	-5
Flordaking	350	peach	cling	yellow	fair	medium	medium	medium	MR	VR	0
April Glo	200	nectarine	semi-free	yellow	good	medium	medium	high	S	MS	0
UFBlaze	300	peach	cling	yellow	good	medium	very firm	high	VR	R	1
Sundollar	350	nectarine	cling	yellow	good	medium	very firm	medium	VR	MR	3
Zee Fire	250	nectarine	cling	yellow	good	medium	firm	high	S	MS	3
Flordastar	225	peach	semi-free	yellow	good	medium	medium	medium	R	R	4
Flordaprince	150	peach	semi-free	yellow	good	medium	medium	high	VS	MS	4
Earliglo	200	nectarine	cling	yellow	good, tart	medium	firm	high	S	MS	5
Suncoast	375	nectarine	semi-free	yellow	very good	medium	very firm	medium	VR	R	5
Flordacrest	350	peach	semi-free	yellow	good	medium	very firm	high	MS	VR	6
Desertgold	300	peach	semi-free	yellow	good	medium	medium	medium	S	S	7
Sungem	425	nectarine	semi-free	yellow	very good	medium	firm	high	MR	R	8
ValleGrande	200	peach	semi-free	yellow	good	medium	medium	medium	MR	MS	10
Sunsplash	400	nectarine	semi-free	yellow	good	medium	medium	medium	MR	VR	10
Red Glo	250	nectarine	cling	yellow	good, tart	med-large	firm	high	S	MS	10
Red Roy	400	nectarine	cling	yellow	good	large	firm	high	S	MS	12
Crimson Gold	450	nectarine	semi-free	yellow	good	small-med	firm	high	MS	MS	12
TropicBeauty	150	peach	semi-free	yellow	very good	medium	very firm	high	MR	VR	14
Sunmist	300	nectarine	semi-free	white	very good	med-large	medium	medium	VR	MR	15
Flordaglo	200	peach	semi-free	white	good	large	firm	high	MR	R	15
Royal Glo	400	nectarine	cling	yellow	good, tart	large	firm	high	S	MS	15
Gulfprince	400	peach	cling	yellow	good	medium	firm	high	VR	VR	16
Sunraycer	275	nectarine	semi-free	yellow	good	large	firm	medium	VR	VR	17
Brittany Lane	400	peach	cling	yellow	very good	med-large	firm	high	S	MS	18

Sunfre	450	nectarine	semi-free	yellow	good	large	firm	medium	S	MR	18
UF2000	300	peach	semi-free	yellow	good	medium	very firm	high	R	R	19
TropicSweet	175	peach	free	yellow	best	large	medium	high	MS	R	22
Flordagold	325	peach	semi-free	yellow	good	med-large	very firm	high	VS	MR	23
TropicSnow	200	peach	free	white	best	large	medium	high	R	VR	24
FlordaRio	400	peach	semi-free	yellow	very good	large	very firm	medium	VR	VR	28

Notes on Cultivars:

Flordadawn is the earliest-ripening peach in North America. Long bloom period spreads out the frost risk.

UFGold has fruit with non-melting flesh that can be harvested tree-ripe and still firm.

Earlitreat has full red blush, tips tend to soften rapidly.

Flordaglobe blooms later than other early-ripening, low-chill peaches.

UFBeauty has non-melting flesh, needs heavy thinning if no frosts. Dense trees need summer pruning for light per

Flordaking is the standard early peach in north FL and south GA due to fairly large size, but has many split pits.

April Glo fruit have a tip and pronounced suture, mottled red over yellow skin.

UFBlaze has non-melting flesh, needs heavy thinning if no frosts. Dense trees need summer pruning for light pene

Sundollar needs regular irrigation to maintain uniform soil moisture during fruit growth, to prevent fruit cracking.

Zee Fire has 85% red blush. Tree can be girdled to increase fruit size without increasing split pits.

Flordastar has 60% red blush, some red in flesh, few split pits except if crop is light.

Flordaprince blooms very early due to very low chill needs. Thin early to achieve 2-inch fruit.

Earliglo has 90% red blush, prone to split pits. Very vigorous tree, needs heavy thinning for good fruit size.

Suncoast has attractive fruit with 80% red blush, few split pits.

Flordacrest has 70% red blush, many split pits when crop is light.

Desertgold is grown in the deserts of CA, AZ and Mexico.

Sungem sets many fruit buds, needs heavy thinning if no frosts occur. Fruit may crack and russet in humid areas.

ValleGrande has 60% red blush; grown in south TX.

Sunsplash has 60% red blush, suffers skin russet and cracking some years.

Red Glo has 85% red blush, prone to split pits.

Red Roy has been widely planted recently in CA. Do not girdle tree; that causes fruit cracking.

Crimson Gold has good hardiness for a CA nectarine, but is VS to brown rot.

TropicBeauty bears young, has very attractive fruit with 80% red blush.

Sunmist has attractive purple skin, some red in flesh, few split pits.

Flordaglo has thin skin, 50% red blush, few split pits even when crop is light. It needs heavy thinning.

Continued on next page

Royal Glo has attractive red-over-yellow fruit.
Gulfprince has long bloom period so resists frost. Fruit have 50% red blush, non-melting flesh.
Sunraycer has 90% red blush, few split pits.
Brittany Lane is rapidly becoming an important cultivar in CA. It's highly flavored, with high sugars and high acidity
Sunfre fruit may russet but are otherwise very attractive. Flower buds are very cold-tender.
UF2000 has fruit with non-melting flesh that can be harvested tree-ripe and still firm.
TropicSweet has very high-quality fruit with sugar speckles if weather is hot and dry during harvest.
Flordagold has very high fruit bud set, so needs heavy thinning if crop not reduced by frosts.
TropicSnow sets many buds but blooms very early due to low chill; needs heavy thinning if no frost.
FlordaRio crops more reliably than Flodaking, due to later bloom. Fruit have 75% red blush.
FlordaGrande is the lowest-chill peach cultivar, grown in south TX and Mexico. Fruit have 50% red blush.
Rio Grande has a large suture bulge and 60% red blush.
La Festival is good for local sales in medium-low chill areas.

Table 18-6.

MEDIUM- TO HIGH-CHILL NECTARINES WITH DISEASE RESISTANCE, FOR HUMID AREAS, ZONES 5 OR 6 TO 8

In approximate order of ripening.

	Chill	Flower Bud Hardiness		Fruit						Resistance to		Days
	Hours			Stone	Flesh					Bact.	Fruit	After
Cultivar	Needed	Winter	Spring	Freeness	Color	Flavor	Size	Firmness	Yields	Spot	Cracking	Redhaven
Westbrook	750	medium	hardy	cling	yellow	very good	medium	medium	high	VR	MS	-21
Carolina Red	800	medium	hardy	cling	yellow	good	medium	firm	high	MR	R	-16
Eastern Glo	750	medium	medium	free	yellow	best	medium	med-firm	high	MR	MS	-11
Juneprincess	850	medium	medium	free	yellow	good	large	firm	high	MR	MR	-7
Arrington	750	medium	medium	cling	yellow	good	medium	very firm	very high	MR	MR	-6
Earliscarlet	800	hardy	hardy	semi-free	yellow	good	medium	firm	high	R	MR	-5
Harblaze	850	hardy	hardy	semi-free	yellow	good	medium	med-firm	medium	MR	MR	-3
Bradley	750	medium	medium	cling	yellow	good	med-large	very firm	very high	VR	MR	0
Redbud	850	hardy	hardy	free	white	good	medium	med-firm	med-high	MR	MR	1
Hardired	950	very hardy	hardy	semi-free	yellow	fair	medium	medium	medium	R	MR	5
Roseprincess	850	medium	medium	free	white	very good	medium	med-firm	med-high	MR	MR	7
Mericrest	1000	very hardy	hardy	free	yellow	fair	medium	medium	medium	MR	MR	10
Redchief	750	medium	medium	free	white	good	small	medium	high	MR	MR	16
Cavalier	900	hardy	very hardy	free	yellow	good	medium	medium	medium	MR	MR	20

Disease resistance ratings: VR = very resistant, R = resistant, MR = moderately resistant, MS = moderately susceptible, S = susceptible, VS = very susceptible

Day of first harvest, in average days before (-) or after 'Redhaven' peach, which ripens about June 20 in southern GA and central CA, July 2 in southern MO and central NC, Aug. 2 in central MI and central NJ, and Aug. 20 in western WA and northern NY.

Notes on Cultivars:

Westbrook blooms 4-6 days later than most peaches/nectarines. Few split pits, too soft to ship.
Carolina Red blooms a few days earlier than most peaches/nectarines, so needs the best sites. Very few split pits, VR to flesh browning.
Eastern Glo has full red blush, smooth skin finish.
Juneprincess has 85% red blush.
Arrington has non-melting flesh, giving it high firmness, ships well. Fruit have mild flavor, 60% red blush.
Earliscarlet sets many hardy flower buds, so usually needs heavy thinning for good fruit size. Fruit have 85% red blush.
Harblaze is MR to brown rot and powdery mildew. Fruit have 90% red blush, soften quickly near harvest, so pick promptly.
Bradley has non-melting flesh, giving it high firmness, ships well. Fruit have 60% red blush.
Redbud sets well so often needs heavy thinning. Fruit have nearly full red blush.
Hardired has very hardy buds, so normally needs heavy fruit thinning to develop good size. Fruit MR to brown rot, have very dark red skin.
Roseprincess sets fruit well, has good red blush but may russet, higher flavor (more acid) than most nectarines.
Mericrest blooms late but has fewer buds and often doesn't set well in the Southeast. Fruit have dull red color.
Redchief has bright red skin, mild flavor, stores well.
Cavalier fruit has slightly bitter taste, like 'Elberta' peach, but very good quality.

Table 18-7.

COMMON COMMERCIAL NECTARINE CULTIVARS

In approximate order of ripening.

These cultivars are susceptible to bacterial spot and brown rot and have little flower bud hardiness, so aren't generally suited to the East.

Cultivar	Chill Hours Needed	Fruit						Avg.Days After Rich Lady
		Stone Freeness	Flesh Color	Flavor	Size	Firmness	Yields	
Mayfire	600	cling	yellow	fair, sweet-tart	small-med	firm	high	-30
Rose Diamond	750	cling	yellow	good, sweet-tart	medium	firm	high	-20
Kay Glo	950	cling	yellow	fair, sweet-tart	large	very firm	medium	-16
Arctic Star	500	semi-free	white	low-acid, mild	large	firm	high	-12
Diamond Bright	700	cling	yellow	best, sweet-tart	medium	firm	high	-9
Red Diamond	650	free	yellow	good, sweet-tart	small	very firm	high	-7
June Pearl	800	cling	white	low-acid, sweet	medium	firm	high	-6
Arctic Sweet	850	cling	white	low-acid, sweet	med-large	firm	high	-6
Zee Grand	850	cling	yellow	good, sweet-tart	small-med	firm	high	-6
Spring Bright	800	cling	yellow	good, sweet-tart	medium	very firm	very high	-3
Honey Blaze	850	semi-free	yellow	low-acid, sweet	med-large	very firm	high	-1
Honey Kist	950	semi-free	yellow	low-acid, sweet	medium	very firm	very high	1
Ruby Sweet	800	cling	yellow	low-acid, sweet	medium	firm	high	8
Ruby Diamond	650	free	yellow	good, sweet-tart	large	very firm	very high	13
Ruby Pearl	800	cling	white	low-acid, sweet	med-large	firm	high	13
Diamond Ray	750	cling	yellow	very good	medium	very firm	high	18
Grand Pearl	800	free	white	best, low-acid	med-large	firm	high	18
Kay Pearl	800	free	white	best, low-acid	med-large	firm	high	18
Bright Pearl	800	cling	white	low-acid, sweet	medium	very firm	medium	22
Summer Bright	850	free	yellow	good, sweet-tart	large	firm	high	26
Summer Grand	750	free	yellow	good, sweet-tart	large	firm	high	27
Fire Pearl	800	cling	white	low-acid, sweet	medium	very firm	high	34
Fire Sweet	800	cling	yellow	low-acid, sweet	large	firm	high	34
Arctic Queen	700	free	white	low-acid, sweet	med-large	firm	high	36
Summer Fire	700	cling	yellow	best, sweet-tart	large	very firm	high	40
July Red	850	cling	yellow	good, sweet-tart	large	firm	high	47
Arctic Blaze	800	cling	white	low-acid, sweet	large	firm	high	49
August Pearl	800	cling	white	low-acid, sweet	large	very firm	high	57
Red Jim	750	cling	yellow	good, sweet-tart	large	firm	high	60
Arctic Pride	850	cling	white	low-acid, sweet	large	firm	high	60
August Red	850	cling	yellow	good, sweet-tart	large	firm	very high	67
Sept. Red	850	cling	yellow	best, sweet-tart	medium	med-firm	very high	68
August Fire	950	cling	yellow	good, sweet-tart	large	firm	high	59
Arctic Snow	950	free	white	low-acid, sweet	very large	firm	high	81
Arctic Mist	850	free	white	low-acid, sweet	large	firm	high	86

Day of first harvest, in average days before (-) or after 'Rich Lady' peach, which starts ripening about June 8 in Fresno, CA, June 22 in Modesto, CA and southern GA, July 2 in Sacramento, CA and central NC, July 22 in western OR, and Aug. 5 in western WA. Days between cultivars increases where nights are cool, decreases where nights are hot.

Notes on Cultivars:
Mayfire sets heavily, sizes erratically, so needs heavy thinning.
Rose Diamond, one of the main early nectarines, has few split pits and full red skin color.
Kay Glo crops poorly some years.
Arctic Star blooms early due to low chill needs, fruit hang well but crack, russet, have some split pits.
Diamond Bright has 75-85% red blush, high fruit quality.
Red Diamond crops inconsistently, needs heavy fruit thining when it crops, VS to brown rot.
June Pearl, with dark red skin, has been widely planted as an early white cultivar.
Arctic Sweet is low-acid but has high flavor, very sweet.
Zee Grand has some split pits; do not girdle.
Spring Bright, the leading cultivar, has very sweet, crisp fruit with dark red skin.
Honey Blaze has full red skin color.
Honey Kist, a top-scored fruit in blind taste tests, may reach 30% sugar if tree-ripened.
Ruby Sweet has mild, low-acid flavor and crisp flesh. Tree is dense, upright, medium vigor.
Ruby Diamond, a seedling of Red Diamond, is hardier so more consistently productive, with larger fruit.
Ruby Pearl has deep red skin, crisp flesh. Tree is dense, upright, medium vigor.
Diamond Ray has full red skin color, bears regularly.
Grand Pearl, a newer white nectarine, has been widely planted.
Kay Pearl, a new white cultivar with full maroon skin, has been heavily planted recently.
Bright Pearl has 80-90% red blush, crops erratically.
Summer Bright is widely grown. Fruit hang well on tree, have full red blush.
Summer Grand, an older cultivar, is no longer planted but still producing.
Fire Pearl has 80-90% red blush, crops consistently.
Fire Sweet has dark red skin and very sweet, tough flesh.
Arctic Queen develops corky flesh in some years.
Summer Fire fruit have dark red skin and crisp flesh.
July Red has nearly full red blush nad very good quality.
Arctic Blaze has nearly full red blush and excellent flavor.
August Pearl has been the most widely-planted late white nectarine in recent years.
Red Jim has almost full red blush.
Arctic Pride is susceptible to powdery mildew.
August Red has consistent yields, full red color.
September Red has full red color, excellent flavor. It blooms very late; may avoid frosts.
August Fire has full red skin color.
Arctic Snow, a widely planted white cultivar, is susceptible to powdery mildew, ships well.
Arctic Mist is very susceptible to powdery mildew.

CHOOSING CULTIVARS

Nurseries offer hundreds of peach cultivars, and new ones debut each year. Unlike many apple and pear cultivars which have unique flavors and coloration, many peach cultivars are very similar in appearance and taste. Consumers distinguish only 3 factors: peach vs. nectarine, yellow vs. white flesh, and sweet/tart vs. low-acid and very sweet. Growers choose among these categories based on the desires of their customers, whether wholesale or retail, and how well such cultivars do in their area.

The number of chill hours (between 32 and 45 F) needed to break winter dormancy is one of the biggest determinants of where each cultivar will do well. (See Fig. 1-2, the Chilling Zone Map.) Peach and nectarine cultivars vary greatly in their needs for winter chilling, and growers should choose cultivars that require about as many chill hours as their area normally provides. Many commercial peach growing districts in California, Georgia, South Carolina and other Southern states usually receive between 650 and 850 chilling hours, so growers in these areas choose from among the many cultivars that need that much chilling. Peach growers in northern Florida, southern Texas, and southern California need lower-chill cultivars with chill needs of 300 to 500 hours, and growers in southern Florida and Mexico need cultivars with very low chill needs of 100 to 250 hours. If such growers plant cultivars with chill needs that are too high for their area, such trees will be very slow to break dormancy and will set few fruit, and many fruit that do set will abort.

Most parts of the U.S., as well as Canada, receive plenty of winter chilling and growers should choose cultivars with the highest chill needs of 950 to 1050 hours, since these will bloom later than cultivars with lower chill needs and be less prone to lose a crop to spring frosts. Peach growers in high-chill regions should also choose cultivars with high levels of bud hardiness in both winter and spring, to avoid losing crops to either winter cold or spring frosts.

Peach and nectarine growers generally plant a succession of cultivars with ripening dates spaced about 10 days apart, to spread harvest over the entire season or the part of the season they're targeting. The main California peach and nectarine shipping season lasts from early May until late August, with a few cultivars ripening in late April, and others ripening in September and early October. Low-chill peach and nectarine growers plant mainly for the very early shipping season from early April to early May. Peaches for processing, grown chiefly in California but also in a few other areas, ripen mainly in July and August. Growers who sell fruit locally usually want peaches ripening over the longest possible season for that area.

Cultivars with firm fruit are desirable to reduce bruising for all markets, especially shipping (which poses more bruising hazards) and canning (to ensure that the canned product remains firm). Fresh-market customers prefer larger fruit, so growers choose cultivars that deliver large fruit size more consistently, although fruit thinning, irrigation and other factors greatly influence fruit size. High yield potential is important for good economic returns, though it's less important than high chill needs and fruit bud hardiness in high-chill areas. Cultivars with a nearly full red blush on the fruit skin are more desirable than those with less skin color, and ideally cultivars should have few fruits with shape defects such as pointed tips or enlarged sutures. More than a few split pits in fruit are undesirable both for fresh-market and processing, since consumers may bite into sharp, projecting split pits or pit fragments, and split pits encourage the growth of fruit-rotting organisms.

Peach and nectarine growers in California and other arid areas do not have to choose cultivars based on disease resistance, but growers in humid eastern North America should. Eastern growers need cultivars with at least moderate resistance to bacterial spot, a disease that may defoliate susceptible cultivars and severely reduce yields. Resistance to peach leaf curl can minimize problems from that disease, especially in very humid areas such as the Pacific Northwest coast, where it can be very severe. Nectarines are prone to fruit cracking and brown rot in humid areas so few growers raise them in such areas, but those who do should choose cultivars that offer the most resistance to cracking and brown rot and also resist bacterial spot.

Peach and nectarine cultivars also vary in how fast the flesh turns brown once cut. Consumers appreciate cultivars which are less susceptible to flesh browning once cut, so retail growers should consider this factor too when choosing cultivars.

Table 18-2 details traits of many of the most widely-planted peach cultivars grown for fresh market in the western U.S., mainly California. Some of these cultivars are also grown in the Southeast, although they lose more crops to cold there and need bacteriocide sprays due to high susceptibility to bacterial spot. Table 18-3 lists the most common processing peach cultivars grown in the western U.S., where over 90% of processing peaches are grown. Table 18-4 describes peach cultivars with disease resistance and medium to high chill needs, widely grown in eastern North America. Table 18-5 presents commonly-grown low-chill peaches and nectarines for the Southeast and Southwest. Table 18-6 details nectarines with disease resistance and medium to high chill needs for eastern North America and Table 18-7 describes nectarine cultivars that are widely planted in the western U.S..

ORCHARD LAYOUT AND TRAINING SYSTEMS

Most growers space rows 18 to 20 feet apart, and trees within rows 12 to 16 feet apart. Higher densities have been tried and in most cases cause excessive crowding and reduced yields after a few years, unless a disciplined summer pruning system is followed. Growers who don't plan to irrigate should space trees farther apart, up to 24 x 24 feet, to allow tree roots to draw from a larger soil area during drought. When the distance between rows is greater than the distance between trees, orienting rows north-south will maximize light interception. Plant rows on the contour if the orchard site slopes moderately or more. All modern peach cultivars are self-fruitful, so plant each cultivar in a block to make harvesting easier.

Medium-low density, open center system. Most commercial growers continue to plant peaches and nectarines at medium-low densities and to train them to the open center system. Spacings of 18 x 16 feet (151 trees per acre) and 18 x 14 feet (172 trees per acre) are common in California, while Eastern peach orchards range from 20 x 18 feet (121 trees per acre) to 16 x 12 feet (226 trees per acre). Growers in Northern areas can plant at higher densities because trees don't grow as large with the shorter growing season.

Planting costs for trees and labor for this system are about $1000 to $1400 per acre, much less than those of higher-density systems. Growers train each tree to 3 or 4 main scaffold limbs, following open center training detailed in chapter 4. Western growers have traditionally allowed scaffold limb to grow fairly upright, at about a 60-degree angle, but many have recently begun to tie branches down to a more horizontal, 50-degree angle as most Eastern growers do, so trees will stay shorter. Trees should start bearing 2 years after planting, and fill their allotted space within 4 to 6 years.

Medium-high density, quad V system. Many new Western peach plantings follow this system, which spaces trees 9 to 12 feet apart within rows that are 16 to 18 feet apart, for a density of 226 to 302 trees per acre. Each tree is trained to a "quad" of 4 scaffold limbs which are selected to grow down the row and perpendicular to it, forming a V shape down the row and another V across the row. This system increases planting costs for trees and labor to about $2200 per acre.

This system produces higher yields during the first 2 to 4 cropping seasons, but after that yields drop below those of the open center system, especially at closer in-row spacings, as trees grow too close and shade one another. At that point growers could remove every other tree or remove the 2 scaffold limbs from each tree that grow down the row, leaving trees with just 2 scaffolds each that form a V shape looking down the row. Tree or limb removal is expensive, however, and removal of large scaffolds creates entry points for bacterial canker in the Southeast or fungal cankers in Northern areas, so limb removal is only practical in warm arid areas such as California. Alternatively, growers can prune back limbs that crowd each other each year. This system also requires heavier fruit thinning to produce fruit of adequate size.

High-density, V-hedgerow system. One large peach-growing firm in California and a few Southeastern growers use this system, which others have tried and abandoned because it is difficult to manage as trees mature. This system spaces trees 6 to 8 feet apart within rows that are 14 to 16 feet apart, for a density of 340 to 518 trees per acre. Each tree is trained to just 2 scaffolds that grow perpendicular to the row, forming a V shape when looking down the row. This system increases planting costs for trees and labor to about $3600 per acre.

This system is very productive during the first 2 to 4 cropping seasons, but no more productive than open-center trees after that, and requires heavier fruit thinning and hand pruning 4 to 5 times each summer to maintain productivity. Machine hedging of such hedgerows caused too much tree damage and regrowth and has been discontinued. Keep nitrogen (N) levels low to avoid excessive vegetative growth. Having just 2 scaffold limbs per tree makes an inherently weaker tree than an open-center tree with 3 or 4 scaffolds (Photo 18-6), so this system is not suited to colder climates where scaffold could suffer winter injury.

High-density, V-trellis system. This system is similar to the V-hedgerow system but uses a V-shaped trellis to aid in training scaffold limbs (Photo 18-7). The trellis adds $3000 to $5000 per acre to establishment costs, so very few commercial growers in North America have adopted it. California growers replant to change cultivars

Photo 18-6. Peach trees trained to 2 scaffolds, as for the V hedgerow system, are more prone to splitting under heavy crop loads than those trained to 3 or 4 scaffolds, so this system shouldn't be used in colder regions. (Photo by Ernest Christ, Rutgers Univ., NJ)

Photo 18-7. A V-trained system, whether freestanding hedgerow or trellis-supported, as here, allows trees to be spaced just 6 feet apart, since only 2 scaffold limbs are chosen and trained to the V shape. The trellis aids in training scaffold limbs, but adds $3000 to $5000 per acre to establishment costs, so North American growers who try high-density peaches use the V hedgerow rather than the V trellis.

every 8 to 12 years, and found that keeping the trellis made soil preparation for replanting very difficult, as they couldn't subsoil, level or fumigate the soil. Some peach growers in Australia and Italy use this system.

Planting and care of new trees. Plant peaches and nectarines from late winter to early spring in Southern areas, and in spring in Northern areas. Order trees from a nursery 10 to 12 months in advance, to ensure getting the cultivars and rootstocks you want, and specify certified virus-free trees. Trees budded the previous summer and 30 to 36 inches tall are best, since taller trees will just be cut back more after planting.

When nursery trees arrive, keep roots moist until planting, and soak roots in water for an hour before planting. Dig holes just large enough to accommodate roots, cutting back any extra long or broken roots, and plant trees at the same depth that they grew in the nursery. In windy areas, plant trees so that each scion curves into the direction of the prevailing wind. Firm soil around the newly-planted roots, and water trees in or irrigate immediately after planting.

Cut back newly-planted trees to a height of 24 to 36". Remove any branches arising less than 18 inches off the ground, and wrap trunks with white tree guards, growth tubes or aluminum foil to protect against sunscald. California growers often apply both white paint to prevent sunscald, and white milk cartons to protect against above-ground rodents. Follow directions for open center or perpendicular V training outlined in chapter 4.

ORCHARD FLOOR MANAGEMENT, IRRIGATION AND FERTILIZERS

Peach growers generally use herbicides to control all vegetation within tree rows, especially when peaches are planted on raised beds. Peaches planted on flat ground might need herbicide treatment only around each tree for the first couple years, which reduces chemical costs but may increase application costs. Generally, growers apply a pre-emergent herbicide in fall after trees have gone dormant, at half strength the first year and full strength in subsequent years. They use 2 in-season sprays of contact herbicides to kill weeds that escaped the preventative treatment.

Controlling all grass and weeds around trees in a wider area, up to a 4-foot radius from each tree, induces young peach trees to develop limbs at wider angles than if vegetation grows nearer to the trees, which aids tree training. Grass sod close to trees especially inhibits root growth, which may reduce cytokinen levels in trees and lead to narrower branch angles.

Groundcovers and irrigation. Most Western peach growers cultivate 4 to 5 times between tree rows in summer (Photo 18-8), and grow winter cover crops (usually a mixture of vetch, winter peas, bell beans or oats) to build soil organic matter and soil N. Those using flood irrigation till soil between trees into long mounds in summer, to create furrows for the water near the trees but not around the trunk (Photo 18-9). Trickle irrigation costs more to establish, but cuts water use and pumping costs in half as compared to flood irrigation, so many growers install trickle irrigation in new plantings. Design trickle irrigation systems to apply up to 50 gallons of water per tree per day.

A few Western peach growers grow a permanent sod of clovers or grasses between peach rows, but this increases irrigation needs by 25%. Clover sods encroach on tree rows and are very difficult to control with herbicides, and can also lead to gopher or stinkbug problems. Grass sods reduce stinkbug problems but increase N fertilizer needs. Permanent sods reduce dust and thus reduce scale insects, because dust hurts scale parasites.

Peach growers in humid areas generally grow a permanent grass sod between tree rows, to reduce soil erosion and to protect tree roots from cultivation damage. This is especially important in the Southeast, where cultivation can leave roots prone to a tree-killing disease complex dubbed peach tree short life, caused by several nematodes and canker diseases.

Photo 18-8. Most Western peach growers cultivate 4 to 5 times between tree rows in summer, and grow winter cover crops to build soil organic matter.

Photo 18-9. Western growers who practice flood irrigation create ridges in summer to direct water around the trees. Trees were planted on mounds or berms so water doesn't stand around the trunks.

Peaches in Eastern areas often lose crops to spring frosts if not protected, so many growers use sprinkler irrigation to provide frost protection. Traveling gun sprinklers and under-tree sprinklers are used the most, since they cost much less than over-tree sprinklers, and can also be utilized on vegetables or row crops than many peach growers raise. Peaches are fairly low trees, and protecting just the lower 2/3 of the tree canopy often provides an adequate peach crop.

Young peach trees require regular irrigation during the growing season in arid areas, and irrigation in humid areas whenever no rain has fallen during the past 2 weeks. Mature peach trees need 42 acre-inches of water to mature a crop and to set fruit buds for next year during late summer and early fall. Slight water stress during Stage I of fruit growth, the first 4 to 6 weeks after bloom when cells divide and fruit grow rapidly, does not hurt yields or fruit size, except in very early-season cultivars. Peaches can withstand moderate water stress (deficit irrigation) during Stage II of fruit growth, the lag phase when fruit size increases only slowly, again except for very early-ripening cultivars which have little or no lag phase.

Adequate water is especially important during Stage III of fruit growth, the "final swell" period of about 6 weeks before harvest, when fruit gain rapidly in size. Inadequate water during the final swell period not only reduces fruit size and thus price received, but also causes about 20% of fruit to be unmarketable, compared to about 5% unmarketable fruit with adequate water during final swell. Also avoid water stress after harvest of most cultivars, during late August and early September when carpels for next year's crop are differentiating, to greatly reduce the number of unmarketable double fruit or beaked fruit that form next year.

Fertilizer applications. Peach trees respond well to heavy N applications during the first few years, so trees will rapidly fill their allotted space. Apply N in a ring around each tree or down each side of the tree rows, to avoid stimulating excess groundcover or weed growth between the trees. Sandy soils need higher N rates than loams or silt loams. Generally, apply 0.05 to 0.07 pounds actual N per tree a few days after planting trees and again in late spring, for a total of 0.1 to 0.14 pounds actual N per tree the first year. Apply 0.1 to 0.14 pounds actual N per tree during the early spring growth flush and again in early summer of the second growing seasons, for a total of 0.2 to 0.28 pounds actual N per tree. The third season, apply 0.16 to 0.20 pounds actual N per tree in early spring and again in either early summer or after harvest of early cultivars, for a total of 0.32 to 0.40 pounds actual N per tree. Monitor shoot growth to adjust N rates (Photo 18-10). If shoots grow more than 2 feet in the first or second year, reduce N the next year, and increase N if shoots grow less than 1 foot in the first or second year.

Bearing peach trees should make 1 to 2 feet of terminal growth each season, and have leaf N levels of 2.6 to 3.0% in midsummer. Many Western growers overfertilize, producing excessive growth which causes some interior branches to die from shading. Trees generally need 0.4 to 0.6 pounds actual N per tree per season when 3 to 5 years old, and 0.6 to 0.7 pounds actual N per tree when 6 years old or older. Western growers apply 30 to 70% of the fertilizer in early spring, and the rest after harvest, since they usually don't have a problem with winter damage. Early-season cultivars receive the larger amount of N after harvest, while later-season cultivars receive the large amount of N in early spring. Eastern growers apply half the fertilizer in early spring, and the other half in late spring if a crop has set, because applying N after harvest of all but the earliest Eastern cultivars could stimulate vegetative growth too late into fall, and subsequent winter damage. If Eastern growers lose a crop to frosts, they skip the second N application that year.

Test soil for P, K, Ca, Mg and pH levels every few years, and apply these nutrients as needed. Use leaf analysis if you suspect any nutrient deficiency. Iron (Fe) and zinc (Zn) deficiencies usually develop if soil pH goes above 7.8, so if this occurs add sulfur to lower the pH, and apply iron sulfate and/or zinc sulfate sprays during the first season to meet immediate Fe and Zn needs.

PRUNING, THINNING AND MANAGING CROPS

Pruning peach trees helps to reduce excess crop loads, allowing remaining fruit to grow larger, and improving return bloom for next year's crop. Pruning also helps control tree size, improves light penetration into trees for better fruit color and sugars, and helps maintain tree health. When crops aren't reduced by frosts, growers must thin excess crops during flower bud initiation or development, bloom and/or early fruit development. Some growers also girdle trees to increase fruit size, and support branches with heavy crops.

Dormant pruning. Peaches are more cold-tender than most fruit trees, so wait until late winter to start dormant pruning. Where winters are cold, delay dormant pruning of peaches until pruning of all other fruit trees is finished, and peach buds start to swell or even bloom (Photo 18-11). This makes it easier to spot dead branches that need to be removed, and to adjust the pruning of fruiting branches based on the percentage of buds that are alive and those that were winter-killed. When most flower buds are still alive and severe frosts during bloom aren't expected, as often occurs in the main peach-growing regions with mild winters, growers can inexpensively reduce crop load by thinning out fruiting shoots, and heading back those that remain, during dormant pruning.

Photo 18-10. Adjust nitrogen (N) fertilizer rates based on shoot growth. If shoots grow less than 1 foot long or have yellow leaves, as here, increase N rates. If shoots grow more than 2 feet long, reduce N rates.

First, remove all dead, diseased or damaged branches, and old gray wood. Also remove any rootstock suckers, trunk watersprouts or branches hanging below 3 feet off the ground, to allow air circulation and herbicide applications. Remove completely all upright, vigorous 1-year-old branches arising from the upper sides of main branches (Photos 18-12, 18-13), since these would produce shoots that would shade the tree interior. Also remove 1-year-old branches that hang below horizontal, and branches that cross and rub against other branches. Remove fruiting branches less than 6 inches long, since these short branches won't have enough leaves to produce good-sized fruit.

This will leave a herringbone pattern of fruiting branches arising mainly from the sides of older branches. Remove up to half of these fruiting branches, until the remaining ones are spaced 6 to 8 inches apart. Head these back to 12 to 15 inches long for fresh market peaches, to increase fruit size. Growers of processing peaches should only cut back branches longer than 18 inches, since processors

Photo 18-11. Delay dormant pruning of peaches until cold weather has passed. In northern areas, wait until budswell or bloom to prune, when cold-damaged buds and branches can be easily distinguished from healthy ones.

don't want huge fruit and total yields are greater if fruiting branches are cut back less. If frosts may occur during bloom, wait until after bloom to head back the fruiting branches.

Many growers have workers use pneumatic pruners to make cuts faster (18-14). Some growers use mechanical cutting bars to mow the tops of peach trees while dormant, which heads back the upper fruiting branches (Photo 18-15). Dormant-season mowing generally produces far fewer damaged branches than does summer mowing, but workers must still do follow-up hand pruning, to remove or cut back branches with jagged ends.

Thinning flower buds or blossoms. Peach flowers are self-fruitful, and wild bees pollinate them well enough that growers do not need to bring in beehives. Only 10% of flowers need to set to produce a full crop, so unless winter cold kills flower buds or frosts kill many flowers, growers need to thin flower buds, open blossoms or newly-set fruit.

Reducing the number of flower buds that set, or thinning during bloom, increases the size of remaining fruit more than fruit thinning, and is widely practiced in California, but is too risky in many Eastern and Northern areas where spring frosts threaten during bloom. Most growers in colder regions choose cultivars that set a large number of hardy flower buds, and so need heavy fruit thinning if no frosts reduce the crop, but growers in California or other regions with rare frosts may choose cultivars that set fewer buds. California growers may also apply the gibberellic acid Ralex™ (a 4% solution of GA3, also called Release™) at 20 to 40 ppm in mid-summer, just before flower buds

Photo 18-12. Healthy peaches trees grow vigorously and need extensive pruning once they begin to bear, to reduce heavy crop loads and admit enough light to produce large, well-colored fruit. This 3-year-old tree row before dormant pruning has many branches that need removal.

Photo 18-13. The same tree row, with upright branches removed and competing branches thinned out, admits more light, especially to tree centers.

Photo 18-14. Many growers have workers use pneumatic pruners to speed the process.

Photo 18-15. Some growers use a mechanical cutting bar to mow off tops of peach trees and lower tree height. Mowing produces less branch damage when done in the dormant season than during the summer, but still requires follow-up hand pruning to remove injured branches so they don't become infected with canker diseases. (Photo by Ernest Christ, Rutgers Univ., NJ)

initiate for next year's crop, to reduce the number of buds that form and bloom the following spring. This compound also improves fruit size, firmness and storage life during the season it's applied, but results vary with cultivar and environmental conditions, so growers should test it on a few trees of each cultivar before treating entire blocks. Generally, trees treated with Ralex™ need 50 to 60% less hand thinning of fruit the next spring.

Another method of reducing fruit buds, that may be safer in areas with spring frosts, is to apply soybean oil late in the dormant season, about 7 to 8 weeks before bloom. The oil suffocates weak flower buds while dormant, but has no effect if applied too late in the dormant season, such as 3 to 4 weeks before bloom. Use 6 to 12% food-grade soybean oil mixed with Latron™ surfactant, in 100 gallons of water per acre. Dormant soybean oil has reduced peach blooms by 60%, increased final fruit size by ¼ inch, delayed bloom by up to 7 days, and yet advanced fruit maturity and allowed earlier harvest. This delayed bloom may reduce the risk of frost injury enough to allow such early thinning.

Mechanical thinning during bloom is possible. Western growers may thin peach blossoms with strong sprays of water, which knock off some of the flowers, at 60 to 80% bloom. Hand-gun attachments to spray tanks must be used to get the spray force strong enough, making this method impractical on large acreages. Some growers use a mechanical rope thinner, a tractor-pulled device that rotates over the tree row, with thick ropes hanging down that drag through the branches. The ropes rub off some flower buds or flowers while not hurting tree bark or branches, but can damage new shoots so this can't be used later as a fruit thinner. A new rope thinner costs about $4200 so requires 30 acres of peaches to pay for itself in a year (or 10 acres of peaches to pay for itself in 3 years), as compared to chemical and hand thinning.

Two surfactants, Entry™ (Armothin™) and Wilthin™, desiccants that destroys flowers' stigmas and styles, are labeled as **blossom thinners** for peaches. Since these compounds inhibit pollination, spraying too early could reduce the crop too much, while spraying too late would not provide enough thinning. Western growers apply them at 60 to 80% bloom, but Eastern growers should generally wait until 85 to 90% bloom. Select 5 representative branches from each of 10 trees in a block, count the total number of flower buds and the number of open flowers on each branch, and calculate (total open blooms/total flower buds x 100) to accurately determine the percent bloom. During cool seasons when bloom is more drawn out, two applications (at mid- and late bloom) at lower rates (such as 1.5% Entry™) may do a better job than one spray at a higher rate (such as 3% Entry™). Another dessicating surfactant, WK™ (dodecyl ether of polyethylene glycol), is sometimes used as a peach bloom thinner in the Southeast, but may cause overthinning and damage to leaves at concentrations of 2% or higher. Some Southern growers use ammonium thiosulfate fertilizer sprays during bloom to thin peaches, but this causes many fruit to stop growing but remain on the tree, making it difficult to distinguish these fruit from normal fruit during follow-up hand thinning.

The thinning effect of these bloom thinners varies greatly with cultivar, so always test a few trees of each cultivar before spraying an entire block. Spraying on cool days or at night produces greater thinning than mid-day sprays on warm days, but do not apply these bloom thinners if foliage is wet or rain is expected within 2 hours. Use at least 100 and preferably 200 gallons of water per acre, since low-volume (even electrostatic low-volume) applications don't work well. Use an air blast sprayer, directing most of the spray into the top 2/3 of the tree. These bloom thinners generally reduce the need for follow-up hand thinning by 33 to 50%. Use of both Ralex™ and Entry™ can reduce hand thinning by 60 to 70%, in areas where spring frosts seldom occur.

Fruit thinning. No safe post-bloom chemical thinners have been developed for peaches, as those tested have all caused defoliation or gummosis of trunks and limbs. Western growers rely on mechanical and/or hand thinning as a follow-up to bloom thinning, and Eastern growers frequently utilize mechanical and/or hand thinning alone.

Thin fruit within 2 to 3 weeks after bloom for the greatest increase in size and sugars of remaining fruit. Many growers use pneumatic limb shakers to knock off lightly-attached fruit (Photo 18-16). A few growers who mechanically harvest peaches for processing use trunk shakers to thin fruit, but these can damage tree roots and shorten tree life. Mechanical thinning leaves many fruit still too closely spaced (Photo 18-17), so most growers have workers remove additional fruit by hand, to a spacing of 5 to 6 inches apart on the branch (Photo 18-18). Thin cultivars in order of ripening. Workers should keep the largest fruit and selectively remove fruit with any insect scars or damage. Thinned fruit on the orchard floor provide a significant source of brown rot spores, so destroy as many of them as possible after thinning by raking them into piles and burning them, or by disking a cultivated orchard, or by applying high-N urea to the orchard floor soon after thinning is done.

Tree girdling. Some growers girdle early-season peach and nectarine trees, to increase fruit size, advance fruit maturity, and allow earlier harvest and fewer pickings. Gaining adequate fruit size for market is difficult during the short, 6- to 9-week fruit development period of very early-ripening cultivars, and earlier harvest of better-sized fruit can greatly increase the price received. Girdling involves removing a thin strip of bark, containing the phloem which transports sugars from leaves down to roots, from around the upper trunk or base of each scaffold limb. This stops downward movement of sugars and growth hormones at that point, allowing more sugars and growth-stimulating compounds to flow into developing fruits, until the girdle heals a few weeks later. Xylem tissues, which transport water up from roots, are not disrupted, so adequate water still flows up to the fruits to increase their size.

Girdling increases the growth rate of fruit only until the girdle heals, so timing is important. Girdle at the end of the first rapid increase in fruit size (stage I in fruit growth), just before the onset of the lag phase (stage II) that limits fruit growth until final swell in the last 6 weeks before harvest (stage III). Determine this stage by cutting open a

Photo 18-16. Many growers use pneumatic limb shakers to mechanically thin excess fruit.

Photo 18-17. Follow-up hand thinning is needed where fruit are still too closely spaced even after chemical or mechanical thinning.

Photo 18-18. Selectively remove smaller fruit or fruit with insect or disease damage, and keep the largest fruit spaced 5 to 6 inches apart on the branch.

few sample peach pits, and girdle when seeds are 9 to 12 mm long.

Girdling requires experienced workers, since girdles that are too deep or wide can kill trees. Western growers practice girdling more than Eastern growers, since even properly-made girdles can serve as an entryway for bacterial canker or other diseases that are worse in humid areas. Girdling also increases split pits on some cultivars, when fruits grow too quickly just as pits are hardening, so do not girdle cultivars that are prone to split pits. Some split pits also split the pointed end of the fruit, leaving it open to decay organisms so that it rots much more quickly. Internal split pits can't be seen from the outside, but such fruit often ripen too quickly or erratically, making harvest and packing very difficult.

Summer pruning. Summer pruning can limit peach trees to a desired height, since it stimulates far less regrowth than dormant pruning. Summer pruning also allows more light to enter tree canopies, which can improve fruit color and size, and increases development of fruiting wood in the lower tree canopy.

Western peach and nectarine growers have traditionally let scaffold limbs grow upright at about a 60-degree angle, and allowed trees to grow to 12 or 13 feet tall so their canopies touch. This enables tractors to drive under tree canopies and almost completely utilizes all light falling on the orchard, so increases yields, but also increases labor costs since workers must use ladders to prune, hand-thin, and harvest. Eastern growers generally train scaffold limbs to grow at wider angles, keep peach trees to a height of 8 or 9 feet, and drive tractors between rather than under the trees. This reduces total yields per acre but also greatly reduces production costs per box of fruit produced, since use of ladders is minimized. Some California growers are now training peach limbs at wider angles, and lowering tree heights with summer pruning.

Some growers practice mechanical summer hedging of tops and sides of peach trees, but this practice leaves jagged cuts that serve as openings for diseases and for insects that bore into branches. Consequently, mechanical pruning must be followed by detailed hand pruning to remove broken branches, so it doesn't reduce hand labor needs very much. Western growers who practice summer topping generally do it after harvest, to avoid causing some of the fruit to fall before harvest or to suddenly let in too much light and cause sunburn on fruit.

Many growers have stopped mechanical hedging in summer, which produces many heading cuts, and instead have workers use pneumatic or hand pruners to make mainly thinning cuts, which produce far less regrowth. Workers cut back upright branches to outward-growing branches or shoots, which allows better light penetration and increases fruiting wood in lower parts of the tree, in addition to lowering tree height. Medium-low density orchards usually need pruning only twice each summer, but higher-density peach orchards do better with 4 or 5 lighter pruning operations per summer. Removing vigorous upright shoots in the tree interior with hand pruners, about 40 days after bloom, improves light penetration and thus fruit size, color and sugar content.

Supporting heavily-laden branches and preventing pre-harvest drop. Western growers of processing peaches commonly support branches with heavy crops by tying rope or wire around the perimeter of young trees, or stretching it down each side of each row of older trees, about a month before harvest. This prevents most branches and limbs from breaking under the crop's weight as fruit gain size. Fresh market peaches are generally thinned more heavily to develop greater fruit size, and fewer fruit produce lower total yields that don't endanger branches as much, but some growers may prop up heavily-laden limbs with boards to prevent breakage (Photo 18-19).

Processing peaches may drop just a few days or weeks before harvest, especially if the crop is heavy and light penetration into the tree is low, causing growers to wait for adequate groundcolor development for the processor to accept the fruit. Fresh market peaches may occasionally suffer pre-harvest drop also. Summer pruning at least a month before harvest can reduce this problem by allowing more

Photo 18-19. Peach limbs loaded with heavy crops may need to be propped up with boards, as here, a few weeks before harvest to prevent limb breakage. Processing peaches produce especially heavy crops so generally need support from rope or wire stretched down the row or around trees a month before harvest.

light into the tree canopy and accelerating color development. Reducing N levels should also help, since excess N causes excessive vegetative growth and reduces fruit color development. Keeping soil water levels steady also reduces pre-harvest drop, and drip-irrigated orchards usually suffer less from it than do flood-irrigated orchards.

Increasing red skin color of fruit. Summer pruning a month or more before harvest also increases red skin color of fruit, which has become more important for fresh-market peaches. Another way to increase red skin color is to lay 5-foot-wide strips of reflective polyethylene mulch down both sides of each tree row, 2 to 4 weeks before the first expected harvest date of each cultivar. The reflective mulch increases light to the trees, increases daytime air temperatures in tree canopies, and reduces daytime relative humidity. This improves red skin color between 16 and 44%, advances harvest by a few days, and increases fruit sugars, although it also causes fruit to be softer. Reflective mulch does not influence fruit size.

PEST MANAGEMENT

Peaches and nectarines suffer from many pests that can kill trees or decimate fruit crops, so growers must work diligently to prevent or control them. However, newer pest control techniques are enabling peach growers to greatly reduce pesticide use as they do this.

Pests that can kill trees. Prevent most of these pests by choosing the orchard site carefully, modifying the site as needed, and protecting roots and trunks from damage.

Phytophthora root rot, incited on peaches and nectarines by any of 11 species in the water-mold fungus genus *Phytophthora*, spreads in wet soils, causing death of roots and eventually death of trees. Prevent it by building raised beds on soils with any drainage problems, and by firmly packing soil around trees when planting to avoid the soil settling into a depression that would hold water around each tree. Schedule irrigations to minimize the duration of soil saturation. Chemicals provide only limited control of Phytophthora root rot in commercial orchards.

Oak root rot, also called mushroom root rot or Armillaria root rot, is caused by the mushroom fungus *Armillaria mellea*. This fungus grows in roots of many tree species, including oaks, willows and all fruit and nut trees, and survives in roots of removed trees for up to 10 years. Some strains of this fungus kill trees rapidly upon infection, while others cause a slower death over 2 to 5 years of decline. Avoid this disease by planting on land that was in grass and/or legume crops, pasture or hay for the past few years if possible. If planting on land cleared from oak forest in the past few years, or replanting a previously removed orchard, remove as much of the old roots as possible, wait several years, and fumigate before planting peaches.

Verticillium wilt, caused by the soilborne fungus *Verticillium dahliae*, penetrates the roots of young trees and plugs up the xylem tubes, causing one or more limbs to wilt. Such limbs may sprout new growth and recover, but sometimes limbs or entire trees die. This fungus also attacks cotton, tomatoes, peppers, melons, and cucumbers, so avoid planting after these crops, especially if Verticillium-susceptible cultivars were grown. Fumigation of infested soil does little to prevent this disease, so choosing sites previously in grass or legume crops is more effective.

Ring nematodes (*Criconemella xenoplax*) feed on peach roots from the outside, greatly reducing feeder roots and predisposing trees to infection by **bacterial canker** (*Pseudomonas syringae*). This can kill many trees on sandy soils in warm regions during the first year after planting peaches on a site formerly planted to stone fruits, grapes or nut trees. Avoid such sites, or grow nematode-reducing crops for 2 years, fumigate before planting, or use soil solarization followed by introduction of the beneficial bacterium, *Pseudomonas* spp. isolate BG33R. Use Guardian rootstock, which resists ring nematodes and bacterial canker as well as root knot nematodes. Avoid using Nemaguard or Nemared rootstocks, which resist root knot nematodes but are more susceptible than most peach rootstocks to ring nematodes and bacterial canker. Avoid cultivation, which damages roots and exacerbates ring nematodes and bacterial canker.

Perennial canker may invade cold-damaged wood, and frequently causes limb loss and tree death in colder peach-growing regions. This fungal canker, also called **Leucostoma canker**, **Cytospora canker**, or **Valsa canker**, is caused by either *Leucostoma cincta* or *L. persoonii*. Planting cold-hardy cultivars, and painting tree trunks white by early winter to reduce cold damage, help to prevent this disease. Train trees during the first growing season to produce wide-angled branches, since narrow-angled branches are more susceptible to winter injury, invasion by borers, and splitting under heavy crop loads. Control rodents, borers and other insects and diseases that create infection sites for cankers. Don't fertilize with N any later than early summer in colder peach-growing regions, and delay pruning until growth begins in spring. Remove cankers as soon as they're seen, and burn all prunings from cankered trees.

Fungal gummosis, caused by *Botryosphaeria dothidea*, attacks water-stressed or poorly managed trees and contributes to general decline of trees. Blisters appear on young bark and enlarge over the next several years if not removed, killing branches or entire trees. Removing and destroying infected branches and limbs is the only control. Where fungal gummosis is rampant, avoid highly susceptible cultivars such as 'O'Henry' and 'Flameprince'.

Peach tree borer (*Syanthedon exitiosa*) and **lesser peach tree borer** (*S. pictipes*) larvae drill into trunks and limbs and can kill trees. Adults of both are clear-winged moths with yellow-orange and black bands on their bodies, and look somewhat like wasps. Peach tree borer (Photo 18-20) generally attacks the main trunk a few inches above or below ground, and is mainly a pest of young trees, since a single borer can kill a tree. Lesser peach tree borer attacks mainly mature trees, anywhere on the trunk or limbs where larvae can get under dead bark. Fully grown larvae pupate under tree bark or in soil near a trunk, then emerge as adults, mate, and lay eggs on trees. Each female moth lays an average of 400 eggs, so even a small moth population can decimate a peach orchard. Peach tree borer has one generation per year with moths emerging from early to late summer, while lesser peach tree borer has two generations per year and moths emerge in late spring and again in late summer.

Mating disruption using pheromone twist ties can eliminate the need for insecticides to control borers, in orchards of 5 acres or more. Place twist-tie dispensers for lesser peach tree borer pheromone before first moth flight, at about shuck fall, at a rate of 100 dispensers per acre. Use 200-250 lesser peach tree borer dispensers per acre to also control peach tree borer, or place additional pheromone dispensers specific for peach tree borer, at a rate of 100 per acre, about 6 weeks after shuck split. Smaller growers protect peach plantings from lesser peach tree borer with a pyrethroid insecticide to control adults during the first cover spray about 2 weeks after shuck split, timed by pheromone traps. They also kill larvae of both borers with a summer trunk spray of a long-residual insecticide such as chlorprifos (Lorsban™), best applied with a handgun either after harvest or well before the pre-harvest interval allowed.

Meadow voles (*Microtus* species), also called meadow mice or field mice, and **pocket gophers** (*Thomomys* species), may girdle trunks just below

Photo 18-20. Peach tree borer tunnels into trunks, and one borer can kill a young tree. Prevent this pest with mating disruption or trunk insecticide sprays.

the soil line, killing trees. Using herbicides to control vegetation in the tree row, and cultivating or mowing the groundcover short between tree rows, will reduce meadow vole populations. Check for burrows of both of these pests in late fall, winter and early spring, and set traps or poison baits if many are found.

Indirect pests that weaken trees, reducing yields and tree longevity. Prevent most of these diseases before planting by choosing virus-free trees and rootstocks with resistance to local soil pests, and by treating the soil if needed. Annual dormant sprays of trees prevent recurring diseases, indirect insects and mites of foliage and buds.

Virus diseases can seriously reduce vigor and yield of peach and nectarine trees. Prevent these by planting only certified virus-free trees. **Peach stunt**, the most serious virus disease in California, is caused by simultaneous infection with two viruses, ***Prunus* necrotic ringspot virus** and **prune dwarf virus**. Each virus by itself causes mild stunting in peach and nectarine trees; together they cause severe stunting of spring shoots and smaller fruits that reduce yields by an average of 18%. About ¼ of non-virus-certified nursery trees are infected with one or both of these viruses, so buying virus-free trees prevents much loss.

Yellow bud mosaic, caused by a strain of **tomato ringspot virus**, produces yellow blotches along main leaf veins that later fall out, and yellow leaf buds that fail to grow from the second year of infection on. After several years, only the tips of branches produce fruit. This virus is spread by the **dagger nematode** (*Xiphinema americanum*) and by cultivation equipment and water in irrigation ditches, so growing nematode-reducing crops or fumigating before planting, and avoiding cultivation and furrow irrigation, can reduce this disease. Also plant virus-free trees.

Phytoplasma diseases of peaches and nectarines include **peach yellow leaf roll** and **X disease**. X disease often infects one or two scaffolds for several years before slowly spreading through a tree, causing leaves to turn yellow, stay smaller, roll inward, and often develop red spots. Peach yellow leaf roll also causes yellow leaves but they have enlarged midveins, and the disease spread more rapidly within trees so that they usually die within 3 years of infection. Leafhoppers spread both diseases, so controlling leafhoppers and weeds that harbor them reduces these diseases. Also remove any infected trees as soon as they're spotted.

Root knot nematodes (several *Meloidogyne* species) enter peach roots, feed from the inside, and form galls. High populations can develop on sandy soils in warm regions, when soils are not fumigated and susceptible rootstocks are used. Such high populations stunt trees, reducing yields enough that orchards are not economical. Always test sandy soils for nematodes before planting, and use rootstocks that resist root knot nematodes if high populations occur. Also avoid using legume cover crops or sod where root knot nematodes are a problem, since populations can build up on legumes.

Root lesion nematodes (*Pratylenchus vulnus*) migrate through the root cortex of most woody perennials, including stone fruits. Lesion nematodes can cause severe stunting in new plantings, and reduce fruit size and yields in older plantings. Prevent damage by testing soil before planting, and if lesion nematodes occur, grow nematode-reducing crops or fumigate before planting. Choose rootstocks with some resistance to lesion nematodes (Bailey, Rutgers Redleaf or Tzim Pee Tao) if root knot nematodes are not also a problem.

Peach leaf curl, caused by the fungus *Taphrina deformans*, induces newly-emerging peach leaves to pucker, redden (Photo 18-21), and eventually fall off, which greatly weakens the tree. Secondary leaves grow, but this reduces the tree's carbohydrate reserves for fruit development, so fruit size and yields suffer. Some peach cultivars have moderate resistance to leaf curl, but none are strongly resistant. The fungus overwinters on bud scales, and must be killed with fungicides during the dormant season to prevent damage. Prevent leaf curl with a dormant fungicide such as a copper compound or lime sulfur, applied after leaves drop in fall or just before buds start to grow in spring, or at both times in very wet years or areas.

Shot hole or **Coryneum blight**, caused by the fungus *Stigmina carpophila*, kills buds and girdles twigs as the fungus invades these tissues during rainy weather from fall to spring. Shot hole lesions may also develop on leaves, flowers or fruits, but these usually do little damage. A late fall fungicide spray for leaf curl will also usually control shot hole, but a second spray in mid-winter may be needed during very rainy winters in warmer areas.

San Jose scale (*Quadraspidiotus perniciosus*) may attack peach and nectarine trees wherever they're grown, killing twigs, branches and sometimes entire trees, and infesting fruit when populations are high. Prevent it with a dormant oil spray, and by pruning out and burning heavily infested branches. Cultivated peach orchards may suffer more scale than those with a permanent sod groundcover, since dust interferes with beneficial parasites that help control scale insects, and such orchards may need summer sprays for scale as described in the pear chapter.

Two-spotted spider mite (*Tetranychus urticae*) occurs throughout North America and **Pacific mite** (*T. pacificus*) occurs on the West Coast, and both may damage peaches. Predator mites that feed on these pests usually keep their populations below damaging thresholds, unless insecticides or fungicides used to control other pests also kill predator mites. Avoid using wettable sulfur fungicide or pyrethroid insecticides during the growing season when predator mites are active, as both are highly toxic to predator mites but not so toxic to pest mites. The dormant oil spray controls pest mites overwintering in or under trees without hurting predator mites, which overwinter in the groundcover between tree rows or in nearby woods or fields.

Photo 18-21. Peach leaf curl can defoliate trees, reducing their yields and longevity. Prevent it with a dormant fungicide spray in late fall, early spring or both times.

Keeping trees well-irrigated reduces mite damage, and sprinkler irrigation can wash mites off trees. A sod groundcover and oiling roads to reduce dust will also reduce mite damage. If summer miticides are needed, use selective compounds that won't hurt predators much, such as cinnamaldehyde (Valero ™) or hexythiazox (Savey ™).

Diseases that attack flowers, fruits and foliage. Prevent hard-to-control bacterial diseases by choosing resistant cultivars in humid regions, and prevent fungal diseases with cultural practices and timely fungicide sprays during the growing season.

Bacterial spot, caused by the bacterium *Xanthomonas campestris* pathovar *pruni*, is potentially the most devastating disease of peaches and nectarines in humid regions. Bacterial spot causes small lesion on leaves, with centers that turn brown and later fall out leaving a "shot-hole" appearance (Photo 18-22), and causes pits and cracks on fruit. Disease severity varies greatly from year to year, but it can cause complete defoliation and crop loss on susceptible cultivars during years with long periods of warm, wet weather. Breeders have produced many fine cultivars with moderate to high resistance to bacterial spot, and growers in humid regions should plant only resistant cultivars. Fixed copper sprays during the dormant season, which may be applied to control peach leaf curl, also reduce bacterial populations.

Brown rot, caused by the fungus *Monolinia fruticola*, is the main fruit-rotting disease of peaches and nectarines before harvest, and also causes much fruit loss after harvest. Brown rot can cause severe crop losses in all peach-growing regions, and is especially devastating on nectarines, since the lack of fuzz to protect the fruit leaves them very susceptible to brown rot. Crack-resistant nectarine cultivars suffer less from brown rot than other nectarines, since any fruit cracks will immediately become diseased. Infected fruit (Photo 18-23) shrivel into mummies (Photo 18-24), which may remain on the tree or fall to the ground, and provide spores that invade next year's flowers and fruit. Brown rot spores also attack small branches, and such infections provide openings for perennial canker. Reduce brown rot by having workers remove all mummies still in trees during dormant pruning, by raking up and destroying thinned fruit or applying urea over them to stimulate faster decomposition, and by avoiding excess N during the final swell phase of fruit growth. Apply protectant fungicides at first bloom, full bloom and petal fall, and 2 to 3 times before harvest starting about 3 weeks before har-

Photo 18-22. Bacterial spot, generally the worst disease of peaches and nectarines in eastern North America, causes spots on fruits and leaves. Centers of leaf spots often fall out, creating "shot-hole" symptoms as seen here, and severe infection can defoliate trees and greatly reduce yields and tree longevity. Choose cultivars resistant to bacterial spot where it's a problem.

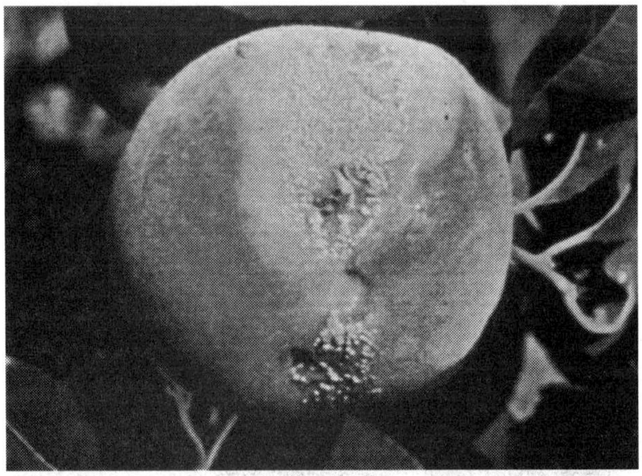

Photo 18-23. Brown rot is a problem on peaches and especially nectarines wherever they are grown, and is particularly serious in humid areas. Prevent it with cultural practices and fungicide sprays during bloom and before harvest.

Photo 18-24. Remove brown rot mummies from last year that still hang on trees during dormant pruning, to greatly reduce the disease this year.

vest or when fruit start to color. Rotate fungicides used among several with different modes of action, to avoid developing resistance to any particular fungicide, and choose fungicides that also control any other diseases that are a problem during each spray period.

Peach scab, caused by the fungus *Cladosporium carpophilum*, produces dark velvety spots on fruit (Photo 18-25), and severe infections cause stunted, misshapen and cracked fruit. Fruit cracks are very susceptible to secondary brown rot infection. The fungus also infects twigs, where it overwinters. Prune trees to keep canopies open, so fruit and leaves dry quickly and fungicides applied will spread evenly over the fruiting surfaces. Apply fungicides at shuck split (when sepals and other flower parts separate from fruit and fall off) and every 7 to 10 days thereafter until 40 days before harvest, as long as weather is warm and humid. Rotate fungicides used among several with different modes of action.

Powdery mildew, caused by the fungus *Sphaerotheca pannosa*, produces white, powdery mycelial growth on young shoots, leaves and fruit, and scars or distorts the fruit. Powdery mildew develops most under warm temperatures and high relative humidity but not on wet surfaces, so is a problem mainly in California and other arid areas with warm, humid conditions in spring. Most nectarine cultivars are quite susceptible to powdery mildew, but peach cultivars differ in susceptibility, so growers can minimize damage by choosing resistant cultivars. The biocontrol agent *Ampelomyces quisalis* (AQ10™) controls powdery mildew, as do the natural materials potassium bicarbonate (Kaligreen™) and chitosan (Elexa 4™). Sterol inhibitors such as myclobutanil (Nova™) or fenbuconazole (Indar™) control brown rot blossom blight in addition to powdery mildew. Apply such materials at bloom, and if conditions are highly favorable for powdery mildew or cultivars are very susceptible, again at petal fall and shuck split.

Rusty spot of peaches is caused by the fungus *Podosphaeria leucostricha*, which causes powdery mildew in apples. Most problems in peaches arise in blocks that are next to blocks of mildew-susceptible apple cultivars. The fungus causes orange-brown lesion on peach fruit, which can render fruit unmarketable on very susceptible cultivars such as 'Loring' and 'Jefferson'. Where rusty spot is a problem, prevent it with a fungicide that controls powdery mildew at petal fall, shuck split, first cover and second cover.

Fruit-attacking insects and birds. These pests feed directly on fruit, leaving it unmarketable and opening fruit to attack by brown rot and other fruit-rotting fungal diseases.

Oriental fruit moth (*Grapholitha molesta*) is generally the worst pest of peach fruit (Photo 18-26), and larvae also burrow into terminals of new shoots (Photo 18-27), causing dieback and creating openings for bacterial or fungal cankers. This pest occurs wherever peaches are grown in North America, has 5 to 6 generations per year in

Photo 18-25. Peach scab disfigures fruit, rendering it unmarketable. Prevent it with fungicide sprays from shuck split until 40 days before harvest, if weather is warm and humid.

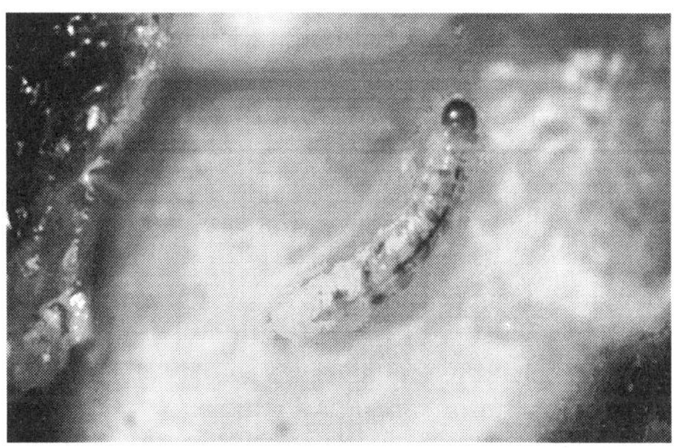

Photo 18-26. Oriental fruit moth, generally the worst pest of peach fruit, has 3 to 6 generations per year. Growers with more than 5 acres of peaches can control it with mating disruption. (Photo by D. Polk, Rutgers Univ., NJ)

Photo 18-27. Oriental fruit moth also attacks peach shoots, causing dieback, and leaves them vulnerable to canker diseases.

Southern peach districts with long growing seasons, and 3 to 4 generations per year in Northern peach areas. First-generation larvae mainly attack shoots, second- and third generation larvae do the most damage to fruit, and fourth- to sixth-generation larvae may attack late-ripening fruit or late shoot growth.

Many peach and nectarine growers with 5 acres or more use mating disruption, either with pheromone twist ties or sprayable pheromone, to control Oriental fruit moth with fewer or no pesticides. California researchers found that the most economical arthropod pest management program included a dormant oil spray for mites and scale, one Oriental fruit moth mating disruption application at first moth catch in pheromone traps (usually during bloom), and one pyrethroid insecticide spray in early summer.

A second mating disruption application costs more, but can eliminate the need for insecticides for this pest. Southeastern growers who must also control plum curculio use an insecticide to control first-generation Oriental fruit moth and plum curculio, and some apply mating disruption to control second- and third-generation moths. Growers who use several pyrethroid sprays per season to control Oriental fruit moth often must apply a special miticide because pyrethroids kill predator mites, and some Oriental fruit moth populations are developing resistance to pyrethroids. Where Oriental fruit moth has caused damage to fruit and shoots during the season, growers should apply sprayable pheromone or an insecticide after harvest to control later generations and prevent population buildup for next season.

Peach twig borer (*Anarsia lineatella*) is another moth pest, whose larvae also feed in shoots and on fruit. It may occur throughout North America but is mainly a pest in California. Shoot feeding kills terminal growth and can severely damage young trees and increase pruning costs. Fruit are most susceptible to borer feeding from the onset of color development until harvest.

Two bloom sprays of Bacillus thuringiensis (Bt) or the insect growth regulator spinosad (Success ™) control overwintering larvae on trees as they emerge, and also reduce populations of other moth pests in the larval stage such as leafrollers, but don't help control Oriental fruit moth since it's in the adult stage during bloom. Apply the first spray when 20 to 40% of larvae have emerged from their hibernating structures, and the second when 80 to 100% of hibernating structures are vacant. Pheromone dispensers for peach twig borer mating disruption have been available since 1995, and many growers apply these at first moth catch soon after bloom and again 60 days later.

Plum curculio (*Conotrachelus nenuphar*) may cause great damage to peaches where it occurs, east of the Rocky Mountains, and needs control every year. It is even more of a pest on stone fruits than on apples and pears, since the beetle larvae can tunnel more extensively through the softer flesh and often remain in the fruit at harvest (Photo 18-28), in addition to deforming fruit and causing many small fruit to drop. Control plum curculio on peaches as described under apples, with insecticide sprays at petal fall, 10 days later (about shuck split in peaches), and in Southern areas where 2 generations occur per year, in mid-summer when monitoring shows that adult beetles have emerged. Very early ripening peach cultivars, harvested before the

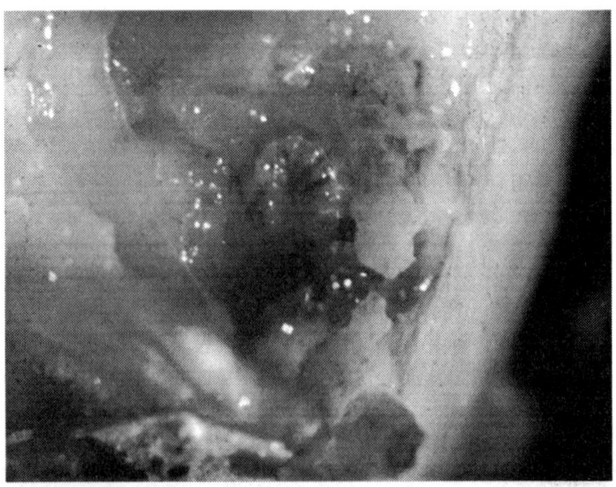

Photo 18-28. Plum curculio is a serious pest of peach fruit in eastern North America. Control it with insecticide sprays at petal fall, shuck split and mid-summer.

second curculio generation emerges in Southern areas, won't need the third spray.

Insecticides such as indoxacarb (Avaunt™) and phosmet (Imidan ™) are effective against curculio but have minimal toxicity to beneficial arthropods. Several pyrethroid insecticides labeled for use on peaches are also effective against plum curculio, and could be used for the first 2 sprays without stimulating pest mite problems if mite predators are not yet present in the orchard.

"**Cat-facing insects**" include **consperse stinkbug** (*Euschistus conspersus*), **Southern green stinkbug** (*Nezera viridula*), **tarnished plant bug** (*Lygus lineolaris*), **lygus bug** (*Lygus hesperus* and *L. elisus*), and other surface feeders that cause dimpled, misshapen or scarred fruit (Photo 18-29). Tarnished plant bug also feeds on flower buds, causing blossom drop and early fruit drop, and lygus bugs also feed on new shoot tips, causing dieback. Monitor these plant bugs with white sticky traps, hanging 1 or 2 traps for every 5 acres in border rows near wooded areas or legume crops, where they build up.

Photo 18-29. Several true bugs feed on the surface of young peach fruit and cause "cat-facing": dimpled, misshapen or scarred fruit. Insecticide sprays at petal fall and shuck split reduce cat-facing injury. A mowed grass sod between peach trees produces far fewer cat-facing insects than does periodic disking, and mowing when most plant bugs are nymphs will destroy many of them.

If excess blossom drop is a problem, spray an insecticide at the pink stage just before bloom to control tarnished plant bug, although if the crop is heavy this insect can actually provide beneficial thinning at this time. Insecticide sprays at petal fall and shuck split, which may be applied anyway for plum curculio, will reduce cat-facing injury to fruit. A mowed grass sod between peach trees produces far fewer cat-facing insects than does a weedy groundcover resulting from periodic disking, and mowing the groundcover when the majority of plant bugs are nymphs will destroy many of them because they do not yet have wings to fly to new food sources.

Omnivorous leafroller (*Platynota stultana*) feeds on leaves and fruit surfaces, but foliage damage is usually minor. Fruit feeding is shallow and extensive, and provides openings for brown rot and other diseases. A bloom spray of Bt reduces populations of this moth pest without hurting predators and at least 12 species of parasites that feed on leafrollers, and can provide significant control if they are not disrupted. Insect-eating birds such as bluebirds and orioles also help control leafrollers.

Birds that peck peaches include **starlings** (*Sturnus vulgaris*), **house finches** (*Carpodacus mexicanus*), and **robins** (*Turdus migratorius*). Large flocks of such birds can move into a peach orchard as fruit begin to ripen, and peck many fruit to find the ripest (Photo 18-30). Pecked fruit can not be marketed, and attract wasps and develop brown rot, which may spread to sound fruit. House finches may also invade orchards in winter and early spring and eat fruit buds, which can reduce the crop if flocks are large and concentrate on particular trees rather than eating a few buds from all trees. Growers use several different frightening devices, rotating them and moving them around so the birds don't get used to them, to repel birds. Modified Australian crow traps with poison baits

Photo 18-30. Birds such as starlings, house finches and robins may peck many peaches just before harvest, leaving them open to infection by brown rot or other fruit rots. Traps with posion baits may be used for starlings and house finches in most areas. Rotate bird scare devices often so birds don't get used to them.

have also been used to trap large numbers of starlings and house finches, and plans for these are available from California farm advisors.

HARVEST, PACKING AND MARKETING

Harvest usually begins in the third year after planting peaches or nectarines, and trees reach full bearing potential between the fifth and seventh years after planting. Some California growers now try to induce some cultivars to bear in the second growing season, by fertilizing and irrigating heavily and pruning minimally during the first growing season.

Most peaches and all nectarines are harvested by hand (Photo 18-31). Wholesale fresh market growers hire workers to harvest peaches and nectarines, as do most retail growers. Very few peaches are sold pick-your-own to customers, since they are large fruits that can be harvested fairly quickly by workers, and people need to be trained to pick peaches at the correct maturity stage without bruising them. Most growers of processing peaches also hire workers to harvest them, since processors pay lower prices for machine-harvested peaches. Machine harvesters utilize a tree shaker and a catching frame.

Photo 18-31. All fresh-market peaches and nectarines, and most processing peaches, are harvested by hand.

Peach and nectarine growers determine fruit maturity on the basis of groundcolor change from green to yellow-green to yellow (Photo 18-32). Groundcolor of white-fleshed fruit changes to a lighter green. Many wholesale growers and processors use color chips to compare with fruit, and harvest when a large number of fruit reach the acceptable yellow-green color stage. Retail growers who sell locally can allow a full yellow groundcolor to develop before harvesting, so fruit accumulate maximum sugars and flavor. Fruit picked too early will soften after harvest but will not develop full flavor or sweetness, and have a mealy texture with less juice.

Photo 18-32. Groundcolor change from green to yellow-green to yellow determines maturity of peaches and nectarines. Groundcolor of white-fleshed fruit changes to a lighter green.

Most cultivars will require 3 or 4 pickings spaced several days apart, over a period of 8 to 14 days, to completely harvest all fruit, but some cultivars ripen over a shorter period and require just 2 pickings. A short harvest period is desirable for processing peaches that are mechanically harvested. Fruit located near the top and outside of each tree normally ripen first because they receive more heat and light, while fruit in the shaded tree interior take longest to ripen.

Photo 18-33. Workers will gently dump fruit from their harvest bags into these sanitized plastic field bins, which should be lined with padding before harvest. Bins should be located in shade while they are filled, and rapidly transported to the packing house or processor for cooling and packing or processing.

Post-harvest handling of peaches. Hand- or machine harvested fruit are often dumped as gently as possible from picking containers into sanitized, padded field bins (Photo 18-34) for transport to the grower's packing shed or a processor's plant. Smaller growers often have workers pick directly into wide buckets, and load these onto trailers for transport to the packing shed. Keep field bins or trailers in the shade while loading, and avoid harvesting during the heat of the day. Use air-suspension systems rather than spring-suspension systems on transport vehicles to minimize fruit bruising.

Fruit must be cooled immediately upon reaching the packing shed or processing plant. Many wholesale growers use hydrocoolers (Photo 18-34), which cool fruit very rapidly but require 50 to 200 ppm chlorine to avoid spreading post-harvest rot diseases such as brown rot, rhizopus rot (caused by the fungi *Rhizopus stolonifer* or *R. arrhizus*) and sour rot (caused by the yeast *Geotrichum candidum*). Many processors and retail growers place peaches in room coolers to get out the field heat before packing, and some small growers have workers rapidly pack fruit into retail or wholesale containers before using a forced-air or room cooler.

Large wholesale growers run peaches and nectarines through a packing line to remove diseased or damaged fruit, wet-brush peaches to remove fuzz, and apply wax and a fungicide to nectarines and peaches. Fruit are then separated into size groups mechanically by weight, hand- or machine-packed into molded trays in 22- to 35-pound, waxed cardboard boxes, and shipped immediately in refrigerated trucks to grocery stores (Photo 18-35). Small growers usually just have workers sort out damaged fruit and pack fruit into retail or wholesale containers by hand, then cool containers in a room or forced-air cooler. Leaving fuzz on peaches helps them to retain water and firmness longer after harvest so wax isn't needed, and avoids damage which leads to fruit rots, so post-harvest fungicide treatment isn't needed. Peach fuzz consists of many fine hairs that are extensions or epidermal cells, so removing fuzz ruptures these cells and causes greater water loss from fruit and greater fruit rots, even when fruit are then waxed and treated with fungicides.

Most fresh market peach cultivars store well under refrigeration for only 2 to 3 weeks, so they must be sold quickly. Controlled-atmosphere storage does not increase storage life of peaches or nectarines very much, and has only limited use for ocean transport of exported fruit. Keep storage temperatures at 31 to 32 F, as temperatures of 36 to 46 F can cause chilling injury, also called internal breakdown. This causes mealy, dry texture and increases susceptibility to fruit-rotting organisms. Early-season peaches and nectarines are least susceptible to chilling injury, while late-season cultivars are most susceptible to it.

Photo 18-34. Many wholesale peach growers use hydrocoolers to rapidly cool fruit after harvest. Hydrocoolers require 50 to 200 ppm chlorine to avoid spreading post-harvest rot diseases such as brown rot and rhizopus rot. (Photo by R.G. Bellinger, Clemson Univ., SC.)

Processing peaches. During the two-month harvest season for canning and freezing peaches, processing plants run 24 hours a day, 7 days a week. Food and Agriculture Dept. inspectors check samples of each load as it arrives at the plant, looking for defects such as green fruit, bruises, other damage and brown rot, and adjust payments accordingly. Bins of peaches go into room coolers for a few hours or if already cool, are unloaded directly to be processed. A conveyor belt passes the peaches through a sizer, where designated rollers direct each size class to the appropriate-sized pitting machine. Workers inspect peaches as they roll by, removing any with defects. Belts position the peaches stem

Photo 18-35. After cooling, wholesale peaches are separated into size groups mechanically by weight, and hand- or machine-packed into molded trays in waxed cardboard boxes. Packed peaches are shipped immediately in refrigerated trucks to grocery stores. (Photo by R.G. Bellinger, Clemson Univ., SC.)

downward, suture forward in line with a machine's knife which cuts peaches in half. A pitting machine removes each pit by twisting peach halves away from the pit, and a second pitter cuts the stringy pit cavity from each peach half.

Peaches next travel in water through the peeling station, where they pass under sprays of a solution that removes the skin. A vibrating conveyor belt picks up the peeled peaches, shaking them so they turn cup side up. Electric eyes and workers inspect peach halves for split pits or pit fragments, which are removed, and for poorly peeled fruit, which are returned to the peeler. Peaches then move onto a size grader, are turned to cup down position, and some enter machines which slice or dice them. The peach pieces move into slots that measure the amount of fruit for each particular can size, and empty cans pass underneath them so the correct amount of fruit falls into each can at the correct time. Cans travel down the line, syrup is added to each can, and filled cans pass through a seaming machine which seals them. A jet of steam blows across tops of containers to remove air before lids are tightened. Cans are heated to 190 to 195 F, then pass through an area at 100-110 F to cool. Finally, cans pass through quality inspection, are labeled, and are stored in the processors warehouse until they are trucked out to grocery stores.

Further Reading

Baugher, T.A., K.C. Elliott, and D.W. Leach. 1991. Improved methods of mechanically thinning peaches at full bloom. *J. Amer. Soc. Hort. Sci.* 116:766-769.

Chalmers, D.J., P.D. Mitchell, and L.van Heek. 1981. Control of peach tree growth and productivity by regulated water supply, tree density, and summer pruning. *J. Amer. Soc. Hort. Sci.* 106:307-312.

Greene, D.W., K.I. Hauschild, and J. Krupa. 2001. Effect of blossom thinners on fruit set and fruit size of peaches. *Hort. Technology* 11:179-183.

Layne, D.R., Z. Jiang, and J.W. Rushing. 2001. Tree fruit reflective film improves red skin coloration and advances maturity in peach. *Hort. Technology* 11:234-242.

Marini, R.P. and D.L. Sowers. 2000. Peach tree growth, yield, and profitability as influenced by tree form and tree density. *HortScience* 35: 837-842.

Ogawa, J.M., E.I. Zehr, G.W. Bird, D.F. Ritchie, K. Uriu, and J.K. Uyemoto. 1995. *Compendium of stone fruit diseases.* APS Press, St. Paul, MN.

Pendergrass, R., R.K. Roberts, D.E. Deyton, and C.E. Sams. 2000. Economics of using soybean oil to reduce peach freeze damage and thin fruit. *Hort. Technology* 10:211-217.

Pickel, C., J. Hasey, W. Bentley, W.H. Olson, and J. Grant. 2002. Pheromones control oriental fruit moth and peach twig borer in cling peaches. *Cal. Agric.* 56:170-176.

Williamson, J.G., P.C. Andersen and W.B. Sherman. 2002. Peaches and nectarines for central and north Florida. Florida Coop. Ext. Serv. Circ. 1159, available online at: http://edis.ifas.ufl.edu/BODY_MG374

South Carolina's web site (http://www.clemson.edu/hort/peach/peachhome.htm) provides much information and links on peach culture, pest management, market information.

Chapter 19
Japanese, European, and American Plums

Plums belong to the genus *Prunus* in the rose family or Rosaceae, like peaches, cherries, apricots, and almonds. These are all stone fruits or drupes, with a smooth (hairless) skin and a fleshy mesocarp surrounding a stony endocarp or pit, which encases a single seed. Plums include several dozen different species that originated in Asia, Europe and North America, many of which have been crossed by breeders to produce cultivars derived from 3 or more different species. In contrast, peaches and nectarines, sweet cherries, tart cherries, apricots and almonds each comprise a single species. Basically, any stone fruit that's not in one of those species is called a plum.

TYPES OF PLUMS, ADAPTATION AND USES

Three types of plums, and a new plum-apricot hybrid species, are grown in North America. **Japanese plums** derived mainly from *P. salicina* (which actually came from China) but include some native American plum germplasm, chiefly *P. angustifolium*, to provide disease resistance. Luther Burbank first introduced these to California and developed cultivars which are still grown today, such as 'Santa Rosa'. Japanese plums are large, round, and very juicy, and usually have red, yellow or almost black skin color (Photo 19-1). Flesh color is usually amber-yellow or red, with red-fleshed plums sometimes called "blood plums". Japanese plums are mainly eaten fresh, with some processed into preserves or baby food.

European plums are mainly *P. domestica* (originally from western Asia), and prunes are a type of European plum with sugar content high enough to dry well without molding. The prune-type European plums typically have an oval shape, a blue-purple skin color, and a dry, mealy, yellow-amber flesh that's not good for eating fresh (Photo 19-2). Other European plums may be round with a red or yellow skin and flesh color, and offer unique

Photo 19-1. Large, juicy, round to heart-shaped Japanese plums, such as this 'Shiro', are mainly sold fresh, with some processed into preserves or baby food.

Photo 19-2. Prunes, a type of European plum with sugar content high enough to dry well without molding, typically have an oval shape, a blue-purple skin color, and a dry, mealy, yellow-amber flesh. This cultivar, 'Stanley', has been the standard in Eastern North America.

flavors for fresh eating, but are usually smaller than Japanese plums and not as juicy.

American plum cultivars have been selected from the wild or resulted from recent breeding programs of several species (*P. americanum*, *P. angustifolium*, *P. munsoniana*), so haven't had as much time for selection of larger fruit, and are generally quite small and round, with red or yellow skin and flesh (Photo 19-3). Wild stands of some American plums are harvested in small areas and processed into jams.

California fruit breeders, the Zaiger family, have recently introduced **pluots™**, their term for **plumcots** that are ¾ Japanese plum and ¼ apricot. Pluots™ and other plumcots offer the tangy taste of apricots and the high sugar content and hairless skin of plums, and are catching on with consumers.

Japanese plums and plumcots have about the same cold hardiness and heat tolerance as peaches, so grow well in most peach growing regions in USDA hardiness zones 6 to 9, if growers choose cultivars with appropriate chill needs and resistance to diseases of the region. European plums are more cold-hardy but less heat tolerant, doing best in USDA hardiness zones 5 to 7. Most American plum cultivars were bred for the Northern Plains region of the U.S. and Canada, within USDA hardiness zones 3 to 5, and survive consistent cold but not fluctuating winter temperatures. A few hybrids of Japanese and American plums were selected from the Southeast, and offer heat tolerance and resistance to diseases which are most serious there.

California produces 94% of the commercial plums grown in the U.S., about 768 million pounds per year, which is about 30% of the amount of U.S. peach production. Oregon, Washington, Idaho and Michigan produce most of the other 6%, while New York and Ontario, Canada also grow a few plums. Japanese plums account for about 55% of total plum production, and most of these are sold on the fresh market, with a small amount processed into jams, preserves and baby food. European plums make up about 45% of plum production, almost entirely dried into prunes. Non-prune-type European plums, plumcots and American plums contribute less than 1% of plum production, and are mainly sold fresh at small retail markets or processed into jams.

Most fresh-market plums are clingstones, but growers may want to plant freestone cultivars if they are available during a particular ripening season, since many customers prefer freestones. Prune-plums are generally semi-freestone to nearly freestone, which aids pit removal during processing into dried prunes.

Photo 19-3. Native American plums, such as this Chickasaw or sandhill plum (*Prunus angustifolia*), produce small, tart fruit that make excellent jam and preserves.

TREE GROWTH AND LIMITING FACTORS

Plums are small trees, growing 15 to 20 feet tall and wide without pruning. Japanese plums and plumcots tend to be more spreading like peaches, while European plums can grow very upright, though cultivars within each group vary greatly. Wild American plums often sucker profusely and form thickets of large bushes, but can be grafted onto plum or peach rootstocks to grow as trees, with rigorous sucker removal.

Well-adapted plum trees last longer than peaches, often 20 to 30 years, although poorly-adapted plums that are susceptible to diseases of the region often die within a few years, before producing much of a crop. Japanese plums and plumcots usually take about a year longer to come into bearing than do peaches, producing a good crop in their third or fourth growing season. European and American plums grow more slowly, and tend to produce their first good crop in the fourth or fifth growing season.

Plums bear most fruit on short spurs arising from 2-year-old or older wood, but also bear fruit from lateral buds on 1-year-old branches like peaches (Photo 19-4). Japanese plums, plumcots and American plums bloom very early but tend to produce huge numbers of flowers, so often produce a partial crop even when frosts occur during bloom. European plums produce fewer blossoms but bloom 1 to 2 weeks later than other plums, usually escaping spring frosts. When frosts do not reduce plum crops, they must be thinned by hand. Fresh-market plums must also be harvested by hand, so labor needs for fresh plums are about the same as for peaches.

Prunes for drying are most often mechanically harvested using the shake-and-catch method, but this requires large acreages to justify the expensive equipment. Demand for dried prunes has declined in North America in recent years, until the cost of growing them exceeded the price growers received for them. A few years ago California, which grows 99% of prunes, initiated a program to pay prune growers to remove some of their acreage, to reduce supply and hopefully lead to increased prices. Labels for prune products were also changed from "prunes" to "dried plums" to update their image and hopefully increase demand.

SITE SELECTION AND PREPARATION

Japanese plums and plumcots need a site similar to that for peaches. They bloom early so need an area with few spring frosts, or a site such as a north-facing slope that will delay bloom. In California, Japanese plums do best at elevations of 500 to 1000 feet, on ridge tops to avoid frost problems. Many Japanese plums and plumcots are grafted on peach rootstocks and these need excellent soil drainage, but those grafted on plum rootstocks need only moderately well-drained soil.

European plums need a site similar to that for apples. They bloom later so frosts usually cause few problems. In California, optimum elevations for European plums are 1500 to 2000 feet so they get enough winter chilling and to minimize fruit flesh injury from summer heat. European plums are grafted on plum rootstocks that tolerate moderately well-drained soil.

American plum cultivars bred for the Northern Plains do best in that region, where winters are steadily cold. If planted in the lower Midwest with its fluctuating winter temperatures, they often bloom very early and lose crops to frost.

Photo 19-4. Plums bear most fruit on short spurs arising from 2-year old or older wood, as here, but also bear fruit from lateral buds on 1-year-old branches like peaches.

Test soil for pH and nutrients at least 6 months before planting, as for other fruit trees, and modify as needed. Make raised beds if planting plums on peach rootstocks or semi-dwarfing rootstocks, but flat planting is fine if plum rootstocks are used.

SELECTING ROOTSTOCKS

Nurserymen graft plum cultivars onto several kinds of plum, peach, and peach-plum hybrid rootstocks. 'Citation' peach-plum hybrid rootstock is often considered the best rootstock for Japanese plums and plumcots in California, as it has good compatibility with them and provides a semi-dwarf tree. It requires excellent soil drainage, however, since it's very susceptible to crown rot. Produced from cuttings, 'Citation' may have a shallower root system than seedling rootstocks, so it may also need more irrigation.

Peach seedling rootstocks are often used for Japanese plums and plumcots, producing a full-size tree with good compatibility. These are susceptible to crown rot so need good soil drainage or raised beds, and need a site free of crown gall or oak root rot. Choose a peach rootstock adapted to your climate and pests (see peach chapter, table 18-1), for example, 'Nemaguard' in California or 'Guardian' in the Southeastern U.S. Peach rootstocks will need greater pest control for peach tree borers than do plum rootstocks.

Many European plums are not as compatible with peach rootstocks and are often propagated on plum rootstocks, which may also be used for Japanese and American plums. Table 19-1 details the attributes of common plum rootstocks. Seedlings of 'Myrobalan' European cherry plum (*Prunus cerasifera*) have been widely used and offer excellent tree anchorage and hardiness to USDA zone 4a, but produce large trees and are susceptible to root knot nematodes, oak root fungus and bacterial canker. 'Myrobalan 29c', a selection grown from cuttings, offers strong resistance to root knot nematodes and a bit less susceptibility to oak root fungus and bacterial canker, but has a shallower root system since it's grown from cuttings, and trees may lean. However, neither Myrobalan rootstock can be used for prunes or many other European plums, since they are highly susceptible to tomato ringspot virus. This virus causes prune brown line disease, necrotic tissue at the rootstock-scion junction that eventually kills the tree. Peach rootstocks are also susceptible to tomato ringspot virus, so can't be used for prunes.

'Marianna 2624', a hybrid of cherry plum and the native American Munson plum (*P. cerasifera* x *P. munsoniana*), resists tomato ringspot virus, so is the most commonly used rootstock for prunes, especially when replanting a site formerly in prunes, apples, tomatoes or other crops susceptible to this virus. 'Marianna 2624' also resists root knot nematodes and crown gall, but is very susceptible to bacterial canker and somewhat shallow-rooted since it's propagated by cuttings, and suckers heavily. 'Marianna M-40', a newer clone, has better anchorage even though it's also grown from cuttings, produces far fewer suckers, and is not quite as susceptible to bacterial canker.

'Damas GF1869', a hybrid of European plum and the European blackthorn plum (*P. domestica* x *P. spinosa*), offers resistance to alkaline soil, bacterial canker and several other diseases, but suckers badly. 'St. Julien A', a semi-dwarfing selection of the European Damson plum (*P. insititia*) also offers resistance to alkaline soil, bacterial canker, several other diseases, and cold. 'Deep Purple', a 1965 University of Minnesota cultivar now recommended by University of California as a rootstock that's produced by micropropagation, also resists bacterial canker, root knot and lesion nematodes, and cold damage.

CHOOSING CULTIVARS

Choose Japanese plums, plumcots and other fresh-market plums to provide a succession of cultivars ripening during each week of the harvest season. This is similar to peaches, except that various plum cultivars offer more distinguishing characteristics, such as skin color or markings or unique shapes, that enable consumers to more readily distinguish the various cultivars and ask for certain ones by name. Also choose cultivars that need about as many chill hours as your area normally provides.

Outside of the arid Southwest (mainly California) where most fresh-market plums are grown, growers should also choose cultivars that bloom late or resist frosts, and have resistance to serious diseases of the area. Growers throughout the eastern half of North America and the Pacific Northwest need resistance to bacterial spot and bacterial canker, and Southeastern growers also need resistance to plum leaf scald.

European plum cultivars bloom later, so frost resistance isn't needed, but cold hardiness is important in Northern areas, and resistance to diseases is important when it's available. Cold hardiness is the most important trait for American plums in Northern areas, while later bloom and disease resistance are more important for American plums in Southern areas. Prune cultivars must dry well and fit industry standards, which are set by 'Improved French', which accounts for 95% of prune production.

Table 19-2 describes traits of many of the most widely-planted Japanese plum cultivars in the western U.S. that have low chill needs, while Table 19-3 details western Japanese plum cultivars with medium to high chill needs. Table 19-4 offers medium- to high-chill Japanese plum cultivars with resistance to some serious diseases in the Southeast in USDA zones 6 to 9, while Table 19-5 lists low-chill Japanese plum cultivars with disease resistance for USDA zones 9 and 10 of the Gulf Coast. Table 19-6 lists the common prune cultivars grown in the western U.S., and Table 19-7 details European plums with disease resistance or extra hardiness for Northeastern and North Central North America. Table 19-8 describes American plums, mainly bred for the extremely cold winters of North Central North America. Table 19-9 covers plumcots, including one cultivar with disease resistance.

ORCHARD DENSITY AND LAYOUT

Most plum orchards, of all types, are laid out in rows 16 to 18 feet apart, with trees spaced 9 to 16 feet apart within rows. This creates a medium-high density planting. Orient rows north-south to gain maximum light interception, though where sunlight is intense and fruit sunburning can be a problem, east-west rows may be better.

Japanese plum and plumcot orchards are laid out like those of peaches. Most new plantings, on 'Citation' or other semi-dwarfing rootstocks, are planted at a medium-high density of 194 to 302 trees per acre, using the quad V system as described for peaches. Such trees are spaced 9 to 14 feet apart within rows that are generally 16 feet apart. Older orchards on standard-sized peach or plum rootstocks often employ the open center system at a medium-low density of 151 to 172 trees per acre, with trees spaced 14 to 16 feet apart within rows that are 16 to18 feet apart.

European plums and many American plums grow more upright, but most often trees are on full-size rootstocks so need slightly wider spacing. Orchards are usually laid out like those of semi-dwarf apples, from 172 to 272 trees per acre. Such trees are spaced 10 to 14 feet apart within rows that are 16 to 18 feet apart.

Table 19-1.

COMMON PLUM ROOTSTOCKS

In approximate order of tree height.

Peach rootstocks described in Table 18-1 are also used for Japanese plums.

Rootstock	Hardy to zone	Tree size	Anchor-age	Resistance/susceptibility to: Wet soil	Dry soil	Alkaline soil	Bacterial canker	Oak root rot	Crown gall	Brown line	Nematodes: Root knot	Nematodes: Lesion	Root suckers grown
Citation	6b	12-14'	good	MR	S	S	MS	S	S	S	R	S	very few
St. Julien A	4b	12-14'	very good	R	MS	R	MS	MS	R	R	S	S	few
Marianna 2624	5a	14-15'	good	VR	MS	S	MS	MR	MR	R	VR	S	many
Marianna M-40	5a	14-15'	very good	VR	MS	S	MR	MR	R	R	R	S	few
Damas	5a	16-18'	good	R	S	R	R	MR	R	R	R	S	many
Deep Purple	4a	16-18'	good	R	MS	S	R	MS	S	S	R	R	few
Myrobalan 29c	4a	16-18'	good	R	MS	S	MS	MS	MR	VS	VR	MR	few
Myrobalan seedling	4a	16-20'	best	R	MR	S	S	S	R	VS	S	MR	very few

Disease resistance ratings: VR = very resistant, R = resistant, MR = moderately resistant, MS = moderately susceptible, S = susceptible, VS = very susceptible.

Comments:

Citation, a semi-dwarfing peach-plum hybrid, is the most common rootstock for new Japanese plum plantings in CA.
St. Julien A, a cold-hardy semi-dwarfing plum rootstock, also offers resistance to alkaline soil.
Marianna 2624 is used with prune-plums to prevent prune brown line, and with Japanese plums for a smaller tree.
Marianna M-40 has better anchorage and less bacterial canker than Marianna 2624.
Damas offers resistance to bacterial canker, root knot nematodes, alkaline soil.
Deep Purple, micropropagated by CA nurseries, is a cold-hardy MN plum clone R to canker, nematodes.
Myrobalan 29c, grown from cuttings, is more uniform than Myrobalan seedling but less well-anchored; trees may lean.
Myrobalan seedling has better anchorage and drought tolerance than cutting-grown plum rootstocks.

Table 19-2.

LOW-CHILL COMMERCIAL JAPANESE PLUM CULTIVARS IN ARID WESTERN U.S.

In approximate order of ripening (varies with area).

These cultivars are susceptible to bacterial spot, canker and leaf scald and have little flower bud hardiness, so aren't generally suited to the East.

Cultivar	Chill Hours Needed	Fruit							Tree		Needs pollen-izer	Bloom time	Days after Santa Rosa
		Stone Freeness	Skin Color	Flesh Color	Flavor	Size	Firmness	Yields	Size	Growth			
Red Beaut	250	semi-free	dark red	yellow	mild, sweet	medium	medium	high	medium	spreads	yes	early	-30
Rubirosa	300	semi-free	dark red	yellow	very sweet	med-large	firm	high	large	upright	yes	early-mid	-14
Black Beaut	250	semi-free	black	amber	sweet-tart	medium	medium	high	large	upright	yes	early	-8
Showtime	300	free	purple	amber	sweet-tart	large	firm	high	large	upright	yes	early-mid	-4
Santa Rosa	300	cling	purple	amber	sweet-tart	medium	medium	medium	large	spreads	no	early-mid	0
Blackamber	400	cling	black	cream	fair, tart	very large	firm	high	medium	upright	yes	middle	3
Queen Rosa	400	cling	red	yellow	mild, sweet	very large	medium	medium	large	spreads	yes	early-mid	10
Laroda	400	cling	purple	yellow	very sweet	medium	firm	high	large	upright	yes	middle	17
Catalina	350	semi-free	black	amber	sweet	med-large	very firm	very high	very large	spreads	yes	early-mid	18
Prime Time	300	free	purple	amber	sweet-tart	large	firm	high	large	upright	yes	early-mid	22
Kelsey	400	free	green	yellow	good, sweet	med-large	medium	high	small	spreads	yes	middle	24
Friar	400	free	black	amber	mild, sweet	medium	firm	high	very large	upright	yes	middle	33
Grand Rosa	400	free	dark red	yellow	sweet-tart	large	firm	high	large	spreads	no	mid-late	42
Rosemary	400	semi-free	red	yellow	sweet-tart	med-large	firm	very high	large	upright	yes	middle	53
Howard Sun	400	free	purple	amber	mild, sweet	large	firm	high	large	upright	yes	middle	56
Casselman	350	cling	dark red	amber	very sweet	med-large	very firm	medium	large	spreads	no	middle	57
John W	450	cling	dark red	orange	sweet	very large	firm	high	medium	upright	no	late	62
Angeleno	450	free	purple	amber	very sweet	med-large	firm	high	medium	spreads	yes	mid-late	72

Day of first harvest, in average days before (-) or after 'Santa Rosa', which starts ripening about June 10 at Fresno, CA, June 24 at Modesto, CA, July 4 in Sacramento, CA, July 24 in western OR, and Aug. 8 in western WA.

Notes on Cultivars:
Red Beaut ships and stores well, but may crop irregularly, VS to bacterial spot.
Rubirosa is increasingly planted as a very early cultivar.
Black Beaut ships and stores well.
Showtime, a seedling of 'Santa Rosa', offers larger fruit and ripens a few days earlier.
Santa Rosa, one of the oldest cultivars, suffers pre-harvest drop and uneven ripening.
Blackamber, the third most-produced plum in California, needs heavy thinning to prevent biennial bearing, has a long storage life.
Queen Rosa has light crops when young and some years after mature.
Laroda is quite tart until fully tree-ripened, ships very well.
Catalina, develops sugar early on the tree, so can be picked early when ships well.
Primetime, a newer cultivar, is being widely planted.
Kelsey is R to bacterial spot, so might be grown in some eastern areas without plum leaf scald or severe bacterial canker.
Friar, an older cultivar, is the most-produced plum in California. Fruit resist cracking, but VS to bacterial spot.
Grand Rosa fruit resist cracking, ship and store well.
Rosemary bears regularly.
Howard Sun ships and stores very well.
Casselman, a bud mutation of 'Late Santa Rosa', resists fruit cracking, has light crops some years and suffers pre-harvest drop.
John W is a newer late cultivar from CA with very large fruit, distinctive speckles on skin.
Angeleno, a standard very late Japanese plum, bears regularly, ships and stores very well.

Table 19-3.

MEDIUM- TO HIGH-CHILL COMMERCIAL JAPANESE PLUM CULTIVARS IN ARID WESTERN U.S.

In approximate order of ripening (varies with area).

These cultivars are susceptible to bacterial canker and leaf scald and have little flower bud hardiness, so most aren't suited to the East.

	Chill	Fruit							Tree		Needs		Days after
Cultivar	Hours Needed	Stone Freeness	Skin Color	Flesh Color	Flavor	Size	Firmness	Yields	Size	Growth	pollen-izer	Bloom time	Santa Rosa
Earliqueen	550	cling	dark red	yellow	sweet-tart	med-large	firm	high	large	upright	yes	early-mid	-22
Murietta	650	free	black	amber	low-acid	medium	firm	high	medium	spreads	yes	early-mid	0
Hiromi Red	850	cling	red	amber	sweet, mild	large	firm	high	large	upright	yes	middle	1
Wickson	500	semi-free	yellow	yellow	very sweet	med-large	firm	high	medium	upright	no	early-mid	12
Simka	550	free	purple	yellow	sweet, mild	med-large	firm	high	medium	spreads	no	late	14
Fortune	500	semi-free	red	yellow	good	large	firm	high	medium	upright	yes	middle	18
Royal Diamond	550	semi-free	black	amber	sweet	med-large	firm	high	medium	upright	yes	late	37
Joanna Red	800	semi-free	red	yellow	sweet-tart	very large	firm	med-high	large	upright	yes	middle	38
Emerald Beaut	700	free	green	yellow	very sweet	medium	medium	high	large	upright	yes	very late	60
Betty Anne	800	cling	red	yellow	sweet, mild	large	firm	very high	medium	upright	yes	middle	61
Autumn Beaut	850	cling	dark red	yellow	sweet-tart	medium	firm	very high	large	upright	yes	late	88

Day of first harvest, in average days before (-) or after 'Santa Rosa', which starts ripening about June 10 at Fresno, CA, June 24 at Modesto, CA, July 4 in Sacramento, CA, July 24 in western OR, and Aug. 8 in western WA.

Notes on Cultivars:

Earliqueen is being planted widely due to earliness and high yields.

Murietta has low-acid fruit which is gaining in popularity.

Hiromi Red has been heavily planted recently, 25% of all new plum plantings in 2001.

Wickson, an older cultivar, is little planted now but still producing, R to bacterial canker but VS to brown rot, bacterial spot.

Simka blooms late, is partly self-thinning, R to bacterial spot. Fruit holds well on tree, ships and stores well.

Fortune, a newer cultivar, has been quickly planted and is now the fourth largest-producing plum in CA.

Royal Diamond's late bloom can avoid frost damage, needs another late-blooming pollenizer.

Joanna Red has lower yields than most newer cultivars, but very large fruit.

Emerald Beaut blooms very late, needs late pollenizer. Fruit holds very well on tree but bruises easily, so handle carefully.

Betty Anne is S to genetic leaf shot-hole.

Autumn Beaut blooms late, needs another late-blooming pollenizer. Fruit resists cracking.

Table 19-4.

JAPANESE PLUMS WITH DISEASE RESISTANCE, FOR HUMID AREAS, USDA ZONES 6 TO 9A.

All are self-unfruitful trees and need a pollenizer.

Cultivar	Hardy to Zone	Chill Hours Needed	Fruit							Resistance to:				Tree		Bloom time	Days after Santa Rosa
			Stone Freeness	Skin Color	Flesh Color	Flavor	Size	Firmness	Yields	Bact. Spot	Bact. Canker	Black Knot	Leaf Scald	Size	Growth		
AU-Amber	7b	650	cling	purple	amber	very good	medium	medium	very high	VR	VR	VS	MR	large	spreads	early	-21
AU-Rubrum	6b	700	cling	maroon	red	gooc	large	medium	high	R	MR	R	R	large	spreads	early-mid	-10
Rubysweet	6b	650	free	bronze	red	best	large	firm	high	R	R	S	MR	large	upright	early-mid	-6
AU-Roadside	6b	650	semi-free	magenta	red	very good	medium	medium	medium	VR	VR	VR	VR	large	spreads	early-mid	1
AU-Producer	6b	600	semi-free	dark red	red	gooc	med-small	firm	very high	VR	VR	MR	R	large	spreads	mid-late	2
AU-Rosa	6b	750	cling	dark red	red	very good	large	firm	medium	R	R	R	R	large	upright	early-mid	4
Shiro	5b	600	cling	yellow	yellow	very good	medium	firm	very high	MR	S	MR	S	large	spreads	middle	5
Starking Delicious	5b	700	semi-free	red	red	good	large	firm	high	VR	MR	MR	S	large	spreads	middle	7
Ozark Premier	5b	750	semi-free	magenta	yellow	good	large	firm	medium	MR	R	MS	S	large	spreads	early	9
Black Ruby	6b	650	free	dark red	yellow	gooc	medium	firm	high	MR	MR	S	MR	medium	upright	early-mid	10
Byrongold	6b	650	semi-free	yellow	yellow	good	medium	med-firm	high	VR	VR	S	MR	very large	spreads	early	12
Explorer	7b	650	semi-free	purple	amber	good	large	med-firm	low	R	MR	R	MR	medium	spreads	early	15
Crimson	6b	850	cling	red	red	good	medium	firm	very high	VR	VR	VR	S	medium	spreads	middle	20

Disease resistance ratings: VR = very resistant, R = resistant, MR = moderately resistant, MS = moderately susceptible, S = susceptible, VS = very susceptible.

Day of first harvest, in average days before (-) or after 'Santa Rosa', which begins ripening about June 18 at Byron, GA, about July 2 in Raleigh, NC, and about July 24 in Mountain Grove, MO.

Notes on Cultivars:
AU-Amber has much better flavor and disease resistance than other early plums, but blooms early due to low chill needs.
AU-Rubrum stores well.
Rubysweet has very high-quality fruit.
AU-Roadside is not firm enough for shipping but sells well in local markets.
AU-Producer needs heavy thinning for good fruit size, stores well.
AU-Rosa has fruit similar to 'Santa Rosa' that keep well, much greater disease resistance.
Shiro tree and fruit buds are very cold-hardy. Fruit may crack in humid areas so best picked a little green and ripened indoors.
Starking Delicious has cold-hardy buds so produces consistently. Fruit flesh is tart, rather tough.
Ozark Premier is an old standard Eastern plum due to hardiness, consistent production. Fruit hang well on tree but don't store well.
Black Ruby, from a cross of Queen Ann x Santa Rosa, produces high-quality fruit.
Byrongold colors well before ripening, so don't harvest too early.
Explorer has cold-tender fruit buds, so is not for zone 7a or colder.
Crimson needs heavy thinning for good fruit size, stores well for 3 weeks or more, ships well.

Table 19-5.

LOW-CHILL JAPANESE PLUMS WITH DISEASE RESISTANCE

For humid areas, USDA zones 9 to 10.

All are self-unfruitful trees and need a pollenizer. All bloom early in lower-chill areas but late in higher chill areas where they're adapted, due to high heat needs after chill units are met. All produce high yields and need heavy fruit thinning to size well. All have good flavor, firm flesh.

	Chill	Fruit				Resistance to:		Tree		Days
	Hours	Stone	Skin	Flesh		Bact.	Leaf			after
Cultivar	Needed	Freeness	Color	Color	Size	Spot	Scald	Size	Growth	Gulfruby
Gulfbeauty	250	cling	dark red	yellow	small	VR	MR	large	spreads	-7
Gulfruby	275	cling	dark red	yellow	small-med	MS	VR	large	spreads	0
Gulfgold	350	cling	yellow	yellow	small-med	R	S	small	spreads	5
Gulfrose	275	semi-free	dark red	red	medium	R	R	large	upright	7
Gulfblaze	250	semi-free	red	orange	medium	MR	VR	large	spreads	7

Disease resistance ratings: VR = very resistant, R = resistant, MR = moderately resistant, MS = moderately susceptible, S = susceptible, VS = very susceptible.

Day of first harvest, in average days before (-) or after 'Gulfruby', which starts ripening about May 18 at Gainsville, FL.

Notes on Cultivars:
Gulfbeauty skin is sour but flesh is sweet.
Gulfruby skin is sour but flesh is sweet, fruit skin sunburns some years.
Gulfgold is a genetic semi-dwarf tree.
Gulfrose has the best fruit quality, greatest commercial potential of these low-chill, disease-resistant plums.
Gulfblaze trees come into bearing early. Skin is sour but flesh is sweet.

Table 19-6.

COMMON PRUNE-PLUM CULTIVARS FOR THE ARID WEST.

All have egg-shaped fruit with firm yellow flesh, and are semi-freestone. All are self-fertile, upright trees that need about 800 hours chilling. These cultivars are susceptible to bacterial canker and bacterial spot and have little flower bud hardiness, so aren't suited to the East.

	Fruit							Days
Cultivar	Skin Color	Flavor	Size	Use	Yields	Tree Size	Bloom time	before French
Tulare Giant	blue	good	very large	fresh	very high	large	middle	-30
Gerrans	red-purple	very good	small	drying	high	medium	early	-12
Sutter	purple	very good	large	drying	high	large	middle	-8
Improved French	red-purple	good	medium	drying	medium	large	middle	0

Day of first harvest, in average days before (-) 'Improved French', which ripens about Aug. 1 in Fresno, CA and Aug. 15 in Modesto, CA.

Notes on Cultivars:

Tulare Giant is used for fresh market due to fruit twice the size of French prune, which would take much longer to dry and have a large pit.

Gerrans fruit have higher sugar content than French prune, and a drying ratio of 2.5 to 2.75 to 1.

Sutter, a newer CA cultivar, has higher sugar, lighter flavor and more complex fruity taste than French prune.

Improved French is the standard prune cultivar, with a drying ratio of 3 to 1 and a small pit. Trees have moderate tendency to alternate bear.

Table 19-7.

EUROPEAN PLUMS WITH DISEASE RESISTANCE OR GREATER HARDINESS

For humid areas, USDA zones 4 to 7.

All need about 800 chill hours, bloom late and produce high yields.

Cultivar	Hardy to USDA Zone	Use	Fruit: Stone Freeness	Skin Color	Flesh Color	Flavor	Size	Firmness	Resistance: Bact. Spot	Black Knot	Tree: Size	Growth	Needs pollenizer	Harvest time
Demontfort	5b	fresh	free	blue	yellow	very good	medium	medium	MR	MR	large	spreads	yes	early
Mont Royal	4a	fresh	free	blue	yellow	good	medium	firm	MR	MR	small	upright	no	early
Seneca	5a	fresh	cling	purple	amber	very good	large	firm	R	MR	large	upright	yes	early
Count Althann's Gage	5b	fresh	cling	purple	amber	best	med-large	firm	R	MR	medium	upright	no	early
Castleton	5a	prunes	free	purple	yellow	good	large	firm	R	S	medium	spreads	no	early
Golden Transparent	5b	fresh	cling	yellow	yellow	best	large	medium	R	MR	medium	spreads	no	early-mid
Stark Blue Ribbon	5a	prunes	free	blue	yellow	very good	very large	firm	MR	MR	large	spreads	no	early-mid
Stanley	4b	prunes	free	blue	yellow	good	large	firm	R	VS	large	upright	no	middle
Longjohn	5a	prunes	free	blue	amber	good	very large	firm	R	S	large	upright	yes	middle
Shropshire Damson	5a	preserves	cling	purple	yellow	best	small	medium	VR	VS	medium	spreads	no	middle
Valor	5a	prunes	semi-free	purple	amber	very good	med-large	firm	R	MS	large	upright	yes	middle
Oneida	5a	prunes	semi-free	purple	amber	very good	large	firm	R	MR	large	upright	no	mid-late
President	5a	fresh	free	blue	orange	good	large	firm	VR	R	medium	upright	yes	late

Disease resistance ratings: VR = very resistant, R = resistant, MR = moderately resistant, MS = moderately susceptible, S = susceptible, VS = very susceptible.

Average ripening times in central MI and northern NJ (about 2 weeks later in northern NY):
very early: early Aug., early: mid-Aug., early-mid: late Aug., middle: early Sept., mid-late: mid-Sept., late: late Sept.

Notes on Cultivars:
Demontfort has a rich flavor.
Mont Royal is a genetic semi-dwarf tree, very cold-hardy.
Seneca fruit may crack slightly during heavy rains and develop brown rot, but much less than 'Stanley' or other prunes.
Count Althann's Gage, also called **'Reine Claude Conducta'**, has a rich flavor.
Castleton cracks much less than 'Stanley', pits well, produces consistently high yields.
Golden Transparent, in the 'Green Gage' group, is one of the finest European desert plums.
Stark Blue Ribbon, a seedling of 'Stanley', ripens earlier with larger fruit.
Stanley, the standard prune-plum grown in the East, is sour around pit, VS to fruit cracking & brown rot.
Longjohn crops at a young age and consistently, and has larger and better quality fruit than 'Stanley'.
Shropshire Damson, also called **'French Damson'**, has a tart, spicy flavor great for preserves. Fruit hang well on tree, can be mechanically harvested. R to bacterial canker.
Valor ripens in 'Italian Prune' season, with much large fruit, better flavor, disease resistance.
Oneida is top-ranked for flavor and size, but has light crops some years.
President has a rich flavor, R to bacterial canker plus bacterial spot, black knot.

Table 19-8.

AMERICAN HYBRID PLUMS FOR NORTHERN PLAINS STATES

In approximate order of ripening (varies with area).

All are spreading trees with yellow-fleshed fruit and need a pollenizer.

Cultivar	Best in Zones	Fruit							Tree Size	Harvest time
		Stone Freeness	Skin Color	Fresh Flavor	Jam Quality	Size	Firmness	Yields		
Pipestone	3b-4	cling	red	very good	very good	large	medium	high	large	early
Underwood	3 & 4	cling	red	good	good	med-large	medium	medium	large	early
Tecumseh	3	cling	dark red	good	good	medium	firm	medium	large	early
La Crescent	3 & 4	free	yellow	best	very good	small-med	firm	medium	large	early-mid
Superior	4 & 5	cling	dark red	very good	very good	large	firm	high	large	middle
Alderman	3b-5	cling	red	best	good	large	medium	high	large	middle
Monitor	3	cling	bronze	good	good	med-large	firm	very high	medium	middle
Toka	4 & 5	free	red	good	very good	small-med	firm	high	small	middle
Kaga	3 & 4	cling	red	good	very good	small-med	firm	high	small	middle
Waneta	3	cling	yellow/purple	good	fair	med-large	firm	very high	medium	mid-late
Ember	3 & 4	cling	yellow/red	very good	very good	med-large	firm	medium	small	mid-late
Redglow	3b-5	semi-free	red	very good	best	large	medium	medium	large	late

Average ripening times in central MN:

early: early Aug., early-mid: mid-Aug., middle: late Aug., mid-late: early Sept., late: mid-Sept.

Notes on Cultivars:

Pipestone fruit has thin, tough skin that peels easily, stringy flesh.
Underwood has a long harvest season, tart skin that peels easily, stringy flesh.
Tecumseh is a hybrid of 'Shiro' Japanese plum and 'Surprise' American plum.
La Crescent fruit are very sweet, with an apricot flavor and thin skin.
Superior bears young, often 2nd season, flowers profusely, needs heavy thinning to get good fruit size.
Alderman bears young, often 2nd season, flowers profusely, needs heavy thinning to get good fruit size.
Monitor has very sweet fruit, but tend to crack in rainy weather.
Toka, a native plum (*P. simonii*), is a good pollenizer for other cultivars. Fruit have a strong, tangy flavor.
Kaga, a native plum (*P. simonii*), is a good pollenizer for other cultivars. Fruit have a strong, tangy flavor.
Waneta is good for fresh eating but doesn't cook as well as others for jam.
Ember is a hybrid of 'Shiro' Japanese plum and an American plum. Fruit have a rich flavor.
Redglow fruit have a rich flavor, thick tart skin. Fruit become acidic when cooked.

Table 19-9.

MEDIUM- TO HIGH-CHILL COMMERCIAL PLUMCOT CULTIVARS IN ARID WESTERN U.S.

In approximate order of ripening (varies with area).

All are high-yielding and need a pollenizer. All these cultivars except 'Spring Satin' are susceptible to bacterial canker, bacterial spot and plum leaf scald and have little flower bud hardiness, so most aren't suited to the East.

	Chill Hours	Fruit						Tree		Bloom	Days after
Cultivar	Needed	Skin Color	Flesh Color	Flavor	Size	Stone freeness	Firmness	Size	Growth	time	Santa Rosa
Flavorosa	450	purple	red	very good	medium	cling	firm	medium	spreads	very early	-20
Spring Satin	750	red-purple	orange	good	medium	cling	firm	medium	upright	early	-17
Flavor Supreme	550	red/green	red	best	medium	cling	medium	medium	upright	early	-2
Flavorite	550	red	yellow	very good	medium	free	firm	large	spreads	early	20
Flavor Queen	500	yellow-green	yellow	very good	large	cling	firm	large	upright	early	35
Dapple Dandy	450	red/yellow	pink	best	large	cling	firm	large	spreads	early	44
Flavor Gold	550	yellow	yellow	best	large	cling	firm	large	upright	very early	46
Flavor King	700	red-purple	red	best	medium	cling	firm	small	spreads	mid-late	50
Flavor Grenade	550	green/red	yellow	best	medium	free	very firm	small	spreads	middle	52
Flavor Heart	550	blue-purple	yellow	very good	large	cling	very firm	large	spreads	mid-late	63
Flavorich	850	purple	orange	very good	large	cling	very firm	large	upright	mid-late	71
Flavorfall	550	maroon	yellow	good	medium	cling	firm	large	upright	middle	92

Day of first harvest, in average days before (-) or after 'Santa Rosa' plum, which starts ripening about June 10 at Fresno, CA, June 24 at Modesto, CA, July 4 in Sacramento, CA, July 24 in western OR, and Aug. 8 in western WA.

Notes on Cultivars:
Flavorosa bears young; very early blooms are prone to frost damage.
Spring Satin, from GA, is MR to bacterial spot, bacterial canker and plum leaf scald.
Flavor Supreme is prone to fruit cracking and subsequent brown rot; don't overirrigate or overthin.
Flavorite can develop 20% soluble solids when allowed to tree-ripen.
Flavor Queen has a prolonged harvest period of 2 to 3 weeks, is prone to split pits, skin bruising.
Dapple Dandy has unique, attractive speckled fruit that's low-acid, very sweet.
Flavor Gold can develop 22% soluble solids when allowed to tree-ripen. It often sets light crops due to very early bloom.
Flavor King has a spicy flavor. Fruit hang well on tree, but prone to severe cracking, brown rot.
Flavor Grenade can develop 20% soluble solids, low acid when allowed to tree-ripen, but fruit may crack.
Flavor Heart has heart-shaped fruit, needs heavy thinning for good fruit size.
Flavorich colors well before fruit maturity, so test soluble solids, which can get to 18 to 20%.
Flavor Fall extends the plumcot season into late fall. Fruit resist cracking, hang well on tree.

Providing for cross-pollination. Most Japanese plum and all plumcot cultivars need cross-pollination by a different cultivar in order to set commercial crops. Some pairs of cultivars, noted in the tables, are cross-incompatible and should not be used as pollenizers for each other. Pairs of cultivars should have overlapping bloom times in order to serve as pollenizers for each other. Choosing compatible pairs of cultivars is especially important for plumcots, which tend to set light crops. Most Japanese plum and plumcot growers plant 2 border rows of one cultivar, 4 rows of a second cultivar to serve as a pollenizer, then 4 rows of the first cultivar or a third cultivar, then keep alternating 4 rows of each cultivar until the final 2 border rows.

Most American plum cultivars and some European plum cultivars also need cross-pollination by a different cultivar, as only 1 to 2% of self-pollinated flowers will set fruit. Some European plum cultivars, including most prunes, are self-fruitful enough to allow 30% of self-pollinated flowers to set fruit, which is plenty for a commercial crop. Such self-fruitful prune cultivars can be planted in solid blocks for ease of harvest, and to avoid excess fruit set and the need for fruit thinning.

Planting and care of new trees. Specify virus-free trees when ordering plums from a nursery, especially with the new threat of plum pox virus. Plant plums in fall or spring, following general fruit tree planting procedures. Cut back newly-planted trees to 24 to 30 inches for Japanese plums and plumcots, and to 36 to 42 inches for European plums. Remove any branches arising less than 18 inches off the ground, and wrap trees with white trunk guards or paint trunks white and place white milk cartons around them, to prevent damage from sunscald and rodents (Photo 19-5). Remove weak or narrow-angled branches, keeping any strong wide-angled branches that may develop into good scaffolds.

Irrigate trees immediately after planting and at least weekly during the first growing season in the absence of rain. Wait to fertilize until tree growth begins. Control weeds, being careful to keep herbicides well away from the new trees.

GROUNDCOVER MANAGEMENT, FERTILIZERS AND IRRIGATION

Fresh-market plums in arid Western North America are generally grown on sandier soils, and growers often sow winter cover crops, then disk them under in spring and keep orchards vegetation-free in summer to reduce competition for water and nutrients. Fresh-market plum growers in more humid areas, such as the Southeast U.S. and Pacific Northwest, generally use a permanent sod between trees to prevent soil erosion. Prune growers most often use a permanent sod, because the main prune area in California is on heavier clay soils that have slow water infiltration after rains and limit equipment access unless sod is used, and Northeastern prune growers need to prevent soil erosion. Almost all growers use herbicides to control vegetation in the trees rows, using a pre-emergent herbicide in fall after trees have gone dormant and two in-season contact herbicide sprays as for peaches.

Permanent sod vs. winter cover crops. Many plum growers use fine fescues and dwarf ryegrass for a permanent sod system, planting from late September to early November depending on the area. Such a sod costs about $80 to $100 per acre to establish, needs mowing just 3 to 4 times per year in arid areas and 5 to 6 times per year in humid areas, and should last at least 10 years. Sod reduces fruit sunburn and mite problems due to dust, but increases irrigation needs by about 25% and requires about 30 to 40 pounds more nitrogen (N) per year than an orchard kept vegetation-free in summer. Some growers add Ladino, New Zealand white and/or strawberry clovers to the sod to help provide N, but clovers in a permanent sod can increase problems with cat-facing insects on plums. Other growers simply mow the resident vegetation (weeds) that grows between tree rows.

Organic Western plum growers often sow legume cover crops to provide N. Vetch sown between the herbicide strips in the tree rows, covering ¾ of the orchard floor, planted in late summer after plum

Photo 19-5. A young plum orchard was planted on slightly raised beds to ensure good soil drainage around tree trunks. A cover crop was planted between tree rows, which are kept weed-free with herbicides. Branches arising less than 18 inches off the ground were removed, and trunks were painted white to prevent sunscald.

harvest and disked under in spring, can provide 100 pounds of N to the orchard if trees are irrigated in summer to release the N. This may be enough N that no other N fertilizer is needed.

Growers who use winter cover crops should disk them under before trees bloom in spring, since cover crops increase the frost damage over bare soil or closely mown sod. Early disking also preserves the most soil water for trees in Western areas, where late disking of cover crops can increase irrigation needs during the growing season. Growers often disk the orchard one or two more times in spring, but most growers have now switched to mowing whatever vegetation comes up during the summer, rather than disking all summer.

Fertilizer applications. Most plum growers apply 40 to 45 pounds of actual N per acre during the first growing season, and 60 to 65 pounds actual N per acre during the second season. Prune growers apply 70 to 75 pounds actual N per acre during the third season and start N rates for bearing trees in the fourth year, but Japanese plums bear in the third season so receive full N rates for bearing trees from the third year on. A typical California Japanese plum or prune crop needs 100 to 150 pounds actual N per year. A lighter crop needs less N, and vigorous trees may not need N applications every year.

Apply half the N in spring and half in early summer or after harvest. Summer N applications to prune trees might be in the form of potassium nitrate added to 4 pesticide sprays, since prunes are high in potassium (K) and often use more K than N, requiring annual additions of 200 to 300 pounds K_2O per acre. An alternative way of applying K is to inject potassium sulfate into trickle irrigation lines. Growers using well water should check it for nitrates and calculate how much N they're applying by normal irrigation, as some wells contain significant nitrate and growers should reduce N applications accordingly. If using potassium chloride (KCl) to supply just K, band it 4 to 5 feet from tree rows at leaf drop just before fall irrigation; do not apply KCl during the growing season as it can cause chloride toxicity.

Sample leaves for nutrient analysis every 2 to 3 years, and adjust N and K rates as needed. Phosphorous (P) fertilizers are seldom needed in Western plum-growing regions, but may be needed in Eastern areas. Plums and prunes on alkaline Western soils often need zinc (Zn) sprays every other year. Take leaf samples in mid-summer, and if needed apply zinc sulfate at leaf fall or 2 to 4 foliar Zn sprays the following spring.

Irrigation methods, timing and amounts. Irrigation is essential for Western plum growers, who have traditionally used flood or furrow irrigation (Photo 19-6) but are increasingly switching to trickle irrigation or microsprinklers to reduce water use. Those who grow Japanese plums in the East generally need sprinkler irrigation to provide frost protection, and most often use traveling guns or under-tree sprinklers as for peaches. Since European plums bloom later, Northeastern and Midwestern growers may not use any irrigation, although many install trickle irrigation for newer orchards.

Irrigate Japanese plums and plumcots like peaches, providing enough water to keep young trees growing and 38 to 42 acre-inches of water per season to mature trees, assuming trickle irrigation with 90% water-use efficiency. (Flood irrigation would require 48 to 52 acre-inches of water per season for mature trees.) Slight water stress during Stage I of fruit growth, the first 4 to 6 weeks after bloom when cells divide and fruit grow rapidly, does not reduce yields or fruit size except in very early cultivars. Most Japanese plum cultivars can withstand moderate water stress (deficit irrigation) during Stage II of fruit growth, the lag phase when fruit size increases only slowly, again except for very early cultivars which have little or no lag phase. All Japanese plums need full irrigation during Stage III of fruit growth, the "final swell" period of about 6 weeks before harvest, when fruit rapidly increase in size. They also need adequate water for several weeks after harvest when fruit buds are setting for next year's crop, usually during late August and early September.

European plums, especially those to be dried into prunes, can withstand more deficit irrigation, and many growers have reduced the amount of irrigation water they apply by 20 to 40% in the past few years, to about 26 to 32 acre-inches of water per year. This reduces excess tree growth and pruning needs as well as water costs. Like Japanese plums, European plums can withstand slight water stress during Stage I and moderate water stress during Stage II of fruit growth, but European plums need less water during the last part of Stage III, which lasts about 8 weeks because they ripen later. Stop irrigating prunes 4 to 6 weeks before harvest, to reduce costs of both irrigation and post-harvest drying. University of California research has shown this does not increase pre-harvest drop.

California Cooperative Extension farm advisors recommend purchasing a pressure chamber, commonly called a "pressure bomb", and using it to check water stress of prune trees at least once a

Photo 19-6. Western plum growers have traditionally tilled furrows along both sides of each tree row, in order to channel water to the trees. Increasingly, growers are switching to trickle irrigation to reduce water costs.

Photo 19-7. Branches of this upright-growing plum should have been spread during the first few growing seasons in order to develop scaffolds with wider branch angles that can bear heavy crops without breaking.

week. Sample about 10 flagged trees per block (so the same trees are sampled each week) between 1 and 3 p.m., when trees typically undergo the most water stress. Choose a shaded leaf from the lower half of each tree's canopy for placement in the pressure bomb to read water pressure. Readings of 0 to -10 bars pressure indicate no to light water stress, -10 to -20 bars show mild to moderate stress, and readings of -20 to -30 bars show moderate to severe stress. Adjust irrigation rates to keep trees in the -10 to -14 bar range during Stage I of fruit growth, in the -14 to -18 bar range during Stage II, and in the -10 to -14 bar range during the first 2 to 4 weeks of Stage III. Then stop irrigating 4 to 6 weeks before harvest.

TREE TRAINING AND PRUNING

Train Japanese plums and plumcots like peaches, to the open-center system for medium-density plantings or to the quad-V system for higher-density plantings. Train European plums like apples, to a central leader system for medium-density plantings or to the vertical axis system for higher-density plantings. American plum cultivars selected for Northern areas are generally trained like European plums, while those selected for Southern areas are usually trained like Japanese plums, although some American plums are grown as bushes since they sucker freely.

Upright-growing plum cultivars of all types benefit from branch spreading during the first few growing seasons, to develop scaffolds with wider branch angles that can bear heavy crops without breaking (Photo 19-7). Plum branches are generally strong enough to support crops without propping limbs or tying them with rope as is common in peaches. However, prune limbs are encircled

Photo 19-8. Prune limbs are encircled with twine before harvesting with mechanical tree shakers, so fruit will fall onto the catch frames and not outside them.

with twine (Photo 19-8) before harvesting with mechanical tree shakers, so fruit will fall onto the catch frames and not outside them.

Since European and American plums are cold hardy, dormant pruning can be done in early to midwinter. Wait until late winter to practice dormant pruning on Japanese plums and plumcots, as for peaches. European plums often need little summer pruning unless they are on vigorous plum rootstocks or are over-fertilized, while Japanese plums need annual summer pruning once they reach the desired height to keep them low enough to do most work without ladders, and to allow adequate light penetration to produce good fruit color.

Pruning bearing trees. Fruiting spurs of all types of plums live for 5 to 8 years, but produce fruit of good size and quality only for the first 3 or 4 years. Therefore, workers prune out 20 to 40% of the older spurs during the dormant season each year, leaving younger spurs and one-year-old branches that also produce fruit on Japanese plums. Some Japanese plums set very heavy crops and need extensive spur pruning to reduce those crops, while others set lighter crops and need less spur pruning. European plums, including prunes, need only light spur thinning.

Older plum orchards may need renewal pruning to admit enough light for good fruit production. Check the shadows cast be tree scaffolds in midwinter, and if they shade a large percentage of the orchard floor, remove one or two large secondary or tertiary upper scaffolds on each tree. This admits more light to middle and lower scaffolds and increases their fruit set, reduces growing points for watersprouts, and reduces future pruning costs.

Prune growers often conduct dormant pruning only every other year to reduce costs. They often use a mechanical mower during the dormant season to lower tree height, then follow up with hand pruning to remove damaged and dead branches and thin out crowded or crossing branches. Eliminating use of ladders, by mechanical topping and using pole-mounted saws and shears for follow-up hand pruning, is essential for reducing pruning costs. Semi-dwarfing rootstocks, regulated deficit irrigation and reduced N applications can reduce the need for height-control pruning. Alternate-year pruning does not reduce yields but may reduce fruit size if trees crop too heavily, so growers may need to thin fruit more heavily during the seasons after they skip dormant pruning.

Some Japanese plum growers mechanically top trees in late summer after harvest to control tree height. Workers must do follow-up hand pruning in late winter to remove vigorous shoots that proliferated from branch stubs created by the mower. Japanese plums on 'Citation' rootstock that receive limited N and regulated deficit irrigation generally don't need summer mowing, but still require hand pruning of overly vigorous branches and water sprouts in summer to reduce shading in the centers of trees.

POLLINATION AND FRUIT THINNING

Growers of fresh-market plums and plumcots generally place hives of honeybees in their orchards during bloom to provide adequate pollination. Bring in beehives when the first plum flowers open, just after mowing or disking the groundcover to reduce any competing blooms. Place one to two hives per acre in groups of 6 to 12 hives at each location. Choose open, weed-free spots protected from winds, where the morning sun will quickly dry off dew and warm the hives, since honeybees don't fly until the air temperature gets to 60 F and winds are less than 15 mph. Some growers place hives on black plastic to absorb sunlight and increase hive temperature more rapidly each morning.

Bloom thinning, hand fruit thinning and girdling in fresh-market plums. Fresh-market plum need fruit thinning every year in Western commercial areas, and most years in Eastern North America, unless the crop is greatly reduced by frosts during bloom. Markets require larger fruit, and fruit must be spaced 3 to 4 inches apart on branches to avoid spreading brown rot. Thinning methods used for peaches are also used for plums. Western growers use a combination of bloom-thinning chemicals or high-pressure water sprays at bloom, mechanical fruit thinning, and follow-up hand thinning. Most growers apply a single spray of a bloom thinner such as 3% Entry™ (Armothin™) at 70 to 80% bloom (Photo 19-9), but a more conservative approach is to apply a lower dose of 2% Entry at 70 to 80% bloom and a second 2% spray 5 days later if needed, after carefully monitoring bee activity and weather conditions. Bloom thinning is generally too risky for Eastern growers, who use mechanical fruit thinning with limb shakers and follow-up hand thinning. Since plums bear fruit in clusters, costs of hand thinning may exceed $1000 per acre.

Fresh-market growers may occasionally girdle trees of early-season Japanese plum cultivars, as

Photo 19-9. Apply a bloom thinning chemical at 70 to 80% bloom, as shown here. To avoid overthinning, use a low dose and then apply a second spray 5 days later if needed.

described for peaches, to increase fruit size and induce ripening a few days earlier. Many plum cultivars do not respond well to girdling, but those that do rarely have problems with split pits that some peach cultivars have. Girdling must be done during or just after petal-fall; girdling later than 3 weeks after bloom is ineffective on plums. Do not girdle weak trees since that could debilitate them, and do not girdle if a light crop is expected, since girdling won't increase fruit size or hasten maturity if the crop is light. Do not girdle cultivars such as 'Santa Rosa' that may soften rapidly, since girdling may cause fruit to ripen very fast.

Mechanical fruit thinning in prunes. Prune orchards often set heavy to extremely heavy crops, which produce a significant amount of small worthless fruit if not thinned, and cause limb breakage, limb and shoot dieback, and leaf scorch from potassium deficiency. Prune growers usually mechanically harvest their crops, and those with tree shakers use these to mechanically thin the fruit. Shake trees as pits within fruit begin to harden at the tip end, during a 3-week period from 3 to 6 weeks after bloom.

Thinning done later than this does not increase size of remaining fruit as much, and removes larger fruit. Before thinning, determine your orchard's optimum crop yield per acre based on previous crop records. Some prune orchards can produce 4 dry tons per acre averaging 65 fruit per pound, while other orchards may only produce 2 or 3 dry tons per acre of fruit this size.

To determine the amount of thinning to do over a prune orchard, follow this method developed by University of California Cooperative Extension Advisors G.S. Sibbett, H. Andris, M. Norton and B. Holtz. Choose a tree several trees in from the end of a row and with a crop load typical of the orchard. Lay 2 tarps (at least 10-foot by 20-foot each) under each half of the tree, and shake as much fruit off as possible onto the tarps by shaking for 7 full seconds. Place all fruit, leaves, and twigs shaken off into 5-gallon buckets, weigh each bucket, and record the total weight. Next, take several 1-pound samples from the buckets, and count all sound fruit in each sample, without counting any yellow or shriveled fruit that would have fallen off on its own without shaking. Multiply the average number of fruit per pound by the total pounds of material removed, to determine the number of sound fruit removed from the tree. Estimate the number of fruit remaining on the tree, and add this number to the number of fruit removed to get the total number of fruit per tree.

Determine the crop load each tree should support by multiplying the desired yield (in dry tons per acre) by the desired fruit size (number of dry fruit per pound) and dividing this by the number of trees per acre. For example, (3 dry tons per acre) multiplied by (65 dry fruit per pound) divided by (109 trees per acre for trees spaced 20 feet by 20 feet) = 3578 fruit per tree. Multiply this number by 1.2 to determine the number of fruit per tree that should remain right after thinning, since about 20% additional natural fruit drop will occur between thinning and harvest. Subtract the number of fruit desired on the tree right after thinning from the total number of fruit per tree to get the number of fruit that should be removed per tree by shaking.

Now place tarps under 3 other typical trees and do trial shakings, one for 4 seconds, one for 5 seconds, and one for 6 seconds. Use a stopwatch to accurately time each shaking. Collect material shaken from each tree and weigh it, and multiply this weight by the number of sound fruit per pound to determine the number of fruit removed. Choose the shaking time that removes the amount of fruit closest to that desired, and use that for the rest of the orchard. Do not shake a tree longer than 6 or 7 seconds or shake a tree more than once, since multiple shakes or overly long shakes can cause excessive fruit injury and later drop.

PEST MANAGEMENT

Most pests that attack peaches also attack plums, so pest management is similar, although some peach pests only rarely attack plums and a few other pests attack only or mainly plums.

Pests that can kill trees. Plums on peach rootstocks suffer from all the "pests that can kill trees" described for peaches, and need the same prevention methods. Plums on certain plum rootstocks (see table 19-1) are resistant to **Phytophthora root rot**, **oak root rot**, **bacterial canker**, **root knot nematodes**, **lesion nematodes** and **crown gall**, and are less susceptible than peach rootstocks to **peach tree borer** and **American plum borer**, which can greatly simplify pest management for plums. Choosing plum rootstocks and cultivars hardy enough for your area, and painting trunks white or using white tree guards, prevents problems with **perennial canker** and **meadow voles** as described in the peach chapter.

Indirect pests that weaken trees, reducing yields and tree longevity. A new virus disease, **plum pox** or sharka, entered Canada and the U.S. from Europe a few years ago and poses a great threat to all stone fruits and especially to plums. Eradication and quarantine efforts have so far confined this disease to Ontario and southeast Pennsylvania and may have eliminated it, but growers should learn about it and be prepared to begin prevention efforts if it should reoccur and spread. The virus can infect over 134 species of plants in 16 families, and is spread by aphids from infected orchards or weeds to new plantings. Plum pox virus causes green fruit to drop prematurely, and any remaining fruit remain small and do not ripen properly, so are unfit even for processing.

Growers of infected stone fruit plantings in Ontario and PA have been forced to bulldoze out and burn their trees. Officials hope to eradicate plum pox from North America. Growers should not order stone fruits from nurseries in the areas where plum pox has been found, and should order certified virus-free trees of all stone fruits. Growers who propagate their own trees should have all bud wood sources tested for viruses. Any suspicious-looking trees should be reported immediately to an Extension Agent for virus testing. 'Myrobalan' seedling and 'Myrobalan 29C' rootstocks are both very resistant to plum pox virus, but no other stone fruit rootstocks have resistance. No stone fruit cultivars grown commercially in North America have resistance, but several European cultivars do, and the USDA's Dr. Ralph Scorza has genetically engineered a European plum tested as "C5" with resistance, by inserting a gene into the 'Bluebyrd' cultivar that keeps the virus from multiplying.

Yellow bud mosaic, caused by a strain of **tomato ringspot virus** and described in the peach chapter, also infects European plums. Prunes are hypersensitive to this virus and will die if grafted on peach seedlings or 'Myrobalan' plum rootstock and planted in soil infested with dagger nematodes, which spread the virus. This is called **prune brownline disease**. Use other plum rootstocks for European plums.

Other peach viruses are not often a problem on plums. The phytoplasma diseases of peach, peach yellow leaf roll and X disease, do not infect plums in the field so needn't be controlled. Plums are seldom attacked by shot hole, so that peach disease also needn't be controlled.

Prune leaf rust fungus, caused by the fungus *Tranzchelia discolor*, can cause substantial defoliation in prune orchards and sometimes in Japanese plums, but the disease only causes economic damage if defoliation occurs before harvest. Monitor plum orchards at least every 2 weeks for this disease, starting in late spring and continuing until mid-July. Check at least 40 trees per block, looking especially at trees with vigorous growth, younger replanted trees, and known rust "hot spots". Treat with a fungicide such as wettable sulfur as soon as a single leaf with symptoms is seen. In California with no summer rainfall, one sulfur application can protect trees for up to 2 months, so the best treatment time is generally mid-June. No rust treatments are needed if no rain falls from mid-May to mid-June.

Several **aphids** attack plums in spring, including **mealy plum aphid** (*Hyalopterus pruni*) and **leaf curl plum aphid** (*Brachycaudus helichryse*). Both cause leaves to curl, and secrete honeydew that may cause fruit cracking and results in fungal growth on the honeydew and split fruit. A dormant oil spray the previous winter can kill overwintering aphid eggs and prevent damage. Monitor 20 trees per orchard for aphids in fall, looking at up to 100 leaves per tree, and count a tree as aphid-infested if any leaves on the tree have aphids. If aphids are scattered across the orchard, treat the entire orchard with a dormant oil spray; if they are concentrated in certain spots, treat only those spots. Dormant oil also prevents most problems with **San Jose scale, two-spotted spider mite** and **Pacific mite**, all of which may damage plums. Start scouting for mites in early summer, in known mite "hot spots" or places where water stress is common, and treat as for peaches if needed.

Diseases that attack flowers, fruits and foliage. Growers in the Southeastern or Midwestern U.S. should choose plum cultivars resistant to **bacterial spot.** It causes the same damage to plums as it does to peaches.

Brown rot (Photo 19-10) is the worst fruit-rotting disease of plums, just like peaches, and is

Photo 19-10. Control brown rot, the worst fruit-rotting disease of plums, with the same combination of sanitation and fungicides used for peaches.

controlled with the same combination of sanitation and fungicides. Prune growers may sometimes not get all tree blocks harvested, and if so should shake off the fruit of any unharvested blocks as soon as possible, before fruit turn into mummies that would create a brown rot epidemic the next year. Hand removal and destruction of all mummies from Japanese plum trees after harvest will also greatly reduce brown rot the following year.

Powdery mildew may attack Japanese plums but usually does less damage than on peaches, and does not attack European plums. It is not usually a problem outside of California, and there only during springs with high humidity, when the fungicide program outlined for peaches should be followed. Japanese plum cultivars differ in resistance to powdery mildew but no information has been quantified. Peach scab only rarely occurs on plums, so doesn't need control.

Plum pockets, caused by the fungus *Taphrina communis* (a cousin of the peach leaf curl fungus *T. deformans*, which does not attack plums), occurs only sporadically in Piedmont orchards of the Southeastern U.S. Infected leaves and new shoots become twisted and puckered like peach leaf curl, but foliar symptoms are less conspicuous and not always present on plums. This disease also attacks plum fruit, distorting and enlarging them. Deformed fruit turn brown-black with age and drop from the tree, and when cut open reveal empty "pockets" instead of flesh under the skin. Most European plums and pure Japanese plums resist plum pockets, since the disease attacks mainly cultivars with some American plum germplasm, especially *P. angustifolia*. Removing wild plums near orchards greatly reduces the disease. A single copper or lime-sulfur fungicide spray applied during dormancy in late fall or early spring, as for peach leaf curl, controls plum pockets if the disease is present.

Fruit-attacking insects and birds. These direct pests are the same for plums as for peaches, and control methods outlined for peaches are followed. **Oriental fruit moth** and **cat-facing insects** are major pests wherever plums are grown, and **pest birds** can become severe pests. **Peach twig borer** and **omnivorous leafroller** are mainly problems in California, and **plum curculio** (Photo 19-11) only occurs east of the Rocky Mountains.

HARVEST AND PROCESSING

Determine the optimum time to harvest plums by fruit groundcolor, soluble solids and flesh firmness. The groundcolor of fresh market plums and plumcots should turn from green to light green or yellow-green. Soluble solids should reach 14 to 18% for Japanese plums, 16 to 20% for plumcots, 12 to 16% for fresh-market European and American plums, and 22 to 24% for European prune-plums for drying. Fruit firmness should be 6 to 8 pounds pressure for Japanese plums and plumcots, and 4 to 6 pounds for prunes for drying.

Avoid harvesting plums too early, even though some fruit will drop before the optimum harvest time. Fruit can gain 3% in size per day as harvest nears, and fruit harvested too soon often receive a much lower price for smaller size. Soluble solids also increase rapidly in the last few days before harvest. Fresh-market fruit harvested too early will never develop full flavor, and prune-plums harvested too early will require more drying time and may not achieve the sugar level needed to prevent spoilage in dried fruit.

Harvesting, ripening and storing fresh-market plums. Hand-harvest fresh-market plums like peaches, going over each cultivar block 2 to 4 times during an 8- to 14-day period and picking only ripe fruit during each harvest. Handle and cool fresh-market plums like peaches, except plums have no fuzz to remove. Most fresh-market plums ripen naturally at the retail outlet, but some cultivars ripen very slowly and need exposure to ethylene gas to ripen. This may come from placing them near apples or pears that give off ethylene naturally, or by exposing them to 10 to 100 ppm ethylene for 3 to 4 days at room temperature of 68 to 77 F.

Japanese plums keep well in refrigerated storage at 30 to 31 F for about 4 weeks, although some cultivars that are prone to internal breakdown can only be stored for 2 to 3 weeks. Controlled-atmosphere storage has only limited use for some plum cultivars that aren't prone to internal breakdown.

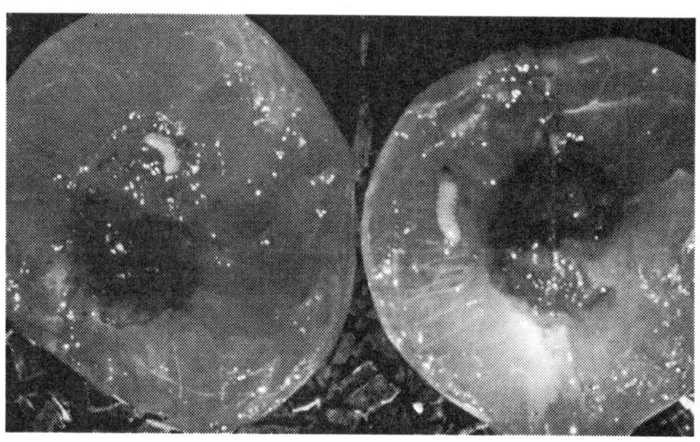

Photo 19-11. Plum curculio is a serious direct pest of plum fruit east of the Rockies. Control it as for peaches.

Harvesting, drying and processing prunes. Most prune growers use shake-and-catch tree shakers to mechanically harvest the fruit. Start harvest earliest in blocks with older trees but lighter crops, because fewer fruit means higher soluble solids and larger fruit size. Next harvest older blocks with moderate crops and then those with heavy crops. Harvest younger, strong-growing trees that have high potassium levels last, since such trees will retain fruit the longest without dropping them.

Many prune growers use field sizing chains with their harvesters, since small or very small prunes reduce the price received. Harvested fruit pass over the sizing chains and small ones drop out in the field, so the fruit going to the dryer have a higher proportion of large sizes. Fruit are deposited into field bins, which are transported to dehydrators.

At the dehydrator, prunes are washed and mechanically spread on large trays, which are stacked 24 to 28 trays high on carts and rolled into a 40-foot-long dehydrator tunnel. Prunes stay in the dehydrator, subjected to a constant flow of dry air heated to 180 to 185 F, for 18 to 24 hours. Dehydrating reduces 3 pounds of fresh fruit to about 1 pound of prunes containing about 21% moisture, the moisture content at which they store best. The dried prunes are then delivered to the processing plant, where they remain in refrigerated storage until further processing, which takes place all year until 2 weeks before the next harvest season.

During processing, prunes are pitted, inspected for quality, graded for size by moving them over large shaker screens, and packaged. Machines can shove pits from whole prunes without cutting them, and follow-up machines blow away prunes that still have pits. Some prunes are cut for packaging as diced prunes. Automatically weighed fruit flows into bags, cartons, bulk cases and cans in packs ranging in size from 1 ounce to 50 pounds. Prune juice is made by cooking prunes in tanks of hot water, draining the juice from the tanks, and blending different lots to provide a uniform taste and composition that is at least 18.5% prune solids. About one third of the prune crop is processed into juice, juice concentrate, puree and baby food.

Since prunes are naturally dark, they are not treated with sulfur dioxide to prevent darkening as apricots, apples, and other light-colored dried fruits are. The only preservative used is potassium sorbate, which prevents mold and yeast growth.

Prune consumption in the U.S. has declined slightly in recent years, but consumption in Europe, Japan and other Asian countries has increased, so that 45% of U.S. prune products are now exported. Prune processing plants label packages in at least 20 different languages.

Further Reading

Johnson, R.S., H. Andris and K. Day. 2002. Experiences with chemical thinning of plums. Available online at: http://www.utahhort.org/talks/2002/ScottJ04.htm

LaRue, J.H. and R.S. Johnson. 1989. *Peaches, plums and nectarines: growing and handling for fresh market*. Univ. of California pub. 3331.

Sibbett, G.S., H. Andris, M. Norton, and B. Holtz. 1998. Mechanically shake thinning prunes in the San Joaquin Valley. *UCCE Tulare Co. Prune News*, May 1998. Available online at: http://cetulare.ucdavis.edu/pub/pru0598.htm

Chapter 20
Sweet and Tart Cherries

Cherries are stone fruits in the genus *Prunus* of the rose family (Rosaceae), and the sub-genus *Cerasus* that produces smaller fruit than most stone fruits. Two cherry species native to eastern Europe and western Asia, around the Caspian and Black Seas, have been developed as commercial crops: sweet cherries derived from *Prunus avium*, and tart cherries from *P. cerasus*. **Sweet cherries** produce larger, meatier and sweeter fruit than **tart cherries** do, and 58% are grown for fresh market, while 31% are processed into maraschino cherries and 11% are canned. Sweet cherries grown for fresh market must be hand harvested, while those grown for processing are machine harvested. Tart cherries (Photo 20-1), sometimes called sour cherries, are too tart for fresh eating but have more juice, so 99.5% are grown for processing and mechanically harvested. About 53% of tart cherries are frozen, 34% are canned for pie fillings, and 14% are made into juice, wine or maraschino cherries.

The total North American cherry crop of about 670 million pounds per year, consisting of about 2/3 sweet cherries and 1/3 tart cherries, exceeds production of plums and is about 30% that of peaches. The U.S. is the world's largest cherry producer, growing about 22% of the total production, followed by Iran and Turkey. European countries also produce many cherries.

ADAPTATION AND GROWTH HABITS OF CHERRY SPECIES

Sweet cherry trees are sensitive to both winter cold and summer heat, and are difficult to grow outside of areas with relatively mild winters (to USDA hardiness zones 6a to 5a) and cool summers. They also bloom very early, so need areas with little danger of spring frosts. Most sweet cherry cultivars suffer from bacterial canker of trunks, fruit cracking and subsequent brown rot, and other diseases in humid areas, so do best under irrigation in arid areas. The West Coast region produces 90% of the sweet cherries grown in North America. Washington produces about 41% of the total crop, California raises about 27%, Oregon grows about 18% and British Columbia grows about 4% of the total North American sweet cherry crop of about 440 million pounds per year.

Tart cherry trees are more cold-hardy, to USDA zones 4a to 3a, but still suffer from summer heat, limiting cherry production to about the northern half of the U.S. and southern Canada. Tart cher-

Photo 20-1. Tart cherries are grown almost entirely for processing into pie fillings, jams, juice and other products.

ries bloom later and generally suffer much less bacterial canker, fruit cracking and brown rot than do sweet cherries, so are mainly grown in the Upper Midwest and Northeastern North America. Michigan produces over 68% of U.S. tart cherries, about 152 million pounds per year, along with 18 million pounds or 4% of sweet cherries, making it the second largest cherry-producing state after Washington. Utah grows about 11% of U.S. tart cherries, Washington produces about 9%, Wisconsin grows about 6%, New York yields about 3%, Pennsylvania raises about 2% and Oregon produces about 1%. Ontario also grows many tart cherries, and the total North American tart cherry crop is about 230 million pounds per year.

Sweet cherries are naturally large trees, reaching 50 feet or greater height in nature, and many cultivars do not begin to bear until 5 to 7 years after planting. Tart cherries are smaller trees, growing about 20 feet tall in nature, and begin bearing earlier, often just 3 to 4 years after planting. Sweet cherries develop fruit-bearing spurs from 1-year-old wood, which bear the next season on 2-year-old wood, and continue producing for 10 to 12 years. Tart cherries produce 55 to 65% of their crop on shorter-lived spurs that produce for 3 to 5 years, and the rest of the crop from lateral fruit buds on 1-year-old branches like those of peaches. Fruit buds of both species produce clusters of 2 to 5 flowers, but fewer fruit per cluster usually set for tart cherries than for sweet cherries, causing fewer brown rot problems on tart cherries.

Sweet cherry orchards often produce well for 25 to 30 years, up to an age of 30 to 35 years, while tart cherry orchards remain productive for 20 to 25 years, until 24 to 28 years old. Sweet cherries generally produce 9000 to 11,000 pounds per acre in the Northwest and 6000 to 9000 pounds per acre in California and other states, while tart cherries produce 7000 to 9000 pounds per acre. A few cultivars of both species are genetic dwarfs, staying smaller and beginning bearing sooner, and recently-developed semi-dwarfing rootstocks now can be used to keep trees smaller and induce earlier bearing.

Several cherry species native to North America are harvested from the wild to a limited extent, mainly for processing into jams and jellies. Western sand cherry (*P. besseyi*), native to the Northern Plains, is a small, spreading shrub growing just 4 to 5 feet tall, with dark purple fruit. Sand cherries are very cold-hardy and drought-tolerant, and several cultivars have been named and planted in the north-central U.S. and south-central Canada. Sand cherry has also been crossed with Japanese plum (*P. salicinia*) and native American plums (*P. hortulana* or *P. munsoniana*) to produce cherry-plum cultivars, also planted in the Northern Plains. Cherry-plums are small trees, commonly budded on hardy *P. americana* rootstock, and grown similarly to the American hybrid plums discussed in chapter 19.

Several cherry species native to eastern North America could be domesticated and grown using the techniques outlined below for sweet and tart cherries, although no improved selections from these species are currently available. Wild black cherry (*P. serotina*), a common woodland tree east of the Great Plains, often reaches heights of 80 feet or more, making it difficult to harvest although the fruit are quite flavorful. It could benefit from dwarfing rootstocks and training techniques discussed below. Pin cherry (*P. pennsylvanica*) and chokecherry (*P. virginiana*), also native to Eastern woods and meadows, are smaller trees growing to 30 feet tall, but fruit are more acid with higher tannins. Wild chokecherries commonly harbor a phytoplasma that causes X-disease, a devastating disease of cultivated cherries, so Eastern cherry growers should remove all chokecherries on their land. Western cherry growers should remove any wild bitter cherry (*P. emarginata*) on their land, as these can harbor X-disease and cherry viruses.

ROOTSTOCKS AND ORCHARD DENSITY

For many years, only two rootstocks were used for cherries. 'Mazzard', grown from seed of wild sweet cherry (*P. avium*) selections, was used for almost all sweet cherries and sometimes used for tart cherries. It's still used for 30 to 40% of new sweet cherry plantings. 'Mazzard' offers good adaptation to heavier soils and some resistance to *Phytophthora* and oak root rots, but produces very large trees that are slow to begin bearing. Several clonal selections of 'Mazzard' with greater disease resistance were made, but are not often grown.

'Mahaleb', grown from seed of *P. mahaleb*, another cherry species native to southwest Europe, was and still is the most common tart cherry rootstock. It induces earlier and heavier bearing, keeps trees about 30% smaller, resists bacterial canker, and tolerates drought and sandy soils better than 'Mazzard'. 'Mahaleb' has sometimes been used for certain sweet cherry cultivars where bacterial canker is a problem, but many sweet cherry cultivars show a delayed incompatibility with 'Mahaleb'.

Several other semi-dwarfing rootstocks have been developed in recent years, clonally propagated, and are now being widely used for new cherry plantings. 'Colt', a hybrid of *P. avium* x *P. pseudocerasus* developed in England, was initially reported to be semi-dwarfing but later found to produce a semi-standard tree only 10% smaller than standard. 'Colt' is also more sensitive to cold injury and drought than other cherry rootstocks, so hasn't been used much in North America.

Several of the Gisela series of rootstocks, developed in Giessen, Germany from crosses of tart cherry (*P. cerasus*) cultivars with *P. canescens*, have been widely propagated and planted in the last few years. 'Gisela 5' produces trees about 50% the size of standard and 'Gisela 12' produces trees about 60% of standard size, but both these rootstocks have poor anchorage, so trees need support by stakes or trellises. 'Gisela 5' is also more prone to disease problems on replant sites, so avoid it if replanting cherries on old orchard land. 'Gisela 6' has better anchorage and produces a tree about 80% of standard size, and has become the most popular sweet cherry rootstock for new plantings in North America during the last few years.

Table 20-1 compares the traits of each of these cherry rootstocks.

Low density orchards. Most older sweet cherry orchards are low density plantings on 'Mazzard' standard sweet cherry rootstock. About 101 to 110 trees per acre were generally planted 18 feet apart within rows spaced 22 to 24 feet apart, or 20 feet by 20 feet apart. Annual pruning keeps trees 18 feet wide and 14 feet tall. Such low density orchards do not bear a commercial crop until 5 to 7 years after planting, and do not become profitable until 10 to 12 years after planting, but are very productive and generally profitable for the next 20 years or so. Such orchards are only removed when cold damage and wood rots start reducing production, usually 30 to 35 years after planting.

Sweet cherries on 'Gisela 6' or 'Colt' rootstocks begin bearing a year sooner and are planted a couple feet closer, commonly 16 feet apart within rows 20 to 22 feet apart, to produce a density of 124 to 136 trees per acre. Tart cherry orchards have usually been planted at this same density because trees begin bearing 3 years after planting and stay smaller than sweet cherries do. 'Mahaleb', the most common rootstock for tart cherries, has about a 30% dwarfing effect, but the most common cultivar, 'Montmorency', along with the newer 'Balaton' and 'Jubileum', are vigorous for tart cherries. If these cultivars are planted on 'Mazzard' rootstock, they don't begin bearing until 4 years after planting and need spacings like sweet cherries on 'Mazzard'.

Medium density orchards. Sweet cherries on 'Mahaleb' (when compatible) or 'Gisela 12' rootstocks can be planted 14 to 15 feet apart within rows spaced 18 to 20 feet apart, producing a density of 155 to 172 trees per acre (Photo 20-2). Genetic semi-dwarf tart cherries, such as 'Danube', 'Meteor' and 'Surefire', on 'Mahaleb' can be planted 12 feet apart in rows 16 to 18 feet apart, creating a density of 201 to 226 trees per acre. Trees on 'Gisela 12' will need staking, but those on 'Mahaleb' are free-standing.

Photo 20-2. Cherries on semi-dwarfing rootstocks can be planted in hedgerows at medium densities, increasing yields in early years of the planting.

Some growers have also used these medium densities for sweet cherries on 'Mazzard' and the more vigorous tart cherry cultivars on 'Mahaleb'. They prune the trees more intensively to control tree spread as well as height.

Medium-high density orchards. Both sweet and tart cherries on 'Gisela 5' rootstock can be planted as close as 8 to10 feet apart within rows spaced 14 to 16 feet apart, producing a density of 272 to 388 trees per acre. Such medium-high density orchards need support, most often trellises, and intensive training and pruning. This system is becoming more common for fresh-market sweet cherries that require hand harvesting, but trellises may not fit into the mechanical harvesting system used for tart cherries or sweet cherries grown for processing.

Some growers plant freestanding cherries on semi-dwarfing rootstocks at medium-high densities to increase early returns from the orchard, but treat every other tree within each row as a temporary tree to be removed after several years of fruiting. Thus an orchard planted at 9- by 14-foot spacing would, after removing temporary trees, leave a tree spacing of 14 by 18 feet, with rows running perpendicular to the original rows. Temporary trees must be pruned

Table 20-1.

COMMON CHERRY ROOTSTOCKS

In approximate order of tree height.

Rootstock	Hardy to Zone	Height in Feet		Yr. Begins Bearing		Anchor-age	Resistance/susceptibility to:									Nematodes		Root Suckers Grown
		Sweet	Tart	Sweet	Tart		Dry Soil	Wet Soil	Alkaline Soil	Bact. Canker	Oak Root Rot	Crown Gall	Stem Pitting	Voles	Peach Tree Borer	Root knot	Lesion	
Gisela 5	4b	20	10	3rd	2nd	fair	S	MR	MS	MR	MS	MR	MS	S	S	S	S	few
Gisela 12	4b	24	12	4th	3rd	fair	MS	MR	MS	MR	MS	MR	R	S	S	S	S	none
Mahaleb	4a	28	14	4th	3rd	very good	MR	VS	MR	R	S	MR	VS	VS	S	R	MR	many
Gisela 6	4b	32	16	4th	3rd	good	MS	MR	MS	MR	MS	MR	MR	S	S	S	S	none
Colt	6b	36	18	4th	3rd	very good	VS	MS	S	R	S	VS	MR	MR	MS	MS	S	none
Mazzard	5b	40	20	5th	4th	very good	MS	MS	S	S	MR	S	MR	MR	MS	R	S	many

Disease resistance ratings: VR = very resistant, R = resistant, MR = moderately resistant, MS = moderately susceptible, S = susceptible, VS = very susceptible.

Comments:

Gisela 5, grown from cuttings, needs staking, timely specialized pruning. Don't use with precocious cultivars.

Gisela 12, grown from cuttings, needs staking, timely specialized pruning, is gaining popularity.

Mahaleb, seed-grown, needs sandy loam soils, attracts gophers, MS to southern root knot nematode though R to northern RKN. Incompatible with sweet cherry.

Gisela 6, grown from cuttings, induces heavy bearing, high fruit quality, needs timely specialized pruning.

Colt, grown from cuttings, is suited only to heavier soils, low-vigor sites in warmer regions with irrigation.

Mazzard, seed-grown, produces a large tree, slow to begin bearing.

differently from permanent trees, with very little pruning while young to encourage rapid growth and early fruiting, then rather severe pruning for a few years before removal so they don't compete with permanent trees.

High and ultra-high density orchards. Several European and Australian researchers have developed systems in which cherries are planted 4 to 6 feet apart in rows 12 to 15 feet apart, producing trees densities of 518 to 605 trees per acre. One version of the Tatura trellis system from Australia even gets cherry tree densities up to 1162 trees per acre. European growers, and a few growers in British Columbia where land prices are very high, have begun to plant such high density cherry orchards. Most North American growers are waiting for test results of such systems in their areas before committing to the extensive training, tree support, and growth control required such systems.

SELECTING CULTIVARS

Many newer cultivars of sweet cherries are self-fruitful, but growers must choose several different cross-compatible cultivars to provide cross-pollination for most sweet cherries. All tart cherry cultivars are self-pollinating, and solid blocks of just one cultivar are often grown.

Sweet cherry cultivars. Sweet cherries grown for fresh market are usually dark red, large, firm-fleshed cultivars with high flavor. Many of these cultivars are very susceptible to fruit cracking or doubling, brown rot, bacterial canker and cold damage, so are grown in arid, irrigated areas with mild winters and few spring frosts. 'Bing' has long been the predominant fresh sweet cherry cultivar due to its high fruit quality, but growers are starting to take advantage of newer cultivars from breeding programs in their new plantings. Some newer cultivars offer better resistance to cold, fruit cracking, fruit doubling due to early summer heat, or diseases such as bacterial canker or powdery mildew. Cultivars that ripen earlier or later than 'Bing' extend the ripening season, making better use of harvest labor.

Sweet cherries grown for processing most often have clear yellow skin or are yellow with a red blush, since most are grown for maraschino cherries and those are first bleached and then dyed bright red. Cultivars grown for canning may be dark red. Processing cherries produce heavy to very heavy yields of firm, medium-sized fruit, and generally resist at least one important disease or fruit problem.

Table 20-2 details the traits of the most commonly-grown sweet cherries in Western North America, both for fresh market and processing. Most of these cultivars do not do well and consequently are not grown much in Eastern North America, although some newer cultivars from Europe, such as 'Attika' and 'Regina', have resistance to fruit cracking, bacterial canker and cherry leaf spot and may do well in the East. Cherry leaf spot is a serious disease in the East but not much problem in the West, while powdery mildew is often a problem in arid areas but not in humid areas, so Western growers can minimize fungicide sprays by choosing cultivars with mildew resistance, while Eastern growers can reduce sprays more with cultivars resistant to leaf spot.

Table 20-3 describes many commonly-grown or new sweet cherry cultivars for Eastern North America, both for fresh market and processing. Some resistance to fruit cracking and bacterial canker is essential for most Eastern locations, and late blooming is also important unless frost protection is available.

Cultivars that bear young will produce crops sooner in both West and East, but some of these cultivars are too naturally precocious for the most dwarfing Gisela rootstocks such as 'Gisela 5.' Pair rootstocks and cultivars that together produce a good cropping level. Cultivars that yield less, such as 'Tieton' or 'Regina' sweet cherries, or come into bearing later, such as 'Bing' or 'Lapins', do best on rootstocks such as 'Gisela 5' that induce greater spur formation. However, cultivars which naturally bear heavy yields starting at a young age, such as 'Chelan' or 'Sweetheart' sweet cherries, can set excessive fruit on 'Gisela 5', causing fruit to remain small and receive much lower market prices. Such precocious, heavy-yielding cultivars do better on rootstocks that do not increase yields as much, such as 'Gisela 6'. Cultivars with a more spreading tree form need less tree training and corrective pruning.

Tart cherry cultivars. Tart cherry cultivars all have bright red to dark red fruit skin, and are classified into two fruit types based on flesh and juice color. Morellos have red flesh and juice, while amarelles have white flesh and clear juice.

The high-yielding amarelle 'Montmorency' (Photo 20-3) has long dominated the tart cherry industry in North America, accounting for 99% of production. In recent years, processors have expressed interest in morello-type cultivars, and growers have begun to plant them. Some of these tart cherry cultivars also offer greater cold hardi-

Table 20-2.

COMMON SWEET CHERRY CULTIVARS IN ARID WESTERN NORTH AMERICA

In approximate order of ripening (varies with area).

All need about 700 to 800 hours chilling. All except Regina are S to cherry leaf spot, and most others don't do well in northeastern North America.

		Fruit					Susceptibility to								
Cultivar	Best in Zones	Color	Flavor	Size	Firmness	Yields	Crack	Double	Bact. Canker	Powdery Mildew	Tree Form	Bears Young	Needs Pollen-izer	Bloom Time	Harvest Time
Early Burlat	5b-7	dark red	very good	large	medium	high	S	VS	R	S	upright	no	yes	early-mid	very early
Black Tartarian	5b-7	dark red	good	small-med	soft	high	S	R	S	MR	upright	no	yes	early	early
Chelan	5b-7	dark red	good	med-large	firm	very high	MS	MS	MS	VR	spreads	yes	yes	early	early
Santina	5b-7	dark red	good	large	med-firm	medium	MR	MS	S	S	spreads	no	no	middle	early
Brooks	5a-7	bright red	best	large	firm	medium	VS	MR	S	S	upright	no	yes	early	early
Bada	5a-7	yellow/red	good	large	firm	high	R	MR	MR	S	spreads	yes	yes	middle	early-mid
Tulare	5b-7	bright red	best	large	firm	high	MR	R	S	S	upright	no	yes	early	early-mid
Royal Ann	6a-7	yellow/red	good	large	firm	high	S	VS	VS	S	upright	no	yes	middle	early-mid
Tieton	6a-7	dark red	good	very large	firm	medium	S	VS	S	MR	upright	no	yes	early-mid	early-mid
Index	5b-7	dark red	very good	med-large	firm	very high	MS	MS	R	S	upright	no	no	middle	early-mid
Cristalina	5a-7	dark red	good	med-large	firm	high	MR	MS	S	S	upright	yes	yes	middle	early-mid
Benton	5b-7	dark red	best	large	firm	high	MR	MS	S	S	upright	no	no	mid-late	middle
Bing	6a-7	dark red	best	med-large	med-firm	med-high	VS	VS	S	VS	upright	no	yes	middle	middle
Hardy Giant	5a-7	dark red	good	medium	medium	very high	R	MS	VS	S	upright	no	yes	middle	middle
Ranier	5b-7	yellow/red	very good	large	firm	high	MS	R	MR	MS	spreads	yes	yes	early-mid	middle
Van	5a-7	dark red	good	medium	very firm	very high	S	S	VS	MR	upright	yes	yes	early-mid	middle
Sandra Rose	5a-7	dark red	best	very large	medium	very high	MR	MS	S	S	spreads	no	no	middle	mid-late
Lambert	5b-7	dark red	very good	med-large	firm	high	S	MR	MS	S	upright	no	yes	mid-late	mid-late
Sonata	5b-7	dark red	good	very large	firm	medium	MS	MS	S	S	upright	yes	no	mid-late	mid-late
Selah	5b-7	dark red	best	large	med-firm	high	S	MS	S	S	upright	no	no	early-mid	late
Attika	5a-7	dark red	best	large	very firm	very high	MR	MS	R	MS	spreads	yes	yes	late	late
Lapins	6a-7	dark red	best	large	medium	very high	MR	MS	S	MR	upright	no	no	early	late
Skeena	5b-7	dark red	good	large	very firm	very high	R	MS	S	S	spreads	yes	no	early-mid	late
Regina	5a-7	dark red	very good	med-large	very firm	medium	R	MS	MR	MS	spreads	yes	yes	late	very late
Sweetheart	6a-7	bright red	good	medium	very firm	very high	MS	S	S	S	spreads	yes	no	early-mid	very late

Disease and disorder susceptibility ratings: VR = very resistant, R = resistant, MR = moderately resistant, MS = moderately susceptible, S = susceptible, VS = very susceptible.

Average ripening times in Sacramento, CA: very early: May 10-20 early: May 15-25, early-mid: May 20-30, middle: May 25-June 5, mid-late: June 1-10, late: June 5-15, very late: June 10-20. Ripening times are about 3 weeks later in western OR and 4 weeks later in western WA.

Notes on Cultivars:
Early Burlat, an old English cultivar, offers very early fruit of high quality, but is VS to deep suture disorder, cherry leaf spot.
Black Tartarian, an older cultivar, is VS to crinkle-leaf and deep suture disorders, but MR to brown rot.
Chelan, the main early cultivar in the Pacific NW, fruits young and heavily, needs heavy annual pruning for good fruit size; don't use precocious rootstocks.
Santina does best on Gisela rootstocks, which increase cropping potential, as does gibberellin.
Brooks ripens evenly; fruit hang well on tree for 7 days, but prone to crack in final week.
Bada is a genetic semi-dwarf tree about 60% of standard size. Fruit are used for maraschino cherries.
Tulare has been widely planted in CA in recent years, due to good resistance to rain-induced cracking and heat-induced doubling of fruit.
Royal Ann, also called **Napolean**, is an old French cultivar grown for maraschino cherries, R to crinkle-leaf disorder, cross-incompatible with Bing, Lambert.
Tieton does best on Gisela rootstocks, which increase cropping potential. R to crinkle-leaf disorder, cross-incompatible with Chelan.
Index has low acidity and high sugars so can be harvested early with good quality.
Cristalina tends to overset, needs heavy pruning each year or gibberellin to maintain fruit size, firmness and flavor.
Benton, also called **Columbia**, has been picked over Bing in consumer taste panels.
Bing is the standard sweet cherry for the Pacific Northwest. VS to crinkle-leaf and deep suture disorders. Use Gisela rootstocks to hasten bearing.
Hardy Giant does well only in arid areas where bacterial canker isn't a problem.
Ranier does well on dwarfing rootstocks to increase light penetration, blush development.
Van bears crops 1-3 years ahead of Bing, often oversets, prune hard to prevent this. VS to deep suture disorder, *Cytospora* canker.
Sandra Rose has very large fruit with great flavor but is prone to fruit marking from frost and wind damage.
Lambert, an older cultivar, is little planted now but older plantings stil produce, bearing large, loose clusters. Cross-incompatible with Bing, Royal Ann.
Sonata also has very large fruit but is acidic and prone to fruit marking from frost and wind damage, fruit dimples where water collects, leading to cracks.
Selah, also called **Liberty Bell**, fruit grow in loose clusters so have less brown rot.
Attika buds are VS to frost so it needs best sites though it blooms late. Does well on Gisela rootstocks. MR to brown rot.
Lapins trees are very upright and vigorous, hard to manage when young. Fruit grow in tight clusters VS to brown rot, may overcrop. VS to cherry leaf spot.
Skeena trees are open, easier to grow than Lapins. Fruit grow in loose clusters but may overset on Gisela rootstocks.
Regina from Germany is R to cherry leaf spot, cracking, bacterial canker. Use Gisela rootstocks to increase yields, cross-incompatible with Van.
Sweetheart fruits young and heavily, needs heavy annual pruning for good fruit size; don't use precocious Gisela rootstocks or trees stunted.

Table 20-3.

COMMON SWEET CHERRY CULTIVARS IN NORTHEASTERN NORTH AMERICA

In approximate order of ripening (varies with area).

All need about 700 to 800 hours chilling.

Cultivar	Best in Zones	Fruit					Susceptibility to			Tree Form	Bears Young	Needs Pollen-izer	Bloom Time	Harvest Time
		Color	Flavor	Size	Firmness	Yields	Crack	Bact. Canker	Leaf Spot					
Cavalier	5a-7	dark red	good	med-large	firm	medium	MR	MR	S	spreads	no	yes	early-mid	very early
Viva	5a-7	dark red	very good	medium	medium	medium	MS	S	MR	upright	no	yes	late	early
Hartland	5b-7	dark red	good	medium	medium	high	MS	R	MS	spreads	yes	yes	middle	early
Viscount	5a-7	dark red	good	large	firm	high	MS	S	S	upright	yes	yes	early	early-mid
Sam	5a-7	dark red	fair	large	medium	high	R	R	R	upright	no	yes	late	early-mid
Valera	6a-7	dark red	very good	medium	firm	high	MS	S	MR	upright	yes	yes	middle	early-mid
White Gold	5a-7	yellow/red	good	medium	firm	very high	MR	MR	R	upright	no	no	mid-late	early-mid
Kristin	4b-6	dark red	very good	med-large	firm	high	MR	R	S	upright	yes	yes	early	middle
Summit	6a-7	bright red	good	very large	medium	medium	MS	MS	S	upright	no	yes	middle	middle
Emperor Francis	5b-7	yellow/red	good	large	firm	very high	MR	S	MR	upright	no	yes	early-mid	middle
Nelson	5a-7	dark red	best	large	firm	high	R	MR	S	upright	no	yes	mid-late	middle
Black York	5a-7	dark red	very good	med-large	firm	very high	MR	R	S	spreads	no	yes	middle	middle
Angela	5a-7	dark red	good	med-large	medium	very high	R	R	S	upright	no	yes	late	mid-late
Blushing Gold	5a-7	yellow/red	good	medium	firm	high	MR	MR	S	upright	no	yes	middle	mid-late
Ulster	5b-7	dark red	good	large	firm	high	MR	R	MR	upright	no	yes	middle	mid-late
Black Gold	5a-7	dark red	best	large	firm	high	MR	MR	S	spreads	no	no	late	mid-late
Hedelfingen	5b-7	dark red	best	medium	medium	very high	MR	S	MS	upright	yes	yes	mid-late	mid-late
Somerset	5a-7	dark red	very good	medium	very firm	high	VS	MR	S	spreads	yes	yes	early	late
Gold	4b-7	yellow	fair	small	firm	very high	R	MR	MS	upright	no	yes	late	late
Hudson	5a-7	dark red	good	med-large	firm	high	MR	R	S	upright	no	yes	late	very late

Disease and disorder susceptibility ratings: VR = very resistant, R = resistant, MR = moderately resistant, MS = moderately susceptible, S = susceptible, VS = very susceptible.

Average ripening times in SW MI: very early: June 20-30, early: June 25-July 5, early-mid: July 1-10, middle: July 5-15, mid-late: July 10-20, late: July 15-25, very late: July 20-Aug. 1. Ripening times are about 3 weeks earlier in SW VA, and about 1 week later in Geneva, NY.

Notes on Cultivars:
Cavalier yields best on Gisela 5 rootstock; often takes 10 years to bear well on Mazzard. Fruit store well.
Viva is cross-incompatible with Cavalier, so needs a different pollenizer. Fruit grow in loose clusters, so only MS to brown rot.
Hartland fruits heavily on 1-year-old wood, and is well-suited to mechanical harvest. Cross-incompatible with Summit.
Viscount may crack badly on young trees, less cracking on full crops of mature trees, may alternate bear. Cross-incompatible with Ranier, Hudso
Sam has tight fruit clusters VS to brown rot, and VS to powdery mildew, so needs extra sprays. Allow to fully ripen or flavor is poor. R to fruit doub
Valera bears loose clusters with few fruit in each, so only MS to brown rot.
White Gold is being planted for maraschino cherries and as a self-fertile pollenizer for Emperor Francis.
Kristin does best in cool-summer areas. It's cross-incompatible with Emperor Francis, Ulster and Somerset, so needs a different pollenizer.
Summit may crack badly on young trees, less cracking on full crops of mature trees. Cross-incompatible with Hartland.
Emperor Francis has better flavor, larger fruit, less cracking than Napolean, another yellow/red cultivar. Cross-incomatible with Kristin, Ulster, So
Nelson, an older cultivar, has among the best crack resistance. Virus-free budwood is newly available, so it's offered again.
Black York is a new canning cultivar, with heavy yields of smaller fruit.
Angela is R to X-disease and bacterial spot, but is slow to begin bearing. Use Gisela rootstocks to help this.
Blushing Gold, a new cultivar for maraschino cherries, has fruit similar to Emperor Francis and is cross-compatible with it.
Ulster is cross-incompatible with Kristen, Emperor Francis and Somerset. Cracking may be bad on young trees, but gets less on mature trees.
Black Gold has good frost tolerance, and fruit stay firm despite harvest-season heat.
Hedelfingen, a very old German cultivar, tends to overbear, causing small fruit some years. Prune hard to stop this. R to powdery mildew.
Somerset has wide branch angles, low vigor, so needs less training, pruning. Don't use Gisela 5 or 7. Incompatible with Kristin, Ulster, Emperor F
Gold is a main cultivar used for maraschino cherries, R to bird damage since birds don't see yellow fruit, MR to brown rot.
Hudson fruit hang well on tree, store well. Tree resists sunscald damage. Use Gisela rootstocks to increase early yields. Incompatible with Viscou

Table 20-4.

COMMON TART CHERRY CULTIVARS IN NORTH AMERICA

In approximate order of ripening (varies with area).

All have bright red skin, are self-fertile, and need 850 to 950 hours chilling.

Cultivar	Best in Zones	Fruit					Susceptibility to			Tree		Bears Young	Bloom Time	Harvest Time
		Flesh Color	Flavor	Size	Firmness	Yields	Leaf Spot	Bact. Canker	Brown Rot	Height	Form			
Jubileum	4b-7	red	sweet-tart	large	firm	high	S	MS	R	15-18'	spreads	no	early-mid	early
Danube	4b-7	red	sweet-tart	large	firm	very high	S	MR	MR	12-14'	spreads	no	early-mid	early
Montmorency	4a-7	white	tart	medium	med-firm	very high	VS	MR	S	15-18'	spreads	no	early	middle
North Star	3b-7	red	very tart	small	medium	high	MR	VS	S	6-8'	upright	yes	middle	mid-late
Surefire	4a-7	red	very tart	medium	firm	high	S	MR	S	12-14'	upright	yes	late	mid-late
Balaton	4b-7	red	sweet-tart	large	very firm	very high	S	R	S	18-20'	spreads	yes	early	late
Meteor	3a-7	white	tart	medium	med-firm	high	MS	MR	MR	10-12'	upright	yes	late	late

Disease and disorder susceptibility ratings: VR = very resistant, R = resistant, MR = moderately resistant, MS = moderately susceptible, S = susceptible, VS = very susceptible.

Average ripening times in SW MI: early: June 25-July 5, middle: July 5-15, mid-late: July 10-20, late: July 15-25. Ripening times are about 3 weeks earlier in SW VA and about 1 week later in Geneva, NY.

Notes on Cultivars:
Jubileum, from Hungary, fruit hang well on tree, has high sugars, can be sold fresh. Juice is red, flesh pulls off pit easily.
Danube, also called **Erdi Botermo** from Hungary, has high sugars, can be sold for fresh eating. Juice is red.
Montmorency, long the chief cultivar for pie fillings, has clear juice.
North Star is little grown except where extreme hardiness is needed, or as a small tree for home plantings. Juice is red.
Surefire crops reliably due to late bloom. MS to bacterial spot, S to powdery mildew. Juice is red.
Balaton from Hungary has high sugars, can be sold fresh. Juice is red. Large pits may jam processing equipment. Buds VS to frost.
Meteor has irregularly-shaped pits, may cause problems in processing equipment. Juice is pink.

Photo 20-3. 'Montmorency', a high-yielding amarelle, accounts for 99% of tart cherry production in North America.

ness, later bloom, smaller tree size, earlier bearing, sweeter fruit, or better resistance to brown rot than 'Montmorency'. However, cultivars such as 'Balaton' and 'Meteor' have larger or irregularly-shaped pits that may cause problems in processing equipment designed for 'Montmorency'. The Hungarian tart cherry cultivars new to North America — 'Balaton', 'Danube' and 'Jubileum' — have fruit sweet enough to sell fresh at retail markets. Table 20-4 describes the most commonly-grown tart cherry cultivars in North America.

SITE SELECTION, PREPARATION AND PLANTING

Sweet cherries need sites with excellent air drainage, preferably a site near a temperature-moderating body of water or a north-facing slope to delay bloom. Tart cherries need good air drainage but not necessarily a bloom-delaying site since they bloom later. Tart cherries on 'Mahaleb' rootstock do require excellent soil drainage, generally sandy loam soils or loams high in organic matter. Cherries on 'Mazzard' or 'Colt' rootstocks do well on silt loams or clay loams with moderate to good drainage, and cherries on 'Gisela' rootstocks adapt well to many soil types as long as they have moderate to good soil drainage. An irrigation system will be needed for cherries on 'Colt' or 'Gisela 5' rootstocks even in high-rainfall areas, and trees on 'Gisela 5' or 'Gisela 12' will require staking or trellises.

In recent years, many older low-density apple orchards have been removed in the Northwest, and the land has been planted to cherries. If planting an old orchard site, or a site cleared from oak forest within the past few years, dig up and burn as much of the old roots as possible to avoid oak root rot and other diseases, and sow grass crops for a couple of years before planting cherries. Do not follow any crops susceptible to Verticillium wilt. Nematodes are not usually a problem on heavier soils, but test sandy soils for nematodes. The Gisela rootstocks and 'Colt' are susceptible to nematodes, so if nematodes are present, soil must be fumigated or sown to nematode-reducing crops before planting. 'Mahaleb' offers resistance to the most common nematodes in Northern cherry areas, and 'Mazzard' each offer resistance to the most common nematodes in Western cherry areas.

Test soil pH and nutrients at least 6 months before planting, and adjust soil pH to within a range of 6.0 to 7.5 if needed. Add phosphorous (P) or potassium (K) if these test low. Few cherry growers make raised planting beds, but these might be needed for 'Mahaleb' rootstock if soil drainage is less than ideal, or for 'Mazzard' or 'Colt' rootstocks if soil is somewhat poorly drained.

Order certified virus-free trees, to avoid numerous cherry viruses. Where bacterial canker is a serious disease, it may girdle and kills low-budded sweet cherry trees, so order high-budded sweet cherries or plant canker-resistant rootstocks, allow them to grow for a year, and graft cultivars 12 to 18 inches out on side limbs. Plant cherry trees in spring, since fall planting risks winter injury. Plant as for other fruit trees, and irrigate right after planting. If planting rootstocks susceptible to crown gall on soil that may harbor that disease, dip tree roots in the biocontrol product Galltrol™ just before planting.

Orchard layout for pollination. Tart cherries are self-pollinating, and many growers produce only one cultivar. Growers who produce several cultivars generally grow each one in a solid block for ease of harvesting and spraying, since some cultivars need fewer or more pesticide sprays than others.

Most sweet cherry growers produce several cultivars, some of which are self-infertile and must be interplanted with self-fertile or cross-compatible pollenizer cultivars to assure good fruit set. A common layout is to plant a row of each cultivar, with self-fertile or pollenizer cultivar rows placed between cultivar rows that need cross-pollination, and repeat that pattern of rows across the orchard. Where one cultivar is the main crop and the pollenizer cultivar is desired in lesser quantities, pollenizer trees can be planted in every third row to produce 33% of the crop, or as every third tree in every third row to produce 11% of the crop.

TREE TRAINING SYSTEMS

Many different training systems have been used for cherries, and in recent years researchers have developed many new systems that are now being tested. No single training system has become dominant, but higher-density systems are starting to replace low-density training systems.

Low density, open center system. Sweet cherries planted at wide spacings have generally been trained to the open-center system described in chapter 4 (Photo 20-4). This system fights the natural tendency of cherries to produce a strong central leader, which must be repeatedly cut back. Most cultivars produce few branches and these tend to grow upright, so they also need extensive heading of main and lateral branches to induce branching, and branch spreading to produce wider branch angles.

Such heavy pruning of young trees delays the onset of fruiting, making this system uneconomical for the first 6 to 8 years or so. As trees age, the fruiting zone moves upward and outward, requiring ladders for hand harvest and reducing harvest efficiency by 30 to 40%. Few new plantings use open-center training.

Low to medium density, modified leader system. Tart cherries have traditionally been trained to the modified leader system, described in chapter 4, and sweet cherries planted at medium densities have also utilized this system. This system requires little pruning during the first two years, so encourages earlier fruiting. The leader is allowed to grow up to about 5 feet off the ground, then cut back during the second dormant season to a strong lateral branch that grows upward and outward. Choose 5 or 6 lateral branches with wide angles, arising between 2 feet and 5 feet off the ground and well-distributed around the trunk, as scaffold limbs, and remove other branches. Pruning is minimal for the next few years, mainly heading back any overly vigorous scaffolds to keep them in balance with the others. As trees age, remove weak interior branches to keep trees open enough for good light and spray penetration.

The modified leader system has low pruning labor requirements and can produce high yields, especially for tart cherries and for sweet cherry cultivars that grow rapidly to fill allotted space, are naturally spreading, and begin to bear early. This system can also be used for high-density plantings where half the trees are temporary and will be removed after about 10 years for tart cherries or 12 to 14 years for sweet cherries. This system eventually produces large trees, which work well for mechanical harvest for processing, but require ladders for hand harvest of fresh-market sweet cherries.

Photo 20-4. Older low-density plantings of sweet cherries and some tart cherries were trained to open-center trees. This training method delays bearing and is less used today.

Medium-high density, slender spindle system. This system, described for apples in chapter 16, is increasingly used for sweet cherries on Gisela rootstocks in Europe and the Pacific Northwest of North America, producing trees a few feet taller and wider than for apples. Trees are planted 8 to 10 feet by 15 to 16 feet apart, as whips that are headed at 30 to 32 inches tall to induce side branches. The central leader is not headed again for the next 3 to 4 years, but is instead encouraged to grow rapidly to a height of 12 to 16 feet. Leaders should grow 3 to 4 feet each season, producing a whorl of 4 to 5 lateral branches at the end of each growing season, for a total of 3 or 4 whorls per tree. Remove any narrow-angled laterals, spread remaining lateral branches to horizontal and pinch them in midsummer to stimulate side branching, but do not cut back laterals during dormant pruning until after they begin to bear fruit. When the central leader reaches the desired height, maintain that height by cutting it back to a weak lateral branch each dormant season. Maintain tree width similarly, by heading back limbs to a lateral branch each dormant season.

The slender spindle system, also called the Vogel system in cherries, encourages early fruit production by almost eliminating pruning during the first 4 years, and early fruiting reduces vegetative growth and keeps trees smaller. Pruning cuts on bearing trees are made mostly into older wood, which produces less regrowth. Most fruit produced by this system can be hand-harvested from the ground since fruit-laden branches bend down, although ladders or picking platforms will still be needed if central leaders are allowed to grow more than 12 feet tall.

Medium-high density, angle system. Trees trained to this system are planted at about the same spacing as for slender spindle, but on a 45-degree angle, and developing lateral shoots are trained to a post and wire support system at right angles to the trunk. Keeping the main trunks angled weakens trees, enabling them to be kept below 8 or 9 feet tall, which allows all harvesting, pruning and other labor to be done from the ground. However, this system fights the natural growth habit of cherries, and requires much additional labor to develop and maintain tree structure. Workers must induce lateral bud development by applying Promalin™ growth regulator, pinch laterals once or twice per growing season to induce side branches, thin laterals to 12 to 16 inches apart on the trunk, and tie them to wires at the correct angle. When the leader reaches the desired length, and when lateral branches become too strong after several years, they are cut back to weak side branches.

Few North American cherry growers have adopted the angle system, also called the Marchand trellis system, because of the detailed tree training and trellis system required. Some European growers use it because it can induce heavy early yields.

Medium-high density, freestanding V-hedgerow system. This system is described as the perpendicular V system in chapter 4 and as the V-hedgerow system for peaches in chapter 18. It produces a V-shaped hedgerow of open-centered trees, each containing 2 scaffolds perpendicular to the row, without a trellis. Cherries trained to this system are planted about 6 feet apart in rows 16 to 18 feet apart, and headed at 8 to 12 inches off the ground to produce low-arising scaffolds. Two branches, one on each side, are chosen as scaffolds and encouraged to grow rapidly without heading during the next 3 years. Treat with Promalin™ growth regulator in early spring to encourage lateral branching, and minimize pruning of lateral branches to induce fruiting as early as possible.

The open center of this system allows high light interception, enabling this system to produce yields almost as high as the V-shaped Tatura trellis described below, but without the expense of the trellis. However, training naturally upright trees into two spreading scaffolds, without a trellis to tie them to, can be as difficult as training cherries to the open-center system. Once scaffolds are trained, this system also requires more labor for to keep trees free of upright shoots, so few North American growers have used it for cherries yet.

High density, close spindle system. This system, also described for apples in chapter 16, is similar to the slender spindle system except planted closer, at 4 to 5 feet by 12 to 14 feet, and whips are not headed at planting. All strong laterals are removed or cut back to 6- to 8-inch long stubs, leaving only weak laterals. As trees age, new weak laterals grow and laterals older than 4 or 5 years old are removed, leaving the most fruiting branches in the least amount of space. This system requires extensive summer pruning, cutting vigorous branches back to 6- to 8-inch long stubs 3 to 4 times per growing season, starting about 6 weeks after bloom and again every 4 weeks. Tree height is maintained at 10 to 12 feet tall.

The close spindle system can be managed without stakes or trellises if the dwarfing rootstock chosen has good enough anchorage on the available soil type, or if genetic dwarf cultivars are grown on semi-dwarfing rootstocks, but in most cases tree support is needed. This system, also called the Zahn vertical axis system in cherries, can produce very high yields early in the planting's life, so is increasingly popular in Europe and parts of the Pacific Northwest. It requires extensive hand pruning to maintain trees in such limited space, but almost all pruning and harvesting can be done from the ground.

High density, Spanish bush system. The goal of the Spanish bush system is to harvest all fruit from the ground by converting cherry trees into open-centered bushes. Trees are planted 6 to 7 feet apart in rows 12 to 14 feet apart, and headed at planting at just 8 to 12 inches off the ground. During the first growing season, 5 or 6 lateral branches are allowed to grow 12 to 16 inches long, then tipped or cut back to about 10 to 12 inches long in early summer. If sufficient growth occurs, branches are tipped again in mid- to late summer. Remove all upright or crossing branches during either summer or dormant pruning. Continue to tip back lateral branches during the second and third growing seasons, creating a bush that should begin to bear during the third season. After harvest during the fourth and subsequent seasons, top bushes at a height of 8 feet and remove growth in the top center of the bush, keeping only a few weak shoots there for the duration of the planting.

This system requires the most dwarfing rootstocks and further vigor control from less fertile soils with limited irrigation and fertilization. It works best with naturally spreading cultivars that fruit early, and is very difficult to manage with strongly upright cultivars.

High or ultra-high density, Tatura V-trellis system. This system, described in chapters 4 and 18 as the perpendicular V- trellis, requires the most dwarfing rootstocks and a V-shaped trellis with 5 wires on each side. Trellis arms are 8 to 9 feet long, extending 2 to 3 feet into the ground and 6 to 6.5 off the ground, and wires are spaced one foot apart starting about 2 feet off the ground. Whips are planted about 5 feet apart within rows 15 feet apart (581 trees per acre), and headed at 8 to 12 inches off the ground. Two branches, one on each side, are chosen as scaffolds and trained along the wires up each side of the trellis. Do not head these scaffolds during the next 3 years, treat with Promalin™ growth regulator in early spring to encourage lateral branching, and minimize pruning of lateral branches to induce fruiting as early as possible. Bending lateral branches down and tying them to the wires also stimulates early fruiting and reduces tree vigor. Once trees have borne crops for a couple of years, excess lateral branches can be thinned out.

A Tatura trellis system can instead be planted with twice as many trees just 2.5 feet apart. Do not head the trees, but allow the leader to grow and train it to just one side of the trellis, so every other tree is trained to the left side and alternate trees are trained to the right side. By practicing minimal pruning for the next few years, this system can begin bearing just 2 or 3 years after planting, which is a year or two sooner than the double-armed V-trellis system. Such a planting requires 1162 trees per acre, which doubles the tree costs of planting.

The Tatura trellis can produce the highest yield per acre of all these cherry training systems, because it allows maximum light interception and distribution within trees, but the elaborate trellis adds $3000 to $5000 per acre to establishment costs and few North American growers have adopted this system. This system also requires more labor for training and maintenance to keep trees free of upright shoots.

GROUNDCOVER MANAGEMENT AND FERTILIZERS

Most cherry growers raise a permanent sod between trees to minimize soil erosion and cultivation damage to cherry roots, but some Western growers use winter cover crops and summer cultivation to reduce irrigation needs. Almost all growers maintain a 3- to 4-foot wide weed-free strip under trees rows with herbicides.

Apply enough nitrogen (N) to young trees so they produce about 24 to 30 inches of terminal growth per season, with some training systems requiring slightly more growth than this and other systems requiring less. During years 3 to 5 after planting, trees on Gisela rootstocks will need more N than trees on 'Mazzard' or 'Colt' rootstocks on the same soil type, to stimulate adequate vegetative growth and support a fruit crop. Once trees start to bear, they should produce only 6 to 12 inches of terminal growth per season, and mature bearing trees should produce just 4 to 8 inches of terminal growth per season.

Growers who irrigate cherries often add N to irrigation water. This spreads out N treatments over the growing season, helping to avoid N leaching from soil, and excess N near harvest which can increase brown rot susceptibility. If broadcasting N over the orchard floor rather than fertigating, apply half the N in early spring and half after harvest.

Usually, N is the only fertilizer element that cherry growers need to add each year. If phosphorous (P) is needed, apply it in early spring with a broadcast N application or in an irrigation treatment. Apply potassium (K) if mid-summer leaf samples have less than 1.5% K, adding it to irrigation water just before harvest or with the broadcast N application after harvest. Use potassium nitrate or a mulch of mushroom compost rather than potassium chloride, which can burn foliage if rainfall and irrigation are not sufficient to leach out the chloride. Do not apply K unless leaf samples indicate it is deficient, since excess K reduces magnesium (Mg) uptake, and that is more commonly deficient. A mild Mg deficiency can be corrected with a single summer spray of magnesium sulfate (Epsom salts).

Boron (B) deficiencies often occur in Western cherry regions, reducing yields and causing stunted shoot growth called rosette. Do not apply B to non-bearing trees, but sample leaf B content the year before the first crop is expected, and apply about 0.8 ounce B per tree to the soil as borax if leaves have less than 35 ppm B. About every 3 years thereafter, apply about 1.5 ounce B per tree to the soil as borax as a maintenance dose to prevent deficiency. If B deficiency does occur, spray foliage twice with 5 to 8 pound sodium pentaborate per acre each time, since the foliar spray will alleviate the deficiency faster than soil treatments. Spray in fall before leaves drop, and again in spring before bloom.

Zinc (Zn) deficiencies may also cause rosette in Western cherry regions, and non-bearing as well as bearing trees may need Zn applications. If Zn levels fall below 14 ppm in mid-summer leaf samples, apply a single spray of zinc sulfate or zinc chelate, and if rosette symptoms also appear, apply two sprays. Treat non-bearing trees as soon as the deficiency is noticed. Apply zinc sulfate to bearing trees either late in the dormant season before any buds start to grow, or soon after harvest, or apply zinc chelate in the first cover spray 10 to 14 days after petal fall. Do not apply Zn within 3 days before or after an oil spray, or tree damage can occur.

TRAINING TECHNIQUES FOR YOUNG TREES

Sweet cherries in all training systems benefit from training techniques that induce earlier bearing, and high-density systems require such techniques to help recoup the increased planting costs. Sweet cherries grow naturally as understory forest trees, branching mainly just below the termination of each year's previous growth, to produce a canopy that extends high and wide to compete for sunlight and to shade out undergrowth. Traditional low-density training systems modify this growth habit to control tree height and induce more fruiting wood to form on lower branches, and high-density systems have even fewer branches and need every branch to produce an optimum fruit load. Tart cherries are smaller trees, start to bear earlier naturally and generally have less upright branches, so don't need the training techniques below as often, but may need them to improve tree structure.

Summer heading cuts on 2- to 3-foot-long current-season shoots or one-year-old branches, removing about 1/3 of the length, can induce lateral branching without delaying cropping much. **Dormant heading cuts** induce lateral branching below the cuts but reduce overall tree growth and delay the onset of cropping, so use dormant heading only sparingly to get trees into bearing as early as possible. Also avoid **dormant thinning cuts** as much as possible until trees begin to bear.

Branch spreading, described for training apple trees in chapter 16, reduces terminal growth, strengthens branches, and induces more lateral branches, spur formation and flowering on young trees. Use clothespins early in the growing season and branch spreaders later in the season for free-standing trees, and tie branches of trellised trees to the trellis wires. Do not bend branches so much as to cause bark cracks, as these provide openings for bacterial canker.

Promalin™ growth regulator, also described for training apple trees in chapter 16, is used even more extensively for training sweet cherry trees. Apply one part Promalin™ mixed with 3 parts white paint, at the green tip stage of bud development in early spring, to induce more lateral branching. Temperatures during bud swell and bud break greatly affect response to Promalin™, as cool temperatures may cause little branching while overly warm temperatures may cause excessive branching and require hand-thinning of lateral shoots.

Scoring above buds can induce them to grow into branches, and works from first bud swell until bud break, a longer period than does Promalin™ application. Remove a wide strip of bark so healing doesn't occur until after new shoot growth begins, but do not cut into wood or strong winds can break the new shoots. Scoring risks bacterial canker infection, so do not practice scoring where the disease is serious.

Bud removal prevents trees from wasting energy on growing shoots in unwanted locations, such as along the trunk below the point where the lowest scaffold is desired. Removing buds in undesirable locations channels more of the tree's resources into producing desired shoots and branches, enabling faster tree growth and earlier fruiting. Where bacterial canker is serious, do not remove buds during cool, humid weather that favors the disease, or apply copper sprays right after bud removal.

TREE PRUNING AND OPTIMIZING CROPPING LEVELS

Serious overcropping can occur 4 or 5 years after planting on many sweet cherry cultivars on 'Gisela 5' rootstock. Pruning strategies during the second and third years after planting should be modified to prevent excess cropping a couple years later, because most fruiting spurs arise from 2-year-old or older branches. Summer heading of longer shoots or dormant-season heading of longer one-year-old branches, during years 2 and 3 after planting, removes some potential fruiting sites that would otherwise bear excess fruit in years 4 and 5, as well as stimulating lateral branching. Fruit spurs form at greater density near the ends of branches, so removing just the tip 10 to 25% of each branch can eliminate 25 to 50% of the future crop load.

Mature cherry trees on Gisela rootstocks also need more pruning than trees on 'Mazzard', 'Colt' or 'Mahaleb'. During dormant pruning, remove most of the fine, shaded branches that develop on such trees, since these branches tend to set excessive fruit. Avoid making large pruning cuts on all cherry trees, as these provide entry sites for bacterial canker. If large cuts must be made, wait until a few weeks before bloom so they will be less subject to cold injury and will heal more rapidly as temperatures start to warm.

Tart cherry trees are quite cold-hardy, so dormant pruning (other than large cuts) can be done starting in January, as for apples. Where winters are mild, sweet cherries may also be pruned all winter. Delay sweet cherry pruning until later in winter in colder regions, since they are less cold hardy. However, complete all dormant cherry pruning several weeks before buds start to swell in spring, since bacterial canker starts spreading then and will attack any recent pruning wounds that haven't yet healed over.

Mature cherry trees may be mechanically hedged on tops and sides between 6 and 7 weeks after full bloom, to keep trees in bounds, promote spur development inside the tree, and increase coloring of ripening fruit. Set the cutting bar so that 10 to 50% of the current season's shoot growth is removed, trying to never cut into one-year-old wood. Taper side cuts at a 60-degree angle so trees are narrower at top and wider at bottom, to admit more light and avoid shading lower branches. Trees that were hedged in summer will require more detailed follow-up hand pruning later in summer or the next dormant season, to remove stubs and thin out excess shoots growing just below cuts.

Pollination and crop load adjustment. Almost all cherry growers rent honeybee hives during bloom to provide adequate pollination. They place one to two hives per acre in orchards once 30 to 40% of flowers are open, and keep hives in the orchard for 4 to 5 days. Since sweet cherries bloom early, native mason bees, described in chapter 10 and also called blue orchard bees or orchard mason bees, do a better job of pollinating because they work at lower temperatures than honeybees. Recent research in commercial sweet cherry orchards in northern Utah found that cherry yields were twice as high during some years when mason bees were used in place of honey bees.

Bees can spread several cherry viruses by gathering pollen from infected trees and bringing it to healthy trees, so remove any virus-infected cherry trees from within or around the orchard before renting or setting out bees. When contracting with a beekeeper who brings bees in from other areas which may harbor cherry viruses, wait at least 48 hours after the bees last foraged before placing them in your orchard, since bees will store all pollen gathered earlier by that time. For example, growers in Oregon and Washington rent bees transported from California, where cherry rugose mosaic virus occurs, but have excluded the virus from the Pacific Northwest by waiting at least 48 hours after bees leave California before releasing bees into Northwest orchards.

Cherry growers do not thin fruit, since fruit are small compared to other tree fruits, and a higher percentage of fruit can set without reducing fruit size. Growers adjust crop load during dormant pruning, doing more branch thinning if a heavy crop is expected and less if a lighter crops is expected due to a heavy crop the previous year, frost predictions, or other problems.

Gibberellic acid to adjust cropping or fruit size. Gibberellic acid (GA) may be applied to young trees of precocious sweet or tart cherry cultivars to prevent them from flowering heavily before trees have grown enough vegetatively to become established. Tart cherry growers may also use GA on mature trees to stimulate development of lateral shoots and fruiting spurs, and increase long-term production capacity. GA to adjust cropping must be applied the year before the desired result, when temperatures are predicted to stay above 70F for several days, usually when terminal shoots have 5 to 7 leaves or about 3 to 4 weeks after normal bloom time. Do not use a surfactant with GA.

Treat precocious young cherry trees during their second year after planting, if deemed necessary to prevent flowering during the third year. Apply two sprays of 50 ppm GA each, 2.5 to 3 weeks apart, or a single application of 100 ppm GA if trees are vigorous and less control of flowering is needed. Over the next couple of years, reduce GA rates gradually, as complete elimination of GA would cause excessive fruit set in years 4 and 5, resulting in stunted trees. During the third season apply 30 to 40 ppm GA in one spray, and during the fourth year apply 15 to 20 ppm GA in one spray, which will enable trees to flower and fruit at reduced rates during years 4 and 5. Thereafter, discontinue GA use or continue it at the rate of 10 to 15 ppm GA per season if tree growth is weak, to provide a moderate cropping level so fruit will grow larger.

Apply 10 to 20 ppm GA to mature tart cherry trees, using the lower rate on more vigorous orchards and the higher rate on low-vigor orchards, with 15 ppm the most common rate. This stimulates growth of lateral shoots and fruiting spurs, reducing the occurrence of long branches without fruit buds. Such GA applications may be needed for trees with mild infections of Prunus necrotic ringspot virus, which reduces vigor and yields only slightly.

Mature sweet cherries may occasionally receive 10-30 ppm GA about 3 weeks before harvest, to induce larger, firmer and sweeter fruit. This GA treatment slows fruit coloring so delays harvest by 3 to 5 days, and is also used to lengthen harvest and better utilize harvest labor.

IRRIGATION, FROST PROTECTION AND PREVENTING FRUIT DISORDERS

Irrigation is necessary for cherry production in arid Western areas, where most growers use trickle systems or micro-sprinklers. Researchers in British Columbia found that drip-irrigated trees grew less vegetatively and produced higher yields than trees irrigated with micro-sprinklers, so trickle irrigation can be another tool to keep cherry trees smaller. The average Western cherry orchard requires 42 inches of water per season, but trickle systems might cut that amount to 32 to 36 inches per season.

Young cherry trees need regular irrigation during the entire growing season. Young trees on Gisela rootstocks can stop growing during dry periods in late spring and early summer, causing them to "runt out" and produce excessive fruit the following year. Established trees need regular irrigation as fruit gain size and as fruit buds for next year form in late summer, but can withstand drier soils just before and after harvest. Regulated deficit irrigation would seem to offer a way to reduce irrigation costs by reducing water applied from harvest until fruit buds set in late summer, but this has not been studied much yet.

Eastern cherry-growing areas generally provide enough rainfall for cherry production on standard and semi-dwarfing rootstocks, but dwarfing rootstocks require irrigation. Sprinkler irrigation systems also offer means to reduce frost and heat damage.

Protecting against frost and fruit doubling. Growers may install sprinkler irrigation to provide frost protection, especially for earlier-blooming sweet cherries. Under-tree sprinklers can provide enough protection for dwarf cherry trees, but standard-size cherries need over-tree sprinklers, which cost much more to install. Consequently, many Eastern sweet cherry growers use orchard heaters (Photo 20-5) or wind machines (Photo 20-6) for frost protection instead. Western growers sometimes hire helicopters to hover over their orchards when frosts threaten during bloom, which provides the same protection as wind machines without the permanent investment, since frosts during bloom don't usually occur every year there.

Where summers are hot, as in certain California sweet cherry-growing districts, growers may install over-tree irrigation to provide both spring frost protection and summer evaporative cooling. Temperatures above 86 F cause many fruit buds to develop two pistils and produce double fruit or fruit with spur-like appendages or deep sutures, which are unmarketable. Evaporative cooling from late June through early September reduces the average daily temperature in the tree canopy during bud formation for next year's crop, and this reduces fruit doubling, spurred fruit, and deep sutures in fruit. Irrigation water for evaporative cooling must be low in dissolved sodium, chlorine, carbonates and boron. Begin sprinkling each afternoon when air temperatures reach 95 F, and continue sprinkling at intervals until sunset.

Reducing fruit cracking. Cherry fruit may crack, or split open, shortly before harvest, render-

Photo 20-5. Orchard heaters can also protect against frost damage of sweet cherries.

Photo 20-6. Some sweet cherry growers use wind machines to protect against frosts during bloom, while other growers hire helicopters for the same purpose.

ing it unfit for sale and very prone to brown rot, which easily spreads to surrounding fruit even if they haven't cracked. Tart cherries rarely suffer from cracking so growers usually don't need to take special precautions for them, but these might be needed if heavy rainfall is predicted just before harvest. Sweet cherries are much more susceptible to cracking, so growers should choose the most crack-resistant cultivars of sweet cherries where cracking is often a problem.

Irrigation management can also minimize fruit cracking unless very heavy rains occur just before harvest. Keep trees well-irrigated as fruit gain in size, but reduce irrigation about 10 days before harvest to provide slight water stress as harvest approaches. Cherries with reduced fruit water content can take in more water after a rain without splitting.

Calcium chloride sprays, applied during heavy rains if they occur as cherries are gaining size, can reduce fruit cracking by an average of 18%. The calcium chloride increases the ion content of the rainwater, which slows fruit water uptake, thus reducing cracking. Calcium chloride is easiest to apply during rains with an over-tree sprinkler irrigation system, but a sprayer can be used if deployed quickly after rain starts, before the ground becomes too wet. With a sprinkler system, program controllers to begin applying a 0.75% calcium chloride solution when 1 mm of rain accumulates, and continue applying it for 4 minutes. Then the system automatically resets to wait for accumulation of an additional 1 mm of rain, when it will apply more calcium chloride for another 4 minutes. A positive injection pump system with back check and flow control valves ensures even application of calcium chloride across the orchard. Application of calcium chloride does not affect cherry fruit size, maturity date or storage life, nor does it influence soil pH because the amounts applied are small. However, calcium residues may persist on fruit until harvest, requiring fruit washing before marketing.

Surfactant sprays can also reduce fruit water uptake, and have reduced fruit cracking by up to 50% in several European trials. A Belgian trial applied a 0.1% solution of the non-ionic spreader-sticker Citowett™, 18 days and 8 days before harvest, and reduced cracking on six cultivars to very low levels compared with unsprayed controls. German, Austrian and Dutch research with Citowett™and Pril™, another spreader-sticker, also showed good results when applied about a week before harvest.

Tree covers help some New Zealand and European cherry growers reduce fruit cracking as well as bird damage and need for fungicides to control brown rot. Many New Zealand cherry orchards, even some of standard-size trees, are covered by large polyethylene structures similar to greenhouses. The high expense of such structures can be recouped as the fruit is exported to Japan during the Christmas season, where demand and prices are high. In Europe, growers use dwarf trees to minimize the area to be covered, erect metal hoops connected with wires or lightweight metal rods over each tree row, and drape polyethylene covers and bird netting over the structures. English researchers have found that short polyethylene covers, extending just over the tops of trees, reduce cracking by about 50%. Longer polyethylene covers, extending down tree sides to within 3 feet of the ground, provide no further cracking protection and cause humidity to build up in the tree canopy, which may increase brown rot and other diseases. A Norwegian study used wires to suspend plastic sheets in a flat plane 1.5 to 5 feet above the tops of cherry trees, and found that covering reduced fruit decay to 6 to 11% from 48% in uncovered plots. One U.S. cherry grower hangs thick polyethylene sheets from wires above the trees, which has also reduced cracking. However, covering is less effective on some sweet cherry cultivars very prone to cracking, such as 'Bing' and 'Ranier'.

PEST MANAGEMENT

Cherries require fewer pesticide sprays than most fruit trees, partly because they ripen earlier in the season, but several pests can be quite severe if not controlled. Only about 1% of North American cherry growers raise certified "organic" cherries, and a key pesticide used by such organic growers, rotenone to control cherry fruit fly, was recently removed from use.

Pests that can kill trees. Young cherry trees often suffer more premature death than many other fruit trees, so careful management is needed to prevent these threats.

Bacterial canker, caused by *Pseudomonas syringae*, often causes death of cherry trees, and is generally the worst pest in most cherry-growing areas (Photo 20-7). Cool, wet weather during bloom spreads the bacteria, which invade frost-damaged flower parts or branch wounds and cause blossom blast and cankers in wood (Photo 20-8). Choosing cultivars and rootstocks with at least moderate resistance to bacterial canker can minimize the disease. So can ordering high-budded trees or planting canker-resistant rootstocks, allowing them to grow for a year, and grafting cultivars 12 to 18 inches out on side limbs. Avoid training systems that require limb bending, and complete dormant pruning of cherries well before the canker-susceptible period during bud swell, so pruning wounds will have healed by then. Copper materials applied preferably at about 75% leaf drop in fall, or during early bud swell in spring, can reduce bacterial canker. Copper materials may also be applied during bloom to prevent blossom blight, but at 25 to 35% the rate to avoid injury to trees, and do not apply copper any later in the growing season.

Collar rot, also called ***Phytophthora* root rot** and caused by several species of that fungus, thrives in wet soils and can kill cherry trees just as described for apples and other fruit plants. Do not plant cherries on 'Mahaleb', a very susceptible rootstock, on heavier soils, and build raised beds before planting a moderately susceptible rootstock, such as 'Mazzard', on somewhat poorly drained soils. Gisela rootstocks have moderate resistance to collar rot so don't often need raised beds.

Oak root rot, also called mushroom root rot or Armillaria root rot and caused by the mushroom fungus *Armillaria mellea*, may attack cherry trees as described for peaches. If planting on land cleared from oak forests in the last few years, or on old orchard land from which apples or other fruit trees were recently removed, dig up and burn as much of the old roots as possible, and sow grass crops for a couple of years before planting cherries. Choose 'Mazzard' rootstock if oak root rot presents more of a threat than bacterial canker, and if orchard density and training system allow for larger trees.

Verticillium wilt, caused by the soil-borne fungus *Verticillium dahliae*, attacks roots of cherries and many other fruit plants, plugging the water-carrying xylem tubes. Vigorous trees may produce enough new xylem to survive infection for several years, but individual limbs or entire trees may suddenly wilt in mid-summer, then lose their leaves. Soil infested with *Verticillium* must be rotated out of susceptible crops for several years or solarized for a summer before planting. Chemical fumigation temporarily reduces this pathogen, but it quickly rebounds to even higher populations once competing microorganisms are killed.

X-disease or **cherry buckskin**, caused by a phytoplasma, is spread by leafhoppers from diseased wild chokecherries or bitter cherries and cultivated cherries. Infected trees develop sparse, light green foliage and fail to mature their fruits, which remain small and pink. X-disease infected trees on 'Mahaleb' rootstock tend to die quickly, so if promptly removed do not serve as reservoirs of the disease, but infected trees on 'Mazzard' and other cherry rootstocks die back over several years and can provide significant inoculum for further infections. Inspect cherry orchards for X-disease regularly during the growing season, remove and burn any suspicious trees as soon as they're seen, and also destroy any wild chokecherries or bitter cherries within 500 feet of the orchard.

Shothole borer (*Scolytus rugulosus*) and **ambrosia beetle** (*Xyleborus dispar*) are important borer pests of cherries in the Pacific Northwest,

Photo 20-7. Bacterial canker may invade frost-damaged flower parts or branch wounds, as here, especially during cool, wet weather at bloom time.

Photo 20-8. Bacterial canker infections lead to cankers that can kill trees, so choose cultivars with resistance where the disease is serious, and apply copper sprays in late fall or early spring.

with shothole borer more common. These borers tunnel under the bark of limbs or the trunk, which can kill major limbs or entire trees. These pests prefer weak or dying trees, so irrigate and fertilize to keep trees vigorously growing, and promptly remove dead or diseased portions of trees. These borers can build up in pruning woodpiles, so promptly burn or destroy all prunings. If borer populations become high, apply endosulfan (Thiodan™) to trunks after harvest.

Peach tree borer (*Syanthedon exitiosa*) and American plum borer (*Euzophera semifuneralis*) may also attack cherry trees. Prevent both as described for peach tree borer in chapter 18.

Voles (*Microtus* species) and pocket gophers (*Thomomys* species) may girdle cherry trunks and kill trees, especially on 'Mahaleb' rootstock which is very susceptible to them. Avoid this rootstock where these pests are populous, and control them as described for apples in chapter 16.

Indirect pests that weaken trees, reducing yields and tree longevity. Planting only certified virus-free trees on rootstocks with resistance to any soil nematodes present will prevent most virus and nematode diseases of cherries. Most of the indirect arthropod pests are normally controlled by natural enemies, so choose pesticides to control major pests wisely, to avoid disrupting natural biological control.

Cherry mottle leaf virus, transmitted by a bud mite from wild bitter cherry trees, may attack trees in Western North America, causing leaves to develop an irregular shape and a mottled chlorosis between veins. Mild new infections reduce yields and fruit size and delay ripening, while long-established infections can cause severe tree stunting and crop failure. Remove any wild bitter cherry trees within 500 feet of the orchard, and any orchard trees with symptoms.

Cherry stem pitting virus occurs in parts of California, where it has infected up to 44% of trees in certain orchards. The virus causes delayed bud break and smaller leaves, with buds never developing on lower parts of trees, and fruit too small and poorly colored to market. The virus is soilborne and especially serious on replant sites, which should be fumigated or sown to grass crops for several years before replanting. 'Mahaleb' rootstock is very susceptible to this virus, so don't use it on problem sites.

Cherry leaf roll virus causes chlorotic leaf spots, leaf curling, shoot dieback, bark splitting, gumming, and decline of trees, which may die within 5 years of infection. Decline is more rapid if trees are also infected with **Prunus necrotic ringspot virus**, which by itself retards growth by about 5% and reduces yields but often shows few symptoms. **Cherry rugose mosaic virus** is a strain of Prunus necrotic ringspot virus that affects cherries much more seriously, but has been confined mainly to California by bee transport agreements that specify waiting 48 hours before setting hives into orchards outside the state. Plant only certified virus-free trees, and promptly remove any trees with virus symptoms.

Sour cherry yellows, caused by prune dwarf virus and made worse if Prunus necrotic ringspot virus is also present, causes chlorotic rings on expanding leaves of newly-infected tart cherry trees. A year or more later, leaves on infected trees develop a yellow and green mottle and drop early. This disease reduces yields by up to 50%, but remaining fruit are larger and firmer than fruit from healthy trees. Again, plant only certified virus-free trees and destroy any infected trees.

Nematodes usually don't build to damaging numbers on the heavier soils that most cherry orchards occupy, but can present a problem on sandy soils. Fortunately, 'Mahaleb' rootstock, which is favored for sandy soils, has resistance to **lesion nematode** (*Pratylenchus vulnus*) and **northern root knot nematode** (*Meloidogyne hapla*). 'Mahaleb' is susceptible to **southern root knot nematode** (*M. incognita*), but among cherry-growing areas that species is only a problem in California. 'Mazzard' resists both northern and southern root knot nematodes but is susceptible to lesion nematodes, so has been favored in California, where lesion nematodes present less of a problem. The Gisela rootstocks are susceptible to nematodes, so if nematodes are present, soil must be fumigated or sown to nematode-reducing crops before planting.

Crown gall, caused by *Agrobacterium tumefaciens*, may attack cherry roots, producing galls that weaken and may eventually kill trees. Most cherry rootstocks have moderate resistance to crown gall, but 'Mazzard' is susceptible and 'Colt' is very susceptible. Dip tree roots in the biocontrol product Galltrol™ when planting susceptible rootstocks on infested soil.

Mite pests of cherries include **European red mite** (*Panonychus ulmi*) and **two-spotted spider mite** (*Tetranychus urticae*) in all cherry regions, plus **McDaniel mite** (*T. mcdanieli*) in the Pacific Northwest. Cherry rust mite (*Aculus fockeui*) also occurs on cherries but is not a pest unless populations become excessive, and in fact is needed as an alternate food source for predator mites that help control the pest mites. Apply a dormant oil spray to reduce European red mites without hurting cherry rust mites or predators, and choose insecticides that do the least harm to cherry rust mites and predator mites. Over-tree irrigation also reduces mite problems, as does paving or oiling farm roads to reduce dust. Controlling weeds in fall rather than spring greatly reduces migration of two-spotted spider mites into trees.

Scale pests of cherries include **San Jose scale** (*Quadraspidiotus perniciosus*) and **European fruit leucanium scale** (*Parthenolecacium corni*). A dormant oil spray reduces scale populations, but if scale are abundant, mix an insecticide (usually diazinon) with the oil to increase control. Organic growers use fish emulsion oil, which is more expensive but works.

Black cherry aphid, *Myzus cerasi*, is the main aphid pest of cherries. Young, vigorously growing trees may develop high aphid populations, possibly reducing the development of future fruiting wood. Dormant oil sprays for mites or scale help to reduce aphid populations, but growers should monitor young trees early in the season and apply an insecticide if needed.

Diseases attacking flowers, fruit and foliage. At least one fungal disease that attacks fruits or causes defoliation occurs in every cherry-growing area of North America, generally requiring fungicide applications.

Brown rot (generally **American brown rot** caused by *Monilinia fruticola*, occasionally **European brown rot** caused by *M. laxa* on European cherry cultivars) can cause blossom blight and a severe fruit rot (Photo 20-19) in areas where rains occur during bloom or as harvest approaches. The desert climate of most Western cherry production areas means that these growers do not spray for brown rot, and many have never seen it in their orchards. However, cherry growers close to the Pacific Coast or in Northeastern North America must apply fungicides, at pink bloom stage if blossom blight is a problem, and from petal fall to close to harvest to prevent fruit rots. Tight-clustered cultivars are especially prone to brown rot, so avoiding such cultivars can help reduce the disease. Harvest cherries as completely as possible, or fruit left on trees can become mummies that hang on until next year and spread brown rot.

Powdery mildew, caused by the fungus *Podosphaera clandestina*, becomes very destructive during years of unusually high spring rainfall in desert climates where most commercial cherries grow, and may cause damage in humid climates during dry years. Powdery mildew first infects leaves, on which powdery white mycelia of the fungus grow, then spreads to fruit, which can become covered by the fungus. To protect fruit, growers must prevent leaf infections with fungicide sprays, starting when powdery mildew first appears or is predicted to appear by a model based on temperature and rainfall data. If brown rot is not also a problem, growers can control powdery mildew with the biocontrol product *Ampelomyces quisqualis* (AQ10™) or with summer-weight horticultural mineral oil. However, oil sprays followed by insecticides (for cherry fruit fly) can cause leaf drop on a few sweet cherry cultivars, such as Lapins and Sweetheart. Sterol inhibitor fungicides, such as myclobutanil (Rally™ or Nova™), fenarimol (Rubigan™) or propiconazole (Orbit™), or sulfur fungicides will control both powdery mildew and brown rot.

Cherry leaf spot, caused by the fungus *Blumeriella jaapii*, is the major disease of tart cherries and can cause serious damage to sweet cherries in Eastern areas, but rarely occurs in Western cherry production areas. The fungus overwinters on fallen leaves, then at about petal fall, ejects spores that infect new leaves, causing leaf spots and defoliation if infection is severe. This loss of leaves reduces yields, fruit size and fruit sugar content. Where cherry leaf spot is serious, growers can minimize fungicide applications by choosing cultivars with the most resistance to the disease. Monitor temperature and leaf wetness periods from petal fall until close to late summer, to determine when fungicides are needed (Table 20-5). For ex-

Photo 20-9. Brown rot attacks cherry orchards close to the Pacific Coast and in Northeastern North America, where growers must apply fungicides from petal fall until close to harvest to prevent it. Harvest cherries as completely as possible, or fruit left on trees can become brown rot mummies, as here, that hang on until next year and spread the disease.

Table 20-5.

APPROXIMATE MINIMUM NUMBER OF LEAF WETNESS HOURS NEEDED AT VARIOUS TEMPERATURES FOR CHERRY LEAF SPOT INFECTIONS

Avg. F Temp.	Hours Wetness
46	28
47	25
48	23
49	20
50	19
51	17
52	15
53	14
54	12
55	11
56	10
57	9
58	8
59-60	7
61-62	6
63-68	5
69-70	6
71-72	7
73	8
74	9
75	11
76	12
77	14
78	16
79	18
80	21
81	28

ample, a wetting period of just 5 hours calls for a fungicide spray if average daily temperatures fall between 63 and 68 F, while average temperatures of either 59 to 60 F or 71 to 72 F require 7 hours wetting before a spray is needed. When average temperatures are either below 47 F or above 80 F, no fungicides are needed until leaves are wet for 28 continuous hours.

Fruit-attacking insects and birds. These direct pests can cause severe damage on cherries, and are the key pests that must be controlled.

Cherry fruit flies, native to North America, are the worst fruit-attacking insects in most cherry-growing regions, and the cause of most insecticide applications to cherries. Flies have one generation per year, emerging from the soil in late spring to lay eggs on developing fruit, which often appear normal until the maggots inside are almost fully grown. **Western cherry fruit fly** (*Rhagoletis indifferens*), found in the Pacific Northwest since the 1940's, does little actual damage and is rarely found in commercial orchards due to intensive spraying to meet a zero tolerance restriction for shipments to California (where it doesn't occur) or export. Inspectors at each cherry packinghouse check each fruit load coming in before it is accepted, and if a single maggot is found in a cherry, the entire load is rejected, and all other fruit loads produced from that orchard are intensively inspected. Inspectors check cherries again after packing.

Northwestern cherry growers begin spraying for fruit flies when pest management advisors detect first fly emergence on monitored sentinel trees, or when temperature-based models predict emergence. Growers continue spraying every 7 to 10 days until harvest, abiding by each chemical's days-to-harvest restriction interval, then apply another spray or two after harvest to control flies on unharvested fruit that remain on trees, for a total of 3 to 5 sprays per season. Spray materials are rotated between knockdown chemicals that kill adult flies by contact only during or very soon after application, and residual-action chemicals that remain on tree surfaces, killing adult flies that feed on them. Knockdown chemicals include pyrethroids, malathion and the insect growth regulator spinosad (Success™ or Spintor™) the last considered organic. Residual-action chemicals include carbaryl (Sevin™), diazinon, phosmet (Imidan™), azinphos-methyl (Guthion™), and dimethoate (Digon™, Dimate™). The only product that controls larvae developing inside unharvested fruit is dimethoate, so it is used for the post-harvest spray, considered the most important spray of the year since it allows growers to enter the next cherry season with a very low cherry fruit fly population.

Two other fruit fly species, the **cherry fruit fly** (*R. cingulata*) and the **black cherry fruit fly** (*R. fausta*), attack the cherry crop in Northeastern North America. Growers monitor fruit flies with yellow sticky cards hung in trees in late spring, and begin spraying 5 to 6 days after first adult flies are caught, using the same spray materials and including a post-harvest spray.

Plum curculio (*Conotrachelus nenuphar*) also attacks cherries in Eastern North America, and there is zero tolerance for larvae in fruit (Photo 20-10). Most sprays for cherry fruit flies in the Northeast (except for spinosad) will also control plum curculio, but in the lower Midwest and mid-South where temperatures are thankfully too warm for cherry fruit flies, plum curculio becomes the main fruit-infesting pest of cherries. Control it as described for apples in chapter 16, with an insecticide spray at petal fall and a second spray 10 days later.

Leafrollers, including the **pandemis leafroller** (*Pandemis pyrusana*) and the **obliquebanded leafroller** (*Choristoneura rosaceana*), have become important secondary pests in Pacific Northwest cherry orchards, probably because they've developed resistance to many chemicals commonly used to control cherry fruit flies. The new insect growth regulator spinosad (Success™) does help control second-generation leafrollers when applied for cherry fruit flies, and does not hurt beneficial predators and parasites that help control leafrollers, so use of spinosad is increasing. Also apply the biocontrol product Bacillus thuringeinsis (Bt; Dipel™, Javelin™, etc.) pre-bloom, at petal fall and 10 days later to reduce first-generation leafrollers.

Birds can eat many whole cherries and peck at others, leaving them open to brown rot, and can decimate a crop if not controlled. Many European growers erect bird netting over dwarf cherry trees. North American growers generally use frightening devices or the chemical bird repellent derived from 'Concord grapes', methyl anthranilate (Bird Shield™ or ReJex-it™). Spray this repellent when birds first appear and once a week thereafter until a week before harvest, to allow time for sunlight and high temperatures to break down the compound so no off-flavors remain at harvest.

HARVEST, HANDLING, AND MARKETING

Fruit color change and soluble solids have traditionally been used to determine harvest time for cherries, along with fruit acidity for tart cherries. Tart cherries turn bright red, dark sweet cherries turn dark red, and light sweet cherries turn solid yellow or yellow with a red blush. Soluble solids should reach 15 to 16% for tart cherries, and 18 to 20% for sweet cherries. Fruit removal force, measured by a pull gauge, has become the most reliable maturity index in recent years.

Sweet cherries for fresh market are hand harvested, usually with stems (pedicels) left on, to avoid creating openings for brown rot where stems were removed. Cherries for processing, including about 42% of sweet cherries and almost all tart cherries, are mechanically harvested using trunk shakers. Fruit fall onto padded catch frames, roll past fans that blow out leaves and many stems, and fall into large tanks of cold water that immediately cool and wash the fruit. Growers of processing cherries often apply the growth regulator ethephon to loosen fruit from pedicels, so many cherries shake off without stems attached (Photo 20-11).

Ethephon use to facilitate mechanical harvest. Ethephon may be sprayed on tart and sweet cherries, 8 to 14 days before expected harvest, to loosen fruit for mechanical harvest. Ethephon enables more fruit to be shaken loose, increasing yields and leaving very few unharvested fruit on trees in which cherry fruit flies could build up. Fruit should have reached full size and be starting to turn yellow or a tinge of red when ethephon is applied, as fruit may fail to enlarge further after

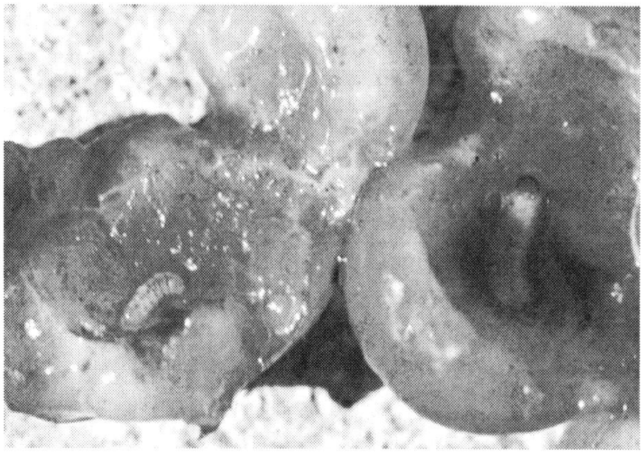

Photo 20-10. Plum curculio attacks cherry fruit east of the Rocky Mountains. Most sprays for cherry fruit flies in the Northeast will also control plum curculio, but where it is the main fruit-infesting pest, control it with an insecticide at petal fall and a second spray 10 days later.

Photo 20-11. Tart cherry growers often apply the growth regulator ethephon to loosen fruit from pedicels before mechanical harvest, so many cherries shake off without stems attached, and loose stems are blown away. Processing cherries are cooled in water tanks for 6 to 8 hours, then immediately frozen or canned.

ethephon is applied. Do not apply ethephon if temperatures are predicted to exceed 85 F during the next 3 days, as such hot temperatures produce excessive ethephon activity that can cause pre-harvest drop. Do not spray ethephon if temperatures are below 60 F either, as such low temperatures greatly reduce ethephon activity.

Try to apply ethephon 10 to 14 days before harvest, so a lower rate can be used to prevent tree injury. If ethephon application is delayed until 8 or 9 days before harvest, a higher rate must be used. If temperatures are predicted to be above average (but not over 85 F) during the next 3 days, the ethephon rate can be reduced, while a higher rate must be used if cooler temperatures are predicted. Ethephon may be applied with fungicides or insecticides, but do not tank-mix ethephon with crack inhibitors, bird repellents, buffers or surfactants which could reduce its effectiveness.

Concentrate spraying of ethephon, using 20 to 80 gallons of water per acre, produces the same effect as does dilute spraying, so growers generally spray concentrated ethephon. Full-size trees generally receive 0.5 to 1.0 pints ethephon per acre for tart cherries, 1.0 to 2.0 pints ethephon per acre for yellow or yellow-red sweet cherries, and 1.5 to 2.5 pints ethephon per acre for dark red sweet cherries. Reduce these rates according to tree-row volume for younger or dwarf trees.

Cooling, processing and packing cherries. Most cherries are hydrocooled soon after harvest, and growers should keep chlorine levels in the hydrocooler at 50 to 200 ppm to avoid spreading brown rot or other post-harvest diseases. Mechanically-harvested cherries are shaken onto padded catch frames, inclined so the cherries roll directly into large water tanks, holding 1000 pounds of cherries and kept at 48 F. Within 30 minutes, water tanks are placed on pallets in cooling stations in the orchard, where ice is added. Cherries remain in the tanks for 6 to 8 hours, constantly flushed with cold water to wash and cool the fruit, and to minimize bruising. Processing cherries are then brought to plants where they are de-stemmed, pitted and packed for freezing or canning. Many of the frozen cherries are later canned, processed into juice or jam, or processed into maraschino cherries.

Maraschino cherries originally were preserved by soaking in a 70-proof liqueur distilled from fermented juice of the wild marasca cherry, native to the Dalmatian Coast of Italy and Croatia. Today, maraschino or "cocktail"cherries are first bleached of natural color with sulfur dioxide, then soaked in a brining solution of sodium metabisulfate, calcium chloride, and citric acid to preserve them. The brined cherries are then sweetened by soaking in solution of corn syrup and fructose, and finally artificially flavored and dyed a bright red or green.

Fresh-market sweet cherries are machine-sorted by size and color, then packed into shallow flats and shipped immediately in refrigerated trucks or ships. Cherries must be sold within 2 weeks. They are prone to the same post-harvest diseases as peaches, apples and other tree fruits, mainly brown rot, Rhizopus rot, gray mold and blue mold.

Further Reading

Arndt, H. 1996. High density cherry systems. Tree Fruit Leader 5(1):1-5. Available online at http://www.agf.gov.bc.ca/treefrt/newlett/high-density_cherry.htm

Borve, J. and A. Stensvand. 2003. Use of a plastic rain shield reduces fruit decay and need for fungicides in sweet cherry. *Plant Dis.* 87:523-528.

Gutzwiler, J. and G. Lang. 2001. Sweet cherry crop load and vigor management on Gisela rootstocks. *Acta Hort.* 557:321-325.

Lang, G.A. 2001. Intensive sweet cherry orchard systems--rootstocks, vigor, precocity, production and management. *Compact Fruit Tree* 34(3):23-26.

Lang, G. and J. Flore. 1999. Reducing rain cracking in cherries. *Good Fruit Grower* 50(4):34-38.

Michigan State University's cherry websites have much information on cultivars, training systems, growth regulators, pests, etc. Available online at http://www.msue.msu.edu/fruit/cherry.htm and http://www.maes.msu.edu/nwmihort

Chapter 21
Apricots

Apricots are the least widely-grown of stone fruits in North America, due to their very early bloom time that requires special regions or sites to avoid crop loss to frost. Apricots originated on mountain slopes in China where winters stayed continuously cold until spring, and many cultivars are hardy to USDA zone 5 in midwinter, but warmer temperatures in late winter induce them to bloom far too early. Traders carried apricot cuttings across the Persian Empire to the Mediterranean region, where they flourished in the mild winters and dry summers and acquired their botanical name, *Prunus armeniaca*. The Mediterranean region still produces most of the world's apricots, with Turkey raising about 20% and Spain, Italy, France and Greece each growing about 10%. The United States is the sixth-largest apricot producing country, producing 170 to 200 million pounds per year, which is about 4% of the world's apricots and about 8% of the size of the U.S. peach crop.

Spaniards first planted apricots in California mission gardens during the 1700's, and California recorded its first major apricot production in 1792. California now produces about 94% of U.S. apricots, Washington grows another 5%, and Utah, Oregon, Michigan and Ontario raise small amounts. Only about 21% of apricots are sold for fresh market during their 6-week-long ripening and storage period, from mid-May to early July in California. The remaining 79% of the crop is processed, with about 53% of that canned, 20% made into juice, 16% frozen and 11% dried. Growers receive the highest prices for fresh apricots, followed by canned, frozen and dried apricots, and the least for juice.

Yields for apricots in California average 3.5 to 4.5 tons per acre, just $^1/_3$ to $^1/_2$ of yields for fresh-market peaches. U.S. apricot growers have faced severe competition from lower-priced imports of dried and canned apricots in recent years. This caused bankruptcy of a California apricot growers' cooperative processing facility in 2000, and the company that bought the facility now buys about 40% fewer apricots than before. Some apricot growers have been forced out of business, and the crop shows little potential for expansion except as a specialty fresh fruit for direct markets outside of traditional apricot-growing areas.

FRUIT AND TREE GROWTH HABITS

Apricot fruits are smaller than those of most peach cultivars, and have a light pubescence or "fuzz" that is less than that of peaches (Photo 21-1). The flesh is less juicy and sweet than peaches but has a unique, tangy flavor that many people

Photo 21-1. Apricot fruits are smaller than those of most peaches, and have a light pubescence or "fuzz" that is less than that of peaches. (photo from Michigan State Univ., E. Lansing, MI)

have never tasted fresh, and retail growers of fresh apricots can readily stimulate demand by offering samples. All commercial apricots have yellow non-melting flesh, although some white-fleshed apricot cultivars exist, and commercial cultivars are freestone.

Breeding efforts in Ontario, New Jersey and Washington have produced a few apricot cultivars which bloom later or have more frost-hardy flowers, and some of these cultivars offer resistance to common diseases of humid areas. Later-blooming apricots have both a higher chilling requirement and a higher heat requirement after chilling is met before they'll bloom. Most apricot cultivars have poisonous seeds, as do most peaches, but a few apricot cultivars bear edible seeds like almonds, and these are sometimes called "alpricots". Retail growers in areas outside of California may find niche markets with frost- and disease-resistant apricots and alpricots. Also, the cancer drug laetrile is derived from bitter apricot seeds.

Zaiger Genetics in California has recently released several "aprium™" cultivars, which are ¾ apricot and ¼ Japanese plum (Photo 21-2). These offer larger and sweeter fruit than apricots, and can be pollinized by apricots or early-blooming Japanese plums. These aprium™ cultivars have low chill requirements and grow well where apricots are grown commercially, but do not have frost or disease resistance for other areas.

Apricot trees are moderately small, growing to about 25 feet tall in nature, and can be kept smaller by pruning. Apricots are long-lived where adapted, often living 30 to 40 years. Consequently, growers maintain orchards 2 to 3 times as long as peach orchards, change cultivars less frequently, and have not yet converted many orchards to higher tree densities.

Trees take several years to begin bearing, usually producing the first commercial crop 4 years after planting. Apricots bear fruit on spurs on 2-year-old or older branches and also bear fruit from lateral buds on 1-year-old branches. Spurs produce larger fruit but bloom 4 to 7 days before lateral buds, so frosts during spur bloom can eliminate most large fruit.

SITE SELECTION AND PREPARATION

Choose the most frost-free site available for apricots. Outside of frost-free parts of California, north-facing slopes are preferred, as they can delay bloom by up to 10 days over a south-facing slope. Areas moderated by a large body of water also have less frost damage.

Photo 21-2. Apriums, such as this 'Flavor Anne', are hybrids that are ¾ apricot and ¼ Japanese plum. (photo from Dave Wilson Nursery, Hickman, CA)

Apricots are very adaptable as to soil type, since they may be propagated on many different *Prunus* rootstocks including peaches, plums, peach-plum hybrids, and peach-almond hybrids as well as on apricot rootstocks. Apricots are the most tolerant of the stone fruits to alkaline and somewhat dry soils. Ideally, choose a site that was planted to grass crops for the last several years. Avoid sites cleared from oak forest in the past few years, to avoid oak root rot, and avoid fields where crops susceptible to Verticillium wilt were grown. Sow old orchard sites to grass for several years before planting apricots.

Test sandy soils for nematodes the summer before planting, and solarize or fumigate the soil if high levels of harmful nematodes occur. Solarization also reduces Verticillium wilt, and is best done with large sheets of clear plastic over the entire site that will be removed before planting, rather than applying clear plastic mulch down future tree rows. Test soil pH, and add lime if it's below 6.0 or sulfur if it's above 8.5. Also test phosphorous and potassium, and add these several months before planting if levels test low.

Western apricot growers prepare the site the year before planting. A custom operator first subsoils 2 to 3 feet deep in two directions to break up any hardpan, then the grower disks twice to break up clods, levels twice, and creates raised beds about 12 inches high. In humid areas when planting into a grass sod, growers can preserve most of the sod and plant trees in rows of glyphosate-killed sod, generally on flat ground, but raised beds might be needed on somewhat poorly drained soil for certain rootstocks (Photo 21-3).

SELECTING ROOTSTOCKS

California apricot growers generally choose peach or plum rootstocks adapted to their soil type, detailed in rootstock charts 18-1 and 19-1 for the peach and plum chapters. 'Citation' peach-plum hybrid is a popular choice for new plantings since it is semi-dwarfing, keeping apricot trees 14 to 18 feet tall, and induces them to bear about a year earlier, usually 3 years after planting. 'Citation' resists root knot and ring nematodes and tolerates wet soil, but needs extra irrigation in dry soil, so don't plant this rootstock on raised beds. It is susceptible to oak root rot, crown gall, and peach tree borer, and moderately susceptible to bacterial canker.

Plum rootstocks are commonly used for apricots on somewhat wet soils. 'Marianna 26-24' and 'Myrobalan 29c' are common choices, since both have moderate resistance to oak root rot and crown gall and strong resistance to root knot nematodes. Both these plum rootstocks produce full-size trees and are moderately susceptible to bacterial canker and peach tree borer. However, the graft union between apricot cultivars and myrobalan plum rootstocks is weak and may break under strong winds 3 to 4 years after planting, and trees that survive start bearing a year later than trees on other rootstocks. 'Torinel' rootstock, a hybrid of two European plum cultivars (*Prunus domestica*) that has better compatibility with apricots, produces a slightly smaller tree and induces precocity, has resistance to root knot nematodes and bacterial canker, and is good for replant sites.

'Nemaguard' and 'Lovell' peach rootstocks produce full-size trees, and 'Nemaguard' resists root knot nematodes while 'Lovell' has moderate resistance to lesion nematodes, but both need ideal soil drainage and are susceptible to peach tree borer and bacterial canker. On alkaline soil, apricot growers usually choose one of the peach-almond hybrids adapted to alkaline sites. 'Atlas' may delay bloom slightly, delays fruit maturity, and increases fruit size.

Photo 21-3. Apricots benefit from planting on raised beds in somewhat poorly drained soils. Intercropping a small grain crop between the young trees provides some income before the trees start bearing.

Where dwarfing isn't desired and soils factors don't require special resistance, apricot trees generally live longest on apricot rootstocks. However, since far fewer apricots are propagated than other stone fruits, some apricot cultivars may only be available from nurseries on peach or plum rootstocks. Apricot seedling rootstocks produce full-size trees tolerant of both somewhat wet and somewhat dry or alkaline soil, with moderate resistance to peach tree borer and strong resistance to root knot and lesion nematodes, but are susceptible to bacterial canker and more susceptible to Verticillium wilt than other *Prunus* species. Apricot growers in colder climates should choose Manchurian apricot seedling rootstocks, which are hardy to USDA zone 4 and also keep trees somewhat smaller.

CHOOSING CULTIVARS

Since most apricots are processed but fresh apricots fetch a higher price, growers choose cultivars that will be accepted by canners or other processors but can also be sold fresh. Growers who concentrate on the fresh market choose a succession of cultivars that ripen over the entire apricot season, often growing very early- and late-ripening cultivars that may not be suitable for processing. Apricot cultivars vary greatly in chill requirements from about 300 to 1000 hours, and growers should choose cultivars that need about as many hours of chilling as their area normally provides. Where frosts are sometimes a problem (usually in high-chill areas) choose cultivars with chill needs on the high side of average for the area, but if frosts generally don't occur (usually in low-chill areas) choose cultivars with chill needs on the low side of average for the area.

Table 21-1 details the uses and traits of many apricot cultivars commonly grown in the arid West of North America. Western cultivars generally have little resistance to frosts which are seldom a problem there, and are usually very susceptible to diseases which again aren't serious in the irrigated dry climate.

Apricot growers in the rainy parts of the Northwest, northern Midwest and Northeast need to choose cultivars that bloom later or have more frost-hardy flowers. They also need resistance to fungal canker, fruit cracking and subsequent brown rot in all those areas, and to bacterial spot if east of

Table 21-1.

COMMON COMMERCIAL APRICOT CULTIVARS FOR ARID WESTERN NORTH AMERICA

In approximate order of ripening (varies with area).

All are freestone (except as noted) with yellow-orange skin and flesh, and all are S to bacterial spot, so don't do well east of the Rocky Mountains.

Cultivar	Chill Hours Needed	Fruit: Uses	Fruit: Flavor	Fruit: Size	Fruit: Firmness	Fruit: Yields	Needs pollen-izer	Bloom time	Harvest time
Poppy	350	fresh	fair	medium	firm	medium	no	early	very early
Apache	300	fresh	good	medium	firm	high	yes	early	very early
Royal Rosa	500	fresh	very good	medium	firm	high	no	early	very early
Earlicot	450	fresh	very good	large	firm	medium	yes	early-mid	early
Lorna	300	fresh	very good	large	firm	high	no	early-mid	early
Orangered	500	fresh	very good	large	firm	high	yes	early	early
Robada	500	fresh	very good	medium	firm	high	no	middle	early
Goldensweet	600	fresh, dried	very good	large	firm	high	no	mid-late	early
Jordanne	400	fresh	very good	very large	firm	high	yes	early	early-mid
Ambercot	500	fresh	good	medium	firm	high	yes	early	early-mid
Goldbrink	800	fresh	good	very large	firm	medium	yes	early	early-mid
Nicole	500	canned, preserves	best	medium	firm	high	yes	early-mid	early-mid
Castlebrite	500	fresh	good	medium	firm	high	no	early-mid	early-mid
Flavor Giant	600	fresh, dried	very good	very large	firm	high	no	early-mid	early-mid
Katy	400	fresh, canned	very good	large	firm	high	no	early	middle
Helena	500	fresh	good	very large	firm	high	no	middle	middle
Impr. Flaming Gold	400	fresh, canned	good	large	firm	med-high	no	early	middle
Tri Gem	600	canned	good	med-large	very firm	med-high	no	middle	middle
Tomcot	600	fresh, dried	very good	med-large	med-firm	high	no	early	middle
Goldstrike	800	fresh, dried	very good	medium	very firm	medium	yes	early	mid-late
Goldbar	800	fresh,dried	very good	med-large	very firm	medium	yes	early	mid-late
Goldrich	800	fresh, canned	good	large	firm	high	yes	middle	mid-late
Rival	800	fresh	good	large	med-firm	medium	yes	early-mid	mid-late
Wenatchee	700	canned, fresh, dried	good	large	firm	med-high	no	early	late
Perfection	600	canned, fresh, dried	best	large	firm	med-high	yes	early	late
Patterson	600	canned, fresh, dried	good	medium	firm	high	no	middle	late
Royal Blenheim	500	canned, dried, fresh	best	medium	medium	high	no	early	late
Tilton	800	canned, fresh	good	med-large	firm	very high	no	mid-late	very late
Fantasme	500	fresh, canned	very good	medium	firm	high	no	early-mid	very late
Westley	600	dried	very good	large	very firm	high	no	middle	very late
Autumn Royal	500	fresh, canned, dried	very good	medium	medium	high	no	very late	extra late
Earli Autumn	800	fresh	very good	medium	firm	medium	no	very late	extra late
Autumn Glo	700	fresh, dried	very good	large	firm	medium	no	very late	extra late

Average ripening times in Fresno, CA: very early: May 15-30, early: May 20-June 5, early-mid: May 25-June 15, middle: June 1-20, mid-late: June 8-25, late: June 15-30, very late: June 22-July 5, extra late: July 25- Aug. 10. Ripening times are about 3 weeks later in western OR and 4 weeks later in western WA.

Notes on Cultivars:
Poppy is being planted for very early fresh market.
Apache needs early, diligent thinning for good fruit size, responds well to branch girdling. Fruit ship and store well.
Royal Rosa is a precocious and semi-dwarf tree, has low-acid fruit.
Earlicot sets few fruit so needs little thinning, is S to fruit cracking if rains fall just before harvest.
Lorna has fruit much bigger than Castlebrite, but S to fruit cracking and pre-harvest drop, incompatible with Marianna 2624 rootstock.
Orangered from France hangs well on tree, ships well, has MR to bacterial spot but is not winter-hardy enough for most Eastern areas.
Robada has deep orange skin and edible kernels, handles and ships well.
Goldensweet fruit have high sugars, tree blooms later than other early apricots.
Jordanne fruit have a slight orange-red blush, ship well.
Ambercot is a good pollenizer for other cultivars.
Goldbrink is an early-ripening cultivar suited to high-chill areas.
Nicole fruit have high sugars, low acid, strong apricot aroma suited for canned halves, pulp, marmalade.
Castlebrite, a main early cultivar, colors early but is very tart until fully ripe, so don't pick too soon. It needs fruit thinning to increase size.
Flavor Giant has very large, sweet-tart fruit.
Katy, a major chipping cultivar in low-chill areas, is not for drying as fruit get too dark after sulfuring.
Helena bears regularly, fruit hold well on tree, can be harvested over a long season.
Improved Flaming Gold colors 2-3 weeks before ripe so don't pick too soon, has strong red blush, not for drying as fruit get too dark.
Tri Gem was developed for mechanical harvest for canning. Fruit need thinning for good size.
Tomcot is one of the most consistent producing apricots in the West.
Goldstrike often has light to medium fruit set, reducing need to thin fruit, is not suited for canning.
Goldbar often has light to medium fruit set, reducing need to thin fruit, is not suited for canning.
Goldrich, also called 'Sungiant', colors early but is acid till fully ripe, hangs well on tree, so don't pick too soon. R to fruit cracking.
Rival fruit have mild, low-acid flavor, are S to cracking, color early so don't pick too soon, may crop erraticly.
Wenatchee, or '**Wenatchee Moorpark**', an older cultivar grown in WA, is VS to frost, fruit cracking, but bears heavily where few frosts.
Perfection, an older cultivar, is a very winter-hardy tree, grown in WA. Fruit soften quickly after harvest, don't store well.
Patterson, a main late cultivar, bears consistently heavy crops with little pre-harvest drop, stores well.
Royal Blenheim, an old mixture of 2 very similar English cultivars, does best in cool suumers, VS to fruit cracking, tends to alternate bear.
Tilton, the main late canning cultivar, colors well before ripe, so don't pick early. Poor flavor when dried. MR to brown rot, VS to peach scab.
Fantasme, a new cultivar from France, has high quality for late-season fresh market.
Westley may set light crops, has very good drying quality and dry-away ratio.
Autumn Royal, a sport of 'Royal Blenheim', extends fresh apricot season into late summer in medium-chill areas.
Earli Autumn has semi-clingstone fruit, handles and ships very well, is not for drying.
Autumn Glo extends fresh apricot season into late summer in high-chill areas.

Table 21-2.

FROST- AND DISEASE-RESISTANT APRICOT CULTIVARS FOR HUMID AREAS, USDA ZONES 5 TO 8A

In approximate order of ripening (varies with area).

All are freestone with yellow-orange skin and flesh, and all need 700-850 hours chilling.

Cultivar	Hardy to Zone	Fruit				Resistance to:			Spring Bud Hardiness	Needs Pollen-izer	Bloom Time	Harvest Time
		Flavor	Size	Firmness	Yields	Bact. Spot	Fungal Canker	Fruit Cracks				
Earlyblush	5b	good	medium	med-firm	high	MR	MR	R	hardy	no	early-mid	very early
Jerseycot	5a	good	medium	medium	medium	R	R	MS	very hardy	no	middle	very early
SunGem	5a	very good	medium	very firm	high	VR	MR	R	very hardy	no	mid-late	early
Harcot	6a	very good	med-large	firm	medium	VR	R	MS	medium	no	mid-late	early
Wilson's Delicious	6b	good	large	firm	medium	R	S	R	medium	no	mid-late	middle
Haroblush	5b	good	med-large	med-firm	medium	VR	VR	MR	very hardy	no	middle	middle
Harojoy	5b	good	med-large	firm	medium	VR	VR	R	very hardy	no	middle	middle
Goldcot	5b	fair	medium	medium	high	S	S	MS	very hardy	no	early-mid	middle
Sundrop	5a	good	medium	med-firm	high	MR	S	R	hardy	yes	early-mid	middle
Harostar	5b	good	med-large	firm	med-high	MR	R	R	hardy	no	middle	mid-late
Hargrand	5b	good	large	firm	medium	MR	R	MS	hardy	no	early-mid	mid-late
Harglow	6b	very good	medium	firm	high	VR	MR	MS	tender	no	late	late
Vivagold	6a	good	large	firm	med-high	MR	S	R	medium	yes	middle	late
Harval	5a	very good	large	firm	med-high	R	R	R	very hardy	no	late	late

Disease and disorder susceptibility ratings: VR = very resistant, R = resistant, MR = moderately resistant, MS = moderately susceptible, S = susceptible, VS = very susceptible.

Average ripening times in SW MI: very early: June 25-July 5, early: July 5-15, middle: July 15-25, mid-late: July 20-30, late: July 25-Aug. 5. Ripening times are about 3 weeks earlier in SW VA, about 2 weeks later in Geneva, NY, and about 3 weeks later in Harrow, Ontario.

Notes on Cultivars:
Earlyblush is 3 weeks earlier than 'Goldcot', has a 40% orange blush, softens quickly so pick promptly.
Jerseycot had at least a partial crop during 20 of 22 years in New Jersey despite frosts. Fruit soften and drop quickly, so pick promptly
SunGem is a precocious tree, fruit have 30% orange blush, MR to brown rot.
Harcot has edible kernels, tends to overset, needs fruit thinning for size.
Wilson's Delicious, an older cultivar, produces consistently.
Haroblush has high flower bud density, mainly on spurs, so usually sets a crop.
Harojoy has atttractive, red-blushed fruit and edible kernels.
Goldcot is a genetic semi-dwarf tree, has staggered heavy bloom so usually sets fruit, widely grown in MI for baby food, S to peach sc
Sundrop colors early so don't harvest too soon, sets heavily some years and needs detailed fruit thinning.
Harostar has uniform-ripening fruit with a good orange blush.
Hargrand has much larger fruit and resists bacterial spot much better than 'Veecot', which ripens with it.
Harglow is a genetic semi-dwarf tree, resists bacterial spot much better than 'Harogem', which ripens with it.
Vivagold has excellent processing quality, edible kernels. Fruit color early, so don't harvest too soon.
Harval has larger and better-flavored fruit that resist cracking much better than 'Harlayne', which ripens with it.

Table 21-3.

COMMON COMMERCIAL APRIUM CULTIVARS FOR ARID WESTERN NORTH AMERICA

In approximate order of ripening (varies with area).

All are partly self-fruitful, set larger crops if cross-pollinated, and bloom early. All are freestone, with yellow-orange skin and flesh. All are susceptible to bacterial spot, so don't do well east of the Rocky Mountains.

Cultivar	Chill Hours Needed	Fruit			Harvest time
		Flavor	Size	Firmness	
Flavor Anne	400	good	large	firm	very early
Tasty Rich	500	best	med-large	firm	early
Honey Rich	300	best	med-large	very firm	early-mid
Flavor Delight	550	very good	medium	firm	middle

Average ripening times in Fresno, CA: very early: May 15-30, early: May 20-June 5, early-mid: May 25-June 15, middle: June 1-20. Ripening times are about 3 weeks later in western OR and 4 weeks later in western WA.

Notes on Cultivars:
Flavor Anne has very sweet, orange fruit.
Tasty Rich, released in 2001,has a spicy-sweet flavor.
Honey Rich fruit hang well on tree, but delicate skin discolors when handled.
Flavor Delight, the first aprium released in 1989, has sweet yellow fruit that scores highly in taste tests.

the Rocky Mountains. Table 21-2 details cultivars bred to resist frost and diseases, which all have fairly high chill requirements of 700 to 850 hours. Later-blooming apricots generally have a higher heat requirement after chilling has been satisfied. No apricot cultivars survive long in the hot, humid Southeastern U.S.

Aprium™ cultivars are described in Table 21-3. Cultivars developed so far have low to medium chill needs and no frost or disease resistance, so do well in low-chill areas of the West where commercial apricots are grown. Few growers raise apriums™ yet, but as more cultivars are developed and growers gain experience with them, more commercial apricot growers in the West will add apriums™.

ORCHARD LAYOUT AND PLANTING

Most apricot growers still space trees widely, about 18 to 20 feet apart within rows and 22 to 24 feet between rows, producing 90 to 110 trees per acre. Growers using 'Citation' rootstock space trees closer, usually 14 to 16 feet apart within rows spaced 18 to 20 to feet apart, for 136 to 172 trees per acre. Apricot growers have not yet tried the high-density V-hedgerow system used for peaches, because apricot trees cost more than peaches and are prone to fungal and bacterial cankers if trained to only 2 limbs. Apricot research plots trained to a V-hedgerow with 427 trees per acre produced yields 15 to 25% higher than did low-density plantings, and apricots trained to a cordon system with 363 trees per acre produced yields 67 to 100% higher, so growers may adopt such high-density systems in the future.

Most apricot cultivars are self-fruitful or partially self-fruitful, but some require cross-pollination. Growers generally raise several different cultivars, and may alternate single rows of each cultivar to provide greatest cross-pollination. Alternatively, they may plant several rows of a self-fruitful cultivar, then a single row of a second cultivar needing cross-pollination, then several rows of a third self-fruitful cultivar, a single row of the second cultivar, etc. across the block.

Western apricot growers generally plant trees in late winter or early spring, but may plant in fall. Growers in USDA zones 6a or north should plant only in spring. If crown gall infests the soil, dip roots of trees in the biocontrol agent Galltrol™ before planting. Young apricot trees are very sensitive to sunscald, so wrap trees in white plastic tree guards or paint trunks white.

ORCHARD FLOOR MANAGEMENT, IRRIGATION AND FERTILIZERS

Most apricot growers in arid Western areas raise winter cover crops between tree rows, and cultivate about 4 times each summer to reduce groundcover competition for water and nutrients. Most of these growers use furrow irrigation, so they create irrigation furrows during cultivation operations. Water is applied as fruit gain size in May, and again in July and August as fruit buds for next year form, at a yearly application rate of about 18 inches per acre for a mature orchard. Weeds in tree rows are controlled with one application of a pre-emergence herbicide plus glyphosate contact herbicide during the dormant season, and one application of glyphosate contact herbicide during summer.

New Western apricot plantings may utilize trickle irrigation, laid under black plastic mulch down tree rows. This has reduced water requirements during the first 3 years of orchard establishment by up to 90%. Black plastic mulch in tree rows controls most weeds, requiring only limited spot treatment with glyphosate for stray weeds that develop, and increases soil temperatures which increases flowering and fruiting of young trees. In some areas, black plastic mulch may also cause earlier flowering which would increase frost risk, and could cause ripening fruit to soften too quickly or develop sunscald. Opaque white or white-over-black plastic mulch would do better in such areas. Previous attempts at planting trees in clear plastic mulch, laid down tree rows to solarize soil and left in place, injured or killed many young trees due to excessive heat buildup unless the mulch was subsequently painted white, which is very labor-intensive.

Apricot growers in humid areas raise permanent sods between trees to avoid soil erosion, and generally control weeds in tree rows with herbicides, although some may use plastic mulch. Where frosts are a problem, apricot growers rely on overtree sprinklers for frost protection as well as irrigation, or use orchard heaters to protect against frosts and generally no irrigation system since apricots need little water compared to other fruit trees.

Nitrogen (N) is the only nutrient that needs to be added to most apricot orchards. About 6 weeks after planting, workers apply about 12 pounds of actual N per acre by hand around each tree. Most growers apply N 2 to 4 times per season through the irrigation system in subsequent years, adding about 20 pounds total N per acre in the second year, and 40 pounds total N per acre in the third year. Young apricot trees should make 24 to 30 inches of new terminal growth per season, and mature bearing trees should make 16 to 24 inches of terminal growth per season, so adjust N fertilization rates to produce such growth. Do not over-fertilize, as trees 2 to 5 years old often grow excessively and may not mature their wood early enough to avoid winter injury. Do not add N between stone hardening and harvest, as that could cause overly soft fruit and greater incidence of fruit rots. Excess N levels can cause corky spots in apricot fruit.

During the third summer and every third year thereafter, growers take leaf samples to help determine how much N to add the following year. Leaves should have 2.3 to 3.5% N by dry weight. If deficiencies are suspected for any other nutrients, most commonly potassium (K) and sometimes zinc (Zn), the leaf samples are tested for those nutrients also. K may be added through the irrigation system or broadcast on the soil. Zn is most economically applied as a dormant zinc sulfate spray at least 2 weeks before applying delayed-dormant oil. Do not spray zinc sulfate to apricots during the growing season since it can cause phytotoxicity, although zinc chelate sprays may be used on apricots in leaf.

TRAINING AND PRUNING

Most apricot growers train trees to a modified central leader, as described in chapter 4. The first scaffold branch should arise 18 to 24 inches off the ground if trees are to be hand-harvested, and at least 30 inches off the ground if trees will be machine-harvested with a trunk shaker. Spreading or removing any narrow-angled branches is very important, since these would be very susceptible to winter injury and subsequent fungal or bacterial canker or Eutypa dieback, all of which plague apricots.

Traditionally, scaffold branches have been allowed to grow at about a 60-degree angle until canopies touch over the aisles, and trees have been allowed to reach heights of 14 to 18 feet tall. Growers generally train new plantings to wider angles and lower tree heights of 10 to 12 feet, to minimize the amount of work that must be done with ladders.

Both young and mature apricot trees need summer or dormant heading of long, slender branches to induce more side branching, with the bulk of pruning done in late summer. Delay dormant pruning until after bloom, and adjust the amount of pruning to the degree of crop loss from frost and subsequent crop load. Mature trees with heavy crop loads need considerable thinning out of branches to reduce some of the crop, and to keep trees open to light to induce growth of new shoots for future fruiting wood, resulting in annual rather than biennial cropping. During years with light crops, reduce the amount of delayed-dormant thinning-out pruning, and many trees will need no pruning at this time.

Most apricot growers do not use mechanical hedgers for summer pruning, although some larger growers utilize them after harvest to reduce labor costs. Mechanical summer hedging, as described for peaches, should only cut into current season's growth and must be followed by detailed hand pruning to remove stubs and broken branches. Growers who don't use mechanical hedging often find that skilled workers using hand or pneumatic pruners can complete summer pruning in less time than that required by mechanical hedging plus follow-up hand pruning.

Routine prunings are placed in row middles, shredded, and incorporated into the soil, delayed dormant prunings along with the winter cover crop, and summer prunings during the last disking for weed control before planting the next winter cover crop. Dead or diseased branches are placed in piles and burned to avoid spreading diseases.

POLLINATION, CROP LOADS AND FRUIT SIZE

Apricot growers generally rent beehives to ensure good pollination, placing them at a rate of 1 hive per acre in groups of 6 to 12 hives on black plastic at each location, where the morning sun will quickly warm the hives. Disk the winter cover crop before bud break or mow the sod between trees just before bringing in bees, both to eliminate competing blooms and to reduce the frost risk to apricots. Since apricots bloom early when the weather is often too cool for honeybees, growers might consider using mason bees (also called blue orchard bees) instead of honey bees, as described in chapter 10 and for cherries.

Compacting the soil between apricot trees with a light roller just before bloom will reduce the frost risk somewhat. After compacting, apply a light irrigation to wet the top 8 to 12 inches of soil, maintain moisture in this area throughout the frost danger period, and avoid soil cultivation during that period.

Fruit thinning. Where apricots are grown commercially, fruit set is usually adequate enough to require some fruit thinning almost every year. Blossom thinning is too risky for apricots, since frosts may still occur all during and after their bloom period. Growers who mechanically harvest fruit for processing may use tree shakers to mechanically thin the fruit as described for prunes, but generally tree shaking would remove too many fruit for apricots. Some growers use pneumatic limb shakers to mechanically remove some fruit before follow-up hand thinning.

Commercial apricot growers must generally thin most fruit by hand. Workers first use poles with rubber hoses attached to the ends to knock off loosely-held fruit from branches, starting 3 to 4 weeks after full bloom. A week or so later, workers come back and twist off additional fruit by hand, to break up fruit clusters and achieve a spacing of 3 to 4 inches between fruit (Photo 21-4). Start thinning cultivars with the heaviest fruit set first, and selectively remove small and insect-damaged fruit. Thin the bottoms of trees more heavily, since lower fruit does not get as large, and remove fruit from tips of shoots which would swing in the wind. When fruit set is good, apricots bear fruit in clusters like plums, and hand thinning costs may approach $1000 per acre like that of plums.

Limb girdling and propping. Early-season apricots may benefit from limb girdling, as described for peaches, to increase fruit size and advance the harvest date by about 5 days. Make girdling cuts just below fruiting branches on limbs rather than trunks of apricots, and leave at least one major branch per tree ungirdled. Start girdling at about petal fall and finish within 10 days, being careful not to cut too deeply or widely. Do not girdle if a light crop is expected. Do not practice girdling in areas where bacterial canker is serious, since the cuts can serve as entry points for the disease.

About 1/3 of late-season apricots need limb propping to prevent limb breakage as fruit size increases in the final 3 weeks before harvest. Prop limbs with boards as for fresh-market peaches.

PEST MANAGEMENT

Most pests that attack peaches also attack apricots, so pest management is very similar. A few pests that attack apples, pears and grapes also attack apricots, while apricots do avoid some peach pests.

Pests that can kill trees. Apricots suffer from all the "pests that can kill trees" described for peaches, and need the same prevention methods. Borers that attack apples may also harm apricots,

Photo 21-4. Thin apricots to 3 or 4 inches apart, starting 3 to 4 weeks after full bloom. Fruit will grow to fill in the space, and when mature will look like these 'Earlicot' apricots. (photo from Dave Wilson Nursery, Hickman, CA)

such as the **Pacific flatheaded borer** (*Chrysobothris mali*); control it as described for apples. Preparing the site to reduce tree-killing pests, and choosing rootstocks adapted to your climate, soil type and pests will prevent many tree deaths.

Eutypa dieback, caused by the fungus *Eutypa lata*, infects fresh pruning wounds when rain falls within 6 weeks after pruning, as described for grapes. Control it as detailed for grapes, by minimizing large pruning cuts and doing most pruning in late summer, at least 6 weeks before fall rains begin. Coat any pruning wounds larger than 2 inches in diameter with a fungicide, and remove diseased limbs at least one foot below the visible disease margin, sterilizing pruning tools between cuts and burning prunings.

Indirect pests that weaken trees, reducing yields and tree longevity. Apricots suffer from far fewer **virus and phytoplasma diseases** than do other stone fruits. **Prunus necrotic ringspot virus** can infect apricots but causes little yield loss. Apricots infected with the phytoplasma causing **X-disease** showed no leaf symptoms and no apparent reduction in fruit quantity or quality. Most apricot cultivars are susceptible to **plum pox virus**, described in the plum chapter as a potential threat in North America though apparently eradicated from this continent as of now. If it does become established, some apricot cultivars such as 'Goldrich', 'Harcot' and 'Harlayne' are resistant to it. Several other virus diseases hurt apricot yields in Europe, western Asia and New Zealand, but have thankfully been kept out of North America.

The same **nematodes** attack apricots as peaches, including **root knot nematodes** and **root lesion nematodes**. Control these as described for peaches, with soil solarization and/or grass crops before planting apricots, and resistant rootstocks if needed.

San Jose scale and **spider mites** may attack apricots just as they do peaches, and are again controlled with a delayed-dormant oil spray, by pruning out and burning scale-infested branches, and by sprinkler irrigation or summer miticide sprays if mites build to damaging populations.

Diseases that attack flowers, fruits and foliage. Just as it does peaches, **bacterial spot** attacks apricots in humid areas, so apricot growers east of the Rocky Mountains also need cultivars with resistance to this disease. **Brown rot** is again the worst disease of flowers and fruits in all apricot-growing areas, and is controlled with protectant fungicides during bloom and before harvest as described for peaches. Hand removal of mummies from trees, raking up and destroying thinned fruits, and avoiding excess N during the final swell phase of fruit growth again reduce brown rot of apricot fruit.

The pathogens that cause brown rot (*Monilinia fruticola* or *M. laxa*), along with the gray mold pathogen (*Botrytis cinerea*) or *Sclerotinia sclerotiorum*, may also cause **jacket rot** on apricots, which occurs during the jacket stage early in fruit growth, when flower part remnants are still attached to the fruit before shuck split. Jacket rot causes a brown discoloration of the fruit under the jacket, and the young fruit wither and fall off the tree within a few weeks. One fungicide application at full bloom usually prevents jacket rot.

Powdery mildew, caused by the same fungus that attacks peach and nectarine (*Sphaerotheca pannosa*) or a related species (*P. tridactyla*), produces the same white, powdery mycelial growth on young shoots, leaves and fruits, and older lesions on fruits turn scabby. Apricots have little fuzz so have about the same susceptibility to powdery mildew as nectarines, and again this disease is worst during humid seasons in warm, arid climates such as California. Control it as described for peaches, with biocontrol agents when it's the only disease problem, or with sterol inhibitor fungicides which also control brown rot blossom blight. Rusty spot, a peach disease caused by the apple powdery mildew fungus, does not attack apricots.

Shot hole disease, caused by the fungus *Wilsonomyces carpophilus*, may kill apricot buds over the winter and cause leaf and fruit spots in spring. Centers of leaf spots eventually fall out, causing

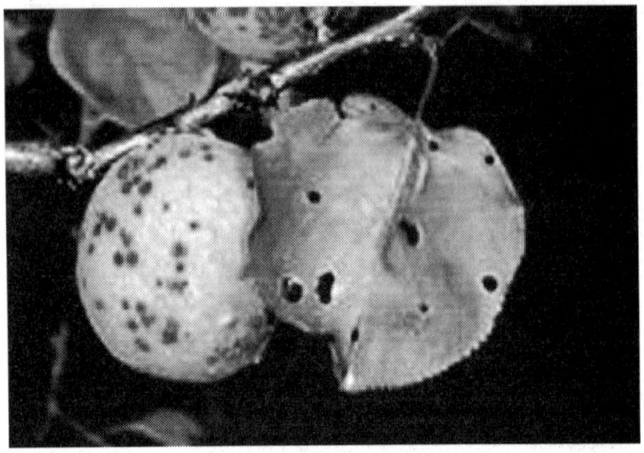

Photo 21-5. Shot hole disease may defoliate apricot leaves by late spring and cause scab-like spots on fruit. Prevent it with a dormant-season fungicide. (photo from Univ. of California.)

"shot holes" (Photo 21-5). This disease may become severe on apricots, causing defoliation by late spring and scab-like spots on fruit, in contrast to its mainly foliar damage on peaches. Prevent shot hole with a dormant-season fungicide applied between mid-November and mid-December before winter rains begin, similar to the fungicide needed to control leaf curl in peaches. Leaf curl does not attack apricots, and peach scab occurs only rarely on apricots and usually doesn't cause economic damage.

Fruit-attacking insects and birds. The same fruit-feeding pests attack apricots as peaches, and are controlled as described for peaches. These include **Oriental fruit moth** wherever apricots are grown in North America, **peach twig borer** in California, **plum curculio** east of the Rocky Mountains, and various **"cat-facing" insects** everywhere. The first two pests may be controlled with mating disruption, while the last two require insecticide sprays. **Oblique-banded leafroller** and **green fruitworm**, described for pears, also attack apricot fruits and again may be controlled with a bloom spray of the biocontrol agent *Bacillus thuringiensis* (Bt). Pest birds include those mentioned for peaches, though apricots are usually not as bothered by birds because they are less juicy.

HARVEST AND PROCESSING

Skin color changes from green to yellow generally determine apricot harvest dates, and the desired shade of yellow-green varies with the cultivar based on how fast each one ripens. Most cultivars soften very quickly once skin turns fully yellow, making them very susceptible to bruising and subsequent decay. Some newer cultivars, however, develop an orange-yellow color well before they reach full maturity, and such cultivars should be tested for soluble solids and flesh firmness to determine the best harvest date. Soluble solids should be greater than 10% and firmness 9 to 10 pounds force.

Mechanical and semi-mechanical harvest and processing. Apricots for processing are harvested mechanically with tree shakers, or semi-mechanically with limb shakers. Fruit must be mature enough to shake off easily but not fully ripe or it would bruise easily.

Most apricot cultivars ripen their fruit over a 10-day to 2-week period, so that some fruit on each tree are ripe while others are still quite immature, making mechanical harvest by trunk shaking more difficult. A few newer cultivars, such as 'Tri Gem', were selected for mechanical harvest, and ripen most fruit at the same time. The earliest-ripening fruits of other cultivars must be hand-harvested before mechanical harvest when 80% of the fruit on the tree are at peak maturity, and less-mature fruit must be sorted out before processing. Using impact rather than inertia tree shakers, and well-padded catching frames, reduces fruit damage to only 3% greater levels than that caused by hand harvesting. Trunk-shaking harvesting equipment is very expensive, requiring about 50 acres of fruit trees that will be mechanically harvested to justify the equipment.

A few smaller apricot growers use pneumatic limb shakers, hooked metal rods, or rubber mallets to shake fruit loose from individual limbs. Such limb shakers are much less expensive, and experienced workers can shake just long enough to remove mature fruit while leaving immature fruit on, so trees can be harvested multiple times over 2 weeks or so, just as in hand harvesting. Two long, sloping, padded platforms are mounted on a trailer to form a V-shaped catching frame that is parked under each limb as it's shaken. Shallow trays are placed on the trailer to catch fruit that rolls down the sloped pads.

Over half the apricots processed in the U.S. are canned in a manner similar to that described for peaches. About 20% are made into juice called "apricot nectar", about 16% are frozen, and some of the frozen or canned apricots are further processed into jams, marmalades and baby foods.

Drying has declined in the U.S. because cheaper, easy-to-transport dried apricots are available from Turkey and other countries. Apricots that are dried are usually laid on paper trays between the rows of the orchard to dry in the sun (Photo 21-6), then are treated with sulfur dioxide gas to inhibit browning. Apricots may also be dried in electric dehydrator tunnels as described for prunes, except that sulfur dioxide gas is added to inhibit browning of apricots.

Hand harvest and handling for fresh market. Fresh-market apricots must be hand-harvested. Many growers who will sell most apricots for processing have workers hand-harvest the earliest fruit for fresh market. Workers generally pick into stonefruit picking bags with shoulder straps, enabling them to use both hands to harvest, and empty these picking bags from the bottom into shallow transport containers, which should be filled with no more than 4 or 5 layers of fruit. When ladders must be used, instruct workers to start picking from the top of the ladder when the bag is empty, moving down the ladder as the bag fills. Also instruct workers to pick all fruit of acceptable maturity from each tree before moving on to the next. Each tree will need to be harvested at least twice and often 3 times, with pickings spaced 5 to 7 days apart.

Transport harvested fruit to the packing shed on a springed trailer at moderate speed to avoid bruising. Handle and cool fresh-market apricots like peaches, except that apricots' minimial fuzz need not be removed. Post-harvest fruit rots of fresh apricots include brown rot and Rhizopus rot. Control these as for peaches, with orchard fungicides and sanitation, rapid cooling after harvest, and chlorine in hydrocoolers.

Transport and store apricots at a temperature of 31 to 32 F, with a relative humidity of 90 to 95%. Fruit stored at temperatures between 36 and 46 F have a very short market life and lose flavor. Most cultivars store well for 1 to 3 weeks at 32 F. Controlled-atmosphere does not appreciably increase storage life so is little used, though it does help retain fruit firmness and ground color. Apricots left out at room temperature will soften very rapidly, at a rate of 3 pounds force per day at 68 F, so must be sold that day. Apricots with firmness readings of 2 to 3 pounds force are ready to eat.

Photo 21-6. Sun-dried apricots are laid on paper trays between the rows of the orchard. (Photo from Stemilt Growers, Stockton, CA.)

Further Reading

Anderson, K.K., K.R. Day, H.L. Andris, W.B. Coates, W.C. Ferriera, S.M. Southwick, K.M. Klonsky, and R.L. DeMoura. 2003. Sample costs to establish an apricot orchard and produce apricots. Univ. of Calif. Coop. Exten. Pub. AC-SJ-03-1 fresh market, AC-SJ-01-2 for processing. Available online at: http://www.agecon.ucdavis.edu/outreach/crop/cost-studies/ApricotsFreshSJV.pdf and /ApricotsProcSJV.pdf

Lamb, R.C. and W.C. Stiles. 1983. *Apricots for New York state*. New York Food and Life Sci. Bull. No. 100. Available online at: http://www.cce.cornell.edu/suffolk/grownet/TREFRUIT/apricot.htm

Stapleton, J.J., E.J. Paplomatas, R.J. Wakeman, and J.E. DeVay. 1993. Establishment of apricot and almond trees using soil mulching with transparent (solarization) and black polyethylene film: Effect on Verticillium wilt and tree health. *Plant Pathology* 42:333-338.

University of California. 2002. UC pest management guidelines: Pests of apricot. Available online at: http://www.ipm.ucdavis.edu/PMG/selectnewpest.apricots.html

Chapter 22
Almonds and Pistachios

Almonds, another of the stone fruits, produce the largest nut crop in North America. Closely related to peaches, almonds belong to the species *Prunus dulcis* in the Prunoideae subfamily of the Rosaceae family. Almonds originated in what is now Iran, spread through the Middle East and around the Mediterranean Sea, and were brought to California by the Spanish in the 1700's. Commercial planting of almonds began in California in the mid-1800's, increasing until the state now supplies all U.S. production and over half of world almond production. Countries around the Mediterranean Sea and in the Middle East together produce almost half of the world's almonds, with Spain producing about 16%, Syria growing about 8%, Italy raising about 6%, Iran producing about 5% and Morocco growing about 4%.

Pistachios (*Pistacia vera*) belong to a different family (the Anacardiaceae or cashew family) but also originated in Iran, need the same Mediterranean climate as almonds, and in North America are grown almost solely in California using similar management techniques. Pistachios are a much more recent crop in the U.S. than almonds, as negligible amounts of psitachios were grown just 30 years ago. Today the U.S. is the second-largest pistachio producer after Iran, which produces 48% of the world crop, while the U.S. grows 31% of the crop and Turkey produces 9%.

Planting of both almonds and pistachios increased greatly in the U.S. over the last 20 years, due to increased demand for the crops and the fact that harvesting and many other operations can be mechanized, in contrast to most fresh-market fruits such as peaches. More than 70% of the U.S. almond crop is exported, with over half of that going to Europe and a quarter going to Asia, making almonds California's largest export crop in dollar value. Between 22 and 30% of the U.S. pistachio crop is exported, primarily to Europe, with more exported when crops are heavy and less exported in years with a light crop.

About 80% of U.S. almonds and 98% of U.S. pistachios grow in California's San Joaquin Valley, and the other 20% of almonds grow in the Sacramento Valley of California. Southern Arizona grows about 1500 acres of pistachios, about 2% of the U.S. crop. Over 600,000 acres of almonds have been planted, the most land devoted to any tree crop in California. Currently, 530,000 bearing acres produce over 1.5 billion pounds of in-shell nuts per year. About 88,000 acres of pistachios produce about 250 million pounds of in-shell nuts per year, in a biennial production cycle with over 300 million pounds in heavy years and about 180 million pounds in light years. Another 23,000 acres of young pistachio orchards have not yet begun bearing.

TREE GROWTH AND LIMITING FACTORS

Almonds are small trees, growing slightly larger than peach trees to 25 or 30 feet tall (Photo 22-1), but can live much longer, and remain productive for

Photo 22-1. Almond trees grow similarly to peaches and also begin bearing 3 to 4 years after planting, but can live much longer, and remain productive for up to 50 years. They bloom much earlier so are only adapted where late winter or spring frosts don't occur. (Photo by Bill Kenney, BK International, LLC, Lodi, CA.)

up to 50 years. They begin bearing about a year later than peaches, 3 to 4 years after planting. They have long, linear leaves similar to those of peaches, but somewhat shorter and thicker. Perfect flowers are nearly identical to peach flowers in structure, but are borne mainly on short lateral spurs from 2-year-old or older wood, and sometimes laterally on long one-year-old branches as on peach trees. Like peaches, almond flowers bear heavy, sticky pollen that is transferred by insects. Unlike peaches, all commercial almond cultivars are self-sterile, so growers must plant 2 or 3 cultivars together to ensure cross-pollination.

Pistachios are also small trees, about the same size as almonds, but grow slower and take longer to reach bearing age, usually 7 years (Photo 22-2). They live about as long as almonds and also remain productive for up to 50 years. They have pinnately compound leaves similar to those of walnuts and pecans. Pistachio trees are dioecious, with each tree bearing only female flowers or only male flowers, so growers plant orchards mainly with a nut-producing female cultivar, interspersing a few male trees to provide pollen. Female flowers occur in clusters on new lateral shoots, and male flowers are borne on pendulous catkins and produce light, dry pollen disseminated by wind, like those of other tree nuts.

Photo 22-2. Pistachios also grow as small trees that remain productive for up to 50 years. They grow slowly and take longer than almonds to reach bearing age, about 7 years after planting.

Photo 22-3. Young pistachio fruit occur in grape-like clusters. Each fruit matures into a shuck that dries and splits open, exposing the nut inside.

Young almond fruits look very much like young peaches, but the fleshy fruit (mesocarp) does not undergo a final swell period like peaches do. Instead, it begins to dry up into a hull or shuck as nutrients flow to the seed (kernel) inside the pit (endocarp or shell), then the shuck cracks open before harvest to reveal the nut inside. Nuts can be shaken free from the tree when mature with mechanical tree shakers, then harvested off the ground with mechanical harvesters. Harvested nuts are transported to hulling plants to remove the hulls, then 92% are cracked and sold as whole almond kernels or pieces. About 8% of almonds are sold uncracked, in-shell.

Young pistachio fruit occur in grape-like clusters (Photo 22-3). Like almonds and other nuts, the fruit matures into a shuck that dries and splits open, revealing the nut consisting of kernel (seed) and shell (endocarp). Unlike other nuts, a high percentage of pistachio shells also split open by harvest time. This allows consumers to easily remove kernels from unshelled nuts, but precludes harvesting pistachios from the ground so kernels won't get dirty. U.S. growers harvest pistachios using mechanical tree shakers and catching frames to avoid having nuts fall on the ground. The split shells also leave kernels vulnerable to mold fungi if humid conditions occur. Pistachios are grown in dry climates to prevent this, and must be kept dry after harvest. Hulls are mechanically removed and in-shell nuts are washed and dried within 24 hours after harvest, leaving unblemished, ivory-colored shells. Iranian and Turkish pistachio exporters dye shells red to cover up surface blemishes that develop due to slower processing.

Almonds have a very low chilling requirement of 300 to 500 hours and a very low heat requirement after chilling, so they bloom extremely early, even earlier than apricots. They bloom in late February in California, and require production areas with little chance of frost at that time. Pistachios have a higher chill requirement of 850 to 1000 hours, so are planted in the northern part of the San Joaquin Valley and other areas that usually provide enough chilling. Growers must sometimes bypass that chill requirement with an oil spray during seasons when natural chilling isn't enough in the areas where pistachios are grown.

Both almonds and pistachios need a Mediterranean climate of cool, moist winters and hot, dry summers to thrive and avoid many diseases which attack them. Almonds are cold-hardy to about 0 F (USDA zone 7) and pistachios tolerate temperatures down to about 6 F (USDA zone 7b). Almonds require large amounts of irrigation water, typically 48 to 52 acre-inches per year for mature trees, for good annual cropping. Pistachios are more drought-tolerant and need 40 to 44 acre-inches per year for mature trees, but even well-irrigated trees still tend to alternate bear.

Heavy planting of almonds in recent years has led to California marketing orders that restrict the amount of crop that each grower can sell during certain high-production years, which can make almond operations unprofitable during those years. Pistachio acreage has not yet increased enough to saturate demand, and pistachios fetch much higher prices per pound than other nuts. Pistachio growers typically receive $1.00 to $1.40 per pound for in-shell nuts (equivalent to $2.00 to $2.80 per pound wholesale for shelled kernels), while almond and walnut growers get about 45% of those prices, receiving $0.90 to $1.30 per pound for nutmeats wholesale. Almonds and pistachios produce about the same amount of nuts per acre, but production swings more widely from year to year with pistachios than with almonds. Pistachios produce 1700 to 3000 pounds per acre of in-shell nuts, while almonds produce 1400 to 2100 pounds of kernels per acre, which is about 2100 to 3000 pounds per acre of nuts in shells.

SITE SELECTION AND PREPARATION

Choose an area and site with minimal chance of spring frosts, especially for almonds. Deep, well-drained, uniform loamy soils induce greatest production from almonds, while pistachios produce best on deep, alkaline sandy loams and tolerate more salinity than most fruit and nut crops. Both almond and pistachio growers generally subsoil 5 to 6 feet deep a year before planting to break up hardpans, then disk twice and smooth the soil surface with a floater.

Adjust the soil pH, if needed, to between 6.2 and 7.5 for almonds, and between 6.8 and 8.5 for pistachios. Test soil for phosphorous (P) and potassium (K) and add these before planting if needed. Avoid planting after crops susceptible to Verticillium wilt, such as cotton or melons, since the pathogen may have built up in such soils. Also avoid land cleared from woodlands within the past 10 years, to avoid oak root rot.

Almonds are often planted on old orchard ground, after pushing over old trees in fall and chipping trunks and branches for burning at cogeneration plants or sale as mulch. Remove stumps and as many roots as possible, then subsoil and disk. Ideally the land should be planted to grass crops for several years to reduce nematodes and fungal pathogens, but growers often fumigate the soil and replant early the next spring.

Both almond and pistachio growers generally make raised beds for planting, since both species are very sensitive to wet soil and Phytophthora root rot. Just before planting trees, growers apply a pre-emergence herbicide to these planting beds.

SELECTING ROOTSTOCKS

The two main rootstocks used for almonds are 'Nemaguard' and 'Lovell' peach rootstocks, detailed in Table 18-1 in the peach chapter. 'Nemaguard' is used when root knot nematodes are a problem but lesion and ring nematodes are not, while 'Lovell' is used in the opposite situation since it is less susceptible to lesion and ring nematodes but is susceptible to root knot nematodes. 'Lovell' is also less susceptible to peach tree borer and bacterial canker, but trees on 'Nemaguard' may bloom 1 to 3 days later than those on 'Lovell'.

Several peach-almond hybrids, also detailed in Table 18-1 in the peach chapter, are sometimes used as rootstocks for almonds, especially on alkaline soils. 'Hansen 536' has low chill needs so induces even earlier bloom in almonds, while 'Nickels' has higher chill needs and delays bloom, making it the better choice if spring frosts may occur. Both these peach-almond hybrids strongly resist root knot nematodes but are susceptible to lesion and ring nematodes and bacterial canker, like 'Nemaguard'. Two other peach-almond hybrids, 'Viking' and 'Atlas', have only moderate resistance to root knot nematodes but also have moderate resistance to bacterial canker. Almond seedlings could also be used as rootstocks for almonds, but are not usually available from nurseries.

A semi-dwarfing rootstock occasionally used for almonds is 'Mariana 2624' plum, detailed in Table 19-1 in the plum chapter. Some almond cultivars exhibit delayed incompatibility with this rootstock, but cultivars such as 'Mission', 'Padre' and 'Ruby' show excellent compatibility and can be planted much closer on 'Mariana 2624', resulting in greater yields per acre. This rootstock may sucker, but deeper planting can reduce suckering.

Pistachio cultivars are budded onto seedlings of several *Pistacia* species, including *P. integerrima*, *P. atlantica* and *P. terebinthis*. All these species resist nematodes better than *P. vera* does. *P. integerrima* is usually preferred because it resists Verticillium wilt, to which the others are susceptible. Two new rootstocks developed by University of California, 'Pioneer Gold II' and 'UCB-1', are hybrids of *P. integerrima* and *P. lentiscus* and are very vigorous, cold tolerant, and resistant to Verticillium wilt.

CHOOSING CULTIVARS

U.S. production of both almonds and pistachios is dominated by a single cultivar in each industry. Although about 50 almonds cultivars are grown in the U.S., 'Nonpareil' accounts for 60% of U.S. almond acreage, due to its thin shell and high crackout rate of up to 70% kernels to 30% shells. Many almond plantings include 'Nonpareil' as the main cultivar, plus 2 or sometimes 3 other pollinizer cultivars with bloom times at the beginning to middle and middle to end of the bloom period of 'Nonpareil'. However, some growers now choose 'Butte' as the main cultivar for at least part of their plantings, since it has quality similar to 'Nonpareil' but blooms later and has a better chance of escaping frost damage. 'Butte' should also be planted with 2 or sometimes 3 other pollinizer cultivars with bloom times at the beginning to middle and middle to end of its bloom period. Choosing several cultivars that ripen at different times (although bloom times overlap) can spread out the harvest season, reducing the number of laborers needed.

Table 22-1 details traits of the most commonly-planted almond cultivars. Thin shells are desirable when nuts will be cracked, as most are, but thick hard shells are better for nuts sold in-shell. Nuts with a well-sealed shell suffer from fewer insect and fungus problems. Medium-large kernels are generally more desirable, but cultivars with small to medium-sized kernels are in demand for processing and may have higher overall yields. Smaller trees can be planted at higher densities. Cultivars that are less susceptible to anthracnose need fewer fungicides, and fortunately 'Nonpareil' has moderate resistance to this disease. Some cultivars, such as 'Carmel' and some strains of 'Nonpareil', are very susceptible to "crazy top", a non-infectious disorder that causes bud failure in trees, often not showing up until the second or third year in the orchard. Avoid this disorder by choosing less susceptible cultivars or resistant strains of 'Nonpareil'.

Pistachio production in the U.S. is even more concentrated, as the female cultivar 'Kerman' accounts for 95% of production. 'Kerman' was grown in California from seed bought in a marketplace in Iran, and was introduced in the 1950's. It produces high yields of large nuts, but suffers more severe alternate bearing than many other pistachios, produces a rather high amount of blank nuts with aborted kernels, and about 20% of its nuts do not split their shells. Other female cultivars grown on limited acreage include 'Red Aleppo', an older cultivar, and 'West Hills', the first new female cultivar from the University of California pistachio breeding program. 'West Hills' offers higher yields and fewer stained nuts and loose kernels than 'Kerman', and matures with 'Kerman' although it blooms a week earlier. 'Peters' is the main male pollinizer cultivar planted. A new male cultivar from the University of California pistachio breeding program, 'Randy', begins blooming a year before 'Peters', so may be planted as a pollinizer instead of 'Peters' in the future.

ORCHARD LAYOUT AND PLANTING

Most almond orchards have been planted at low densities of 71 to 91 trees per acre, spaced 24 to 28 feet between rows and 20 to 24 feet apart within rows. Newer orchards using 'Lovell' peach rootstock may be planted closer, often 15 to 16 feet apart in rows spaced 20 feet apart, for a density of 136 to 145 trees per acre. Some almonds on 'Mariana 2624' rootstock have been planted as close as 10 feet apart within rows spaced 18 to 20 feet apart, for a density of 217 to 242 trees per acre.

Almond growers generally plant single rows to each cultivar, alternating 2, 3, or 4 cultivars across the orchard to provide for cross-pollination. However studies have shown that alternating 2 cultivars every other tree down each row produces about 9% greater yields than planting solid rows of each cultivar. Closer tree spacing creates hedgerows that bees tend to fly down rather than across, so higher-density plantings especially benefit from planting 2 or more cultivars in each row. Cultivars to be combined in a row either should mature at the same time and be ones processors are willing to buy mixed, or should mature at different times and be harvested in 2 separate operations.

Pistachios are commonly planted 16 to 18 feet apart within rows spaced 18 to 20 feet apart, for a density of 121 to 151 trees per acre. Growers plant 1 male tree for every 24 female trees, so male trees are planted (or grafted) as every fifth tree in every fifth row. Since pistachios are slow-growing trees, some growers plant temporary trees to be removed later, to increase yields during the early years. Under this system, growers space trees 11 to 12 feet apart within rows spaced 22 to 24 feet apart, then after 15 or 16 years, every other tree within each row is removed, leaving trees spaced 22 to 24 feet apart in both directions.

Both almond and pistachio growers commonly hire custom planting crews to plant trees in January and February. Almonds are ordered from a nursery as bare-root grafted or budded trees, like those of other tree fruits and nuts. Dipping roots in the biocontrol agent Galltrol™, just before planting, can prevent problems with crown gall. After planting, workers place white plastic tree guards around trees to protect them from sunscald and rodents.

Pistachios do not transplant well bare-root, so are commonly planted as container-grown seedling rootstocks ordered from a nursery, then field-budded to cultivars the following spring or summer by a custom grafter. When temperatures are high during budding, grafters should place the bud on the north side of each rootstock. Workers cut a shallow notch in the rootstock about 1 inch above the bud, 5 to 7 days after budding, to force bud growth, and may head back the rootstock by about 50% to further induce bud growth. A 5-foot-tall stake is placed by each tree and the developing shoot from the bud is tied to it every 10 days from mid- to late summer during the first 2 years. The rootstock will be cut off just above the forced bud during the winter after budding.

ORCHARD FLOOR MANAGEMENT, IRRIGATION AND FERTILIZERS

Almond and pistachio growers generally disk between young trees 4 to 5 times per growing season, to save water and nutrients for the trees and help them become established faster (Photo 22-4). Tree rows are kept weed-free by applying a preemergence herbicide each winter to prevent most summer annual weeds from growing. Workers spot-treat weeds that do develop with a contact herbicide three times during spring and summer.

Once trees begin to bear, growers disk and roll the soil between the trees before bloom each spring, so the soil is weed-free and compacted to retain the most heat and help prevent frost damage. Pistachio growers often disk 2 or 3 more times in summer, which minimizes weed competition for water and nutrients. Most almond growers allow resident vegetation to grow and mow it 4 to 6 times from late spring until mid-summer. About 3 to 4 weeks before harvest, some almond growers spray the entire orchard floor with glyphosate to eliminate vegetation that would clog mechanical harvesters, and compact the ground to facilitate mechanical harvest. After harvest, both almond and pistachio growers often allow resident vegetation to grow over winter, or they may sow a winter annual legume as a cover crop to provide nitrogen (N) when it is disked under in late winter.

Irrigation methods and water quality. Most almond and pistachio growers use micro-sprinklers or other forms of trickle irrigation to reduce water needs (Photo 22-5). Some almond growers use under-tree sprinklers so they can provide frost protection if temperatures drop below 28 F during bloom or below 32 F once the jackets are

Photo 22-4. Many almond and pistachio growers control all orchard floor vegetation between young trees in order to minimize competition for water and nutrients. This orchard uses porous tubing, a type of drip irrigation.

Photo 22-5. Most almond and pistachio growers use drip irrigation, as here, or micro-sprinklers because these systems offer the greatest water-use efficiency. Herbicides control weeds in the tree rows, and resident vegetation is mowed from late spring to mid-summer.

Table 22-1.

COMMON COMMERCIAL ALMOND CULTIVARS FOR ARID WESTERN NORTH AMERICA

In approximate order of ripening (varies with area).

	Nut				Tree				
Cultivar	Kernel size	Shell	Seal	Yield	Size	Growth	Anthrac-nose suscep.	Bloom time	Days after Nonpareil
Nonpareil	med-large	thin	poor	medium	large	spreads	MR	middle	0
Price	small-med	thin	poor	high	medium	upright	VS	middle	5
Livingston	medium	thin, soft	good	very high	med-large	upright	MS	late	7
Sonora	med-large	thin	poor	very high	medium	spreads	S	very early	7
Jenette	medium	thin, soft	fair	high	small	upright	S	middle	10
Aldrich	medium	semi-hard	good	high	large	upright	MS	middle	14
Durango	med-large	semi-hard	very good	high	large	upright	S	early	14
Ne Plus Ultra	large	soft	good	low-med	medium	spreads	S	very early	14
Butte	small-med	semi-hard	fair	very high	large	spreads	VS	late	15
Carmel	medium	soft	good	high	medium	upright	S	middle	15
Wood Colony	medium	semi-soft	good	high	medium	spreads	MS	middle	16
Peerless	medium	very hard	very good	medium	medium	upright	S	early	17
Padre	small-med	hard	good	high	large	upright	S	late	24
Winters	med-large	thin	good	very high	large	upright	VS	early	25
Monterey	large	soft	good	very high	medium	spreads	VS	middle	26
Alamo	small-med	hard	very good	very high	large	upright	S	mid-late	26
Mission	small-med	hard	very good	high	large	upright	S	late	28
Ruby	small-med	semi-hard	good	high	large	upright	S	late	30
Fritz	small-med	soft	poor	high	medium	upright	S	middle	40

Anthracnose susceptibility ratings: MR = moderately resistant, MS = moderately susceptible, S = susceptible, VS = very susceptible.

Harvest times listed in average days after 'Nonpareil', which is harvested in late August in Sacramento, CA.

Notes on Cultivars:
Nonpareil, the industry standard, has 70% crackout. Incompatible with Marianna 2624 rootstock. Some strair
Price may alternate bear so needs more thinning to prevent this. S to crazy top.
Livingston blooms late so may avoid frosts, good pollenizer for 'Butte', not for 'Nonpareil'.
Sonora may alternate bear due to very early bloom and very high yield potential, so needs annual pruning, fe
Jenette has good yields from a compact tree that can be planted closer.
Aldrich nuts have 62% crackout. Excellent pollenizer for 'Nonpareil'. Needs tip pruning for first 3 years.
Durango has early but prolonged bloom, attracts bees, excellent pollenizer for 'Nonpareil', shell resists insect
Ne Plus Ultra is prone to frost damage due to very early bloom, VS to brown rot blossom blight.
Butte has had highest yields in regional cultivar trials. Late bloom reduces frost risk, but may not pollenize 'N
Carmel bears young, most popular pollenizer for 'Nonpareil', but VS to crazy top, brown rot blossom blight, sc
Wood Colony has a prolonged bloom period, good pollenizer for 'Nonpareil', consistent production.
Peerless blooms early so risks frost, best cultivar for in-shell market due to light-colored, hard shell. S to craz
Padre consistently yields heavily. Late bloom reduces frost risk, may not pollenize 'Nonpareil', good for 'Butte'
Winters blooms early so risks frosts. Thin shell gives 65% crackout, but higher worm and ant damage.
Monterey bears young, good pollenizer for 'Nonpareil', but has high tendency to form double nuts.
Alamo has mid-late bloom, good pollenizer for both 'Nonpareil' and 'Butte', forms a few double nuts.
Mission, also called **Texas**, blooms late so may miss frosts, can't pollinize 'Nonpareil', good for 'Butte'. S to c
Ruby blooms latest, 7 days after Nonpareil, so may miss frosts, can't pollinize 'Nonpareil', good for 'Butte'.
Fritz is good pollenizer for 'Nonpareil', extends harvest to 6 weeks later than 'Nonpareil'.

off young nuts. Some older almond and pistachio orchards use furrow irrigation, but many of these orchards have been converted to trickle irrigation. In addition to reducing water use while reducing water stress, micro-sprinklers and other forms of trickle irrigation enable almonds to begin bearing a year earlier, in the third year, and do not wet the foliage so reduce diseases such as brown rot, shot hole and scab. Place trickle irrigation emitters several feet away from tree trunks to provide water where need most and to avoid excess water around trunks, which can aggravate Verticillium wilt or Phytophthora root rot.

More and more almond growers are trying regulated deficit irrigation, because during drought years, water districts limit growers to between 30 and 75% of their normal water allocations, depending on drought severity. When water is limited, Kern County Extension irrigation specialist Blake Sanden recommends irrigating at 50% deficit levels until about July 1, or about 3 weeks before the normal irrigation cut-off before harvest. Apply 4 to 6 inches of water during the next 3 weeks to fill the soil water profile, then stop irrigating 2 to 3 weeks before harvest so soil is firm enough for harvest equipment and less hull rot occurs. Orchards irrigated by micro-sprinklers need to have irrigation cut off about 3 weeks before harvest, while drip-irrigated orchards don't need to have water cut off until 2 weeks before harvest. After harvest, apply 6 to 9 inches of water as fruit buds are forming for next year's crop. This approach may cause a grower to lose 10 to 15% of the current crop in kernel shrinkage, but ensures next year's crop. The worst thing a grower can do is to use up the entire water allocation before harvest, and lose next year's crop due to bud failure.

Pistachios are very salt-tolerant and can utilize irrigation water with total soluble salts of up to 2500 ppm, which is 2.5 times the tolerance level of almonds, peaches and most other tree crops. Irrigation water with about 10 ppm boron (B) salts is beneficial, since pistachios have a high B requirement, while most tree crops can not tolerate more than 1 ppm B in irrigation water. Using saline irrigation water for pistachios reduces the crop's overall need for water, allowing regulated deficit irrigation at 45 to 75% the normally recommended amounts of 40 to 44 inches per year. Pistachios irrigated with water containing 2500 ppm total soluble salts used only 18 inches of water, so the grower would only need to apply 20 inches per year to supply this when accounting for irrigation system losses.

Pistachios need about 4 inches of water from early spring until mid-May, when nuts attain full shell size. Growers should avoid stressing pistachios with too little water in spring, and also avoid excess water which can aggravate Verticillium wilt. From mid-May until early July when rapid kernel growth begins, 50% deficit irrigation can be practiced with no reduction in yields. During kernel expansion from early July until mid-August, irrigation needs increase and only a mild, 25% deficit irrigation schedule should be tried. During final kernel filling and shell splitting from mid-August to harvest in mid-September, pistachios need full irrigation. Unlike almonds, pistachios tolerate severe post-harvest water stress.

Fertilizer additions. Nitrogen (N) is the main fertilizer nutrient that almond and pistachio growers add. Almond growers typically apply 20 pounds actual N per acre the first year, 40 pounds the second, 100 pounds the third, 140 pounds the fourth, 200 pounds the fifth, and 240 pounds actual N per acre from year 6 on. Pistachio growers typically apply no N the first year, 33 pounds actual N per acre the second, 66 pounds the third, 100 pounds the fourth, 135 pounds the fifth and sixth years, and 170 pounds actual N per acre from year 7 on. However, growers should take leaf samples in mid-summer of the second year, fourth year and at least every third year thereafter, and test for N levels to guide them in fertilizing. The ideal range for leaf N concentration, both to maximize yields and prevent N leaching loss, is 2.2 to 2.5% N by dry weight for both almonds and pistachios. Researchers have found that many growers apply too much N, and some mature orchards produce well with only 100 pounds of N added per year.

Growers should also test irrigation water for nitrate content, as researchers have found that some irrigation water contains enough nitrates that almond trees which received no N fertilizer for 3 to 4 years still had adequate leaf N levels for top yields. Many growers apply N through irrigation lines throughout the growing season, resulting in more efficient uptake than soil application just twice a year, and can reduce the total amount of N applied per year. Growers who apply N to the soil add half in early spring and half in early summer.

Almonds use more potassium (K) than N, and if K drops below 1.0 to 1.2% by dry weight in leaves, it hurts yields the following year, so test leaf samples for K also when testing for N. Many growers apply 400 to 500 pounds of potassium sulfate (165 to 200 pounds actual K) per acre per year to keep soil K levels up enough to prevent deficiency, but during years of low almond prices such K additions may not be cost-effective. K can be added through the irrigation system for greater tree uptake, using potassium nitrate or potassium sulfate rather than the cheaper potassium chloride if salinity or chlorine toxicity is a problem.

Pistachio K uptake is also very high during kernel filling, and mid-summer leaf levels should be about 1.7% for optimum yields. Some soil types provide plenty of K and don't need any K fertilizer, but adding up to 200 pounds actual K per acre has significantly increased yields and reduced blank and stained nuts on soils high in kaolinite clays which do not fix K. Adding K may reduce incidence of panicle and shoot blight, and foliar sprays of both N and K may be needed during heavy crops of pistachios' "on" year, to get these nutrients into trees. Both almonds and pistachios rarely need phosphorous (P) fertilizer.

Almonds and pistachios usually need zinc (Zn) additions on alkaline soils, so test leaf samples for Zn also. Leaf levels of less than 15 ppm Zn for almonds or 7 ppm Zn for pistachios require Zn fertilizer additions. If needed, add zinc sulfate to pesticide sprays the next spring and summer, or spray zinc sulfate in late November to defoliate trees if old leaves still hang on which would spread diseases the next season.

Pistachios use more boron (B) than any other tree crop, and thrive on saline soils that are high in B. Mid-summer pistachio leaf samples should have at least 80 ppm B. Soils low in B need annual applications of 20 to 40 pounds B per acre, often in the form of borax. Almond leaf samples should have 30 to 65 ppm B, but soils high in B can cause B toxicity in almonds, which causes terminal shoot to lose all leaves up to 12 inches from the tips. Almonds on soils low in B often receive a spray of soluble B in October, which defoliates trees and helps them harden for winter in addition to providing B. Some almond growers add B to the late winter pre-emergent herbicide spray down the tree row.

TRAINING, PRUNING AND MANAGING CROPS

Almond growers train trees to the open center system (outlined in chapter 4) like lower-density peaches, choosing 3 main scaffold branches per tree during the first growing season and removing lower branches and root suckers a couple of times per season. Pistachio growers train trees to the modified leader system (also outlined in chapter 4), choosing 4 or 5 main scaffold branches per tree during the first two growing seasons and also removing lower branches and root suckers a couple of times per season. The lowest scaffold branch should arise at least 2.5 feet off the ground for almonds and 3 to 3.5 feet off the ground for pistachios, to leave room for the trunk shaker when harvesting.

Almond trunks eventually grow too wide for trunk shaking on most rootstocks, so growers must switch to limb shaking after 15 or 16 years. Plan ahead for that by carefully selecting scaffold branches, choosing ones spaced at least 3 to 6 inches apart on the trunk, to allow the shaker access to each limb. Almond growers may practice long, short or intermediate pruning of scaffolds. Long pruning makes no major cuts on scaffolds, allowing the tree to develop quickly and bear nuts earlier, but trees usually need rope stretched around the branches' perimeter to support fruit, and limbs may break even when roped. Short pruning involves heading each of the 3 scaffolds back to 18 to 24 inches to stimulate growth of secondary branches, then thinning out these branches, and delays bearing but keeps trees shorter with no need to support the branches with rope. Intermediate pruning, the most common method, cuts scaffolds back less severely, at 42 to 48 inches, reducing the number of secondary branches stimulated and the need to thin these branches. Intermediate-pruned trees do not require roping unless trees are especially vigorous.

Select pistachio scaffold branches that are spaced 6 to 8 inches apart on the trunk. Head these scaffolds back to about 30 inches during the dormant season, and during the next dormant season, head secondary branches that arise from the scaffolds back to about 30 inches. Continue heading back new growth to about 30 inches each dormant season for several years until the tree framework is established. It's very important to establish a good bearing framework during the first 5 years, while heading cuts still stimulate side branching, because once trees get older branches have very few lateral vegetative buds, and heading cuts simply cause branches to die back without stimulating lateral branching. Once pistachio trees start to bear, they need little pruning for several years.

Almond fruiting spurs live about 5 years, so on mature trees from age 6 on, remove older, minimally productive wood to stimulate growth of new fruiting wood. Pruning every other year reduces costs over annual pruning but does not reduce yields, so many growers prune half their almond trees each year. Mature almond tree pruning consists mostly of thinning cuts.

When almond trees are 9 or 10 years old, permanently mark the secondary limbs to be removed when switching from trunk shaking to scaffold shaking, including branches at the bases of the scaffolds that would interfere with shaking equipment. Then instruct pruning crews to remove branches mainly from the marked limbs over the next few years, so when branches cross or compete, branches arising from temporary limbs are removed and branches from permanent limbs remain. When trees reach 14 to 16 years of age, removed the marked limbs completely.

Mechanical tree topping and hedging. Some almond growers use dormant mechanical topping in January and February once trees mature, to keep trees open to light and more productive. Workers follow this up with hand pruning to remove stubs and broken or diseased branches. However, mechanical pruning reduces yields in the year following pruning, and results in excessive regrowth that requires increased hand thinning the following year.

Dormant mechanical topping does benefit older pistachio trees, which do not respond to conventional pruning practices used for other trees, and topping also reduces alternate bearing. Bearing wood on pistachios spreads further out each year, until eventually branches bend down and are sunburned. Thinning-out pruning does not stimulate new growth on remaining branches, but mechanically cutting tops of trees, before branches get so long they begin to bend down, does stimulate needed new growth.

Topping during the winter before the "off year" of the biennial bearing cycle stimulates new shoot growth, then topping the following winter heads back these new branches and stimulates them to develop lateral fruiting buds, which will produce a crop during the next "off year". Two successive winters of topping mitigated alternate bearing through the next 3 cycles (6 years) without any additional topping in years 3 through 8, and increased cumulative yields. Topping would be needed again in years 9 and 10.

Hedging older pistachio trees on the sides also stimulates growth of new fruiting wood closer to the trunk, but can reduce current-year yields too much if both sides are done at once. Hedging only one side of each row over 2 years, and the other side over the next 2 years, can maximize future benefits while retaining current yields. Practice hedging during years between topping operations, for example years 3 though 6 or 5 through 8 after 2 years of topping.

Pollination, frost protection and fruit set. Almond growers rent beehives from commercial beekeepers, placing them in the orchard in early February, before bloom begins. They place 1 hive per 2 acres in the third year when the orchard begins blooming, 2 hives per acre in the fourth year, and 3 hives per acre from the fifth year on. The almond industry needs about 1.2 million beehives to pollinate the crop each spring, and about half these colonies come from outside California. The beekeeper should remove hives from the orchard when 90% of blooms on the latest-blooming cultivar have shed their pollen, as the grower will soon need to apply insecticides to control pests.

Pistachios are wind-pollinated like other nut crops so no bees need be rented, and bloom late enough that they usually need no frost protection. During some mild winters pistachios may not receive enough winter chilling in southern parts of California's San Joaquin Valley. After such mild winters, an oil spray applied in late January or early February for scale and mite control will also stimulate earlier and more uniform flowering of pistachios.

Almond growers may use orchard heaters, sprinkler irrigation or wind machines for frost protection, or may hire helicopters during years that bad frosts are predicted. Combining wind machines or helicopters with orchard heaters or irrigating the orchard floor (especially with under-tree sprinklers, but even with micro-sprinklers) can provide greater frost protection than one method alone.

Researchers have increased fruit set in almonds by 45% with sprays of 72 ppm of the gibberellin-cytokinen mixture Promalin™ at the early fruit set stage, and have increased fruit set by 77% when 200 ppm n-m-tolilphtalamic acid was added to the Promalin™. Neither growth regulator is currently registered for use on almonds, but may be in the future.

Other researchers have reduced alternate bearing in pistachios by applying a combination of cytokinen and urea nitrogen fertilizer during the "on" year of the cycle. This increases flower bud retention during the subsequent "off" year. Currently, the only cytokinens that may be applied to pistachios are seaweed extracts, which have low cytokinen concentrations, but several companies are pursuing registration of higher-concentration cytokinens.

PEST MANAGEMENT

Pest management for both almonds and pistachios is similar to that for peaches, with a few less total pests but a couple additional serious pests, notably navel orangeworm and several foliar and fruit diseases. An almond integrated pest management program begun in 1998 has tested alternative pest control methods, and has significantly reduced pesticide use in almonds. Pistachios received few pesticide sprays in the early years of the industry, but more pests have attacked as plantings increase.

Pests that can kill trees. Almonds and pistachios are susceptible to all the "pests that can kill trees" described for peaches, and need the same prevention methods. These include Phytophthora root rot, oak root rot, Verticillium wilt, ring nematodes, bacterial canker, perennial canker, peach tree borer, and meadow voles. Choosing sites that were in grass crops for several years before planting, building raised beds, choosing rootstocks resistant to pests of the region and soil type, and using white plastic tree guards on young trees can prevent many tree deaths.

Verticillium wilt, caused by the soil fungus *Verticillium dahliae*, is especially serious on pistachios. Soil fumigation reduces this fungus for a couple of years but then it rebounds to higher levels than before with fewer beneficial soil fungi to compete with it, so crop rotation and soil solarization are better pre-plant treatments long-term than fumigation. Do not plant pistachios within a few years of cotton, which builds up high soil levels of *V. dahliae*. Choosing Verticillium-resistant rootstocks such as *Pistacia integerrima*, 'Pioneer Gold II' or 'UCB-1' will greatly reduce tree loss on problem soils as compared to orchards on susceptible rootstocks such as *P. atlantica*.

Indirect pests that weaken trees, reducing yields and tree longevity

Almonds and pistachios have far fewer virus and phytoplasma diseases than do tree fruits such as peaches. None are generally serious on pistachios, while almonds may face **yellow bud mosaic** caused by tomato ringspot virus, which only seriously damages the cultivar 'Mission'. If this virus occurs in your area, don't plant 'Mission'. **Root knot** and/or **lesion nematodes** can be serious pests, so sample for them before planting and solarize or fumigate the soil if needed, and choose rootstocks resistant to the predominant nematode pest.

Several **canker diseases** attack almond and pistachio trunks or branches that were damaged by mechanical shakers during harvesting or suffered sunburn. **Ceratocystis canker**, caused by the fungus *Ceratocystis fimbriata*, is the most common of these in almonds, causing water-soaked areas with amber-colored gum balls at canker margins. Cankers grow slowly from year to year, and can girdle and kill infected limbs or trunks after 3 to 4 years, especially trees of the most susceptible almond cultivars such as 'Nonpareil', 'Mission' and 'Ne Plus Ultra'. **Band canker**, caused by the fungus *Botryosphaeria dothidea*, mainly attacks 3- to 5-year-old trees of these same almond cultivars or 'Carmel', as well as pistachios. It often girdles a limb within a year but doesn't recur the next year. **Foamy canker**, caused by several *Zyomonas* species, mainly attacks 2- or 3-year-old 'Carmel' trees but occasionally infects other almond cultivars, usually just a few scattered trees in an orchard.

Prevent canker diseases by repainting tree trunks white each fall to avoid sunburn. Also stop irrigation a couple weeks before harvest so bark is less likely to tear during tree shaking, and limit shaking time to 2 to 3 seconds so as not to injure the bark. If bark does become damaged, shave off rough portions to promote rapid callus formation and healing. Inspect trees over winter for cankers and remove infected bark plus an inch of healthy-looking bark around the canker, and ½ inch of woody tissue below the bark. Do not apply any wound dressing to the area; that would slow healing and actually promote fungal growth.

San Jose scale (*Quadraspidiotus perniciosus*) and **spider mites** (*Tetranychus* species) may attack almonds just as they do peaches. **European fruit lecanium** (*Parthenolecanium corni*) is another scale insect that attacks both almonds and pistachios, and **frosted scale** (*P. pruinosum*) and **black scale** (*Saissetia oleae*) also attack pistachios. **Citrus flat mite** (*Brevipalpus lewisi*) may also infest pistachios, and **European red mite** (*Panonychus ulmi*), **citrus red mite** (*P. citri*), and **brown mite** (*Bryobia rubrioculus*) may be pests of almonds. Control all these scales and mites with a delayed-dormant oil spray, by pruning out and burning scale-infested branches, and by sprinkler irrigation or summer miticide sprays if mites build to damaging populations.

Diseases that attack flowers, fruits and foliage. Both almonds and pistachios suffer from several fungal and bacterial pathogens that can decimate crops by killing flowers or young fruit surrounding the nuts, or by defoliating the leaves. Both crops regularly need fungicide sprays to prevent some of these diseases.

Bacterial blast, caused by *Pseudomonas syringae*, the same bacterium that causes bacterial canker, can be severe on almonds during cold, rainy springs. Rain spreads the bacteria and frosts during bloom damage flowers, leaving them vulnerable to the disease, which kills flowers and young shoots. Cultivars that bloom early, such as 'Ne Plus Ultra' or 'Sonora', are especially susceptible to bacterial blast, but it can also be serious on mid-blooming cultivars such as 'Nonpareil', 'Carmel', 'Fritz', 'Price' and 'Wood Colony'. Choosing later-blooming cultivars and protecting against frost damage are the best ways to prevent this disease.

Shot hole disease, incited by the fungus *Wilsonomyces carpophilus*, attacks almonds but not pistachios. It causes early-spring leaf spots whose centers later fall out, resulting in "shot holes". This disease may become severe on almonds during wet springs, causing defoliation by late spring and scab-like spots on fruit which can induce young fruit to drop, in contrast to its mainly foliar damage on peaches. Sprinkler irrigation worsens this disease, so in sprinkler-irrigated orchards, shortening the time that trees are wet can reduce infections. Warm winters also worsen shot hole by allowing infected old leaves to remain on trees over winter until new leaves appear, which rapidly get infected. If a mild winter is predicted, spraying zinc sulfate on trees in late November to defoliate them will reduce the disease next spring. Two contact fungicide applications, at full bloom and the jacket stage, usually control shot hole. Soft-shelled cultivars are much more susceptible to shot hole than hard-shelled cultivars, and growers who have sprayed only rows of soft-shelled cultivars have gotten good control. If rains continue, apply additional fungicide sprays to soft-shelled cultivars 1 to 2 weeks apart. Sprays for shot hole should also control the less-common **leaf blight** of almonds, caused by the fungus *Hendersonia rubi*.

Brown rot on almonds is caused mainly by *Monilinia laxa* and secondarily by *M. fruticola*; it leads to blossom blight and jacket rot as described for apricots. Most years a single fungicide at pink bud will prevent it on most cultivars, but during rainy springs a second spray at full bloom or petal fall is needed. Cultivars with tight clusters, such as 'Butte', 'Carmel', 'Mission', and 'Ne Plus Ultra', are very susceptible and need the second spray every year. The second spray should also control shot hole and **jacket rot** caused by either *Monilinia* species or by *Botrytis cinerea* or *Sclerotinia sclerotiorum*.

Botrytis shoot blight in pistachios is also caused by *Botrytis cinerea*, which infects male and female flowers, then spreads into shoots, girdling and killing them. Cool, rainy springs favor this disease also, so it is mainly a problem in orchards in the northern Sacramento Valley and is not usually serious in the San Joaquin Valley, where most pistachios are grown. One or two fungicide sprays at bloom can prevent it when cool, wet weather is predicted.

Alternaria leaf spot, caused by the fungus *Alternaria alternata*, attacks both almonds and pistachios. It causes large lesions on leaves in early summer, and can defoliate them within a few weeks. Sprays for almond scab should control this disease also on almonds, but pistachios may require special sprays for this disease. 'Butte', 'Carmel' and 'Wood Colony' almonds are very susceptible to alternaria leaf spot and may also need extra sprays.

Almond scab, caused by the same fungus that incites peach scab (*Cladosporium carpophilum*), has become a much greater threat to almonds in recent years. The disease produces fuzzy black spots on young fruit and leaves (Photo 22-6) just like those on peaches. Fruit lesions do little damage, but leaf scab can defoliate trees in a short time. All almond cultivars are susceptible to scab, but 'Carmel' is very susceptible. Control scab with a dormant lime sulfur spray plus 2 to 3 fungicide sprays between 2 and 5 weeks after bloom.

Anthracnose, caused by the fungus *Colletotrichum gloeosporioides*, has also become a severe pest of almonds in the last few years. Tan-colored leaf lesions cause some leaves to drop, but deep, rust-colored fruit lesions do the most damage, destroying nut kernels inside. Control anthracnose with 4 to 6 sprays in spring starting at pink bud, rotating a strobilurin fungicide such as azoxystrobin (Abound™) or myclobutanil (Rally™) with a contact fungicide such as captan, maneb or ziram to avoid developing fungal resistance to the products. Planting less susceptible cultivars, which fortunately include 'Nonpareil', can reduce the number of sprays needed.

Almond rust (*Tranzschelia discolor* f. sp. *dulcis*) causes rust-colored lesions on leaves and can defoliate trees during warm, humid weather in late spring and summer. This disease has also gotten worse in recent years, because warm winters have allowed infected old leaves to remain on trees over winter until new leaves appear, which get rapidly infected. Break this disease cycle by spraying zinc sulfate on trees in late November to defoliate them. One or two sulfur sprays in late spring also reduce rust. 'Carmel' is one of the most susceptible cultivars.

Photo 22-6. Almond scab can defoliate trees. Control it with 3 to 4 fungicide sprays. (Photo from Univ. of California.)

Photo 22-7. Botryosphaeria blight is the worst pistachio disease, blighting pistachio fruit panicles and shoots during wet weather and warm temperatures of 80 to 86 F. (Photo from Univ. of California.)

Hull rot of almonds, caused by either *Monilinia* species that causes brown rot or by *Rhizopus stolonifer*, can occur several weeks before harvest cause, so 2 or 3 fungicide sprays may be needed in the weeks leading up to harvest. Hull rot frequently occurs in sprinkler-irrigated orchards on lower limbs that are directly hit by water, and spreads into the limbs and eventually kills them, so lowering sprinkler trajectories or switching to micro-sprinklers can reduce the need for fungicides and avoid lower limb losses. Withholding irrigation at first signs of hull split, then providing a full irrigation before reducing water again 2 to 3 weeks before harvest will also reduce hull rot. Keep leaf N levels between 2.5 to 2.7%; higher N levels induce more hull rot. 'Nonpareil', 'Butte' and 'Sonora' are among the most susceptible cultivars.

Botryosphaeria blight is the worst pistachio disease, blighting pistachio fruit panicles and shoots (Photo 22-7) during wet weather and warm temperatures of 80 to 86 F. It's caused by the fungus *Botryosphaeria dothidea* that also causes band canker in almonds. Sprinkler irrigation spreads this disease, so limiting the time that trees are wet, lowering the projection angle of sprinklers, or using micro-sprinklers instead will reduce the disease. Removing last year's fruit stems and leaf petioles, which remain on trees over winter and spread the disease in spring, greatly reduces the disease the next year. A fungicide spray at bloom and from late spring to close to harvest may also be needed, alternating a strobilurin fungicide such as azoxystrobin (Abound™) with a triazole fungicide such as tebuconazole (Elite™) to avoid development of fungal resistance.

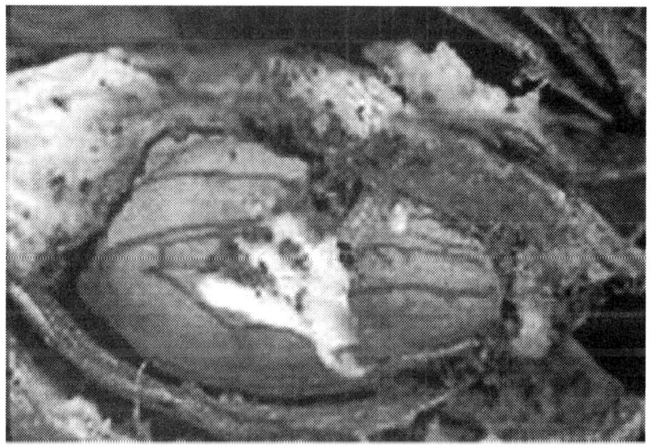

Photo 22-8. Navel orangeworm larvae, the most serious direct pest of both almonds and pistachios, tunnel into nuts to feed on the kernels. (Photo from Univ. of California.)

Bacterial spot is not a problem in the arid areas where almonds are grown, nor does it attack pistachios. Powdery mildew and peach leaf curl do not infect almonds or pistachios.

Insects and mammals that attack fruits and nuts. Few insects and vertebrate pests directly attack almond and pistachio fruits and nuts, but those that do can cause extensive damage if not prevented or controlled.

Navel orangeworm (*Amyelois transitella*) is the most serious direct pest of both almonds and pistachios. This moth pest lays eggs in split hulls or shells, on nut surfaces, or on stems close to fruit, and larvae hatch and tunnel into nuts to feed on the kernels (Photo 22-8). Such damage predisposes the kernels to infection by mold fungi *Aspergillus flavus* or *A. parasiticus*, which produce a chemical that's highly toxic to people, aflatoxin. Navel orangeworms overwinter in various larval stages or as pupae in mummy nuts on the tree or on the ground, and has 4 generations per growing season. Remove mummy nuts from trees by shaking them in January or February, and destroy mummy nuts on the ground by March 15 by disking or flail mowing. Such sanitation can eliminate the need for navel orangeworm sprays in an orchard at least 1/4 miles from infested trees. Harvesting, picking up and hulling nuts as quickly as possible, and getting them quickly to a handler where they'll be fumigated after harvest, will also reduce navel orangeworm infestation.

Almond cultivars with soft shells or poor seals, such as 'Ne Plus Ultra' and 'Nonpareil', are much more susceptible to navel orangeworm than hard-shelled cultivars with good seals such as 'Mission', 'Alamo' and 'Padre'. Susceptible cultivars generally need one or two sprays of the biological insecticide *Bacillus thuringiensis* (Bt) or the soybean oil product Stealth™ at the onset of hull split, while resistant cultivars need no sprays if sanitation has been practiced. If sources of infestation outside the orchard occur within 1/4 mile of the orchard, another Bt spray will be needed during spring egg-laying, timed with egg traps baited with almond press cake and placed in the orchard in early April. The spring spray will also help control peach twig borer.

Peach twig borer (*Anarsia linatella*) larvae bore into twigs of almonds, making it difficult to train young trees, and also bore into nuts, damaging them directly and creating preferred sites for navel orangeworm egg-laying. Soft-shelled cultivars are again most susceptible. Growers have traditionally applied a dormant oil spray combined with an organophosphate insecticide to reduce peach twig borer populations by up to 95%, but many growers today have switched to applying just dormant oil and then spraying Bt in spring. Mating disruption for peach twig borer is not cost-effective on almonds, but pheromone traps are widely used to monitor the pest during the growing season and time any in-season sprays if needed.

The same **cat-facing bugs** that attack peaches also attack almonds and pistachios. Early-season feeding may cause nut abortion, while late-season feeding causes dark spots on almond kernels and stained shells in pistachios, both of which are culled during sorting of harvested nuts. Many plant bugs move into nut orchards after early spring rains produce heavy weed growth, after cutting of nearby alfalfa fields, or during sudden hot weather in summer as annual crops and native vegetation dry up. Frequent mowing of the orchard to prevent rank weed growth, especially of legumes and mustard weeds, will reduce plant bug problems. Monitor populations of these pests in fields or weeds that harbor them, and apply an insecticide such as a pyrethroid when agronomic crops are cut or weeds dry up if many bugs are found.

Obliquebanded leafroller (*Choristoneura rosaceana*) may become abundant in pistachio orchards in early summer, especially if broad-spectrum insecticides are used to control plant bugs and also harm the beneficial insects that usually keep obliquebanded leafroller in check. These moth larvae web leaves together and feed in the webs, protected from bird predators and many insecticides. Outbreaks of this pest may cause defoliation, and the larvae may also feed on developing nut kernels. Monitor this pest with pheromone traps and also by scouting orchards for larvae and damage. If needed, apply an insect growth regulator that targets moth pests, such as tebufenozide (Confirm™), or a bio-insecticide that doesn't harm beneficial insects, such as spinosad (Success™ or SpinTor™). These have proved more effective in penetrating webs and controlling these leafrollers than Bt has.

Pavement ant (*Tetrameorium caespitum*) and **southern fire ant** (*Solenopsis xyloni*) can damage almonds on the ground after tree shaking and before nut harvest, and damage increases the longer nuts are left on the ground. Harvesting nuts within 4 days of shaking minimizes ant damage. Shaking trees too early increases ant damage, since nuts must dry on the ground before they can be harvested, and that takes longer than drying on the tree. Monitor ant colonies in the orchard in spring, and if ants are numerous, apply an insecticide bait in mid-summer, up to 2 weeks before harvest. Cultivars with poorly-sealed shells suffer more ant damage than well-sealed almond cultivars. Hard-shelled, well-sealed cultivars such as 'Alamo', 'Mission', 'Padre' and 'Peerless' need no ant treatments.

Ground squirrels (*Spermophilus beecheyi*) can carry away many nuts during harvest, in addition to feeding on bark, buds and blossoms over winter. Ground squirrels can be severe pests if nearby wooded areas provide habitat for them. Reduce populations by setting up poisoned bait stations in late summer from which the squirrels can regularly feed, and keeping them stocked with fresh bait.

HARVEST, HULLING, SHELLING AND MARKETING

Almonds and pistachios may be hand-harvested during their first light crop, in the third growing season after planting trees for almonds and the fifth season for pistachios, to prevent damage to young trunks from tree shakers. This involves laying tarps under trees and shaking limbs by hand or striking limbs with rubber mallets, so nuts fall on the tarps. Mechanical harvest begins in the fourth growing season for almonds and the sixth season for pistachios. Almond growers knock nuts to the ground with tree shakers, then use mechanical sweepers and harvesters to pick up the nuts. Pistachio growers use shake-and-catch mechanical harvesters.

Almond and pistachio growers prepare for harvest by mowing grass and weeds in orchard row middles, controlling weeds within tree rows with herbicides, and filling any ruts and tamping the soil down. Some almond growers spray the entire orchard floor with glyphosate herbicide, 3 to 4 weeks before harvest, to eliminate vegetation that would clog mechanical harvesters, and compact the ground to facilitate mechanical harvest. Growers stop irrigating 2 to 3 weeks before the anticipated harvest date, so tree bark will tighten and be less prone to injury by mechanical tree shakers.

Determining harvest time. Almond hulls, the fruit around the nuts, begin to split open 30 to 45 days before optimal harvest time. Almond kernels have achieved 85% of their dry weight by hull split, so harvesting slightly early may not reduce yields much and can avoid navel orangeworm infestation if the crop is knocked to the ground before egg-laying. Navel orangeworm egg-laying usually increases greatly 30 to 40 days after hull split of 'Nonpareil', so many growers aim to harvest 28 to 30 days after hull split, especially with 'Nonpareil' and other early-ripening cultivars. However, almond growers should generally wait until both kernels and hulls have dried to below 10% moisture content before shaking trees. Nuts shaken at higher moisture contents take longer to dry on the ground than on the tree, worsening problems with ants, squirrels and hull rot.

As harvest time approaches, growers shake limbs by hand to see how easily the nuts fall off. Once conditions seem about right, they bring in a mechanical shaker and shake a test tree. If more than 5% of the nuts remain on the tree after shaking, they wait a day or two and repeat the test on another tree. When shaking removes 99% of the nuts on the test tree, harvest begins.

Almond growers shake and harvest most cultivars in separate batches, since the processor will pay up to 15 cents a pound less if a cultivar such as 'Nonpareil' contains 10% or more of other cultivars. 'Carmel', 'Ne Plus Ultra' and 'Wood Colony' must also be harvested separately. Certain similar cultivars may be harvested together, such as 'Butte' and 'Padre', and 'Mission' and 'Alamo'. Several different cultivars may be harvested together as "California-type" walnuts, including 'Aldrich', 'Fritz', 'Merced', and 'Price'.

Pistachios are mature when hulls have turned yellow or reddish and separate easily from the tree. By that time many hulls have split and so have the shells inside the hulls, leaving kernels exposed, so pistachio growers do not shake nuts onto the ground where kernels would get dirty. Instead, they use catching frames like those used for mechanically-harvested tree fruits. Growers generally harvest within 7 to 10 days after nuts first mature, although nuts will hang on the trees for up to a month, because later harvest increases navel orangeworm infestation and results in more stained shells.

Orchard drying, sweeping, pickup and stockpiling of almonds. After tree shaking, almonds are left on the orchard floor to dry further. Early-harvested nuts may need 1 to 2 weeks on the ground to finish drying, while nuts shaken at about 10% moisture content usually need just 3 to 4 days on the ground. Dry conditions generally prevail during harvest in central California, where most almonds are grown, but growers in northern California may experience high humidity or rain during almond harvest. Such growers must harvest nuts at higher moisture contents, so hulling plants in northern California often dry the crop by blowing heated air onto harvested nuts.

Once nuts are adequately dry, a self-propelled or tractor-pulled sweeper blows nuts from around trees and sweeps them into windrows. A mechanical nut harvester then uses rubber paddles to flip nuts up from the ground onto a conveyor belt that carries them to a trailer. The harvester blows out leaves and other light debris, and a chain conveyor removes large sticks. Harvested nuts are then transported to a huller, or stockpiled and fumigated.

Stockpiles of nuts are best placed on a paved or cement surface, in pyramids or long triangular stacks. Cover piles with 6-mil-thick polyethylene sheets, anchoring the sheets to the floor with tubes filled with sand or soil, then fumigate to kill any navel orangeworms present. After fumigation, remove polyethylene sheets if rain is unlikely, to prevent moisture accumulation under the plastic and reduce the chance of spontaneous combustion. Almonds with less than 10% moisture content can be stockpiled for several weeks before hulling. Nuts with higher moisture content have much higher risk of developing mold and should not be stockpiled, but hulled immediately.

Almond cleaning, drying, hulling and shelling. A few almond growers operate their own hulling facilities, but most growers haul freshly-harvested or stockpiled almonds to a large commercial or cooperative hulling plant. Nuts first go through a cleaning machine to remove debris. An aspirator blows out leaves and dust, then the nuts pass over a moving screen which removes large sticks. Rotating or vibrating finer screens sift out soil particles, and a gravity table removes dirt clods, rocks and other heavy contaminants.

Nuts with less than 8 to 12% moisture, depending on the type of hulling machinery, go directly to hulling machines, but those with higher moisture content must go through a heated-air dryer. Most processing plants use continuous-flow belt dryers, which reach high temperatures of 200 to 250 F and rapidly move a shallow layer of nuts through the dryer within 20 minutes to 1 hour. Such high temperatures can kill any navel orangeworms present, eliminating the need for fumigation if nuts are processed immediately after harvest. Some older hulling facilities use stationary batch dryers of large loads at 130 F for 6 to 10 hours, but batch drying causes nuts at the bottom of the bin to become too dry and brittle, and the lower temperature will not control navel orangeworm.

A hulling cylinder, the most common type of huller, can handle hulls with up to 12% moisture and will crack some almond shells. Each hulling cylinder contains a rotating shaft with metal rods called "hulling fingers" welded perpendicularly to its axis, so the rods rotate inside the cylinder and beat the hulls from almonds that are fed into it. Hulls, in-shell almonds, shelled almonds and shells flow out through a grate of metal bars to a screen which separates them. By adjusting rotation speed, clearance between rods and bars, and bar spacing, the operator can control the amount of shelling that occurs.

A shear roll, the other type of almond hulling machine, works well only with nuts of 8% moisture or less, but can remove hulls that adhere more tightly to nuts, so is used in areas which produce a high percentage of nuts with tightly adhering hulls. Almonds move along a belt underneath a rotating rubber cylinder that removes hulls without cracking nuts, so shear rolls are desirable when processing nuts for the in-shell market. Processors that produce shelled kernels have hulled nuts flow on to a cracking roll, consisting of two counter-rotating steel cylinders, which crack shells as nuts pass between them.

About 92% of the total almond crop is sold shelled, while 8% is sold in-shell. About 80% of the in-shell nuts are exported and make up 9% of total almond exports, while 91% of exports are sold shelled. About 95% of almonds sold within the U.S. are shelled, and about 5% are sold in-shell.

Pistachio hulling and processing. Freshly-harvested pistachios are loaded into trailers furnished by a processor and hauled to the processing plant. There they are weighed, hulled in a machine similar to the shear roll used for almonds, washed, dried to a moisture content of 5%, and stored in large bins until they are roasted. Pistachios must be hulled and dried within 24 hours of harvest to avoid shell staining and aflatoxin contamination.

Electric-eye sorters remove pistachios with unsplit or stained shells and loose kernels. Nuts with unsplit shells will be sent to a pecan or almond cracking plant, loose kernels will be utilized as shelled nuts, and nuts with stained shells are usually discarded in the U.S. but may be routed to an area where they are dyed red to hide the stains. Split-shelled nuts on stainless steel racks pass under spray nozzles of salt solution, then enter a roasting oven for 20 minutes at 250 F. Ventilators cool the nuts, which stay at room temperature for 24 hours before bagging.

Processing plants report pistachio yields to growers in 3 categories: clean in-shell split nuts, closed nuts to be shelled, and loose kernels. Growers receive more than twice the price per pound for in-shell split nuts as they do for closed nuts or loose kernels. Fortunately, growers usually get about 80% split nuts and about 20% closed nuts, with less than 0.5% loose kernels. About 75% of pistachios are sold in-shell as roasted, split nuts for fresh consumption, while 25% are sold as chopped kernels to food processors that produce ice cream, candy and baked goods.

Storage and marketing. Dried almonds and pistachios store well for 1 to 2 years at 0 F, at 60 to 75% humidity for almonds and 55 to 65% humidity for pistachios. Nuts are generally marketed at room temperature and have a shelf life of several months. Most growers sell almonds and pistachios wholesale to commercial or cooperative processors, who store the nuts and market them to food products manufacturers and through supermarkets and warehouse clubs. A few large growers do their own processing and market their products directly to consumers over the Internet.

Further Reading

Crane, J.C. and J. Maranto. 1987. Pistachio production. Univ. of Cal. pub. 2279. 15 pp.

Ferguson, L. and D. Kester. 1998. Proceedings of the 2nd International Symposium on Pistachios and Almonds. *Acta Hort.* 470.

Ferguson, L., J. Maranto, and R.H. Beede. 1995. Mechanical topping mitigates alternate bearing of 'Kerman' pistachio (Pictacia vera L.). *HortScience* 30:1369-1372.

Freeman, M.W., M.A. Viveros, K.M. Klonsky, and R.L. DeMoura. 2003. Sample costs to establish and produce almonds, San Joaquin Valley South, micro-sprinkler irrigation. Univ. of Cal. Coop. Ext. AM-VS-03-2. Available on-line at: http://www.agecon.ucdavis.edu/outreach/crop/cost-studies/AlmondSprinkVS2003.pdf

Integrated Pest Management for Almonds, rev. ed. 2002. Univ. of Cal. pub. 3308. 199 pp.

Kallsen, C.E., B.A. Holtz, L.C. Hendricks, L. Ferguson, S.G. Sibbett, R.H. Beede, K.M. Klonsky, and R.L. DeMoura. 2000. Sample costs to establish and produce pistachios. Univ. of Cal. Coop. Ext. Available on-line at: http://www.agecon.ucdavis.edu/outreach/crop/cost-studies/pistachio2000.pdf

Micke, W.C., ed. 1996. Almond production manual. Univ. of Cal. pub. 3364. 289 pp.

Rice, R.E., W.J. Bentley, and R.H. Beede. 1988. Insect and mite pests of pistachio in California. Univ. of Cal. pub. 21452. 26 pp.

Teviotdale, B.L., D.A. Goldhamer, and M. Viveros. 2001. Effects of deficit irrigation on hull rot disease of almond trees caused by *Monilinia fruticola* and *Rhizopus stolonifer*. *Plant Dis.* 85:399-403.

Chapter 23
Walnuts

Walnuts, the second largest nut crop in North America and the world, belong to the genus *Juglans* in the walnut family, the Juglandaceae. The Persian or English walnut, *J. regia*, accounts for almost all commercial nut production due to the mild flavor and thin shells of these walnuts (Photo 23-1). Like almonds and pistachios, this species originated in Persia (now Iran). It spread throughout Asia and Europe, and arrived in North America via English ships, thus giving rise to the oft-used common name "English walnuts". Over 200,000 acres of bearing Persian walnut orchards in North America, 99% in California, produce over 600 million pounds of in-shell walnuts per year. California produces 38% of the world's walnut crop and is the second largest producer after China, which grows 44% of the world crop. About 40% of the U.S. walnut crop is exported, primarily to Japan, Germany, Spain, Italy and Israel.

Several other *Juglans* species native to North America are utilized for their wood and as rootstocks for Persian walnuts, but only the eastern black walnut, *J. nigra* is commercially harvested for its edible kernels. Native to the eastern half of North America, eastern black walnuts have very thick shells and a much stronger flavor than Persian walnuts, that some consumers like but many feel is too intense. Over 750 eastern black walnut cultivars have been selected from the wild or from seedlings of wild selections, and some are grown in small commercial orchards, but 99% of black walnuts are collected and sold from native stands. Each year, 25 million pounds of eastern black walnuts are collected, yielding only 2 million pounds of edible kernels due to their thick shells.

TREE GROWTH, ADAPTATION AND LIMITING FACTORS

Persian walnuts grow into medium-sized trees, 40 to 50 feet tall (Photo 23-2), but newer cultivars have been selected to be genetically semi-dwarf and reach 30 to 40 feet tall. They are long-lived trees, and orchards generally remain productive for at least 40 to 50 years. Older cultivars generally began bearing 5 to 7 years after planting, but newer cultivars have been selected for precocity and begin bearing 3 to 4 years after planting. Like other tree nuts, walnuts tend toward a biennial bearing

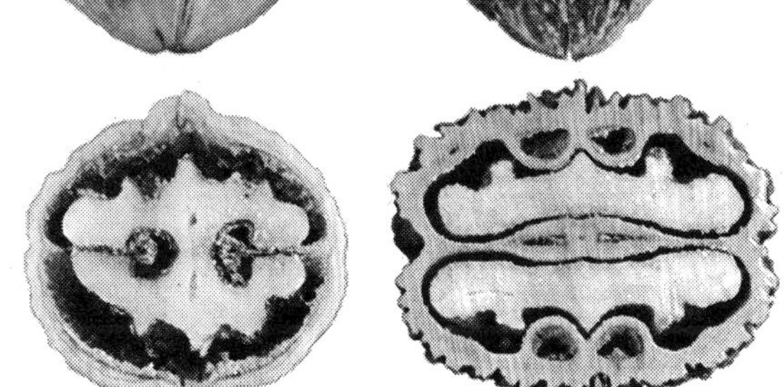

Photo 23-1. Persian walnuts (left) have thin shells and a mild flavor, while eastern black walnuts (right) have very thick shells and a much stronger flavor.

Photo 23-2. Persian walnuts grow into medium-sized trees, 40 to 50 feet tall, but newer cultivars have been selected to stay more compact.

cycle of a heavy crop one year followed by a light crop the next, although biennial bearing is not as pronounced in walnuts as in some other nuts such as pecans and pistachios.

Most Persian walnut cultivars are only cold-hardy to about 10 F (USDA zone 8) but have rather high winter chill needs of about 800 hours below 45 F, and need warm but not hot summers and no early fall or late spring freezes. Such exacting climate needs limit commercial production to the valleys of central and northern California. Winters are too mild and summers too hot for Persian walnuts in southern California desert areas and the Southeastern U.S., and winters are too cold or spring or fall freezes too common in most other parts of North America.

Some Persian walnuts, known as 'Carpathian walnuts' because they were selected from the wild in the Carpathian Mountain region of Poland, have greater winter cold hardiness, to about -15 F (USDA zone 5b). However, most 'Carpathian walnuts' sold are seedlings with highly variable characteristics; most have poor nut size and quality. A few superior Carpathian walnut cultivars have been selected and are propagated by nurseries, but most are still very susceptible to late spring or early fall freezes. Improved cultivars of Carpathian walnuts need to be tested in replicated trials before Persian walnut production can expand to colder areas.

Black walnuts grow into very large trees 80 to 100 feet tall (Photo 23-3), one factor that has limited their planting for nut production. Like Persian walnuts, older cultivars don't begin bearing until 5 to 7 years after planting, but a few newer cultivars have been selected for precocity and begin bearing 3 to 4 years after planting. Heavy early production can help keep trees smaller. Orchards of well-chosen cultivars can remain productive for 60 to 80 years or more.

Walnuts have pinnately compound leaves and are monoecious, bearing separate staminate (male) and pistillate (female) flowers on the same tree (Photo 23-4). Staminate flowers are borne on pendulous catkins and produce light, dry pollen disseminated by wind. Pistillate flowers, which produce the fruit, occur on terminal and lateral shoots of new growth. Since they are wind pollinated, they have no showy petals to attract insects. Older Persian walnut cultivars produce pistillate flowers mainly on shoot terminals, but newer cultivars have been selected for lateral bearing. Greater

Photo 23-3. Black walnuts grow into very large trees that may reach 80 to 100 feet tall.

Photo 23-4. Walnuts bear pistillate (female) flowers at the tips of new growth, and staminate (male) flowers in pendulous catkins that develop from one-year old branches. These catkins are just starting to arise from the sides of this branch, while the pistillate flowers will develop later from the tip of the terminal growth, so this cultivar is protandrous.

lateral bearing increases yields and distributes nut bearing throughout the tree canopy, rather than concentrating it on the tree periphery. Cultivars with increased lateral bearing also tend to come into bearing at a younger age, which increases overall yields and helps keep trees smaller.

Most often, staminate and pistillate flowers of each walnut cultivar open at different times, so a cultivar can not pollinize itself. Cultivars whose staminate flowers open before the pistillate flowers are called protandrous, while cultivars whose pistillate flowers open before the staminate flowers are called protogynous. Some Persian walnut cultivars yield best without cross-pollination so should be planted in solid blocks of a single cultivar each, as excess pollen reduces fruit set. Orchards of other Persian walnut cultivars and black walnut orchards must contain at least two cultivars to provide cross-pollination.

Persian walnut growers generally plant the majority of the orchard to protandrous cultivars, since later blooming of female flowers reduce frost risk and susceptibility to walnut blight and codling moth. If the main cultivars need cross-pollination, they also plant at least one cultivar that sheds pollen during the time that the main cultivars' pistillate flowers are receptive. This pollenizer cultivar may be protogynous, or a protandrous cultivar that sheds pollen earlier than the main cultivars. A black walnut orchard should contain at least one protandrous cultivar and at least one protogynous cultivar.

Like other tree nuts, walnut kernels are seeds that grow within shells (endocarp) within fleshy fruit (mesocarp) called hulls or shucks. Persian walnut hulls (Photo 23-5) split slightly as the nuts mature, but Eastern black walnut hulls (Photo 23-6) do not, and hulls of both types of walnuts must be removed after harvest. Walnuts are shaken to the ground with mechanical trunk shakers and mechanically harvested off the ground in the same way as almonds. Persian walnuts in California yield 2500 to 3000 pounds of in-shell nuts per acre each year on average. Black walnuts have traditionally produced much less, but higher-density plantings of high-yielding cultivars may produce up to 2000 pounds of in-shell nuts per acre per year.

SITE SELECTION AND PREPARATION

Walnuts produce best on fertile, well-drained soils at least 10 feet deep. Avoid soils with a seasonally high water table and saline soils. Since walnuts leaf out later in spring than many other trees, air drainage is not as critical as it is with most fruit plants, except for Carpathian walnuts in eastern North America.

The ideal site was planted to grass crops for several years before planting walnuts, but the most common walnut rootstocks have some resistance to Verticillium wilt, oak root rot and root knot nematodes, so walnuts may not suffer as much from planting after other fruit crops. Soils that were previously planted to tree or vine crops should be tested for nematodes, especially lesion nematodes, and treated if needed. Soil fumigation or solarization is not usually needed, unless lesion nematodes are numerous or when using Persian walnut rootstock to avoid black line disease, because that rootstock is very susceptible to several soil pests as described below.

Photo 23-5. Walnut kernels (seeds) develop inside shells (endocarp) inside fleshy fruit (mesocarp) called hulls or shucks. These Persian walnut hulls will split slightly when nuts are mature.

Photo 23-6. Black walnut shucks do not split open at maturity.

Soils with a plowpan layer may need deep ripping, to a depth of at least 3 feet, before planting. Orchards to be irrigated by surface flooding or furrows must be graded to a uniform slope, but most new walnut plantings utilize micro-sprinkler or trickle irrigation so do not need grading. Walnut growers generally plant on flat ground, but some make slightly raised beds (Photo 23-7) if a rootstock susceptible to wet soil is chosen. A non-saline irrigation water source is essential.

Test soil samples for pH, phosphorous (P) and potassium (K) well before planting. Add lime if pH falls below 5.5, and sulfur if it tests above 7.5. Most walnut soils don't need P additions, but some may need K. Soils with slow water infiltration and permeability rates benefit from adding organic materials and gypsum, which gives off calcium (Ca) that replaces excess soil sodium, and in some cases, excess magnesium (Mg) or K.

Photo 23-7. Walnut growers may plant trees on slightly raised beds if soil is somewhat poorly drained and a rootstock susceptible to Phytophthora root rot is chosen.

SELECTING ROOTSTOCKS

Nurserymen graft Persian walnuts onto 3 different seed-grown rootstocks, and some Persian walnuts are also grown as self-rooted cultivars propagated by cuttings or micro-propagation. Each rootstock or self-rooted cultivar offers resistance to certain diseases and environmental stress but is susceptible to others (Table 23-1), so growers should choose the root system that best suits their soil type and disease problems.

Northern California black walnut (*J. hindsii*) seedlings have very good, deep anchorage on good soils (Photo 23-8), but grow poorly on heavy soils prone to waterlogging. This rootstock withstands high soil boron (B) and resists Verticillium wilt, oak root fungus, and root-knot nematode. It has moderate susceptibility to crown gall but is very susceptible to lesion nematodes and black line disease. Black line occurs when a Persian walnut scion

Photo 23-8. Older Persian walnuts on California black walnut rootstock clearly show the graft unions between the cultivar with lighter-colored bark and the rootstock with darker bark.

Table 23-1.

COMMON ROOTSTOCKS FOR PERSIAN WALNUTS

	Susceptibility to									
									Nematodes	
Rootstock	Wet soil	Saline soil	High soil B	Black line	Deep bark canker	Vert. wilt	Oak root rot	Crown gall	Root knot	Lesion
N. California black walnut	S	MS	R	S	VR	R	MR	MS	MR	VS
Paradox hybrid seedling	MR	S	S	S	VR	R	R	VS	MR	MR
English walnut seedling	S	S	S	VR	S	S	VS	S	S	VS

Pest susceptibility ratings: VR = very resistant, R = resistant, MR = moderately resistant, MS = moderately susceptible, S = susceptible, VS = very susceptible.

infected with cherry leafroll virus transmits the virus to a black walnut rootstock, which is highly susceptible to it. The virus kills cells at the graft union, causing a black line there and tree death.

'Paradox hybrid' seedlings result from a cross between northern California black walnut and Persian walnut. Paradox rootstocks perform well on less fertile soils and on heavy clay soils with somewhat poor drainage, but develop toxicity in soils high in B or salts. Paradox rootstocks resist Verticillium wilt and oak root fungus and have moderate resistance to root knot and lesion nematodes, but are susceptible to black line disease and very susceptible to crown gall.

Persian (English) walnut seedling rootstocks or self-rooted cultivars are increasingly used in areas where black line is prevalent, since this species tolerates cherry leafroll virus. Using self-rooted cultivars can reduce the cost of buying trees, as rooting cuttings uses much less labor than grafting, but cutting-propagated root systems tend to be shallower and less well-anchored. Using just one Persian walnut cultivar as a rootstock creates greater tree uniformity in the orchard, and a cultivar with certain traits can be chosen, such as 'Manregian' English walnut from Manchuria which has greater cold hardiness. Generally, Persian walnut seedlings are susceptible to most soil pests including root-knot and lesion nematodes, Verticillium wilt, oak root rot, crown gall and crown rot. Therefore, using English walnut seedling rootstocks or self-rooted cultivars may require soil fumigation or solarization or other special pre-plant treatments.

Hardier Persian walnuts (Carpathian walnuts) and Eastern black walnut cultivars are generally grafted onto Eastern black walnut seedling rootstocks.

CHOOSING CULTIVARS

Over the last 10 years, Persian walnut growers have chosen mainly lateral-bearing cultivars that come into bearing quickly and produce high yields of thin-shelled nuts with a high percentage of kernels as opposed to shells. Almost 82% of new walnut plantings have been to three cultivars that meet those criteria: 'Chandler', 'Howard' and 'Tulare'. These cultivars receive a premium price from shellers per pound of harvested nuts, due to their higher percent kernel than other cultivars, and their higher yields also increase grower returns.

Late bloom time is another desirable trait that these three cultivars share. Protandrous cultivars that bloom later than other protandrous cultivars have reduced risk of crop reduction from spring frosts (Photo 23-9), and escape most damage from walnut blight and codling moth, the most serious direct pests of walnuts. Table 23-2 compares these cultivars and other Persian walnuts that are commonly grown in western North America.

Photo 23-9. Late spring frosts may kill new shoot growth on walnuts and reduce the number of pistillate flowers that develop, hurting nut yields. Cultivars that leaf out and bloom later generally avoid frost damage.

Table 23-3 describes some of the most widely-available cultivars of hardier Persian walnuts (Carpathian walnuts) for parts of eastern North America. No commercial plantings of these hardier Persian walnuts have yet been made, but they may offer potential on some favorable sites.

Eastern black walnut cultivars that bear on spurs, start bearing at a young age, produce high yields of thinner-shelled nuts, and leaf out and bloom late are also desirable. Eastern black walnuts do not exhibit lateral bearing from side shoots of current season growth as do Persian walnuts, but spur-bearing black walnut cultivars bear nuts from long-lived short spurs as well as from shoot terminals (Photo 23-10). Black walnuts should also have resistance to anthracnose, a common disease in humid areas that defoliates susceptible cultivars early and greatly reduces kernel filling. Cultivars with a low tendency toward alternate bearing and a high percent kernel are also desirable. Table 23-4 details the traits of some common Eastern black walnut cultivars.

ORCHARD LAYOUT AND PLANTING

Persian walnuts grow into large trees but take many years to do so, so traditionally trees have been planted twice as close within rows as between rows, planning that every other tree in each row will be removed after 10 to 12 years. This leaves a square pattern of trees the same distance apart within

Table 23-2.

COMMON PERSIAN WALNUT CULTIVARS IN WESTERN NORTH AMERICA,

in approximate order of harvest (varies with area).

All are protandrous except Chico, which is protogynous.

	Nut					Tree			Lateral	Days after Payne		Needs	Susceptibility to		
Cultivar	Kernel size	% Kernel	Light kernels	Shell seal	Yield	Size	Growth	Bears young	bud fruiting	Leafing	Female flowering	pollen-izer	Walnut blight	Codling moth	Harvest time
Chico	small	47%	60%	good	very high	small	upright	yes	95%	2	-10	yes	VS	VS	early
Serr	large	57%	75%	good	high	large	spreads	yes	60%	-1	-3	no	S	VS	early
Payne	small-med	50%	60%	very good	high	medium	spreads	yes	90%	0	0	no	VS	VS	early
Vina	medium	49%	75%	good	very high	medium	spreads	yes	72%	8	4	yes	VS	S	early-mid
Howard	med-large	49%	92%	good	very high	small	upright	yes	90%	15	11	yes	MR	MR	middle
Tulare	large	53%	86%	good	very high	medium	upright	yes	72%	12	11	no	MS	MS	middle
Hartley	medium	46%	82%	good	high	med-large	spreads	no	8%	17	13	yes	MR	MR	mid-late
Chandler	med-large	49%	95%	fair	very high	medium	upright	yes	90%	17	13	yes	MR	MR	mid-late
Cisco	small-med	46%	86%	very good	medium	medium	upright	yes	77%	25	19	yes	R	R	late
Franquette	small	47%	90%	good	medium	large	upright	no	5%	26	19	yes	R	R	late

Pest susceptibility ratings: R = resistant, MR = moderately resistant, MS = moderately susceptible, S = susceptible, VS = very susceptible.

Average harvest times in Davis, CA: early: Sept. 15-19, early-mid: Sept. 20-24, middle: Sept. 25-29, mid-late: Sept. 30-Oct. 4, late: Oct. 5-9.

Notes on Cultivars:

Chico needs a pollenizer with early catkin bloom, such as Payne. Good for high-density plantings, but small, hard-to-remove kernels undesirable.
Serr suffers pistillate flower abscission from excess pollen, so plant alone, shake trees when catkins form to remove pollen. Little planted now.
Payne, found in 1898, was standard cultivar and still planted, but less now since early leafing makes it prone to frost, walnut blight, codling moth.
Vina was heavily planted but rarely planted now, since fairly early leafing makes it prone to frost, walnut blight, codling moth.
Howard, one of top 3 cultivars planted now, is good for high-density plantings. Use Cisco or Franquette as pollenizer.
Tulare, top cultivar planted now, is good for high-density plantings, solid block fine.
Hartley is slow to begin bearing, has few lateral fruit buds, and S to deep bark canker, so less planted now.
Chandler, one of top 3 cultivars planted now, is good for high-density plantings. Use Cisco or Franquette as pollenizer.
Cisco has lower yields but is planted as a pollenizer for Chandler or Howard due to late pollen shed.
Franquette is another pollenizer for Chandler or Howard, but slow to begin bearing, has few lateral fruit buds, so less planted now.

Table 23-3.

HARDIER PERSIAN WALNUT CULTIVARS

for parts of USDA zone 5b to 7b in eastern North America, in approximate order of harvest (varies with area)

Cultivar	Nut			Tree bears young	Lateral bud fruiting	Alternate bearing tendancy	Time of Leafing	First flowers	Needs pollen-izer	Pest resistance	Harvest time
	Kernel size	% Kernel	Yield								
Somers	medium	52%	high	yes	no	low	early	female	yes	R to husk maggot	early
Shiawassee	large	58%	high	yes	yes	medium	early	male	yes	R to anthracnose	early
Colby	small-med	53%	medium	no	no	low	early	male	no		early
Hansen	small-med	60%	high	yes	yes	low	early	male	no	R to anthracnose	early
Kentucky Giant	large	44%	high	yes	yes	medium	late	male	no		early
Sauber Giant	very large	48%	medium	yes	no	high	early	male	yes		early
Greenhaven	med-large	46%	medium	no	no	low	middle	male	yes		early
Lake	large	50%	medium	yes	no	medium	late	female	yes	very good shell seal	middle
Perry	large	56%	high	yes	yes	medium	middle	male	yes	R to anthracnose	middle
Coble #2	large	45%	high	yes	yes	medium	middle	male	yes		middle
Kaiser	very large	48%	medium	no	no	high	middle	male	yes		late
Schaefer	med-large	48%	high	yes	yes	low	middle	male	yes	very good shell seal	late
Allegheny	medium	52%	high	no	no	low	late	male	no	very good shell seal	late

Pest resistance ratings: R = resistant.

Average harvest times in central MO: early: Sept. 1-14, middle: Sept. 15-28, late: Sept. 29 - Oct. 12.

Notes on Cultivars:

Somers bears consistently and nuts crack easily, with light-colored kernels.
Greenhaven nuts fill well, have light-colored kernels, crack easily.
Shiawassee bears large clusters of nuts with a light kernel color.
Colby may not be as hardy to winter cold as other Carpathians, but produces well where adapted.
Hansen, one of the oldest culitvars, is a small tree that can be planted closer.
Kentucky Giant avoids frost due to late bloom, has a thin shell, but often doesn't fill well.
Sauber Giant is frost-prone due to early leafing, so plant it only on the best sites.
Lake bears some nuts even during bad frost years, and is an important pollenizer for most other cultivars.
Perry bears large clusters of nuts with a light kernel color.
Coble #2 fills nuts well.
Kaiser nuts fill well, have light-colored kernels.
Schaefer has a thin shell.
Allegheny cracks out in whole halves, misses frosts due to late leafing.

Table 23-4.

EASTERN BLACK WALNUT CULTIVARS

for nut production in eastern North America
In approximate order of harvest (varies with area).

Cultivar	Nut			Alternate bearing tendancy	Lateral spur fruiting	Leafing, days after Davidson	First flowers	Anthracnose suscep.	Harvest time
	In-shell weight	% Kernel	Yield						
Davidson	22 g	27%	high	medium	yes	0	female	S	early
Football	20 g	33%	medium	severe	yes	11	male	VS	early
Sparrow	19 g	32%	high	medium	no	17	female	MR	early
Emma K	19 g	34%	low	high	yes	5	female	S	middle
Mintle	16 g	31%	medium	high	no	7	female	VS	middle
McGinnis	17 g	32%	medium	high	yes	5	male	S	middle
Drake	19 g	31%	high	medium	no	17	female	S	middle
Kwik Krop	17 g	31%	low	severe	yes	18	female	S	middle
Rupert	18 g	26%	high	low	yes	8	female	MR	late
Surprise	20 g	33%	medium	low	yes	13	female	VS	late
Scrimger	19 g	31%	low	medium	no	17	female	MR	late
Thomas	22 g	24%	low	medium	no	21	female	MR	late
Thomas-Meyers	23 g	34%	high	low	no	23	female	MS	late

Anthracnose susceptibility ratings: MR = moderately resistant, MS = moderately susceptible, S = susceptible, VS = very susceptible.

Average harvest times in central MO: early: Sept. 1-14, middle: Sept. 15-28, late: Sept. 29 - Oct. 12.

Notes on Cultivars:
Davidson bears young and heavily, but early-leafing branches are prone to spring frosts.
Football bears young and has high % kernel, but severe alternate bearing and anthracnose limit its use.
Sparrow is the best overall cultivar with high yield, high % kernel, resists anthracnose, but doesn't bear young.
Emma K has a thinner shell and high % kernel, but alternate bears and nuts may not fill well in heavy-crop years.
Mintle has top flavor but small kernels, alternate bearing and anthracnose pose problems.
McGinnis has a thinner shell and light-colored but small kernels that crack out easily, but alternate bears.
Drake has high yields but doesn't bear young.
Kwik Krop bears young and cracks easily, but severe alternate bearing, low yields, anthracnose limit its use.
Rupert bears young and annually, yields well, resists anthracnose, but kernels are small.
Surprise has large kernels that crack out in quarters, but is anthracnose-prone.
Scrimger has anthracnose resistance but doesn't bear young and has low yields.
Thomas was the first cultivar selected, in 1881. Yield and % kernel are too low to plant it today.
Thomas-Meyers avoids late frosts due to very late leafing, large nut with high % kernel, but doesn't bear young.

Photo 23-10. Eastern black walnuts do not exhibit lateral bearing on new shoots as do some Persian walnut cultivars, but some black walnut cultivars bear on spurs as well as on shoot terminals.

Photo 23-11. Conventional Persian walnut orchards have been planted in a square pattern, as here, or in a rectangular pattern where every other tree would be removed after 10 to 12 years, leaving a square pattern. Many newer orchards have been planted to hedgerows to achieve higher tree densities.

rows as between rows. The distance between and within rows depends on the cultivar's ultimate tree size. Medium-sized trees like 'Chandler' and 'Tulare' should be planted 28 to 35 feet apart between rows, and 14 to 18 feet apart within rows. Small trees like 'Howard' and 'Chico' can be planted 24 to 26 feet between rows, and 12 to 13 feet between trees within rows. A medium-large cultivar like 'Hartley' requires 30 to 36 feet between rows, and large cultivars like 'Serr' and 'Franquette' need 34 to 40 feet between rows, but few large cultivars are being planted today. When growers do not want to remove trees, they may plant them in the square pattern but slightly closer, for example 24 to 28 feet apart in both directions for medium-sized trees (Photo 23-11).

Recently most walnut growers have planted trees in hedgerows, with rows spaced twice as far apart as trees within rows, but all trees are kept for the life of the orchard, and are maintained at this tighter spacing with dormant mechanical pruning. Hedgerow plantings have increased early yields since trees are trained to fruit as soon as possible. Most growers plant medium-sized cultivars such as 'Chandler' and 'Tulare' 15 to 16 feet apart within rows that are 30 to 32 feet apart. Some growers have planted such medium-sized trees a few feet closer, at 24 to 26 feet between rows, and hedged them more severely. Small trees like 'Howard' and 'Chico' are planted 11 to 12 feet apart within rows 20 to 24 feet apart. Medium-large to large cultivars have not performed as well in hedgerows as in conventional plantings.

Most walnut cultivars yield more when some cross-pollination is provided. If a grower wants to maximize the main cultivar and minimize a pollenizer, as little as 3 to 5% of the total orchard can be planted to the pollenizer. Plant pollenizer trees at regular intervals such as every 10th row or 400 feet, and perpendicular to the direction of the prevailing wind. A few walnut cultivars are self-pollinating and are planted in solid blocks. Some Persian walnuts, especially 'Serr' but possibly also 'Chandler' and 'Vina', have the unique problem of aborting many newly-pollinated female flowers if orchard pollen loads are high. Plant these cultivars only in solid blocks if they are-self-pollinating like 'Serr', or plant them with no more than 3 to 5% pollinizers if they need some cross-pollination, as higher percentages of pollinizers reduce yields.

Plant walnuts as bare-root trees in late winter to early spring in California, and in spring in other parts of North America. Dig holes as for other trees, just a few inches deeper than needed to accommodate the root system so trees don't settle too deeply, and break up sides of auger-dug holes to allow roots to penetrate. Dip roots in a crown gall biocontrol solution (Galltrol™) to avoid that disease. Drive an 8-foot-long stake 2 feet into the ground next to each newly planted tree, to aid in training the leader (Photo 23-12).

Irrigate after planting if soil is dry, but do not irrigate soil that is already moist, to avoid problems with root rot. Cut newly-planted trees back to about 4 live buds above the graft union, or about 28 to 30 inches tall. Use white plastic tree guards or paint trunks white and install a wire cage around each tree to prevent sunscald and rodent damage.

ORCHARD FLOOR MANAGEMENT, IRRIGATION AND FERTILIZERS

Young walnut orchards may be intercropped to soybeans, dry beans or vegetable crops for the first 3 to 4 growing seasons if these annual crops are sown well away from the tree rows. Legumes such as beans can benefit tree growth by adding N to the soil. Corn, milo and other fast-growing annual grass crops compete too much with the trees for water and nutrients. By the time walnut trees start bearing a good crop in year 4 or 5, their roots will extend over most of the orchard floor and should not be damaged by the extensive cultivation needed for intercrops.

Most walnut growers plant a winter cover crop between bearing walnut rows after harvest in fall, or allow a former cover crop to reseed itself or native vegetation to grow over the winter. Most mow this vegetation once at about 4 to 5 inches in spring, then some disk it into the soil once to minimize water competition, and mow the resulting regrowth periodically. Some just mow the vegetation without disking, and others mow only strips along both sides of tree rows, leaving a center strip to flower and reseed itself. Such strips can provide habitat for beneficial arthropods, but legumes may also attract pests such as plant bugs.

All growers mow the groundcover closely or herbicide it just before harvest. Cover crops add organic matter to the soil, improving water penetration and infiltration, and legume crops add N. When orchards are irrigated by micro-sprinklers or trickle systems that are placed within tree rows, cover crops do not take much of that irrigation water. One irrigation after walnut harvest is generally sufficient to germinate cover crop seed.

Photo 23-12. Drive a stake into the ground next to each newly planted tree, to aid in training the leader, and paint trunks white to avoid sunscald.

If nematode pests build up in the orchard, planting certain cover crops can reduce them. For example, if lesion nematodes occur, planting sudan grass or barley will greatly reduce them. Sudan grass will also reduce ring nematodes and Northern root knot nematodes, although it may increase Southern root knot nematodes. Barley, sweet clover and white vetch are poor hosts for Southern root knot nematodes but good hosts for Northern root knot nematodes.

Older conventional walnut orchards in a square pattern may be cross-mowed in two directions if irrigated by movable sprinklers or the flood system. Such orchards use herbicides to control weeds in a square around each tree. Hedgerow orchards use herbicides in a wide strip down each row, so mowing and other operations are done in only one direction, and they can utilize any type of irrigation. Northern California black walnut and Eastern black walnut rootstocks secrete a chemical called juglone into the soil, which inhibits the growth of many other plants but nut grasses or legumes, so acts as a natural herbicide.

Irrigation systems and water quality. Walnuts need about 42 to 45 inches of water per year, mainly during the hot summer months from early May until late August. Winter rains generally wet the soil enough to provide for spring moisture needs, so irrigation isn't needed until trees use up most available soil water in the top 2 to 3 feet of soil.

This can be determined by placing deep soil tensiometers, that measure moisture content at a depth of 30 inches, at several locations in the orchard. Nuts grow most rapidly during the 5 or 6 weeks from the end of bloom until shell hardening in mid-June, and moisture stress during this time will reduce nut size. Moisture stress during summer can reduce kernel filling and crack-out percentage and cause undesirable darkening of kernels.

If drought forces cutbacks in irrigation allotments, irrigate at 25% to 50% deficit levels throughout the season, rather than using up all water allotted early and cutting back severely later. Heat-sensitive cultivars need near-normal irrigation levels during the 2 weeks before harvest to prevent sunscald of the fruit (Photo 23-13).

Many newer walnut orchards use micro-sprinklers, since they provide a broader wetting pattern than other forms of trickle irrigation, and cut water use considerably from the amounts used by large sprinklers or furrow irrigation. Micro-sprinklers and other forms of trickle irrigation must cover at least 50% of the orchard ground surface to provide enough water for walnuts. Place micro-sprinkler emitters or sprinkler heads so they do not wet tree trunks or soil right around the trunks, to avoid inducing Phytophthora collar rot or Armillaria root rot (Photo 23-14). Since walnuts leaf out later than most fruit and nut crops, sprinkler irrigation for frost protection isn't needed.

Check irrigation water for sodium, chloride, nitrogen (N) and boron (B) content, since much water used for irrigation contains these chemicals. Water high in N can enable growers to reduce N fertilizer rates, but water high in sodium, chloride, or B is toxic to walnuts. Well water chemical levels stay fairly constant year-round, so test irrigation wells and do not use those with a sodium absorption ratio greater than 3, or with more than 4 ppm chloride or 2 ppm B. Canal water may be high in salts or B early and late in the growing season but is lower in summer, so test such water before each planned irrigation and use it only when salt and B levels are below those just mentioned.

If salts have already built up in the soil, they may be pushed below the root zone with several leaching irrigation treatments that apply more water than field capacity for that soil. Winter rainfall may accomplish some leaching, and winter irrigations may be needed to provide more leaching, since it's easiest to apply excess water when evapotranspiration is lowest. Such leaching may be needed every few years, and growers should routinely test soil for salinity to determine when leaching is needed.

Fertilizer additions. A 4000-pound walnut crop removes about 100 pounds of actual N from the soil, so growers need to add back at least that amount each year, from a legume cover crop and/or fertilizer additions. Since soils tie up some of the added N, growers usually apply 140 to 200 pounds of actual N per year. Take mid-summer leaf samples and test irrigation water before deciding how much N to add. Leaf samples should have 2.2 to 2.8% N, and trees with higher N levels need no additional N that year. A legume cover crop plus N in irrigation water may supply 80 to 120 pounds of actual N per year, so reduce N fertilizer accordingly.

Photo 23-13. Many Persian walnut cultivars are heat-sensitive and need near-normal irrigation levels during the 2 weeks before harvest to prevent sunscald of the fruit.

Photo 23-14. Walnut orchards that receive sprinkler irrigation should have sprinkler heads placed so they do not wet tree trunks or soil right around the trunks, to avoid inducing Phytophthora collar rot or Armillaria root rot.

Make at least 2 N applications per year, in early spring and early summer, and on sandier soils 3 or 4 smaller applications per year are better, adding one in late spring and one in late summer. Young trees should receive no more than 1 ounce of actual N per tree per year of tree age at each application, so may need more frequent applications of smaller amounts of N. Irrigate after each N application, or apply N through the irrigation system.

Walnut trees may develop K deficiencies on soils high in Mg. Leaves turn pale, then edges of leaflets curl upward and bronze. Take leaf samples to confirm a deficiency. If K is needed, apply it through the irrigation system or band it along tree rows in fall and irrigate it in.

Low-ph soils may develop Mg deficiency, as basal leaves on vigorous shoots develop chlorosis at the margins and the base of the midrib. Raising soil pH with dolomitic lime will alleviate this, and while waiting for this to take effect, growers can spray magnesium sulfate or Epsom salts on the foliage.

Zinc (Zn) deficiencies are common in walnut orchards, especially those on sandy soils. Zn deficiency delays opening of vegetative and flower buds, and causes leaves to remain small and chlorotic and appear in tufts, giving rise to the name "little leaf". Deficiency symptoms may be confined to several branches while the rest of the tree looks healthy. Where Zn deficiency may be a problem, test midsummer leaf samples for Zn. If below 18 ppm, apply foliar sprays of zinc sulfate or zinc chelate the next spring after some bloom has occurred.

TRAINING, PRUNING AND MANAGING CROPS

Train walnut trees based on the system desired and each cultivar's growth habit. In conventional plantings, where orchardists plan to eventually remove half the trees, growers usually train trees to the modified leader system, often using minimum pruning for lateral-bearing cultivars that begin to bear at a young age. Closer-planted orchards, where all trees will remain, are trained to the hedgerow system to become a solid wall of growth down each tree row. Tree training takes about a year longer for the modified leader system than for the hedgerow system, which reduces yields during the early years of the planting, but the hedgerow system requires expensive mechanical hedging equipment.

Lateral-bearing cultivars need mainly heading cuts to encourage scaffold development, while terminal-bearing cultivars usually need few heading cuts but instead need more thinning of unnecessary branches. Lateral-bearing cultivars produce greater yields, especially in the first few years of production.

Training young trees for the first 3 years. Encourage trees to grow rapidly during the first growing season. Select the strongest and straightest shoot to become the trunk, and tie it loosely to the stake every few weeks so it doesn't snap off in high winds. Pinch the tips off of other shoots after they've grown 10 to 12 inches long, and remove any root suckers. For hedgerow training, head the main shoot when it reaches 4.5 to 6 feet long to stimulate branching. For modified leader training, allow the main shoot to grow 7½ to 9 feet long before heading it.

During the first dormant season, remove any side branches arising from the trunk less than 3 feet off the ground for hedgerow training or less than 5 feet off the ground for modified leader training, to allow room for the trunk shaker during mechanical harvest. Head the leader at 3.5 to 5 feet for hedgerow training, and at 6.5 to 8 feet for modified leader training.

Lower scaffold limbs should develop during the second growing season, ideally arising at a wide, 75- to 90-degree angle and then growing outward at a 45- to 60-degree angle. Remove any narrow-angled branches that arise, and those that grow horizontally or below. Tip scaffold branches back slightly if they grow more than 4 feet long, so they don't bend, break or grow horizontally. Allow small non-scaffold branches to grow without pruning, so these can bear as soon as possible; they will be removed in a few years as scaffolds begin to bear.

Select lower scaffolds during the second dormant season. Retain 2 to 3 scaffold limbs plus the leader, with scaffolds spaced at least 1 foot apart vertically and 120 degrees apart horizontally, and remove other vigorous lateral limbs at their bases. Lightly head back scaffolds and the leader of lateral-bearing cultivars, but do not head the scaffolds of terminal-bearing cultivars. Allow less vigorous lateral branches to grow without pruning to provide early fruiting.

Additional scaffolds will develop during the third growing season, and should be tipped back if they grow more than 4 feet long. During this season, allow the leader to bend and become the uppermost scaffold for modified leader training. Again, allow small non-scaffold branches to grow without pruning.

In orchards where half the trees will be removed 10 to 12 years after planting, prune the temporary trees as little as possible during the first 3 years to encourage early fruiting.

Mechanical pruning of hedgerows from year 3 on. Hedgerow pruning during the third and subsequent dormant seasons consists almost entirely of heading cuts, so it is generally mechanized. Top hedgerows at a height that will remove ⅓ to ½ of last season's shoot length, and hedge sides to remove about 1/3 of last season's shoot growth. Cut by hand any upright shoots missed by hedging equipment, such as those growing between trees in the row.

From the fourth dormant season on, many growers hedge sides every other year, before the expected heavy crop of the alternate bearing cycle.

This helps reduce the heavy crop and avoids hurting the light crop the following year. Side hedging should remove about 1/3 of last season's shoot growth, but should remain within 4 to 5 feet of trunks so the total hedgerow canopy is maintained at 8 to 10 feet wide after each hedging. Wider canopies will produce too much shading of interior wood in such densely planted walnut orchards. Open alleyways between hedges should remain at least 12 to 14 feet wide to accommodate equipment.

Top hedging isn't needed once trees have grown to a height approaching the distance between rows, and is usually done only once during the third dormant season. If trees have not yet neared this height by the fourth or fifth dormant season, they may be topped once more to stimulate further upward growth by cutting off fruiting branches.

If hedgerow trees begin to crowd each other, hand pruning will be needed to remove some competing branches and admit more light. If trees grow taller than the distance between rows, workers must selectively cut back excessively long branches to a lateral branch by hand, using a pruning tower. Such hand pruning of tops results in much less undesirable regrowth than would machine topping.

Pruning conventional plantings from year 3 on. Choose additional scaffold limbs during the third dormant season for modified leader training, for a total of 4 to 6 primary scaffolds. Also select secondary scaffolds from among the branches growing from the primary scaffolds. Remove other vigorous limbs, and allow less vigorous lateral branches to grow without pruning. Again, lightly head back the scaffolds and the leader of lateral-bearing cultivars, but do not head the scaffolds of terminal-bearing cultivars.

Continue to lightly head the scaffolds each dormant season thereafter, as long as they produce 3 feet or more of new growth. Once trees reach the desired size or terminal growth of scaffold slows to less than 3 feet per year, stop heading them. Continue to leave small branches unpruned for maximum fruiting.

Most growers prune mature trees every other year, before the expected heavy crop of the alternate bearing cycle. Workers must prune trees by hand, but they remove much less wood than machine hedgers do, and growers avoid the expense of side hedgers. Workers mainly remove any crossing or competing branches to maintain good light penetration, and shorten any excessively long limbs.

In orchards where half the trees will be removed 10 to 12 years after planting, continue to prune the temporary trees as little as possible until 5 or 6 years before they will be removed. Then start to remove a large limb from each temporary tree each year, to allow permanent trees room to expand. After 4 or 5 years of this severe pruning, each temporary tree should have only a single upright limb or a narrow cluster of limbs left, and permanent trees will have grown to fill most of the orchard. Final removal of temporary trees should then cause little reduction in yield. Walnut wood is very valuable, and large limbs and trunks from removed trees may be sold for production of gun stocks or furniture.

Managing pollination to prevent pistillate flower abortion. When orchard pollen loads are high, 'Serr' and some other cultivars drop many newly-pollinated female flowers (Photo 23-15). 'Serr' can lose up to 90% of female flowers, while other cultivars are less prone to this problem but can lose up to 40% of female flowers. In established orchards planted primarily to 'Serr' that contain pollenizer trees, removing all pollenizer trees, or top-grafting pollenizer trees to 'Serr', has improved yields between 20 and 86%. Orchards of other cultivars that suffer from this disorder but still need some cross-pollination may benefit from removing or grafting over enough pollenizer trees to leave just 3 to 5% pollenizers.

Most years, 'Serr' finishes shedding pollen and catkins drop from trees before 'Serr' female flowers become receptive to pollen. During years when 'Serr' female flowers look like they'll become receptive while 'Serr' is still shedding pollen, using a tree shaker to shake off catkins before female flowers

Photo 23-15. Some Persian walnut cultivars, especially 'Serr', may drop many newly-pollinated female flowers, shown here, if orchard pollen loads are high. Such cultivars should be planted in solid blocks of a single cultivar each.

become receptive will also reduce pistillate flower abortion. Other cultivars that do not self-pollinate need no shaking to remove catkins.

PEST MANAGEMENT

Walnuts can be grown with fewer pesticides than many other fruits and nuts, particularly if pest-resistant cultivars and rootstocks and a site with few soil pests are chosen.

Pests that can kill trees. Walnuts suffer from many of these same pests mentioned for other fruit trees such as peaches, and again site selection and modification and rootstock choice are the main ways to prevent them. **Phytophthora root rot** (caused by several *Phytophthora* species) can be a problem on wet soils with slow water infiltration, and adding organic matter to such soils, building raised beds and choosing 'Paradox hybrid' rootstock when possible will prevent it. 'Paradox hybrid' or northern California black walnut rootstocks also resist **oak root rot** (caused by *Armillaria mellea*) and **Verticillium wilt** (caused by *Verticillium dahliae*).

Unfortunately, Persian walnuts on these two rootstocks may also suffer from **black line disease**, caused by **cherry leafroll virus**. Where this occurs, mainly in coastal walnut-growing areas and in the lower Sacramento and upper San Joaquin valleys, growers plant own-rooted Persian walnuts or cultivars grafted to Persian walnut seedling rootstocks. These are susceptible to all the diseases listed above, so need extra care to prevent them, such as fumigating or solarizing the soil and building raised beds.

All walnut rootstocks are less attractive to voles than other fruit and nut trees, so usually suffer little vole damage. Deer and rabbits also avoid walnuts, so prevention is only needed for new plantings located near natural habitats such as streamside thickets or woods.

Indirect pests that weaken trees, reducing yields and tree longevity. Both 'Paradox hybrid' and northern California black walnut rootstocks resist **root knot nematodes** (*Meloidogyne* species), and 'Paradox hybrid' rootstock also resists **lesion nematodes** (*Pratylenchus* species) but Persian walnut seedling rootstock or own-rooted trees are susceptible to both. Removing old roots and waiting a year before replanting walnuts reduces root knot nematodes by 90%, and keeping the site in grass crops for a few years before planting walnuts reduces all nematodes. Again, Persian walnut seedling rootstock or own-rooted trees probably need soil fumigation or solarization before planting.

Both 'Paradox hybrid' and Persian walnut roots are very susceptible to **crown gall**, caused by the soil bacterium *Agrobacterium tumefaciens*, so dip them in a biocontrol solution of antagonistic bacterium *A. radiobacter* (Galltrol™) before planting. Northern California black walnut rootstocks are less susceptible to crown gall but still benefit from this treatment. If crown gall is found in the planting, growers surgically remove gall tissues and then paint cuts with Gallex™, a crown gall eradicant derived from petroleum compounds 2,4 xylenol and M-cresol.

Two bacterial bark canker diseases may afflict walnuts that have been weakened by other stresses such as drought, excess soil moisture, root rot diseases, or wounds from tree shaking equipment. **Deep bark canker**, caused by *Erwinia rubrifaciens*, produces deep cracks running down scaffold and trunks (Photo 23-16), from which a dark brown fluid oozes from late spring through early fall. This canker usually enters trees through wounds made by tree shakers, and shakers can spread the disease from tree to tree. Minimize tree shaking time to avoid hurting trunks, and avoid introducing the pathogen into the orchard or spreading it by drenching shaker pads in a 10% bleach solution before shaking trees. Improving soil drainage where needed, and installing soil tensiometers to aid in keeping soil moisture levels adequate, will also reduce deep bark canker. The cultivar 'Hartley' is very susceptible to deep bark canker, and is being planted less for this reason. **Shallow bark canker**, caused by *E. nigrifluens*, affects many walnut cultivars but the dark, superficial cankers seem to do little damage to trees. Preventing tree stress will also prevent shallow bark canker.

Walnut aphid (*Chromaphis juglandicola*) used to be a major pest in California walnut orchards, but a parasitic wasp was introduced from France and Iran in the 1960's and now controls it so well that pesticide treatments are rarely needed. **Dusky-veined aphid** (*Callaphis juglandis*) has now become a pest in some California orchards (Photo 23-17). Monitor aphids on mature leaflet samples every other week from early May until populations start to build up, then once a week until late August. Check the upper surface of each leaflet for dusky-veined aphid and the lower surface for walnut aphid, taking 5 leaflets from each of 20 interior trees per 20 acres. Monitor border rows separately, because pesticide drift onto these edge trees sometimes disrupts natural control, necessitating treatment of just the border rows. Choose pesticides that do the least damage to aphid

Photo 23-16. Prevent deep bark canker by avoiding injuring trees during tree shaking for harvest. The Persian walnut cultivar 'Hartley' is very susceptible, so is planted less now.

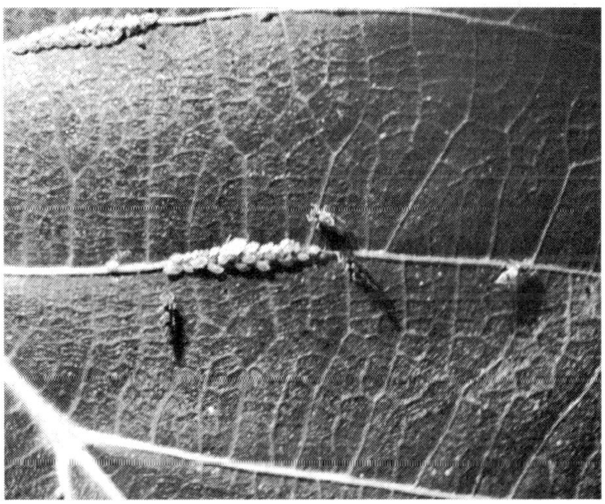

Photo 23-17. Monitor the upper surface of walnut leaflets for dusky-veined aphid and the lower surface for walnut aphid, and treat with a selective insecticide if necessary.

parasites as well as to natural enemies of mites and scales. Several aphids may occur on Eastern walnuts but rarely need control.

The same mite pests attack walnuts as most other fruit and nut trees: **two-spotted spider mite** (*Tetranychus urticae*), **Pacific mite** (*T. pacificus*) and **European red mite** (*Panonychus ulmi*). These mites frequently build to damaging numbers in walnut orchards when broad-spectrum pesticides, applied to control other pests, also kill mite predators. Choose pesticides that do the least damage to mite predators, and apply a dormant oil spray to reduce overwintering pest mites but not predators. Avoid treating a low to moderate population of European red mite in spring, since it can serve as an important food source for predator mite populations and enable them to build up enough to control spider mites in summer. Minimizing dust by adequate irrigation, maintaining a groundcover, and oiling roads also reduces pest mite buildups. Use selective miticides if needed.

A large variety of scale insects may attack walnut branches, sucking sap from the inner bark and causing dieback of inside fruiting wood on lateral-bearing cultivars. These include **walnut scale** (*Quadraspidiotus juglansregiae*), **San Jose scale** (*Q. perniciosus*), **frosted scale** (*Lecanium pruinosum*), and several others that are only sporadic pests. Again avoid using broad-spectrum pesticides that would kill natural enemies of scales, which can be very effective in controlling them. A dormant oil spray will prevent most scale outbreaks. Prune out and burn scale-infested branches as soon as they're seen.

Walnut caterpillar (*Datana integerrima*) can defoliate trees in the eastern half of North America during some years when populations climb to high levels. Walnut caterpillars form large colonies and all crawl down at the same time to molt on lower tree trunks, where they can be effectively controlled with directed sprays. Birds, *Trichogramma* wasps and other natural enemies keep populations of this pest low during most years, but during outbreaks growers must spray with Bt or other selective insecticides that target moth pests, such as the insect growth regulators tebufenozide (Confirm™) or diflubenzuron (Dimilin™). Spot treatments with these materials will also control other moth pests that may defoliate walnuts, such as **fall webworm** (*Hyphantria cunea*, photo 23-18), **fruit-tree leafroller** (*Archips argyrospilus*) or **red-humped caterpillar** (*Schizura concinna*). All three of these pests occur everywhere that walnuts are grown in North America.

Diseases that attack flowers, fruits and foliage. Walnuts suffer from few foliar and fruit diseases as compared to other fruits and nuts, and the need for pesticides to control these diseases can be minimized by choosing late-leafing cultivars that avoid them.

Walnut blight, caused by the bacterium *Xanthomonas campestris* pv. *juglandis*, is the most serious above-ground disease of Persian walnuts in California. This disease may also attack Eastern black walnuts, but Persian walnuts are much more susceptible. The pathogen attacks succulent young shoots and leaves during rains in early spring, causing lesions that may girdle shoots, leaf spots, and malformations of leaves. Bacteria also infect

Photo 23-18. Control fall webworm and other sporadic moth pests of walnuts with a directed spray of the biological insecticide Bt or an insect growth regulator.

Photo 23-19. Walnut anthracnose is the major foliage disease of eastern black walnuts and Persian walnuts grown in eastern North America.

catkins, causing them to shrivel and blacken, and infect pollen and are spread to female flowers. There bacteria invade the flowers and young fruit, causing water-soaked spots that develop into black, sunken lesions. Kernels of nuts attacked before shell hardening shrivel, while those attacked after shell hardening have discolored spots.

Apply 1 to 3 copper or biocontrol sprays before rains, starting at first pistillate bloom and continuing every 7 to 14 days until the rainy season ends in about mid-May, to prevent blight. Late-blooming cultivars such as 'Chandler' and 'Howard' avoid most of the disease period so need the fewest sprays, often just 1 or 2. Early-blooming cultivars such as 'Chico' and 'Payne' usually need 3 to 6 sprays, and growers should rotate materials to avoid inducing bacterial resistance to them. The biocontrol agent *Bacillus subtilis* (Serenade™) is effective against walnut blight, as is Sporan™, a concentrated blend of plant essential oils; both are considered organic. Some strains of walnut blight bacteria have developed resistance to copper, and if these occur, adding maneb to the copper material (allowed only in Tulare County) improves control.

Walnut anthracnose, caused by the fungus *Gnomonia leptostyla*, is the major disease of eastern black walnuts and Persian walnuts grown in the eastern half of North America, where this disease occurs. The fungus attacks leaves during rainy periods soon after full leaf expansion, causing small dark spots that later enlarge and merge to form large dead areas (Photo 23-19). The disease spreads rapidly in early summer, and can completely defoliate trees by mid- to late summer. This premature leaf drop is a major cause of poor kernel fill, and the fungus can also infect nut husks, causing sunken, necrotic spots which also hurt kernel development. Apply 2 to 4 fungicide sprays, when leaves reach ½ to ¾ mature size and at two-week intervals thereafter, to prevent most anthracnose. Walnut cultivars that leaf out late miss most of the anthracnose infection period so need only 2 sprays, while earlier-leafing cultivars need 3 or 4 sprays.

Branch wilt, caused by the fungus *Hendersonula toruloidea*, attacks sunburned portions of outer branches in Persian walnuts, inducing leaves to yellow, wither and turn brown. Wilting spreads to larger limbs, which retain dead leaves in autumn well after healthy leaves have fallen, making diseased limbs easy to recognize. Prune out such limbs as soon as they're seen, cutting at least a foot below the visible disease margin, and burn all prunings. Trees weakened by drought, nutrient deficiencies, or mite outbreaks are prone to sunburn damage that leads to branch wilt, so provide sufficient irrigation, nutrients and mite control during hot weather to prevent this disease.

Insects, birds and mammals that attack fruits and nuts. Choosing late-blooming and mid- to late-ripening walnut cultivars greatly reduces the number of insect pests that need control.

Codling moth (*Cydia pomonella*), described in chapter 16 for apples, is the key insect pest of early- and mid-blooming Persian walnut cultivars, often requiring insecticide treatments that disrupt natural enemies of other pests and lead to outbreaks of aphids, mites and scales, which then require more pesticides. Very late-blooming cultivars such as 'Cisco' and 'Franquette' generally escape this pest and need no treatments for it, while late-blooming cultivars such as 'Chandler' and 'Hartley' often need fewer insecticide treatments. Codling moth is a less serious pest of Persian walnuts in eastern North America, although choosing late-blooming cultivars is still wise. This pest does not attack eastern black walnut.

Growers time insecticide sprays with pheromone traps or the newly-available kairomone lure that attracts both sexes of codling moth, placed in mid-March in known codling moth "hot spots" in the orchard. For the least disruption of beneficial arthropods, use insect growth regulators such as tebufenozide (Confirm™) or diflubenzuron (Dimilin™). Dimilin™ must be applied before egg laying (50 to 100 days from the first 2 consecutive moth catches on traps) and Confirm™ must be applied before egg hatch (200 days from the first 2 consecutive moth catches on traps), so trap monitoring is critical to success with these products. Starting in May, scout for nuts infested with codling moth larvae; more than 3% infestation indicates a second insecticide spray is needed to control the second generation. Where codling moth populations are lower, the second generation can be reduced by releasing *Trichogramma* wasp parasites. Mating disruption has generally proven too expensive for walnut producers, but a trial using it allowed 83% of walnut growers to eliminate use of organophosphate insecticides.

Navel orangeworm (*Amyelois transitella*) causes the same damage to Persian walnuts as described for almonds and pistachios, and is prevented in the same ways with orchard sanitation, one or two insecticide sprays at husk split if needed, prompt nut harvest and fumigation after harvest by the processor. This pest is a scavenger that only attacks nuts after husk split or nuts damaged by codling moth or walnut blight, so late-ripening walnut cultivars that avoid those two pests also suffer less navel orangeworm damage, and usually need no sprays for it. Navel orangeworm does not occur in eastern North America.

Walnut husk fly (*Rhagoletis completa*), a cousin of apple maggot, seriously attacks mid- to late-season walnut cultivars in most areas of California and other Western states, except certain parts of the central and southern San Joaquin Valley, where hot summers preclude this pest. Flies emerge from the soil in mid- to late summer, lay eggs on walnut fruit, and the resulting larvae feed in the husks for 3 to 5 weeks. Larval feeding causes husks to decay and stain walnut shells, and early-season feeding may cause shriveled or darkened kernels. Damaged husks do not split normally to release the nuts, so are difficult to harvest. Control this pest with insecticides timed 7 to 10 days after flies start appearing on yellow panel and green sphere sticky traps (Photo 23-20) and a few nuts show evidence of egg-laying. The bioinsecticide spinosad (Success™ or SpinTor™) is a safer material, considered organic, that controls walnut husk fly. Early-ripening cultivars often mature before most husk fly larvae develop, so don't need insecticide treatments for this pest, though they need more codling moth, walnut blight and navel orangeworm control.

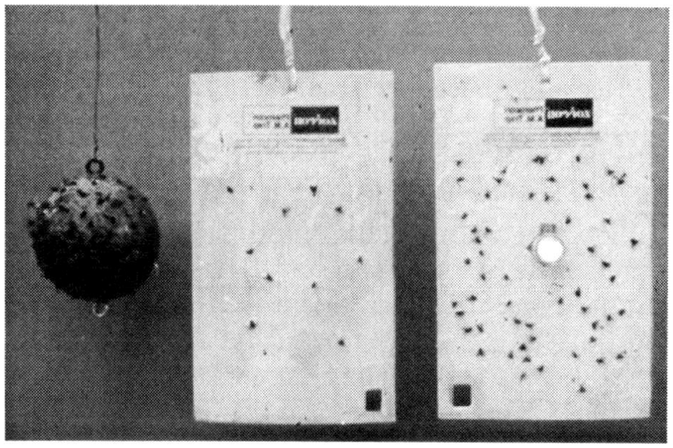

Photo 23-20. Control walnut husk fly with insecticides timed 7 to 10 days after flies start appearing on yellow panel and green sphere sticky traps, shown here.

Eastern walnut husk fly (*R. suavis*), a different species in the same genus, occurs in the eastern half of North America. Late-maturing black or Persian walnut cultivars avoid this pest if harvested as soon as hulls soften, because eggs laid after late September are unable to mature before harvest. If control is necessary, time insecticides based on sticky trap catches and egg-laying as above.

Two weevils may attack walnuts in eastern North America. **Butternut curculio** (*Conotrachelus juglandis*) feeds on Persian walnuts, eastern black walnuts, and butternuts. **Black walnut curculio** (*C. retentus*) attacks eastern black walnuts, and butternuts. Larvae of both species burrow into nuts, causing up to 50% to drop shortly after pollination and infesting some nuts at harvest, and also tunnel into new shoots, which may be killed back to old wood. Apply an insecticide just after pollination to prevent these weevils if they are problems.

Ground squirrels (*Spermophilus beecheyi* in California and related species elsewhere) feed on young and mature nuts in trees and on the ground, and may also severely damage trickle irrigation tubing by gnawing on it. Control them with poison baits as described for almonds and pistachios and in Chapter 9, Pest Management. Several species of **tree squirrels** (*Sciurus* species) and **pocket gophers** (*Thomomys* species) may occasionally invade walnut orchards, and can be controlled by trapping. **Crows** (*Corvus brachyrhynchos*) may take large numbers of walnuts in localized areas where large flocks congregate, and growers use frightening devices and toxic baits to control them.

GROWTH REGULATORS, HARVEST, PROCESSING AND MARKETING

Persian walnut hulls split open a few weeks before nuts are ready to harvest, and hulls of a few cultivars peel back to expose the nut within, but hulls of most Persian walnut cultivars and all black walnut hulls merely loosen contact with the nut shell. Persian walnut growers use mechanical tree shakers to knock nuts to the ground, then use mechanical sweepers and harvesters to pick up the nuts as described for almonds. Black walnuts could also be mechanically harvested if orchards are planted, but currently the crop is hand-harvested from native stands. Both types of walnuts must be hulled and shelled after harvest.

Both Persian walnut growers and those who harvest black walnuts will receive higher prices if the kernels are large in size and lighter in color, with little internal damage from insects or mold and little external damage from broken shells or adhering hulls once the nuts are hulled. Once nuts reach maximum size and the packing tissue between and around the kernels halves turns brown, harvesting as soon as possible will produce the lightest-colored kernels with the least internal damage (Photo 23-21).

However, hulls must soften enough to produce an indentation when a thumb is pressed into them (Photo 23-22) before harvest can take place, or the hulls will be too difficult to remove. Hulls may not mature to this stage until 2 to 3 weeks after kernels are mature, and waiting for hull maturation increases kernel darkening and damage from walnut husk fly, navel orangeworm and kernel mold. Many Persian walnut growers use a growth regulator to allow earlier harvest and preserve quality, and best black walnut quality is achieved if nuts are harvested as early as possible.

Use of ethephon to hasten harvest. Persian walnut growers often apply ethephon, the growth regulator that releases ethylene gas, as harvest approaches. Ethephon loosens walnut fruit, enabling greater nut removal and hullability at an earlier date. This can allow harvest 7 to 10 days earlier, resulting in a much greater proportion of light-colored kernels and much less damage from walnut husk fly, navel orangeworm and kernel mold. Ethephon is especially used on the earliest-maturing cultivars such as 'Payne' and 'Chico', but has also been used on mid- and late-ripening cultivars.

To promote earlier harvest, growers may apply ethephon when the packing tissue inside sampled nuts has just turned brown, to promote harvest about 14 days later, which is 7 to 10 days earlier than normal. This allows about 90% of nuts to be harvested in the first shaking, and a second shaking a week or so later is often economical. Alternatively, growers may wait about 10 days after the packing tissue of sampled nuts has turned brown to apply ethephon, at about 10 days before normal harvest. Harvest will then occur at the normal time but just one shaking will remove almost all nuts, eliminating the need for a second shaking, and almost all hulls will be easy to remove.

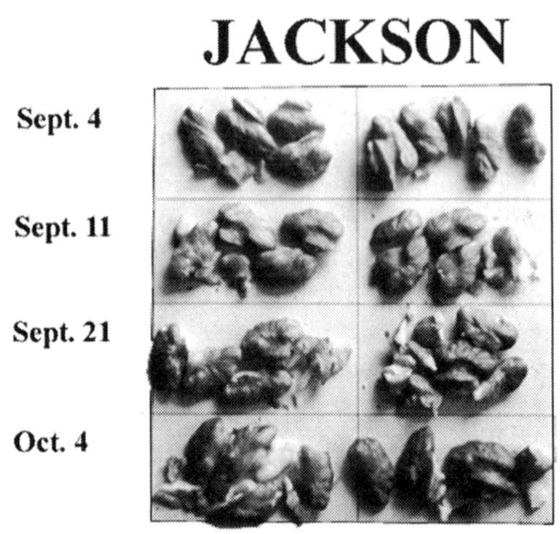

Photo 23-21. Harvesting as soon as possible after nuts mature produces kernels with the lightest color and least defects. This Eastern black walnut cultivar was mature at the first harvest date, and kernels darkened with delayed harvest.

Photo 23-22. Before harvest can take place, walnut hulls must soften enough to produce an indentation when a thumb is pressed into them, as in these Eastern black walnuts, or the hulls will be too difficult to remove from the nuts.

Growers must be careful to achieve thorough spray coverage when applying ethephon, since it is not translocated through leaves to nuts, but must be sprayed on nuts directly to affect them. Large growers must apply ethephon to different sections of the orchard in a sequence over a period of several days, applying ethephon to no more acreage during one day than can be harvested, picked up, hulled and dried in a day.

Mechanical tree shaking, nut pickup, hulling and drying. Walnut growers begin tree shaking about 2 weeks after ethephon application or when a test shaking removes at least 90% of the nuts from a tree. Unlike almonds, walnuts are not left on the ground to dry for several days to weeks before harvesting, but are windrowed and picked up with mechanical harvesters as soon as possible after shaking, and dried after harvest. Kernels can rapidly darken in walnuts left on the orchard floor as stains from the hulls seep into the nuts, especially in sun-exposed nuts when temperatures exceed 90 F or in shaded nuts when temperatures exceed 100 F, and most quality loss occurs within 9 hours after tree shaking.

Growers run newly-harvested walnuts through a hulling machine, then dry them before bringing them to a processing plant to be cracked. Stationary bin, pallet bin, trailer, or grain trailer dryers are used. Air heated to 110 F is blown through the nuts until they reach an average moisture content of 8%. Nuts harvested early in the season may start at 30 to 40% moisture and need 24 to 36 hours in the dryer to reach the desired moisture content. As the season progresses, moisture content of newly-harvested nuts drops, until at the end of the season, nuts may be at 8% moisture when harvested and not need drying.

Black walnuts harvested by hand from the wild are brought to local buying stations in the Midwest, where they are hulled, dried and bagged for shipment to a commercial sheller.

Walnut processing, storage and marketing. One large California grower processor, Diamond Walnut Foods, cracks, packages and markets about 50% of Persian walnuts, and other private commercial processors handle the rest. One commercial company in Missouri, Hammons Products, processes black walnuts. Shellers recover over 40% kernels from Persian walnuts, packaging them as kernel halves or pieces. Black walnuts harvested from the wild yield only 6 to 10% useable kernel pieces, so processors have developed markets for the shells. Ground black walnut shells are used to clean and polish metals and to aid in oil well drilling, and are ingredients in paint, explosives and make-up. Some Persian walnut shells are also ground and used as an industrial abrasive or in tire manufacturing to improve the grip of tires. Some Persian walnut processors burn the shells to produce heat and electricity for the processing plant.

About 60% of U.S. Persian walnuts are sold in the U.S. and about 40% are exported. Over 94% of nuts for the U.S. market are shelled and sold to grocery stores and warehouse clubs for retail sales and to food product manufacturers, while 6% are sold in-shell to grocery stores. About 58% of exported walnuts are sold shelled, and 42% are sold in-shell. Black walnuts are all sold shelled in the U.S., mainly to manufacturers of food products such as ice cream and baked goods.

Commercial packages of walnut kernel halves or pieces are often treated with synthetic antioxidants before packaging to retard development of rancidity, a common problem in nuts with high oil content. Shelled walnuts should be stored under refrigeration at 34 to 35 F for up to several months, or at 0 F for up to 2 years, to prevent kernels from turning rancid. Unfortunately, many supermarkets display shelled walnuts at room temperature, and if such products do not move quickly, some may develop rancidity before being sold. Many consumers do not know that walnuts, pecans and other nuts should be refrigerated or frozen when stored at home, and the industry should do a better job of making this fact known to increase consumer satisfaction and demand.

Further Reading

Jones, J. E., R. Mueller and J. W. Van Sambeek, eds. 1998. *Nut production handbook for eastern black walnut.* Southwest Missouri Resource Cons. and Dev. Inc. 147 pp.

Ramos, D.E., ed. 1998. *Walnut production manual.* Univ. of Cal. pub. 3373. 317 pp.

Reid, W.R., M.V. Coggeshall, and K.L. Hunt. 2004. Cultivar evaluation and development for black walnut orchards. In: *Sixth National Black Walnut Symposium.* West Lafayette, IN.

Reil, W. 2002. Walnut production in Yolo and Solano Counties of California. Available on-line at http://ceyolo.ucdavis.edu/Fruits_and_Nuts/2768.pdf

Strand, L.L., ed. 2003. *Integrated pest management for walnuts.* Univ. of Cal. pub. 3270. 136 pp.

Chapter 24
Pecans

Pecan trees are native to North America, and produce the largest commercial crop of native fruit or nut trees. The United States is the world's largest pecan producer, harvesting 200 to 400 million pounds of in-shell nuts per year, about 75% of the world's production. Mexico grows about 22% of the world's pecans, and Australia raises about 2%. Most U.S. pecans are sold in the U.S., as only 5 to 10% of the crop is exported, mainly to Canada and Mexico.

Pecans are native to floodplains of the Mississippi River and its tributaries, from central Illinois and southeastern Iowa south to Texas and parts of Mexico. About 25% of pecan production is harvested from managed native stands, while the other 75% comes from planted orchards. Georgia produces about 30% of U.S. pecans, Texas grows about 22%, and other Southern states provide most of the rest: New Mexico about 18%, Oklahoma about 8%, Arizona about 6%, Alabama and Louisiana each about 4%, and about 1% each from California, Mississippi, Kansas, North Carolina, South Carolina and Florida.

The pecan (*Carya illinoinensis*) belongs to the hickory genus in the walnut family, the Juglandaceae. Pecans have a much thinner shell that is far easier to crack than other hickories, and yield a greater proportion of nutmeats to shells. Consequently, much more selection and breeding of cultivars has taken place among pecans than among other hickories. A few cultivars have been selected and grown on a very limited basis from two other hickory species, shagbark hickory (*C. ovata*) and shellbark hickory (*C. lacinosa*). Both of these hickory species have also been crossed with pecans to create "hicans", but hicans have very low yields so are little grown.

ADAPTATION, TREE GROWTH AND BEARING HABIT

Pecans require a long, warm to hot growing season of at least 160 days to produce good crops of nuts with well-filled kernels. Cooling degree days (the sum of each day's mean temperature minus 65 F) must exceed 1200 per year, and solar radiation levels should usually be high during the growing season. Some pecan cultivars are winter-hardy to about -20 F (USDA zone 5) so can survive in most parts of the U.S., but where summers are cool, cloudy or short, nut kernels will not fully develop and will be unmarketable. U.S. pecan production can be divided into 3 main regions: the Southeast, the Southwest, and the lower Midwest (often called "Northern pecans", though this region includes the Texas panhandle, northern Oklahoma, and southern Missouri). Pecans are very deep-rooted trees and can be grown without irrigation in the Southeast and lower Midwest, but produce better with irrigation in those regions and require irrigation in the Southwest.

Pecans grow into very large trees, 70 to100 feet tall and 80 to 120 feet wide, and can live for 200 years or more (Photo 24-1). Native trees and those grown from seed often do not begin to bear until 8 to 12 years of age. Breeders have selected cultivars that begin bearing 4 to 6 years after transplanting grafted trees and produce a more compact tree, and growers have planted such cultivars at closer spacings in recent years in an effort to increase production in the early years of a pecan planting. Since pecan orchards do not produce a profit until the 13th to 15th year, many growers intercrop agronomic crops between trees to provide additional income.

Many pecan cultivars have a strong alternate bearing habit, producing a heavy crop one year and a light crop the next. Winter damage, drought, pest outbreaks or other factors that significantly reduce the crop can trigger this cycle, so growers manage these factors as much as possible to avoid biennial bearing. An overly heavy crop can also cause reduced nut set the following year and start an alternate bearing cycle, so growers must thin very heavy crops to prevent this. Despite such management efforts, the pecan industry still tends to produce much heavier crops every other year and lighter crops during the "off" years in between.

Like walnuts, pecans have pinnately compound leaves and are monoecious, producing pistillate (female) flowers at the tips of new shoots and staminate (male) flowers in pendulous catkins that hang from 1-year-old branches (Figure 24-1). Pecans are completely dichogamous, with staminate and pistillate flowers opening at different times so a cultivar can not pollinize itself. Protandrous cultivars shed pollen before their pistillate flowers are receptive to it, while pistillate flowers of protogynous cultivars become receptive to pollen before it is shed from that cultivar. Pecan orchards need at least one protandrous cultivar and at least one protogynous cultivar to provide good cross-pollination.

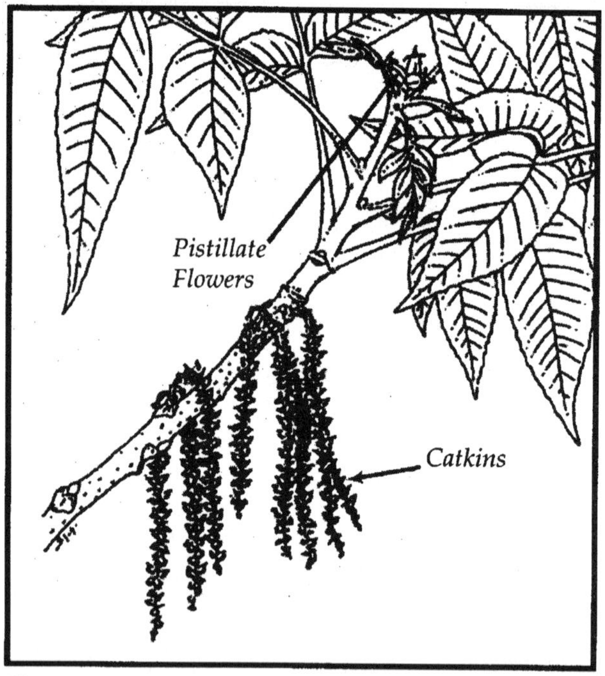

Figure 24-1. Pecans are monoecious, producing pistillate (female) flowers at the tips of new shoots and staminate (male) flowers in pendulous catkins that hang from 1-year-old branches.

SITE SELECTION AND PREPARATION

When planting pecans, choose a soil with moderate to good internal drainage, at least 3 feet of topsoil, and subsoil that allows free penetration of roots, water and air. Native pecans grow mainly in deep alluvial soils in floodplains along rivers and streams, and pecans can withstand several weeks of surface flooding (Photo 24-2), but the water table in such soils usually ranges from 10 to 25 feet

Photo 24-1. Pecans grow slowly into very large trees that can live for 200 years or more.

Photo 24-2. Pecans are native to river bottoms and can withstand several weeks of surface flooding.

below the soil surface during summer. The water table should not be closer than 6 feet from the soil surface in summer. Pecans will not perform well on an upland soil with a subsoil impervious to root growth, but can do well on a deep upland soil with a penetrable subsoil.

Pecans leaf out and bloom late so spring frosts are not usually a problem. Early fall frosts can threaten crops in areas with shorter growing seasons, so in those areas growers should choose early-ripening cultivars.

Many riparian soils where pecans once grew naturally have been cleared for agronomic crop production, but in recent years government payments have encouraged farmers to replant highly erodable floodplain lands to trees which hold the soil. Pecan plantings can qualify for conservation payments if planted with at least two other tree species which can be removed at the end of the contract period 10 to 15 years later, and if no pecans are harvested until the contract period ends. At least 300 trees per acre must be planted in such programs, so growers should weigh the cost of maintaining so many trees, plus the eventual costs of removing 90% of them, with the benefits of conservation payments before making such a contract. Alternatively, some agronomic crop farmers are simply diversifying into pecans on their land that's subject to periodic flooding.

Where riparian woods containing pecans exist, a native pecan grove can be created by removing all other tree and shrub species and thinning out the pecan trees. Native pecan production is centered in eastern Texas, Oklahoma, Louisiana and Kansas.

Modifying a site for planting. About 6 months before planting pecans, take soil samples from a depth of 8 to 15 inches. Modify soil pH if it falls outside the range of 6.0 to 7.5, and add phosphorous (P) or potassium (K) if these test low. Eradicate any perennial weeds such as johnsongrass or bermudagrass with glyphosate (Roundup™) herbicide. Remove any trees or underbrush, and burn brush piles away from where trees will be planted, as wood ashes on burning sites can raise soil pH and give off excess K, creating nutrient imbalances.

If the soil contains a plowpan or a clay hardpan, subsoil to a depth of 24 to 30 inches, both in the direction that rows will be planted, and perpendicular to that direction. Do this when soil is slightly dry to avoid compacting it. Planting a winter legume crop in fall, and disking it under in spring before planting pecans, helps loosen the soil and provides needed nitrogen (N).

Pecan orchards require smooth, level soil for efficient mechanical nut harvest, so growers must take soil from high spots and use it to fill in low spots before planting. Where furrow or flood irrigation will be practiced, land must be graded and water movement structures must be created. In some bottomland soils with moderately good drainage, constructing drainage ditches before planting will improve the site enough for good pecan production.

CHOOSING CULTIVARS

Pecan growers should choose cultivars that are adapted to their area's growing season (Table 24-1) and produce larger nuts, yet cultivars must ripen nuts before the first fall freeze (Photo 24-3). Cultivars whose shucks do not split open before the average date of the first fall freeze in the area will produce "stick-tights," whose shucks stick to the nuts and are unmarketable (Photo 24-4). Relatively early nut maturity also allows nuts to be harvested before Thanksgiving or Christmas, when demand and prices are higher, and when the weather is relatively dry. If wet winter weather delays harvest into January, nut quality can deteriorate.

Cultivars with a high percent kernel, the proportion of nut weight that is kernel rather than shell, fetch a higher price from shellers than cultivars with lower percent kernel. High yields also contribute to higher returns, if the nuts are good quality with a high percent kernel. Large nut size is especially important for growers who sell retail pecans.

Precocious cultivars, which come into bearing by 5 or 6 years after planting, are desirable as long

Photo 24-3. Pecan growers should choose cultivars whose shucks split open before the average date of the first fall freeze in the area, allowing nuts to drop and be harvested.

Table 24-1.

COMMON PECAN CULTIVARS

In approximate order of ripening (varies with area).

Cultivar	Regions where adapted	Nut			First flowers	Tree bears at age	Alternate bearing tendancy	Resistance to					Days before(-) or after Stuart	
		Kernel size	% Kernel	Yield				Pecan scab	Aphids		Phylloxera			
									Yellow	Black	Stem	Leaf	Leafing	Harvest
Peruque	MW	small	59%	medium	male	5-7	low	MS	R	MS	R	R	-1	-32
Colby	MW	medium	45%	medium	female	5-7	low	S	MR	MS	R	R	9	-30
Posey	MW	medium	53%	medium	female	6-8	low	R	VS	MS	R	R	7	-26
Kanza	MW	small-med	52%	med-high	male	5-7	medium	R	R	MS	R	R	8	-23
Pawnee	SW, MW	large	57%	high	male	4-6	high	S	R	MS	R	R	1	-21
Major	MW	small	50%	medium	male	7-9	low	R	R	MS	R	R	4	-21
Giles	MW	small-med	53%	high	female	5-7	medium	S	MS	MS	R	R	12	-12
Chetopa	MW	med-large	55%	medium	female	6-8	low	MR	MS	MS	R	R	10	-10
Candy	SE	small-med	50%	med-high	female	5-7	low	MS	MR	MR	MR	R	12	-10
Caddo	SE	small-med	56%	very high	male	5-7	low	R	MR	S	S	S	-6	-8
Shawnee	SE, SW	med-large	59%	med-high	female	6-8	medium	R	R	MS	R	R	2	-7
Gloria Grande	SE	large	48%	high	female	6-8	low	VR	S	VS	R	R	0	-6
GraTex	SW	med-large	59%	medium	female	7-9	low	R	S	S	MS	S	2	-5
Wichita	SW	large	60%	high	female	4-6	high	VS	MS	MR	MS	R	-3	-4
Forkert	SE, SW	med-large	62%	med-high	female	5-7	low	MS	S	S	R	R	-6	-4
Elliot	SE	small-med	55%	medium	female	5-7	high	VR	S	S	R	R	-15	-3
Maramec	SE	large	56%	low-med	female	4-6	low	MS	S	S	MS	S	-1	-3
Melrose	SE	med-large	54%	med-high	female	6-8	low	VR	R	R	R	R	-4	-1
Cheyenne	SE, SW	med-large	59%	high	male	4-6	med-high	MR	VS	S	MS	R	-2	-1
Stuart	SE	large	48%	low-med	female	8-10	low	MS	VS	MS	S	MS	0	0
Cape Fear	SE	large	55%	high	male	4-6	high	MR	S	VR	MR	R	-9	3
Kiowa	SE, SW	very large	58%	high	female	5-7	high	MR	MS	MS	R	R	-6	3
Western Schley	SW	large	57%	high	male	4-6	low	VS	VS	MS	S	S	-1	3
Desirable	SE	large	53%	medium	male	7-9	low	MS	MS	MR	S	R	-5	4
Choctaw	SW	very large	58%	high	female	6-8	medium	MR	MS	VS	VS	R	2	5
Sumner	SE	large	53%	med-high	female	4-6	low	R	S	S	R	R	0	11

Regions: MW = Midwest, from southern IL to northern OK and TX panhandle, USDA zones 5b to 7a
SE = Southeast, from NC and southern OK to FL and east TX, USDA zones 7b to 9b
SW = Southwest, from west TX to CA, USDA zones 7b to 9b

Pest resistance ratings: VR = very resistant, R = resistant, MR = moderately resistant, MS = moderately susceptible, S = susceptible, VS = very susceptible.

Average maturity date of 'Stuart' in central GA and AL: Oct. 25. 'Cheyenne' (1 day after 'Stuart') is mature about Oct. 8 in Brownwood, TX and about Oct. 21 in Tulare, CA. 'Colby' (30 days before 'Stuart) is mature about Sept. 25 in Chetopa, KS and about Sept. 28 in Columbia, MO.

Notes on Cultivars:
Peruque is very cold-hardy but VS to bird predation due to early ripening, small size and thin shell, which may crack during harvest.
Colby is very hardy to winter cold and leafs out late, avoiding most spring frosts.
Posey has dark kernels so harvest as early as possible to prevent excessive darkening. It leafs out late.
Kanza is very hardy to winter cold and leafs out late, avoiding most spring frosts.
Pawnee trees are upright, suited to high density, MR to low zinc, R to fall freezes, winter damage.
Major nuts are small, round, and don't crack well, limiting use. VS to leaf scorch, potato leafhopper, low zinc.
Giles breaks bud late, usually avoiding spring frosts; matures too late for most northern pecan areas, but does well in southern Midwest.
Chetopa breaks bud late; matures too late for most northern pecan areas, but does well in southern Midwest.
Candy breaks bud late; produces small nuts sold like natives for candy making, VR to bunch disease.
Caddo nuts are VS to bird predation due to early ripening, small size and thin shell. R to bunch disease; VS to leaf scorch, drought.
Shawnee is VS to stink bugs, mites, poor soil drainage; R to winter damage where it matures.
Gloria Grande, a 'Stuart' seedling, is R to stink bugs, birds, bunch disease, leaf scorch, downy spot, fall freezes, winter damage.
GraTex thins naturally so does not alternate bear, is R to leaf scorch but S to powdery mildew.
Wichita has narrow branch angles that need spreading, VS to low zinc, stink bugs, mites, hickory shuckworm; needs fruit thinning.
Forkert nuts may crack during mechanical harvest; VS to crow predation, fall freezes; MR to bunch disease.
Elliot leafs out very early so needs best sites; needs fruit thinning; R to drought, excess water; S to bunch disease, shuck disease.
Maramec is VR to winter injury in areas where it matures; strong limbs are R to breakage.
Melrose trees are upright but have strong branches, suited to high density; R to mites and low zinc, but VS to early fall freezes.
Cheyenne trees are about 2/3 the size of most cultivars, so can be planted closer but need excellent management. VS to low zinc.
Stuart, one of the first pecan cultivars selected in 1874, still provides over 20% of the crop though it's late to bear; VR to bunch disease.
Cape Fear is VS to leaf scorch, spring and fall freezes, so needs good water and air drainage. VS to stink bugs; needs fruit thinning.
Kiowa is VS to hickory shuckworm, low zinc; needs fruit thinning to prevent alternate bearing.
Western Schley is widely planted in SW as it's R to salt, low zinc, fall frosts, winter damage; VS to potato leafhopper, birds, bunch disease.
Desirable thins naturally so does not alternate bear, but weak limbs break easily; S to bunch disease, drought, poor soil drainage.
Choctaw is R to fall frosts, but nuts may crack and split sutures during harvest, so harvest and store promptly. VS to hickory shuckworm.
Sumner has strong branches R to breakage; VS to downy spot, stink bugs, fall freezes, winter injury.

Photo 24-4. Cultivars whose shucks do not split open before the first fall freeze produce "stick-tights", whose shucks stick to the nuts and can't be removed, so are unmarketable.

as they do not overbear as mature trees. Some precocious cultivars do tend to overbear in maturity, creating a severe alternate bearing cycle unless the heavy crop is reduced.

Cultivars with severe alternate bearing that were planted in the past have mostly been top-worked to other cultivars or removed in recent years, and growers now prefer to plant cultivars with lower alternate bearing tendencies.

Resistance to pecan scab is extremely important for pecan growers in the Southeast, where susceptible cultivars require 10 to 15 fungicide sprays per season to control this disease. In certain parts of the Southeast, the pecan scab fungus has overcome the resistance of some cultivars, so growers should check with their state's Extension Service to be sure cultivars listed as resistant to scab are still resistant in their area.

Midwestern pecan growers should also choose cultivars resistant to scab to minimize the need for fungicides, though usually only 2 to 3 sprays are needed per season there. Pecan growers in the arid Southwest can grow cultivars susceptible to scab since the low humidity prevents scab from being a problem there.

Resistance to both yellow and black pecan aphids is important to Southwestern pecan growers. Resistance to yellow pecan aphids can reduce pesticide use in the Southeast and Midwest, but black pecan aphids are only occasionally a problem in those regions. Pecan stem phylloxera is not a problem in the Southwest, but resistance to this pest can minimize insecticide use in the Southeast and Midwest. Most pecan cultivars offer resistance to pecan leaf phylloxera, a related insect that occurs in most areas where pecans are grown.

Pecan orchards need at least one protandrous cultivar and at least one protogynous cultivar to ensure cross-pollination. Cultivars that leaf out late have less risk of frost damage to new shoots and developing pistillate flowers.

SELECTING ROOTSTOCKS

Pecans are propagated on seedling pecan rootstocks. If a nursery does not specify the seed source of its pecan rootstocks, ask them before ordering. Seedlings of certain cultivars or selections have specific adaptation to the soil problems of certain areas. Native pecan seedstocks should generally be used within a 200-mile radius of their origin.

Seedlings of the Mexican pecan selection 'Borrado' make good rootstocks in south Texas where cotton root rot is a problem, since they tend to resist that disease. In west Texas where soils are alkaline, seedlings of native pecan trees selected for enhanced zinc uptake make better rootstocks. Seedlings of 'Moore', selected in Florida, are widely used in the Southeast and proved most productive in a northwest Louisiana rootstock trial in the 1930's. Seedlings of 'Giles', a native pecan selection from southeast Kansas, have greater cold hardiness and are widely used as rootstocks by midwestern nurseries.

Several seedling rootstocks induce earlier spring growth which is susceptible to spring frost damage. These include 'Curtis' and 'Elliott', both selected in Florida, and 'VC-168', selected in southern Arizona. These rootstocks should only be used in areas where spring frosts aren't a problem, such as southern California or near the Gulf Coast.

PROPAGATING PECANS AND ESTABLISHING AN ORCHARD

Pecan orchards can be very long-lived, and trees eventually grow very large, so pecan growers plan on removing some trees as the orchard matures. Pecan orchards with more than 50 trees per acre are considered high density, and mature pecan orchards can have just 4 to 9 trees per acre. Pecan is a valuable hardwood tree, and temporary trees can be pruned to at least 10 feet of straight trunk so that removed trees can be sold to timber buyers.

Orchard layout. Most growers plant pecans 30 to 35 feet apart in both directions, with trees aligned in a square pattern and 36 to 48 trees planted per acre. Such orchards require tree thinning within 12 to 15 years after planting. Some pecan growers do not want to remove trees until 20 to 24 years after planting, so plant trees at a 50- by 50-foot spacing (Photo 24-5), which is only 17 trees per acre. Other growers plant rows 60 to 70 feet apart and trees within rows 30 to 35 feet apart, using 18 to 24 trees per acre, to make intercropping agronomic crops between the rows easier.

Another way to make intercropping easier but plant more trees per acre is to plant in beds that are 2 trees wide and offset, so trees are 40 feet apart within rows but 30 feet apart between rows, and the 2 rows are just 22 feet apart (Photo 24-6). Distances between double rows are 58 feet, allowing easier intercropping without disturbing pecan roots within the beds. This system allows planting of about 27 trees per acre.

Some southwestern pecan growers have tried hedgerow plantings like those used for walnuts, with trees spaced 17½ to 35 feet apart within rows spaced 35 to 40 feet apart. When trees begin to crowd, limbs are mechanically hedged with large circle-saw hedgers instead of removing trees (Photo 24-7). Hedging may be done in one direction only as for walnuts, or if trees are spaced the same distance within as between rows, trees may be hedged in one direction one year and in the perpendicular direction a year or two later. Hedging must be repeated every 2 to 4 years throughout the life of the orchard, to encourage nut production in tree interiors.

This hedging system only works where light intensity is high and on cultivars that can fruit well in tree interiors, such as the southwestern cultivars 'Western Schley' and 'Wichita'; most southeastern and midwestern cultivars do not bear well inside the tree canopy. Hedging is also limited to arid areas where wood rots in pruning wounds are not a problem, and the high cost of hedging equipment, and

Photo 24-6. Planting pecans in offset double rows makes intercropping the land between the double rows easier, without disturbing pecan roots within the double-row beds.

Photo 24-7. Higher-density pecan plantings that require hedging only work where light intensity is high, the climate is arid so wood rots in pruning wounds aren't a problem, and on cultivars that can fruit well in tree interiors, such as the southwestern cultivars 'Western Schley' and 'Wichita'. The high cost of hedging equipment, and resulting tree damage, have limited this production method even in the Southwest.

Photo 24-5. Most growers plant pecans in a square pattern, with trees 30 to 35 feet apart in both directions, but some plant trees 50 feet by 50 feet apart as here, to reduce later tree thinning and to make intercropping easier, as with these soybeans.

resulting tree damage, have limited this production method even there. One large southwestern pecan orchard uses hedging, and an Australian grower has produced crops averaging over 3200 pounds per acre for 6 consecutive years using top and side hedging on one side every year.

Growers must include both protandrous and protogynous cultivars in plantings, and should arrange them in a way that ensures some of both types will remain as permanent trees.

Planting pecan trees. Most pecan growers buy bare-root, grafted trees as with other fruit and nut tree crops. Plant bare-root trees in early spring in most areas, or in late fall or winter in the Southwest. Young pecan trees have a strong taproot and few lateral roots so are difficult to transplant, and growers should always keep roots moist and plant as soon as possible after receiving trees. Prune off broken roots and cut the taproot back to 24 inches before planting each tree, and prune off about 1/3 of the top growth of each tree to reduce water demands on the root system.

Some growers plant container-grown, grafted pecan trees to preserve more of the young trees' root systems and enable faster orchard establishment. Plant container-grown trees in either mid-fall or early spring. Dig holes twice as wide as the container but no deeper

A few growers transplant pecan seedling trees and plan on grafting them to cultivars the following year. This method takes a year or two longer for a pecan orchard to begin bearing, but grafted trees of cultivars the grower wants may be unavailable.

Growing or buying rootstocks and grafting trees. Growers who propagate their own pecan trees often grow rootstock trees from seed. Choose medium-sized, well-filled nuts in the fall from a cultivar or native tree adapated to your area, and stratify them in moist sand at 30 to 40 F for 90 to 120 days before planting in spring, after frost danger has passed. Plant nuts in a nursery row, and transplant the most vigorous seedlings to the orchard the following spring. Seedling pecan trees may also be purchased from many states' Department of Conservation, or from commercial nurseries, and can be planted directly in the orchard.

Allow transplanted rootstock trees to grow until they exhibit strong annual growth before grafting to cultivars. Many trees will be too small to graft after just one year in the orchard, so the grower must wait to graft until subsequent years. The three-flap graft is commonly used.

CREATING A NATIVE PECAN GROVE

Where pecan trees grow natively in mixed hardwood tree stands, landowners often convert such land into native pecan groves. A well-managed native pecan grove can yield an average of 1000 pounds of pecans per acre per year, almost as much as many planted pecan orchards yield. The food industry buys many high-quality native pecans due to their high oil content and excellent flavor. It takes about 4 years after competing trees are removed and management begins before a native pecan grove reaches good commercial production, about the half the time required after planting an orchard of grafted trees.

Landowners can receive income from timber sales, hay crops or cattle pasturing in addition to nut harvests as they work through the following steps to a native pecan grove.

Removing trees other than pecans and brush. Native pecans grow primarily along creek and river bottoms where they compete with various species of oaks, ash, elm and other trees for light and space. The ground is usually covered with low shrubs and vines, which compete for water and nutrients. Pecan trees under such competition have narrow tree canopies and very low nut yields (Photo 24-8). Nut production increases immediately after removing all trees other than pecans and clearing the underbrush (Photo 24-9).

Some trees removed in this initial clean-up operation may have value as sawlogs, or can be chopped and sold for firewood. Following understory removal, growers should establish a permanent grass/legume groundcover to hold the soil. This groundcover can be utilized for cattle grazing in summer. Cattle grazing may reduce pecan yields through soil compaction, but the combined income will be greater than income from pecans only. Remove cattle in fall, at least 2 months before pecan harvest, to avoid contamination from cattle droppings at harvest.

Pecan tree evaluation and thinning. Most native pecan stands will still be too crowded even after all other trees are removed. Remaining pecan trees usually have inverted pyramid shapes since they developed under intense competition for light. Pecan trees bear nuts only at the tips of new growth, so only the very tops of such trees are productive. Removing some of the pecan trees allows light to penetrate into the grove and stimulates remaining trees to "bush out" and develop lower branches. Nut production increases as tree canopies become larger and well-rounded.

Thin native pecan groves gradually, over several years, until neighboring trees' branches are 10 feet apart. Gradual thinning avoids sun scald and wind breakage that could result from removing too many trees all at once. Some of the removed trees will probably be worth selling to timber companies (Photo 24-10), since pecan wood is the third most valuable hardwood, after black walnut and black cherry. Culled trees and tops may be sold as fuel wood, as pecan makes an excellent barbecue wood and a good fireplace wood.

Landowners should inspect pecan trees during initial grove clean-up and again during the next few growing and harvest seasons to select the best trees to keep. The most important pecan trees to save are the small ones, with diameters of less than 4 inches, since these trees can be grafted to superior cultivars. The next most desirable trees to keep are those with diameters of 4 to 12 inches, as these will develop a good nut-bearing canopy faster than larger trees. Trees with a diameter greater than 24 inches take longer to begin bearing, require too much space, would shade adjacent younger trees and inhibit their growth, and are usually more valuable as sawlogs.

Photo 24-8. Native pecan trees in river bottoms have narrow tree canopies due to competition with many other trees, and very low nut yields.

Photo 24-9. Nut production from native pecan groves increases immediately after removing all trees other than pecans, clearing the underbrush and establishing a ground-cover that can be grazed by cattle.

Photo 24-10. Pecan trees removed when thinning native groves or planted orchards may be sold to timber companies, as pecan wood is the third most valuable hardwood.

Among medium-sized trees that won't be grafted, retain trees with the largest nuts, the best production, and the least damage from pecan scab, pecan phylloxera or other pests. These traits can be recorded on a metal tag attached to each tree. Remove trees that bear few nuts over several years, or nuts so small that more than 140 are needed to weigh a pound. Also remove trees whose shucks do not split open before the average date of the first fall freeze, as such shucks stick to the nuts, rendering them unmarketable.

Multi-trunked, badly forked or leaning trees are difficult to mechanically shake for nut harvest, so are commonly removed when straight, single-trunked trees occur in the vicinity. Also remove damaged and decaying trees. Mark trees to be kept, with paint or plastic tape tied around the trunks, before removing other trees or contracting with a timber company or other operator to remove trees. Fortunately, the ideal timber tree is tall and straight with a narrow crown, just the opposite of the traits desired for a nut-producing tree.

Top-working young native trees. Most native pecan stands will contain several to many young trees that can be top-worked to improved pecan cultivars with larger nuts and higher yields. Choose cultivars adapted to the area, as discussed above, obtain scionwood, and top-work in early spring. The three-flap graft (Photo 24-11) works well on trees with a diameter up to 2 inches, while the bark inlay graft (Photo 24-12) works well on trees with a diameter of 2 to 4 inches.

Attach a stake to top-worked trees to help train the new leader arising from the scion, and to ensure

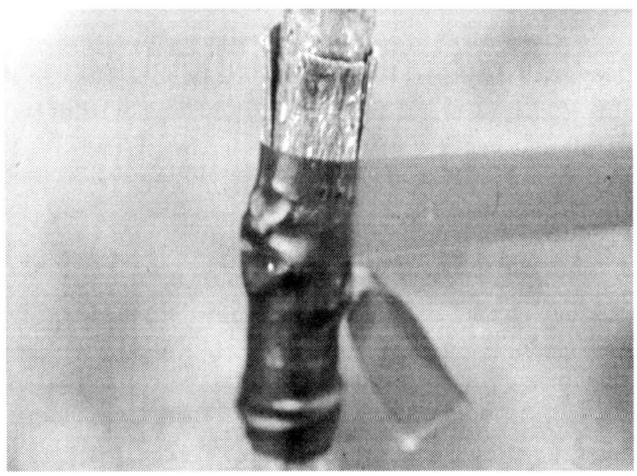

Photo 24-11. The three-flap graft works well for top-working small pecan trees with a diameter up to 2 inches.

Photo 24-12. The bark inlay graft is used to top-work trees with a diameter of 2 to 4 inches.

Photo 24-13. Attach a stake to top-worked trees to help train the new leader arising from the scion, and to ensure that strong winds do not blow the graft off the tree.

that strong winds do not blow the graft off the tree (Photo 24-13). Train top-worked trees like other grafted trees as described below. Top-worked trees should begin to bear 3 to 4 years after grafting.

Pruning medium to large native trees. Retained medium to large trees with multiple trunks should be pruned to a single trunk, removing no more than one trunk per year from each tree, to make tree shaking easier. Also remove limbs and branches which arise from trunks or limbs at a narrow angle, since these are prone to snapping off in high winds, ice storms, and under heavy crop loads. Cut back major limbs over several years rather than shocking trees by making large cuts all in one year.

TREE THINNING IN PLANTED PECAN ORCHARDS

Pecan trees in planted orchards must also be gradually thinned to avoid crowding and maintain nut production. Most pecan orchards will undergo 3 tree thinnings over 40 years. For an orchard planted at a spacing of 30 feet by 30 feet or 35 feet by 35 feet, the first thinning takes place about 15 to 20 years after planting, when half the trees will be removed. Trees that were planted in a square pattern can be thinned so rows now run diagonally, as shown in Figure. 24-2, by removing every other tree in each row in an offset pattern, so that trees planted 35 feet apart are now 49½ feet apart and trees planted 30 feet apart are now 42½ feet apart. This leaves 18 to 24 trees per acre.

Trees planted in double-row beds will have one entire row of each bed removed, leaving trees at a spacing of 40 feet by 80 feet and 14 trees per acre. Trees planted at 30 feet by 60 feet or 35 feet by 70 feet, and those planted at 50 feet by 50 feet, do not need this first thinning.

The second tree thinning takes place 10 to 12 years after the first, about 26 to 32 years after planting, and half the remaining trees will be removed. Orchards that were planted at 50 feet by 50 feet, and those planted with trees 30 to 35 feet apart and that received a first thinning so trees are now 43 to 50 feet apart, are thinned by removing every other tree in each row in an offset pattern. This again creates new rows that run diagonally to the old rows, and trees will now be spaced 60 to 70 feet apart, with 9 to 12 trees per acre. Orchards that were planted 30 by 60 feet apart or 35 by 70 feet apart to allow intercropping will now have every other tree in each row removed but not in an offset pattern, leaving trees spaced 60 to 70 feet apart in both directions, so again a square pattern of 9 to 12 trees per acre remains. Orchards originally planted to double rows and now spaced 40 feet by 80 feet will have every other tree removed, leaving 7 trees per acre spaced 80 feet by 80 feet.

Some pecan orchards may be maintained at a spacing of 60 to 80 feet between trees without additional tree thinning, if less vigorous cultivars were chosen, but many will need an additional tree thinning when trees are about 40 to 50 years old. The third thinning again removes every other tree in each row in an offset pattern, again creating new rows that run diagonally to the old rows, and trees will now be spaced 85 to 99 feet apart. This leaves 4.5 to 6 trees per acre.

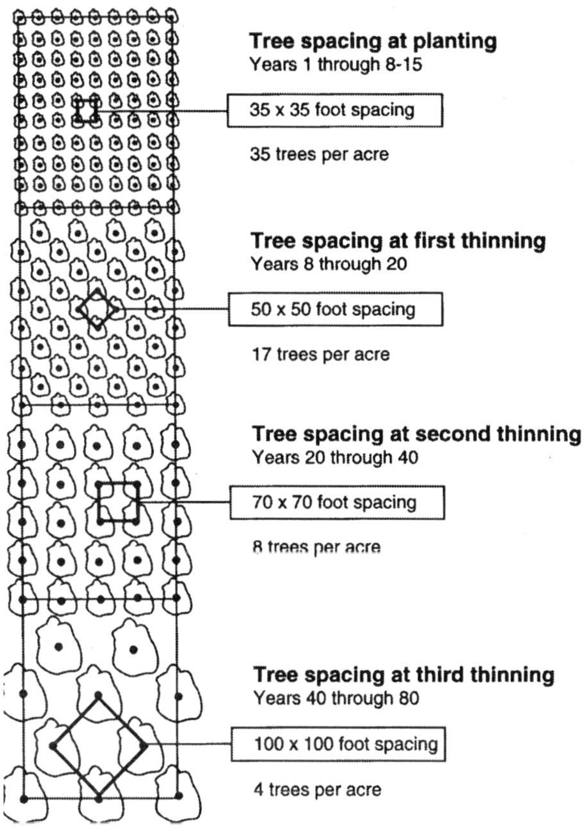

Fig. 24-2. Pecan orchards must be gradually thinned to avoid tree crowding and maintain nut production. Most undergo 3 tree thinnings over 40 years, as shown. (Figure from G.R. McEachern and L. A. Stein, eds., Texas pecan handbook.)

ORCHARD FLOOR MANAGEMENT, IRRIGATION AND FERTILIZERS

Pecan growers frequently plant intercrops to generate income during the first 4 or 5 years after orchard establishment. Summer annual soybeans and biennial clovers work well because they add nitrogen (N) to the soil and are less competitive for moisture than summer annual grasses, such as corn and sorghum. Vegetable crops, wheat, dry beans and cotton also have been intercropped with pecans for the first few years after planting the orchard. Some pecan growers plant peaches between pecan trees, planning that the peaches will be removed within 10 years since peach trees are short-lived anyway, but peach intercrops reduce the growth of pecan trees.

Southeastern and Midwestern pecan growers should plant a permanent groundcover by the time pecan trees are 5 years old (Photo 24-14). Cool-season perennial pasture grasses such as bluegrass and ryegrass make good groundcovers. Avoid warm-season perennial grasses such as Bermudagrass as they are too competitive, and legumes such as clovers and alfalfa because they attract stinkbugs. Growers should mow this groundcover during the entire growing season to conserve water for the pecan trees. Southwestern pecan growers generally keep the orchard floor free of vegetation to conserve water after the pecan orchard begins bearing, since the dry climate does not cause soil erosion.

Irrigation systems. Most native pecan groves, and many planted pecan orchards in the Southeastern and Midwestern U.S., receive no irrigation, since pecans are deep-rooted trees that usually mature good crops with natural rainfall in those areas. An irrigation system is essential for pecans in the Southwest, and may be installed to improve crop quality and consistency in planted Southeastern orchards. The water source should have less than 350 to 500 ppm dissolved salts, a sodium absorption ratio of less than 3 to 5, and less than 1 to 2 ppm boron, although sandy soils can withstand slightly more saline irrigation water.

Flood irrigation (Photo 24-15) was the most common method for Southwestern pecans due to low initial cost to build dikes and water control

Photo 24-14. Southeastern and Midwestern pecan growers should plant a permanent groundcover, such as this sod, by the time pecan trees are 5 years old.

Photo 24-15. Flood irrigation was the most common irrigation method for Southwestern pecans due to low initial cost, but increasing water costs and low efficiency have reduced its use in recent years.

structures, but increasing water costs and low efficiency of this method have led many growers to switch to sprinkler or drip irrigation in recent years. Sprinkler systems work well for mature pecan plantings, but waste water on young orchards unless the intercrop also needs irrigation. Pecan trees leaf out and bloom late so do not need sprinkler systems for frost protection, which is fortunate since sprinklers couldn't cover many of the blooms on such tall trees. Pecan growers who use sprinkler systems install under-tree systems (Photo 24-16) to avoid wetting the foliage and increasing disease problems.

Trickle irrigation systems (Photo 24-17) are the most efficient, but capacity must increase as trees grow. Many growers start with a single lateral line down one side of each tree row, either on the soil surface or buried about 6 inches in the soil, and place a single emitter about 1½ feet away from each tree. As trees grow and water requirements increase, an additional emitter per tree is punched into the line every couple of years until there are 4 emitters per tree providing water evenly around the trunk from 1½ to 20 feet away. Then a second lateral line is added on the other side of each tree row, and additional emitters are added up to 4 per tree from the second line, for a total of 8 emitters per tree. When trees come into production, micro-sprinkler emitters can replace drip emitters to wet a greater area. As trees are removed from the orchard, their emitters will then wet the soil for the expanding root systems of remaining trees. The system must be designed from the beginning to provide enough water for a mature orchard, about 55 inches per year. This is difficult to do, so drip irrigation is declining in popularity for pecans, especially in the Southwest where inadequate water application can exacerbate soil salinity.

Pecan water needs. During a pecan orchard's first two growing seasons, irrigate once a week in the absence of rain, applying seven gallons per tree during each irrigation. From the third year on, check soil water-holding capacity in the top 32 inches of soil once a week using tensiometers (Photo 24-18), neutron probes, or other methods, and irrigate before available water is depleted. Calculate the amount of water to apply per tree from pan evaporation measurements multiplied by a water-use factor based on temperature and multiplied by tree rooting area calculated from canopy diameter.

Mature pecan trees need water during 4 critical periods: initial shoot growth and flowering in mid-spring, fruit sizing in mid-summer, kernel filling in

Photo 24-16. Solid-set, under-tree sprinkler irrigation systems avoid wetting the foliage and increasing disease problems.

Photo 24-17. Trickle irrigation is the most efficient type of irrigation, but capacity must increase as trees grow.

Photo 24-18. Soil tensiometers can measure soil water-holding capacity in the top 32 inches of soil. Check these once a week to determine when to irrigate.

late summer, and shuck split in fall. Lack of water during March or April can reduce the number of nuts that develop, and water shortage during June or July will reduce nut size. Drought during August or September will cause kernels to shrivel and fail to fill the shells. Finally, nuts need water in October so shucks will split properly and nuts will shake free from shucks during tree shaking.

Since pecan trees need water from late March until early October, deficit irrigation does not usually work well on pecans. However, the trees are deep-rooted, and non-irrigated pecans can withstand 3 to 4 weeks of no rain if the soil has a high water-holding capacity that was full before the drought started. Drip-irrigated orchards may receive water 2 to 3 times per week during spring and fall, and every other day or daily during summer heat, while flood-irrigated orchards receive much larger amounts of water at 2- to 3-week intervals. Each mature pecan tree needs 120 to 200 gallons of water per day.

Fertilizing. All pecan orchards and native pecan groves need N fertilizer for good tree growth and yields. Many soils need no other added nutrients, but high-rainfall areas may also need P and K, and alkaline soils in arid areas need foliar zinc (Zn).

Apply N three times per year, in early March, early May and early October (Photo 24-19). Apply enough N to stimulate 3 to 4 feet of terminal growth per season on non-bearing trees, 1 to 2 feet of terminal growth on young bearing trees 6 to 12 years old, and 6 to 12 inches of terminal growth on mature trees over 12 years old. Too little N reduces tree growth and delays the onset of bearing, while too much N produces excess vegetative growth that is prone to wind breakage and winter damage. Generally, each tree should receive $^1/_3$ to $^1/_2$ pound of actual N per inch of trunk diameter per season. River-bottom orchards or native groves usually need 90 to 100 pounds of actual N per year. Upland orchards on sandy soils need 120 to 140 pounds of actual N per year.

Adjust N applications to reduce biennial bearing if that becomes a problem. Apply more N during "on" years of heavy crops, and less N during "off" years of light crops. Extra N during the heavy-crop year will ensure that trees have adequate N to both ripen the heavy crop and set fruit buds for next year, while less N during a light-crop year will reduce vegetative growth and fruit set for next year.

Where soil pH is above 7.0, zinc becomes unavailable. Trees on such soils need 4 to 5 foliar applications per year of zinc sulfate or zinc nitrate. Apply zinc sprays at 2-week intervals starting at leaf burst.

TRAINING, PRUNING AND MANAGING CROPS

Train pecan trees to a central leader as described in chapter 4. Retain as many lower branches as possible while trees are young to maximize photosynthesis. Remove lower branches when they grow to 1 inch in diameter to allow equipment access.

Pruning low-density and high-density orchards. Prune mature trees only enough to permit free air movement and admit light throughout the tree canopy. Temporary trees that will be removed later during tree thinning are often "wisped" in the Southeast by removing entire limbs as they start to compete with limbs of permanent trees.

High-density orchards must start mechanical hedging by the time trees are 8 to 12 years old, so no large cuts are made into limbs. Thereafter, hedgerow plantings require mechanical pruning of sides every 2 to 3 years. Successful growers have hedged one side one year and the opposite side the next year, cutting sides 10 to 12 feet from trunks and tops at 33 feet.

Mechanical tree shakers used for harvesting do a good job of shaking most dead or damaged branches from pecan trees, so little detail pruning of mature trees is needed. If any limbs or branches become diseased or broken by storms, growers should remove them as soon as they're noticed. Most such pruning is done with a chainsaw on pecan trees, and often requires a hydraulic lift or platform to reach the branches in need of pruning.

Mechanical fruit thinning. Many pecan cultivars tend to alternate bear and benefit from fruit thinning during heavy crop years. Thinning

Photo 24-19. Apply enough nitrogen (N) to mature pecan trees (over 12 years old) to stimulate 6 to 12 inches of terminal growth per season. Apply N several times between March, as here, and September.

the young fruit, before kernels reach full water stage in late July to mid-August, enables remaining nuts to fill better and induces much greater return bloom the following year, evening out production from year to year.

Chemical thinning agents have not proven effective on pecans, and often cause defoliation, so mechanical thinning using trunk shakers is the best method. Since thinning is done in late summer when sap is still actively flowing, tree shaking for fruit thinning could potentially cause more bark damage than trunk shaking at harvest time. Growers can prevent most damage by using softer donut pads rather than hard shaker pads, by lubricating the pads with silicone grease, and by using the shortest shaking time that produces effective fruit thinning.

Determine whether a tree needs thinning by examining 100 terminal shoots in mid-canopy, counting the ones which have at least one nut as fruiting shoots, and determining the percentage of shoots that are fruiting. Cultivars with larger nuts can support a lower percentage of fruiting shoots and thus need thinning more often than cultivars with smaller nuts.

Generally, the optimum crop load for cultivars with very large nuts, such as Kiowa or Choctaw, is 45 to 50% fruiting shoots. The optimum crop load for cultivars with large nuts, such as Wichita or Pawnee, is 50 to 60% fruiting shoots, while the optimum crop load for cultivars with medium-sized nuts, such as Cheyenne or Giles, is 60 to 70% fruiting shoots. If higher fruiting percentages than these occur, plan on shaking trees to thin them.

Cut open several sample nuts starting in late July or early August to determine the best time to thin. After each female flower was pollinated and fertilized in late spring, the ovule (the tissue that becomes the kernel) expanded and lengthened from the nut tip toward the peduncle or stem. Nuts enlarge during early summer, reaching their final size by late July to early August, while ovules are still growing. The space within the nut is filled with liquid endosperm until the ovule extends to the stem end of the nut, when the kernel begins to fill and nuts pass from the water stage to the dough stage.

Thin when ovules are about 50% expanded (Photo 24-20) and endosperms are still liquid so nuts are still in the water stage. Thinning before this stage requires more shaking time and can damage trees, while thinning after this stage will not improve nut quality or return bloom the following year as much.

Mechanical thinning usually removes individual fruits within clusters rather than entire clusters, and greater thinning occurs on large, stiff branches than on small, limber branches. Growers should shake trees in spurts of 2 to 3 seconds each until thinning is complete (Photo 24-21), leaving the desired percentage of fruiting shoots and 2½ to 3 nuts per cluster on average. This may require removing more than half the young fruit. Nut distribution after thinning may be enhanced by shaking trees in two directions, a north/south orientation and an east/west orientation.

PEST MANAGEMENT

Irrigated pecans planted in arid areas suffer from few diseases and insect pests, although they require greater inputs for water and zinc. Native pecans have fewer diseases than pecans planted in the Southeast, because humidity is lower where pecan trees are native. Southeastern pecans need the most pest management, and growers should choose cultivars that resist pecan scab and other major diseases in their area.

Pests that can kill trees. Pecans suffer from fewer pests that can kill trees than do most other fruit and nut crops. Since pecans withstand wetter soils, Phytophthora root rot rarely causes problems, and most areas need no special soil modifications or rootstock choices.

Cotton root rot, caused by the fungus *Phymatrotrichum omnivorum*, may occur in heavy clay, high pH soils in the southwestern U.S. and Mexico. Where this disease occurs, do not plant pecans within 2 years of growing cotton, and plant trees on slight mounds to avoid water standing around trunks. Seedlings of the Mexican pecan selection 'Borrado', which tend to resist cotton root rot, make

Photo 24-20. Thin excess pecans when ovules are about 50% expanded as shown, and endosperms are still liquid so nuts have not yet reached full water stage.

Photo 24-21. Pecan growers use a tree shaker to mechanically thin pecans. Shake trees in spurts of 2 to 3 seconds each until thinning is complete.

good rootstocks. Adding two parts cedar shavings to one part soil around trees when planting greatly reduces tree mortality from cotton root rot.

Deer (*Odocoileus* species) may browse on foliage or bark of young pecan trees, killing them or preventing them from developing well, where woods occur near pecan plantings or groves and deer populations are high. Electric or conventional deer fencing is often needed around young orchards where deer populations are high. Leasing hunting rights may adequately protect older orchards where deer populations are low to medium.

Indirect pests that weaken trees, reducing yields and tree longevity. Ask other local growers or Extension workers which of these indirect pests are troublesome in your area before planting, and choose cultivars with resistance to these pests, modify the soil if needed or treat with a biocontrol agent to prevent problems.

Crown gall, caused by the bacterium *Agrobacterium tumefaciens*, commonly infects pecan trees as well as many other tree species. It very rarely kills pecan trees, but reduces their vigor and yields. Before planting a new orchard, dip tree roots in the bicontrol solution *A. radiobacter* (Galltrol™) to prevent most infections for several years. Older trees that became infected with crown gall but received fertilizer and good control of other pests have produced well for over 50 years, so removing older trees with crown gall is not economically justified. Growers may surgically remove galls from older trees and paint them with Gallex™, a crown gall eradicant derived from petroleum compounds 2,4 xylenol and M-cresol.

Root-knot nematodes (*Meloidogyne* species) may cause leaf yellowing and dieback of new growth on young pecan trees in sandy soils, reducing tree growth and delaying bearing. Test sandy soils for nematodes before planting, and if nematodes occur at moderate to high levels, grow cover crops to reduce them as outlined in Chapter 2. **Ring nematodes** (*Criconemella xenoplax*) may also damage young pecan trees in soils deficient in magnesium (Mg), so apply dolomitic lime to increase Mg levels in such soils.

Bunch disease, caused by a phytoplasma that is spread by a leafhopper, can infect pecan trees in the Southeast and Midwest. Infected trees have a proliferation of shoots on lower scaffolds, with larger than normal leaves (Photo 24-22), in contrast to the smaller and chlorotic leaves of shoot bunches on zinc-deficient trees. Bunch disease reduces yields and causes inferior nut quality. Where bunch disease is serious, avoid very susceptible cultivars such as 'Elliot' and 'Desirable', and choose resistant cultivars such as 'Candy' and 'Stuart'. Native pecan producers should remove any trees with bunch disease. Remove and burn or otherwise destroy infected trees as soon as symptoms are seen. If only a few scattered bunches occur on large trees, pruning out infected branches at least a foot below the bunched shoots can be enough to prevent disease spread.

Aphids may attack pecans in all regions, and are often the most serious insect pests in the Southwest, where few other insects attack pecans. Three aphid species are commonly pecan pests. **Yellow pecan aphid** (*Monelliopsis pecanis*) and

Photo 24-22. Bunch disease, a problem in the Southeast, causes a proliferation of shoots on lower scaffolds, with larger than normal leaves.

black-margined aphid (*Monellia caryella*) are similar and often grouped together as "yellow aphids" (Photo 24-23). Both secrete honeydew which serves as a growth substrate for sooty mold which reduces photosynthesis and yields, and very high populations of yellow pecan aphid can defoliate trees. **Black pecan aphid** (*Melanocallis caryaefoliae*) causes much more damage than yellow aphids, as just 3 aphids per compound leaf can cause severe leaf damage, defoliation, reduced kernel filling and decreased fruit bud formation for next year's crop.

Water stress and low leaf N levels increases damage from black aphids, so irrigated trees that receive adequate N fertilizer suffer less damage. During the growing season, growers should monitor aphid populations and apply selective insecticides if populations exceed thresholds of 25 yellow aphids per compound leaf or 3 black aphids per compound leaf. Avoid using pyrethroid insecticides for control of other insects in early to mid-summer, as these can kill aphid predators and lead to an outbreak of yellow aphids.

Phylloxera are aphid-like insects that create galls on leaves or shoots. Five species are common on pecans, including one that attacks stems and fruit, and four that attack leaf blades. **Pecan stem phylloxera** (*Phylloxera devastratix*) attack shoots, fruit, and leaf midribs (Photo 24-24), and is the most damaging species, inhibiting flower production.

Pecan leaf phylloxera (*P. notabilis*) is the most prevalent species attacking the leaves, and can cause significant defoliation. Many pecan cultivars resist one or more species of phylloxera, but each species may develop biotypes that sometimes overcome that resistance in local areas. Growers can generally avoid phylloxera damage by choosing resistant cultivars or by applying an insecticide at leaf burst to orchards that had problems with phylloxera the previous year.

Pecan leafscorch mite (*Eotetranycus hicoriae*) is the worst spider mite attacking pecans, and large infestations can cause defoliation. They prefer shady interior parts of trees, feeding on the undersides of leaves, so growers should monitor these locations when weather is hot and dry. Mite outbreaks may occur after use of certain insecticides for other pests, such as carbaryl (Sevin™), so use other insecticides which don't hurt mite predators whenever possible. If mites have been problems in past years, apply a selective miticide if mites increase to more than 10 per compound leaf during the growing season.

Photo 24-23. Yellow pecan aphids, such as these black-margined aphids, reduce pecan yields, and very high populations can defoliate trees.

Photo 24-24. Pecan stem phylloxera attacks shoots and fruit as well as leaves, forming galls. Here, a single gall is sliced open to reveal the adult phylloxera inside. Choose cultivars resistant to this pest to avoid the need for special pesticide treatments.

Fall webworm (*Hyphantria cunea*) outbreaks occur some years in all pecan-growing regions. These gregarious caterpillars build webs which contain 100 or more webworms, where they feed on pecan leaves. Predators and parasites control webworms most years, but when outbreaks occur, spot-treat webs with a selective insecticide such as tebufenozide (Confirm™) or the biological insecticide *Bacillus thuringiensis* (Bt; Dipel™ or Javelin™).

Diseases that attack foliage, flowers and fruit. These key diseases must be avoided by planting in regions where they don't occur, or choosing cultivars resistant to them where they do occur.

Pecan scab, caused by the fungus *Cladosporium caryigenum*, is the worst disease of pecans in the Southeast and Midwest, where growers should choose cultivars resistant to scab. The Southwest is too arid for scab development, so susceptible cultivars may be grown there. Since pecan orchards may continue to produce for over 100 years and pecan scab pressure in the Southeast is heavy, strains of the fungus have developed that now overcome the resistance of many Southeastern cultivars which formerly escaped scab, and such trees require 6 to 8 fungicide sprays per season. Growers must rotate among different fungicide groups, using both protective and systemic fungicides, to avoid developing fungal strains resistant to certain fungicides. Scab pressure is less in the Midwest and many trees in native groves resist pecan scab, and where scab sprays are needed, usually just 1 or 2 per season suffice.

Scab causes olive-green lesions on leaves and fruit that darken to brown as the season progresses (Photo 24-25). Leaves are susceptible only in early spring while they expand, and fruit are susceptible from nut set until the water stage. Severe leaf infections can cause defoliation, early fruit infections cause nuts to abort, and later fruit infections cause kernels to fill poorly or shucks to fail to open at harvest. In addition to choosing resistant cultivars and using fungicides if needed, growers should plant trees no closer than 35 by 35 feet apart to allow good air circulation to dry trees quickly after rains, and should thin trees when they begin to crowd. Irrigated orchards should utilize trickle systems, micro-sprinklers or low-angle sprinklers that do not wet the foliage.

Fungal leaf scorch may occur in the Southeast, especially in late summer on trees suffering from water stress. The disease may be caused by any of several fungi in the genera *Phomopsis, Glomerella, Fusarium, Pestalotia, Alternaria*, or *Gleosporium*. Infected leaflets die from the tips or margins inward, and severe infections can attack shucks and cause defoliation that reduces yields. Irrigation reduces fungal leaf scorch, as do fungicide sprays for scab or other diseases. Where fungal leaf scorch causes problems during most years, growers should avoid very susceptible cultivars such as 'Cape Fear' and 'Choctaw'.

Insects, birds and mammals that attack fruits and nuts. These are the key pests that must be controlled in many pecan-growing regions. Generally 3 to 5 insecticide sprays per season prevent damage from the insect pests, while vertebrate pests are controlled with habitat management, hunting and scare tactics.

Pecan nut casebearer (*Acrobasis nuxvorella*) caterpillars feed on newly-set nut clusters in May to mid-June, tunneling into young fruit and causing them to abort (Photo 24-26). This moth pest occurs throughout the Southeast and Midwest and west to New Mexico. During some years when crop set is heavy and casebearer populations are low, feeding actually provides some beneficial fruit thinning of an excess crop. When pest populations are heavy, however, casebearers may damage 40 to 65% of nuts if not controlled, which can greatly reduce yields. Pecan growers should monitor this pest with pheromone traps and sample nut clus-

Photo 24-25. Pecan scab, the worst disease of pecans in the Southeast and Midwest, causes olive-green lesions on leaves and fruit that darken to brown as the season progresses. Choose cultivars that resist scab in these regions.

Photo 24-26. Pecan nut casebearer caterpillars feed on newly-set nut clusters, tunneling into young fruit and causing them to abort. If nut set is heavy and pest populations are low, this can actually provide beneficial fruit thinning, but heavy pest populations require insecticide treatment.

ters for damage. When damage exceeds 2 to 4% of sampled fruit, apply a single spray of a selective insecticide such as tebufenozide (Confirm™) or a pyrethroid.

Hickory shuckworm (*Cydia caryana*) feeding before shell hardening may also cause premature nut drop, and feeding after shell hardening causes stick-tights with poorly filled kernels that may contain larvae. This moth pest occurs throughout the Southeast and Midwest and west to New Mexico, and has 3 to 5 generations. The overwintering generation emerges in spring before fruit set and does little damage, mainly infesting phylloxera galls. Subsequent generations before shell hardening cause nut drop, and generations emerging after shell hardening feed on the shucks, interrupting the flow of sugars to the seeds.

Insecticide treatments for hickory shuckworm before shell hardening are only necessary if the orchard has had a previous history of heavy nut drop. Hickory shuckworm after shell hardening is often controlled by insecticides applied for pecan weevil. In regions where pecan weevil does not occur, a single insecticide applied at shell hardening will usually prevent damage.

Pecan weevil (*Curculio caryae*) is generally the worst fruit-attacking insect of pecan where it occurs, throughout the Southeast and Midwest. This snout beetle drills holes through both shuck and shell (Photo 24-27) to lay eggs in the pecan kernel, where larvae hatch and devour the kernels (Photo 24-28). Early-ripening cultivars are most susceptible, and weevils may infest almost every nut on such trees if not controlled. Infested nuts drop to the ground and larvae remain in the soil for 1 to 2 years, then pupate and remain in the soil as adults for 1 more year before emerging to mate and lay eggs, usually on the same tree. Certain trees may be infested year after year while adjacent trees suffer little damage, so growers should identify those trees and place cone-shaped weevil traps there to determine when adult weevils emerge from the soil. Spray infested trees thoroughly with carbaryl (Sevin™) or with a pyrethroid insecticide such as lambda-cyhalothrin (Warrior™) when weevils begin to emerge, or at the gel stage and again 10 to 14 days later.

Squirrels (*Sciurus* species) can be serious pests of pecans, as a single squirrel can carry off 50 to 60 pounds of pecans per year. Squirrels may also damage the bark on small trees or new grafts of top-worked trees, which can kill the tree or the graft. Reduce squirrel populations by destroying any squirrel nests in the pecan orchard, hunting well before harvest, and placing traps on the trunks of trees in the periphery of the orchard.

Photo 24-27. Pecan weevil, the worst insect pest of pecan in the Southeast and Midwest, drills holes through shucks to lay eggs in pecan kernels.

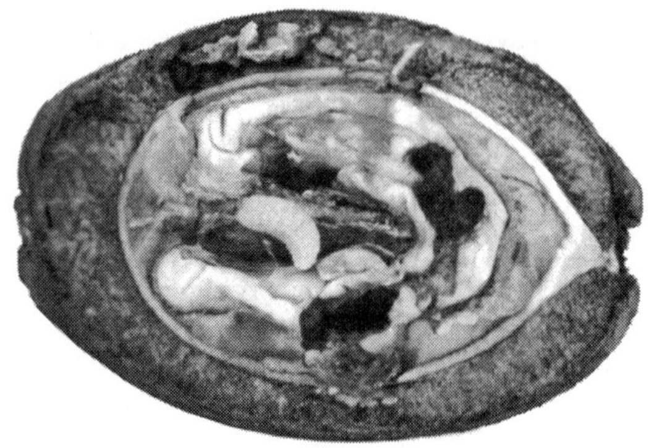

Photo 24-28. Pecan weevil larvae hatch inside pecan nuts and devour the kernels. Two insecticide sprays in late summer or early fall prevent weevil damage.

Crows (*Corvus brachyrhynchos*) can also carry away many pecans, especially native pecans and smaller cultivars such as 'Peruque' or 'Candy'. Shooting, poison baits or rotating scare devices may be necessary if a flock of crows descends on a pecan area.

HARVEST, PROCESSING AND MARKETING

Pecan shucks split open and peel back from the nuts when they are mature, allowing the nuts to be shaken free to the ground and harvested. Shucks must split before the first killing frost, which kills the shucks and prevents further opening, or the nuts will remain as "stick-tights" enclosed in the shucks, which can't be removed with common processing equipment. Choosing cultivars that normally split their shucks before the first killing frost helps ensure a minimum of stick-tights.

All commercial growers harvest pecans mechanically. Most use 3 pieces of equipment: trunk shakers, nut harvesters, and nut cleaners. Before harvest can begin, growers must prepare the orchard floor and determine the best time to harvest.

Preparation for harvest. Remove any cattle grazing in the pecan orchard or grove at least 2 months before harvest, to allow manure droppings time to decompose so they don't contaminate harvested nuts. Mow the groundcover a week to 10 days before harvest to keep it short enough for best harvester operation and allow time for grass clippings to decompose. Clean-cultivated orchards must have the surface packed down with heavy rollers following the last disking operation before harvest. Rake up and burn sticks and other debris that would otherwise clog the harvester.

Wait until 70 to 80% of nut clusters have open shucks, and kernel moisture is below 10%, before beginning harvest. Shake a test tree, and if at least 80% of nuts drop after 10 seconds or less of shaking, proceed to harvest other trees of that cultivar, or trees at a similar development stage in a native grove. Harvesting too early can damage tree trunks because sap is still flowing and bark slips easily. However, waiting too long to begin harvest allows nutmeats to darken and lose quality.

Tree shaking, nut pickup, and cleaning. Trunk shakers used for pecans come in a range of sizes and price ranges, and even small growers with just a few acres find that a mechanical shaker quickly pays for itself. Most growers shake nuts onto the ground, then pick them up with nut harvesting machines. Smaller growers may shake nuts onto large tarps, then transfer the nuts into sacks, avoiding the need to purchase a much more expensive mechanical harvester. Shakers may be tractor-mounted or self-propelled, with shaker heads ranging from 32 inches to 46 inches in diameter.

Some pecan cultivars will drop almost all their nuts with just one tree shaking, while other cultivars require a second shaking a couple of weeks after the first. The shaker operator should maintain the padded clamp to avoid causing trunk damage, and should be sure the clamp is tight enough on each tree to avoid any slipping. Vary the length of shaking each tree based on tree size and crop load, shaking larger trees longer than smaller ones and trees with heavy crops longer than those with light crops. Generally, shake each tree for 6 to 10 seconds (Photo 24-29).

Most tree shakers have sweeps in front to push fallen nuts out of the path of the equipment. If the shaker does not, a worker with a hand rake must clear a path for the shaker before the operator proceeds to the next tree, to avoid crushing nuts on the ground.

Pecan growers harvest nuts from the ground immediately after tree shaking, instead of leaving nuts to dry on the orchard floor as for almonds. Most planted pecan orchards use both a windrower and a mechanical harvester designed to pick up nuts in windrows (Photo 24-30), as described for almonds and walnuts. Native pecan growers use mechanical harvesters designed to pick up pecans from the entire orchard floor (Photo 24-31), so they needn't purchase sweeper machines to windrow the nuts. Harvester operators must adjust the machines to

Photo 24-29. Vary the shaking time for each pecan tree based on tree size and crop load, generally shaking each tree for 6 to 10 seconds.

Photo 24-30. Most planted pecan orchards use both a windrower and a mechanical harvester designed to pick up nuts in windrows.

Photo 24-31. Mechanical harvesters designed for irregularly-spaced native groves pick up nuts from the entire ground surface, immediately after tree shaking.

Photo 24-32. Workers must pass harvested pecans through nut cleaning machines to remove dirt, leaves, cracked and damaged pecans.

pick up as little dirt and clods as possible, yet not crack the nuts or blow many nuts out with leaves and twigs.

Harvested nuts are generally placed into sacks or super-sacks, large polypropylene or nylon sacks that hold up to 1500 pounds of nuts and are moved with tractors or fork-lifts. After tree shaking and harvesting operations are completed, workers dump sacks of nuts onto nut cleaning machines (Photo 24-32). Such machines first pass nuts through a "squirrel cage" tumbler which removes loose dirt, small sticks and other debris. Nuts next pass over a series of chains to remove clods larger than the pecans, then go through hullers to remove any shucks sticking to some nuts. Next, an air stream blows out leaves and light nuts that didn't fill well or were devoured by insects. Finally, pecans pass over an inspection table where workers remove cracked nuts, stick-tights, or other damaged nuts.

After cleaning, growers may sell their pecans wholesale to a shelling plant, sell nuts retail themselves, or store some or all of the nuts in cold storage. Retail nuts are sold in-shell, cracked, or as nutmeats and processed pecan products.

Drying and storing nuts. Traditionally, pecan growers have waited to begin harvest until kernel moisture content is below 10% and do not store nuts long on the farm, so haven't needed to invest in nut drying equipment. Some larger pecan growers now begin harvest earlier, with nuts at moisture contents up to 20%, and use drying bins or trailers to reduce moisture content of nuts that are harvested early. Temperatures in drying bins should not exceed 90F. Nuts for cold storage should be dried to 4% moisture content, which takes about 40 hours of drying if nuts start at 20% moisture, or about 12 hours of drying if nuts start at 8% moisture. Shellers dry most nuts for storage, with growers drying only those nuts harvested at moisture contents greater than 10% or nuts they plan to store themselves.

Some pecans are stored in-shell, and cracked just before they are marketed. This enables the unbroken shells to protect nutmeats from discoloration, mold or insect damage, although unshelled pecans require twice as much cold storage space as shelled nutmeats do. Other pecans are first cracked and shelled, then stored as nutmeats in 30-pound boxes until they are bagged, shipped and marketed.

In-shell or shelled pecans that were dried to 4% moisture content store well for up to 2 years, at a temperature of 0 F and relative humidity of 65 to 70%. In-shell pecans last 18 months and shelled pecans keep for 12 months at a temperature of 32 F. Pecan nutmeats readily absorb odors and flavors from nearby materials, so store them apart from odoriferous vegetables or fruits and those that give off ethylene, such as apples and pears.

Cracking, shelling, sizing and marketing nutmeats. Pecan processors want as many pecan kernels as possible to crack out into "whole halves", since the two halves of each kernel are attached by a narrow piece that readily separates, rather than cracking into smaller pieces. Pecan halves fetch higher prices than pieces and also store better since fewer cells were broken. Certain cultivars produce a higher percentage of halves than others, and shellers may pay more for such cultivars.

Conditioning pecans before cracking will increase the yield of halves from about 50% to up to 87%. Condition pecans by soaking them in cold water containing chlorine for 1 to 2 hours, or in hot water for 3 to 5 minutes, then hold them for 12 to 24 hours before cracking.

Pecan crackers feed nuts into cracking chambers and crack them one at a time. Adjust cracking machinery to produce the greatest amount of whole halves and the least amount of crushed nuts. After cracking, nuts pass into shellers, where shell fragments and the middle partitions in pecans, between the two halves, are removed. Remaining nutmeat halves and pieces pass over screens to remove shell and partition fragments, then pieces are separated from halves. Pecan halves move down conveyor belts and are sorted into 8 size classes, ranging from small topper to mammoth, based on number of halves per pound.

Sized nutmeats next pass through a dryer, which reduces kernel moisture level to 3-4%, then are cooled and packaged. Packaged pecans should be kept under refrigeration with 65-70% relative humidity, preferably at 0 F until they are marketed and at 32 F during display at stores.

Most larger-sized pecan halves are sold to consumers, while smaller-sized halves and pieces are sold to food processors for use in baked goods, ice cream, and other products.

Further Reading

Goff, W.D., J.R. McVay, and W.S. Gazaway, eds. 1989. Pecan production in the Southeast. Alabama Coop. Ext. Service Circular ANR-459. Auburn, AL. 222 pp.

McEachern, G.R. and L. A. Stein, eds. 1997. Texas pecan handbook. Texas Agric. Ext. Serv. Hort. Handbook 105.

Reid, W.R. and K.L. Hunt. 2000. Pecan production in the northern United States. HortTechnology 10: 298-301.

Worley, R.E. 2003. Compendium of pecan research and production. Edwards Bros. Inc., Ann Arbor, MI. 479 pp.

Chapter 25
Hazelnuts and Chestnuts

Hazelnuts and chestnuts are minor nut crops in North America, and both suffer from serious blight diseases that have historically limited production to a few areas. Hazelnuts, also called filberts, are grown on about 30,000 acres, with 99% of that acreage in Oregon. Chestnuts are grown on about 1000 acres, with about 66% of that in California and most of the rest in Oregon, Washington and British Columbia. The U.S. produces only about 4% of the world's hazelnuts and less than 1% of the world's chestnuts. Turkey raises 71% of the world's hazelnuts, and Italy grows about 12%. China produces 40% of the world's chestnuts, Korea grows 15%, and Italy, Turkey and Japan each raise about 10% of the world's chestnuts.

Hazelnuts have an oil content of 58 to 66%, about the same as almonds and lower than the 70 to 75% oil content of walnuts and pecans, so taste drier. Chestnuts have an oil content of just 3%, much lower than that of other nuts. Chestnuts, sometimes called "the potato of nut crops," have a starchy texture that resembles a potato with a mildly nutty flavor. Both nuts are used mainly in cooked or baked products, or ground into flour for pastries.

Hazelnuts belong to the genus *Corylus* in the birch family, the Betulaceae. All commercial production utilizes the European hazelnut, *C. avellana*, although most older cultivars are susceptible to eastern filbert blight that occurs in eastern North America and has invaded parts of the Pacific Northwest. The American hazelnut, *C. americana*, has resistance to eastern filbert blight, but nuts are much smaller and lower in quality. A few small plantings of hybrids between European and American hazelnuts have been made in eastern North America.

Chestnuts belong to the genus *Castanea* in the beech and oak family, the Fagaceae. American chestnuts (*C. dentata*) once were dominant trees in the Appalachian hardwood forest of eastern North America, but the chestnut blight fungus, introduced from Asia in the early 1900's, destroyed almost all of them. European chestnuts (*C. sativa*) are also susceptible to chestnut blight, as are most Japanese chestnuts (*C. crenata*), but these species and their hybrids have been planted in West Coast areas where chestnut blight does not occur. Chinese chestnuts (*C. mollissima*) offer strong resistance to chestnut blight and have been planted in recent years. Breeders have also hybridized the various chestnut species, and some blight-resistant hybrid cultivars have been planted commercially.

ADAPTATION, TREE GROWTH AND BEARING HABITS

Hazelnuts have a chilling requirement of 800 to 1200 hours so require areas that provide that much winter chilling. However, male flowers shed pollen and female flowers are receptive very early (Photo 25-1), usually in late January and February, so the best areas for hazelnuts have mild winters with consistent temperatures that rarely drop below 15 F in late winter. Hazelnuts also cannot tolerate dry summer heat. These climate requirements and eastern filbert blight effectively limit hazelnut production to the Pacific Northwest of North America.

Photo 25-1. Pendulous hazelnut catkins (male flowers, at bottom) shed pollen and female flowers (at top) are receptive in late winter, limiting hazelnut production to areas without late winter frosts but that receive at least 800 hours winter chilling.

Photo 25-2. Chestnuts bloom in early summer, so do not suffer from frost damage and are widely adapted.

Chestnuts have a much lower chilling requirement of about 300 to 700 hours, but don't bloom until early summer (Photo 25-2), so have a much wider adaptation. Chestnuts are cold-hardy to about -20 F, so grow well throughout much of USDA hardiness zones 5 to 8. Although most commercial plantings have been made in the West Coast region, chestnuts have also been planted in Eastern states such as Ohio, Michigan, and Georgia, and chestnuts have great potential to expand further. The U.S. exports no chestnuts but imports 10 million pounds per year, mainly from Italy, and chestnuts could be planted on 4000 additional acres in the U.S. just to displace imports. Promotion could also increase chestnut consumption, which averages just ½ pound per person per year in the U.S., but 1 pound in Europe and 2 pounds in China.

Hazelnuts grow naturally as small, multi-stemmed trees to a height of 15 to 25 feet, but are trained to single trunks to make mechanical harvest easier (Photo 25-3). American and European chestnuts grew into very large trees that reach 60 to 100 feet tall, but Chinese and Japanese chestnuts are medium-sized trees that reach a height of 30 to 50 feet.

Both hazelnuts and Chinese and hybrid chestnuts begin to bear 3 to 4 years after planting, and reach full production by 7 years after planting. Like other nut crops, both are wind-pollinated and monoecious and require pollenizer cultivars for best production. Both are long-lived trees where blight diseases do not occur or blight-resistant cultivars are chosen, and orchards can remain productive for 40 to 80 years.

Hazelnuts are borne singly or in clusters of up to 12 nuts, enclosed in leafy husks that are modified whorls of bracts (Photo 25-4). Chestnuts are borne

Photo 25-3. Hazelnuts grow naturally as small, multi-stemmed trees but are trained as single-trunk trees to make mechanical harvest and other operations easier.

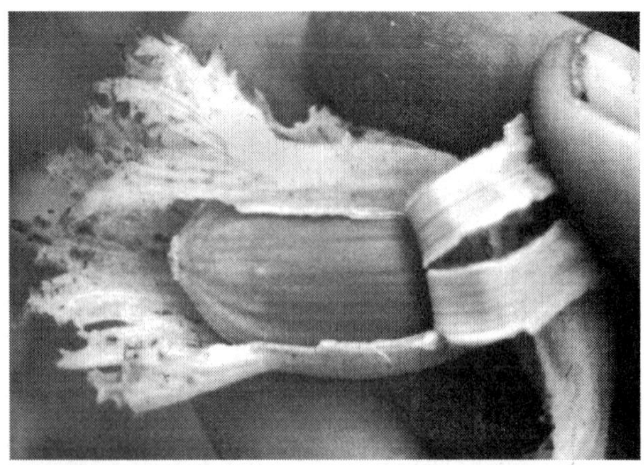

Photo 25-4. Hazelnuts are borne singly or in clusters of up to 12 nuts, enclosed in leafy husks that are modified whorls of bracts.

in spiny burs that are modified whorls of bracts (Photo 25-5), and have 1 to 3 nuts per bur. Both hazelnuts and chestnuts ripen in early fall, usually between early September and late October.

SITE SELECTION AND PREPARATION

Hazelnuts and chestnuts both do best on well-drained soils rich in organic matter, with a pH of 6.5 to 7.0. Air drainage is not as big a factor as it is with most other fruit and nut crops, since hazelnuts bloom so early and chestnuts bloom so late. However, avoid low-lying frost pockets for hazelnuts, as spring frosts may reduce cluster buds and shoots even two months after flowering.

Test soil for pH, phosphorous (P), potassium (K), calcium (Ca) and magnesium (Mg) at least several months before planting, and modify pH or add nutrients as needed. Compacted soils or those with a plowpan should be ripped well before planting to break this up. If soil is somewhat poorly drained, construct berms to improve drainage, especially for chestnuts which are very susceptible to Phytophthora root rot. Hazelnuts withstand a fairly high soil water table in early spring as long as it becomes lower by late spring and soil below 4 feet deep drains freely.

Avoid planting hazelnuts or chestnuts, which are prone to oak root rot, on soil that may contain this pathogen, or else fumigate or solarize the soil the year before planting. Control any perennial weeds before planting.

CHOOSING CULTIVARS

Production of both hazelnuts and chestnuts in the U.S. currently relies on one main cultivar each, with pollenizer cultivars interplanted to provide cross-pollination. However, breeders and growers have selected new blight-resistant cultivars of both crops, and growers have started planting them. (Tables 25-1 and 25-2)

'Barcelona' has accounted for over 67% of hazelnut production in Oregon for the last 110 years. Its medium-large kernels of good quality and strong resistance to big bud mite have contributed to its popularity, but 'Barcelona' is moderately susceptible to eastern filbert blight. 'Ennis' supplies 14% of hazelnut production, mainly for the in-shell market due to its very large nuts, but is very susceptible to both eastern filbert blight and big bud mite. The main cultivar planted as a pollenizer for 'Barcelona' or 'Ennis' before 1980 was 'Daviana', but many growers removed these trees due to high susceptibility to eastern filbert blight and big bud mite, and replaced them with 'Butler',

Photo 25-5. Chestnuts are borne in spiny burs that are modified whorls of bracts.

which is moderately susceptible to blight. 'Gamma' and the VR series of pollenizers, released by Oregon State University, are very resistant to eastern filbert blight and are good pollenizers for 'Barcelona', since they shed pollen in midseason. Some growers are replacing pollenizers in older 'Barcelona' orchards with 'Gamma' or VR pollenizers, either by topworking existing pollenizer trees or removing pollenizers and replanting.

Four cultivars released by Oregon State University in the last few years offer resistance to eastern filbert blight and to big bud mite, and are now being planted the most. 'Sacajawea' and 'Lewis' have nuts almost as large as 'Barcelona', with higher percent kernel, fewer nuts with defects, and higher marketable yields. 'Santiam' and 'Clark' have smaller nuts but less defects than 'Barcelona', and similar yields. 'Lewis' trees are 25% smaller than 'Barcelona' and 'Clark' trees are 30% smaller, yet yields per tree are higher than 'Barcelona', so planting these new cultivars allows more trees to be planted per acre and increases yields per acre. Female flowers of 'Santiam' emerge late, while those of 'Clark' emerge very late. This may reduce cold damage to flowers, but requires pollenizers that shed pollen late. 'Gamma' is a good pollenizer for 'Lewis', and 'Gamma' or 'Lewis' are good pollenizers for 'Sacajawea'. 'Delta', 'Epsilon' and 'Zeta' are more pollenizers from Oregon State University that are very resistant to eastern filbert blight. 'Delta' sheds pollen in late midseason, so is a good pollenizer for 'Lewis'. 'Epsilon' sheds pollen late and 'Zeta' sheds pollen very late, and both are good pollenizers for 'Santiam' and 'Clark'.

'Colossal', a European-Japanese hybrid chestnut, has been the main chestnut cultivar planted in the U.S. during the last 15 years. It produces high yields of very large, sweet nuts, and is susceptible

Table 25-1.

COMMERCIAL HAZELNUT CULTIVARS AND POLLENIZER CULTIVARS FOR THE PACIFIC NORTHWEST OF NORTH AMERICA

In approximate order of flowering for main cultivars and pollen shed for pollenizers.

Cultivar	Cultivar type	Nut			Tree size, % of Barcelona	Resistance to		Days before (-) or after Barcelona	
		Size	% Kernel	Yield		Eastern filbert blight	Big bud mite	Flowering	Harvest
Sacajawea	main	med-large	52%	med-high	85%	VR	VR	2	-14
Santiam	main	medium	50%	medium	80%	VR	R	8	-14
Clark	main	medium	50%	medium	70%	R	R	10	-8
Lewis	main	med-large	48%	med-high	75%	R	MR	4	-6
Barcelona	main	large	42%	medium	100%	MS	VR	0	0
Ennis	main	very large	47%	high	85%	VS	S	8	7
Butler	pollenizer	med-large	48%	medium	95%	MS	MR	0	-2
Gamma	pollenizer	medium	52%	low-med	95%	VR	MR	2	-3
Delta	pollenizer	medium	49%	low-med	85%	VR	R	6	2
Epsilon	pollenizer	medium	52%	low-med	70%	VR	VR	8	-3
Zeta	pollenizer	medium	50%	low-med	95%	VR	MR	10	-3

Pest resistance ratings: VR = very resistant, R = resistant, MR = moderately resistant, MS =

Notes on cultivars:
Sacajawea has excellent-flavored nuts with few defects. Use 'Lewis' or 'Gamma' as pollenizer.
Santiam nuts are too small for the in-shell market but suited to the kernel market.
Clark trees can be planted closer than 'Barcelona'. It needs a late pollenizer, 'Epsilon' or 'Zeta'.
Lewis trees can be planted closer than 'Barcelona', and trees yield more in early years.
Barcelona, the standard hazelnut cultivar for over 100 years, accounts for 67% of production.
Ennis produces very large nuts but is little planted now due to filbert blight susceptibility.
Butler, widely planted as a pollenizer for 'Barcelona', tends to alternate-bear.
Gamma is a mid-season pollenizer for 'Barcelona', 'Sacajawea' and 'Lewis'.
Delta, a mid-late pollenizer for 'Lewis', is incompatible with 'Barcelona' or 'Ennis'.
Epsilon is a late pollenizer for 'Clark' and 'Santiam'; incompatible with 'Barcelona' and 'Ennis'.
Zeta is a very late pollenizer for 'Clark' and 'Santiam'; incompatible with 'Barcelona' and 'Ennis'.

Table 25-2.

HYBRID CHESTNUT CULTIVARS FOR NORTH AMERICA

In approximate order of ripening (varies with area).

Cultivar	Species	Nut			Tree		Resistance to chestnut blight	Pollen	Harvest time
		Size	Yield	Flavor	Size	Growth			
Colossal	J x E	very large	high	good	large	spreads	S	sterile	early
Doyle	J x E	very large	high	good	large	spreads	S	sterile	early
Eaton	C x (J x A)	med-large	medium	best	large	spreads	R	fertile	early
Sleeping Giant	C x (J x A)	med-large	high	best	very large	upright	R	fertile	early
Okei	E	very large	low	good	large	spreads	S	fertile	early-mid
Skookum	C x E	medium	medium	very good	large	upright	S	sterile	early-mid
Myoka	C x E	med-large	high	good	large	upright	R	fertile	middle
Peach	C	large	high	good	large	upright	R	fertile	middle
Qing	C	very large	high	best	large	spreads	R	fertile	middle
Jack's	C	large	high	best	small	spreads	R	fertile	middle
Carolina	C x C(C x A)	medium	high	good	large	spreads	R	fertile	middle
Willamette	C x C(C x A)	very large	medium	good	large	upright	R	fertile	mid-late
Revival	C x C(C x A)	large	high	very good	large	spreads	R	fertile	late

Species: J = Japanese, E = European, C = Chinese, A = American.

Blight resistance ratings: R = resistant, S = susceptible.

Average harvest times in central CA and southern MO: early: Sept. 12-22, early-mid: Sept. 16-26, middle: Sept. 20-30, mid-late: Sept. 24-Oct. 4, late: Sept. 28-Oct. 10. Harvest is about 3 weeks later in western OR and central OH.

Notes on cultivars:

Colossal, the industry standard in California, bears young and heavily; too blight-susceptible for the East.
Doyle, a seedling of 'Colossal', is also too blight-susceptible for the East.
Eaton nuts are very sweet, store well.
Sleeping Giant has very sweet nuts that peel easily, but very large trees need wider spacing.
Okei produces very large nuts and abundant pollen so is planted as a pollenizer for 'Colossal'.
Skookum is grown in the West but is too blight-susceptible for the East.
Myoka is a good pollenizer with much pollen. Nuts must be separated from burs, but shell peels easily.
Peach has sweet nuts with shells that peel easily.
Qing (pronounced Ching) trees have good branch angles, very sweet nuts peel easily and store well.
Jack's trees grow just 20-25 feet tall and wide rather than 35 feet, so can be planted closer.
Carolina nuts are mildly sweet, best cooked.
Willamette has sweet nuts with shells that peel easily.
Revival bears young and has very dense, crunchy, sweet nuts that peel easily and store well.

to chestnut blight but survives well on the blight-free West Coast. Many seedlings of 'Colossal', including a cultivar named 'Doyle', have also been planted. Blight-resistant "Dunstan hybrids" — third-generation offspring of Chinese-American chestnut crosses such as 'Revival', 'Carolina', and 'Willamette' — have large to very large nuts and have also been planted. So have Chinese-European hybrids with medium-large nuts and some blight resistance, 'Skookum' and 'Myoka', and blight-resistant hybrids of Chinese and Japanese-American chestnuts, 'Eaton' and 'Sleeping Giant'. Several pure Chinese chestnut cultivars, including 'Peach', 'Qing' and 'Jack's', have been planted in a few trial plantings in the eastern U.S.

Photo 25-6. Hazelnuts are propagated mainly by mound layering, as in this stool bed.

PROPAGATION AND PLANTING

Hazelnuts are propagated mainly by mound layering, creating "stool beds" to propagate more plants (Photo 25-6). Tissue culture is also used to propagate newer cultivars faster.

Some chestnut growers plant grafted trees, but many have planted seedling trees of cultivars, such as 'Colossal' seedlings. When chestnuts are grafted, the rootstocks are generally seedlings of the same species or hybrid, and often seedlings of the cultivar to be grafted. For example, 'Colossal' scions are most often grafted onto 'Colossal' seedlings. This avoids delayed graft incompatibility, which often plagues chestnuts if cultivars are grafted onto rootstocks of a genetically distant genotype.

Tree spacing and pollenizer layout. Most hazelnut and chestnut growers plant trees at moderately high densities and either prune trees as hedgerows to limit their height and spread, or plan on removing half the trees about 10 years after planting, when trees start to crowd. Hazelnut growers often plant trees 10 feet apart within rows and 18 feet between rows, using 242 trees per acre. Chestnut growers may similarly plant trees in a hedgerow system of 26 feet between rows and 13 feet between trees within rows, creating a density of 129 trees per acre. Other chestnut growers use a lower-density system, planting trees 22 to 25 feet apart in rows that are 25 to 28 feet apart, for tree densities of 70 to 80 trees per acre.

Usually, hazelnut and chestnut growers plant one main cultivar in each tree block, and intersperse pollenizer trees which produce nuts of lower quality and yields, so they maximize the number of main cultivar trees planted and minimize the number of pollenizer trees. Hazelnut growers have planted one pollenizer tree for every 18 to 20 trees of the chosen cultivar in recent plantings, although older orchards often contain one pollenizer for every 8 or 9 trees of the desired cultivar. Chestnut growers plant one pollenizer for every 8 to 10 trees of the desired cultivar. Planting a pollenizer as every 6th tree in every 3rd row in hazelnut orchards, and as every 3rd tree in every 3rd row in chestnut orchards, achieves these ratios. If pollenizer cultivars produce good-quality nuts and high yields themselves, more pollenizers can be planted.

Planting and care of new trees. Most hazelnut growers plant in early winter, as soon as possible after receiving trees from the nursery. Chestnuts may also be planted in early winter where winters are mild, or in fall or early spring where winters are more severe. Plant hazelnuts so the top root is 2 to 3 inches below the soil surface, and plant chestnuts at the same depth that they grew in the nursery. Prune the top of each tree to a single stem and lightly head it back to a height to 30 to 36 inches. Water newly-planted trees well, and irrigate them again whenever the soil starts to dry out during the first year after planting.

Place plastic trees guards around each tree soon after planting to prevent sunscald and damage from voles or rabbits, pressing the guard into the soil to prevent voles from digging under them. Placing woodchip mulch around young hazelnut and chestnut trees, 3 to 4 inches deep, helps preserve soil moisture and increases tree growth almost as much as irrigation, without attracting voles or rabbits the way straw mulch would.

ORCHARD FLOOR MANAGEMENT, IRRIGATION AND FERTILIZERS

Most hazelnut and chestnut growers mow the resident vegetation (weeds or whatever comes up) from early spring until harvest in mid-fall, then allow it to grow over the winter to increase soil organic matter. Some growers sow legume cover crops in fall to provide nitrogen (N), especially subterranean clover or other clovers which will self-seed each fall thereafter, but orchards with clover cover crops attract more gophers (a serious pest of both crops) and usually require more irrigation. Growers use pre-emergent herbicides and spot-treat with glyphosate to control weeds in tree rows.

Irrigation. Many older hazelnut orchards do not receive irrigation, but recent plantings have usually been established with trickle irrigation. Young orchards that receive irrigation come into bearing a year or two sooner, and generally pay for the investment in irrigation within a few years.

Chestnut trees do not often need irrigation when planted in humid eastern North America, but require irrigation in arid western North America. Trickle irrigation is the method most often used (Photo 25-7). Adequate water is critical during July and August when nuts are gaining size. Water shortages during nut sizing will reduce nut size by up to 50% and severely limit the price the chestnuts receive. Chestnuts add the greatest amount of nut weight in the last 2 weeks before harvest, so irrigation just before harvest ensures this weight gain and promotes proper bur split.

Fertilizing. Nitrogen (N) is the main fertilizer element needed, as with other crops. Concentrated N applications to newly-planted hazelnuts can kill trees, so apply only 1 to 2 ounces actual N per tree the first year, and be sure to spread it out well and keep it at least a foot away from tree trunks. Chestnut growers apply much more N, 8 to 9 ounces per tree, during the first year.

Hazelnut growers generally apply 2 ounces actual N per tree the second year, and increase N rates by 2 to 3 ounces per year, until they apply full rates of 1.5 to 2.0 pounds actual N per tree from year 12 on. This means they apply 240 to 300 pounds of actual N per acre to mature trees. Chestnut growers also increase N rates by 2 to 3 ounces per year until they apply full rates of 2 to 3 pounds actual N per tree from year 12 on. Since mature chestnut orchards have far fewer trees per acre than hazelnuts, growers apply only 100 pounds of actual N per acre to mature trees. Most growers of both crops apply fertilizer N every year, but clover cover crops greatly reduce the need for added N although they require more irrigation and gopher control.

Photo 25-7. Trickle irrigation benefits this chestnut planting in arid western North America, and most new hazelnut plantings also use trickle irrigation.

Potassium (K) is not generally needed each year, but some soils may need it every few years. Take soil and leaf samples for K at least every 3 years. When needed, band K, usually as potassium sulfate, just inside the dripline of tree rows. Large amounts of K need to be applied to a concentrated area, because smaller amounts that are spread out are quickly tied up in the soil and become unavailable to plant roots.

Most hazelnut and chestnut soils do not need applications of phosphorous (P) or other macronutrients. Hazelnut trees over 6 years old often benefit from boron (B) applications. Some growers broadcast borax on the soil at a rate of 5 to 7 pounds actual B per acre; it can be toxic if concentrated so should always be broadcast, not banded. Most hazelnut growers instead apply B as foliar sprays in late May to increase nut set, as described below.

TRAINING, PRUNING, TREE THINNING AND MANAGING CROPS

Hazelnut and chestnut orchards that will be grown as hedgerows, retaining all planted trees for the life of the orchards, are trained and pruned very differently from traditional orchards, which will receive one or two tree thinnings 10 to 20 years after planting. The hedgerow plantings offer higher yields of larger nuts, but require more intensive management and investment in mechanical pruning equipment.

Training young trees. Train hazelnuts to an open-center system by selecting 3 to 5 major scaffold limbs during the first 2 to 3 years, all arising at about 3 feet off the ground to allow equipment to run under them, and cutting the leader off above them. In a hedgerow system, growers generally choose 4 scaffold limbs growing in an X-pattern across the row, so limbs from closely-planted adjacent trees don't grow into each other as much. Where every other tree will be removed from the orchard later, scaffold positions do not matter.

Train hedgerow chestnut trees to a V-system like that described for high-density peaches, with two main scaffold limbs. During the first 4 years, encourage growth of the leader and 2 main branches that grow perpendicular to the row, by cutting the leader back and other branches off while allowing the 2 chosen branches to grow unchecked. About 5 years after planting, remove the leader completely just above the 2 scaffold branches, leaving a V-shaped hedgerow that admits plenty of light for maximum production of large nuts.

Train lower-density chestnuts, where temporary trees will be removed, to a modified leader system. The central leader may need to be staked to keep it upright, so it will develop buds in a spiral fashion, from which scaffold limbs will grow. Select 4 to 6 scaffolds that arise in a spiral fashion around the trunk, starting 3 to 4 feet off the ground and with at least 8 to 12 vertical inches between sequential branches. After all desired scaffolds have been selected, cut off the leader above the top scaffold.

Hedging mature trees. Once trained, hazelnut and chestnut trees need little pruning for the first few years of cropping. Mature trees need rejuvenation pruning to cut back overly dense growth and admit more light. Increasingly, this is accomplished by mechanical hedging.

About 70% of a hazelnut crop is produced on shoots from 1-year-old branches that have grown longer than 6 inches, while the rest is produced from shorter branches. Dormant pruning stimulates growth of new branches in mature orchards, so increases yields. Heavy pruning (removing up to half the bearing surface) of about 20% of the trees each year, rather than light pruning (removing 5 to 8% of the bearing surface) of all trees each year, takes much less labor time so is the most common method.

Mechanical side hedging of 20% of the trees each year, using gang saws provided by a custom operator, has become the main hazelnut pruning method in recent years. Workers should do follow-up hand pruning to remove stubs, to avoid limb dieback which could aggravate both fungal and bacterial filbert blights.

Hedgerow chestnuts also may be mechanically hedged to limit tree height to 12 to 13 feet tall, and to trim sides so branches from adjoining rows do not overlap. Again, follow-up hand pruning is needed to remove stubs.

Pruning and removing temporary trees in low-density orchards. Once trees begin to crowd, after 7 to 8 years, prune back the temporary trees more severely than the permanent trees for the next few years, so they compete less with the permanent trees. Do this by removing one main limb from each temporary tree each year, which grows toward a permanent tree. About 10 years after planting, remove every other tree in each row for the first tree thinning.

If the original planting had trees closer within the row than between rows, such as 10 feet apart within rows and 18 feet between rows for hazelnuts or 13 feet by 26 feet for chestnuts, growers remove the second, fourth, sixth, etc. tree in each row. This leaves hazelnut trees spaced 20 feet by 18 feet (Photo 25-8) and chestnut trees spaced 26 feet by 26 feet.

Generally growers remove trees on the diagonal (in a staggered pattern from row to row) if the original planting was made in a square pattern with equal distances between and within rows. For example, chestnut trees planted 25 feet by 25 feet apart would be thinned diagonally as shown in the pecan chapter, Figure 24-2, to create a 35-foot by 35-foot spacing.

Hazelnut orchards are generally planted so they need only one tree thinning, while chestnut orchards that are not managed as hedgerows need a second tree thinning about 10 years later, or about 20 years after planting. This second tree thinning is again done diagonally, to create the final tree spacing of 50 feet by 50 feet.

Sucker control and increasing nut set in hazelnuts. Hazelnut growers control root suckers to keep trees to a single trunk, rather than letting them develop multiple trunks. Suckers used to be pulled or cut off by hand, but now are controlled by a herbicide spray in spring and summer, when suckers are 6 to 9 inches tall. Most orchards need 3 or 4 treatments per year. Herbicides for sucker control include paraquat, 2,4-D amine formulation, and glufonisate-ammonium (Rely™).

Hazelnuts drop many female flower clusters in late spring, which reduces potential yields. A boron (B) spray applied to foliage in late May reduces this cluster drop and increases nut cluster set by up to 33% in orchards with low B levels. Use up to 1 pound of actual B per acre, usually in the form of 2 to 5 pounds sodium pentaborate (Solubor™) per 100 gallons water. Do not apply B to trees younger

Photo 25-8. Mature hazelnut trees are generally spaced 18 to 20 feet apart after removing any temporary trees. Tree canopies touch to capture almost all light falling on the orchard.

Photo 25-9. Eastern filbert blight, the worst disease of hazelnuts in North America, produces cankers that enlarge each year until they girdle scaffold limbs and eventually kill the tree. (Photo by J.W. Pscheidt, Oregon State Univ.)

than 5 years old, or to trees whose midsummer leaf samples show B levels over 150 ppm.

PEST MANAGEMENT

Choosing blight-resistant cultivars is the most important pest management strategy for both hazelnuts and chestnuts. Hazelnuts require several sprays per season to control key pests, but some pesticides considered organic can be used. Spray programs are not currently required in western chestnut orchards, so chestnuts are a good crop for organic production there. Eastern chestnut orchards require minimal pesticide applications.

Pests that can kill trees. Plan ahead for these serious pests before planting trees. Choose blight-resistant cultivars, modify the soil if needed to prevent soil-borne diseases, and protect young trees from diseases, insects and mammals that can kill them.

Eastern filbert blight, caused by the fungus *Anisogramma anomala* that is native to eastern North America on American hazelnuts, is the worst disease of commercial hazelnuts on this continent. Surviving American hazelnuts have developed resistance to this disease so suffer little damage, but all European hazelnuts are at least moderately susceptible to it, and some cultivars are highly susceptible. The disease produces cankers on twigs and branches that first appear as long, sunken brown streaks, then enlarge each year until they girdle scaffold limbs and eventually kill the tree (Photo 25-9).

Eastern filbert blight has now spread into much of the Pacific Northwest, so growers should avoid highly susceptible cultivars such as 'Ennis' and 'Daviana'. 'Barcelona' is less susceptible, and can survive for many years if orchards are rigorously pruned to remove diseased branches at least a foot below visible cankers, and sprayed with a copper or strobulurin fungicide at budbreak and every 2 weeks for the next 6 weeks. Growers who wish to reduce fungicide use should plant cultivars with at least moderate resistance to Eastern filbert blight, such as 'Lewis', 'Clark' 'Sacajawea,' or 'Santiam.' Also replace susceptible pollenizer cultivars with very resistant ones, such as 'Gamma', 'Delta', 'Epsilon' or 'Zeta'.

Bacterial blight of hazelnuts, caused by the bacterium *Xanthomonas campestris* pv. *corylina*, may also girdle and kill trees up to 5 years old, and causes dieback of nut-bearing branches in trees up to 10 years old. Start with disease-free planting material and plant in early winter, as trees planted in late spring are especially susceptible to bacterial blight the next fall. Most seasons, one copper spray in late August or early September, before the first heavy fall rains, will prevent bacterial blight on young hazelnut trees. Seasons with heavy, prolonged winter rains require a second spray when about 75% of the leaves have fallen from the trees.

Chestnut blight, caused by the fungus *Cryphonectria parasitica* which was introduced from Asia in the early 1900's, is the chief disease that kills chestnut trees east of the Rocky Mountains. Some susceptible trees die from cankers at a very young age (Photo 25-10), while others reach a diameter of a foot or more and bear a few crops before trees are killed. No American chestnuts have demonstrated adequate resistance to chestnut blight, and most that do not succumb to the disease have merely escaped it as the fungus does not occur in the im-

Photo 25-10. Chestnut blight, the chief disease that kills chestnut trees east of the Rocky Mountains, produces cankers that kill some trees at a young age, while other infected trees reach large sizes and bear a few crops before they die. (Photo from Univ. of Guelph, Ontario.)

mediate area. Some genotypes of Chinese and Japanese chestnuts and their hybrids do possess true resistance to chestnut blight, and those who want to grow chestnuts for nut production should plant such resistant cultivars. Since chestnut blight is still rare along the West Coast, growers there have been able to plant susceptible cultivars such as 'Colossal'.

Phytophthora root rot, caused by several *Phytophthora* fungi, can also kill chestnut trees on poorly-drained soils. This is the most common chestnut disease in California. Do not plant on poorly-drained soils, and build raised beds before planting soils with somewhat poor drainage to chestnuts.

Oak root rot, caused by the mushroom fungus *Armillaria mellea*, has also killed hazelnut and chestnut trees on sites infested with this pathogen. Avoid old orchard sites or sites where oak trees grew during the past 10 years, or fumigate or solarize the soil before planting.

Shothole borer (*Xyleborus dispar*), a bark and timber beetle, may kill young chestnut trees in the Pacific Northwest. This pest, introduced from Europe in the early 1800's, has become ubiquitous in forests of the Western Cascades and attacks all broadleaf tree species, but chestnut appears to be a preferred host. Two generations per year of adult borers tunnel into sapwood of apparently healthy trees showing no signs of stress, although many attacked trees have actually suffered the stress of recent transplanting, sunburn, drought, excess soil water, mechanical or cold damage. Prevent some shothole borer damage by transplanting carefully, using tree guards, irrigating, constructing raised beds if needed, and avoiding mechanical damage. Shothole borers can also be trapped, either to time insecticide sprays or to mass-trap many beetles and remove them from the orchard. Some growers make traps from 2-foot lengths of PVC (polyvinylchloride) pipe, capping each end and drilling small holes in the upper side, filling them with alcohol and coating the outside with the sticky material Tanglefoot™. The alcohol attracts the beetles, and the Tanglefoot™ traps them.

Oriental chestnut gall wasp (*Dryocosmus kuriphilus*) was introduced into Georgia in 1974 on chestnut cuttings from Korea and Japan, where it is native. This pest has now spread into McMinnville, Tennessee, a major nursery area that has the potential to spread the gall wasp across the country, and has spread up the Appalachian Mountains and been found in Ohio. This pest turns chestnut buds into galls (Photo 25-11), reducing fruiting and suppressing shoot growth. Infestations that continue for several years weaken trees and usually kill them. Gall wasps have one generation per year. Young larvae overwinter inside chestnut buds, maturing and converting the buds into galls in spring. Adult female wasps emerge from galls in early summer, lay eggs in clusters of buds, and the eggs hatch into larvae in mid-summer, which develop slowly in fall and winter. Since larvae are always protected inside the buds, no effective chemical controls have been developed.

A few years after the gall wasp invaded Georgia, the USDA introduced several of its natural enemies from Japan, and one parasitic wasp has been very effective in controlling this pest. Chestnut growers should inspect all nursery material before planting, to be certain no bud galls are present, and should inspect their orchards each spring for the presence of galls. If any are found, prune off and burn the infested shoots, and contact the USDA office in Athens, GA to obtain parasites of the gall wasp to introduce.

Deer (*Odocoileus* species) may kill many young hazelnut and chestnut trees near wooded areas. An electric deer fence is the most cost-effective control method for large plantings, but it must be maintained regularly and vegetation under the wires must be controlled with herbicides. Small growers may use deer repellents or place a 5-foot tall, welded wire cage around each tree.

Pocket gophers (*Thomomys* species) may kill a dozen or more hazelnut or chestnut trees per season, especially if clover grows in the groundcover. Frequent, short mowing of the groundcover reduces pocket gophers. Large populations must

Photo 25-11. Oriental chestnut gall wasp turns chestnut buds into galls. Infestations that continue for several years weaken trees and usually kill them. (Photo by J.A. Payne, USDA, ARS.)

be controlled by poison baits and by trapping the survivors.

Voles (*Microtus* species) and rabbits (*Sylvilagus* species) are especially attracted to both hazelnuts and chestnuts due to their smooth, thin bark when young. Place plastic tree guards around each tree soon after planting, pressing the guard into the soil to prevent voles from digging under them. Controlling all vegetation in a 3-foot-diameter area around each tree or a 3-foot-wide strip down the tree row greatly reduces vole and rabbit damage.

Indirect pests that weaken trees. Hazelnuts and chestnuts suffer from far fewer indirect pests than do most fruit and nut crops. Starting with certified virus-free plants will prevent virus diseases, and several indirect insect pests can be controlled biologically.

Hazelnut mosaic, caused by **apple mosaic virus**, is the most common virus disease of hazelnuts in Europe and has been found in breeding lines in the U.S., but has not yet been found in commercial plantings or wild hazelnuts in North America. This virus causes chlorotic ringspot, line patterns and flecking on older leaves. Prevent this and two other similar viruses of hazelnuts by buying certified virus-free plants to start a new planting. Chestnuts are rarely infected with viruses, but certified virus-free plants are still a good idea. Hazelnuts and chestnuts are usually free of phytoplasma diseases in North America.

Filbert aphid (*Myzocallis coryli*) feeding on leaves reduces percent fill and size of nuts, which reduces yields. Filbert aphid often required pesticide treatments to hazelnuts until a beneficial parasitic wasp, *Trioxys pallidus*, was imported to provide biological control. This wasp parasite has become well established in the Willamette Valley of Oregon, making aphid sprays unnecessary in most hazelnut orchards. If aphid populations do build up, sample 3 terminal branches per tree and 3 leaves per terminal, counting the number of aphids per leaf, and also look for signs of parasitism. Apply an insecticide if little or no parasitism is seen and aphids per leaf reach a level of 20 or more in April, 30 or more in May, or 40 or more in June or July.

Tent caterpillars (*Malacosoma* species) may sometimes damage hazelnuts or chestnuts, and **yellow neck caterpillar** (*Datana ministra*) is an occasional pest of chestnuts. These gregarious insects feed in large colonies and can completely defoliate young trees during seasons when they are not controlled by natural enemies. Sprays of the biological insecticide *Bacillus thuringiensis* (Bt) or of insect growth regulators such as tebufenozide (Confirm™) can control outbreaks of these pests.

Diseases that attack flowers, fruits and foliage. Hazelnuts and chestnuts also suffer from far fewer fruit and foliar diseases than do most fruit and nut crops. During most years in most locations, no measures beyond those used for blight diseases are needed to control any fruit or foliar diseases.

Kernel decay of hazelnuts may be caused by several different fungi. *Mycosphaerella punctiformis* causes kernel tips to blacken, *Septoria ostryae* causes internal discoloration of kernels, and *Nematospora coryli* causes dark, sunken lesions on kernel surfaces. The fall copper spray used to prevent bacterial blight on young hazelnut trees will also prevent kernel decay. Occasionally, older trees which are no longer receiving copper sprays for bacterial blight will require special sprays to prevent kernel decay. Flail mowing of the orchard floor stirs up disease spores that can infect hazelnut flowers and cause to kernel decay, so delaying mowing until late spring, after hazelnut flowering is over, will reduce kernel decay.

Brown stain of hazelnuts is a physiological disorder of unknown cause. Soft, water-soaked spots develop on shells when fruit are about half-grown, spots turn dark brown, and later both shell and kernel disintegrate into a watery brown mass. No control methods are known.

Insects, birds and mammals that attack flowers, fruits and nuts. Several insects can be serious direct pests of hazelnuts and chestnuts, and birds and squirrels may take many hazelnuts, so these pests require control where they occur.

Big bud mites (primarily *Phytoptus avellanae*) attack hazelnut axillary flower buds, destroying them so they do not develop into nuts. Susceptible cultivars such as 'Daviana' and 'Ennis' require several sprays of an insecticide, such as endosulfan, during the spring mite migration from buds infected the previous year to new buds. Planting cultivars with high resistance to big bud mites, such as 'Barcelona', 'Hall's Giant', and the pollenizers 'Delta' and 'Epsilon', eliminates the need for such insecticide sprays. Cultivars with intermediate mite resistance, such as 'Lewis', 'Clark', and the pollenizers 'Gamma' and 'Zeta', usually need no insecticide treatments but may require them if big bud mite populations build up to high levels.

Filbertworm (*Cyadia latiferreana*) is the most serious insect pest of hazelnuts in Oregon, usually requiring 2 insecticide sprays each year. Growers monitor filbertworm populations with pheromone traps placed in the orchard in mid-June, and apply sprays when trap catches average 2 to 3 moths per trap, usually about July 10 and again 3 weeks later. Filbertworm has also attacked about 10% of chestnut plantings in the Pacific Northwest.

Filbert leafroller (*Archips rosana*) and **omnivorous leafroller** (*Cnephasia longana*) larvae feed on hazelnut buds and developing leaves in early spring but have only one generation per year, so one or occasionally two insecticide sprays are generally sufficient to control these pests. **Obliquebanded leafroller** (*Choristoneura rosaceana*) has 2 generations per year and feeds from May until harvest, rolling leaves together and feeding on the outsides of developing nuts. Start checking for leafroller larvae about mid-March by inspecting 3 terminal branches per tree and 3 leaf clusters per terminal, and treat with an insecticide when 20 to 25% of clusters are infested with leafrollers. Place one pheromone trap per 5 acres in mid-May to monitor adult moth flight, and treat if 40 or more moths are caught per week. The biological insecticides *Bacillus thuringiensis* (Bt) or spinosad (Success™), or insect growth regulators such as methoxyfenozide (Intrepid™), can be used to kill the larvae.

Chestnut weevils (*Curculio* species and close relatives) may develop into serious pests of chestnuts, but have become rare in the wild because most native chestnuts have been killed by chestnut blight. Consequently, it often takes 10 to 15 years before chestnut weevils find new chestnut plantings and build up to economically damaging populations. Adults are "snout beetles" similar to pecan weevil, and the female uses her long proboscis to drill a hole through the shell and then lay eggs in the kernel of each nut. Larvae may devour the kernel and exit the nut to burrow into the soil, leaving an empty shell (Photo 25-12), or remain to infest the nut at harvest. In California, *Curculio elephas* and *Balanius rectus* are the species that may cause damage. East of the Rocky Mountains, **small chestnut weevils** (*Curculio sayi*) emerge after 2 to 3 years in the soil in late May through July, while **large chestnut weevils** (*C. caryatrypes*) emerge after 1-2 years in the soil in late July and August. Minimize weevils by promptly harvesting chestnuts as soon as they ripen, then place nuts in a hot water treatment at 130 F for 20 minutes to kill unhatched weevil eggs, then immediately cool nuts to 32 to 34 F. If weevil populations build up in older chestnut orchards, scout for adults by jarring trees over a white sheet and counting the fallen weevils. Three applications of carbaryl starting in mid-August can prevent damage where weevil populations have built up.

Blue jays (*Cyanocitta cristata*) and **squirrels** (*Sciurus* species) often congregate in hazelnut orchards, consuming large amounts of nuts. Placing steel traps on tall posts for blue jays or short posts for squirrels, and leaning these posts against trees in the outer rows of the orchard, can catch many of these pests. A few growers have erected a 20-inch-tall chicken wire fence, topped with a slanted cover and an electric wire, to keep squirrels out of hazelnut orchards. Chestnuts' spiny burs generally protect their nuts from attack by birds or mammals.

HARVEST, PROCESSING AND MARKETING

Hazelnut and chestnut growers prepare for harvest by flail mowing the orchard just before nuts begin to drop in early September, to grind up dropped blank nuts (which fall earlier than well-filled nuts) and twigs. Clean-cultivated orchards,

Photo 25-12. Chestnut weevil larvae, such as these small chestnut weevils, may devour kernels and exit nuts to burrow into the soil, leaving an empty shell as here, or remain to infest nuts at harvest. (Photo by R. Bessin, Univ. of Kentucky.)

which are rare today, must be leveled, smoothed and rolled in two directions before harvest.

Hazelnut growers may apply ethephon during late August or early September to hasten husk opening and harvest. Researchers have tested ethephon applications on chestnuts, in an attempt to condense the 3-week long harvest, but ethephon has not hastened chestnut ripening.

Mechanical and hand harvest. Hazelnut growers use mechanical trunk shakers and mechanical nut harvesters, like those used for other tree nut crops. Chestnut growers allow nuts to fall naturally soon after burs split open (Photo 25-13) rather than shaking trees, so that harvested nuts have uniform moisture contents. Tree shakers would dislodge many immature nuts whose burs hadn't split yet. Most chestnut growers harvest nuts from the ground by hand, but a few use nut-harvesting machines.

Photo 25-13. Chestnut burs split open and nuts drop naturally when ready to harvest. Chestnut growers do not use tree shakers, as they would dislodge many immature nuts whose burs hadn't split yet.

Hazelnut growers begin tree shaking when some nuts start to fall naturally or after ethephon application. Tractor-pulled mechanical nut sweepers (Photo 25-14) windrow nuts in the center of the row. A pick-up machine follows, lifting nuts and depositing them via a conveyor belt into a trailer or tote box, while blowing out leaves and small twigs and removing husks with hulling cylinders. Tote boxes require a forklift to transport them from the orchard.

Photo 25-14. Tractor-pulled mechanical nut sweepers windrow hazelnuts, then harvesting machines pick up the nuts and deposit them in tote boxes or trailers. Tote boxes, shown here, require a forklift to transport them from the orchard. (Photo by C. Kempler, Agric. Canada, Agassiz, B.C.)

Chestnut growers must harvest at least every other day over a 3-week period to maintain nut quality. Growers who harvest chestnuts by hand often pick nuts from the burs as they harvest. Mechanically-harvested chestnuts, or those that have been raked up, must be separated from burs after harvest. One California chestnut grower has developed a wringer-type separator made from tires that twists the burs so nuts pop out without receiving scratches. Some Italian chestnut growers use a vacuum separator that sucks nuts into a rotating drum that removes the burs and sorts nuts by size.

Processing, storing and marketing hazelnuts. Growers run harvested hazelnuts through a processing line that includes a cleaning machine like that described for pecans, a washer to remove mud, a fresh water rinse, and a chlorine rinse to kill bacteria. Hazelnuts are then dried at 90 to 100 F to a kernel moisture content of 8 to 10%, either by the grower or the buyer. Hazelnuts are sold on a dry-weight basis, so growers are docked if nuts have more than 10% moisture. Hazelnuts store best at 36 to 40 F.

About 60% of U.S.-produced hazelnuts are exported, but the U.S. often imports as many or more hazelnuts as it exports. Most imports, mainly from Turkey, are shelled hazelnuts, whereas 80% of U.S. hazelnuts for export are sold in-shell, with over half of these going to Hong Kong and China. The price of hazelnuts produced in Turkey plummeted between 2001 and 2003 due to very large world supply, and the U.S. added an 8% tariff to shelled Turkish hazelnut imports, which otherwise sell for less than the U.S. cost of production.

The major U.S. cultivars, especially 'Barcelona' and 'Ennis', have been mainly sold in-shell due to their large nut size and thick shells that resist breakage. However, demand for in-shell hazelnuts is declining while demand for shelled hazelnuts is increasing, and most growers are choosing cultivars for new plantings that have thinner shells, crack easily, and are mainly sold as shelled nuts. Most U.S. hazelnuts sold in this country are sold as shelled nuts, mainly to food manufacturers who make chocolate candies, flavored coffees, hazelnut butter, cookies, ice cream, mixed nuts and other products.

Handling, storing and marketing chestnuts. Once removed from burs as soon as possible after harvest, chestnuts should be handled like apples rather than like other nut crops. Wash them, cool them as soon as possible, and store them at 32 to 33 F with as high a humidity as possible, to retain quality and prevent molding.

Chestnuts are primarily sold fresh in the shell, within 3 months of harvest. Fresh chestnuts that mature early and have not been dried command the highest market prices. Chestnuts sold direct to consumers at the farm generally fetch $2.50 to $4 per pound, while those sold at farmers' markets go for $4 to $5 per pound and chestnuts in grocery stores sell for $6 to $7 per pound. Chestnut prices are highest in September, than gradually fall until late December, and demand greatly declines after January 1. Chestnuts are graded into 6 size categories, and should be in the top 3 sizes (greater than 1½ inches diameter) to be sold as fresh nuts.

Smaller chestnuts should be shelled ("peeled") and dried to 8% moisture, which turns them rock-hard, then ground into flour or used to create products such as chestnut chips. Dried chestnuts can be stored at room temperature as they will not mold, but require tightly-sealed containers to exclude grain pests such as Indian meal moth and flour beetles. Dried chestnuts can be rehydrated in boiling water like dried beans.

Further Reading

Growing hazelnuts in Oregon. 1985. Oregon State Univ. Ext. Serv. Ext. Circular 1219.

Hazelnut pest management guide for the Willamette Valley. 1994. Oregon State Unvi. Ext. Serv. EM 8328.

Hunt, K., M. Gold and W. Reid. 2003. *Growing Chinese chestnuts in Missouri*. Univ. of Missouri Center for Agroforestry.

Olsen, J. 2003. *Nut growers handbook.* Available online at http://oregonhazelnuts.org/handbook.htm

Ridley, D. and J. Beaumont, eds. 1999. *The Australian chestnut growers' resource manual.* Dept. of Nat. Res. and Env., Agric. Victoria.

Vossen, P. 2000. *Chestnut culture in California.* Univ. of Calif. Div. of Agric. And Nat. Res. pub. 8010.

Chapter 26
Figs and Mulberries

Figs and mulberries both belong to the mulberry family, the Moraceae. Both types of trees produce fruits that consist of many individual ripened ovaries, arising from separate flowers in a cluster, which have fused together. Botanically, these are called "multiple fruits". Both are minor crops in North America, although figs are an important fruit crop and mulberry leaves are an important feed crop for silkworms and livestock worldwide.

Fig (*Ficus carica*) originated in the fertile semi-desert region of southern Arabia, and was first cultivated in Mesopotamia over 5000 years ago. Figs thrive under the mild winters and hot, dry summers of a Mediterranean climate, and are cold-hardy to about 10 F (USDA zone 8). The world produces over 1 million metric tons of figs per year, with about 28% of them grown in Turkey, about 19% raised in Egypt, about 8% each from Greece and Iran, about 7% from Morocco, about 6% each from Spain and Algeria, and about 4% from the U.S. Figs are grown on 13,000 acres in 14 U.S. states, although 98% of U.S. fig production comes from California. Texas, Louisiana, South Carolina and Mississippi also produce significant quantities of figs.

Mulberries include several species in the genus *Morus*. Red mulberry (*M. rubra*), also called American mulberry, is native to eastern North America from Massachusetts to the Gulf Coast and west to Kansas, and is adapted to USDA zones 5 to 8. White mulberry (*M. alba*), native to China, was introduced to North America during colonial times in an attempt to establish silkworm culture. The silkworm industry failed, but the white mulberry became naturalized and has produced natural hybrids with the native red mulberry. White mulberry and red-white mulberry hybrids vary greatly in hardiness among cultivars; some are hardy to -30 F (zone 4) while others are only hardy to -10 F (zone 6). Black mulberry (*M. nigra*), native to western Asia and grown for its fruit in Europe since before Roman times, was also introduced to North America but is generally cold-hardy only to 0 F (zone 7), so has been grown mainly in the Pacific Northwest and Deep South, zones 7 to 10. Russian mulberry (*M. tatarica*), originated in China but was introduced into eastern Europe about 1500 years ago, and has been widely planted in windbreaks in the U.S. in recent years because it is cold-hardy to -40 F (zone 3) and very drought-resistant, but it produces inferior fruit.

About 95% of U.S.-grown figs are machine-harvested for drying, and just 5% are hand-harvested for fresh market. About 85% of U.S. figs are grown for domestic sales, and about 15% are exported. No commercial mulberry industry exists in North America, but many Asian and European countries grow significant amounts of mulberries, using the leaves of white mulberries as silkworm fodder and the fruit of black and white mulberries for poultry feed and human consumption. Superior mulberry cultivars have been selected in the U.S., and mulberries have great potential for machine harvest for juice, wine and jelly production.

TREE GROWTH AND BEARING HABITS

Figs start bearing crops 2 to 3 years after planting, while mulberries begin to bear 5 to 7 years after planting. Figs and mulberries are both very long-lived. Figs and black mulberries continue to bear for 100 years or so, while red and white mulberries bear for 60 to 70 years.

Figs are small trees, reaching 15 to 30 feet tall, with large palmate leaves (Photo 26-1). Fig flowers are inverted so develop inside the multiple fruit, with both male and female flowers enclosed in stem tissue. At maturity the interior of the fig contains the remnants of these flower structures, including unfertilized ovaries which are commonly but erroneously called seeds (Photo 26-2).

Two botanical types of figs occur, common figs and Smyrna figs. Common figs produce fruit parthenocarpically, without any pollination or fertilization, so have no true seeds. Smyrna figs must be pollinated by a tiny fig wasp that enters the fruit to ensure fertilization and fruit growth, and require male pollenizer trees that produce inedible swollen tissues with male flowers, called "caprifigs". Common figs produce 2 crops per year in California's mild climate and make up 57% of fig acreage, while Smyrna figs produce only 1 crop per year and are grown on 43% of fig acreage. Common figs yield 6000 to 8000 pounds of fresh figs per acre per year, which is 1500 to 2000 pounds dried figs, while Smyrna figs yield 6000 to 7000 pounds of fresh figs per acre per year, which is 1500 to 1750 pounds dried figs.

Red and white mulberries and their hybrids grow into medium-large trees, 50 to 70 feet tall. Black mulberries are smaller trees, 20 to 30 feet tall. Mulberry trees have large leaves that may be lobed or unlobed, and are either dioecious or monoecious. Both male and female flowers occur on catkins that hang from the axils of current season shoots or spurs on older wood (Photo 26-3). Mulberries are wind-pollinated, but many cultivars set fruit without pollination.

SITE SELECTION AND PREPARATION

Figs perform best in areas with less than 800 chilling hours, where winter temperatures do not drop below 10 F. They need full sun, and a south-facing slope is ideal to provide maximum warmth in winter and sunlight in summer. They need well-drained soil but can be grown in many soil types, and are relatively salt-tolerant for fruit crops. Figs are shallow-rooted, so don't need very deep soil but do require irrigation water.

Most black mulberry cultivars are only hardy to 0 F, and do best where summers are hot and dry. They bloom late and need full sun, so a south slope would again be ideal. White and red mulberries and their hybrids are hardier and bloom earlier, so

Photo 26-1. Figs are small trees, reaching 15 to 30 feet tall, with large palmate leaves, and begin bearing crops 2 to 3 years after planting. (Photo by T.S. Gladwin, Passion Fruit Farms, Merced, CA.)

Photo 26-2. Fig fruits contain the remnants of flower structures inside the fruit, since flowers are inverted, but most cultivars, such as these 'Mission' figs, develop parthenocarpically so flowers are never fertilized. (Photo by T.S. Gladwin, Passion Fruit Farms, Merced, CA.)

Photo 26-3. Mulberry fruit develop from catkins that hang from the axils of current season shoots or spurs on older wood. (Photo by Dr. Ed Hager, California Rare Fruit Growers.)

an east or northeast slope might be better to delay bloom a bit. Mulberries are deep-rooted so require deep, well-drained soil. They are drought-tolerant so usually don't need irrigation in Eastern areas, but require it in the arid West.

Soil pH for figs and mulberries should be between 6.0 and 7.4. Figs are especially prone to cotton root rot on alkaline soils, so test soil pH before planting and lower it with sulfur if it is above 7.4. Also test soil for phosphorous (P) and potassium (K), and add these nutrients before planting if they test low. Test sandier soils for root-knot nematodes, and if these are present, fumigate or solarize the soil, or sow nematode-reducing cover crops for 2 years before planting.

Fig growers generally deep-rip the soil to a depth of 5 to 6 feet at least several months before planting, to break up any underlying hardpan that could stop root growth and water drainage. They next disk the soil to break up clods, and float it to smooth the ground. Most fig growers no longer level the site, since they will install drip irrigation rather than using furrow or flood irrigation as was previously done. Mulberry growers in Western areas could follow these same procedures, while mulberries in Eastern areas could be planted directly into herbicide-killed strips in a sod hayfield. Figs and mulberries do not need raised beds.

CHOOSING CULTIVARS

Fig and mulberry cultivars must be cold-hardy enough for the area to be planted, and should need about as many chilling hours as the area normally provides. Only a small percentage of these fruits can be sold fresh, so cultivars should produce fruit that suits the intended processing market.

Three cultivars of common figs are widely grown in California for drying: 'Adriatic,' 'Mission' and 'Kadota'. 'Adriatic' provides 42% of dried figs, 'Mission' accounts for 29%, and 'Kadota' provides 3% of dried figs. One cultivar of Smyrna fig, 'Calimyrna,' provides 26% of dried fig production as well as almost all the fresh-market figs, which constitute 5% of total fig production. Other common figs grown in California and the Pacific Northwest include 'Conadria,' 'Black Jack' and 'California Brown Turkey,' while those grown in the southeast include 'Eastern Brown Turkey,' 'Celeste' and 'Alma'. Table 26-1 describes these fig cultivars.

Some of the most-planted mulberries include 'Illinois Everbearing,' a white mulberry x red mulberry hybrid, and 'Black Persian' and 'Pakistan' black mulberries. Table 26-2 details traits of these and other mulberry cultivars.

PROPAGATION, PLANTING AND CARE OF NEW PLANTS

Figs are most often propagated by hardwood stem cuttings. Growers should mark trees in summer from which to take cutting material next winter, choosing trees that are free of fig mosaic symptoms. Growers collect one-year-old terminal branches, 6 to 8 inches long, in late winter. They wrap these branches in bundles and place the bundles upside down in callusing trenches from mid-January to mid-April, covering the cuttings with 2 to 4 inches of soil. Once calluses have formed, they place individual cuttings 6 to 12 inches apart and right side up in a propagation row, with 1 to 2 inches of the cutting above the soil line and 5 to 6 inches of the cutting below the soil line. The cuttings should develop roots and grow shoots 3 to 4 feet tall during the following growing season, and will be ready to transplant to the new orchard the next dormant season.

Mulberries are generally budded or grafted onto seedling mulberry rootstocks in early spring. Be sure the rootstock mulberry species is hardy enough for your area.

Tree spacing and orchard layout. Figs are generally planted 12 feet apart in rows spaced 20 to 24 feet apart, using 151 to 181 trees per acre. Mulberry trees grow much larger so should be planted 16 to 18 feet apart in rows 24 to 28 feet apart, using 86 to 113 trees per acre. Rows are generally oriented north-south for maximum light interception, but should be planted on the contour on hilly ground.

Common figs and most cultivated mulberry cultivars need no pollenizer trees. Smyrna figs need caprifigs as pollenizers, but caprifig trees are now planted separately some distance away from 'Calimyrna' orchards, since caprifigs were found to harbor overwintering *Fusarium* spores that cause fig endosepsis in 'Calimyrna' fruit. The winter caprifig crop is picked and hung in 'Calimyrna' orchards to provide pollination.

Planting and care of new plants. Growers transplant 1-year-old rooted fig cuttings or grafted mulberry trees into the new orchard in early spring. Plant as described for other fruit trees, and water well after planting. Newly-planted figs and mulberries are prone to sunscald and attack by rodents, so especially benefit from white plastic trees guards placed around the trunks.

Soon after planting, remove any narrow-angled branches and those arising less than 18 inches off the ground, except when training figs to the multi-

Table 26-1.

WIDELY-GROWN FIG CULTIVARS

In approximate order of ripening (varies with area).

Cultivar	Where grown	Chill hours needed	Fruit				Yields		Susceptibility to:			Tree size	Harvest time
			Size	Skin color	Flesh color	Eye	First crop	Second crop	Fig Mosaic	Fig Smut	Sour Rot		
Celeste	SE	550	small	purple	amber	closed	none	high	S	R	R	large	early
Adriatic	CA, NW	300	med-large	yellow	pink	medium	small	high	VS	VS	S	large	middle
Conadria	CA, NW	300	medium	yellow	red	small	medium	very high	MR	MS	MR	very large	middle
California Brown Turkey	CA, NW	300	large	brown	pink	open	small	high	S	MS	S	medium	middle
Eastern Brown Turkey	SE	400	med-large	brown	pink	medium	medium	high	S	MR	MS	very large	mid-late
Mission	CA	300	large	purple	red	closed	medium	high	VS	R	R	large	mid-late
Calimyrna	CA	300	large	yellow	amber	open	none	med-high	MR	VS	VS	large	mid-late
Kadota	CA	600	medium	amber	amber	open	small	high	MR	MR	MS	large	mid-late
Black Jack	CA, SE	400	very large	purple	red	medium	medium	high	S	S	MR	small	late
Alma	SE	500	medium	yellow	amber	closed	medium	high	S	R	R	medium	very late

Where grown: CA = California, NW = Pacific Northwest, SE = Southeast; all in USDA hardiness zone 8-10.

Pest susceptibility/resistance ratings: VS = very susceptible, S = susceptible, MS = moderately susceptible, MR = moderately resistant, R = resistant.

Average ripening times in San Joaquin Valley, CA: early: early May 1st crop, early June 2nd crop; middle: mid-May 1st crop, mid-June 2nd crop; mid-late: late May 1st crop, late June 2nd crop; late: early June 1st crop, mid-July 2nd crop, very late: mid-June 1st crop, late July 2nd crop.

Notes on cultivars:
Celeste, also called **Celestial** and **Tennessee Mountain**, is one of most cold-hardy figs, closed eye R to dried fruit beetle, sour rot.
Adriatic, also called **Verdone**, is the main commercial fig. It leafs out early; frost-prone in SE.
Conadria, also called **Adriatic Hybrid** as it's a seedling of Adriatic, is precocious.
California Brown Turkey, also called **Black Spanish**, has few seeds, best fresh, not for drying or canning.
Eastern Brown Turkey, also called **Texas Everbearing**, is hardier than California B.T., will bear on suckers next year if frost-killed.
Mission, also called **Black Mission**, is R to oak root rot; closed eye R to dried fruit beetle, sour rot.
Calimyrna is the only fig listed that needs pollination by caprifig wasps; many seeds.
Kadota fruit has very few seeds, eye open but filled with honeydew, which stops dried fruit beetle, sour rot.
Black Jack trees are smaller, can be planted closer.
Alma fruit is fine with small seeds, tree is R to fig rust, precocious, but frost-prone as young tree.

Table 26-2.

COMMON MULBERRY CULTIVARS

In approximate order of ripening (varies with area).

Cultivar	Species	Best in zones	Fruit				Tree height	Harvest time
			Size	Flavor	Firmness	Yield		
Shangri-La	black	7-9	large	very good	soft-med.	high	18-22'	early
Collier	white X red	5-8	medium	very good	medium	very high	25-30'	early
Illinois Everbearing	white X red	5-8	med-large	best	medium	very high	25-30'	middle
Silk Hope	white X red	6-9	large	very good	medium	high	26-32'	middle
Black Beauty	black	7-10	large	very good	soft	high	12-16'	middle
Black Persian	black	7-10	large	best	soft-med.	very high	22-28'	late
Pakistan	black	7-10	very large	best	firm	very high	24-30'	late
Wellington	white	4-7	large	good	soft	very high	30-35'	late

Average ripening times in central CA & central OK: early: early June to mid-July, middle: mid-June to late July, late: late June to early Aug. Harvest tlmes are about 1 month later in western OR & northern IL for cultivars adapted there.

Notes on cultivars:
Shangri-La does well in humid Deep South as well as West Coast.
Collier is hardy, adapted to most of U.S.
Illinois Everbearing is hardy and precocious, often bears 3 to 4 years after planting.
Silk Hope is not as precocious as IL Everbearing, but better adapated to humid Southeast.
Black Beauty is a smaller tree than most, but too disease-prone in humid Southeast.
Black Persian does well in humid Deep South as well as West Coast.
Pakistan has firmer fruit than other mulberries, so doesn't stain as much.
Wellington is the hardiest mulberry with good-tasting, though soft, fruit.

trunk system described below. Prune back remaining branches and the leader to 2/3 their length.

SOIL MANAGEMENT, IRRIGATION AND FERTILIZERS

Most fig and mulberry growers allow resident vegetation to grow in the aisles between rows, and control weeds within rows with a fall-applied residual pre-emergent herbicide. During the first 2 years after planting, these herbicides are used at half the rate that is used for older plantings. Each spring, weeds which survive the pre-emergent herbicide are spot-treated with glyphosate or other post-emergent herbicides.

Fig growers generally disk the resident vegetation 4 times per year during the first 2 years, then just once per year from year 3 on, when the orchard is bearing. This disking is performed before the first harvest and followed by a floater to smooth the soil for mechanical harvest. Since drip irrigation is generally used and doesn't wet the row middles, the resident vegetation may need no mowing during the harvest season in arid areas.

Fig and mulberry growers in humid areas will need to mow the row middles every few weeks during the growing season, in both young and bearing orchards. Such orchards should not be disked due to the threat of soil erosion.

Irrigation methods and water use. Drip irrigation is now standard for new fig plantings. Some older fig orchards are still watered by flood or furrow irrigation, especially for fresh-market figs, but many dried fig growers have replaced these methods with drip irrigation as water costs have increased.

Figs are a relatively low-water-use crop when drip irrigation is used. A newly-planted orchard needs only 4 acre-inches of water during the first season, and just 9 acre-inches of water during the second season. The orchard should begin to bear during the third season, when it needs 16 acre-inches of water, and will need 18 acre-inches of water during the fourth season. Mature fig orchards, from the fifth season on, need only 24 acre-inches of water per year if applied through a drip system.

Since figs bear over a long season, from mid-

May to mid-October, irrigation water must be consistently applied from the start of fruit growth in spring until a few weeks before final harvest. Generally growers apply 1 inch of water per week from early April to early September.

Mulberries are usually not irrigated in Eastern areas, but require irrigation in the West. They need about the same amount of water per year as figs do, timed to coincide with maximum shoot growth of young trees and maximum fruit growth of bearing trees, generally from mid-April to early August.

Fertilizer additions. Nitrogen (N) is the main nutrient needed by figs and mulberries, like other fruit crops, but figs and mulberries need less N than most other fruit crops. A newly-planted orchard needs only 10 pounds actual N per acre during the first season, and just 20 pounds actual N per acre during the second season. A fig orchard should begin to bear during the third season, when it needs 30 pounds actual N per acre, and will need 40 pounds actual N per acre during the fourth season. From the fifth season on, fig and mulberry orchards need 50 pounds actual N per acre per year. Apply N after harvest, in early to late summer for mulberries and in fall for figs.

Figs and mulberries generally need no other fertilizer nutrients besides N in California. Where initial soil tests indicated low P or K, test the soil again every 3 to 4 years and add more P or K if needed. Alkaline soils need sulfur (S) applications every 3 to 4 years to keep pH lower.

TRAINING, PRUNING AND MANAGING CROPS

California fig growers generally train trees to the open-center system described in chapter 4 and used for low-density peaches. Growers in zones 9-10 of the Southeast also use the open-center system.

Growers in zone 8 of the Southeast, where frosts may occasionally cause dieback of fig branches and trunks, train figs to the multi-trunk stool system. Several trunks are allowed to grow so if some get killed back, others may live, or if all trunks are killed back, several new trunks will grow from the roots or bases of old trunks. These growers raise figs as bushes rather than as trees. They head back newly-planted trees by one-half, choose 3 to 8 widely-spaced leaders that grow from this stub, and each year head back the selected leaders by one-third to one-half.

Mulberries are generally trained to a modified leader tree, described in chapter 4, like low-density apples.

Crop management and pruning figs. Smyrna figs, including 'Calimyrna,' require male caprifigs as pollenizers, and the pollinator wasps that overwinter in caprifigs as pollinators. Caprifig trees produce 2 crops per year of small, inedible caprifigs containing male flowers. The winter caprifig crop from nearby plantings is picked, placed in paper bags or baskets, and hung in 'Calimyrna' orchards to provide pollination. The summer crop of caprifigs is prevented by spraying a growth regulator or destroyed by disking or burning.

Common figs require no pollination and produce 2 crops per year. The early-summer or "breba" crop began to form the previous fall, at the tips of last season's branches. These figs develop slowly over the winter and ripen in early summer. Many common fig cultivars produce only a very light breba crop. The second common fig crop, borne at the base of the current season's branches, ripens in late summer and is the main crop (Photo 26-4).

Pruning affects the size of the 2 common fig crops. For a larger breba crop, do not cut back branches during the dormant season, since that would remove the develop figs. Conversely, cutting back all branches during the dormant season can eliminate the first fig crop and increase the second fig crop.

Figs grown for drying are allowed to grow to nearly their full height, and pruning consists mainly of removing suckers and crowded branches to thin out dense areas. Some cultivars, such as 'Adriatic,' also need dormant heading back of top growth, at least every other year, to induce vigorous new branches to grow rather than weak short twigs. 'Calimyrna' trees produce long, upright shoots without branches, which need light heading in summer or must be cut back heavily as one-year-old branches the following winter.

Taller fig trees grown for fresh fruit, such as 'Kadota,' are pruned to lower heights and a nearly flat top to facilitate hand harvest, by cutting inside branches shorter than outer branches. Larger growers may use mechanical pruning equipment to mow off the tops of trees. 'Black Jack' is a genetic semi-dwarf cultivar that stays lower naturally, so is good for hand harvest for fresh fruit, but ripens late.

Pruning mulberries. Prune mulberries mainly during the dormant season, since branches can bleed excessive amounts of sap if large cuts are made in summer. However, once trees have developed the desired height and spread, pruning lateral branches to 6 leaves in mid-summer will encourage these branches to become spurs, which bear much fruit while keeping the tree in bounds and open

Photo 26-4. The early-summer or "breba" fig crop begins to form the previous fall, at the tips of last season's branches, and ripens in early summer. Many common fig cultivars produce only a very light breba crop, so branches may be cut back in winter to eliminate the first crop and increase the second crop. The second common fig crop, borne at the base of the current season's branches and shown here, ripens in late summer and is the main crop.

enough for good light penetration. The only other pruning needed is to remove any diseased, damaged, dead or overly dense branches.

PEST MANAGEMENT

Figs and mulberries generally suffer from far fewer pests than do most fruit trees, especially in the regions where they are best adapted.

Pests that can kill trees. The only pest management needed for young fig orchards during most years is to monitor soil pH and to prevent rodent damage. Mulberries often have no serious pests that kill trees.

Cotton root rot, caused by the fungus *Phymatotrichum omnivorum*, is the main killer of fig trees in alkaline soil. Test soil pH before planting and lower it with sulfur if it is above 7.4.

Ground squirrels (*Spermophilus* species) and **pocket gophers** (*Thomomys* species) may feed on fig tree bark, killing young trees. Where populations of these pests are high, growers apply poison bait in fig orchards. Avoid mulching around fig tree trunks, so as not to create a habitat that favors these rodents, and use white plastic tree guards to protect trunks.

Indirect pests that reduce yields and tree longevity. These pests can generally be prevented by physical methods such as choosing disease-free propagating material or modifying the soil, or are usually controlled naturally by biocontrol agents.

Fig mosaic virus causes yellow spots on leaves and fruit, causing susceptible cultivars to produce fewer and smaller fruit, and some fruit to drop prematurely. 'Mission' is the most seriously damaged cultivar, while cultivars such as 'Conadria,' 'Kadota' and 'Eastern Brown Turkey' suffer little damage when infected with fig mosaic virus. Avoid propagating this virus by choosing cutting material from trees that are free of mosaic symptoms. Fig mites transmit fig mosaic virus, so removing and destroying trees with mosaic symptoms and controlling fig mites may reduce disease spread.

Root-knot nematodes (*Meloidogyne* species), **lesion meatodes** (*Pratylenchus vulnus*) and **dagger nematodes** (*Xiphinema index*) attack fig trees, causing a gradual decline and loss of vigor over several years. High nematode populations also cause many premature fruit to drop. Test soil for nematodes before planting, and if present, fumigate or solarize the soil, or sow nematode-reducing cover crops for 2 years before planting.

Fig scale (*Lepidosaphes conchiformis*) may feed on fig leaves, branches or fruit, reducing tree vigor and yields. Scale feeding on fruit causes callus tissue to form on fruit skin, giving fruit a warty appearance, and scale-damaged fruit cannot be sold fresh. A wasp parasite normally gives excellent control of fig scale, so avoid any insecticides that would disrupt this parasite. Dormant oil can control scale if it becomes a problem, without disrupting the parasite, but fig growers do not often need to apply dormant oil.

Spider mites (*Tetranychus* species) may form webs on both fig and mulberry trees, sucking sap from leaves and reducing tree vitality. Heavy spider mite feeding causes leaf bronzing, defoliation and reduced yields. Naturally-occurring predaceous mites and other natural enemies often provide complete control of spider mites if not disrupted, so avoid using any chemicals for other pests that would disrupt these natural enemies, such as sulfur fungicide or pyrethroid insecticides in summer. Keep trees well-irrigated to minimize mite damage. If spider mites build up, summer oil sprays can reduce them without hurting predators much.

Fig rust, caused by the fungus *Physopella fici*, may defoliate fig trees in humid climates if not controlled. Rust lesions on leaves first appear in spring as small, yellow-orange spots (Photo 26-5), which enlarge and become more numerous as the season progresses, and may cause premature defoliation. Figs in humid climates usually require 1 to 4 copper fungicide sprays in spring, starting when the first leaves have reached full size and continuing every 3 to 4 weeks while weather remains humid.

Photo 26-5. Fig rust may defoliate fig trees in humid climates if not controlled by 1 to 4 copper fungicide sprays in spring. (Photo by T. Eaker, Univ. of Georgia.)

Cercospora leaf spot, caused by several species of *Cercospora* fungus, attacks mulberry leaves, producing small spots that later enlarge and coalesce. This disease is not generally damaging in the arid West, but can defoliate mulberries in the humid Southeast. Some mulberry cultivars, noted below Table 26-2, suffer less from this disease so are better adapted to the Southeast, and choosing them can reduce the need for fungicides. Applying 2 or 3 sprays of a fungicide such as mancozeb or the sterol inhibitor mycobutanil (Nova™), when new growth appears in spring and every 2 weeks thereafter if wet weather prevails, can reduce Cercospora leaf spot. Using urea fertilizer as the N source applied after harvest, and applying it to fallen leaves on the ground to hasten their decomposition, will also reduce this disease.

Diseases that attack flowers and fruits. The dry climate where most figs are grown prevents most fruit rots and other diseases, but these diseases may require prevention in humid climates or during occasional rainy harvest seasons in dry climates. Several fig cultivars resist some of these diseases. Mulberries have few fruit diseases.

Fig smut, caused by several species of *Aspergillus* fungus, turns the internal tissues of the fig cavity black and powdery with spores. Fig smut usually starts at fruit eyes when figs are still green, so cultivars with small, closed eyes, or those with honeydew covering the eye so spores can't penetrate, resist the disease much more than cultivars with large, open eyes. The Southeastern cultivars 'Alma' and 'Celeste,' and the California cultivar 'Mission,' generally have the most resistance to fig smut, while 'Adriatic' and 'Calimyrna' are most susceptible, suffering up to 30% losses in California during some years. Removing all culled fruit from the orchard, and controlling dried fruit beetles and vinegar flies as described below, reduces fig smut, as does reducing dust during late summer by avoiding cultivation then. These cultural practices usually control fig smut enough that no fungicide treatments for it are needed.

Sour rot of figs, caused by various yeasts and bacteria, attacks figs that were damaged by dried fruit beetles or vinegar flies, which spread the pathogens. Infected figs turn pink and later become water-soaked, with an odor of fermentation. Again, cultivars with small, closed eyes, or those with honeydew covering the eye so insects can't penetrate, resist sour rot much more than cultivars with large, open eyes. Controlling dried fruit beetles and vinegar flies, as described below, also reduces sour rot.

Fig endosepsis attacks Smyrna-type figs and is also called internal rot, brown rot, eye-end rot, pink rot and soft rot. Caused by several species of *Fusarium* fungus, it starts as internal brown streaks and discolored areas at the base of flowers inside the young fruit, with no outer symptoms. Eventually a water-soaked circular area appears on the fruit skin, usually around the fruit eye. This area later extends down the fruit sides and turns purple, and the infected area has an off-flavor.

Fusarium spores overwinter in the summer crop of caprifigs, which is why caprifig trees are now planted outside of 'Calimyrna' orchards and the winter caprifig crop is picked and brought into the orchards to provide for pollination. Winter caprifigs are split in half, and any with internal discoloration are burned or otherwise destroyed. Healthy caprifig halves are treated with benomyl fungicide, placed in paper bags or baskets, and hung in 'Calimyrna' orchards to provide pollination. The summer crop of caprifigs is prevented by spraying a growth regulator or destroyed by disking or burning.

Surface rot of fig, caused by the fungus *Alternaria alternata*, may attack common and Smyrna-type figs during humid seasons. Small, round, brown or black spots cover the fruit surface. Any skin cracks make the fruit more susceptible to surface rot, so regular irrigation, to prevent cracking due to drought followed by heavy rains, helps prevent this disease. Growers in humid areas may need to apply a fungicide such as maneb when humid conditions prevail about 3 to 4 weeks before harvest begins.

Popcorn disease of mulberry, caused by the fungus *Ciboria carunculoides*, replaces fruit carpels with white fungal sclerotia, which enlarge and extend beyond normal fruit carpels, appearing somewhat like popcorn. White mulberry cultivars and some red-white mulberry hybrids are more susceptible to popcorn disease than black mulberries, so plant less susceptible cultivars where they are hardy enough. Where popcorn disease occurs, monitor fruit for the disease early in the season, and remove and destroy any infected fruit.

Insects and birds that attack flowers and fruits. These are the key pests of figs and mulberries. Controlling these insects on figs prevents most of the fruit diseases that they spread, and controlling birds is also often necessary for both figs and mulberries.

Dried fruit beetles, several *Carpophilus* species, may attack ripe or fermenting figs. The black or brown beetles (Photo 26-6) and white larvae are just 0.1 to 0.2 inches long, and their presence in any fruit will cause an entire load to be rejected by fig processors. Dried fruit beetles transmit yeasts and bacteria that cause fruit souring. All commercial fig cultivars are susceptible to these pests, but cultivars with smaller eyes, such as 'Mission,' are infested less than cultivars with large eyes, such as 'Calimyrna'. The Eastern cultivar 'Celeste' has a tightly closed eye, and the Texas cultivar 'Alma' has a small eye that's sealed with a drop of resin, so dried fruit beetles attack these cultivars much less. Dried fruit beetles overwinter on mummified fruit, so growers attempt to remove all such fruit from orchards to reduce the overwintering population. Early, rapid, complete harvesting and prompt fumigation of fig crops reduces summer generations of dried fruit beetles. Bait trapping of these beetles in large containers filled with cull fruit, water and yeast may effectively lure these pests from fig crops and reduce their populations.

Vinegar flies (*Drosophila* species) also have tiny white larvae just 0.2 inches long that cause the same damage to figs as dried fruit beetles, and also transmit yeasts and bacteria that cause fruit souring. Late-ripening cultivars are especially susceptible to vinegar flies, which breed in fermenting or decaying fruit. Promptly removing all culled fruit from the orchard and destroying it reduces vinegar flies, as does rapid harvesting and prompt fumigation of fig crops.

Birds of several species may peck figs and love mulberries. Birds so prefer mulberries that they may be used as trap crops for the early-summer fig crop or for other fruits that ripen at the same time, such as cherries or early peaches. Mulberries produce so much fruit that there is still plenty to harvest despite the birds, and thankfully birds take entire mulberries rather than just pecking fruit as they do on figs and peaches. Fig and mulberry growers may need to rotate bird scare devices, especially for the later fig crops when birds have less alternate food sources.

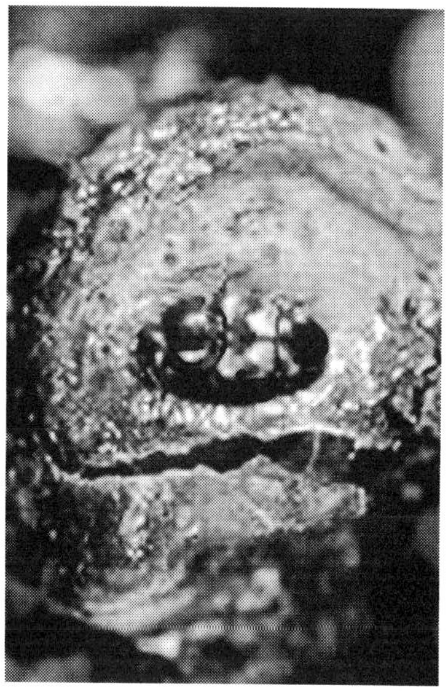

Photo 26-6. Dried fruit beetles transmit yeasts and bacteria that cause fruit souring. Cultivars with smaller eyes are infested less than cultivars with large eyes, so choose the former to reduce this pest. (Photo from Univ. of Calif.)

HARVEST AND PROCESSING

About 95% of U.S. figs are machine harvested for processing, but some growers hand-harvest some figs for fresh market. Larger growers may hand-harvest the earliest figs, before many are ready to machine harvest, and very small growers may hand-harvest all their figs to provide a long season of ripe figs to farmers' markets or other outlets. Mulberries are often harvested semi-mechanically, and harvest could be fully mechanized if a mulberry industry developed.

Hand harvest for fresh market. Start harvesting fresh figs when they begin to soften and the skin color starts to change to its mature shade. Workers must wear gloves when harvesting figs, as the white sap exuded from broken pedicels (fruit stems) can irritate the skin (Photo 26-7). Harvest at least 2 or 3 times per week for up to 4 months per each variety to be sold fresh. Forced-air cooling to 32F soon after harvest is highly recommended for fresh figs.

Black mulberry fruit are difficult to hand harvest, since berries tend to collapse when pulled off, staining hands and clothing with bright red juice. Red and white mulberries and their hybrids pull off easier with less staining, but hand harvest requires ladders and is time consuming. Most growers prefer to spread sheets on the ground and shake

Photo 26-7. Workers wear gloves when harvesting figs, as the sap exuded from broken fruit stems can irritate the skin. (Photo by R. Steinacher, Maywood Farms, Corning, CA.)

Photo 26-8. Most growers spread washed figs on paper trays laid on the orchard floor, and let them dry in the sun as done for raisins. (Photo by R. Steinacher, Maywood Farms, Corning, CA.)

the limbs by hand, so ripe mulberries fall onto the sheets, where they can be gathered for packing into sale containers or for home processing. Mulberries ripen over a long period of several weeks, so 3 to 4 shakings and harvests are needed.

Machine harvest for processing. Figs grown for drying are allowed to partially dry on the tree, until some figs begin to drop. At that point a few growers apply ethephon to speed up ripening, then hire helicopters to fly over the orchard to hasten fruit drop. Some growers use tree shakers to induce the rest of the figs to fall to the ground, but many just repeatedly harvest the figs that drop naturally when fully ripe. Many growers use air-blast sprayers, spraying only air, to induce the latest-ripening figs to drop. Fig growers use sweeper machines to windrow figs in the row middles, then use harvester machines to pick fruit up off the ground.

Harvest occurs at 1- to 2-week intervals for up to 4 months for common figs, from June 10 to Oct. 15 in California. 'Calimyrna' figs require 6 harvests per year for the single crop of mature trees. Harvested fruit are washed, may be stored for a few days at 32 to 34 F, and are then further dried.

Mulberries could be mechanically harvested using shake-and-catch machines used for cherries or other tree fruits.

Drying and marketing figs. Most growers spread washed figs on paper trays laid on the orchard floor, and let them dry in the sun as done for raisins (Photo 26-8), while some use electric dryers set at 60 to 70 F. Fig growers dry figs to a moisture content of 14 to 20%. Sulfur is applied to some yellow fig cultivars, such as 'Adriatic' and 'Kadota,' to keep them from browning, while 'Calimyrna' will naturally dry to a pleasing amber color without sulfur. Purple or brown figs such as 'Mission' or 'Brown Turkey' are allowed to darken to a deep purple-black while drying.

Figs are sorted while spreading and them on drying trays and collecting them after drying, or while loading and unloading them from electric dryers. Any with rots or other defects are removed and destroyed. Dried figs are sold to processors or a coop that rehydrates them to a moisture content of 26 to 31%, further sorts our blemished fruit, and markets them to grocery stores or food manufacturers. Mandatory assessment fees are charged by the California Fig Advisory Board to both growers and processors, to pay for fig marketing and advertising.

Cold storage, canning and freezing of fresh figs and mulberries. Fresh figs remain in good condition for 1 to 2 weeks at 30 to 32 F and 90 to 95% relative humidity, but have a shelf life of only 1 to 2 days at room temperature. Controlled-atmosphere storage with 5-10% oxygen and 12-20% carbon dioxide can increase fig storage life by only 2 weeks, to 3 to 4 weeks total, so isn't generally worth the expense.

Store mulberries under the same conditions as figs, 30 to 32 F and 90 to 95% relative humidity, but mulberries last only 2 to 3 days in cold storage, similar to raspberries and blackberries. Controlled-atmosphere storage offers no benefit.

Figs and mulberries may be promptly canned within hours of harvest, and canned fruit lasts several years. A large fig canning industry existed in the Southeast during the early to mid-1900's, but has greatly declined in recent years as most figs are now dried. Mulberries could be canned for pie fillings as blackberries are, and mulberries may also be frozen in individual quick-freeze chambers used for raspberries, blackberries and blueberries. Frozen mulberries last at least 2 years.

Further Reading

California Rare Fruit Growers. 1997. *Mulberry fruit facts*. Available online at: http://www.uog.edu/cals/PEOPLE/Pubs/mulberry.html

Himelrick, D.G. 1999. *Fig production guide*. Alabama A & M and Auburn Univ. ANR-1145. Available online at: http://www.aces.edu/pubs/docs/A/ANR-1145/

Univ. of Calif. 2003. *IPM pest management guides, fig*. Available online at: http://www.ipm.ucdavis.edu/PMG/selectnewpest.figs.html

Chapter 27
Persimmons and Pawpaws

Persimmons and pawpaws are both temperate-zone, deciduous fruit trees that belong to genera and families with mostly tropical and subtropical members. Both fruits have a more tropical taste and softer texture than most temperate-zone fruits. Both are minor fruits in North America, but have potential to greatly expand.

Persimmons belong to the genus *Diospyros* in the ebony family (Ebenaceae). Commercial persimmons belong to the species *D. kaki* which originated in China and spread to Japan, so are also called kaki, Oriental persimmons or Japanese persimmons. This species is cold-hardy in USDA zones 7 to 10, and the U.S. has about 2200 acres of commercial persimmons, mainly in southern California with a few acres in Florida and other Gulf Coast states. The American persimmon, *D. virginiana*, is native to the eastern U.S. from the Gulf Coast to as far north as New Jersey and Iowa, and is adapted to USDA zones 5 to 9. Some American persimmons are gathered from the wild and sold, and a few cultivars have been selected and planted in small orchards.

Fruit of both types of persimmons turn orange when they ripen in fall, but Oriental persimmons (Photo 27-1) have 4- to 5-inch-diameter fruit while American persimmons have only 1.5- to 2-inch-diameter fruit (Photo 27-2), so the latter have much less pulp. Fruit of most Oriental persimmon cultivars and all American persimmons are very astringent until fully ripe and soft, but some Oriental persimmon cultivars have non-astringent fruit that can be eaten firm-ripe. Many Oriental

Photo 27-1. Oriental persimmons have large, 4- to 5-inch-diameter fruit, and many grow parthenocarpically so are seedless. Some cultivars lose astringency early, so may be eaten crisp.

Photo 27-2. American persimmons have smaller, 1.5- to 2-inch-diameter fruit and most are seeded, so have less pulp. All cultivars must soften before they lose astringency.

persimmon cultivars set fruit parthenocarpically so are seedless, while American persimmons have large seeds.

Pawpaw (*Asimina triloba*) is the only temperate-zone member of the custard-apple family (Annonaceae). Pawpaw is native to much of the eastern U.S., and is adapted to USDA zones 5 to 9. Pawpaw trees produce elongated, 4- to 6-inch-long fruit that turn yellow as they ripen in fall and somewhat resemble a banana (Photo 27-3). Like American persimmons, some pawpaws are gathered from the wild and sold, and a few cultivars have been selected and planted in small orchards. Pawpaw fruits also have large seeds and should be allowed to soften and fully color before harvesting.

China produces about 70% of the world's persimmons, about 3.7 billion pounds. South Korea and Japan each produce about 12% of the world's persimmons, Brazil raises about 3%, Italy grows about 2% and Israel, the U.S., New Zealand and several other countries each grow less than 1%. Only a small, unrecorded amount of American persimmons and pawpaws are gathered from the wild or grown in orchards for sale, and only in the U.S.

GROWTH AND BEARING HABITS

Oriental persimmons grow into medium-size trees 25 to 40 feet tall, while American persimmons grow taller, reaching 35 to 50 feet in height. Pawpaw grows as a small tree or large shrub, 15 to 25 feet tall. Persimmons begin to bear 3 to 4 years after planting, and reach full production 8 to 10 years after planting. Pawpaws are slow-growing and slow to bear, usually beginning fruiting 5 to 7 years after planting and reaching full bearing 10 to 12 years after planting. Both persimmons and pawpaws are long-lived trees, continuing to bear well for 50 to 60 years after fruiting commences.

Persimmons are dioecious, bearing staminate and pistillate flowers on separate trees (Photo 27-4). Growers generally plant 1 staminate pollenizer Oriental persimmon for every 8 to 10 trees of the desired fruit-bearing cultivar if that cultivar requires pollination, but many Oriental persimmon cultivars set fruit parthenocarpically so may be planted in solid blocks. Almost all American persimmons require pollination and need 1 staminate pollenizer for every 5 to 6 trees of the desired fruit-bearing cultivar.

Pawpaws have perfect flowers but they are protogynous, so stigmas are receptive before pollen from the same flower is shed, ensuring that a flower will not self-pollinate (Photo 27-5). Most

Photo 27-3. Pawpaw trees, native to eastern North America, produce elongated, 4- to 6-inch-long fruit that turn light yellow as they ripen in fall and somewhat resemble a banana. (photo by S. Bauer, USDA-ARS)

Photo 27-4. Persimmons are dioecious, bearing staminate and pistillate flowers, shown here, on separate trees. Growers often plant cultivars that set fruit parthenocarpically in solid blocks, so female flowers will not be pollinated and fruit will be seedless.

Photo 27-5. Pawpaws have perfect flowers but they are protogynous, so stigmas are receptive before pollen from the same flower is shed, ensuring that a flower will not self-pollinate. Plant at least 2 different cultivars for cross-pollination.

pawpaw cultivars are also self-incompatible, and require pollen from a different cultivar for flowers to be fertilized. Pawpaw plantings need at least 2 and preferably 3 cultivars, alternated down each row to provide maximum cross-pollination. The earliest-opening flowers on each cultivar will not set fruit, but will provide pollen for later-opening flowers of other cultivars.

Oriental persimmons are hexaploid, with 90 chromosomes. Most American persimmons are tetraploid with 60 chromosomes, but some cultivars are hexaploid, with 90 chromosomes. A few hybrid Oriental-American persimmon cultivars have been developed in Russia, and are now being tested in the U.S. for their cold hardiness and fruit quality.

Mature Oriental persimmon plantings produce 10,000 to 15,000 pounds (5 to 7.5 tons) of fruit per acre. Mature American persimmons produce about 4000 pounds of fruit per acre, which yields about 2000 pounds of persimmon pulp per acre. Pawpaws can produce about 2800 pounds of fruit per acre.

SITE SELECTION AND PREPARATION

Choose a site with good air drainage for Oriental persimmons, which are subject to spring frosts since they have a very low chilling requirement of about 200 hours below 45 F. American persimmons and pawpaws bloom later in spring, so usually don't suffer from frost damage. Oriental persimmons need at least 190 frost-free days and about 3000 total growing degree-days, while American persimmons and pawpaws need at least 160 frost-free days and about 2600 total growing degree-days, to ripen the fruit. None of these fruits produce well in desert regions with very high summer heat, and none tolerates soil salinity.

Persimmons and pawpaws need moderately well-drained to well-drained soil, with a pH of 6.0 to 7.0. Oriental persimmons are especially sensitive to soil pH, and can develop "green blotch", a fruit disorder caused by manganese toxicity, and even tree death at a pH less than 6.0, while they may suffer from manganese and iron deficiencies at a pH over 7.0. Test soil pH, phosphorous (P) and potassium (K) levels at least 6 months before planting. Add soil amendments if needed to get pH within 6.0 to 7.0, P levels to 75 pounds per acre, and K levels to 300 pounds per acre, mixing amendments well into the top 18 inches of soil. Both fruits do well in many soil types as long as drainage is adequate, and benefit from raised beds if drainage is only moderate. Persimmons grown in arid areas require irrigation.

Persimmons need full sun for good fruit production. Pawpaws are understory trees and do best with partial shade for the first 2-3 years and when older where light intensities are high, but may fruit well in full sun in cool, cloudy climates. Persimmons and pawpaws both need protection from desiccating summer winds, so planting them on the north, northeast or east side of a windbreak of deciduous trees will protect the fruit trees from wind, where summer winds blow mainly from the south, southwest or west. Use shorter windbreak trees for persimmons, and plant the persimmons far enough from the windbreak so they do not receive shade from it. Use taller windbreak trees for pawpaws and plant the pawpaws closer to the windbreak, to protect them from excess sunlight.

CHOOSING CULTIVARS

Many Oriental persimmon cultivars set seedless, parthenocarpic fruit when planted in solid blocks of single cultivars. All fruiting cultivars are female trees, but some occasionally sport branches with a few male flowers, so some cross-pollination can occur if cultivars are mixed in a block. Consumers greatly prefer seedless fruit, so most persimmon growers raise just one cultivar, or at most 2 cultivars in separate blocks that are at least ½ mile apart. If pollination is desired to produce large (seeded) fruit of some cultivars, growers plant 'Dai Dai Maru' or 'Gailey' as male pollenizer cultivars.

Oriental persimmon cultivars are divided into 3 groups: cultivars with astringent fruit, cultivars with non-astringent fruit, and pollination-variant cultivars whose fruit is non-astringent when pollinated but astringent when seedless. Astringent fruit must be jelly-soft before they are edible unless astringency is chemically removed, while non-astringent fruit can be eaten while crisp. Astringent cultivars are best adapted to regions with cooler summers and have been most widely planted in cooler parts of California, especially 'Hiratanenashi,' 'Hachiya,' and 'Tanenashi.'

Non-astringent cultivars do best adapted in regions with hot summers and have been increasingly planted both in southern California and in Florida and Texas. Non-astringent cultivars include 'Fuyu' (Photo 27-6), 'Izu,' 'Hana-Fuyu,' 'Jiro,' and 'Gosho.' Few pollination-variant cultivars are grown commercially, but Nishamura Wase has been planted for its earliness, and 'Eureka' has been planted in Florida and Texas because it has frost resistance. Table 27-1 details the traits of the most widely-grown Oriental persimmon cultivars.

Table 27-1.

COMMON ORIENTAL PERSIMMON CULTIVARS

In approximate order of ripening (varies with area).

Cultivar	Hardy to zone	Fruit						Tree size	Harvest time
		Astringent	Size	Flavor	Shape	Cracks	Yields		
Nishamura Wase	8a	yes	med-large	fair	round	none	medium	medium	early
Izu	8a	no	med-large	very good	oblate	none	medium	small	early-mid
Tanenashi	8a	yes	med-large	good	conic	some	high	med-large	early-mid
Hachiya	8a	yes	large	very good	conic	some	high	large	middle
Triumph	8a	yes	medium	good	oblate	none	high	medium	middle
Hiratanenashi	8a	yes	medium	best	oblate	none	high	medium	mid-late
Korean	6b	yes	medium	good	round	none	medium	large	mid-late
Hana-Fuyu	8a	no	med-large	good	oblate	some	high	medium	late
Jiro	7b	no	med-large	very good	oblate	some	high	med-large	late
Eureka	8a	yes	med-large	good	round	none	high	small	late
Sheng	7a	yes	med-large	very good	oblate	none	medium	medium	late
Gosho	8a	no	medium	very good	round	some	low	small	late
Fuyu	8b	no	medium	best	oblate	none	high	medium	very late

Average harvest time in southern CA: early: late Aug. to mid-Sept., early-mid: mid-Sept. to early Oct., middle: late Sept. to mid-Oct., mid-late: early to late Oct., late: mid-Oct. to mid-Nov., very late: late Oct. to late Nov.

Notes on Cultivars:
Nishamura Wase, one of the few pollination-variant cultivars, is one of the earliest-ripening.
Izu, the earliest-ripening non-astringent cultivar, doesn't bear until 5 years but regulates crops well.
Tanenashi tends to alternate bear, so need fruit thinning; sets well parthenocarpically.
Hachiya has been widely planted in CA, sets better when pollinated, parthenocarpic fruit drops early.
Triumph fruit may have astringency chemically removed before sale.
Hiratanenashi fruit have thick peel, long shelf life; widely planted in CA in recent years.
Korean is more hardy to -5 F, leafs out later than other cultivars; best for zones 7-8 of Southeast.
Hana-Fuyu, also called **Giant Fuyu**, regulates crop loads well.
Jiro crops erratically when young, better when older.
Eureka is precocious, resists frosts, does well in zones 8b-9 of Southeast, pollination-variant.
Sheng is more hardy to 0 F, leafs out later than other cultivars; best for zones 7-8 of Southeast.
Gosho has light crops some seasons due to excess shedding of immature fruit.
Fuyu, the most widely-grown persimmon in the world, has 18-19% sugars, needs fruit thinning.

Table 27-2.

COMMON AMERICAN PERSIMMON CULTIVARS

In approximate order of ripening (varies with area).

Cultivar	Hardy to zone	Fruit Size	Flavor	Firmness	Seeds	Yields	Harvest time
Yates	5a	large	best	medium	0 to few	high	very early
Ruby	6a	large	very good	firm	few	med-high	early
Meader	4a	medium	very good	firm	0 to few	high	early
Garretson	4b	small-med	best	soft	small	high	early-mid
Craggs	5a	large	best	medium	few	high	middle
Morris Burton	5a	medium	best	medium	few	high	middle
Killen	5a	medium	very good	firm	few	medium	mid-late
John Rick	4b	med-large	best	firm	2 to 8	high	mid-late
Ennis	5a	medium	best	soft	0 to few	medium	late
Miller	5a	large	good	firm	2 to 8	high	late
William	5a	small	fair	soft	many	none-low	pollenizer

Average harvest time in southern MO: very early: late Aug. to early Sept., early: early to mid-Sept., early-mid: mid- to late Sept., middle: late Sept. to early Oct., mid-late: early to mid-Oct., late: mid- to late Oct.

Notes on Cultivars:
Yates produces seedless fruit (smaller) if no pollenizers grow within ½ mile.
Ruby has low chill needs, fruit loses astringency early.
Meader produces some seedless fruit (smaller), some male flowers.
Garretson has 90 chromosomes and a few male flowers, so may pollenize other cultivars.
Craggs is sometimes called Creggs.
Morris Burton has a rich flavor.
Killen has 90 chromosomes and a few male flowers, so may pollenize other cultivars.
John Rick needs a 90-chromosome pollenizer like 'William', or 'Garretson' or 'Killen' male flowers.
Ennis produces seedless fruit (smaller) if no pollenizers grow within ½ mile.
Miller has high chill needs, so best in zones 5 to 7.
William, a male tree, blooms profusely for a long time; has a few female flowers and fruit some years.

Table 27-3.

COMMON PAWPAW CULTIVARS

In approximate order of ripening (varies with area).

Cultivar	Fruit			Harvest time
	Size	Flavor	Yields	
Overleese	large	best	low-med	early
Rappahannock	med-large	very good	high	early
Pennsylvania Golden	medium	very good	high	early
Davis	med-large	good	medium	early-mid
Prolific	large	best	high	early-mid
Sweet Alice	med-large	good	high	middle
Mitchell	med-large	best	low-med	middle
Taylor	small	very good	low-med	middle
Taytwo	medium	very good	low	middle
Shenandoah	large	best	high	middle
Susquehanna	very large	very good	medium	middle
Mary Foos Johnson	med-large	good	high	mid-late
Sunflower	med-large	good	high	mid-late
Wells	med-large	best	medium	late
Rebecca's Gold	med-large	best	high	late

Average harvest time in central KY: early: early Sept., middle: mid-Sept., late: late Sept. Harvest times are about 3 weeks later in southern MI.

Notes on Cultivars:
Overleese is slow to bear, clusters of 3 to 6 fruit per cluster need thinning, S to vascular decline.
Rappahannock bears fruit singly or in small clusters.
Pennsylvania Golden is precocious and bears reliably, MR to vascular decline.
Davis is slow-growing, slow to bear, keeps well in cold storage.
Prolific is precocious as well as high-yielding.
Sweet Alice has very sweet fruit.
Mitchell is slow to begin bearing, MR to vascular decline.
Taylor is slow to bear, has up to 7 fruit/cluster so needs heavy thinning, MR to vascular decline.
Taytwo, sometimes spelled Taytoo, is very cold-hardy, S to vascular decline.
Shenandoah fruit is often borne singly, so needs little thinning.
Susquehanna has only 3% seeds by weight, since fruit are so large.
Mary Foos Johnson may be the same as Sunflower.
Sunflower is precocious, partly self-fruitful, yields reliably; MS to vascular decline, may not ripen in MI.
Wells is slow to begin bearing, but fruit are very large once it bears; S to vascular decline.
Rebecca's Gold is slow-growing and slow to bear.

Photo 27-6. Some Oriental persimmon cultivars, such as this 'Fuyu', have non-astringent fruit that can be eaten while crisp. (photo by K. Sumner, Gator Persimmons, Plymouth, FL)

Photo 27-7. A few American persimmon cultivars, such as 'Yates' shown here, will set a few very small, parthenocarpic, seedless fruit if not pollinated, but larger fruit like these require cross-pollination by a male cultivar.

All American persimmon cultivars have astringent fruit that must soften before it is edible, and most require male pollenizer cultivars. A few American persimmon cultivars, such as 'Evelyn' and 'Yates' (Photo 27-7), will set a few parthenocarpic, seedless fruit if they are not pollinated, but such fruit will be very small and often drop prematurely.

All pawpaw cultivars also have fruit that must fully soften before it can be eaten, and pawpaw plantings require 2 or more interspersed cultivars to provide cross-pollination. Cultivars of both American persimmons and pawpaws have been selected with larger fruit than most seedlings, and these named cultivars, described in Tables 27-2 and 27-3, are often planted. However, the majority of American persimmons and pawpaws are still gathered from the wild.

ROOTSTOCKS AND PROPAGATION

Seedling rootstocks of a related persimmon species, *Diospyros lotus*, are most widely used as rootstocks for Oriental persimmons, despite the fact that it is susceptible to crown gall. This rootstock keeps trees slightly smaller and induces them to fruit heavier while young, but may cause trees to die sooner. *D. lotus* has delayed incompatibility with some cultivars, such as 'Fuyu,' so they must be propagated on Oriental persimmon rootstock, *D. kaki*. American persimmon seedlings, *D. virginiana*, are used as rootstocks for American persimmon cultivars and for Oriental persimmon cultivars in the Southeast for their cold hardiness and drought tolerance, but are not generally used in California because they sucker badly. Where crown gall or persimmon wilt are serious diseases, use Oriental persimmon seedling rootstocks, since they resist both these diseases, while *D. lotus* is susceptible to crown gall and *D. virginiana* is susceptible to both diseases.

Pawpaw cultivars are grafted onto pawpaw seedling rootstocks. Only a few grafted pawpaw cultivars are available from nurseries, so many growers raise their own seedling rootstocks, plant these in the orchard, and graft or bud desired cultivars onto these rootstocks once they're established (Photo 27-8).

ORCHARD LAYOUT AND PLANTING

Tree spacing of Oriental persimmons depends on the cultivar's tree size, the soil type and training system. Medium-sized trees such as 'Fuyu' can be planted 12 to 14 feet apart in rows that are 18 to 20 feet apart if they are to be trained as individual modified leader trees, for a planting density of 155 to 200 trees/acre. If trained to the perpendicular V system in hedgerows, tree spacing within rows can be cut to 8 to 10 feet apart and rows kept 18 to 20 feet apart, providing 217 to 302 trees/acre. Smaller-growing cultivars, such as 'Izu' or 'Gosho,' can be planted even closer: 7 to 8 feet apart in a hedgerow system or 10 to 12 feet apart for modified leader trees, in both cases in rows 18 feet apart, for a planting density of 200 to 345 trees/acre.

Larger-growing cultivars, such as 'Tanenashi' or 'Hachiya,' are commonly spaced 20 to 24 feet apart both within and between rows when trained as individual modified leader trees, for a planting

Photo 27-8. Many pawpaw growers raise their own seedling rootstocks, plant these in the orchard, and graft desired cultivars onto these rootstocks once they're established. A chip bud was used here. (photo by K. Pomper, Kentucky State Univ.)

density of 76 to 109 trees/acre. Hedgerow-trained trees can be spaced 10 to 12 feet apart within rows and 20 to 22 feet between rows, providing 165 to 217 trees/acre.

Plant American persimmons at the same tree spacings used for these larger-growing Oriental persimmon cultivars. Plant pawpaws at tree spacings used for small-growing Oriental persimmon cultivars, about 18 feet between rows and 8 feet apart within rows, for 302 trees per acre.

When including pollenizer cultivars in a persimmon planting, plant every third tree of every third row to a pollenizer for Oriental persimmons, which places each fruiting tree next to a pollenizer and creates a ratio of 1 pollenizer per 8 fruiting trees. Every third tree of every other row is generally planted to a pollenizer for American persimmons, creating a ratio of 1 pollenizer per 5 fruiting trees. Pawpaw growers should plant equal amounts of at least 2 and preferably 3 cultivars, alternating cultivars down each row to provide maximum cross-pollination.

Planting. Plant Oriental persimmons as early in the dormant season as possible, preferably in late fall or early winter to allow roots time to grow before top growth begins in spring. American persimmons and pawpaws may also be planted in fall in zones 6b to 9, but should be planted in spring in zones 5 and 6a.

Oriental persimmons are usually planted as grafted, bare-root trees, like most other fruit trees. American persimmons and pawpaws are difficult to transplant, so are best planted as small container-grown trees to keep all roots intact. These may be grafted trees or seedling rootstocks to be grafted the next year.

Dip persimmon roots in a biocontrol solution to protect against crown gall just before planting. Pawpaws do not need this treatment.

Care of new plants. After planting persimmon and pawpaw trees, place white plastic tree guards around them to protect against sunscald and rodents. Do not use translucent plastic tree tubes, as these cause trunks to heat to high temperatures and have caused many tree deaths, especially in pawpaws. Water in the new trees thoroughly, and irrigate at least once a week during the growing season if no rain falls during the period.

Many growers prune newly-planted persimmon trees back to 3 feet tall right after planting, but if deer populations are high in the area, do not prune persimmons back as the deer will do it. Deer feeding on unpruned young trees may not damage them much, but trees that were pruned back at planting and then browsed by deer produced little new growth for the next 2 years. Pawpaws do not usually attract deer feeding, but do not need pruning at planting since they grow slowly.

Install a stake next to each persimmon tree by mid-spring after planting, if a trellis will not be used. Young persimmon trees are prone to branch and limb breakage in strong winds, and staking or trellising for the first few years prevents this damage.

Pawpaws grow slowly and do not need staking or trellising, but do need shading for the first 2 to 3 years after planting. If a windbreak does not already provide mid-day shade, erect temporary shadecloth or other shading structures.

SOIL MANAGEMENT, IRRIGATION AND FERTILIZERS

Persimmon and pawpaw growers generally maintain sod row middles between tree rows, and control all vegetation within tree rows with herbicides, as commonly practiced for other tree fruit crops (Photo 27-9). Some Oriental persimmon growers in California grow winter cover crops, then cultivate orchards in summer to reduce water use by the groundcover.

Irrigation needs and methods. Oriental persimmons require irrigation in arid areas, where drip and furrow irrigation are the most common methods. Persimmons need 36 to 48 inches of water per year, applied regularly in spring and summer and at a reduced rate in fall. Hot inland areas of

Photo 27-9. Most persimmon and pawpaw orchards have sod row middles, with all vegetation within tree rows controlled by herbicides, as in this pawpaw orchard. (photo by K. Pomper, Kentucky State Univ.)

Photo 27-10. Drip irrigation can be very beneficial for young pawpaw trees, shown here, and is essential for Oriental persimmons in arid areas. (photo by K. Pomper, Kentucky State Univ.)

California need 2 furrow irrigations or 3 to 4 drip irrigations per week in summer, while cooler coastal regions may need irrigation only once every 2 to 6 weeks.

Oriental persimmons grown on American persimmon rootstocks in the Southeast are quite drought-tolerant once established, but need irrigation at least every 2 weeks during the first 2 to 3 years of establishment. Many growers install drip irrigation systems, especially on sandy soils, which need more irrigation than heavier soils. Micro-sprinkler systems that spray water on the trunks are especially popular, since they can also help protect Oriental persimmon trunks from extreme cold and heat.

Running micro-sprinklers during sunny winter days provides evaporative cooling that prevents trunks from heating to such high temperatures before being subjected to freezing night temperatures, thus reducing sunscald damage. Running micro-sprinklers on cold, still nights causes water to freeze around the trunks, preventing trunk temperatures from dipping below 32 F and preventing freezing damage.

American persimmons and pawpaws are much more cold-hardy and drought-resistant, and are not generally irrigated once established. Drip irrigation can be very beneficial for young trees though, especially pawpaws (Photo 27-10).

Fertilizer nutrients. Oriental persimmons need nitrogen (N) fertilizer and usually potassium (K) also. Apply about 1.5 ounce each of actual N and K per tree (12 to 15 pounds per acre) during the first growing season, and about 3 ounces each per tree (24 to 30 pounds each per acre) during the second growing season. Keep increasing this amount by 1½ ounces each per tree (12 to 15 pounds each per acre) each year until trees reach maturity after 7 to 8 years, when they will receive 100 to 120 pounds per acre (¾ to 1 pound per tree) each of actual N and K during each year thereafter. Divide these rates into 3 applications per year: in March about 2 weeks before leaf burst, in June, and in late August to early September. However, many astringent persimmon cultivars suffer premature fruit drop if N levels are too high when fruit are sizing, so reduce or eliminate the last N application if this problem occurs.

American persimmons and pawpaws benefit from about the same annual rates of N, but apply them only during the first 2 periods to avoid stimulating late growth that would be prone to winter damage in the colder areas where these crops are grown. These crops won't need annual K applications on many soils, but test leaves for phosphorous (P) and K every 3 to 4 years to determine if they need either of these fertilizers.

Oriental persimmons may also need P on some soils, so again test leaves every 3 to 4 years. Magnesium (Mg) deficiencies may occur on some soils, especially those with high amounts of K or calcium, causing stunted and blackened terminal leaf tips. Such soils need dolomitic lime if pH is below 6.5 or magnesium sulfate fertilizer if pH is above 6.5.

Oriental persimmons can suffer from iron or manganese deficiencies on soils with a pH above 7.0. Prevent these deficiencies by lowering the pH to 7.0 before planting and every few years thereafter with sulfur applications. If deficiencies occur, foliar applications of iron sulfate or manganese sulfate can temporarily alleviate symptoms. Soils with a pH less than 6.0 lead to manganese toxicity in Oriental persimmons, which causes green blotch fruit disorder or even sudden death of trees. Lime such soils to raise pH to 6.5 before planting, and every few years thereafter as pH tests dictate.

TRAINING, PRUNING AND MANAGING CROPS

Train low-density Oriental and American persimmons to the modified leader system, as described in chapter 4. Higher density plantings grown in hedgerows are generally trained to the central leader system, also described in chapter 4. Train pawpaws to the central leader system. All these fruit trees need minimal pruning while young, just enough to achieve the desired training and to remove any diseased or damaged branches.

Dormant pruning. Mature Oriental persimmon trees benefit from thinning out some weaker fruiting wood in the interior of trees during the dormant season. This reduces the number of flowers and fruit that set in spring, so helps to prevent overcropping. Do not remove too many interior branches from Oriental persimmons in winter and do not prune them in summer, as fruit are very prone to sunburn. Also head back any overly vigorous 1-year-old branches during dormant pruning, and remove any diseased or damaged branches.

American persimmon trees naturally shed small-diameter branches in fall and winter, so trees don't need interior thinning. Taller limbs may be headed back when trees get too tall to harvest easily, to induce lower side branches to grow.

Mature pawpaw trees need minimal pruning, mainly removing any diseased or damaged branches and some weaker branches in crowded areas within trees. Pawpaws usually suffer from too little rather than too much fruit set, so fruiting branches don't need thinning. Fruit are also prone to sunburn, so summer pruning isn't usually practiced.

Pollination management. Most Oriental persimmon growers want to prevent pollination so that only parthenocarpic, seedless fruit will develop. They do this by planting solid blocks of 1 or 2 fruiting cultivars, and pruning off any branch sports with staminate flowers that develop. Spring frosts usually present no problem when pollination is not desired, although a late frost after fruit set could reduce the crop or leave frost markings on the fruit at harvest.

American persimmon cultivars and a few Oriental persimmon cultivars require pollination for good fruit production, so are planted with staminate trees or with fruiting cultivars that produce some staminate flowers. Native bees provide enough pollination, so no beehives need be introduced. American persimmons bloom late, and the few Oriental persimmons in which pollination is desired do not suffer much from frosts, so frost protection is not often justified.

Pawpaws are pollinated by various species of flies and beetles rather than by bees, and these pollinators are less efficient and may be scarce if weather is cool during bloom. Some growers hand-pollinate pawpaw flowers under such conditions to improve fruit set. Use a small artist's paintbrush to transfer pollen from ripe brown anthers of one cultivar to glossy stigmas of a different cultivar. Hand-pollinate only about half of the flowers on each tree, to avoid creating too heavy a fruit set that will require fruit thinning. Do not hand-pollinate flowers of pawpaw cultivars that often produce clusters of smaller fruit, such as 'Davis' and 'Taylor.'

Pawpaw flowers may be killed by spring frosts (Photo 27-11), so growers may want to use orchard heaters or sprays to reduce fruit damage if a frost is predicted during bloom. Frost damage does not occur often enough to justify installing expensive sprinkler irrigation systems to prevent damage.

Fruit thinning. Many Oriental persimmon cultivars tend to biennial bear, and require fruit thinning to prevent this and to produce larger fruit. Mature trees of large-fruited cultivars should be allowed to ripen 150 to 250 fruit per tree, which means removing about half the fruit that set. Smaller-fruited cultivars may be allowed to ripen 300 to 400 fruit per tree, which means removing about one-third of the young fruit if crop set is heavy, or if crop set is light they may need little or no fruit thinning. Some cultivars, such as 'Izu' and 'Hana-Fuyu,' drop many immature fruit on their own, so need less fruit thinning. A few cultivars,

Photo 27-11. Spring frosts may kill pawpaw flowers, so growers may use orchard heaters or sprays to reduce fruit damage if a frost is predicted during bloom. (photo by K. Pomper, Kentucky State Univ.)

such as 'Gosho,' drop excessive amounts of young fruit during some years, leading to low yields. Such cultivars will probably require fruit thinning the next year to prevent biennial bearing.

The auxin NAA may be used to thin persimmons, but could easily produce too much thinning, so many growers only practice hand thinning. NAA is most often used on cultivars that need heavy fruit thinning, such as 'Fuyu.' If NAA is applied, about 2 weeks after bloom in a manner similar to that described for apples, follow-up hand thinning will still be needed to break up areas of heavy fruit set on branches.

Wait to hand-thin fruit until after the natural post-bloom drop of young fruit, or about 3 weeks after full bloom. Leave 1 to 4 fruit on each fruiting shoot, with fruit spaced 6 inches apart. Some cultivars, such as 'Fuyu' set most fruit in clusters, so require heavy thinning to remove all but 1 fruit per cluster. Keep the largest and best-developed fruit, which are generally the ones toward the base of each fruiting shoot.

American persimmons have much smaller fruit that don't generally increase in size much in response to fruit thinning, and the fruit are borne higher up so they're harder to reach, so fruit thinning is not usually practiced. Pawpaw cultivars that bear smaller fruit and tend to set in clusters of 3 or more fruit (Photo 27-12), such as 'Davis' and 'Taylor,' benefit from thinning to a single fruit per cluster, which increases fruit size and reduces fruit rot. Pawpaw cultivars with larger fruit that sets singly or in doubles usually don't need fruit thinning, especially if hand pollination was not practiced.

Photo 27-12. Pawpaw cultivars that bear smaller fruit and tend to set in clusters of 3 or more fruit, as here, would benefit from thinning to a single fruit per cluster, which increases fruit size and reduces fruit rot. (photo by D. Layne, Kentucky State Univ.)

PEST MANAGEMENT

Persimmons and pawpaws need careful pest control while trees are young to avoid tree deaths or stunting, but once bearing need few pesticides and are easier to grow organically than many other fruit crops.

Pests that can kill trees. Avoid most of these persimmon killers by modifying the soil before planting, and choosing cultivars and rootstocks that resist these problems. Protect young pawpaw trees by providing partial shade for the first few years.

Phytophthora root rot may attack persimmons and pawpaws as described for other fruit trees. Prevent it by building raised planting beds in soils that are only moderately well drained.

Crown gall, caused the by bacterium *Agrobacterium tumefaciens*, attacks persimmons on *D. lotus* or American persimmon rootstocks, so inspect nursery material to be sure none is infected, and dip tree roots in a biocontrol solution of *A. radiobacter* (Galltrol™) just before planting them. Oriental persimmon rootstocks and pawpaws resist crown gall.

Persimmon wilt, caused by the fungus *Cephalosporium diospyri*, attacks Oriental persimmons on American persimmon rootstocks as well as American persimmons. This disease causes severe defoliation and often kills trees after several seasons. Damage from **twig girdler** (*Onsideres cingulatus*), a beetle larvae that cuts off twigs and small branches, affords entry to the wilt fungus, so American persimmons and trees on American persimmon rootstocks may require insecticide control of this beetle in early fall when they lay eggs. Oriental persimmons on Oriental persimmon or *D. lotus* rootstocks resist wilt, so plant them where this disease is serious.

Sudden death syndrome often kills young Oriental persimmon trees in the Southeast. It appears to be caused by mangenese toxicity in soils with a pH less than 6.0, exacerbated by very cold winters or other stresses. Avoid this syndrome by testing soil pH before planting and every 3 years thereafter, adding lime if needed and mixing it into the top 18 inches of soil, and choosing the most cold-hardy cultivars for areas that may experience temperatures below 10 F.

Vascular decline has attacked young pawpaw trees in parts of Oregon, causing many of them to wilt and die within a few years of planting. A pathogen has not yet been identified as the cause of this disease, and so far it has not occurred outside of Oregon, which is outside the pawpaw's natural range. This disease is also called blue stain because it causes a blue and black discoloration beneath the bark of declining trees (Photo 27-13). Survival rates of pawpaw cultivars varied in an Oregon trial exposed to vascular decline. Where vascular decline occurs, plant cultivars with the greatest apparent resistance to this disease, such as 'Wilson,' 'Mitchell' or 'Taylor.'

Persimmon borer (*Sannina uroceriformis*), a clearwing moth larvae, bores into the trunk (Photo 27-14) and taproot of persimmons in southern states, and is especially a problem in Florida where it may kill many young trees. Monitor adult moths by setting out buckets of sugar water laced with an insecticide, and when several moths are caught per trap or in mid-summer, apply a long-residual insecticide such as chlorprifos (Lorsban™) to the lower trunks with a handgun. Once trees start to bear, this treatment should no longer be needed, but if it is, apply the insecticide well before the 21-day pre-harvest interval allowed.

Deer (*Odocoileus* species) may feed on persimmon bark in winter, killing young trees. Protect young trees with cages around each tree or an electric fence around the orchard. Pawpaw bark is not attractive to deer or rabbits.

Sunscald is the main threat to young pawpaw trees, and may hurt persimmons also. Wrap white plastic tree guards around trunks, and provide partial shade for pawpaws for the first 2 to 3 years. Pawpaws resist many soil diseases that kill other fruit trees, such as oak root rot.

Indirect pests that reduce yields and tree longevity. Persimmons and pawpaws do not usually suffer much from virus, phytoplasma or nematode diseases, so generally don't need soil preparation to reduce nematodes or virus-free planting material. Several insects and diseases are indirect pests that may need control, generally with safer pesticides that are considered organic.

Angular leaf spot, caused by *Cercospora kaki*, may prematurely defoliate trees of both Oriental and American persimmons (Photo 27-15). Prevent this with a copper hydroxide fungicide spray at full bloom and 3 to 4 weeks later, and every 14 days thereafter as needed, depending on the degree of leaf wetness and disease severity.

Persimmon psylla (*Trioza diospyri*) attacks new leaves of persimmons in Florida in spring, causing them to crinkle and be misshapen, and stunting the growth of young trees. Nymphs are covered with a powdery white layer and adults have black bodies, and both are protected from pesticides once leaves curl around them. Delayed-dormant oil sprays control most psylla attacks, but if outbreaks still occur, apply a selective insecticide to young, non-blooming persimmons at about the time that older trees bloom. Bearing trees should not need this bloom treatment, although it would not hurt

Photo 27-13. Vascular decline of pawpaws, also called blue stain because it causes a blue and black discoloration beneath the bark of declining trees, kills many young trees in Oregon but hasn't yet occurred in other areas. (photo by K. Pomper, Kentucky State Univ.)

Photo 27-14. Persimmon borer larvae bore into the trunk and taproot of persimmons in southern states. Spray an insecticide on the lower trunks of young trees to prevent it.

Photo 27-15. Angular leaf spot, shown here on Oriental persimmon, also attacks American persimmons and may prematurely defoliate trees. Prevent it with copper hydroxide fungicide sprays at full bloom, 3 to 4 weeks later, and every 14 days thereafter as needed. (photo by K. Sumner, Gator Persimmons, Plymouth, FL)

Oriental persimmons since they have parthenocarpic fruit.

Many **scale** insects may attack persimmons, including **European fruit lecanium** (*Parthenoleucanium corni*) in California and Florida, **olive scale** (*Parlatoria oleae*) in California, **white peach scale** (*Pseudaulacaspos pentagona*) in Florida and Georgia, and **San Jose scale** (*Quadraspidiotus perniciosus*) in all regions, and the last may occasionally attack pawpaws. If scales attack, apply a delayed-dormant oil spray, and if needed spray a light oil in summer. Prune off and burn severely infested branches.

Citrus mealybug (*Panoccus citri*) may sometimes become damaging on Oriental persimmons in southern California and Florida, weakening trees through sap feeding and excreting large amounts of honeydew on the fruit, which serves as a substrate for sooty mold fungus. This pest is usually controlled by natural enemies, especially a lady beetle that was imported for this purpose from Australia in 1891, which is commonly called the mealybug destroyer. Be careful not to apply any pesticides to control other pests that would disrupt this biological control. Monitor the cottony white egg masses of mealybugs in late winter to early spring, and if many are seen, apply a delayed-dormant oil spray as mealybug crawlers start migrating up trunks and branches to the fruit in early spring.

Fall webworm (*Hyphantria cunea*) may congregate on persimmons in early summer and fall during outbreak years in all regions where they are grown. Predators and parasites control these gregarious caterpillars during most years, but when outbreaks occur, growers must physically remove webs with long poles and burn them, or spot-treat webs with a selective moth insecticide such as tebufenozide (Confirm™) or diflubenzuron (Dimilin™).

Red-humped caterpillar (*Schizuva concinna*) may sporadically defoliate persimmon trees in California. Apply a moth-specific insecticide, such as the biocontrol agent *Bacillus thuringiensis* (Dipel™, Javelin™, etc.) or the molting inhibitor tebufenozide (Confirm™), if this pest becomes damaging.

Japanese beetle (*Popillia japonica*) may devour pawpaw leaves where it occurs in eastern North America (Photo 27-16). Control it with milky spore disease, a biocontrol product.

Zebra swallowtail butterfly (*Eurytides marcellus*) larvae feed exclusively on pawpaw leaves, but very rarely in numbers that cause economic damage. The adult butterfly is beautiful, and many gardeners plant pawpaws specifically to provide food for these butterfly larvae. Growers very rarely need to control this caterpillar, but if they do, *Bacillus thuringiensis* (Dipel™, Javelin™, etc.) would work.

Diseases that attack flowers and fruit. Persimmons and pawpaws have few fruit diseases, and careful cultivar choice can often minimize these.

Anthracnose, caused by *Gloeosporium kaki* or *Colletotrichum* species fungi, grows on cracked fruit of Oriental persimmons, and occasionally American persimmons. Prevent anthracnose by choosing cultivars with fruit that rarely cracks, and by irrigating regularly as persimmons ripen to avoid sudden moisture surges that cause fruit cracking.

Fruit spot of pawpaw, caused by *Phyllosticta asiminae*, may disfigure the fruit (Photo 27-17)

Photo 27-16. Japanese beetle may devour pawpaw leaves where it occurs in eastern North America. Control them with a biocontrol product, milky spore disease. (photo by K. Pomper, Kentucky State Univ.)

Photo 27-17. Fruit spot of pawpaw, caused by *Phyllosticta asiminae,* may disfigure the fruit and may also cause a leaf spot that could lead to early defoliation. (photo by K. Pomper, Kentucky State Univ.)

although it doesn't usually reduce yields, and may also cause a leaf spot that could lead to early defoliation. Infected fruit could be used for processed products but not sold fresh. Growers should monitor leaves for signs of infection, and if any appear, apply a fungicide to prevent fruit infection.

Insects that attack flowers and fruit. Persimmons and pawpaws suffer from only a few fruit insects, which are generally easily controlled with very safe insecticides that are considered organic.

Orange tortrix (*Argyrotaenia citrana*) may attack Oriental persimmons in coastal areas of California, feeding on leaves, buds and fruit surfaces. Sample fruit weekly beginning in June, and if more than 1% damage occurs, apply a selective moth insecticide, such as *Bacillus thuringiensis* (Dipel™, Javelin™, etc.) or tebufenozide (Confirm™). Eliminating any mustard-family weeds in the groundcover will reduce populations of orange tortrix, as will conserving natural enemies such as spiders, lacewing larvae and parasitic wasps.

Pawpaw peduncle borer (*Talponia plummeriana*) burrows into the fleshy parts of pawpaw flowers, causing blossoms to wither and drop. This moth larva occurs wherever pawpaws are grown, and may destroy a majority of flowers during some years. Monitor flower damage, and if more than 10% occurs, apply a moth-specific insecticide, such as *Bacillus thuringiensis* (Dipel™, Javelin™, etc.) or tebufenozide (Confirm™), during bloom. These sprays won't hurt pollinating insects.

HARVEST, STORAGE, MARKETING AND PROCESSING

Persimmons and pawpaws must be harvested by hand to prevent fruit bruising, even when fruit is to be processed. Fruit must also be handled very carefully post-harvest.

Harvest time and methods. Growers determine harvest time of persimmons and pawpaws mainly by fruit skin color changes (Photo 27-18), and also by fruit firmness or softening. Many growers of Oriental persimmons use color charts that are published for several groups of cultivars. Non-astringent cultivars generally lose their astringency when fruit color turns from dark yellow to light orange at the calyx, from light orange to dark orange around the fruit middle, and from dark orange to orange-red at the fruit apex. Astringent cultivars of Oriental persimmons do not lose astringency until they turn dark orange-red and soft, but are generally harvested while still firm, orange and astringent, and are treated before or after harvest to remove the astringency.

Photo 27-18. Growers determine harvest time of persimmons and pawpaws mainly by fruit skin color changes, as these Oriental persimmons change from yellow to orange, and also by fruit firmness or softening.

Some Oriental persimmon growers may apply ethephon to increase fruit color and hasten harvest, but ethephon also hastens fruit softening, and treated trees must be harvested quickly to prevent fruit drop. Workers generally use small clippers to cut Oriental persimmons from branches so the calyx remains on the fruit, to avoid tearing off the calyx if fruit were pulled off branches.

Harvest American persimmons when fruit are dark orange-purple, just starting to soften and pull away from the branch easily with the calyx attached to the fruit (Photo 27-19). Clippers are not often used. American persimmons will drop quickly when ripe, so check trees every day as harvest approaches, and if some fruit have started to drop, proceed to harvest the rest of the fruit from that cultivar.

Photo 27-19. Harvest American persimmons, such as this 'John Rick' cultivar, when fruit are dark orange-purple, just starting to soften and pull away from the branch easily with the calyx attached to the fruit.

Pawpaws ripen unevenly on the tree, so fruit must be checked every 2 to 3 days, and only the ripe ones picked (Photo 27-20). Skin color changes to a lighter green or pale yellow and fruit starts to soften when ripe. If cold storage facilities are available, fruit may be harvested just before it starts to change color and stored for up to 4 weeks before marketing, whereas ripe fruit lasts only 1 week in cold storage.

Removing astringency from astringent Oriental persimmons. Astringent cultivars of Oriental persimmons may be treated before or after harvest to remove astringency so they can be picked firm, or may be allowed to soften before harvest so they lose astringency naturally. Asian customers generally prefer firm persimmons so astringent persimmons for those markets must be picked firm and treated, while European and American customers often like soft persimmons so no treatment may be needed. However, persimmons allowed to soften before harvest must be handled very carefully and can not be shipped, so must be sold in local markets.

Astringency may be removed from Oriental persimmons on the tree when fruit first begins to turn yellow, but this method requires a great deal of hand labor. Workers enclose individual fruit in small polyethylene bags containing about 1 ml of 40% ethyl alcohol, and leave bags in place for 3 days. Workers then cut bags in half to allow the alcohol to drain out, leaving bags on fruit to indicate that fruit have been treated. Treated fruit remain on the tree until they mature and turn orange. Fruit treated this way may develop brown or hard flesh, and the heavy labor requirement mostly prohibits its use in the U.S., although it's done on a small scale in Japan.

Alcohol may instead be applied to persimmons after harvest to remove astringency. Sorted fruit are packed into corrugated cardboard cartons lined with heavy paper, and a machine sprays 150 to 200 ml of 40% ethyl alcohol on top of each packed box. A cardboard liner is then placed on top of the fruit, and the box is sealed and shipped or stored. Fruit astringency diminishes to unnoticeable amounts in about 10 days at room temperature, but unfortunately 'Hachiya' and other cultivars soften during that time.

Harvested Oriental persimmons may also be treated with carbon dioxide (CO_2) gas to remove astringency, and this is the most common method used in California. Fruit could be place in a controlled-atmosphere storage room with 80% CO_2 at 68 F or room temperature for 24 hours, or a

Photo 27-20. Pawpaws ripen unevenly on the tree, so fruit must be checked every 2 to 3 days, and only the ripe ones picked. Here, Kirk Pomper harvests pawpaws at Kentucky State University. (photo by C. Pomper)

polyvinyl-chloride film could be placed over boxes of fruit for 24 hours at room temperature, which develops 80% CO_2 levels. Fruit is still astringent when removed from these conditions, but astringency drops to negligible amounts within 3 to 4 days. This method is faster than alcohol treatment and retains fruit firmness, but sometimes causes fruit skin to darken.

Fruit storage. Oriental persimmons have a relatively long shelf life at room temperature. Non-astringent cultivars that ripen early, such as 'Izu,' last about 10 days, while those that ripen late, such as 'Fuyu,' last 20 to 30 days. Astringent cultivars also have a long shelf-life if astringency is not removed, but treatments to remove astringency reduce that shelf life. Such persimmons last about 15 days if astringency was removed on the tree with ethyl alcohol, 7 to 10 days if astringency was removed after harvest with carbon dioxide, and 5 to 6 days if astringency was removed after harvest with alcohol.

Oriental persimmons may be stored up to 3 months at 32 F and 90 to 95% humidity. Fruit skin color gradually darkens, and fruit may soften. Oriental persimmons soften and darken excessively if ethylene gas is present, so do not store them with other fruits that give off ethylene, such as apples, pears or pawpaws. Do not store non-astringent persimmons at temperatures between 41 and 59 F, as they develop flesh browning and softening at those temperatures.

Controlled-atmosphere storage, utilized only for the non-astringent cultivar 'Fuyu,' allows these persimmons to be kept for 5 months. Optimum conditions are 3 to 5% oxygen, 5 to 8% CO_2, and ethylene scubbers to remove all ethylene. Alternatively, 'Fuyu' persimmons may be individually wrapped in polyethylene bags of 0.06 mm thickness, and kept in regular cold storage at 32F for 4 to 5 months due to the modified atmosphere that forms around each fruit within the wrapping.

American persimmons have a shelf life of only 1 to 2 days at room temperature, and last 1-2 weeks in cold storage at 32 F. Customers generally use American persimmons in baking within a few days, or promptly freeze them either whole or as persimmon pulp.

Ripe pawpaws last only 1 to 2 days at room temperature, as the fruit have a higher respiration rate than many other fruits, and give off large amounts of ethylene. Ripe pawpaws last up to a week in cold storage at 40 to 45 F, and unripe pawpaws can be stored up to 4 weeks at those temperatures. Do not store pawpaws at temperatures less than 40 F, or off-flavors could develop.

Marketing and processing. Oriental persimmons are mainly sold fresh. Most consumers eatsthem fresh, often cutting them in half and eating the pulp with a spoon. Some consumers may dry or freeze them at home.

A small amount of Oriental persimmons are commercially dried. Astringent cultivars are generally used, as astringency disappears naturally during the drying process. Fruit are first machine-peeled, then treated with sulfur to prevent darkening, then partially dried for 3 days in heated chambers at about 95 F. Fruit are next spread in the sun to finish drying, and are packaged. Oriental persimmons have such a high sugar content that sugar crystallizes on the fruit surface as they finish drying, giving the fruit a dusty white appearance so that they look like they were dipped in powdered sugar.

American persimmons are most often sold fresh to local markets, and are frozen by consumers and used to make baked goods such as persimmon pudding or persimmon bread. Some growers process persimmons and sell canned persimmon pulp, baked persimmon products, or persimmon wine.

Pawpaws are primarily sold fresh at farmers' markets. Many consumers like to cut pawpaws in half lengthwise (Photo 27-21), remove the seeds, and eat the pulp with a spoon. Many also freeze the pulp for later use in homemade pawpaw bread. A few growers produce products such as pawpaw bread or pawpaw ice cream that they sell to retail markets.

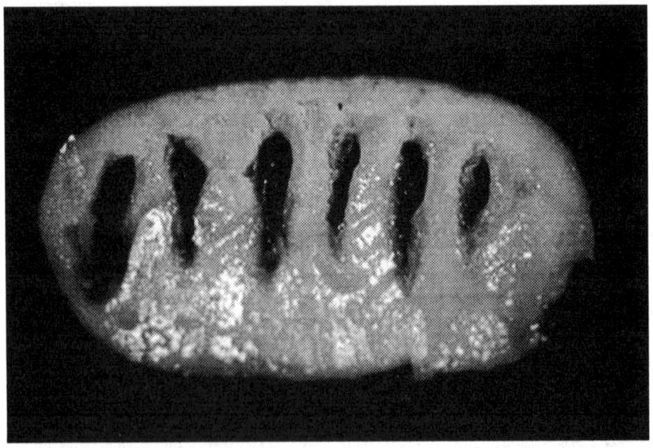

Photo 27-21. Many consumers like to cut pawpaws in half lengthwise, remove the seeds, and eat the pulp with a spoon. (photo by D. Layne, Kentucky State Univ.)

Further Reading

Bordelon, B. 1995. *Growing pawpaws.* Purdue Univ. Coop. Ext. Serv. Pub. HO-220. Available online at http://www.hort.purdue.edu/ext/HO-220.pdf

California Rare Fruit Growers. 1996. *Fruit facts: persimmons.* Available online at http://www.crfg.org/pubs/ff/persimmon.html

Kentucky State Univ. has the national repository for pawpaw germplasm and pawpaw information on its website at http://www.pawpaw.kysu.edu

Kitagawa, H. and P. Glucina. 1984. *Persimmon culture in New Zealand.* Sci. Info. Pub. Cntr., DSIR, P.O. Box 9741, Wellington, New Zealand.

Miller, E.P. and T.E. Crocker. 1994. *Oriental persimmons in Florida.* Fla. Coop Ext. Serv. SP 101. Available online at http://edis.ifas.ufl.edu/MG242

Pomper, K.W., D.R. Layne, R.N. Peterson, and D. Wolfe. 2003. The pawpaw regional variety trial: background and early data. *HortTechnology* 13:412-417.

Chapter 28
Saskatoons and Mayhaws

Saskatoons and mayhaws have both emerged as new North American fruit crops during the last 20 years. Both are members of the rose family (Rosaceae) subfamily Pomoideae, and both produce small pome fruits that are similar to crabapples but ripen much earlier in the season, although saskatoon fruits are often called berries. Both fruits are used primarily for home or commercial production of jellies, jams, syrups and other cooked products. Both plants are native to North America, both were gathered from the wild from prehistoric times until the present, but wild stands of both have declined in recent years due to development. Growers and researchers have selected superior genotypes and crossed these to produce improved cultivars of both saskatoons and mayhaws over the last 30 years or so, and these have been planted. Both crops thrive under conditions where few other fruit crops can grow, and in those regions they have become fruit industries.

Saskatoons (*Amelanchier alnifolia*) are native to the northern plains states, the prairie provinces of Canada, the Northern Rockies, the Northwest, and into the Yukon and Alaska. They are cold-hardy to -45 to -60 F (USDA zones 2b to 1a), grow as bushes to a height of 6 to 14 feet tall, and produce small berry-like pomes that turn blue-purple when ripe. They are sometimes called juneberries or serviceberries, although those common names are more often used for other species in this genus that have a tree form and are not as cold-hardy as *A. alnifolia*, and produce red-purple fruit (Photo 28-1). Other *Amelanchier* species native to North America could be developed as crops, but there is less demand for a new crop in regions where many other fruits grow well, and much less selection for superior fruiting types has been done on other *Amelanchier* species.

Mayhaws include several species of hawthorns that are native to swamps and lowlands in the southern U.S.: *Crataegus aestivalis*, *C. opaca*, and *C. rufus*. These hawthorn species are small trees that grow 20 to 30 feet tall. Fruit generally ripen in early May, thus the name mayhaws, and the early fruiting time helps mayhaw sales, since few other fruits ripen that early (Photo 28-2). Mayhaws are adapted only to the southeastern U.S., USDA zones 7b-9, but can be planted on soil that is too wet for other fruit crops. Other *Crataegus* species native to North

Photo 28-1. Juneberries grow as small trees, are not as cold-hardy as saskatoons. Fruit, such as these *A. x grandiflora*, turn red-purple when ripe.

Photo 28-2. Mayhaws and other hawthorns (*Crataegus* species) also bear pome fruit that can be 1 to 2 inches in diameter and resemble crabapples. Mayhaws are adapted to the Deep South of the U.S. and ripen very early in the season, generally in May.

Photo 28-3. Many mayhaws are gathered from wild stands like these, but planted mayhaw orchards yield about 15 times more fruit per acre.

America are more cold-hardy, ripen later and need well-drained soils. These other hawthorns could be developed as crops, but there is less demand for a new crop that ripens with many other fruits and needs soils on which other fruit crops grow well, so again much less selection for superior fruiting types has been done on other *Crataegus* species.

PRODUCTION AND BEARING HABITS

Over 2000 acres of saskatoons have now been planted, and studies indicate that the present markets for jam and fresh fruit could support another 8000 acres or more. The Canadian province of Alberta alone has over 1000 acres of saskatoons, along with about 500 acres each in Saskatchewan and Manitoba, and about 40 acres in North Dakota. Approximately 400 acres of mayhaws have been planted in Southern states such as east Texas, Louisiana, Mississippi and Georgia, and over 1000 acres of wild mayhaws are harvested. No countries outside of North America have planted saskatoons or mayhaws.

Most fruit of both crops are sold direct to consumers or to processors, or utilized in on-farm processing to sell jam or other products direct to consumers. Larger saskatoon operations use mechanical harvesters for fruit to be processed, while saskatoons and mayhaws to be sold fresh are hand-harvested by workers or customers. Mature saskatoon plantings produce an average of 3000 to 4000 pounds of fruit per acre, while well-managed plantings may produce up to 12,000 pounds of saskatoons in an optimal year. Wild mayhaws yield an average of 132 pounds of fruit per acre (Photo 28-3), while some well-managed planted orchards yield 2000 pounds of mayhaws per acre when mature.

Saskatoons begin to bear 3 to 4 years after planting, and reach full production after 7 to 8 years. Mayhaws start bearing 4 to 6 years after planting, and achieve full production after 9 to 10 years. Both crops are long-lived: saskatoon plantings should remain productive for 30 to 50 years, and mayhaw orchards should bear good crops for 50 to 70 years.

Both crops bear fruit on short spurs that arise from 1-year-old or older branches. Spurs produce clusters of fruit that ripen early in the season for each crop's area of adaptation. Mayhaws also have pointed spurs that act like thorns.

SITE SELECTION AND PREPARATION

Saskatoons need well-drained soil like most fruit crops. Mayhaws tolerate wet sites where water may stand for up to several weeks, similar to pecan sites, but also do well in well-drained soil that will stay moist in spring or can be irrigated. Both crops bloom early for their regions, so both need good air drainage to prevent spring frost damage. Both crops may benefit from a northeast slope to delay bloom, if such a site is available, although some saskatoon growers prefer a southern slope to hasten fruit ripening, and choose later-blooming cultivars. If a pick-your-own operation is planned, locate the planting within ½ mile of a paved road and allow room for a future parking area.

Saskatoons benefit from shelterbelts planted to protect them from desiccating winds in summer and/or winter (Photo 28-4). Where winter winds are strong, plant a row or two of evergreen trees to block them, generally to the northwest of the proposed saskatoon planting site. Strong summer winds call for a windbreak of fast-growing deciduous trees, generally planted south of the proposed saskatoon planting site. Choose tree species for the windbreak that are cold-hardy, well-adapted and need no pesticide sprays. Planting mulberries in the summer windbreak may help to lure birds away from the saskatoon crop.

Saskatoons and mayhaws may be planted into the stubble following wheat or another small grain crop, or planted into glyphosate-killed strips in a hay field without tearing up the sod. Raised beds aren't needed for these crops. Control any perennial weeds on the site before planting. Test soil pH and adjust it if needed so it's between 6.5 and 7.5 for saskatoons, and 6.0 to 6.5 for mayhaws. Also test soil for phosphorous (P) and potassium (K) and add these before planting if they test low.

CHOOSING CULTIVARS

Most saskatoon and mayhaw growers plant at least 3 different cultivars to provide a range of ripening times from early to late season, although the earliest-ripening cultivars also bloom early and suffer the greatest frost risk. Growers in marginal areas should plant only cultivars that bloom in the middle to late bloom period, or plant early-blooming cultivars only on sites that delay bloom.

Saskatoon cultivars must be cold-hardy enough for the zone where they'll be planted. Resistance to serious saskatoon diseases, most often berry and leaf spot and juniper rust but sometimes powdery mildew, can minimize need for fungicides. Cold hardiness is not a concern for mayhaws in the areas where they are grown, and no disease resistance has yet been identified in mayhaw cultivars.

Photo 28-5. Saskatoons begin to bloom and bear 3 to 4 years after planting, and reach full production after 7 to 8 years. Saskatoons benefit from shelterbelts planted to protect them from desiccating winds, as here. (Photo from USDA Natural Resources Conservation Service, MT.)

Many saskatoon and mayhaw cultivars ripen slowly over a period of 2 to 3 weeks for each cultivar, which is fine for pick-your-own operations, but cultivars with more concentrated ripening are better for mechanical harvesting. High yields are desirable for both harvest methods, and larger fruit are important for pick-your-own operations. Shorter saskatoon cultivars are easier to pick in hand-harvested operations, but several of the shorter cultivars have inferior fruit flavor, so are not often grown. Mayhaw cultivars should have firm fruit that do not suffer pre-harvest drop, but are not retained so strongly on the tree that they are difficult to hand-harvest or shake loose with machines. Some mayhaw cultivars have few or no thorns, and these are preferred for hand harvesting.

Table 28-1 details the traits of commonly grown saskatoon cultivars, and Table 28-2 describes the characteristics of some of the most-planted mayhaw cultivars.

PROPAGATION AND PLANTING

Saskatoons and mayhaws are mainly self-pollinated, and many seedlings will be mostly true to type of the parent, which is very unusual among fruit crops. Both fruits also produce many nucellar seedlings which are identical to the parent, which is also very unusual. Therefore, about 90% of saskatoon plantings and many mayhaw orchards have been started with seedlings. Saskatoons may also be clonally propagated from suckers, cuttings, or tissue culture, and mayhaw cultivars may be grown from cuttings or grafted onto rootstocks.

Seed propagation. Collect seeds from ripe fruit of superior selections or cultivars, taking care not to mix in any fruit from other, inferior plants. Separate seeds from the fruit by placing fruit in a blender with water, blending, and pouring off the pulp, capturing the seeds in a strainer. Seeds may be planted immediately in a sterile potting medium of peat and vermiculite and some will germinate, but germination can be increased by stratifying the seeds in moist peat/vermiculite in a refrigerator. Saskatoon seed requires 4 months of stratification, while mayhaw seed requires 7 to 8 months of stratification.

After stratifying seeds, plant them in early spring in individual containers of a sterile potting medium with a slow-release fertilizer. Grow these seedlings in a nursery area with daily irrigation

Table 28-1.

COMMON SASKATOON CULTIVARS

In approximate order of ripening (varies with area).

Cultivar	Hardy to zone	Bush height, feet	Suckers	Fruit Size	Fruit Flavor	Fruit Yield	Resistance to: Powdery mildew	Resistance to: Berry Spot	Resistance to: Juniper rust	Bloom time	Harvest time
Thiessen	1	14-16'	medium	very large	best	high	R	S	S	early	early
Martin	1	8-10'	medium	very large	best	high	R	S	S	early	early
Smoky	1	8-10'	many	large	very good	high	S	S	S	middle	middle
Pembina	1	10-14'	few	large	best	high	S	S	S	middle	middle
Parkhill	2	4-5'	few	large	fair	medium	S	R	S	middle	middle
Forestburg	2	9-12'	few	very large	good	very high	S	S	S	mid-late	mid-late
Nelson	2	4-5'	medium	large	very good	high	S	S	R	late	mid-late
Regent	3	5-6'	few	medium	fair	low-med	S	R	R	mid-late	mid-late
Success	3	5-6'	medium	large	good	med-high	S	R	S	mid-late	mid-late
Northline	1	5-6'	many	medium	best	very high	S	MR	S	late	late
Honeywood	1	10-14'	few	large	best	very high	R	R	S	mid-late	late

Disease resistance ratings: R = resistant, MR = moderately resistant, S = susceptible.

Average harvest times in southern Alberta: early: July 4-14, middle: July 8-18, mid-late: July 12-22, late: July 16-26.

Notes on cultivars:
Thiessen, the second-most-planted cultivar, has large fruit good for pick-your-own, but is frost-prone.
Martin, a seedling of Thiessen, has similar fruit but more concentrated ripening for machine harvest.
Smoky, the most-planted cultivar, suckers profusely to fill rows quickly.
Pembina suckers sparsely but plants get larger.
Parkhill suckers sparsely and plants are compact, so plant closer.
Forestburg has large, sweet fruit and is more drought-tolerant than most cultivars.
Nelson plants are compact, late bloom avoids frost damage.
Regent plants are compact and precocious, fruit are sweet but bland, yields low.
Success plants are compact, fruit are very sweet but bland.
Northline, the third-most-planted cultivar, is precocious and has sweet fruit that resist cracking.
Honeywood is precocious, frost-resistant, has very sweet fruit and resists mildew and berry spot.

Table 28-2.

COMMON MAYHAW CULTIVARS

In approximate order of ripening (varies with area).

	Fruit						Bloom	Harvest
Cultivar	Size	Firmness	Skin Color	Flesh	Retention	Yield	time	time
T.O.'s Superberry	medium	very firm	dark red	pink	ideal	low-high	early	early
Mason's Superberry	large	very firm	dark red	pink	too high	low-high	early	early
Superspur	large	soft	red/yellow	yellow	poor	high	middle	early-mid
Saline	medium	very firm	red	orange	too high	high	mid-late	early-mid
Big Red	large	firm	red	pink	ideal	very high	middle	middle
Crimson	med-large	medium	red	pink	poor	high	late	middle
G-5	very large	soft	red	yellow	poor	high	middle	middle
Big V	med-large	soft	light red	pink	poor	high	mid-late	middle
Turnage 57	small-med	soft	light red	yellow	poor	very high	mid-late	mid-late
Texas Star	very large	medium	red/orange	yellow	too high	high	mid-late	mid-late
Turnage 88	medium	medium	light red	yellow	medium	very high	mid-late	late

Average bloom times in southern Georgia: early: Feb. 12-22, middle: Feb. 20-March 3, mid-late: March 1-10, late: May 10-20.

Average harvest times in southern Georgia: early: April 20-30, early-mid: April 25-May 5, middle: May 1-10, mid-late: May 5-15, late: May 10-20.

Notes on cultivars:
T.O.'s Superberry loses many crops to frosts due to early bloom, but harvests easily when it crops.
Mason's Superberry, also called **Texas Superberry**, also blooms early, has light crops many years.
Superspur is a spur-type tree, has high yields when no late frosts occur, but fruit are soft, drop early.
Saline blooms later, crops more consistently, fruit don't drop but may be hard to remove.
Big Red often loses crops to frosts, but has very high yields when no frosts occur and harvests easily.
Crimson has good crops most years due to late bloom, but fruit drop early.
G-5 is a new selection with very large but soft fruit.
Big V sets good crops even after frosts, but soft fruit drop early.
Turnage 57 is very precocious, sets excellent crops even after frosts, but soft fruit drop early and pollen is sterile.
Texas Star has very large fruit and sets good crops even after frosts.
Turnage 88 is slow to begin bearing, but yields well once it starts bearing, even after frosts.

during the growing season, then transplant them into the field in fall. Saskatoon seedlings begin bearing within 3 to 4 years after transplanting, while mayhaw seedlings take 6 to 7 years to begin bearing.

Vegetative propagation. Clonal propagation methods work well for growers with existing plantings who want to propagate particular superior plants in a new planting, or for new growers who want to start with more uniform plantings.

Saskatoons produce root suckers that may be dug up to start new plantings. Sufficient roots must be dug with each sucker or it will not survive transplanting, and roots must immediately be placed in buckets of water and kept moist and cool until transplanting.

Suckers dug on the farm and kept cool and moist usually transplant well, but suckers ordered from nurseries, that may have gotten warm or dry during shipping, may suffer mortality rates of 60% or more after transplanting. Mid- to late fall is usually the best time to dig and transplant root suckers for direct transplanting into a new field. Growers may alternatively dig suckers in early spring while plants are still dormant, cut back top growth to about 2 inches, and transplant them to a mist-irrigated nursery area until they develop strong new tops, then transplant them to the field in late fall.

Saskatoons may also be clonally propagated by cuttings from etiolated shoots, and this method yields rooting percentages and transplant survival of 90% or more. Growers or nurserymen establish a stool bed by cutting mature bushes to the ground in early spring, then covering them with a black polyethylene tunnel supported by wire hoops. The shoots that develop will be etiolated (yellow-white) due to lack of light, and such shoots that lack chlorophyll will root easily when taken as cuttings. After etiolated shoots have grown for 4 to 6 weeks, the north side of the tunnel is slit to allow them to partly green up for 6 days, without the danger of sunburn that would occur if the south, east or west sides were slit. Then propagators take 5- to 7-inch-long cuttings, dip their bases in a 3000 to 10,000 ppm solution of IBA rooting hormone, and stick them in individual small pots filled with a rooting medium such as 50% sand and 50% peat.

Rooting beds are similar to those used for blueberries, usually raised and set outdoors under shadecloth or a clear polyethylene tunnel, and equipped with automatic misting units. Rooted cuttings are ready to transplant to the field when roots show through the bottom of the pots, usually by mid-fall, and should start bearing within 3 years.

Tissue-cultured saskatoon plants are available but are more expensive than seedlings, rooted cuttings or suckers, and are usually smaller. Tissue-cultured plants need more fertilizer to become established and reach bearing size, but begin bearing at about the same time as saskatoon seedlings, usually 4 years after planting.

Mayhaws may be propagated from softwood cuttings. Take cuttings in summer, dip them in a root-promoting hormone (2000 to 6000 ppm IBA), and place in a rooting medium under intermittent mist as is described for blueberries in chapter 15. Cutting-grown trees will be completely identical to the cultivar from which the cuttings were taken and will bear sooner than seedlings, but tend to be shallow-rooted.

Mayhaw scions may also be cut from desired cultivars and grafted onto seedling mayhaw rootstocks. This creates trees which bear fruit identical to the cultivar from which scionwood was taken, but have strong root systems which are adapted to wet sites. Parsley hawthorn, another hawthorn species, is also considered a good rootstock for mayhaws in east Texas and Mississippi.

Do not graft mayhaws onto 'Washington' hawthorn seedlings, a common ornamental tree that's commercially available, because it can not stand wet soil conditions, and mayhaws grafted onto it are severely dwarfed and perform poorly. Graft when mayhaw bark starts to slip, in early to mid-February where mayhaws are grown. Grafted mayhaws usually start bearing a couple years earlier than seedling-grown trees.

Orchard layout. Saskatoon growers plant rows 16-20 feet apart to accommodate mechanical harvesters, or perhaps 12-15 feet apart if a pick-your-own operation is planned and mechanical harvest is not ever anticipated. Plants are spaced 3 to 5 feet apart within rows, with wider spacing for cultivars that grow larger and sucker more, and narrower spacing for cultivars that stay smaller and sucker less. Generally, 700 to 1365 plants per acre are set out.

Mayhaw growers generally plant trees in rows 18 to 20 feet apart, with 15 to 20 feet between trees within the row, producing an orchard density of 110 to 160 trees per acre.

Solid blocks of individual cultivars may be planted, since saskatoons and mayhaws self-pollinate. A north-south row orientation is generally preferred to allow maximum light penetration, and favors snow drifting in saskatoons. Hilly land should be planted on the contour, especially if mechanical harvest is planned.

Planting and care of new plants. Plant saskatoons in mid- to late fall after planting material has hardened off but before the soil freezes, or in late spring after the threat of spring frosts has passed. Plant mayhaws from late fall through early spring. Plant only vigorous plants with well-developed root masses, 12 to 24 inches tall for saskatoons and 3 to 4 feet tall for mayhaws. Plan on at least 10% mortality of saskatoons and 5% for mayhaws, so have extra plants ready to replace those which die. Water in newly-set plants well.

Saskatoons need no pruning at planting, except to remove any broken branches. Retain a single trunk on mayhaw trees, removing any additional trunks or root suckers at planting time, and remove any branches arising less than 24 to 30 inches off the ground. Place white plastic tree guards around mayhaw trunks to protect against sunscald and rodent damage.

SOIL MANAGEMENT, IRRIGATION AND FERTILIZERS

Most saskatoon and mayhaw growers use herbicides to control all vegetation in the planting rows, and mow a sod groundcover between fruit rows, similar to the methods used for most fruit crops in humid areas. Short, native bunchgrasses are recommended groundcovers for saskatoons since they slow-growing and need little mowing, drought-tolerant so don't require irrigation or take much water from the saskatoons, and shade-tolerant. Sheep's fescue is widely recommended. Mayhaws often require a groundcover that withstands periodic flooding; most orchardists just rely on native vegetation.

Keep groundcovers in both saskatoons and mayhaws mowed to a height of 2 to 4 inches during the harvest season and when water is limited, but the groundcover may be allowed to grow taller when it's too wet to mow and in fall as plants are hardening for winter.

Most saskatoon growers apply pre-emergent herbicides to prevent weeds in planting rows in late fall or early spring, when saskatoons are dormant. Some saskatoon growers apply wood chip mulch in the planting rows to help control weeds and conserve soil moisture, similar to sawdust mulch often used for blueberries and straw mulch sometimes used for blackberries. Sawdust, straw or grass clippings should not be used to mulch saskatoons, however, since these materials attract voles, rabbits and other rodents, while wood chips do not.

When mayhaws grow in floodplains, use herbicides to control vegetation only in a 3-foot-diameter circle around each tree, not in strips down entire rows, to minimize bare soil subject to loss during floods. Mayhaws are not generally mulched.

Irrigation needs and methods. Many saskatoon and mayhaw plantings do not have irrigation systems. Both crops mature early in the growing season, so mid- to late summer droughts do not hurt fruit production. Mulching saskatoons conserves enough soil moisture that irrigation of mature plantings isn't often needed in many areas. Mayhaws are deep-rooted, drought-tolerant trees, and are often planted on wetter sites that don't usually need irrigation.

Saskatoon and mayhaw plantings without irrigation systems must be watered by hand for the first 2 or 3 years during droughts, generally by driving a tractor or truck hauling a water tank through the field, and filling buckets to apply water to each fruit plant. Once plantings are well-established, plants will survive droughts without this watering.

Saskatoon and mayhaw growers who install irrigation systems generally use drip irrigation. Young plantings should be irrigated whenever less than 1 inch of rain fell during the previous week. Bearing plantings may need only 2 or 3 irrigations per year, during fruit sizing in late spring or early summer and when fruit buds form in late summer.

Fertilizers. On many soils, the only fertilizer element needed for saskatoons and mayhaws is nitrogen (N), like many other fruit crops. Neither crop needs a large amount of N, and mayhaw growers should especially avoid applying excess N as it can encourage fire blight.

Fertilize first-year saskatoon plantings with about 20 pounds actual N per acre in early May and another 25 pounds actual N per acre in early July. From the second year on, add about 35 pounds actual N per acre in early May and another 35 pounds actual N per acre in early July or after harvest.

First-year mayhaw plantings need 3 fertilizer applications in rings around each tree, adding about 1 ounce actual N per tree in March and about 0.5 ounce actual N per tree each in May and July. This is only 14 to 20 pounds of actual N per acre in the first year, depending on the number of trees per acre. From the second year on, add about 1.5 ounces actual N per inch of trunk diameter to each tree in March, and the same amount in July if leaves look pale or shoots have not grown at least 12 to 18 inches. This will require 20 to 60 pounds of actual

N per acre per year, depending on the number of trees per acre and tree diameter.

Test soil for P and K every 3 to 4 years, and use a complete fertilizer if these nutrients are needed for either saskatoons or mayhaws. Other nutrients are seldom deficient on these crops, but take leaf samples if any deficiency symptoms are seen.

TRAINING, PRUNING AND MANAGING CROPS

Train saskatoons as open-centered bushes, similar to blueberry bushes but larger. Train mayhaws into modified-leader trees, as for low-density apples and pears. Most training of both types of plants is accomplished by dormant pruning in early spring.

Training and pruning saskatoons. Encourage root suckers to grow during the first couple years after planting, to fill in the planting row as needed. Prune off at the base any excess suckers that develop and crowd other plants. Once sufficient plants have developed, remove additional suckers each year to prevent overcrowding. Suckers that develop between the planting rows will be mown off with the groundcover.

Starting in the second dormant season after planting, remove all diseased, damaged, weak, and low spreading branches in early spring of each year. From year 5 or 6 on, after saskatoons have borne 2 crops, remove an older branch or two, that is 1 to 1.5 inches in diameter at the base, in the center of each bush each year. This will allow better air circulation in the planting, reducing fruit diseases, and will encourage growth of new fruiting wood. The largest fruit are produced on 1- to 2-year-old branches, so removing older branches and encouraging new growth increases fruit size.

Once a saskatoon planting reaches 14 to 16 years of age, machine topping every 3 to 4 years may be helpful to keep hedgerows to the desired height, especially for pick-your-own plantings. Alternatively, pick-your-own growers can limit their plantings to cultivars which stay 6 to 8 feet tall or lower. Plantings that are machine harvested may be up to 12 to 14 feet tall, so may need no topping.

Training and pruning mayhaws. Train mayhaws to the modified leader system described in chapter 4, with the lowest limb arising at least 24 to 30 inches off the ground to allow room for a trunk shaker to aid in harvesting the fruit. Remove any diseased branches as soon as they're seen, cutting at least 12 inches below the visible disease margin of fire blight-infected shoots or branches. Also remove all damaged, weak, and low spreading branches in early spring of each year. Mayhaws also require annual removal of root suckers, which may grow back less if removed in mid-summer.

Older mayhaw trees benefit from thinning out the centers to allow better light penetration, but this is difficult to do by hand because of the thorny branches. However, machine hedging of mayhaws is not advisable because the damaged branches and resulting regrowth would be very prone to fire blight.

Crop management. Saskatoons and mayhaws are both self-pollinating, so no beehives need be brought into orchards, although wild bees and other insects do pollinate both crops. Both crops bloom early for their regions (Photo 28-5) and may lose some fruit to frosts, which helps increase size of the remaining fruit. Commercial growers do not use any frost protection methods for these crops.

Saskatoons and mayhaws both produce small fruit, and no fruit thinning is done. Pruning out excess branches from bush or tree centers effects some fruit removal and allows larger fruit to develop on outer branches, as mentioned.

PEST MANAGEMENT

Saskatoons and mayhaws suffer from some of the same pests described for apples, plus some additional pests. Many pesticides registered for apples and pears are now registered for all pome fruits, including saskatoons and mayhaws.

Pests that can kill plants. Some of these pests may kill many saskatoons or mayhaws in localized areas, so scout for them regularly and apply control measures promptly if the pests occur.

Photo 28-5. Saskatoons and juneberries (such as these *A. x grandiflora*) bloom early for their regions, as do mayhaws in their area of adaptation. Frosts may reduce crops of both fruits, but generally this provides helpful crop reduction. Saskatoon and mayhaw growers do not employ frost protection during bloom, neither do they thin the fruit.

Wooly apple aphids (*Eriosoma americanum*), also called **wooly elm aphids** because elm is their alternate host where they overwinter, may cause substantial transplant losses in saskatoons as they feed on roots of new plants. Scout for damage in mid-summer by digging around roots of a few plants, checking for masses of these aphids with their wooly white coatings. If present, a soil insecticide, acephate (Orthene™) may be used on non-bearing plantings for the first 3 years of establishment. Direct this spray to the plant roots once per year, in mid-July to early August.

Flat-headed apple borers (*Chrysobothris femorata*) may attack trunks of young mayhaw trees as they do apples. Control them as described for apples, by protecting trunks of young trees with white plastic tree guards or screening material from the soil line to above the graft union.

Cytospora canker, caused by the fungi *Leucostoma cincta* or *L. persoonii*, the same fungi that cause perennial or Cytospora canker of peaches in colder climates, does similar damage in saskatoons. Canker fungi infect shoots through cracks caused by injuries or early fall frosts, and the fungi grow down stems to the crowns, causing stem dieback and eventually killing plants. Prevent canker by applying no N fertilizer after mid-summer to avoid stimulating growth late into fall, and by reducing irrigation and mowing less frequently in fall so saskatoons slow their growth sooner and shoots won't crack after fall frosts. Mechanical harvesting equipment may cause injuries that lead to canker, so prune off broken branches soon after harvest is complete. Do not prune in early fall, as fresh pruning wounds could serve as entry site for canker.

Black leaf or **witches' broom**, caused by the fungus *Apiosporina collinsii*, may seriously infect saskatoons. It causes gray-black, felt-like growth on leaf undersides and multiple branching that resembles a broom, and can spread into crowns and kill plants. Prune off infected shoots or branches at least 8 inches below the site of brooming, in late fall soon after leaf drop. Plants with infected root crowns must be dug up and destroyed.

Fire blight, caused by the bacterium *Erwinia amylovora*, is generally a minor problem on saskatoons since they are grown in cold climates, but can be a major problem on mayhaws since they grow in warm, humid climates. Selection for mayhaw cultivars that resist fire blight would be helpful, but has not yet been done. Control fire blight as described for apples and pears, by avoiding excess N fertilizer, pruning out blighted branches as soon as they're seen and minimizing other pruning, and by applying bacteriocides during bloom if weather is especially warm and wet.

Deer (*Odocoileus* species) are often serious pests of both saskatoons and mayhaws, and **elk** (*Cervus elaphus*) may also browse saskatoons. Deer and elk strip bark from bushes or trees, which can kill the plants especially on mayhaws, and eat many fruit buds especially on saskatoons. High deer populations may necessitate an electric fence around the planting, while low to moderate populations may be repelled by monthly applications of deer-repellent materials.

Rabbits (*Sylvilagus* species) and **voles** (*Microtus* species) can girdle and kill saskatoons and mayhaws with their winter feeding, as they do with other fruit trees. Controlling vegetation around the fruit plants with herbicides reduces habitat for these rodents. Where rodent populations are high, spray a rodent repellent on the bases of saskatoons and mayhaws in the fall to prevent damage.

Indirect pests that reduce yields and plant longevity

Saskatoons and mayhaws do not generally have virus or phytoplasma diseases, partly because many saskatoons and some mayhaws are seed propagated. Several insects and mites may feed on leaves of saskatoons and mayhaws, weakening plants and reducing fruit yields. Natural enemies usually keep these pests in check, and when outbreaks occur, biological or selective pesticides can control them.

Leafrollers attacking saskatoons and mayhaws include **fruittree leafroller** (*Archips argyrospilus*) and **oblique-banded leafroller** (*Choristoneura rosaceana*). Leafroller larvae feed on leaves and buds early in the season, webbing or rolling the leaves to protect themselves from predators, which also protects them from insecticide sprays. Scout for leafroller damage early in the season, and if any is seen, apply a bloom spray of the biological insecticide *Bacillus thuringiensis* (Javelin™, Dipel™, etc.), which will not hurt pollinating insects.

Prairie tent caterpillar (*Malacosoma lutescens*) may attack saskatoons (Photo 28-6), while **Eastern tent caterpillar** (*M. americanum*) may attack mayhaws. Both moth pests have one generation per year of gregarious caterpillars that construct silken nests in forks in branches, which protect them from pesticides, and both feed on leaves. Natural enemies, including birds and fungal diseases, control tent caterpillars adequately most years, but pest outbreaks occur about every 10 years and may defoliate saskatoons or mayhaws. Scattered colonies may be physically removed by twisting a pole through each web, and stepping on

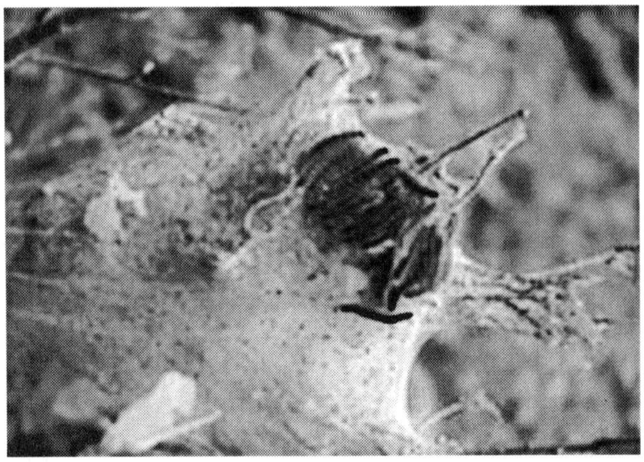

Photo 28-6. Prairie tent caterpillar may attack saskatoons, as here, while Eastern tent caterpillar may attack mayhaws. High-pressure sprays of a molting inhibitor or spinosad, directed at the nests when first formed, will control tent caterpillar outbreaks. (Photo from Agric. Canada.)

the detached webs to kill the caterpillars. High-pressure sprays of a molting inhibitor such as tebufenozide (Confirm™), methoxyfenozide (Intrepid™), or spinosad (SpinTor™), directed at the nests when first formed, will control tent caterpillar outbreaks.

Cherry shoot borer (*Argyresthia oreasella*) larvae may attack saskatoons, tunneling into young shoots in late spring and mid-summer. Attacked shoots wither and die back, which can reduce fruiting the next year if this pest is abundant. Monitor shoots for damage in late spring, and if numerous, apply a molting inhibitor in mid-summer to stop the next generation from developing.

Two-spotted spider mites (*Tetranychus urticae*) are occasional problems in saskatoons and mayhaws, just as in apples and other fruit crops. These pest mites are usually controlled by predator mites, but spider mites may build up to damaging populations in some years, when they can cause foliage bronzing and premature defoliation. Apply a dormant oil spray if mite problems are anticipated, keep plantings adequately irrigated if irrigation is available, and avoid using pesticides that disrupt mite predators, such as sulfur fungicide or pyrethroid insecticides after early summer. If spider mite outbreaks occur, apply a selective miticide that won't hurt predator mites, such as hexythiazox (Savey™) or cinnamaldehyde (Valero™).

Diseases that attack leaves, flowers and fruit. Several severe fungal diseases occur every year on saskatoons and mayhaws, usually requiring about 3 fungicide sprays per year.

Leaf and berry spot, caused by the fungus *Entomosporium mespili*, is the most damaging disease of saskatoons. This disease is favored by high humidity and moderate temperatures of 68-77 F, and causes red-brown spots on leaves and fruit (Photo 28-7). It can defoliate bushes and render all fruit of susceptible cultivars unmarketable. Planting cultivars that resist leaf and berry spot, such as 'Success' and 'Honeywood', can reduce or eliminate the need for fungicides to control this disease. Susceptible cultivars need 3 fungicide sprays per season, at white tip, at petal fall and 2 weeks later at the green fruit stage. Pruning out most suckers, to keep planting rows open to wind currents, also reduces this disease. Canadian researchers have developed an *Entomosporium* disease forecast model based on cultivar susceptibility, saskatoon development stage, temperature, leaf wetness and inoculum potential, that can guide growers in timing sprays.

Rust diseases, caused by several *Gymnosporangium* species, attack both saskatoons and mayhaws. **Saskatoon-juniper rust** is caused by the same fungus that causes **cedar-apple rust** of apples and mayhaws, *G. juniperi-virginianae*, and other species in this genus. Saskatoon-juniper rust is a less frequent problem in saskatoons, occurring mainly when spring weather is unusually cool and the saskatoon crop is delayed. The disease causes the same damage in saskatoons and mayhaws and spreads the same way as described in chapter 16 for apples. **Quince rust**, caused by *G. clavipes*, is the worst disease of mayhaws, infecting mainly the fruit and rarely the leaves. Heavily infected mayhaw fruit does not ripen properly, and has no value. **Hawthorn rust**, caused by *G. globosum*, also attacks mayhaw leaves and fruit (Photo 28-8).

Juniper is required as the alternate host for all these rust diseases, so removing all junipers within a 1-mile radius of the saskatoon or mayhaw planting would prevent them, but is generally impractical. Mayhaw growers need to apply 2 or 3 fungicide sprays to control rust diseases in early spring, from bloom until rust structures on junipers dry up 3 to 4 weeks later. Saskatoon growers should choose a fungicide for leaf and berry spot that also controls saskatoon-juniper rust, such as the sterol inhibitor propiconazole (Topas™ or Engage™).

Mummy berry of saskatoons, also called American brown rot, is caused by the fungus *Monilinia amelancheris*, which is in the same genus as the pathogen that causes brown rot of stone fruits. As in peaches, this disease causes saskatoon fruit

Photo 28-7. Leaf and berry spot, the worst disease of saskatoons, causes red-brown spots on leaves and fruit. Prevent it by choosing more resistant cultivars or by applying 2 or 3 fungicide sprays in early spring. (Photo by P. Fisher, Ontario Min. of Agric. & Food)

Photo 28-8. Hawthorn rust may attack mayhaws. Control it like quince rust, and like saskatoon-juniper rust, with 2 or 3 fungicide sprays in early spring.

third spray for leaf and berry spot also helps control mummy berry, but must be made at least 38 days before harvest.

Brown rot blossom blight (*Monilinia fruticola*), the scourge of stone fruits, may also attack developing mayhaw fruit during some wet springs. Control it as described for peaches, with fungicide sprays at first bloom and full bloom.

Insects and birds that attack flowers and fruit. Several early-season insect pests attack saskatoon and mayhaw flowers and developing fruits, usually requiring 2 or 3 insecticide sprays that can be combined with the 3 early-season fungicide sprays required for fruit disease control.

Saskatoon bud moth (*Epinotia bicordana*) larvae bore into flower buds in early spring, destroying them and preventing flowering and fruiting of the attacked buds. Growers should scout regularly for the tiny, dark gray moths and light green larvae on newly-expanding buds (green tip stage). If the pest occurs, apply 2 or 3 insecticide sprays at 10- to 14-day intervals: the first before bloom, the second at first bloom (if 10 days or more after the first spray), and the last spray at petal fall or a few days later.

Tarnished plant bug (*Lygus lineolaris*), **lygus bug** (*L. hesperus*), and other plant bugs may feed on fruit buds or developing fruit of both saskatoons and mayhaws, causing misshapen fruit. Sprays for saskatoon bud moth also control plant bugs on saskatoons, while mayhaws may require 2 similar sprays just before first bloom and at petal fall. Mowing the groundcover when a majority of plant bugs are still on it and are nymphs will destroy many of them, because they do not yet have wings to fly to the fruit plants.

Apple curculio (*Tachypterellus quadrigibbus*) may also attack saskatoons soon after petal fall, while **plum curculio** (*Conotrachelus nenuphar*) attacks mayhaws, and **hawthorn weevil** (*C. crataegi*) may attack both saskatoons and mayhaws. These snout beetles lay eggs in the fruit, and larvae eat the developing seeds inside the fruit, causing hard, misshapen fruit to develop. The last spray for saskatoon bud moth will help control apple curculio and hawthorn weevil, but another insecticide spray 2 weeks after petal fall may also be needed. Mayhaws usually need 2 insecticide sprays, at petal fall and 10 days later, to control plum curculio and hawthorn weevil.

Saskatoon sawfly (*Hoplocampa montanicola*) larvae feed on developing fruit of saskatoons in late spring and summer, and may cause serious damage. Fruit attacked while young usually drop, while fruit attacked when older often remain attached and consist of empty shells. Sprays for apple curculio and hawthorn weevil will help control saskatoon sawfly.

Many **birds** eat saskatoons, including **starlings** (*Sturnus vulgaris*), **house finches** (*Carpodacus mexicanus*) and **robins** (*Turdus migratorius*). Saskatoon growers may need to erect bird netting or rotate bird scare devices, or use poison baits for large flocks of starlings or other pest birds. Birds usually cause little problems on mayhaws, since the fruit are larger and are too hard to peck.

HARVEST AND PROCESSING

Many saskatoon growers sell their entire crop pick-your-own or hand-harvested for retail sales for the first few years of production, then purchase a machine harvester and sell increasing amounts of fruit to processors as the planting matures and yields increase. Producers with less than 5 acres of saskatoons located within 30 miles of at least 50,000 people may continue to sell their entire crop pick-your-own or hand-harvested for retail sales.

Most mayhaw growers hire workers (often family members) to hand-harvest fruit for retail sales or on-farm jelly production. A few mayhaw growers offer pick-your-own sales, while a few larger growers use tree or limb shakers or shake-and-catch machine harvesters.

Hand harvest. Harvest seasons for saskatoons and mayhaws are short, just 3 to 5 weeks long. Saskatoon harvest generally runs from early to late July in Canada, and mayhaw harvest runs from late April till late May in the Deep South. Producers must attract sufficient numbers of pick-your-own customers and/or hire sufficient workers to harvest the crop during the short season. Many pick-your-own growers also offer harvested fruit, as many customers will want to purchase more fruit than they want to pick themselves.

Saskatoons are ready to harvest when fruit turn red-purple and pull off the branches easily. Workers or customers can often shake entire fruiting clusters over buckets to dislodge ripe fruit quickly. Some growers provide workers with hand-held berry-picking vibrators, which dislodge ripe fruit even more quickly.

Mayhaws are not generally harvested from the trees due to the thorns, although some fruit on young trees or lower, outer limbs of large trees may be harvested from the trees when the fruit turn orange-red. Many growers allow fruit to fall naturally to the ground, where they may be picked up by workers or pick-your-own customers. Some growers spread plastic tarps under the trees to catch falling fruit (Photo 28-9), which are then transferred into containers for sale. Fruit that is allowed to fall before harvesting must be firm enough to resist bruising when it hits the ground; cultivars with softer fruit are best harvested from the tree. Mayhaw fruit grown in swampy areas may fall into water, which cushions the fruit from bruising, but such fruit must be raked up within a day or two of falling so it doesn't rot.

Mechanical harvest. Over-the row harvesters, similar to those used for blueberries and brambles, are used for saskatoons. Because the demand for saskatoons for processing has thus far exceeded supply, processors have paid about the same price per pound for saskatoons as pick-your-own customers have. This makes mechanical harvesting more affordable for saskatoons than for other fruit crops, so that a grower who mechanically harvests as little as 10 acres could pay off the investment in the harvester within a few years of full crops. Each saskatoon cultivar will usually need 2 to 3 mechanical harvests, starting when about $^2/_3$ of the fruit are red-purple. Complete harvesting by 11 a.m. each day to minimize field heat in harvested fruit. Harvested fruit is run across a blower and grading belt to remove leaves and green fruit, and promptly cooled.

Shake-and-catch machine harvesters, similar to those used for peaches and other tree fruits, may be used on larger mayhaw plantings, but require greater acreage to pay back the investment than do saskatoons. Many smaller growers harvest semi-mechanically by spreading tarps under the trees and shaking trees with mechanical trunk shakers or limb shakers, or shaking limbs by hand. Again, finish harvesting by late morning to minimize field heat, run harvested fruit across a blower and grading belt to remove leaves and green fruit, and cool promptly.

Mayhaw cultivars with firm fruit that is retained well on the tree work better for mechanical harvest than cultivars which drop fruit quickly and have softer fruit that bruises easily as it hits the tarp or catch pad. Firm-fruited cultivars may be harvested just 2 or 3 times, starting when at least $^3/_4$ of the fruit have turned orange-red, while cultivars with softer fruit will need 5 or 6 harvests.

Photo 28-9. As harvest time approaches, many mayhaw growers spread plastic tarps under the trees to catch falling fruit which are then gathered in buckets.

Fruit storage and processing. Harvested saskatoons should be promptly cooled or frozen by the grower, customers or processor. Saskatoons that are hand-harvested for retail sale will last about up to 2 weeks at 31-32 F and 90-95% humidity. Machine-harvested saskatoons must be processed or frozen within 24 hours of harvest due to bruising. Many processors have growers bag, freeze and box saskatoons before delivering them to the processor, so prices paid to growers by processors include these services, and growers must have sufficient freezer space for each day's harvest.

Harvested mayhaws can be held at room temperature for only 1 to 2 days, since most fruit are bruised during harvesting. Thus growers must promptly sell, process into jelly, or freeze mayhaws soon after harvest. Many smaller growers freeze mayhaws in 1-gallon, zip-lock plastic freezer bags immediately after sorting them, since most customers use frozen fruit to make jellies, but this requires sufficient freezer space.

Further Reading

Costs and returns for a saskatoon berry orchard. 2004. Available online at: http://www.agr.gov.sk.ca/docs/econ_farm_man/production/fruit/saskatoonberry.asp

Krewer, G. 1998. *Experiments and observations on growing mayhaws as a crop in south Georgia and north Florida.* Available online at http://pubs.caes.uga.edu/caespubs/pubcd/mayhaw.htm

St. Pierre, R.G. 1997. *Growing saskatoons — a manual for orchardists, 5th ed.* Available from Native Fruit Development Program, Dept. of Hort. Sci., Univ. of Saskatchewan, 51 Campus Dr., Saskatoon, Saskatchewan S7N 5A8

Chapter 29
Currants and Gooseberries

Currants and gooseberries are both members of the genus *Ribes* in the gooseberry family (Grossulariaceae). Both grow as small bushes, generally 3 to 6 feet tall and wide, and produce true berries. Both are very minor crops in North America, with less than 400 acres in total currant and gooseberry production on the continent, but are produced in large quantities (over 2 billion pounds per year) in Europe, and are starting to increase in production in North America.

Currants and gooseberries were grown more widely on this continent in the 1700's, 1800's, and early 1900's, and the 1920 census revealed over 7400 acres of commercial currant and gooseberry production in the U.S. alone. Then several *Ribes* species were discovered to be alternate hosts for white pine blister rust, caused by a fungus introduced from Europe several times between 1898 and 1910, which kills five-needled pine trees. Due to the large white pine lumber industry, the federal government and many states prohibited cultivation of *Ribes* in the early 1930's. The federal ban was lifted in 1966, but many states still forbid the planting of European black currants, the most susceptible species, or in some cases any *Ribes* species. Some states have recently modified these laws to allow planting of species or cultivars of *Ribes* that resist white pine blister rust. Since many currant and gooseberry cultivars do resist white pine blister rust, and a market for fresh berries and processed products exists, plantings of these fruits have increased in recent years and have the potential to increase further, perhaps to the level that existed in 1920.

Gooseberry plants have spiny stems, although a few "nearly thornless" cultivars have been selected and propagated, while currants have smooth stems. Gooseberry fruit are much larger than currants, and some large-fruited gooseberry cultivars are mainly sold fresh for eating raw or in desserts. Currants and smaller-fruited gooseberries are processed at home or commercially into jams, juices, wines and other products.

SPECIES AND ADAPTATION

The main commercial currant species all originated in northern Europe. Black currants descend mainly from *R. nigrum*, the widespread European black currant, and have purple-black fruit. Canadian breeders developed rust-resistant black currants by hybridizing the Russian black currant, *R. ussuriense*, with *R. nigrum*. Red currants include cultivars with white fruit (Photo 29-1) as well as those with red fruit (Photo 29-2), and derive mainly from *R. rubrum* (also called *R. sativum*) native to Scandinavia and Siberia. Some red currant cultivars also include germplasm from *R. multiflorum* native to England, *R. petraeum* native to high mountains of Europe and northern Africa, and *R. spicatum* native to Scandinavia. Most red, white and black currant cultivars are very cold hardy, usually to USDA hardiness zone 3 and sometimes to zone 2, but do not withstand the very hot summers of the southern half of the United States.

A few black currant cultivars have also been selected from species native to North America, and these native currants are very ornamental with their large yellow flowers and red fall color (Photo

Photo 29-1. White currants are a color variation of red currants, selected from several European species. (Photo from Cornell Univ., NY)

Photo 29-2. European red currants are mainly processed at home or commercially into jams, sauces and syrups.

29-3). The clove current (*R. aureum* var. *villosum*, formerly *R. odoratum*) is native in the Great Plains states, and the golden currant (*R. aureum*) is native in the Pacific Northwest. These species have greater heat tolerance than other currants. The North American red currant, *R. triste*, has fruit quality similar to European red currants but lower yields, and has not yet been developed for cultivation or used in breeding.

Gooseberry cultivars descend from the European gooseberry (*R. uva-crispa*, formerly *R. grossularia*) and its hybrids with several American gooseberry species. The main native American species used has been *R. hirtellum*, native to most of the continent, but *R. cynobasti* native to the Northeast and *R. missouriensis* native to the U.S. Midwest have also been utilized to a limited extent. Most cultivars grown in North America are hybrids between European and American gooseberries (Photo 29-4), as most pure European cultivars are too disease-susceptible and often cold-tender, although some new cultivars bred in Europe are very promising. Many gooseberry cultivars are cold-hardy to zone 3, while some are only hardy to zone 4, and a few are hardy to zone 2. One cultivar has been selected from the wild northern gooseberry (*R. oxyacanthoides*), native to the Prairie Provinces of Canada, and is hardy to zone 2 as well as highly disease-resistant. Like European currants, gooseberries do not thrive under hot summers.

All currants and gooseberries are diploid, and different species are easily hybridized. European black currants and European gooseberries have been hybridized to create jostaberries, *R. nigridolaria*. Jostaberry fruit is similar to black currants

Photo 29-3. The clove currant, native to the U.S. Great Plains, is very ornamental with its large yellow flowers, shown here, and red-orange fall color. It produces black fruit with a milder taste than European black currants.

but with the size of gooseberries, and is mainly used for juice and wine production, like black currants. Stems are slightly spiny, plants are hardy to zone 3, and again do best with cool summers.

Currants and gooseberries need only 120 to 140 frost-free days to mature fruit and complete their vegetative growth each season, so they do well in short-season areas. Cultivars' winter chilling needs vary from 800 to 1500 hours of temperatures between 32 and 45 F.

PRODUCTION AND BEARING HABITS

Currants and gooseberries generally grow as upright bushes that do not need support, but branches of some high-yielding gooseberry cultivars bend under the weight of the fruit crop, and

Photo 29-4. Most gooseberry cultivars grown in North America are hybrids between European and American gooseberries, such as this 'Pixwell'.

benefit from trellising or staking. Some growers also choose to trellis red or white currants to make hand harvesting easier. Plants usually begin to bear 2 to 3 years after planting, but some precocious cultivars begin bearing in the second season, less than 18 months after planting. Currants reach full production 3 to 4 years after planting, while gooseberries reach full production 4 to 5 years after planting. Currants yield 9000 to 16,000 pounds (4.5 to 8 tons) of fruit per acre, while gooseberries yield 9000 to 12,000 pounds (4.5 to 6 tons) of fruit per acre. Currants and gooseberries will continue producing well for 20 or more years.

Bushes bear flowers on spurs of older stems, and laterally on one year old stems. Currants produce flowers on long, pendulous racemes, so the fruit ripen in long clusters called "strigs" which are harvested as a unit. Gooseberries bear flowers singly, in pairs or in triples, so the fruit ripen this way and grow much larger, and are harvested as individual berries. Both types of fruits contain many large seeds, which are more noticeable in currants since the fleshy fruits are much smaller. Each gooseberry fruit retains a hardened style from the flowers, which becomes a spine, and anthers around the spine. These must be removed after harvest, while currant fruits are smooth.

Currants and gooseberries have perfect flowers, and most cultivars are self-fertile. A few cultivars are partially self-sterile, so benefit from planting with other cultivars to provide cross-pollination. Flowers are insect-pollinated.

Fruit ripen in mid-summer. Each currant cultivar ripens all its fruit over about 2 weeks, although ripe fruit will hang on bushes well for up to 2 weeks on many cultivars. This concentrated ripening allows currants grown for processing to be mechanically harvested. Some gooseberry cultivars also have concentrated ripening or are machine-harvested immature for processing, while other gooseberry cultivars ripen over a longer period and are best sold pick-your-own or hand-harvested for fresh market.

SITE SELECTION AND PREPARATION

Currants and gooseberries do best in cool-summer climates with frequent cloud cover, but may grow well in less cloudy, warm-summer areas if provided a site with partial shade. Gooseberry fruit especially will scald if exposed to direct sunlight under high temperatures. A north-facing slope helps delay spring bloom and also minimizes direct sunlight in summer. Alternatively, provide shade with shadecloth mounted on frames over the planting. The site should also have good air drainage, to avoid spring frost damage and reduce diseases, and should be located at least 1000 feet away from any white pines.

Soil must also be well-drained, but with a high water-holding capacity. Heavier soils are best in warm regions, as they retain more moisture and keep cooler in summer. Choose soil with at least 3 to 5% organic matter, or grow a cover crop to increase soil organic matter to these levels before planting berry bushes. Test soil for pH, phosphorous (P) and potassium (K) well before planting. If necessary, adjust pH to within the range of 5.5 to 7.0, and raise the available soil P level to 30 to 40 pounds per acre and the available soil K level to about 300 pounds per acre. Build raised beds if soil is only moderately well-drained.

Test soil for nematodes before planting *Ribes*, and grow nematode-suppressing crops for a season or two if needed. The dagger nematode (*Xiphinema americanum*) is the most damaging nematode to currants and gooseberries, because it spreads tomato ringspot virus. It's most prevalent on lighter soils in the Pacific Northwest and Northeast of North America.

CHOOSING SPECIES AND CULTIVARS

Before ordering plants, check with your state or province's Department of Agriculture to obtain the latest *Ribes* regulations. Several states still prohibit planting of all black currant cultivars although they allow red currants and gooseberries, and a few states still prohibit planting of all *Ribes*.

If the market will be largely pick-your-own and retail, choose large-fruited gooseberries with very few spines, and possibly several red and white currants with long clusters of large fruit that are easy to pick. Choose several cultivars with early, mid-season and late ripening times to extend the harvest season. European black currants have a very strong, musky flavor that is too intense for many North Americans' tastes, so do not plant them as a retail or pick-your-own crop unless there is known demand in your area. 'Crandall' currants, from a native American species, have black fruit with a milder taste that may work well for pick-your-own markets.

European black currants may be planted for the juice market where a processing plant that wants them is located nearby. Black currant production increased so much during the 1990's that large juice processors are able to dictate to growers the cultivars to plant, production techniques and selling prices, so check with the processor before planting. Red currants may also be planted for juice or jelly, but there is little processing demand for white currants due to the lack of color. A small processing demand exists for canned gooseberries. Plantings for processing must be at least 30 to 40 acres to justify a mechanical harvester, or several smaller farmers might cooperate in purchasing a harvester, or a smaller farmer might contract for custom harvesting with a larger grower who owns a harvesting machine if one lives nearby.

If white pines occur in your region, choose only cultivars with strong resistance to blister rust. Choosing cultivars resistant to powdery mildew, leaf spot and gray mold will minimize the need for fungicides, and choosing currant cultivars resistant to currant borer will reduce the need for insecticides.

Some red and white currant cultivars bloom quite early so are prone to crop loss from spring frosts. Choosing later-blooming cultivars reduces this risk. Most black currants and gooseberries bloom late enough to avoid frosts, but if frosts are often a problem in your area, choose the latest-blooming cultivars of these fruits too. Black currants and gooseberries vary in winter hardiness more than red and white currants, so choose cultivars hardy enough for the area.

Currants are upright bushes that don't need support. Most gooseberry cultivars also do not need support, but a few cultivars tend to sprawl and do better if trellised or staked, so avoid these cultivars if you don't want the expense of a trellis. Finally, choose the highest-yielding cultivars that meet market demands, pest and frost resistance as outlined above.

Table 29-1 details the traits of some common red and white currant cultivars, while Table 29-2 describes common black currants and black currant x gooseberry hybrids. Table 29-3 compares many widely-planted gooseberry cultivars.

PROPAGATION AND PLANTING

Currants and gooseberries are most often propagated by cuttings. Nurseries in mild-winter climates, such as the Pacific Northwest, take dormant stem cuttings in fall, plant them in outdoor propagation beds where they root over winter, and dig and sell the rooted cuttings in early spring or the following fall. Currants and gooseberries may also be rooted from softwood cuttings taken in spring and placed in propagation beds under mist, as described for blueberries. Growers may propagate their own plants from cuttings if bushes are healthy, but it would be wise to have propagating material tested for viruses, as about 12 viruses may infect *Ribes*. Growers should also specify virus-free plants when ordering from nurseries.

Plant dormant rooted cuttings or 1-year-old transplants of currants and gooseberries between October and March. Growers in mild-winter climates usually plant in late fall or early winter, so plants grow more roots over the winter and become well-established before shoot growth begins in spring, but in USDA zone 4 and colder regions, wait until early spring to plant. Do not plant bare-root plants later in spring, since the plants break dormancy early and rapid shoot growth would overwhelm the rudimentary root system. Container-grown plants may be planted later in spring as they have more roots, but cost more than bare-root plants.

Hand-harvested plantings usually space plants 3 to 4 feet apart within rows so they'll grow as bushes, and space rows as far apart as equipment requires, generally 10 to 12 feet apart. This requires 908 to 1452 plants per acre. Plants can be spaced twice as close within rows if they will be trained to cordons for hand harvest, so 1816 to 2904 plants would be needed per acre for cordon training. Space mechanically-harvested plantings closer within rows also, generally 1.5 to 2 feet apart to create a tight hedgerow, but allow more space between rows, usually 12 to 14 feet, to accommodate the harvester. Thus mechanically-harvested plantings require 1556 to 2420 plants per acre. Cultivars that grow more upright can be spaced a bit closer than those that spread more.

Irrigate newly-set plants soon after planting.

SOIL MANAGEMENT, IRRIGATION, AND FERTILIZERS

Most currant and gooseberry growers control weeds in crop rows with herbicides as practiced for other fruit crops, but some growers use mechanical cultivators or flame weeders within rows. Currants and gooseberries benefit greatly from an organic mulch within rows, which keeps the soil cooler in summer and preserves soil moisture. Grass clippings or wheat straw make good mulch in many areas, but where vole or mice populations are high, use wood chips instead to avoid attracting the rodents. Mulch also reduces weed growth, and the mulch is much easier to maintain when weeds that do grow are controlled with herbicides rather than cultivation or flame weeding.

Growers in humid areas, and some in arid areas, sow a less-competitive grass and/or legume sod between rows and keep it mowed during the growing season (Photo 29-5). Growers in arid areas may raise winter cover crops between rows, turn them under in spring, and cultivate between rows during summer to reduce water needs. However, vegetation should be allowed to grow between rows for a few weeks before harvest if rain is at all possible during that time, to avoid mud which annoys pick-your-own customers and can cause harvest machines to get stuck.

Irrigation management. Currants and gooseberries require about 1 inch of water per week during the 3-month period when fruit are growing and ripening, about ¾ inch of water per week during the next month of vegetative growth, and about 1 inch of water per week during the 2 weeks in late summer when fruit buds are setting for next year. This adds up to a water requirement of about 18 inches per year, which means that only 20 inches of water per year need to be applied through a drip irrigation system that is 90% efficient. This low water use requirement is a plus factor for these fruit crops, especially in arid areas of the Pacific Northwest.

Most Western growers use drip irrigation to reduce water use, since currants and gooseberries do not usually suffer from frost damage if later-blooming cultivars are chosen, so sprinkler systems aren't needed. Some larger growers prefer sprinklers, since each sprinkler head covers a large area (Photo 29-6), and sprinklers have far fewer problems with clogging and no problems with rodent damage to irrigation lines. Sprinklers can also provide evaporative cooling in early summer, helpful in areas where summers get a bit hot for currants, and may be used for spring frost protection if needed. Eastern growers generally do not install irrigation systems, since rainfall is usually adequate to meet crop needs.

Fertilizers. During the first year after planting, currants and gooseberries need only about 30 pounds of actual nitrogen (N) per acre, applied in spring, at least 6 weeks after planting, to stimulate shoot growth. During the second growing season they need about 50 pounds of actual N per acre, and they need about 70 pounds of actual N per acre during the third year, which is best applied in both spring and after harvest in summer. If P and K were added before planting, they often won't be needed during the first 3 years.

Mature currant and gooseberry plantings need about 90 pounds of actual N, 20 pounds of P and 40 pounds of K per acre per year. Do not use potas-

Photo 29-5. Many currant and gooseberry growers raise a less-competitive grass sod between rows to prevent soil erosion. This commercial red current field of Dennis Pleasant's in Prosser, WA is trellised for easier machine harvest. (Photo by D.G. James, Washington State Univ.)

Photo 29-6. Some growers, such as Dennis Pleasant, use sprinkler irrigation on currants to provide evaporative cooling in summer, and spring frost control if needed. (Photo by D.G. James, Washington State Univ.)

sium chloride to supply K, as *Ribes* are sensitive to chloride toxicity. Apply about 2/3 of the N fertilizer after harvest in summer to encourage fruit buds for next year, and 1/3 of the N in late fall to promote root growth or early spring to stimulate shoot growth.

TRAINING, PRUNING AND MANAGING CROPS

Most fresh-market currants and gooseberries are trained and pruned as freestanding, individual bushes in a manner similar to blueberries. Some European growers of fresh-market gooseberries and red or white currants train plant to cordons to increase fruit size. Mechanically-harvested plantings are trained to hedgerows, and some growers of processing black currants mechanically prune the plants so they harvest each row only once every other year.

Training and pruning as bushes for hand harvest. When grown as freestanding bushes, currants and gooseberries need little training. Bushes are allowed to grow naturally for the first 3 years, only removing any crossing branches or those that are damaged or diseased.

Red and white currants and gooseberries bear most fruit on short spurs arising from 2- and 3-year-old stems, so from the third dormant season on, older stems are removed each year. During dormant pruning each year, retain 3 or 4 one-year-old stems for future production, plus 3 to 4 each of 2- and 3-year-old stems, for a total of 9 to 12 stems per bush.

Cordon training for hand harvest. This system produces very large, highly-colored fruit that is easier to pick, but requires twice the number of plants per acre, a trellis and stakes, and detailed hand pruning, so it costs more to establish and maintain. Cordon training creates more open plants which can reduce powdery mildew and other diseases, and may be faster to harvest. It is most appropriate for gooseberries and red or white currants in cool-summer climates, as it can greatly increase fruit sunburn in warm-summer climates.

If planning to train plants to cordons, erect a trellis before or soon after planting. The trellis is just to support bamboo or conduit poles on which to train cordons, not to support the plants themselves, so needn't be overly strong. Steel fenceposts about 6½ feet long are generally the least expensive type of post, and should be spaced 12 to 16 feet apart down the row, with anchor posts at each end of each trellis. Drill a hole for the wire near the top of each post, and pound each post about 1 foot into the ground to anchor it well. String number 14 or number 12 wire down each row, through the fencepost holes.

Six-foot lengths of conduit pipe or bamboo work well as stakes to train the cordons. If one cordon is being selected per plant, as is usually done, place one stake next to each plant. If 3 cordons will be chosen per plant of red or white currants, place 3 stakes next to each plant. Pound stakes into the ground about 6 inches, and tie them to the trellis wire to keep them upright.

For cordon training, plant red or white currants about 1.5 feet apart within rows, and gooseberries about 2 feet apart within rows. Early in the first growing season, select 3 stems or branches per plant as cordons for red or white currants, and 1 to 3 stems or branches per plant as cordons for gooseberries, and remove all other branches. Train the cordons up bamboo stakes or conduit pipes, allowing them to grow to a height of 5 to 6 feet before heading the tips. Allow any side shoots on the cordons to grow, but remove other shoots from the base of the plant (suckers) during the growing season.

During the second year, retain all well-spaced, medium-diameter side branches on the cordons, and allow side shoots to grow. Some branches may fruit the second year. Remove smaller side branches and those in poor positions. Also remove all other stems or branches that arise from the base of the bush except one, kept as a potential replacement for any cordons that die, and cut this spare branch back to about 6 inches.

From the second or third season on, immediately after 1-year-old side branches fruit in summer, remove them. Continue to allow well-spaced side shoots to grow into medium-diameter branches that will fruit the following year. This completely renews the fruiting wood each year, enabling plants to produce bigger berries and possibly higher yields per acre.

Hedgerow pruning by hand for mechanical harvest. Plantings to be mechanically harvested are spaced closer within rows, generally 1½ to 2 feet apart to create a tight hedgerow. Plants in hedgerows are trained and pruned similarly to freestanding bushes, except that fewer stems are kept per bush to prevent overcrowding. Since twice as many plants per acre must be pruned, train workers to quickly decide which stems to remove, to minimize pruning time. Keep only upright stems, removing any that droop or that overly entrude on stems from the next plant in the hedgerow.

Black currants bear most fruit on 1-year-old stems or 1-year-old branches from older stems, so remove older stems with few branches. Keep 3 or 4 one-year-old stems plus 2 to 4 older stems with many 1-year-old branches per plant, for a total of 5 to 8 stems per plant.

Red currants or gooseberries grown in hedgerows for mechanical harvest should be pruned to 2 or 3 one-year-old stems for future production, plus 2 to 3 each of 2- and 3-year-old stems, for a total of 6 to 9 stems per bush.

Mechanical pruning for alternate-year mechanical harvest.

Black currants grown for juice may utilize completely mechanized pruning as well as harvest, by mowing half of the planting to the ground each year during the dormant season. The other half of the planting will be mowed to the ground during the following dormant season. Plants that were mown 1½ years earlier will bear fruit and be mechanically harvested each year. Like alternate-year harvest systems for bramble crops, this system reduces total yields, but plants are spaced closer to partially compensate for that yield loss, and this system cuts labor requirements to just 15 hours per acre per year.

Crop Management. Almost all currant and gooseberry cultivars are self-pollinating, and those that need cross-pollination are generally adequately pollinated by wild bees when planted with other cultivars, so no beehives need to be brought into the plantings during bloom. Growers usually do not use any frost protection methods during bloom, other than choosing later-blooming cultivars if spring frosts are a frequent problem in the area.

Like other small fruit crops, no fruit thinning is done on currants or gooseberries. Cordon pruning systems do reduce the number of fruit that develop and thus increase fruit size.

PEST MANAGEMENT

Currants and gooseberries may be attacked by many pests, but most can be prevented without pesticides by choosing resistant cultivars and certified virus-free plants, and by regular pruning out of diseased or damaged stems.

Pests that can kill plants. Most of these pests can be prevented by choosing resistant cultivars and pruning out infested stems.

Currant borer (*Syanthedon tipuliformis*) larvae feed inside the stems of currants but not gooseberries (Photo 29-7). Adults are clear-winged moths with blue backs with yellow markings. Adult females lay eggs on stems in early summer, and larvae develop inside the stems over the winter. Leaves of infested stems wilt and then stems fall over during the next growing season, causing developing fruit to wither. Currant borer adult females exhibit marked non-preference for many currant cultivars, designated as "resistant" or "very resistant" in Tables 29-1 and 29-2, so planting these cultivars can minimize the need to control currant borer. Further reduce damage by pruning out infested canes in fall or early spring, as soon as signs of damage appear, and burning or otherwise destroying them. Mating disruption can also control currant borer if at least 5 acres of currants are planted in one block. Smaller plantings may require treatment with an insecticide in early summer, before larvae enter canes.

Currant stem girdler (*Janus integer*), a sawfly, attacks both currants and gooseberries. Adult female sawflies puncture shoot tips in spring, a few inches from the tip, and insert eggs, then girdle the shoot tips so they droop and wilt (Photo 29-8). Larvae hatch and tunnel through the stems, feeding

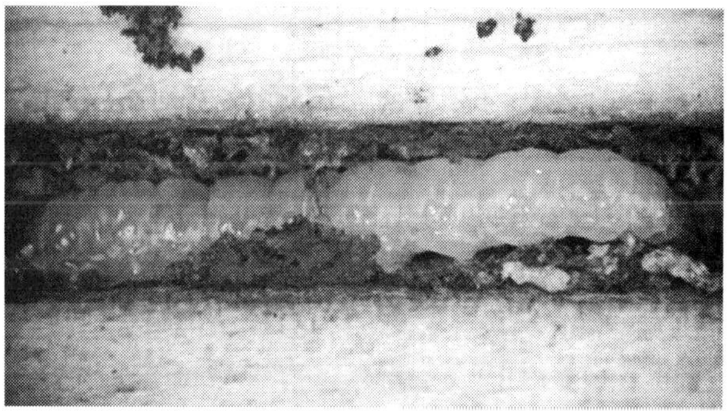

Photo 29-7. Currant borer larvae tunnel inside the stems of currants but not gooseberries. Prevent it by choosing resistant cultivars, or using mating disruption on currant fields of 5 acres or more. (Photo by D.G. James, Washington State Univ.)

Photo 29-8. Currant stem girdler female sawflies insert eggs in shoot tips in spring, then girdle the shoot tips so they droop and wilt. Prune off injured shoots at least 8 inches below the girdle, and burn the prunings. (Photo by D.G. James, Washington State Univ.)

in the pith until they emerge in fall to spin a cocoon in which they overwinter. Larvae pupate in early spring, then emerge as adults in mid-spring. Pruning off injured shoots at least 8 inches below the girdle as soon as they're seen, and burning or otherwise destroying the prunings, helps control this pest.

Cane blight, caused by the fungi *Botryosphaeria ribis* or *Botrytis cinerea*, may kill stems back or to the ground. Prune out infected canes, at least 4 to 6 inches below the visible disease margin, as soon as they're seen, and burn or otherwise destroy the prunings.

Oak root rot, caused by the fungus *Armillaria mellea*, may infect currants and gooseberries if the pathogen is present in the soil, so avoid planting within 10 years of clearing oak woodland or removing other fruit plants in areas where the disease occurs.

Voles (*Microtus* species) and **mice** (*Mus* species) may girdle stems of currants and gooseberries, killing plants. Where these rodents occur in high populations, avoid use of straw mulch which would attract them, and use wood chip mulch instead. Poisoned baits may be necessary.

Indirect pests that reduce yields and plant longevity. Currants and gooseberries may suffer from about 12 different virus and phytoplasma diseases, so specify virus-free plants when ordering planting material. Choosing cultivars that resist foliar diseases, and minimizing the use of pesticides that hurt natural enemies of foliar insect and mite pests, will prevent most other indirect pest problems.

Gooseberry vein banding virus may infect currants as well as gooseberries, and is the most common *Ribes* virus in North America. This virus causes a yellow banding pattern along the main veins of leaves (Photo 29-9) that is most evident in early-spring growth and harder to see by mid-summer. Remove and destroy infected plants as soon as symptoms are seen, and control aphids which spread this virus.

Tomato ringspot virus is another common virus of both currants and gooseberries, especially in the Pacific Northwest and Northeast of North America. This virus causes a yellow mottling or mosaic pattern in leaves (Photo 29-10) that again is most evident in early-spring growth and harder to see by mid-summer. Remove and destroy infected plants as soon as symptoms are seen. This virus is spread by the dagger nematode (*Xiphinema americanum*), so test soil for nematodes before planting *Ribes*, and grow nematode-suppressing crops if needed.

Photo 29-9. Gooseberry vein banding virus, the most common *Ribes* virus in North America, may infect currants as well as gooseberries. Order certified virus-free planting stock, remove and destroy infected plants as soon as symptoms are seen, and control aphids which spread this virus. (Photo by J. Postman, USDA, ARS)

Photo 29-10. Tomato ringspot virus, another common virus of both currants and gooseberries, is spread by the dagger nematode, so test soil for nematodes before planting *Ribes*, and grow nematode-suppressing crops if needed. (Photo by J. Postman, USDA, ARS)

White pine blister rust, caused by the fungus *Cronartium ribicola*, does more damage to pines than to currants and gooseberries. Susceptible *Ribes* cultivars develop orange spores on the undersides of leaves (Photo 29-11), which spread to the pines and can defoliate very susceptible currants. If no 5-needled pines grow anywhere in the area, this disease will not develop, but *Ribes* growers in districts with 5-needled pines should plant only cultivars with strong resistance to white pine blister rust.

Leaf spot or **anthracnose**, caused by the fungus *Drepanopezizia ribis*, produces many dark brown spots on both sides of leaves (Photo 29-12), usually starting early in the growing season. Infected leaves turn yellow and drop, reducing plant vigor and yields, and causing smaller fruit of reduced quality. Prevent leaf spot by choosing cultivars with resistance to this disease. Susceptible cultivars need fungicide sprays, such as a copper fungicide or the strobilurin pyraclostrobin (Cabrio ™) when the first leaf reaches full size and every 10 to 14 days thereafter until dry weather prevails.

Septoria leaf spot, caused by the fungus *Mycospharella ribis*, produces similar symptoms as anthracnose but does not generally appear until late spring, and the leaf spots develop black spores which do not appear on anthracnose leaf spots. Infected leaves again turn yellow and drop, reducing plant vigor but not reducing yields or fruit size as much since the disease appears later in the season. Septoria leaf spot often does not require control, but if it becomes severe a fungicide may be needed at least 2 weeks before harvest or after harvest.

Photo 29-11. White pine blister rust produces orange spores on leaf undersides of susceptible *Ribes* cultivars, which spread to the pines and can defoliate very susceptible currants, such as this 'Ben Alder' black currant. Choose resistant currant and gooseberry cultivars if white pines grow in the region, and plant at least 1000 feet away from any white pines. (Photo by J. Postman, USDA, ARS)

Photo 29-12. Prevent leaf spot, also called anthracnose, of currants and gooseberries by choosing cultivars with resistance to this disease. (Photo by J. Postman, USDA, ARS)

Imported currant worm, *Nematis ribesii*, feeds on foliage of both currants and gooseberries, with gooseberries actually the preferred host. Adults are sawflies, and 2 generations occur per year. Larvae feed in colonies and can consume leaves to the point that only midribs are left (Photo 29-14), so start scouting for damage soon after the leaves have expanded. This is often the most serious insect pest of currants and gooseberries, and may require 1 or 2 insecticide sprays in mid- to late spring.

Currant aphid (*Cryptomyzus ribis*) also feeds on foliage of both currants and gooseberries. Feeding on leaves in early spring causes them to crinkle and turn down at the edges, which results in stunted and distorted new shoot growth. These aphids usually occur in colonies, and secrete honeydew that may fall on developing fruit, serving as a substrate for sooty mold fungi that mar the fruit and leave it unmarketable. Natural predators such as lady beetles usually keep aphids on control, but if currant aphids have become a problem in past years, use a delayed-dormant oil spray to prevent them from reaching damaging levels.

San Jose scale (*Quadraspidiotus pernicious*) may attack currant stems as it does many other fruit crops. Spiny gooseberry cultivars generally escape attack, but scale may damage less spiny

Photo 29-13. Imported currant worm larvae feed in colonies on both currants and gooseberries. Larvae can consume leaves to the point that only midribs are left, so start scouting for damage soon after leaves have expanded, and apply an insecticide as needed. (Photo by J. Postman, USDA, ARS)

gooseberry cultivars. Prevent it with a dormant or delayed-dormant oil spray also.

Diseases that attack stems, flowers and fruit. Currants and gooseberries suffer from few fruit diseases, and many cultivars resist the fruit diseases that do occur, so often need no fungicides to prevent these direct diseases.

Powdery mildew, caused by the fungus *Sphaerotheca mors-uvae*, produces white powdery mycelial growth on shoots and leaves and a gray to dark brown felt-like coating on berries of susceptible currants and gooseberries (Photo 29-14). Damaged leaves may drop during hot weather, infected plants are often stunted, and severely infected plants can die. Many currant and gooseberry cultivars resist powdery mildew, and choosing resistant cultivars is the best management strategy. Susceptible cultivars need fungicide sprays from green tip until late summer, avoiding sprays during the 2 weeks before harvest. Lime sulfur is most often used, but can cause mite outbreaks during warm weather. Potassium bicarbonate (Kaligreen ™) or the powdery mildew biofungicide AQ10™ might be better choices during the summer.

Photo 29-14. Powdery mildew produces white mycelial growth on shoots and leaves and a gray to dark brown felt-like coating on berries of susceptible currants and gooseberries. Avoid it by choosing resistant cultivars or applying selective fungicides. (Photo by J. Postman, USDA, ARS)

Gray mold, caused by the fungus *Botrytis cinerea*, invades currant and gooseberry fruit as described previously for strawberries and raspberries. Infected fruit often drop early, especially in black currants, which is called "run-off". This fungus may also cause cane blight of stems as described above. Choosing currant and gooseberry cultivars that resist gray mold minimizes the need for fungicides, which should be applied at first bloom and petal fall to susceptible cultivars.

Insects and birds that attack flowers and fruit. A few insect pests attack fruit in northern areas where most currants and gooseberries are grown and need control. These direct insect pests but don't occur in regions with warmer summers, so although *Ribes* are harder to grow in such climates, they may need no insecticide sprays for fruit insects.

Gooseberry fruitworm (*Zophodia convolutella*) attacks currants as well as gooseberries. This moth larva is green with dark side stripes, and each one consumes several berries by hollowing them out. Where this pest occurs, apply two insecticide sprays, at petal fall and 10 days later, to prevent it.

Currant fruit fly, also called **gooseberry maggot** (*Epochra canadensis*), infests fruit of all *Ribes* species in northern areas. Larvae feed inside berries, causing premature fruit drop or remaining in harvested fruit. Sprays for gooseberry fruitworm will also control currant fruit fly.

Birds may carry away many red currants, so bird netting over individual rows (Photo 29- 15) or the entire planting may be needed. Birds do not usually bother white currants or yellow-fruited gooseberries because they can't see that color as easily, and the taste of European black currants repels many birds. The chemical bird repellant, methyl anthranilate, has not proven successful on red currants, however. Gooseberry fruit is too large for most birds to carry away, but red-fruited gooseberries may be pecked as they harvest approaches and they color up, so they may also need bird netting.

HARVEST, PROCESSING AND MARKETING

Black currants are grown mainly for processing into juice, so are almost entirely machine-harvested. Some red and white currants and gooseberries are also machine harvested for processing, but many in North America today are sold pick-your-own or hand-harvested for sale at farmers' markets.

Hand harvest. Pick-your-own marketing eliminates about ⅔ of the total labor requirement in producing currants and gooseberries, so most hand-harvested operations aim to have most of their customers pick their own fruit. Most of these operations also hire some workers to harvest fruit for sale on-farm or at farmers' markets, for customers who don't want to pick fruit.

A person can pick about 22 pounds of gooseberries per hour because the fruit are large, and about 18 pounds of red or white currants per hour because

Photo 29-15. Bird netting prevents birds from carrying away red currants, as here, or pecking red-fruited gooseberries.

entire clusters are harvested at once. A person can pick only about 9 pounds of black currants per hour because the fruit and clusters are smaller, which is another reason that most black currants are machine-harvested. Choosing large-fruited gooseberry cultivars and currant cultivars with long, large clusters is important for hand-harvested operations, to increase pick-your-own customer satisfaction and the efficiency of hired harvesters.

Growers may provide customers or workers with hand clippers to cut entire clusters of currants, or pickers may simply pinch off clusters at their bases. Equipping gooseberry pickers with hand-held pneumatic or electric shakers can speed up harvesting, but can damage some fully-ripe fruit which must be sorted out, so the additional sorting may outweigh the gain in harvesting speed. Many currant and gooseberry cultivars can be harvested in 2 pickings over about 2 weeks, since ripe fruit hang well on plants without dropping.

Currants and gooseberries for fresh sale are generally hand-harvested directly into 1/2- or 1-pint containers, as used for raspberries, since these fruits often sell better in smaller containers, and the delicate skin would be easily damaged if placed into larger containers. Avoid plastic mesh baskets, as they may bruise gooseberries and some currants may press through the mesh.

Mechanical harvest. Over-the-row mechanical harvesters designed for currants and gooseberries are manufactured in Europe but are quite expensive (Photo 29-16). Smaller mechanical harvesters built in North America for blueberries or brambles might be adapted to currants and gooseberries to reduce the expense of purchasing a harvester. At least 30 to 40 acres of currants and/or gooseberries are required to justify a mechanical harvester.

Choosing cultivars with a concentrated ripening period is important to machine harvest. European currant growers spray fruit to be machine-harvested with ethephon 10 to 12 days before harvest, to allow once-over harvest of each block. Currants intended for processing into jelly are often harvested slightly immature so that pectin levels will be higher, but are better harvested fully ripe, as fruit have better flavor and still contain plenty of pectin to jell properly. Gooseberries sold to canners are also harvested immature, but again taste much better if allowed to fully ripen.

Fruit storage and processing. Fresh-market red and white currants and gooseberries have delicate skins so must be handled carefully. Fruit that is cooled promptly after harvested fruit may be kept at 34 to 36 F for 3 to 4 days before marketing. Fruit intended for processing may be frozen until utilized.

Black currants have tougher skins, and may be kept in cold storage for 1 to 2 weeks before processing or fresh sales, or may be frozen for future processing.

Marketing. Currants and gooseberries need more extensive marketing to increase consumer demand in North America. Many people in the U.S. especially have never tasted these fruits, while others have negative images of them due to previous efforts at eradicating wild currants and gooseberries. Offering samples of berries, jellies or other fruit products on pick-your-own farms and at

Photo 29-16. This mechanical black currant harvester is commonly used in England. (Photo from Purdue Univ., IN)

farmers' markets should increase demand for fresh fruit. Juice makers can offer samples at grocery stores, warehouse clubs and other retail outlets where currant juice is sold.

Further Reading

Dale, A. 2000. Potential for *Ribes* cultivation in North America. HortTechnology 10:548-554.

Hummer, K.E. and D.L. Barney. 2002. Currants. *HortTechnology* 12:377-387.

Hummer, K.E. and J. Postman. 2000. Currant and gooseberry diagnostic tool. Available online at: http://www.ars-grin.gov/cor/ribes/ribsymp/ribsymp

McKay, S.A. 2002. Gooseberries and red currants: cordon training for fresh fruit production. *The New York Berry News* vol. 1, no. 7:5-6. Available online at: http://www.nysaes.cornell.edu/pp/extension/tfabp/newslett/nybn

Chapter 30

Rewards and Requirements of Home Fruit Growing

Fruit and nut plants offer numerous benefits to home gardeners, so many people are eager to grow them. Yet growing fruits and nuts well requires much more knowledge than growing vegetables or ornamental plants, and many misleading advertisements or articles convince some gardeners that these plants will just take care of themselves, and all the gardener needs to do is reap the bountiful harvest. Extension Service agents, garden center workers, garden writers and others who advise home gardeners should all learn enough about home fruit growing to answer gardeners' many questions.

ECONOMIC BENEFITS AND COSTS OF HOME FRUIT GROWING

If well-adapted species and cultivars are chosen and properly cared for, fruit plants can produce abundant harvests, saving home growers far more money than they would save by investing the same time into growing vegetables. The authors calculated that a 20-foot-long blueberry hedge (Photo 30-1) they planted yielded over $1500.00 worth of blueberries over a 12-year-period that they lived in that house, in addition to beautifying the landscape. The only costs were less than $50 for plants, soil amendments and bird netting, plus the time spent pruning and harvesting the fruit.

Many fruits and nuts freeze well so are quickly and easily stored, while many people enjoy making jams, breads, pies and other homemade fruit and nut products. And of course, what can compare with fresh fruit or berries on the morning cereal or on ice cream (Photo 30-2)?

Some fruit and nut trees are expensive to buy and take 3 to 7 years after planting before they begin to bear, however, so homeowners should investigate this before planting. Strawberries, raspberries and blackberries cost much less for starter plants and begin bearing just 15 months or so after planting, so these fruits should be emphasized to those who may move to a new house in a few years. Those who plan to stay put should definitely be encouraged to plant fruit and nut trees, and can be educated about cultivars that begin bearing sooner and cultural techniques to induce earlier bearing, described in earlier chapters.

Choosing cultivars adapted to the area's climate, and resistant to the main diseases of the area whenever possible, is very important. Many

Photo 30-1. Over a 12-year-period that they lived in this house, the authors harvested over $1500.00 worth of blueberries from this hedge they planted.

cultivars that grow well mainly in arid, mild-winter parts of the western U.S. are sold all over the U.S. each year, and most of these plants die from cold or severe disease attacks, or produce only diseased fruit. Disease-prone fruit plants can require greater costs for pest control than the benefits they provide in fruit. Uninformed gardeners who plant such poorly adapted cultivars not only lose the money spent on the plants, but may give up on fruit growing altogether.

Consult tables 1-1 to 1-3 in chapter 1, and the individual fruit and nut chapters, for the adaptation of each type of plant that you want to grow. Over most of North America where spring frosts threaten almost every year, gardeners would be wise to choose species or cultivars that bloom later than the average last spring frost date, to avoid losing crops most years.

FITTING FRUIT PLANTS INTO THE HOME LANDSCAPE

Many fruit and nut plants make attractive as well as useful additions to a home landscape, and needn't be relegated to a separate area in the backyard. (Photo 30-3). Even those with very little space can often fit in some fruit plants along walkways or walls or in shrub borders.

Ornamental and functional attributes of fruit and nut plants. Blueberries make a beautiful four-season hedge, with white spring flowers, glossy green summer leaves, red fall foliage (Photo 30-4), and red stems in winter. Blueberries and lingonberries can also be interplanted in shrub borders with azaleas, which need the same acid, peaty, sandy soil. Currants and gooseberries look good in shrub borders and thrive under partly shaded conditions, such as along the north side of a house or under tall trees providing light to medium shade, where few other fruit plants produce well.

Many fruit trees have spring bloom as pretty as commonly-planted flowering trees, and can act as specimen trees in a lawn or as a boundary planting along a road or property line (Photo 30-5). Fruit trees can also be underplanted with perennial flowers or low flowering shrubs (Photo 30-6), which compete less with the fruit trees for water than does grass, in a mixed border. Fruit trees can even by planted in very little space by espaliering them against a wall (Photo 30-7), for those willing to prune back branches throughout the growing season.

Nut trees make excellent shade trees (Photo 30-8), although most don't grow as fast as some shade trees such as hybrid red-silver maples. Grapes can be trained on arbors to provide shade as well as fruit (Photo 30-9).

Strawberries make excellent groundcovers or facing plants for perennial flowers, (Photo 30-10) or can edge a walkway (Photo 30-11).

Even spiny raspberries or gooseberries have a landscape function, as a boundary hedge to discourage stray dogs or fruit pilferers from entering the yard.

Placing each type of fruit plant in the best spot. Home situations have the advantage of providing many different microclimates, which can enable gardeners to raise some fruit plants than would not be economical to grow commercially in the area. Early-blooming fruit trees, such as apricots or Japanese plums, can be planted north of the house or garage to delay bloom in spring. Plants that benefit from mid-day shade, such as pawpaws, currants and gooseberries, can be planted under trees that provide filtered shade, such as pecans. Raspberries can be planted on the east side of a garage or garden shed in hot climates, to provide late afternoon shade. Soil in a small area can be modified for blueberries and lingonberries.

Nut trees should be planted where shade will eventually be wanted, such as east or west of the house to provide morning or late afternoon shade for the house. Medium-large trees with fruit that attracts birds, such as mulberries or juneberries, can be planted toward the rear of the yard, where any unharvested fruit won't stain walkways, and may help lure birds from other fruit plantings.

Even a small suburban yard can produce plenty of fruit. The authors raised 65 fruit and nut trees and hundreds of small fruit plants in their 1/3-acre yard, before beginning commercial fruit farming.

CARING FOR HOME FRUIT PLANTINGS

Some fruit and nut plants can be grown with little time devoted to their care, while others need quite a bit of care. Gardeners should learn how much care each type of fruit needs before planting them. Caring for fruit plantings provides needed exercise and can be enjoyable for many gardeners, but some do not like to do certain tasks, such as spraying or thinning fruit, and should avoid types of fruits that need such tasks. Table 30-1 details the care needs for home gardeners of many common fruit and nut plants.

Training and pruning. Fruit and nut trees and grapevines need basic training while young, to avoid developing structural weaknesses that can later cause limbs to split or require severe pruning of older trees or vines. A few hours devoted to train-

Photo 30-2. Gardeners can easily plant several cultivars of small fruits to provide variety, since plants are smaller and more total plants are needed. These late-ripening strawberries, yellow raspberries and blueberries all ripened at the same time, providing a delicious topping for vanilla ice cream.

Photo 30-3. An effective and attractive method of weed control around fruit and nut trees is to plant less competitive perennial flowers, and mulch these with bark chips.

Photo 30-4. Blueberries offer red fall foliage as well as white spring flowers, glossy green summer leaves, and red stems in winter.

Photo 30-5. Fruit trees with pretty spring bloom, such as these dwarf apple trees with two taller pear trees and a pink-petaled crabapple in the background, can serve as a boundary planting along a road or property line as here, or as a specimen trees in a lawn.

Photo 30-6. Fruit trees can also be worked into a mixed border by underplanting them with perennial flowers or low flowering shrubs, such as these rudbeckia, daylilies and others.

Photo 30-7. Fruit trees can by planted in very little space by espaliering them against a wall, as done with this apple tree underplanted with tuberous begonias and white alyssum.

Photo 30-8. Nut trees, such as these pecans, make excellent shade trees.
Photo 30-9. Grapes can be trained on arbors to provide shade as well as fruit
Photo 30-10. Strawberries make excellent facing plants for perennial flowers such as these daylilies, planted under apricot trees along a driveway.

Photo 30-11. Strawberries can also edge walkways if an edging material, such as these boards, prevents the plants from encroaching on the pathway.

ing young trees can often reduce pruning of mature nut trees to nothing except occasional removal of damaged branches, and can reduce pruning of mature fruit trees to mainly heading back overly vigorous shoots in summer and thinning out competing branches in winter. Grapevines, however, will need heavy pruning every year to stay healthy and produce good yields of high-quality fruit.

Blackberries and raspberries need no training the first year except pinching the new canes, but need annual pinching of new canes and removal of old canes after fruiting each year. Blueberries, currants and gooseberries need little or no training, but from the fourth year on, plants also require annual removal of older stems. Strawberries and lingonberries perform best if runners are trained by setting them in desired positions the first year, and plantings are renewed each year after harvest.

Watering, fertilizing and controlling weeds. Choosing types of fruit plants that need about as much water as rainfall normally provides in your area will minimize the need to irrigate once plants are established. However, all fruit and nut plants need some watering during the first couple of seasons, unless rain falls every week during the growing season.

Once established, the fruits and nuts with the lowest water needs (20 to 28 inches per year) include apricots, currants, gooseberries, figs, mulberries, pistachios, and European bunch grapes for raisins or wine. Even these fruits and nuts require timely watering during fruit expansion and fruit bud set, however. Crops with medium water needs (28 to 36 inches per year) include interstem-dwarfed apples, almonds, tart cherries, Chinese chestnuts, hazelnuts, pears, European plums, saskatoons, muscadine grapes, American grapes, and European-American hybrid grapes. Those with high water needs (36 to 42 inches per year) include rootstock-dwarfed apples, sweet cherries, blackberries, mayhaws, pawpaws, American persimmons and black walnuts. Plants with very high water needs (42 to 48 inches per year) include blueberries, lingonberries, raspberries, strawberries, peaches, nectarines, Oriental persimmons, Japanese plums, pecans and Persian walnuts.

Table 30-1.

HOME GARDEN LONGEVITY, YIELDS, AND CARE NEEDS FOR FRUIT AND NUT PLANTS

Plant	Least years to bearing[z]	Years of bearing[y]	# plants for family of 4[x]	Lbs. per plant/yr.	Need for:			
					Water[w]	Pruning	Fruit Thinning	Pest Control[v]
Almonds	3	40-45	2	25-35	medium	medium	none	high
Rootstock-dwarfed Apples	3	20-25	2	120-160	high	medium	high	medium
Interstem-dwarfed Apples	3	30-40	2	160-200	medium	high	high	medium
Apricots, Apriums	4	30-35	2	80-120	low	low	medium	medium
Blueberries, Lingonberries	3	30-35	4-5	10-20	very high	medium	none	low
Blackberries, hybrids	1	15-20	5	3-5	high	very high	none	low
Sweet Cherries	4	25-30	1	30-60	high	high	none	very high
Tart Cherries	3	20-25	1	25-35	medium	low	none	medium
Chinese & hybrid Chestnuts	3	40-60	2	40-60	medium	medium	none	low
Currants	2	18-22	5	4-8	low	low	none	high
Figs	2	80-100	1	30-40	low	low	none	low
Gooseberries	2	18-22	5	6-10	low	low	none	medium
American Bunch Grapes	3	50-60	2	15-20	medium	very high	none	medium
European Bunch Grapes	3	50-60	3-5	10-12	low	high	none	very high
Hybrid Bunch Grapes	3	50-60	3-5	15-20	medium	very high	none	high
Muscadine Grapes	3	50-60	2	25-30	medium	very high	none	high
Hazelnuts	3	40-60	2	10-15	medium	medium	none	medium
Mayhaws	4	50-70	2	10-15	high	medium	none	high
Mulberries	5	60-70	1	20-30	low	low	none	low
Nectarines	2	10-14	1	80-160	very high	high	very high	very high
Pawpaws	5	50-60	2	8-12	high	low	medium	low
Peaches	2	10-14	2	120-200	very high	high	high	high
Asian & European Pears	4	60-80	2	40-120	medium	low	high	low
Pecans	5	80-100	2	40-60	very high	low	none	medium
American Persimmons	4	50-60	1	40-50	high	low	medium	medium
Oriental Persimmons	3	50-60	1	60-70	very high	low	high	high
Pistachios	7	40-45	1	15-25	low	low	none	medium
European Plums	4	25-30	2	40-120	medium	medium	medium	medium
Japanese Plums, Plumcots	2	15-20	1	120-200	very high	high	very high	very high
Raspberries	1	10-15	5-10	2-3	very high	very high	none	medium
Saskatoons	3	30-50	3-5	3-4	medium	medium	none	medium
Strawberries	1	3-5	25	1.5	very high	high	none	medium
Black Walnuts	4	80-100	2	20-30	high	low	none	low
Persian Walnuts	3	60-80	1	60-80	very high	medium	none	medium

[z] if precocious cultivars are chosen

[y] if adapted cultivars are chosen for an appropriate climate and given proper care

[x] at least 2 different cultivars are needed to provide cross-pollination in some cases

[w] low = 18-24"/year, medium = 24-30" /year, high = 30-36"/year, very high = 36-42"/year

[v] if disease-resistant cultivars are chosen

Most fruit and nut plants in a home garden need very little fertilizer, and it's better to err on the side of providing too little rather than too much fertilizer, especially nitrogen (N). Adding too much N is especially dangerous when fruit are gaining size, as that makes fruit more prone to fruit rots. Excess N also stimulates too much vegetative growth, which is more susceptible to diseases such as fire blight, so especially avoid it for apples and pears.

Controlling weeds around the fruit and nut plants conserves both water and nutrients for the desired crops, so is very important. Weed control of most small fruit plants is most easily accomplished with thick organic mulches, which also conserve soil moisture and provide some needed nutrients as they break down.

Straw or grass clipping mulches are often undesirable around fruit and nut trees as they can attract rodents, so an effective and attractive method of weed control is to plant less competitive perennial flowers around fruit and nut trees, and mulch these with wood chips or bark chips. Perennial flowers compete much less with the fruit and nut trees for water and nutrients than does grass, and this also avoids the number one cause of damage to young fruit and nut trees: lawnmowers and weedeaters.

Winter and frost protection, and fruit thinning. Even many well-adapted fruit and nut plants require some winter protection in many areas. Most fruit and nut trees benefit from placing white plastic tree guards around the trunks in fall, or painting trunks white as trees get larger, in most areas where sub-freezing temperatures occur. Strawberries need to be covered with straw or row covers over the winter in the same areas. These jobs do not take long for small home garden plantings.

Small fruit bushes and grapevines generally need no winter protection if well-adapted cultivars were chosen, except that half-high blueberries in USDA zones 3 and 4 benefit from shoveling snow around the bushes to protect them over winter. Gardeners who grow less hardy vines or cane berries, such as tender European grapes or boysenberries, in USDA zone 5 or 6 may lay vines or canes on the ground over winter and cover them with mulch to protect them, then tie up vines or canes to trellises in spring.

Choosing later-blooming species or cultivars is the easiest way to avoid the need for frost protection. Those who want to grow early-blooming fruit trees should plant them on the north side of the house or garage, where the shade will delay spring bloom. Setting automatic sprinklers to run all night when a frost is expected could protect fruit tree blossoms, but is generally not practical for homeowners. Some gardeners protect smaller plants, such as strawberries or lingonberries, by laying sheets of plastic over the plants the night before an expected frost.

Large-fruited tree fruits usually require fruit thinning when crops aren't reduced or eliminated by frosts (Photo 30-12). Otherwise, many tiny fruit will develop and will not ripen properly. Small-fruited tree fruits such as cherries and mulberries, along with small fruits and nuts, do not require any fruit thinning beyond that accomplished by pruning, so gardeners who do not want to thin fruit should choose these crops.

Pest control. Many home gardeners want to completely avoid or minimize the need for pesticides. They must be told about diseases and insect pests that are prevalent in their region on the fruits they want to grow, and the effectiveness or lack of effectiveness of non-chemical pest control methods. For example, peaches, nectarines and plums are almost impossible to grow "organically" east of the Rocky Mountains, where plum curculio and brown rot (plus many other pests) occur. They may be grown successfully in many areas, however, with just 3 or 4 sprays per year. This will necessitate buying a sprayer, a fungicide and an insecticide, which should be included in the projected costs.

Generally, most small fruits--such as strawberries, blackberries, raspberries, blueberries, lingonberries, currants and gooseberries--may be grown without pesticides in most areas if cultivars are chosen that resist the main diseases of the area. Grapes are almost impossible to grow without pesticides, except in arid Western areas.

European and Asian pears are the easiest of the tree fruits to grow without pesticides if cultivars resistant to fire blight are chosen. Mulberries are also an almost pest-free crop except for birds, and many gardeners are willing to share the fruit with birds. Apple cultivars resistant to fire blight, apple scab, cedar apple rust, powdery mildew and fairly resistant to fruit rots may be grown without sprays in many areas, if trapping and other methods are employed to control insects such as codling moth and apple maggot. Tart cherries resistant to leaf spot and brown rot may also be grown without sprays in many areas, if trapping and other methods are employed to control plum curculio, cherry fruit

Photo 30-12. Large-fruited tree fruits, such as these peaches and plums, require fruit thinning to ensure larger fruit like these when crops aren't reduced or eliminated by frosts.

fly and other insects. Figs tend to have few pests in most areas where they can be grown, and pawpaws usually have few pests also.

Most other tree fruits — including disease-susceptible apple cultivars, apricots, sweet cherries, persimmons, peaches, nectarines and plums — require at least 3 or 4 pesticide sprays per year, during years that they bear fruit. Several of these fruits are very prone to losing crops to frost, and need few to no sprays during years with no crops.

Blight-resistant Chinese chestnuts are the easiest nut crops to grow without pesticides. Blight-resistant hazelnuts, blight-resistant Persian walnuts, anthracnose-resistant black walnuts, and scab-resistant pecans need only 2 or 3 insecticide sprays. Almonds bloom so early and pistachios are so cold-sensitive that they're almost impossible to grow in temperate North America outside of the arid southwestern U.S., where they require much pest control.

Many fruit and nut crops are highly attractive to birds, squirrels and other mammals that eat the fruit. Growing blueberries, cherries, red currants and gooseberries, grapes or saskatoons often requires bird netting. Yellow cherry cultivars and white cultivars of currants, gooseberries, and grapes often don't need netting, since those colors look similar to green to birds.

Birds may also cause serious pecking damage on larger fruits, but those fruit trees are usually too big for bird netting, so gardeners may need to employ rotating scare devices, or plant other berry-bearing plants for the birds to hopefully lure them away. Squirrels can carry away many nuts or immature pears, so squirrel guards around tree trunks may be needed.

SIZE AND LONGEVITY OF HOME PLANTINGS AND HARVESTS

Many people are surprised by how much fruit one or two fruit trees can produce during a good cropping year (Photo 30-13). Table 30-1 lists an average yield per plant for each type of fruit crop at maturity, or after 20 years in the case of nut trees, and also provides a suggested number of each type to plant for a family of four. Even by planting these minimal numbers, gardeners will have plenty of extra fruit to give away to family and friends during some years.

Some crops require at least 2 different cultivars to provide cross-pollination, as detailed in Tables 1-1 to 1-3 and the cultivar charts in the individual fruit and nut chapters. If space is limited, some fruit trees can be purchased with 2 or 3 different cultivars grafted onto a single tree, but usually the cultivars available in such combination trees are not disease-resistant. Some gardeners might want to order a disease-resistant cultivar, obtain scionwood of a second disease-resistant cultivar, and learn to graft to create their own 2-in-1 tree.

Other crops are self-pollinating so need only one cultivar, such as peaches and grapes, but gardeners will probably want to plant at least two cultivars that ripen early and late to stretch out the harvest. Bush fruits also self-pollinate but gardeners can easily plant several cultivars since plants are smaller and more total plants are needed.

Photo 30-13. One or two fruit trees, such as this peach tree, can produce large harvests during a good cropping year, which are fun for children to help pick.

Find out when each cultivar you want to plant ripens in your area. If you always take a vacation in late July, don't plant cultivars that ripen then. Most cultivars ripen most of their fruits or nuts within a 10-day period or so, producing much fruit that must be consumed or put up in a short time. If you prefer a less concentrated harvest, choose cultivars that ripen over a longer period.

Properly cared-for fruit and nut plants will keep producing for many years, as Table 30-1 shows. Older nut trees provide shade, and well-maintained older fruit trees or grapevines add dignity to a landscape. Planting fruit and nut plants is an investment that can continue to yield dividends for a lifetime and beyond.

Further Reading

Creasy, R. 1982. *The complete book of edible landscaping*. Sierra Club Books, San Francisco, CA. 379 pp.

Olcott-Reid, B. 1990. A model fruit garden. *Flower and Garden* vol. 34 no. 1: 14-20.

Southwick, L. 1972. *Dwarf fruit trees for the home gardener*. Garden Way Publishing, Charlotte, VT. 118 pp.

Index

Numbers

A

B

C

D

E

F

G

H

L

M

N

O

P

Q

R

S

T

U

V

W

Y

Z